UNITED STATES
DEPARTMENT OF THE INTERIOR
FRED A. SEATON, Secretary
BUREAU OF RECLAMATION
WILBUR A. DEXHEIMER, Commissioner

Hydraulic and Excavation Tables

ELEVENTH EDITION

Reprint of Eleventh Edition, 1957

Hydraulic and Excavation Tables
Eleventh Edition

ISBN-10: 1-930665-93-8
ISBN-13: 978-1-930665-93-4

Library of Congress Control Number: 2004101473

THE BLACKBURN PRESS
P. O. Box 287
Caldwell, New Jersey   07006 U.S.A.
973-228-7077
www.BlackburnPress.com

PREFACE

The first edition of the Hydraulic and Excavation Tables was issued by the Bureau of Reclamation in 1905. It was compiled under the direction of A. P. Davis, and the following engineers were given credit for computing tables in the first edition: Roy H. Bolster, Clarence T. Johnston, C. C. Babb, F. E. Weymouth, E. C. Murphy, Robert Follansbee, and J. C. Hoyt.

The number of printings of this handbook, aggregating over 23,000 copies, is as follows:

Year	Edition	Numbered pages	Numbered tables
1905	First	76	13
1906	Geological Survey	81	18
1910	Second	117	39
1913	Third	147	43
1917	Fourth	168	49
1921 [1]	Fifth	168	49
1934	Sixth	168	49
1935	Seventh	168	49
1940	Eighth	173	49
1946	Ninth	173	49
1950 [2]	Tenth	173	49
1957	Eleventh	350	71

[1] Reprintings of 5th Edition in 1923, 1926, and 1930.
[2] Reprinting of 10th Edition in 1952.

While compiled in the first instance with a view to the requirements of the engineers of the Bureau of Reclamation, the book has had a considerable circulation among other engineers engaged in similar lines of work. Many of the tables have been computed especially for this publication and are available nowhere else in print. In order to make the series complete a few tables taken from other sources have been included, most of which have been extended or modified to conform to the conditions encountered on Reclamation projects.

The present edition constitutes a major revision as compared with the previous editions. Many new useful tables have been

3

added and, in conformance with present Bureau practice, the tables for flow in open channels are now based on Manning's formula rather than the Chezy-Kutter formula. The following tables have been taken, with permission, from Hydraulic Tables, Second Edition, of the War Department, Corps of Engineers: 15, 20, 58, 61, 62, 63, and 64. Part of the information for the following tables was also taken from the above Hydraulic Tables: 1 through 14, 27, 28, 29, 31, 33, 34, and 35. Most of the information in tables 18, 19, and 21 was taken, with permission, from Hydraulics of Steady Flow in Open Channels, by S. M. Woodward and C. J. Posey, copyright 1941, John Wiley & Sons, Inc.

The introductory notes giving typical values of n, in so far as they are applicable to flumes, conform with those given by Fred C. Scobey in Technical Bulletin No. 393, United States Department of Agriculture.

All reported errors have been checked and corrections made, and the thanks of the Bureau are extended to all those who by reporting errors or by making suggestions for improvement have cooperated in this revision. A continuance of this interest is earnestly solicited; all errors reported or suggestions made in the line of constructive criticism are welcomed, since it is only by the continued cooperation of users and publishers that the goal of an entirely reliable and adequate handbook can be attained.

WILBUR A. DEXHEIMER, *Commissioner.*

LIST OF TABLES

HYDRAULIC AND EXCAVATION TABLES

Explanation of Tables

Tables 1 to 14.—Tables 1 to 14 give the value of the mean velocity of water in open channels computed from Manning's formula:

$$v = \frac{1.486 r^{2/3} s^{1/2}}{n}$$

The value of n, the coefficient of roughness, to be used in finding v, depends on the roughness of the materials forming the bed and banks of the channel, irregularities and imperfections in the bed or banks, the load of silt or detritus, aquatic plants, curves, eddies, velocity of water, and various other considerations. Many experiments have been made which have determined the value of n for existing installations. The assigning of a value of n for a proposed conduit is, however, subject to considerable uncertainty, small differences in conditions of apparent minor importance often causing a surprising change in the measured values of n. On this account, where it is important that a channel carry a certain definite quantity of water, a factor of safety must be used. This may be done either by using a value of n greater than the most probable value or preferably the channel may be designed for a discharge greater than that actually desired.

The values of n given in the following list, with typical descriptions of the open channels from which the values were determined, are for the most part abridged from the work of Mr. Fred C. Scobey presented in Technical Bulletins Nos. 150, 393, 652, and in Bulletins 376 and 852 of the United States Department of Agriculture. The original bulletins should be referred to before selecting a value of n for any important project.

$n = 0.008$–0.010; for extremely smooth channels of glass or polished wood used in model work; for the computation of high velocities on steep chutes and spillways where air is entrained. These values correspond to about two-thirds the normal values found in canals at usual veloci-

9

ties and their use gives an approximation of the velocity of the air-water mixture. Depths must be computed by the use of normal value of n.

$n=0.009$; smooth flumes of polished lumber as used in models; should not be used for design of prototype.

$n=0.010$; this value can be taken as the lowest one attained in field tests where conditions approaching the ideal are maintained. It is too low to be anticipated in the design of any flume.

$n=0.011$; for new, straight, untreated semicircular stave flumes; for new, straight, smooth, unpainted semicircular metal flumes free from internal obstructions; for straight, best quality, large-diameter concrete pipe used as a flow line; only applicable to straight reaches; should not be expected to hold for life of structure.

$n=0.012$; for surfaced, untreated lumber flumes in excellent condition; for short, straight, smooth flumes of unpainted metal; for hand-poured concrete of the highest grade of workmanship with surfaces as smooth as a troweled sidewalk with masked expansion joints; practically no moss, larvae, or gravel ravelings; alignment long, straight tangents connected with long radius curves; field conditions seldom make this value applicable.

$n=0.013$; minimum conservative value of n for the design of long flumes of all materials of quality described under $n=0.012$; provides for mild curvature or some sand; treated wood stave flumes; covered flumes built of surfaced lumber, with battens included in hydraulic computations and of high-class workmanship; metal flumes painted and with dead smooth interiors; concrete flumes with oiled forms, fins rubbed down with troweled bottom; shot concrete if steel troweled; conduits to be this class should probably attain $n=0.012$ initially.

$n=0.014$; excellent value for conservatively designed structures of wood, painted metal, or concrete under usual conditions; cares for alignment about equal in curve and tangent length; conforms to surfaces as left by smooth-jointed forms or well-broomed shot concrete; will care for slight algae growth or slight deposits of silt or slight deterioration.

$n = 0.015$; rough, plank flumes of unsurfaced lumber with curves made by short length, angular shifts; for metal flumes with shallow compression member projecting into section but otherwise of class $n = 0.013$; for construction with first-class sides but roughly troweled bottom or for class $n = 0.014$ construction with noticeable silt or gravel deposits; value suitable for use with muddy gravel deposits; value suitable for use with muddy water for either poured or shot concrete; smooth concrete that is seasonally roughened by larvae or algae growths take value of $n = 0.015$ or higher; lowest value for highest class rubble and concrete combination.

$n = 0.016$; for lining made with rough board forms conveying clear water with small amount of debris; class $n = 0.014$ linings with reasonably heavy algae; or maximum larvae growth; or large amounts of cobble detritus; or old linings repaired with thin cement mortar coat; or heavy lime encrustations; earth channels in best possible conditions, with slick deposit of silt, free of moss and nearly straight alignment; true to grade and section; not to be used for design.

$n = 0.017$; for clear water on first-class bottom and excellent rubble sides or smooth rock bottom and wooden plank sides; roughly coated, poured lining with uneven expansion joints; basic value for shot concrete against smoothly trimmed earth base; such a surface is distinctly rough and will scratch hand; undulations of the order of 1 inch.

$n = 0.018$; about the upper limit for concrete construction in any workable condition; very rough concrete with sharp curves and deposits of gravel and moss; minimum design value for uniform rubble, or concrete sides and natural channel bed; for volcanic ash soils with no vegetation; minimum value for large high-class canals in very fine silt.

$n = 0.020$; for tuberculated iron; ruined masonry; well-constructed canals in firm earth or fine packed gravel where velocities are such that the silt may fill the interstices in the gravel; alignment straight, banks clean; large canals of class $n = 0.0225$.

$n = 0.0225$; for corrugated pipe with hydraulic functions computed from minimum internal diameter; average; well-constructed canal in material which will eventually have a medium smooth bottom with graded gravel, grass on the edges, and average alignment with silt deposits at both sides of the bed and a few scattered stones in the middle; hardpan in good condition; clay and lava-ash soil. For the largest of canals of this type a value of $n = 0.020$ will be originally applicable.

$n = 0.025$; for canals where moss, dense grass near edges, or scattered cobbles are noticeable. Earth channels with neglected maintenance have this value and up; a good value for small, head ditches serving a couple of farms; for canals wholly in cut and thus subject to rolling debris; minimum value for rock-cut smoothed up with shot concrete.

$n = 0.0275$; cobble-bottom canals, typically occurring near mouths of canyons; value only applicable where cobbles are graded and well packed; can reach 0.040 for large boulders and heavy sand.

$n = 0.030$; canals with heavy growth of moss, banks irregular and overhanging with dense rootlets; bottom covered with large fragments of rock or bed badly pitted by erosion.

$n = 0.035$; for medium large canals about 50 percent choked with moss growth and in bad order and regimen; small channels with considerable variation in wetted cross-section and biennial maintenance; for flood channels not continuously maintained; for untouched rock cuts and tunnels based on "paper" cross section.

$n = 0.040$; for canals badly choked with moss or heavy growth; large canals in which large cobbles and boulders collect, approaching a stream bed in character.

$n = 0.050-0.060$; floodways poorly maintained; canals two-thirds choked with vegetation.

$n = 0.060-0.240$; floodways without channels through timber and underbrush, friction slope 0.4 to 0.8 foot per mile.

Tables 21, 22, 23, and 24.—Tables 21 and 23 give areas and hydraulic radii for partially filled circular and horseshoe conduit sections, respectively.

Where horseshoe or circular cross sections are used the labor of testing for the critical depth or the hydraulic jump is materially reduced by use of tables 22 and 24, which are supplementary to tables 21 and 23.

For example, suppose that it is desired to find the critical depth for 650 second-feet flowing free in a 9-foot diameter circular conduit.

$$\frac{Q}{D^{5/2}} = \frac{650}{243} = 2.675.$$

Entering table 22, column 3, with this value, by interpolation the corresponding value in column 1 is found of

$$\frac{d}{D} = 0.701.$$

The critical depth is therefore

$$d = 9 \times 0.701 \doteq 6.31 \text{ feet.}$$

The critical velocity head is likewise found from column 2 to be

$$h_v = .3214 \times 9 = 2.893 \text{ feet,}$$

which gives a critical velocity of

$$V = 13.63 \text{ feet per second.}$$

This information determines whether flow at a given point is above or below critical depth, and provision may be made for any contingency likely to occur.

Column 4 gives the hydrostatic pressure upon the cross section of the water prism. The tabular values must be multiplied by D^3 to give pressures in cubic units of water. If the pressure is required in pounds, multiply the tabular value by 62.5 D^3.

Table 24 is identical with table 22 except that it is computed for a horseshoe section of the proportions indicated. These two tables are subject to an error of one unit in the last digit.

Table 36.—Table 36 gives the discharges in cubic feet per second over Cipolletti weirs for various lengths and depths of water on the crest. The formula from which this table is computed is $Q = 3.367 LH^{3/2}$, where Q is the discharge in cubic feet per second, L the length in feet of the crest of the weir, and H the depth in feet of water flowing over the weir. Table 36 may also be used, with modification, for rectangular weirs, as discussed after the table.

The Cipolletti weir differs from the rectangular form in having side slopes of 4 vertical to 1 horizontal, instead of vertical sides. Its coefficient of contraction is unity and hence its discharge is more readily computed than that of the rectangular weir.

Since the discharge is proportional to the length of weir, the table may be used for weirs of any length by multiplying some value found in the table by the proper factor, or by moving decimal points and adding, but the tabular values are not accurate in case the head is greater than one-third the length of the weir.

Table 38.—Table 38 gives the discharge per foot of length over sharp-crested vertical weirs, without end contractions, of heights 2, 4, 6, 8, 10, 20, and 30 feet, computed from Bazin's formula. Although this formula is based on data obtained from experiments with heads not greater than 1.64 feet, discharges for heads of 4 feet and less computed thereby agree within 2 percent with those obtained by use of the Fteley and Stearns formula. The discharge given by this table is corrected for velocity of approach and the head to be used is that observed 16 feet or more upstream from the crest of the weir.

Tables 39 to 41.—Tables 39 to 41 give multipliers to be applied to quantities in table 38 to determine the discharge over broad-crested weirs of various types and dimensions.

EXAMPLE: Suppose the discharge is to be computed over a weir of rectangular cross section that is 10 feet long, 12 feet high, 6 feet wide at crest, and has an observed head of 2.4 feet. Table 38 shows that for a height (p) of 12 feet and a head (h) of 2.4, the discharge is 12.42 second-feet. Table 39 shows that for a height (p) of 12 feet, a crest width (c) of 6 feet, and head (h) of 2.4 feet the multiplier is 0.797. Hence, the discharge is $12.42 \times 0.797 \times 10 = 99.0$ second-feet. With two end contractions the discharge would be $9.9 \left(10 - \dfrac{2 \times 2.4}{10} \right) = 94.2$.

Table 44.—Table 44 presents factors which may be multiplied by the square root of the head to give the discharge of concrete pipe by Scobey's formula (see Bulletin No. 852, Department of Agriculture). Thus, for example, the discharge of a 6-inch pipe with rubber-gasket joints for a head loss of 6 feet per 1,000 would be $0.2078 \times 2.4495 = 0.51$ cubic feet per second.

Tables 48 to 52.—Tables 48 to 52 give the volume of excavation in cubic yards per 100 feet of length for various center depths and side slopes, assuming the ground to be level transversely. The volume required is the difference between two triangular prisms.

In figure 1 is shown the cross section of a canal that has a bottom width of 18 feet and side slopes of 1½ to 1. The amount of material in the prism C B F E is equal to the volume of the prism A C E minus the volume of the prism A B F. As A C E has an altitude of 16 feet and A B F has an altitude of 6 feet, the volume of each for a length of 100 feet can be obtained from the table. Opposite 16 in table 50 is 1,422, which is the volume in cubic yards of A C E per 100 linear feet; opposite 6 is 200, which is the volume of A B F.

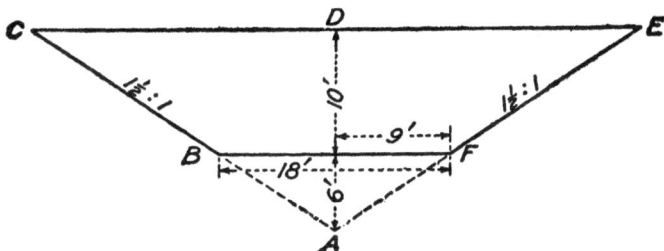

FIG. 1.—Ideal canal section.

As C B F E = A C E − A B F
C B F E = 1,422 − 200
= 1,222

When working up quantities for canal excavation it is only necessary to subtract the volume below the bed once for each mile or for each 10 miles, thus making the use of the table much more rapid.

Tables 53 to 57.—Tables 53 to 57 give the volume of excavation in cubic yards per 100 feet of length, where the surface slopes transversely, for various center depths and side slopes. They differ from tables 48 to 52 only in that the earth surface is sloping ground instead of being level transversely. The surface slope is expressed in per cent, a 10 per cent slope being 10 vertical to 100 horizontal.

Table 69.—Table 69 is designed for use in stadia work and gives the difference in elevation corresponding to specified slant distances for vertical angles of 0° to 20°. The horizontal distances corresponding to the slant distances are also given for various vertical angles.

EXAMPLE: With the instrument at *A* a vertical angle of 3° 10′ is observed on a point *B* which is distant 350 feet by stadia reading; find the difference in elevation of *A* and *B* and the horizontal distance *AB*. Opposite 3° 10′ in the first column of the table, 16.5 is found under a distance of 300 and 22.1 under a distance of 400; and interpolation for a distance of 350 feet gives 19.3 feet for the difference in elevation of *A* and *B*. Interpolation for 350 between the values in the 300 and the 400 distance columns of the horizontal distance lines at 3° and 4° gives, respectively, 349.0 and 348.2; and an additional interpolation gives, for an angle of 3° 10′ and a slant distance of 350, a horizontal distance of 348.9. The horizontal distance of *AB* is therefore 348.9 feet.

Table 1.—*Velocity of water, v, in feet per second, based on Manning's formula* $v = (1.486/n)r^{2/3}s^{1/2}$, **n = .010**

r \ s	.00005	.00010	.00015	.00020	.00025	.00030	.00035	.00040	.00045	.00050
0.2	0.36	0.51	0.62	0.72	0.80	0.88	0.95	1.02	1.08	1.14
0.4	.57	.81	.99	1.14	1.28	1.40	1.51	1.61	1.71	1.80
0.6	.75	1.06	1.29	1.49	1.67	1.83	1.98	2.11	2.24	2.36
0.8	.91	1.28	1.57	1.81	2.02	2.22	2.40	2.56	2.72	2.86
1.0	1.05	1.49	1.82	2.10	2.35	2.57	2.78	2.97	3.15	3.32
1.2	1.19	1.68	2.06	2.37	2.65	2.91	3.14	3.36	3.56	3.75
1.4	1.31	1.86	2.28	2.63	2.94	3.22	3.48	3.72	3.94	4.16
1.6	1.44	2.03	2.49	2.87	3.21	3.52	3.80	4.07	4.31	4.55
1.8	1.55	2.20	2.69	3.11	3.48	3.81	4.11	4.40	4.66	4.92
2.0	1.67	2.36	2.89	3.34	3.73	4.09	4.41	4.72	5.00	5.27
2.2	1.78	2.51	3.08	3.55	3.97	4.35	4.70	5.03	5.33	5.62
2.4	1.88	2.66	3.26	3.77	4.21	4.61	4.98	5.33	5.65	5.96
2.6	1.99	2.81	3.44	3.97	4.44	4.87	5.26	5.62	5.96	6.28
2.8	2.09	2.95	3.62	4.17	4.67	5.11	5.52	5.90	6.26	6.60
3.0	2.19	3.09	3.79	4.37	4.89	5.35	5.78	6.18	6.56	6.91
3.2	2.28	3.23	3.95	4.56	5.10	5.59	6.04	6.45	6.85	7.22
3.4	2.38	3.36	4.12	4.75	5.31	5.82	6.29	6.72	7.13	7.51
3.6	2.47	3.49	4.27	4.94	5.52	6.05	6.53	6.98	7.40	7.80
3.8	2.56	3.62	4.43	5.12	5.72	6.27	6.77	7.24	7.68	8.09
4.0	2.65	3.74	4.59	5.30	5.92	6.49	7.01	7.49	7.94	8.37
4.2	2.74	3.87	4.74	5.47	6.12	6.70	7.24	7.74	8.21	8.65
4.4	2.82	3.99	4.89	5.64	6.31	6.91	7.46	7.98	8.46	8.92
4.6	2.91	4.11	5.03	5.81	6.50	7.12	7.69	8.22	8.72	9.19
4.8	2.99	4.23	5.18	5.98	6.69	7.32	7.91	8.46	8.97	9.46
5.0	3.07	4.35	5.32	6.14	6.87	7.53	8.13	8.69	9.22	9.72
5.2	3.15	4.46	5.46	6.31	7.05	7.73	8.34	8.92	9.46	9.97
5.4	3.23	4.57	5.60	6.47	7.23	7.92	8.56	9.15	9.70	10.23
5.6	3.31	4.69	5.74	6.63	7.41	8.12	8.77	9.37	9.94	10.48
5.8	3.39	4.80	5.88	6.78	7.58	8.31	8.97	9.59	10.18	10.73
6.0	3.47	4.91	6.01	6.94	7.76	8.50	9.18	9.81	10.41	10.97
6.2	3.55	5.02	6.14	7.09	7.93	8.69	9.38	10.03	10.64	11.21
6.4	3.62	5.12	6.27	7.24	8.10	8.87	9.58	10.24	10.87	11.45
6.6	3.70	5.23	6.40	7.39	8.27	9.06	9.78	10.46	11.09	11.69
6.8	3.77	5.33	6.53	7.54	8.43	9.24	9.98	10.67	11.31	11.93
7.0	3.85	5.44	6.66	7.69	8.60	9.42	10.17	10.88	11.54	12.16
7.5	4.03	5.69	6.97	8.05	9.00	9.86	10.65	11.39	12.08	12.73
8.0	4.20	5.94	7.28	8.41	9.40	10.30	11.12	11.89	12.61	13.29
8.5	4.38	6.19	7.58	8.75	9.79	10.72	11.58	12.38	13.13	13.84
9.0	4.55	6.43	7.87	9.09	10.17	11.14	12.03	12.86	13.64	14.38
9.5	4.71	6.67	8.16	9.43	10.54	11.55	12.47	13.33	14.14	14.90
10	4.88	6.90	8.45	9.75	10.91	11.95	12.90	13.79	14.63	15.42
11	5.20	7.35	9.00	10.39	11.62	12.73	13.75	14.70	15.59	16.43
12	5.51	7.79	9.54	11.02	12.32	13.49	14.57	15.58	16.52	17.42
13	5.81	8.22	10.06	11.62	12.99	14.23	15.37	16.43	17.43	18.37
14	6.10	8.63	10.57	12.21	13.65	14.95	16.15	17.26	18.31	19.30
15	6.39	9.04	11.07	12.78	14.29	15.65	16.91	18.08	19.17	20.21
16	6.67	9.44	11.56	13.34	14.92	16.34	17.65	18.87	20.02	21.10
17	6.95	9.82	12.03	13.89	15.53	17.02	18.38	19.65	20.84	21.97
18	7.22	10.21	12.50	14.43	16.14	17.68	19.09	20.41	21.65	22.82
19	7.48	10.58	12.96	14.96	16.73	18.33	19.79	21.16	22.45	23.66
20	7.74	10.95	13.41	15.48	17.31	18.96	20.48	21.90	23.23	24.48

Table 1.—*Velocity of water, v, in feet per second, based on Manning's formula* $v = (1.486/n)r^{2/3}s^{1/2}$, **n** $= .010$—Continued

s / r	.00055	.00060	.00065	.00070	.00075	.00080	.00085	.00090	.00095	.00100
0.2	1.19	1.24	1.30	1.34	1.39	1.44	1.48	1.52	1.57	1.61
0.4	1.89	1.98	2.06	2.13	2.21	2.28	2.35	2.42	2.49	2.55
0.6	2.48	2.59	2.70	2.80	2.90	2.99	3.08	3.17	3.26	3.34
0.8	3.00	3.14	3.26	3.39	3.51	3.62	3.73	3.84	3.95	4.05
1.0	3.48	3.64	3.79	3.93	4.07	4.20	4.33	4.46	4.58	4.70
1.2	3.94	4.11	4.28	4.44	4.60	4.75	4.89	5.03	5.17	5.31
1.4	4.36	4.56	4.74	4.92	5.09	5.26	5.42	5.58	5.73	5.88
1.6	4.77	4.98	5.18	5.38	5.57	5.75	5.93	6.10	6.27	6.43
1.8	5.16	5.39	5.61	5.82	6.02	6.22	6.41	6.60	6.78	6.95
2.0	5.53	5.78	6.01	6.24	6.46	6.67	6.88	7.08	7.27	7.46
2.2	5.89	6.16	6.41	6.65	6.88	7.11	7.33	7.54	7.75	7.95
2.4	6.25	6.52	6.79	7.05	7.29	7.53	7.77	7.99	8.21	8.42
2.6	6.59	6.88	7.16	7.43	7.69	7.95	8.19	8.43	8.66	8.89
2.8	6.92	7.23	7.53	7.81	8.08	8.35	8.61	8.86	9.10	9.34
3.0	7.25	7.57	7.88	8.18	8.47	8.74	9.01	9.27	9.53	9.77
3.2	7.57	7.90	8.23	8.54	8.84	9.13	9.41	9.68	9.95	10.20
3.4	7.88	8.23	8.57	8.89	9.20	9.50	9.80	10.08	10.36	10.63
3.6	8.19	8.55	8.90	9.23	9.56	9.87	10.18	10.47	10.76	11.04
3.8	8.49	8.86	9.23	9.57	9.91	10.23	10.55	10.86	11.15	11.44
4.0	8.78	9.17	9.55	9.91	10.25	10.59	10.92	11.23	11.54	11.84
4.2	9.07	9.48	9.86	10.23	10.59	10.94	11.28	11.60	11.92	12.23
4.4	9.36	9.77	10.17	10.56	10.93	11.29	11.63	11.97	12.30	12.62
4.6	9.64	10.07	10.48	10.87	11.26	11.63	11.98	12.33	12.67	13.00
4.8	9.92	10.36	10.78	11.19	11.58	11.96	12.33	12.69	13.03	13.37
5.0	10.19	10.64	11.08	11.50	11.90	12.29	12.67	13.04	13.39	13.74
5.2	10.46	10.93	11.37	11.80	12.21	12.62	13.00	13.38	13.75	14.10
5.4	10.73	11.20	11.66	12.10	12.53	12.94	13.33	13.72	14.10	14.46
5.6	10.99	11.48	11.95	12.40	12.83	13.25	13.66	14.06	14.44	14.82
5.8	11.25	11.75	12.23	12.69	13.14	13.57	13.99	14.39	14.79	15.17
6.0	11.51	12.02	12.51	12.98	13.44	13.88	14.31	14.72	15.12	15.52
6.2	11.76	12.28	12.79	13.27	13.73	14.18	14.62	15.05	15.46	15.86
6.4	12.01	12.55	13.06	13.55	14.03	14.49	14.93	15.37	15.79	16.20
6.6	12.26	12.81	13.33	13.83	14.32	14.79	15.24	15.69	16.12	16.53
6.8	12.51	13.06	13.60	14.11	14.61	15.09	15.55	16.00	16.44	16.87
7.0	12.75	13.32	13.86	14.39	14.89	15.38	15.85	16.31	16.76	17.20
7.5	13.35	13.95	14.52	15.06	15.59	16.10	16.60	17.08	17.55	18.00
8.0	13.94	14.56	15.15	15.73	16.28	16.81	17.33	17.83	18.32	18.80
8.5	14.51	15.16	15.78	16.37	16.95	17.51	18.04	18.57	19.08	19.57
9.0	15.08	15.75	16.39	17.01	17.61	18.19	18.75	19.29	19.82	20.33
9.5	15.63	16.33	16.99	17.64	18.25	18.85	19.43	20.00	20.54	21.08
10	16.18	16.90	17.58	18.25	18.89	19.51	20.11	20.69	21.26	21.81
11	17.24	18.00	18.74	19.45	20.13	20.79	21.43	22.05	22.65	23.24
12	18.27	19.08	19.86	20.61	21.33	22.03	22.71	23.37	24.01	24.63
13	19.27	20.12	20.95	21.74	22.50	23.24	23.95	24.65	25.32	25.98
14	20.24	21.14	22.01	22.84	23.64	24.41	25.17	25.90	26.61	27.30
15	21.20	22.14	23.04	23.91	24.75	25.56	26.35	27.11	27.86	28.58
16	22.13	23.11	24.06	24.96	25.84	26.69	27.51	28.31	29.08	29.84
17	23.04	24.07	25.05	25.99	26.91	27.79	28.64	29.47	30.28	31.07
18	23.94	25.00	26.02	27.00	27.95	28.87	29.76	30.62	31.46	32.28
19	24.81	25.92	26.98	27.99	28.98	29.93	30.85	31.74	32.61	33.46
20	25.68	26.82	27.91	28.97	29.98	30.97	31.92	32.85	33.75	34.62

Table 1.—*Velocity of water, v, in feet per second, based on Manning's formula* $v = (1.486/n)r^{2/3}s^{1/2}$, **n = .010**—Continued

r \ s	.0010	.0015	.0020	.0025	.0030	.0035	.0040	.0045	.0050
0.2	1.61	1.97	2.27	2.54	2.78	3.01	3.21	3.41	3.59
0.4	2.55	3.12	3.61	4.03	4.42	4.77	5.10	5.41	5.70
0.6	3.34	4.09	4.73	5.29	5.79	6.25	6.69	7.09	7.47
0.8	4.05	4.96	5.73	6.40	7.01	7.58	8.10	8.59	9.06
1.0	4.70	5.76	6.65	7.43	8.14	8.79	9.40	9.97	10.51
1.2	5.31	6.50	7.50	8.39	9.19	9.93	10.61	11.26	11.87
1.4	5.88	7.20	8.32	9.30	10.19	11.00	11.76	12.48	13.15
1.6	6.43	7.87	9.09	10.16	11.13	12.03	12.86	13.64	14.37
1.8	6.95	8.52	9.83	10.99	12.04	13.01	13.91	14.75	15.55
2.0	7.46	9.14	10.55	11.79	12.92	13.96	14.92	15.82	16.68
2.2	7.95	9.74	11.24	12.57	13.77	14.87	15.90	16.86	17.77
2.4	8.42	10.32	11.91	13.32	14.59	15.76	16.85	17.87	18.84
2.6	8.89	10.88	12.57	14.05	15.39	16.62	17.77	18.85	19.87
2.8	9.34	11.43	13.20	14.76	16.17	17.46	18.67	19.80	20.87
3.0	9.77	11.97	13.82	15.46	16.93	18.29	19.55	20.74	21.86
3.2	10.20	12.50	14.43	16.13	17.67	19.09	20.41	21.65	22.82
3.4	10.63	13.01	15.03	16.80	18.40	19.88	21.25	22.54	23.76
3.6	11.04	13.52	15.61	17.45	19.12	20.65	22.08	23.41	24.68
3.8	11.44	14.01	16.18	18.09	19.82	21.41	22.89	24.27	25.59
4.0	11.84	14.50	16.75	18.72	20.51	22.15	23.68	25.12	26.48
4.2	12.23	14.98	17.30	19.34	21.19	22.89	24.47	25.95	27.35
4.4	12.62	15.45	17.84	19.95	21.85	23.61	25.24	26.77	28.21
4.6	13.00	15.92	18.38	20.55	22.51	24.32	25.99	27.57	29.06
4.8	13.37	16.38	18.91	21.14	23.16	25.02	26.74	28.37	29.90
5.0	13.74	16.83	19.43	21.73	23.80	25.71	27.48	29.15	30.72
5.2	14.10	17.27	19.95	22.30	24.43	26.39	28.21	29.92	31.54
5.4	14.46	17.71	20.45	22.87	25.05	27.06	28.93	30.68	32.34
5.6	14.82	18.15	20.96	23.43	25.67	27.72	29.64	31.44	33.14
5.8	15.17	18.58	21.45	23.99	26.27	28.38	30.34	32.18	33.92
6.0	15.52	19.00	21.94	24.53	26.87	29.03	31.03	32.91	34.70
6.2	15.86	19.42	22.43	25.08	27.47	29.67	31.72	33.64	35.46
6.4	16.20	19.84	22.91	25.61	28.06	30.30	32.40	34.36	36.22
6.6	16.53	20.25	23.38	26.14	28.64	30.93	33.07	35.07	36.97
6.8	16.87	20.66	23.85	26.67	29.21	31.55	33.73	35.78	37.71
7.0	17.20	21.06	24.32	27.19	29.78	32.17	34.39	36.48	38.45
7.5	18.00	22.05	25.46	28.47	31.19	33.68	36.01	38.19	40.26
8.0	18.80	23.02	26.58	29.72	32.56	35.17	37.59	39.87	42.03
8.5	19.57	23.97	27.68	30.95	33.90	36.62	39.14	41.52	43.76
9.0	20.33	24.90	28.75	32.15	35.22	38.04	40.66	43.13	45.46
9.5	21.08	25.82	29.81	33.33	36.51	39.43	42.16	44.71	47.13
10	21.81	26.71	30.85	34.49	37.78	40.81	43.62	46.27	48.77
11	23.24	28.47	32.87	36.75	40.26	43.48	46.48	49.30	51.97
12	24.63	30.17	34.83	38.94	42.66	46.08	49.26	52.25	55.08
13	25.98	31.82	36.74	41.08	45.00	48.61	51.96	55.11	58.09
14	27.30	33.43	38.60	43.16	47.28	51.07	54.59	57.90	61.04
15	28.58	35.00	40.42	45.19	49.50	53.47	57.16	60.63	63.91
16	29.84	36.54	42.20	47.18	51.68	55.82	59.68	63.30	66.72
17	31.07	38.05	43.94	49.12	53.81	58.12	62.14	65.91	69.47
18	32.28	39.53	45.64	51.03	55.90	60.38	64.55	68.47	72.17
19	33.46	40.98	47.32	52.90	57.95	62.60	66.92	70.98	74.82
20	34.62	42.41	48.97	54.74	59.97	64.77	69.25	73.45	77.42

Table 1.—*Velocity of water, v, in feet per second, based on Manning's formula* $v = (1.486/n) r^{2/3} s^{1/2}$, **n** = **.010**—Continued

s / r	.0055	.0060	.0065	.0070	.0075	.0080	.0085	.0090	.0095	.0100
0.2	3.77	3.94	4.10	4.25	4.40	4.55	4.69	4.82	4.95	5.08
0.4	5.98	6.25	6.50	6.75	6.99	7.22	7.44	7.65	7.86	8.07
0.6	7.84	8.19	8.52	8.84	9.15	9.46	9.75	10.03	10.30	10.57
0.8	9.50	9.92	10.32	10.71	11.09	11.45	11.81	12.15	12.48	12.81
1.0	11.02	11.51	11.98	12.43	12.87	13.29	13.70	14.10	14.48	14.86
1.2	12.44	13.00	13.53	14.04	14.53	15.01	15.47	15.92	16.36	16.78
1.4	13.79	14.40	14.99	15.56	16.11	16.63	17.15	17.64	18.13	18.60
1.6	15.08	15.75	16.39	17.01	17.60	18.18	18.74	19.29	19.81	20.33
1.8	16.31	17.03	17.73	18.40	19.04	19.67	20.27	20.86	21.43	21.99
2.0	17.49	18.27	19.02	19.74	20.43	21.10	21.75	22.38	22.99	23.59
2.2	18.64	19.47	20.27	21.03	21.77	22.48	23.17	23.85	24.50	25.14
2.4	19.75	20.63	21.48	22.29	23.07	23.83	24.56	25.27	25.96	26.64
2.6	20.84	21.76	22.65	23.51	24.33	25.13	25.90	26.66	27.39	28.10
2.8	21.89	22.87	23.80	24.70	25.57	26.40	27.22	28.01	28.77	29.52
3.0	22.92	23.94	24.92	25.86	26.77	27.65	28.50	29.32	30.13	30.91
3.2	23.93	25.00	26.02	27.00	27.95	28.86	29.75	30.61	31.45	32.27
3.4	24.92	26.03	27.09	28.11	29.10	30.05	30.98	31.88	32.75	33.60
3.6	25.89	27.04	28.14	29.20	30.23	31.22	32.18	33.11	34.02	34.90
3.8	26.84	28.03	29.17	30.28	31.34	32.37	33.36	34.33	35.27	36.19
4.0	27.77	29.00	30.19	31.33	32.43	33.49	34.52	35.52	36.50	37.44
4.2	28.69	29.96	31.19	32.36	33.50	34.60	35.66	36.70	37.70	38.68
4.4	29.59	30.91	32.17	33.38	34.56	35.69	36.79	37.85	38.89	39.90
4.6	30.48	31.84	33.14	34.39	35.59	36.76	37.89	38.99	40.06	41.10
4.8	31.36	32.75	34.09	35.38	36.62	37.82	38.98	40.11	41.21	42.28
5.0	32.22	33.66	35.03	36.35	37.63	38.86	40.06	41.22	42.35	43.45
5.2	33.08	34.55	35.96	37.32	38.63	39.89	41.12	42.31	43.47	44.60
5.4	33.92	35.43	36.88	38.27	39.61	40.91	42.17	43.39	44.58	45.74
5.6	34.75	36.30	37.78	39.21	40.58	41.91	43.20	44.46	45.67	46.86
5.8	35.58	37.16	38.67	40.13	41.54	42.91	44.23	45.51	46.76	47.97
6.0	36.39	38.01	39.56	41.05	42.49	43.89	45.24	46.55	47.82	49.07
6.2	37.19	38.85	40.43	41.96	43.43	44.86	46.24	47.58	48.88	50.15
6.4	37.99	39.68	41.30	42.86	44.36	45.82	47.23	48.60	49.93	51.22
6.6	38.78	40.50	42.15	43.75	45.28	46.77	48.20	49.60	50.96	52.29
6.8	39.56	41.31	43.00	44.62	46.19	47.71	49.17	50.60	51.99	53.34
7.0	40.33	42.12	43.84	45.50	47.09	48.64	50.13	51.59	53.00	54.38
7.5	42.23	44.10	45.90	47.64	49.31	50.93	52.49	54.01	55.50	56.94
8.0	44.08	46.04	47.92	49.73	51.48	53.16	54.80	56.39	57.93	59.44
8.5	45.90	47.94	49.90	51.78	53.60	55.36	57.06	58.72	60.32	61.89
9.0	47.68	49.80	51.84	53.79	55.68	57.51	59.28	61.00	62.67	64.30
9.5	49.43	51.63	53.74	55.77	57.73	59.62	61.45	63.23	64.97	66.66
10	51.15	53.43	55.61	57.71	59.73	61.69	63.59	65.43	67.23	68.97
11	54.51	56.93	59.26	61.49	63.65	65.74	67.76	69.73	71.64	73.50
12	57.76	60.33	62.80	65.17	67.45	69.67	71.81	73.89	75.92	77.89
13	60.93	63.64	66.24	68.74	71.15	73.48	75.74	77.94	80.08	82.16
14	64.02	66.86	69.59	72.22	74.75	77.21	79.58	81.89	84.13	86.32
15	67.03	70.01	72.87	75.62	78.27	80.84	83.33	85.74	88.09	90.38
16	69.98	73.09	76.07	78.94	81.71	84.39	86.99	89.51	91.97	94.36
17	72.86	76.10	79.21	82.20	85.08	87.87	90.58	93.21	95.76	98.25
18	75.69	79.06	82.29	85.39	88.39	91.29	94.10	96.83	99.48	102.06
19	78.47	81.96	85.31	88.53	91.63	94.64	97.55	100.38	103.13	105.81
20	81.20	84.81	88.27	91.61	94.82	97.93	100.94	103.87	106.72	109.49

Table 1.—*Velocity of water, v, in feet per second, based on Manning's formula* $v=(1.486/n)r^{2/3}s^{1/2}$, **n = .010**—Continued

r	.01	.02	.03	.04	.05	.06	.07	.08	.09	.10
0.2	5.08	7.19	8.80	10.16	11.36	12.45	13.45	14.37	15.25	16.07
0.4	8.07	11.41	13.97	16.13	18.04	19.76	21.34	22.82	24.20	25.51
0.6	10.57	14.95	18.31	21.14	23.64	25.89	27.97	29.90	31.71	33.43
0.8	12.81	18.11	22.18	25.61	28.63	31.37	33.88	36.22	38.42	40.50
1.0	14.86	21.02	25.74	29.72	33.23	36.40	39.32	42.03	44.58	46.99
1.2	16.78	23.73	29.06	33.56	37.52	41.10	44.40	47.46	50.34	53.06
1.4	18.60	26.30	32.21	37.19	41.58	45.55	49.20	52.60	55.79	58.81
1.6	20.33	28.75	35.21	40.66	45.46	49.79	53.78	57.50	60.98	64.28
1.8	21.99	31.10	38.09	43.98	49.17	53.86	58.18	62.19	65.97	69.53
2.0	23.59	33.36	40.86	47.18	52.75	57.78	62.41	66.72	70.77	74.59
2.2	25.14	35.55	43.54	50.27	56.21	61.57	66.50	71.10	75.41	79.49
2.4	26.64	37.67	46.14	53.27	59.56	65.25	70.48	75.34	79.91	84.24
2.6	28.10	39.74	48.67	56.20	62.83	68.82	74.34	79.47	84.29	88.85
2.8	29.52	41.75	51.13	59.04	66.01	72.31	78.10	83.50	88.56	93.35
3.0	30.91	43.71	53.54	61.82	69.12	75.71	81.78	87.43	92.73	97.75
3.2	32.27	45.64	55.89	64.54	72.16	79.04	85.38	91.27	96.81	102.04
3.4	33.60	47.52	58.20	67.20	75.13	82.30	88.90	95.03	100.80	106.25
3.6	34.90	49.36	60.46	69.81	78.05	85.50	92.35	98.73	104.71	110.38
3.8	36.19	51.17	62.68	72.37	80.91	88.64	95.74	102.35	108.56	114.43
4.0	37.44	52.96	64.86	74.89	83.73	91.72	99.07	105.91	112.33	118.41
4.2	38.68	54.71	67.00	77.37	86.50	94.75	102.35	109.41	116.05	122.33
4.4	39.90	56.43	69.11	79.80	89.22	97.74	105.57	112.86	119.70	126.18
4.6	41.10	58.13	71.19	82.20	91.91	100.68	108.74	116.25	123.30	129.97
4.8	42.28	59.80	73.24	84.57	94.55	103.58	111.87	119.60	126.85	133.71
5.0	43.45	61.45	75.26	86.90	97.16	106.43	114.96	122.90	130.35	137.40
5.2	44.60	63.08	77.25	89.20	99.73	109.25	118.01	126.15	133.81	141.04
5.4	45.74	64.68	79.22	91.48	102.27	112.04	121.01	129.37	137.22	144.64
5.6	46.86	66.27	81.17	93.72	104.78	114.79	123.98	132.54	140.58	148.19
5.8	47.97	67.84	83.09	95.94	107.26	117.50	126.92	135.68	143.91	151.69
6.0	49.07	69.39	84.99	98.13	109.72	120.19	129.82	138.78	147.20	155.16
6.2	50.15	70.92	86.86	100.30	112.14	122.84	132.69	141.85	150.45	158.59
6.4	51.22	72.44	88.72	102.45	114.54	125.47	135.53	144.88	153.67	161.98
6.6	52.29	73.94	90.56	104.57	116.91	128.07	138.33	147.89	156.86	165.34
6.8	53.34	75.43	92.38	106.67	119.26	130.65	141.12	150.86	160.01	168.66
7.0	54.38	76.90	94.18	108.75	121.59	133.20	143.87	153.80	163.13	171.96
7.5	56.94	80.52	98.62	113.87	127.31	139.47	150.64	161.04	170.81	180.05
8.0	59.44	84.06	102.95	118.88	132.91	145.60	157.26	168.12	178.32	187.97
8.5	61.89	87.53	107.20	123.78	138.39	151.60	163.75	175.06	185.67	195.72
9.0	64.30	90.93	111.36	128.59	143.77	157.49	170.11	181.86	192.89	203.32
9.5	66.66	94.26	115.45	133.31	149.05	163.27	176.35	188.53	199.97	210.78
10	68.97	97.54	119.47	137.95	154.23	168.95	182.49	195.09	206.92	218.11
11	73.50	103.94	127.30	147.00	164.35	180.03	194.46	207.89	220.50	232.42
12	77.89	110.15	134.91	155.78	174.16	190.79	206.07	220.30	233.67	246.30
13	82.16	116.19	142.30	164.32	183.71	201.24	217.37	232.38	246.47	259.81
14	86.32	122.07	149.51	172.64	193.01	211.44	228.38	244.15	258.96	272.96
15	90.38	127.82	156.55	180.76	202.10	221.39	239.13	255.64	271.14	285.81
16	94.36	133.44	163.43	188.71	210.98	231.12	249.64	266.88	283.07	298.38
17	98.25	138.94	170.17	196.49	219.69	240.65	259.94	277.88	294.74	310.68
18	102.06	144.34	176.78	204.13	228.22	250.00	270.03	288.68	306.19	322.75
19	105.81	149.64	183.27	211.62	236.60	259.18	279.94	299.27	317.43	334.60
20	109.49	154.84	189.64	218.98	244.83	268.19	289.68	309.68	328.47	346.24

Table 2.—*Velocity of water, v, in feet per second, based on Manning's formula* $v = (1.486/n) r^{2/3} s^{1/2}$, **n = .011**

r \ s	.00005	.00010	.00015	.00020	.00025	.00030	.00035	.00040	.00045	.00050
0.2	0.33	0.46	0.57	0.65	0.73	0.80	0.86	0.92	0.98	1.03
0.4	.52	.73	.90	1.04	1.16	1.27	1.37	1.47	1.56	1.64
0.6	.68	.96	1.18	1.36	1.52	1.66	1.80	1.92	2.04	2.15
0.8	.82	1.16	1.43	1.65	1.84	2.02	2.18	2.33	2.47	2.60
1.0	.96	1.35	1.65	1.91	2.14	2.34	2.53	2.70	2.87	3.02
1.2	1.08	1.53	1.87	2.16	2.41	2.64	2.85	3.05	3.24	3.41
1.4	1.20	1.69	2.07	2.39	2.67	2.93	3.16	3.38	3.59	3.78
1.6	1.31	1.85	2.26	2.61	2.92	3.20	3.46	3.70	3.92	4.13
1.8	1.41	2.00	2.45	2.83	3.16	3.46	3.74	4.00	4.24	4.47
2.0	1.52	2.14	2.63	3.03	3.39	3.71	4.01	4.29	4.55	4.80
2.2	1.62	2.29	2.80	3.23	3.61	3.96	4.28	4.57	4.85	5.11
2.4	1.71	2.42	2.97	3.42	3.83	4.19	4.53	4.84	5.14	5.41
2.6	1.81	2.55	3.13	3.61	4.04	4.42	4.78	5.11	5.42	5.71
2.8	1.90	2.68	3.29	3.80	4.24	4.65	5.02	5.37	5.69	6.00
3.0	1.99	2.81	3.44	3.97	4.44	4.87	5.26	5.62	5.96	6.28
3.2	2.07	2.93	3.59	4.15	4.64	5.08	5.49	5.87	6.22	6.56
3.4	2.16	3.05	3.74	4.32	4.83	5.29	5.71	6.11	6.48	6.83
3.6	2.24	3.17	3.89	4.49	5.02	5.50	5.94	6.35	6.73	7.10
3.8	2.33	3.29	4.03	4.65	5.20	5.70	6.15	6.58	6.98	7.36
4.0	2.41	3.40	4.17	4.81	5.38	5.90	6.37	6.81	7.22	7.61
4.2	2.49	3.52	4.31	4.97	5.56	6.09	6.58	7.03	7.46	7.86
4.4	2.56	3.63	4.44	5.13	5.74	6.28	6.79	7.25	7.69	8.11
4.6	2.64	3.74	4.58	5.28	5.91	6.47	6.99	7.47	7.93	8.36
4.8	2.72	3.84	4.71	5.44	6.08	6.66	7.19	7.69	8.15	8.60
5.0	2.79	3.95	4.84	5.59	6.25	6.84	7.39	7.90	8.38	8.83
5.2	2.87	4.05	4.97	5.73	6.41	7.02	7.59	8.11	8.60	9.07
5.4	2.94	4.16	5.09	5.88	6.57	7.20	7.78	8.32	8.82	9.30
5.6	3.01	4.26	5.22	6.02	6.74	7.38	7.97	8.52	9.04	9.53
5.8	3.08	4.36	5.34	6.17	6.90	7.55	8.16	8.72	9.25	9.75
6.0	3.15	4.46	5.46	6.31	7.05	7.73	8.35	8.92	9.46	9.97
6.2	3.22	4.56	5.58	6.45	7.21	7.90	8.53	9.12	9.67	10.19
6.4	3.29	4.66	5.70	6.59	7.36	8.07	8.71	9.31	9.88	10.41
6.6	3.36	4.75	5.82	6.72	7.52	8.23	8.89	9.51	10.08	10.63
6.8	3.43	4.85	5.94	6.86	7.67	8.40	9.07	9.70	10.29	10.84
7.0	3.50	4.94	6.05	6.99	7.82	8.56	9.25	9.89	10.49	11.05
7.5	3.66	5.18	6.34	7.32	8.18	8.97	9.68	10.35	10.98	11.57
8.0	3.82	5.40	6.62	7.64	8.54	9.36	10.11	10.81	11.46	12.08
8.5	3.98	5.63	6.89	7.96	8.90	9.75	10.53	11.25	11.94	12.58
9.0	4.13	5.85	7.16	8.27	9.24	10.12	10.94	11.69	12.40	13.07
9.5	4.28	6.06	7.42	8.57	9.58	10.50	11.34	12.12	12.85	13.55
10	4.43	6.27	7.68	8.87	9.91	10.86	11.73	12.54	13.30	14.02
11	4.72	6.68	8.18	9.45	10.56	11.57	12.50	13.36	14.17	14.94
12	5.01	7.08	8.67	10.01	11.20	12.26	13.25	14.16	15.02	15.83
13	5.28	7.47	9.15	10.56	11.81	12.94	13.97	14.94	15.84	16.70
14	5.55	7.85	9.61	11.10	12.41	13.59	14.68	15.69	16.65	17.55
15	5.81	8.22	10.06	11.62	12.99	14.23	15.37	16.43	17.43	18.37
16	6.07	8.58	10.51	12.13	13.56	14.86	16.05	17.16	18.20	19.18
17	6.32	8.93	10.94	12.63	14.12	15.47	16.71	17.86	18.95	19.97
18	6.56	9.28	11.36	13.12	14.67	16.07	17.36	18.56	19.68	20.75
19	6.80	9.62	11.78	13.60	15.21	16.66	18.00	19.24	20.40	21.51
20	7.04	9.95	12.19	14.08	15.74	17.24	18.62	19.91	21.11	22.26

Table 2.—*Velocity of water, v, in feet per second, based on Manning's formula* $v=(1.486/n)r^{2/3}s^{1/2}$, **n = .011**—Continued

r \ s	.00055	.00060	.00065	.00070	.00075	.00080	.00085	.00090	.00095	.00100
0.2	1.08	1.13	1.18	1.22	1.27	1.31	1.35	1.39	1.42	1.46
0.4	1.72	1.80	1.87	1.94	2.01	2.07	2.14	2.20	2.26	2.32
0.6	2.25	2.35	2.45	2.54	2.63	2.72	2.80	2.88	2.96	3.04
0.8	2.73	2.85	2.97	3.08	3.19	3.29	3.39	3.49	3.59	3.68
1.0	3.17	3.31	3.44	3.57	3.70	3.82	3.94	4.05	4.16	4.27
1.2	3.58	3.74	3.89	4.04	4.18	4.31	4.45	4.58	4.70	4.82
1.4	3.96	4.14	4.31	4.47	4.63	4.78	4.93	5.07	5.21	5.35
1.6	4.33	4.53	4.71	4.89	5.06	5.23	5.39	5.54	5.70	5.84
1.8	4.69	4.90	5.10	5.29	5.47	5.65	5.83	6.00	6.16	6.32
2.0	5.03	5.25	5.47	5.67	5.87	6.07	6.25	6.43	6.61	6.78
2.2	5.36	5.60	5.83	6.05	6.26	6.46	6.66	6.86	7.04	7.23
2.4	5.68	5.93	6.17	6.41	6.63	6.85	7.06	7.26	7.46	7.66
2.6	5.99	6.26	6.51	6.76	7.00	7.22	7.45	7.66	7.87	8.08
2.8	6.29	6.57	6.84	7.10	7.35	7.59	7.82	8.05	8.27	8.49
3.0	6.59	6.88	7.16	7.43	7.70	7.95	8.19	8.43	8.66	8.89
3.2	6.88	7.19	7.48	7.76	8.03	8.30	8.55	8.80	9.04	9.28
3.4	7.16	7.48	7.79	8.08	8.37	8.64	8.91	9.16	9.41	9.66
3.6	7.44	7.77	8.09	8.40	8.69	8.98	9.25	9.52	9.78	10.03
3.8	7.71	8.06	8.39	8.70	9.01	9.30	9.59	9.87	10.14	10.40
4.0	7.98	8.34	8.68	9.01	9.32	9.63	9.92	10.21	10.49	10.76
4.2	8.25	8.61	8.97	9.30	9.63	9.95	10.25	10.55	10.84	11.12
4.4	8.51	8.89	9.25	9.60	9.93	10.26	10.58	10.88	11.18	11.47
4.6	8.76	9.15	9.53	9.89	10.23	10.57	10.89	11.21	11.52	11.82
4.8	9.02	9.42	9.80	10.17	10.53	10.87	11.21	11.53	11.85	12.16
5.0	9.26	9.68	10.07	10.45	10.82	11.17	11.52	11.85	12.17	12.49
5.2	9.51	9.93	10.34	10.73	11.10	11.47	11.82	12.16	12.50	12.82
5.4	9.75	10.19	10.60	11.00	11.39	11.76	12.12	12.47	12.82	13.15
5.6	9.99	10.44	10.86	11.27	11.67	12.05	12.42	12.78	13.13	13.47
5.8	10.23	10.68	11.12	11.54	11.94	12.33	12.71	13.08	13.44	13.79
6.0	10.46	10.93	11.37	11.80	12.22	12.62	13.00	13.38	13.75	14.11
6.2	10.69	11.17	11.62	12.06	12.49	12.90	13.29	13.68	14.05	14.42
6.4	10.92	11.41	11.87	12.32	12.75	13.17	13.58	13.97	14.35	14.73
6.6	11.15	11.64	12.12	12.58	13.02	13.44	13.86	14.26	14.65	15.03
6.8	11.37	11.88	12.36	12.83	13.28	13.71	14.14	14.55	14.94	15.33
7.0	11.59	12.11	12.60	13.08	13.54	13.98	14.41	14.83	15.24	15.63
7.5	12.14	12.68	13.20	13.69	14.18	14.64	15.09	15.53	15.95	16.37
8.0	12.67	13.24	13.78	14.30	14.80	15.28	15.75	16.21	16.66	17.09
8.5	13.20	13.78	14.34	14.89	15.41	15.91	16.40	16.88	17.34	17.79
9.0	13.71	14.32	14.90	15.46	16.01	16.53	17.04	17.54	18.02	18.48
9.5	14.21	14.84	15.45	16.03	16.59	17.14	17.67	18.18	18.68	19.16
10	14.71	15.36	15.99	16.59	17.17	17.74	18.28	18.81	19.33	19.83
11	15.67	16.37	17.04	17.68	18.30	18.90	19.48	20.05	20.59	21.13
12	16.61	17.34	18.05	18.73	19.39	20.03	20.64	21.24	21.82	22.39
13	17.52	18.29	19.04	19.76	20.45	21.13	21.78	22.41	23.02	23.62
14	18.40	19.22	20.01	20.76	21.49	22.20	22.88	23.54	24.19	24.81
15	19.27	20.13	20.95	21.74	22.50	23.24	23.96	24.65	25.32	25.98
16	20.12	21.01	21.87	22.69	23.49	24.26	25.01	25.73	26.44	27.13
17	20.95	21.88	22.77	23.63	24.46	25.26	26.04	26.79	27.53	28.24
18	21.76	22.73	23.66	24.55	25.41	26.24	27.05	27.84	28.60	29.34
19	22.56	23.56	24.52	25.45	26.34	27.21	28.04	28.86	29.65	30.42
20	23.34	24.38	25.38	26.33	27.26	28.15	29.02	29.86	30.68	31.48

Table 2.—*Velocity of water, v, in feet per second, based on Manning's formula* $v = (1.486/n)r^{2/3}s^{1/2}$, **n = .011**—Continued

r \ s	.0010	.0015	.0020	.0025	.0030	.0035	.0040	.0045	.0050
0.2	1.46	1.79	2.07	2.31	2.53	2.73	2.92	3.10	3.27
0.4	2.32	2.84	3.28	3.67	4.02	4.34	4.64	4.92	5.19
0.6	3.04	3.72	4.30	4.81	5.26	5.69	6.08	6.45	6.80
0.8	3.68	4.51	5.21	5.82	6.38	6.89	7.36	7.81	8.23
1.0	4.27	5.23	6.04	6.75	7.40	7.99	8.54	9.06	9.55
1.2	4.82	5.91	6.82	7.63	8.36	9.03	9.65	10.23	10.79
1.4	5.35	6.55	7.56	8.45	9.26	10.00	10.69	11.34	11.95
1.6	5.84	7.16	8.26	9.24	10.12	10.93	11.69	12.40	13.07
1.8	6.32	7.74	8.94	9.99	10.95	11.83	12.64	13.41	14.13
2.0	6.78	8.31	9.59	10.72	11.75	12.69	13.56	14.39	15.16
2.2	7.23	8.85	10.22	11.43	12.52	13.52	14.45	15.33	16.16
2.4	7.66	9.38	10.83	12.11	13.26	14.33	15.32	16.24	17.12
2.6	8.08	9.89	11.42	12.77	13.99	15.11	16.15	17.13	18.06
2.8	8.49	10.39	12.00	13.42	14.70	15.88	16.97	18.00	18.98
3.0	8.89	10.88	12.57	14.05	15.39	16.62	17.77	18.85	19.87
3.2	9.28	11.36	13.12	14.67	16.07	17.36	18.55	19.68	20.74
3.4	9.66	11.83	13.66	15.27	16.73	18.07	19.32	20.49	21.60
3.6	10.03	12.29	14.19	15.87	17.38	18.77	20.07	21.29	22.44
3.8	10.40	12.74	14.71	16.45	18.02	19.46	20.81	22.07	23.26
4.0	10.76	13.18	15.22	17.02	18.64	20.14	21.53	22.84	24.07
4.2	11.12	13.62	15.73	17.58	19.26	20.80	22.24	23.59	24.87
4.4	11.47	14.05	16.22	18.14	19.87	21.46	22.94	24.33	25.65
4.6	11.82	14.47	16.71	18.68	20.47	22.11	23.63	25.07	26.42
4.8	12.16	14.89	17.19	19.22	21.05	22.74	24.31	25.79	27.18
5.0	12.49	15.30	17.67	19.75	21.64	23.37	24.98	26.50	27.93
5.2	12.82	15.70	18.13	20.27	22.21	23.99	25.64	27.20	28.67
5.4	13.15	16.10	18.60	20.79	22.77	24.60	26.30	27.89	29.40
5.6	13.47	16.50	19.05	21.30	23.33	25.20	26.94	28.58	30.12
5.8	13.79	16.89	19.50	21.80	23.89	25.80	27.58	29.25	30.84
6.0	14.11	17.28	19.95	22.30	24.43	26.39	28.21	29.92	31.54
6.2	14.42	17.66	20.39	22.80	24.97	26.97	28.83	30.58	32.24
6.4	14.73	18.04	20.83	23.28	25.51	27.55	29.45	31.24	32.93
6.6	15.03	18.41	21.26	23.77	26.03	28.12	30.06	31.89	33.61
6.8	15.33	18.78	21.68	24.24	26.56	28.69	30.67	32.53	34.29
7.0	15.63	19.15	22.11	24.72	27.08	29.25	31.26	33.16	34.96
7.5	16.37	20.05	23.15	25.88	28.35	30.62	32.74	34.72	36.60
8.0	17.09	20.93	24.17	27.02	29.60	31.97	34.18	36.25	38.21
8.5	17.79	21.79	25.16	28.13	30.82	33.29	35.59	37.74	39.79
9.0	18.48	22.64	26.14	29.23	32.01	34.58	36.97	39.21	41.33
9.5	19.16	23.47	27.10	30.30	33.19	35.85	38.32	40.65	42.85
10	19.83	24.29	28.04	31.35	34.34	37.10	39.66	42.06	44.34
11	21.13	25.88	29.88	33.41	36.60	39.53	42.26	44.82	47.25
12	22.39	27.42	31.67	35.40	38.78	41.89	44.78	47.50	50.07
13	23.62	28.93	33.40	37.34	40.91	44.19	47.24	50.10	52.81
14	24.81	30.39	35.09	39.24	42.98	46.42	49.63	52.64	55.49
15	25.98	31.82	36.75	41.08	45.00	48.61	51.97	55.12	58.10
16	27.13	33.22	38.36	42.89	46.98	50.75	54.25	57.54	60.65
17	28.24	34.59	39.94	44.66	48.92	52.84	56.49	59.91	63.16
18	29.34	35.94	41.49	46.39	50.82	54.89	58.68	62.24	65.61
19	30.42	37.25	43.02	48.09	52.69	56.91	60.84	64.53	68.02
20	31.48	38.55	44.51	49.77	54.52	58.89	62.95	66.77	70.38

Table 2.—*Velocity of water, v, in feet per second, based on Manning's formula* $v = (1.486/n) r^{2/3} s^{1/2}$, **n = .011**—Continued

r	.0055	.0060	.0065	.0070	.0075	.0080	.0085	.0090	.0095	.0100
0.2	3.43	3.58	3.72	3.87	4.00	4.13	4.26	4.38	4.50	4.62
0.4	5.44	5.68	5.91	6.14	6.35	6.56	6.76	6.96	7.15	7.33
0.6	7.13	7.44	7.75	8.04	8.32	8.60	8.86	9.12	9.37	9.61
0.8	8.63	9.02	9.39	9.74	10.08	10.41	10.73	11.04	11.35	11.64
1.0	10.02	10.46	10.89	11.30	11.70	12.08	12.45	12.82	13.17	13.51
1.2	11.31	11.82	12.30	12.76	13.21	13.64	14.06	14.47	14.87	15.26
1.4	12.54	13.10	13.63	14.14	14.64	15.12	15.59	16.04	16.48	16.91
1.6	13.71	14.31	14.90	15.46	16.00	16.53	17.04	17.53	18.01	18.48
1.8	14.82	15.48	16.12	16.72	17.31	17.88	18.43	18.96	19.48	19.99
2.0	15.90	16.61	17.29	17.94	18.57	19.18	19.77	20.34	20.90	21.44
2.2	16.95	17.70	18.42	19.12	19.79	20.44	21.07	21.68	22.27	22.85
2.4	17.96	18.76	19.52	20.26	20.97	21.66	22.33	22.97	23.60	24.22
2.6	18.94	19.79	20.59	21.37	22.12	22.85	23.55	24.23	24.90	25.54
2.8	19.90	20.79	21.64	22.45	23.24	24.00	24.74	25.46	26.16	26.84
3.0	20.84	21.77	22.65	23.51	24.34	25.13	25.91	26.66	27.39	28.10
3.2	21.76	22.72	23.65	24.54	25.41	26.24	27.05	27.83	28.59	29.34
3.4	22.65	23.66	24.63	25.56	26.45	27.32	28.16	28.98	29.77	30.55
3.6	23.53	24.58	25.58	26.55	27.48	28.38	29.26	30.10	30.93	31.73
3.8	24.40	25.48	26.52	27.52	28.49	29.42	30.33	31.21	32.06	32.90
4.0	25.25	26.37	27.44	28.48	29.48	30.45	31.38	32.29	33.18	34.04
4.2	26.08	27.24	28.35	29.42	30.45	31.45	32.42	33.36	34.28	35.17
4.4	26.90	28.10	29.24	30.35	31.41	32.44	33.44	34.41	35.36	36.27
4.6	27.71	28.94	30.12	31.26	32.36	33.42	34.45	35.45	36.42	37.36
4.8	28.51	29.78	30.99	32.16	33.29	34.38	35.44	36.47	37.47	38.44
5.0	29.29	30.60	31.85	33.05	34.21	35.33	36.42	37.47	38.50	39.50
5.2	30.07	31.41	32.69	33.92	35.11	36.27	37.38	38.47	39.52	40.55
5.4	30.84	32.21	33.52	34.79	36.01	37.19	38.34	39.45	40.53	41.58
5.6	31.59	33.00	34.35	35.64	36.89	38.10	39.28	40.41	41.52	42.60
5.8	32.34	33.78	35.16	36.49	37.77	39.01	40.21	41.37	42.50	43.61
6.0	33.08	34.55	35.96	37.32	38.63	39.90	41.12	42.32	43.48	44.61
6.2	33.81	35.32	36.76	38.14	39.48	40.78	42.03	43.25	44.44	45.59
6.4	34.54	36.07	37.54	38.96	40.33	41.65	42.93	44.18	45.39	46.57
6.6	35.25	36.82	38.32	39.77	41.16	42.51	43.82	45.09	46.33	47.53
6.8	35.96	37.56	39.09	40.57	41.99	43.37	44.70	46.00	47.26	48.49
7.0	36.66	38.29	39.85	41.36	42.81	44.22	45.58	46.90	48.18	49.43
7.5	38.39	40.09	41.73	43.31	44.83	46.30	47.72	49.10	50.45	51.76
8.0	40.07	41.86	43.57	45.21	46.80	48.33	49.82	51.26	52.67	54.04
8.5	41.73	43.58	45.36	47.07	48.73	50.32	51.87	53.38	54.84	56.27
9.0	43.35	45.28	47.12	48.90	50.62	52.28	53.89	55.45	56.97	58.45
9.5	44.94	46.94	48.85	50.70	52.48	54.20	55.87	57.49	59.06	60.60
10	46.50	48.57	50.55	52.46	54.30	56.08	57.81	59.49	61.12	62.70
11	49.55	51.76	53.87	55.90	57.87	59.76	61.60	63.39	65.13	66.82
12	52.51	54.85	57.09	59.24	61.32	63.33	65.28	67.17	69.01	70.81
13	55.39	57.85	60.22	62.49	64.68	66.80	68.86	70.86	72.80	74.69
14	58.20	60.78	63.27	65.65	67.96	70.19	72.35	74.44	76.48	78.47
15	60.94	63.64	66.24	68.74	71.16	73.49	75.75	77.95	80.08	82.17
16	63.61	66.44	69.16	71.77	74.29	76.72	79.08	81.38	83.61	85.78
17	66.24	69.18	72.01	74.73	77.35	79.89	82.34	84.73	87.05	89.32
18	68.81	71.87	74.81	77.63	80.35	82.99	85.54	88.02	90.43	92.78
19	71.34	74.51	77.55	80.48	83.30	86.03	88.68	91.25	93.75	96.19
20	73.82	77.10	80.25	83.28	86.20	89.03	91.77	94.43	97.02	99.54

Table 2.—*Velocity of water, v, in feet per second, based on Manning's formula* $v = (1.486/n)r^{2/3}s^{1/2}$, **n = .011**—Continued

r \ s	.01	.02	.03	.04	.05	.06	.07	.08	.09	.10
0.2	4.62	6.53	8.00	9.24	10.33	11.32	12.22	13.07	13.86	14.61
0.4	7.33	10.37	12.70	14.67	16.40	17.96	19.40	20.74	22.00	23.19
0.6	9.61	13.59	16.65	19.22	21.49	23.54	25.43	27.18	28.83	30.39
0.8	11.64	16.46	20.16	23.28	26.03	28.52	30.80	32.93	34.93	36.81
1.0	13.51	19.10	23.40	27.02	30.21	33.09	35.74	38.21	40.53	42.72
1.2	15.26	21.57	26.42	30.51	34.11	37.37	40.36	43.15	45.77	48.24
1.4	16.91	23.91	29.28	33.81	37.80	41.41	44.73	47.82	50.72	53.46
1.6	18.48	26.13	32.01	36.96	41.32	45.27	48.89	52.27	55.44	58.44
1.8	19.99	28.27	34.62	39.98	44.70	48.96	52.89	56.54	59.97	63.21
2.0	21.44	30.33	37.14	42.89	47.95	52.53	56.74	60.65	64.33	67.81
2.2	22.85	32.32	39.58	45.70	51.10	55.97	60.46	64.63	68.55	72.26
2.4	24.22	34.25	41.94	48.43	54.15	59.32	64.07	68.49	72.65	76.58
2.6	25.54	36.12	44.24	51.09	57.12	62.57	67.58	72.25	76.63	80.77
2.8	26.84	37.95	46.48	53.67	60.01	65.74	71.00	75.91	80.51	84.87
3.0	28.10	39.74	48.67	56.20	62.83	68.83	74.35	79.48	84.30	88.86
3.2	29.34	41.49	50.81	58.67	65.60	71.86	77.61	82.97	88.01	92.77
3.4	30.55	43.20	52.91	61.09	68.30	74.82	80.82	86.40	91.64	96.59
3.6	31.73	44.88	54.96	63.46	70.95	77.73	83.95	89.75	95.20	100.34
3.8	32.90	46.52	56.98	65.79	73.56	80.58	87.04	93.05	98.69	104.03
4.0	34.04	48.14	58.96	68.08	76.12	83.38	90.06	96.28	102.12	107.65
4.2	35.17	49.73	60.91	70.33	78.63	86.14	93.04	99.47	105.50	111.21
4.4	36.27	51.30	62.83	72.55	81.11	88.85	95.97	102.60	108.82	114.71
4.6	37.36	52.84	64.72	74.73	83.55	91.53	98.86	105.68	112.09	118.16
4.8	38.44	54.36	66.58	76.88	85.96	94.16	101.70	108.73	115.32	121.56
5.0	39.50	55.86	68.42	79.00	88.33	96.76	104.51	111.73	118.50	124.91
5.2	40.55	57.34	70.23	81.09	90.67	99.32	107.28	114.69	121.64	128.22
5.4	41.58	58.80	72.02	83.16	92.98	101.85	110.01	117.61	124.74	131.49
5.6	42.60	60.25	73.79	85.20	95.26	104.35	112.71	120.49	127.80	134.72
5.8	43.61	61.67	75.53	87.22	97.51	106.82	115.38	123.35	130.83	137.90
6.0	44.61	63.08	77.26	89.21	99.74	109.26	118.02	126.16	133.82	141.06
6.2	45.59	64.48	78.97	91.18	101.95	111.68	120.62	128.95	136.78	144.17
6.4	46.57	65.86	80.66	93.13	104.13	114.07	123.21	131.71	139.70	147.26
6.6	47.53	67.22	82.33	95.06	106.29	116.43	125.76	134.44	142.60	150.31
6.8	48.49	68.57	83.98	96.98	108.42	118.77	128.29	137.14	145.46	153.33
7.0	49.43	69.91	85.62	98.87	110.54	121.09	130.79	139.82	148.30	156.32
7.5	51.76	73.20	89.65	103.52	115.74	126.79	136.95	146.40	155.28	163.68
8.0	54.04	76.42	93.59	108.07	120.83	132.36	142.97	152.84	162.11	170.88
8.5	56.27	79.57	97.45	112.53	125.81	137.82	148.86	159.14	168.80	177.93
9.0	58.45	82.66	101.24	116.90	130.70	143.17	154.65	165.32	175.35	184.84
9.5	60.60	85.70	104.95	121.19	135.50	148.43	160.32	171.39	181.79	191.62
10	62.70	88.68	108.61	125.41	140.21	153.59	165.90	177.35	188.11	198.29
11	66.82	94.49	115.73	133.63	149.41	163.67	176.78	188.99	200.45	211.29
12	70.81	100.14	122.64	141.62	158.33	173.44	187.34	200.27	212.42	223.91
13	74.69	105.63	129.36	149.38	167.01	182.95	197.61	211.25	224.07	236.19
14	78.47	110.98	135.92	156.94	175.47	192.21	207.62	221.95	235.41	248.15
15	82.17	116.20	142.31	164.33	183.73	201.26	217.39	232.40	246.50	259.83
16	85.78	121.31	148.57	171.55	191.80	210.11	226.95	242.62	257.33	271.25
17	89.32	126.31	154.70	178.63	199.71	218.78	236.31	252.62	267.95	282.44
18	92.78	131.22	160.71	185.57	207.47	227.27	245.48	262.43	278.35	293.41
19	96.19	136.03	166.61	192.38	215.09	235.62	254.49	272.07	288.57	304.18
20	99.54	140.76	172.40	199.07	222.57	243.81	263.35	281.53	298.61	314.76

Table 3.—*Velocity of water, v, in feet per second, based on Manning's formula* $v = (1.486/n) r^{2/3} s^{1/2}$, **n = .012**

r \ s	.00005	.00010	.00015	.00020	.00025	.00030	.00035	.00040	.00045	.00050
0.2	0.30	0.42	0.52	0.60	0.67	0.73	0.79	0.85	0.90	0.95
0.4	.48	.67	.82	.95	1.06	1.16	1.26	1.34	1.43	1.50
0.6	.62	.88	1.08	1.25	1.39	1.53	1.65	1.76	1.87	1.97
0.8	.75	1.07	1.31	1.51	1.69	1.85	2.00	2.13	2.26	2.39
1.0	.88	1.24	1.52	1.75	1.96	2.14	2.32	2.48	2.63	2.77
1.2	.99	1.40	1.71	1.98	2.21	2.42	2.62	2.80	2.97	3.13
1.4	1.10	1.55	1.90	2.19	2.45	2.68	2.90	3.10	3.29	3.47
1.6	1.20	1.69	2.07	2.40	2.68	2.93	3.17	3.39	3.59	3.79
1.8	1.30	1.83	2.24	2.59	2.90	3.17	3.43	3.66	3.89	4.10
2.0	1.39	1.97	2.41	2.78	3.11	3.40	3.68	3.93	4.17	4.40
2.2	1.48	2.09	2.57	2.96	3.31	3.63	3.92	4.19	4.44	4.68
2.4	1.57	2.22	2.72	3.14	3.51	3.84	4.15	4.44	4.71	4.96
2.6	1.66	2.34	2.87	3.31	3.70	4.06	4.38	4.68	4.97	5.24
2.8	1.74	2.46	3.01	3.48	3.89	4.26	4.60	4.92	5.22	5.50
3.0	1.82	2.58	3.15	3.64	4.07	4.46	4.82	5.15	5.46	5.76
3.2	1.90	2.69	3.29	3.80	4.25	4.66	5.03	5.38	5.70	6.01
3.4	1.98	2.80	3.43	3.96	4.43	4.85	5.24	5.60	5.94	6.26
3.6	2.06	2.91	3.56	4.11	4.60	5.04	5.44	5.82	6.17	6.50
3.8	2.13	3.02	3.69	4.26	4.77	5.22	5.64	6.03	6.40	6.74
4.0	2.21	3.12	3.82	4.41	4.93	5.40	5.84	6.24	6.62	6.98
4.2	2.28	3.22	3.95	4.56	5.10	5.58	6.03	6.45	6.84	7.21
4.4	2.35	3.33	4.07	4.70	5.26	5.76	6.22	6.65	7.05	7.44
4.6	2.42	3.43	4.19	4.84	5.42	5.93	6.41	6.85	7.27	7.66
4.8	2.49	3.52	4.32	4.98	5.57	6.10	6.59	7.05	7.47	7.88
5.0	2.56	3.62	4.43	5.12	5.73	6.27	6.77	7.24	7.68	8.10
5.2	2.63	3.72	4.55	5.26	5.88	6.44	6.95	7.43	7.88	8.31
5.4	2.70	3.81	4.67	5.39	6.03	6.60	7.13	7.62	8.09	8.52
5.6	2.76	3.91	4.78	5.52	6.17	6.76	7.31	7.81	8.28	8.73
5.8	2.83	4.00	4.90	5.65	6.32	6.92	7.48	8.00	8.48	8.94
6.0	2.89	4.09	5.01	5.78	6.47	7.08	7.65	8.18	8.67	9.14
6.2	2.96	4.18	5.12	5.91	6.61	7.24	7.82	8.36	8.87	9.35
6.4	3.02	4.27	5.23	6.04	6.75	7.39	7.99	8.54	9.06	9.54
6.6	3.08	4.36	5.34	6.16	6.89	7.55	8.15	8.71	9.24	9.74
6.8	3.14	4.44	5.44	6.29	7.03	7.70	8.32	8.89	9.43	9.94
7.0	3.20	4.53	5.55	6.41	7.16	7.85	8.48	9.06	9.61	10.13
7.5	3.36	4.74	5.81	6.71	7.50	8.22	8.88	9.49	10.07	10.61
8.0	3.50	4.95	6.07	7.01	7.83	8.58	9.27	9.91	10.51	11.08
8.5	3.65	5.16	6.32	7.29	8.15	8.93	9.65	10.32	10.94	11.53
9.0	3.79	5.36	6.56	7.58	8.47	9.28	10.02	10.72	11.37	11.98
9.5	3.93	5.55	6.80	7.86	8.78	9.62	10.39	11.11	11.78	12.42
10	4.06	5.75	7.04	8.13	9.09	9.96	10.75	11.50	12.19	12.85
11	4.33	6.12	7.50	8.66	9.68	10.61	11.46	12.25	12.99	13.70
12	4.59	6.49	7.95	9.18	10.26	11.24	12.14	12.98	13.77	14.51
13	4.84	6.85	8.39	9.68	10.83	11.86	12.81	13.69	14.52	15.31
14	5.09	7.19	8.81	10.17	11.37	12.46	13.46	14.39	15.26	16.08
15	5.33	7.53	9.22	10.65	11.91	13.05	14.09	15.06	15.98	16.84
16	5.56	7.86	9.63	11.12	12.43	13.62	14.71	15.73	16.68	17.58
17	5.79	8.19	10.03	11.58	12.95	14.18	15.32	16.37	17.37	18.31
18	6.01	8.51	10.42	12.03	13.45	14.73	15.91	17.01	18.04	19.02
19	6.23	8.82	10.80	12.47	13.94	15.27	16.50	17.63	18.70	19.72
20	6.45	9.12	11.17	12.90	14.43	15.80	17.07	18.25	19.36	20.40

Table 3.—*Velocity of water, v, in feet per second, based on Manning's formula* $v = (1.486/n) r^{2/3} s^{1/2}$, **n = .012**—Continued

r \ s	.00055	.00060	.00065	.00070	.00075	.00080	.00085	.00090	.00095	.00100
0.2	0.99	1.04	1.08	1.12	1.16	1.20	1.23	1.27	1.31	1.34
0.4	1.58	1.65	1.71	1.78	1.84	1.90	1.96	2.02	2.07	2.13
0.6	2.07	2.16	2.25	2.33	2.41	2.49	2.57	2.64	2.72	2.79
0.8	2.50	2.61	2.72	2.82	2.92	3.02	3.11	3.20	3.29	3.37
1.0	2.90	3.03	3.16	3.28	3.39	3.50	3.61	3.72	3.82	3.92
1.2	3.28	3.43	3.57	3.70	3.83	3.96	4.08	4.20	4.31	4.42
1.4	3.63	3.80	3.95	4.10	4.24	4.38	4.52	4.65	4.78	4.90
1.6	3.97	4.15	4.32	4.48	4.64	4.79	4.94	5.08	5.22	5.36
1.8	4.30	4.49	4.67	4.85	5.02	5.18	5.34	5.50	5.65	5.79
2.0	4.61	4.82	5.01	5.20	5.38	5.56	5.73	5.90	6.06	6.22
2.2	4.91	5.13	5.34	5.54	5.74	5.92	6.11	6.28	6.46	6.62
2.4	5.21	5.44	5.66	5.87	6.08	6.28	6.47	6.66	6.84	7.02
2.6	5.49	5.74	5.97	6.19	6.41	6.62	6.83	7.02	7.22	7.40
2.8	5.77	6.03	6.27	6.51	6.74	6.96	7.17	7.38	7.58	7.78
3.0	6.04	6.31	6.57	6.82	7.05	7.29	7.51	7.73	7.94	8.15
3.2	6.31	6.59	6.86	7.11	7.36	7.61	7.84	8.07	8.29	8.50
3.4	6.57	6.86	7.14	7.41	7.67	7.92	8.16	8.40	8.63	8.85
3.6	6.82	7.12	7.42	7.70	7.97	8.23	8.48	8.73	8.97	9.20
3.8	7.07	7.39	7.69	7.98	8.26	8.53	8.79	9.05	9.29	9.54
4.0	7.32	7.64	7.96	8.26	8.55	8.83	9.10	9.36	9.62	9.87
4.2	7.56	7.90	8.22	8.53	8.83	9.12	9.40	9.67	9.94	10.19
4.4	7.80	8.14	8.48	8.80	9.11	9.40	9.69	9.98	10.25	10.51
4.6	8.03	8.39	8.73	9.06	9.38	9.69	9.99	10.28	10.56	10.83
4.8	8.26	8.63	8.98	9.32	9.65	9.97	10.27	10.57	10.86	11.14
5.0	8.49	8.87	9.23	9.58	9.92	10.24	10.56	10.86	11.16	11.45
5.2	8.72	9.10	9.48	9.83	10.18	10.51	10.84	11.15	11.46	11.75
5.4	8.94	9.34	9.72	10.08	10.44	10.78	11.11	11.43	11.75	12.05
5.6	9.16	9.57	9.96	10.33	10.69	11.05	11.39	11.72	12.04	12.35
5.8	9.37	9.79	10.19	10.58	10.95	11.31	11.65	11.99	12.32	12.64
6.0	9.59	10.02	10.42	10.82	11.20	11.57	11.92	12.27	12.60	12.93
6.2	9.80	10.24	10.66	11.06	11.45	11.82	12.18	12.54	12.88	13.22
6.4	10.01	10.46	10.88	11.29	11.69	12.07	12.45	12.81	13.16	13.50
6.6	10.22	10.67	11.11	11.53	11.93	12.32	12.70	13.07	13.43	13.78
6.8	10.42	10.89	11.33	11.76	12.17	12.57	12.96	13.33	13.70	14.06
7.0	10.63	11.10	11.55	11.99	12.41	12.82	13.21	13.59	13.97	14.33
7.5	11.13	11.62	12.10	12.55	12.99	13.42	13.83	14.23	14.62	15.00
8.0	11.62	12.13	12.63	13.11	13.57	14.01	14.44	14.86	15.27	15.66
8.5	12.10	12.63	13.15	13.65	14.12	14.59	15.04	15.47	15.90	16.31
9.0	12.57	13.12	13.66	14.18	14.67	15.15	15.62	16.07	16.51	16.94
9.5	13.03	13.61	14.16	14.70	15.21	15.71	16.19	16.66	17.12	17.57
10	13.48	14.08	14.65	15.21	15.74	16.26	16.76	17.24	17.72	18.18
11	14.36	15.00	15.62	16.20	16.77	17.32	17.86	18.37	18.88	19.37
12	15.22	15.90	16.55	17.17	17.78	18.36	18.92	19.47	20.01	20.53
13	16.06	16.77	17.46	18.11	18.75	19.36	19.96	20.54	21.10	21.65
14	16.87	17.62	18.34	19.03	19.70	20.35	20.97	21.58	22.17	22.75
15	17.66	18.45	19.20	19.93	20.63	21.30	21.96	22.60	23.21	23.82
16	18.44	19.26	20.05	20.80	21.53	22.24	22.92	23.59	24.24	24.86
17	19.20	20.05	20.87	21.66	22.42	23.16	23.87	24.56	25.23	25.89
18	19.95	20.83	21.68	22.50	23.29	24.06	24.80	25.52	26.21	26.90
19	20.68	21.60	22.48	23.33	24.15	24.94	25.71	26.45	27.18	27.88
20	21.40	22.35	23.26	24.14	24.99	25.81	26.60	27.37	28.12	28.85

Table 3.—*Velocity of water, v, in feet per second, based on Manning's formula* $v = (1.486/n)r^{2/3}s^{1/2}$, **n = .012**—Continued

r \ s	.0010	.0015	.0020	.0025	.0030	.0035	.0040	.0045	.0050
0.2	1.34	1.64	1.89	2.12	2.32	2.51	2.68	2.84	2.99
0.4	2.13	2.60	3.01	3.36	3.68	3.98	4.25	4.51	4.75
0.6	2.79	3.41	3.94	4.40	4.83	5.21	5.57	5.91	6.23
0.8	3.37	4.13	4.77	5.34	5.85	6.31	6.75	7.16	7.55
1.0	3.92	4.80	5.54	6.19	6.78	7.33	7.83	8.31	8.76
1.2	4.42	5.42	6.25	6.99	7.66	8.27	8.84	9.38	9.89
1.4	4.90	6.00	6.93	7.75	8.49	9.17	9.80	10.40	10.96
1.6	5.36	6.56	7.58	8.47	9.28	10.02	10.71	11.36	11.98
1.8	5.79	7.10	8.19	9.16	10.04	10.84	11.59	12.29	12.96
2.0	6.22	7.61	8.79	9.83	10.77	11.63	12.43	13.19	13.90
2.2	6.62	8.11	9.37	10.47	11.47	12.39	13.25	14.05	14.81
2.4	7.02	8.60	9.93	11.10	12.16	13.13	14.04	14.89	15.70
2.6	7.40	9.07	10.47	11.71	12.82	13.85	14.81	15.71	16.56
2.8	7.78	9.53	11.00	12.30	13.47	14.55	15.56	16.50	17.40
3.0	8.15	9.98	11.52	12.88	14.11	15.24	16.29	17.28	18.21
3.2	8.50	10.41	12.03	13.45	14.73	15.91	17.01	18.04	19.01
3.4	8.85	10.84	12.52	14.00	15.34	16.56	17.71	18.78	19.80
3.6	9.20	11.27	13.01	14.54	15.93	17.21	18.40	19.51	20.57
3.8	9.54	11.68	13.49	15.08	16.52	17.84	19.07	20.23	21.32
4.0	9.87	12.09	13.95	15.60	17.09	18.46	19.74	20.93	22.06
4.2	10.19	12.48	14.42	16.12	17.66	19.07	20.39	21.62	22.79
4.4	10.51	12.88	14.87	16.63	18.21	19.67	21.03	22.31	23.51
4.6	10.83	13.27	15.32	17.13	18.76	20.26	21.66	22.98	24.22
4.8	11.14	13.65	15.76	17.62	19.30	20.85	22.29	23.64	24.92
5.0	11.45	14.02	16.19	18.10	19.83	21.42	22.90	24.29	25.60
5.2	11.75	14.40	16.62	18.58	20.36	21.99	23.51	24.93	26.28
5.4	12.05	14.76	17.05	19.06	20.88	22.55	24.11	25.57	26.95
5.6	12.35	15.12	17.46	19.53	21.39	23.10	24.70	26.20	27.61
5.8	12.64	15.48	17.88	19.99	21.90	23.65	25.28	26.82	28.27
6.0	12.93	15.84	18.29	20.44	22.40	24.19	25.86	27.43	28.91
6.2	13.22	16.19	18.69	20.90	22.89	24.72	26.43	28.04	29.55
6.4	13.50	16.53	19.09	21.34	23.38	25.25	27.00	28.63	30.18
6.6	13.78	16.88	19.49	21.79	23.86	25.78	27.56	29.23	30.81
6.8	14.06	17.21	19.88	22.22	24.34	26.30	28.11	29.82	31.43
7.0	14.33	17.55	20.27	22.66	24.82	26.81	28.66	30.40	32.04
7.5	15.00	18.38	21.22	23.72	25.99	28.07	30.01	31.83	33.55
8.0	15.66	19.18	22.15	24.77	27.13	29.30	31.33	33.23	35.03
8.5	16.31	19.98	23.07	25.79	28.25	30.51	32.62	34.60	36.47
9.0	16.94	20.75	23.96	26.79	29.35	31.70	33.89	35.94	37.89
9.5	17.57	21.51	24.84	27.77	30.42	32.86	35.13	37.26	39.28
10	18.18	22.26	25.71	28.74	31.48	34.00	36.35	38.56	40.64
11	19.37	23.72	27.39	30.62	33.55	36.24	38.74	41.09	43.31
12	20.53	25.14	29.03	32.45	35.55	38.40	41.05	43.54	45.90
13	21.65	26.52	30.62	34.23	37.50	40.50	43.30	45.93	48.41
14	22.75	27.86	32.17	35.97	39.40	42.56	45.49	48.25	50.86
15	23.82	29.17	33.68	37.66	41.25	44.56	47.64	50.52	53.26
16	24.86	30.45	35.16	39.31	43.07	46.52	49.73	52.75	55.60
17	25.89	31.71	36.61	40.94	44.84	48.44	51.78	54.92	57.89
18	26.90	32.94	38.04	42.53	46.59	50.32	53.79	57.05	60.14
19	27.88	34.15	39.43	44.09	48.29	52.16	55.77	59.15	62.35
20	28.85	35.34	40.80	45.62	49.97	53.98	57.71	61.21	64.52

Table 3.—*Velocity of water, v, in feet per second, based on Manning's formula* $v = (1.486/n)r^{2/3}s^{1/2}$, **n = .012**—Continued

r \ s	.0055	.0060	.0065	.0070	.0075	.0080	.0085	.0090	.0095	.0100
0.2	3.14	3.28	3.41	3.54	3.67	3.79	3.90	4.02	4.13	4.24
0.4	4.99	5.21	5.42	5.62	5.82	6.01	6.20	6.38	6.55	6.72
0.6	6.53	6.82	7.10	7.37	7.63	7.88	8.12	8.36	8.59	8.81
0.8	7.91	8.27	8.60	8.93	9.24	9.54	9.84	10.12	10.40	10.67
1.0	9.18	9.59	9.98	10.36	10.72	11.08	11.42	11.75	12.07	12.38
1.2	10.37	10.83	11.27	11.70	12.11	12.51	12.89	13.27	13.63	13.98
1.4	11.49	12.00	12.49	12.97	13.42	13.86	14.29	14.70	15.10	15.50
1.6	12.56	13.12	13.66	14.17	14.67	15.15	15.62	16.07	16.51	16.94
1.8	13.59	14.19	14.77	15.33	15.87	16.39	16.89	17.38	17.86	18.32
2.0	14.58	15.23	15.85	16.45	17.02	17.58	18.12	18.65	19.16	19.66
2.2	15.53	16.23	16.89	17.53	18.14	18.74	19.31	19.87	20.42	20.95
2.4	16.46	17.19	17.90	18.57	19.22	19.85	20.47	21.06	21.64	22.20
2.6	17.36	18.14	18.88	19.59	20.28	20.94	21.59	22.21	22.82	23.41
2.8	18.24	19.06	19.83	20.58	21.30	22.00	22.68	23.34	23.98	24.60
3.0	19.10	19.95	20.77	21.55	22.31	23.04	23.75	24.44	25.11	25.76
3.2	19.94	20.83	21.68	22.50	23.29	24.05	24.79	25.51	26.21	26.89
3.4	20.77	21.69	22.57	23.43	24.25	25.04	25.81	26.56	27.29	28.00
3.6	21.57	22.53	23.45	24.34	25.19	26.02	26.82	27.59	28.35	29.09
3.8	22.36	23.36	24.31	25.23	26.12	26.97	27.80	28.61	29.39	30.16
4.0	23.14	24.17	25.16	26.11	27.02	27.91	28.77	29.60	30.41	31.20
4.2	23.91	24.97	25.99	26.97	27.92	28.83	29.72	30.58	31.42	32.24
4.4	24.66	25.76	26.81	27.82	28.80	29.74	30.66	31.54	32.41	33.25
4.6	25.40	26.53	27.61	28.66	29.66	30.64	31.58	32.49	33.38	34.25
4.8	26.13	27.29	28.41	29.48	30.52	31.52	32.49	33.43	34.34	35.24
5.0	26.85	28.05	29.19	30.29	31.36	32.39	33.38	34.35	35.29	36.21
5.2	27.56	28.79	29.97	31.10	32.19	33.24	34.27	35.26	36.23	37.17
5.4	28.27	29.52	30.73	31.89	33.01	34.09	35.14	36.16	37.15	38.12
5.6	28.96	30.25	31.48	32.67	33.82	34.93	36.00	37.05	38.06	39.05
5.8	29.65	30.96	32.23	33.45	34.62	35.75	36.86	37.92	38.96	39.98
6.0	30.32	31.67	32.97	34.21	35.41	36.57	37.70	38.79	39.85	40.89
6.2	30.99	32.37	33.69	34.97	36.19	37.38	38.53	39.65	40.73	41.79
6.4	31.66	33.06	34.41	35.71	36.97	38.18	39.36	40.50	41.61	42.69
6.6	32.31	33.75	35.13	36.45	37.73	38.97	40.17	41.34	42.47	43.57
6.8	32.96	34.43	35.83	37.19	38.49	39.75	40.98	42.17	43.32	44.45
7.0	33.61	35.10	36.53	37.91	39.24	40.53	41.78	42.99	44.17	45.31
7.5	35.19	36.75	38.25	39.70	41.09	42.44	43.74	45.01	46.25	47.45
8.0	36.73	38.37	39.94	41.44	42.90	44.30	45.67	46.99	48.28	49.53
8.5	38.25	39.95	41.58	43.15	44.67	46.13	47.55	48.93	50.27	51.58
9.0	39.74	41.50	43.20	44.83	46.40	47.92	49.40	50.83	52.22	53.58
9.5	41.19	43.03	44.78	46.47	48.10	49.68	51.21	52.70	54.14	55.55
10	42.63	44.52	46.34	48.09	49.78	51.41	52.99	54.53	56.02	57.48
11	45.42	47.44	49.38	51.24	53.04	54.78	56.47	58.11	59.70	61.25
12	48.14	50.28	52.33	54.31	56.21	58.05	59.84	61.58	63.26	64.91
13	50.77	53.03	55.20	57.28	59.29	61.24	63.12	64.95	66.73	68.46
14	53.35	55.72	57.99	60.18	62.30	64.34	66.32	68.24	70.11	71.93
15	55.86	58.34	60.72	63.02	65.23	67.37	69.44	71.45	73.41	75.32
16	58.31	60.91	63.39	65.79	68.09	70.33	72.49	74.59	76.64	78.63
17	60.72	63.42	66.01	68.50	70.90	73.23	75.48	77.67	79.80	81.87
18	63.08	65.88	68.57	71.16	73.66	76.07	78.41	80.69	82.90	85.05
19	65.39	68.30	71.09	73.77	76.36	78.87	81.29	83.65	85.94	88.17
20	67.67	70.68	73.56	76.34	79.02	81.61	84.12	86.56	88.93	91.24

Table 3.—*Velocity of water, v, in feet per second, based on Manning's formula* $v = (1.486/n)r^{2/3}s^{1/2}$, **n** = .012—Continued

r \ s	.01	.02	.03	.04	.05	.06	.07	.08	.09	.10
0.2	4.24	5.99	7.34	8.47	9.47	10.37	11.20	11.98	12.71	13.39
0.4	6.72	9.51	11.64	13.45	15.03	16.47	17.79	19.01	20.17	21.26
0.6	8.81	12.46	15.26	17.62	19.70	21.58	23.31	24.92	26.43	27.86
0.8	10.67	15.09	18.48	21.34	23.86	26.14	28.23	30.18	32.01	33.75
1.0	12.38	17.51	21.45	24.77	27.69	30.33	32.76	35.03	37.15	39.16
1.2	13.98	19.78	24.22	27.97	31.27	34.25	37.00	39.55	41.95	44.22
1.4	15.50	21.92	26.84	30.99	34.65	37.96	41.00	43.83	46.49	49.01
1.6	16.94	23.96	29.34	33.88	37.88	41.49	44.82	47.91	50.82	53.57
1.8	18.32	25.91	31.74	36.65	40.97	44.88	48.48	51.83	54.97	57.95
2.0	19.66	27.80	34.05	39.31	43.96	48.15	52.01	55.60	58.97	62.16
2.2	20.95	29.62	36.28	41.89	46.84	51.31	55.42	59.25	62.84	66.24
2.4	22.20	31.39	38.45	44.40	49.64	54.37	58.73	62.79	66.59	70.20
2.6	23.41	33.11	40.56	46.83	52.36	57.35	61.95	66.23	70.24	74.04
2.8	24.60	34.79	42.61	49.20	55.01	60.26	65.09	69.58	73.80	77.79
3.0	25.76	36.43	44.61	51.52	57.60	63.09	68.15	72.86	77.28	81.46
3.2	26.89	38.03	46.58	53.78	60.13	65.87	71.15	76.06	80.67	85.04
3.4	28.00	39.60	48.50	56.00	62.61	68.59	74.08	79.20	84.00	88.54
3.6	29.09	41.14	50.38	58.17	65.04	71.25	76.96	82.27	87.26	91.98
3.8	30.16	42.65	52.23	60.31	67.43	73.86	79.78	85.29	90.47	95.36
4.0	31.20	44.13	54.05	62.41	69.77	76.43	82.56	88.26	93.61	98.68
4.2	32.24	45.59	55.83	64.47	72.08	78.96	85.29	91.18	96.71	101.94
4.4	33.25	47.02	57.59	66.50	74.35	81.45	87.97	94.05	99.75	105.15
4.6	34.25	48.44	59.32	68.50	76.59	83.90	90.62	96.88	102.75	108.31
4.8	35.24	49.83	61.03	70.47	78.79	86.31	93.23	99.67	105.71	111.43
5.0	36.21	51.21	62.72	72.42	80.97	88.69	95.80	102.41	108.63	114.50
5.2	37.17	52.56	64.38	74.34	83.11	91.04	98.34	105.13	111.51	117.54
5.4	38.12	53.90	66.02	76.23	85.23	93.36	100.84	107.81	114.35	120.53
5.6	39.05	55.23	67.64	78.10	87.32	95.65	103.32	110.45	117.15	123.49
5.8	39.98	56.53	69.24	79.95	89.39	97.92	105.76	113.07	119.93	126.41
6.0	40.89	57.83	70.82	81.78	91.43	100.16	108.18	115.65	122.67	129.30
6.2	41.79	59.10	72.39	83.59	93.45	102.37	110.57	118.21	125.38	132.16
6.4	42.69	60.37	73.94	85.37	95.45	104.56	112.94	120.74	128.06	134.99
6.6	43.57	61.62	75.47	87.14	97.43	106.73	115.28	123.24	130.71	137.78
6.8	44.45	62.86	76.98	88.89	99.39	108.87	117.60	125.72	133.34	140.55
7.0	45.31	64.08	78.49	90.63	101.33	111.00	119.89	128.17	135.94	143.30
7.5	47.45	67.10	82.18	94.89	106.10	116.22	125.53	134.20	142.34	150.04
8.0	49.53	70.05	85.79	99.07	110.76	121.33	131.05	140.10	148.60	156.64
8.5	51.58	72.94	89.33	103.15	115.33	126.34	136.46	145.88	154.73	163.10
9.0	53.58	75.77	92.80	107.16	119.81	131.24	141.76	151.55	160.74	169.43
9.5	55.55	78.55	96.21	111.09	124.20	136.06	146.96	157.11	166.64	175.65
10	57.48	81.29	99.56	114.96	128.53	140.79	152.07	162.57	172.44	181.76
11	61.25	86.62	106.09	122.50	136.96	150.03	162.05	173.24	183.75	193.69
12	64.91	91.79	112.42	129.81	145.14	158.99	171.73	183.58	194.72	205.25
13	68.46	96.82	118.58	136.93	153.09	167.70	181.14	193.65	205.39	216.50
14	71.93	101.73	124.59	143.86	160.85	176.20	190.31	203.45	215.80	227.47
15	75.32	106.52	130.45	150.64	168.42	184.49	199.27	213.03	225.95	238.18
16	78.63	111.20	136.19	157.26	175.82	192.60	208.03	222.40	235.89	248.65
17	81.87	115.78	141.81	163.74	183.07	200.55	216.61	231.57	245.62	258.90
18	85.05	120.28	147.31	170.10	190.18	208.33	225.03	240.56	255.16	268.96
19	88.17	124.70	152.72	176.35	197.16	215.98	233.29	249.39	264.52	278.83
20	91.24	129.03	158.03	182.48	204.02	223.49	241.40	258.07	273.72	288.53

Table 4.—*Velocity of water, v, in feet per second, based on Manning's formula* $v=(1.486/n)r^{2/3}s^{1/2}$, **n = .013**

r	.00005	.00010	.00015	.00020	.00025	.00030	.00035	.00040	.00045	.00050
0.2	0.28	0.39	0.48	0.55	0.62	0.68	0.73	0.78	0.83	0.87
0.4	.44	.62	.76	.88	.98	1.07	1.16	1.24	1.32	1.39
0.6	.57	.81	1.00	1.15	1.29	1.41	1.52	1.63	1.72	1.82
0.8	.70	.99	1.21	1.39	1.56	1.71	1.84	1.97	2.09	2.20
1.0	.81	1.14	1.40	1.62	1.81	1.98	2.14	2.29	2.42	2.56
1.2	.91	1.29	1.58	1.83	2.04	2.24	2.41	2.58	2.74	2.89
1.4	1.01	1.43	1.75	2.02	2.26	2.48	2.68	2.86	3.03	3.20
1.6	1.11	1.56	1.92	2.21	2.47	2.71	2.93	3.13	3.32	3.50
1.8	1.20	1.69	2.07	2.39	2.67	2.93	3.16	3.38	3.59	3.78
2.0	1.28	1.81	2.22	2.57	2.87	3.14	3.39	3.63	3.85	4.06
2.2	1.37	1.93	2.37	2.73	3.06	3.35	3.62	3.87	4.10	4.32
2.4	1.45	2.05	2.51	2.90	3.24	3.55	3.83	4.10	4.35	4.58
2.6	1.53	2.16	2.65	3.06	3.42	3.74	4.04	4.32	4.58	4.83
2.8	1.61	2.27	2.78	3.21	3.59	3.93	4.25	4.54	4.82	5.08
3.0	1.68	2.38	2.91	3.36	3.76	4.12	4.45	4.76	5.04	5.32
3.2	1.76	2.48	3.04	3.51	3.92	4.30	4.64	4.96	5.27	5.55
3.4	1.83	2.58	3.17	3.66	4.09	4.48	4.84	5.17	5.48	5.78
3.6	1.90	2.68	3.29	3.80	4.25	4.65	5.02	5.37	5.70	6.00
3.8	1.97	2.78	3.41	3.94	4.40	4.82	5.21	5.57	5.90	6.22
4.0	2.04	2.88	3.53	4.07	4.55	4.99	5.39	5.76	6.11	6.44
4.2	2.10	2.98	3.64	4.21	4.70	5.15	5.57	5.95	6.31	6.65
4.4	2.17	3.07	3.76	4.34	4.85	5.32	5.74	6.14	6.51	6.86
4.6	2.24	3.16	3.87	4.47	5.00	5.48	5.91	6.32	6.71	7.07
4.8	2.30	3.25	3.98	4.60	5.14	5.63	6.09	6.51	6.90	7.27
5.0	2.36	3.34	4.09	4.73	5.28	5.79	6.25	6.68	7.09	7.47
5.2	2.43	3.43	4.20	4.85	5.42	5.94	6.42	6.86	7.28	7.67
5.4	2.49	3.52	4.31	4.98	5.56	6.09	6.58	7.04	7.46	7.87
5.6	2.55	3.60	4.41	5.10	5.70	6.24	6.74	7.21	7.65	8.06
5.8	2.61	3.69	4.52	5.22	5.83	6.39	6.90	7.38	7.83	8.25
6.0	2.67	3.77	4.62	5.34	5.97	6.54	7.06	7.55	8.01	8.44
6.2	2.73	3.86	4.72	5.46	6.10	6.68	7.22	7.72	8.18	8.63
6.4	2.79	3.94	4.83	5.57	6.23	6.82	7.37	7.88	8.36	8.81
6.6	2.84	4.02	4.93	5.69	6.36	6.97	7.52	8.04	8.53	8.99
6.8	2.90	4.10	5.02	5.80	6.49	7.11	7.68	8.21	8.70	9.17
7.0	2.96	4.18	5.12	5.92	6.61	7.24	7.83	8.37	8.87	9.35
7.5	3.10	4.38	5.36	6.19	6.92	7.59	8.19	8.76	9.29	9.79
8.0	3.23	4.57	5.60	6.47	7.23	7.92	8.55	9.14	9.70	10.22
8.5	3.37	4.76	5.83	6.73	7.53	8.25	8.91	9.52	10.10	10.65
9.0	3.50	4.95	6.06	6.99	7.82	8.57	9.25	9.89	10.49	11.06
9.5	3.63	5.13	6.28	7.25	8.11	8.88	9.59	10.25	10.88	11.47
10	3.75	5.31	6.50	7.50	8.39	9.19	9.93	10.61	11.26	11.86
11	4.00	5.65	6.92	8.00	8.94	9.79	10.58	11.31	11.99	12.64
12	4.24	5.99	7.34	8.47	9.47	10.38	11.21	11.98	12.71	13.40
13	4.47	6.32	7.74	8.94	9.99	10.95	11.82	12.64	13.41	14.13
14	4.70	6.64	8.13	9.39	10.50	11.50	12.42	13.28	14.09	14.85
15	4.92	6.95	8.51	9.83	10.99	12.04	13.01	13.90	14.75	15.55
16	5.13	7.26	8.89	10.26	11.48	12.57	13.58	14.52	15.40	16.23
17	5.34	7.56	9.26	10.69	11.95	13.09	14.14	15.11	16.03	16.90
18	5.55	7.85	9.62	11.10	12.41	13.60	14.69	15.70	16.65	17.56
19	5.76	8.14	9.97	11.51	12.87	14.10	15.23	16.28	17.27	18.20
20	5.96	8.42	10.32	11.91	13.32	14.59	15.76	16.84	17.87	18.83

Table 4.—*Velocity of water, v, in feet per second, based on Manning's formula* $v = (1.486/n)r^{2/3}s^{1/2}$, **n = .013**—Continued

r \ s	.00055	.00060	.00065	.00070	.00075	.00080	.00085	.00090	.00095	.00100
0.2	0.92	0.96	1.00	1.03	1.07	1.11	1.14	1.17	1.20	1.24
0.4	1.46	1.52	1.58	1.64	1.70	1.76	1.81	1.86	1.91	1.96
0.6	1.91	1.99	2.07	2.15	2.23	2.30	2.37	2.44	2.51	2.57
0.8	2.31	2.41	2.51	2.61	2.70	2.79	2.87	2.96	3.04	3.12
1.0	2.68	2.80	2.91	3.02	3.13	3.23	3.33	3.43	3.52	3.61
1.2	3.03	3.16	3.29	3.42	3.54	3.65	3.76	3.87	3.98	4.08
1.4	3.35	3.50	3.65	3.78	3.92	4.05	4.17	4.29	4.41	4.52
1.6	3.67	3.83	3.99	4.14	4.28	4.42	4.56	4.69	4.82	4.94
1.8	3.97	4.14	4.31	4.48	4.63	4.78	4.93	5.07	5.21	5.35
2.0	4.26	4.44	4.63	4.80	4.97	5.13	5.29	5.44	5.59	5.74
2.2	4.53	4.74	4.93	5.12	5.30	5.47	5.64	5.80	5.96	6.11
2.4	4.81	5.02	5.22	5.42	5.61	5.80	5.97	6.15	6.32	6.48
2.6	5.07	5.29	5.51	5.72	5.92	6.11	6.30	6.48	6.66	6.83
2.8	5.33	5.56	5.79	6.01	6.22	6.42	6.62	6.81	7.00	7.18
3.0	5.58	5.82	6.06	6.29	6.51	6.73	6.93	7.13	7.33	7.52
3.2	5.82	6.08	6.33	6.57	6.80	7.02	7.24	7.45	7.65	7.85
3.4	6.06	6.33	6.59	6.84	7.08	7.31	7.54	7.75	7.97	8.17
3.6	6.30	6.58	6.85	7.10	7.35	7.59	7.83	8.05	8.28	8.49
3.8	6.53	6.82	7.10	7.36	7.62	7.87	8.12	8.35	8.58	8.80
4.0	6.76	7.06	7.34	7.62	7.89	8.15	8.40	8.64	8.88	9.11
4.2	6.98	7.29	7.59	7.87	8.15	8.42	8.68	8.93	9.17	9.41
4.4	7.20	7.52	7.83	8.12	8.41	8.68	8.95	9.21	9.46	9.71
4.6	7.41	7.74	8.06	8.36	8.66	8.94	9.22	9.48	9.74	10.00
4.8	7.63	7.97	8.29	8.61	8.91	9.20	9.48	9.76	10.03	10.29
5.0	7.84	8.19	8.52	8.84	9.15	9.45	9.74	10.03	10.30	10.57
5.2	8.05	8.40	8.75	9.08	9.40	9.70	10.00	10.29	10.57	10.85
5.4	8.25	8.62	8.97	9.31	9.64	9.95	10.26	10.56	10.84	11.13
5.6	8.45	8.83	9.19	9.54	9.87	10.20	10.51	10.81	11.11	11.40
5.8	8.65	9.04	9.41	9.76	10.11	10.44	10.76	11.07	11.37	11.67
6.0	8.85	9.25	9.62	9.99	10.34	10.68	11.00	11.32	11.63	11.94
6.2	9.05	9.45	9.84	10.21	10.56	10.91	11.25	11.57	11.89	12.20
6.4	9.24	9.65	10.05	10.43	10.79	11.14	11.49	11.82	12.14	12.46
6.6	9.43	9.85	10.25	10.64	11.01	11.38	11.73	12.07	12.40	12.72
6.8	9.62	10.05	10.46	10.86	11.24	11.60	11.96	12.31	12.65	12.97
7.0	9.81	10.25	10.66	11.07	11.46	11.83	12.20	12.55	12.89	13.23
7.5	10.27	10.73	11.17	11.59	11.99	12.39	12.77	13.14	13.50	13.85
8.0	10.72	11.20	11.66	12.10	12.52	12.93	13.33	13.72	14.09	14.46
8.5	11.17	11.66	12.14	12.60	13.04	13.47	13.88	14.28	14.67	15.06
9.0	11.60	12.11	12.61	13.09	13.54	13.99	14.42	14.84	15.24	15.64
9.5	12.02	12.56	13.07	13.57	14.04	14.50	14.95	15.38	15.80	16.21
10	12.44	13.00	13.53	14.04	14.53	15.01	15.47	15.92	16.35	16.78
11	13.26	13.85	14.41	14.96	15.48	15.99	16.48	16.96	17.43	17.88
12	14.05	14.68	15.28	15.85	16.41	16.95	17.47	17.97	18.47	18.95
13	14.82	15.48	16.11	16.72	17.31	17.88	18.43	18.96	19.48	19.99
14	15.57	16.26	16.93	17.57	18.18	18.78	19.36	19.92	20.47	21.00
15	16.30	17.03	17.73	18.39	19.04	19.66	20.27	20.86	21.43	21.99
16	17.02	17.78	18.50	19.20	19.88	20.53	21.16	21.77	22.37	22.95
17	17.72	18.51	19.27	20.00	20.70	21.38	22.03	22.67	23.29	23.90
18	18.41	19.23	20.02	20.77	21.50	22.21	22.89	23.55	24.20	24.83
19	19.09	19.94	20.75	21.53	22.29	23.02	23.73	24.42	25.09	25.74
20	19.75	20.63	21.47	22.28	23.07	23.82	24.55	25.27	25.96	26.63

Table 4.—*Velocity of water, v, in feet per second, based on Manning's formula* $v = (1.486/n) r^{2/3} s^{1/2}$, **n = .013**—Continued

r	.0010	.0015	.0020	.0025	.0030	.0035	.0040	.0045	.0050
0.2	1.24	1.51	1.75	1.95	2.14	2.31	2.47	2.62	2.76
0.4	1.96	2.40	2.78	3.10	3.40	3.67	3.92	4.16	4.39
0.6	2.57	3.15	3.64	4.07	4.45	4.81	5.14	5.45	5.75
0.8	3.12	3.82	4.41	4.93	5.40	5.83	6.23	6.61	6.97
1.0	3.61	4.43	5.11	5.72	6.26	6.76	7.23	7.67	8.08
1.2	4.08	5.00	5.77	6.45	7.07	7.64	8.16	8.66	9.13
1.4	4.52	5.54	6.40	7.15	7.84	8.46	9.05	9.60	10.12
1.6	4.94	6.06	6.99	7.82	8.56	9.25	9.89	10.49	11.06
1.8	5.35	6.55	7.56	8.46	9.26	10.01	10.70	11.35	11.96
2.0	5.74	7.03	8.11	9.07	9.94	10.73	11.48	12.17	12.83
2.2	6.11	7.49	8.65	9.67	10.59	11.44	12.23	12.97	13.67
2.4	6.48	7.94	9.16	10.25	11.22	12.12	12.96	13.75	14.49
2.6	6.83	8.37	9.67	10.81	11.84	12.79	13.67	14.50	15.28
2.8	7.18	8.79	10.16	11.35	12.44	13.43	14.36	15.23	16.06
3.0	7.52	9.21	10.63	11.89	13.02	14.07	15.04	15.95	16.81
3.2	7.85	9.61	11.10	12.41	13.60	14.69	15.70	16.65	17.55
3.4	8.17	10.01	11.56	12.92	14.16	15.29	16.35	17.34	18.28
3.6	8.49	10.40	12.01	13.42	14.71	15.88	16.98	18.01	18.99
3.8	8.80	10.78	12.45	13.92	15.25	16.47	17.60	18.67	19.68
4.0	9.11	11.16	12.88	14.40	15.78	17.04	18.22	19.32	20.37
4.2	9.41	11.52	13.31	14.88	16.30	17.60	18.82	19.96	21.04
4.4	9.71	11.89	13.73	15.35	16.81	18.16	19.41	20.59	21.70
4.6	10.00	12.25	14.14	15.81	17.32	18.70	20.00	21.21	22.36
4.8	10.29	12.60	14.55	16.26	17.82	19.24	20.57	21.82	23.00
5.0	10.57	12.94	14.95	16.71	18.31	19.77	21.14	22.42	23.63
5.2	10.85	13.29	15.34	17.15	18.79	20.30	21.70	23.02	24.26
5.4	11.13	13.63	15.73	17.59	19.27	20.81	22.25	23.60	24.88
5.6	11.40	13.96	16.12	18.02	19.74	21.33	22.80	24.18	25.49
5.8	11.67	14.29	16.50	18.45	20.21	21.83	23.34	24.75	26.09
6.0	11.94	14.62	16.88	18.87	20.67	22.33	23.87	25.32	26.69
6.2	12.20	14.94	17.25	19.29	21.13	22.82	24.40	25.88	27.28
6.4	12.46	15.26	17.62	19.70	21.58	23.31	24.92	26.43	27.86
6.6	12.72	15.58	17.99	20.11	22.03	23.79	25.44	26.98	28.44
6.8	12.97	15.89	18.35	20.51	22.47	24.27	25.95	27.52	29.01
7.0	13.23	16.20	18.71	20.91	22.91	24.75	26.45	28.06	29.58
7.5	13.85	16.96	19.59	21.90	23.99	25.91	27.70	29.38	30.97
8.0	14.46	17.71	20.45	22.86	25.04	27.05	28.92	30.67	32.33
8.5	15.06	18.44	21.29	23.80	26.08	28.17	30.11	31.94	33.66
9.0	15.64	19.16	22.12	24.73	27.09	29.26	31.28	33.18	34.97
9.5	16.21	19.86	22.93	25.64	28.08	30.33	32.43	34.40	36.26
10	16.78	20.55	23.73	26.53	29.06	31.39	33.56	35.59	37.52
11	17.88	21.90	25.28	28.27	30.97	33.45	35.76	37.93	39.98
12	18.95	23.20	26.79	29.96	32.82	35.45	37.89	40.19	42.37
13	19.99	24.48	28.26	31.60	34.62	37.39	39.97	42.39	44.69
14	21.00	25.72	29.69	33.20	36.37	39.28	41.99	44.54	46.95
15	21.99	26.93	31.09	34.76	38.08	41.13	43.97	46.64	49.16
16	22.95	28.11	32.46	36.29	39.75	42.94	45.90	48.69	51.32
17	23.90	29.27	33.80	37.79	41.39	44.71	47.80	50.70	53.44
18	24.83	30.41	35.11	39.25	43.00	46.45	49.65	52.67	55.51
19	25.74	31.52	36.40	40.70	44.58	48.15	51.48	54.60	57.55
20	26.63	32.62	37.67	42.11	46.13	49.83	53.27	56.50	59.55

Table 4.—*Velocity of water, v, in feet per second, based on Manning's formula* $v=(1.486/n)r^{2/3}s^{1/2}$, **n = .013**—Continued

r \ s	.0055	.0060	.0065	.0070	.0075	.0080	.0085	.0090	.0095	.0100
0.2	2.90	3.03	3.15	3.27	3.39	3.50	3.60	3.71	3.81	3.91
0.4	4.60	4.81	5.00	5.19	5.37	5.55	5.72	5.89	6.05	6.21
0.6	6.03	6.30	6.56	6.80	7.04	7.27	7.50	7.71	7.93	8.13
0.8	7.31	7.63	7.94	8.24	8.53	8.81	9.08	9.35	9.60	9.85
1.0	8.48	8.85	9.22	9.56	9.90	10.22	10.54	10.84	11.14	11.43
1.2	9.57	10.00	10.41	10.80	11.18	11.55	11.90	12.25	12.58	12.91
1.4	10.61	11.08	11.53	11.97	12.39	12.79	13.19	13.57	13.94	14.31
1.6	11.60	12.11	12.61	13.08	13.54	13.99	14.42	14.83	15.24	15.64
1.8	12.54	13.10	13.64	14.15	14.65	15.13	15.59	16.05	16.49	16.91
2.0	13.46	14.06	14.63	15.18	15.71	16.23	16.73	17.21	17.69	18.15
2.2	14.34	14.98	15.59	16.18	16.75	17.29	17.83	18.34	18.85	19.34
2.4	15.20	15.87	16.52	17.14	17.75	18.33	18.89	19.44	19.97	20.49
2.6	16.03	16.74	17.43	18.08	18.72	19.33	19.93	20.50	21.07	21.61
2.8	16.84	17.59	18.31	19.00	19.67	20.31	20.94	21.54	22.13	22.71
3.0	17.63	18.42	19.17	19.89	20.59	21.27	21.92	22.56	23.17	23.78
3.2	18.41	19.23	20.01	20.77	21.50	22.20	22.89	23.55	24.19	24.82
3.4	19.17	20.02	20.84	21.62	22.38	23.12	23.83	24.52	25.19	25.85
3.6	19.91	20.80	21.65	22.46	23.25	24.02	24.75	25.47	26.17	26.85
3.8	20.64	21.56	22.44	23.29	24.11	24.90	25.66	26.41	27.13	27.84
4.0	21.36	22.31	23.22	24.10	24.94	25.76	26.56	27.33	28.07	28.80
4.2	22.07	23.05	23.99	24.90	25.77	26.61	27.43	28.23	29.00	29.76
4.4	22.76	23.77	24.75	25.68	26.58	27.45	28.30	29.12	29.92	30.69
4.6	23.45	24.49	25.49	26.45	27.38	28.28	29.15	29.99	30.82	31.62
4.8	24.12	25.19	26.22	27.21	28.17	29.09	29.99	30.86	31.70	32.53
5.0	24.79	25.89	26.95	27.96	28.95	29.90	30.82	31.71	32.58	33.42
5.2	25.44	26.58	27.66	28.71	29.71	30.69	31.63	32.55	33.44	34.31
5.4	26.09	27.25	28.37	29.44	30.47	31.47	32.44	33.38	34.29	35.18
5.6	26.73	27.92	29.06	30.16	31.22	32.24	33.23	34.20	35.13	36.04
5.8	27.37	28.58	29.75	30.87	31.96	33.00	34.02	35.01	35.97	36.90
6.0	27.99	29.24	30.43	31.58	32.69	33.76	34.80	35.81	36.79	37.74
6.2	28.61	29.88	31.10	32.28	33.41	34.50	35.57	36.60	37.60	38.58
6.4	29.22	30.52	31.77	32.97	34.12	35.24	36.33	37.38	38.41	39.40
6.6	29.83	31.15	32.43	33.65	34.83	35.97	37.08	38.16	39.20	40.22
6.8	30.43	31.78	33.08	34.33	35.53	36.70	37.83	38.92	39.99	41.03
7.0	31.02	32.40	33.72	35.00	36.22	37.41	38.56	39.68	40.77	41.83
7.5	32.48	33.93	35.31	36.64	37.93	39.17	40.38	41.55	42.69	43.80
8.0	33.91	35.42	36.86	38.25	39.60	40.90	42.15	43.38	44.57	45.72
8.5	35.31	36.88	38.38	39.83	41.23	42.58	43.89	45.17	46.40	47.61
9.0	36.68	38.31	39.87	41.38	42.83	44.24	45.60	46.92	48.21	49.46
9.5	38.03	39.72	41.34	42.90	44.40	45.86	47.27	48.64	49.98	51.27
10	39.35	41.10	42.78	44.39	45.95	47.46	48.92	50.33	51.71	53.06
11	41.93	43.79	45.58	47.30	48.96	50.57	52.13	53.64	55.11	56.54
12	44.43	46.41	48.30	50.13	51.89	53.59	55.24	56.84	58.40	59.91
13	46.87	48.95	50.95	52.88	54.73	56.53	58.27	59.96	61.60	63.20
14	49.24	51.43	53.53	55.55	57.50	59.39	61.22	62.99	64.72	66.40
15	51.56	53.85	56.05	58.17	60.21	62.18	64.10	65.96	67.76	69.52
16	53.83	56.22	58.52	60.73	62.86	64.92	66.92	68.86	70.74	72.58
17	56.05	58.54	60.93	63.23	65.45	67.60	69.68	71.70	73.66	75.57
18	58.22	60.81	63.30	65.69	67.99	70.22	72.38	74.48	76.52	78.51
19	60.36	63.05	65.62	68.10	70.49	72.80	75.04	77.21	79.33	81.39
20	62.46	65.24	67.90	70.47	72.94	75.33	77.65	79.90	82.09	84.22

Table 4.—*Velocity of water, v, in feet per second, based on Manning's formula* $v=(1.486/n)r^{2/3}s^{1/2}$, **n** $=.013$—Continued

r \ s	.01	.02	.03	.04	.05	.06	.07	.08	.09	.10
0.2	3.91	5.53	6.77	7.82	8.74	9.58	10.34	11.06	11.73	12.36
0.4	6.21	8.78	10.75	12.41	13.88	15.20	16.42	17.55	18.62	19.62
0.6	8.13	11.50	14.08	16.26	18.18	19.92	21.51	23.00	24.39	25.71
0.8	9.85	13.93	17.06	19.70	22.03	24.13	26.06	27.86	29.55	31.15
1.0	11.43	16.17	19.80	22.86	25.56	28.00	30.24	32.33	34.29	36.15
1.2	12.91	18.25	22.36	25.82	28.86	31.62	34.15	36.51	38.72	40.82
1.4	14.31	20.23	24.78	28.61	31.99	35.04	37.85	40.46	42.92	45.24
1.6	15.64	22.11	27.08	31.27	34.97	38.30	41.37	44.23	46.91	49.45
1.8	16.91	23.92	29.30	33.83	37.82	41.43	44.75	47.84	50.74	53.49
2.0	18.15	25.66	31.43	36.29	40.57	44.45	48.01	51.32	54.44	57.38
2.2	19.34	27.34	33.49	38.67	43.24	47.36	51.16	54.69	58.01	61.14
2.4	20.49	28.98	35.49	40.98	45.82	50.19	54.21	57.96	61.47	64.80
2.6	21.61	30.57	37.44	43.23	48.33	52.94	57.18	61.13	64.84	68.35
2.8	22.71	32.11	39.33	45.42	50.78	55.62	60.08	64.23	68.12	71.81
3.0	23.78	33.63	41.18	47.55	53.17	58.24	62.91	67.25	71.33	75.19
3.2	24.82	35.10	42.99	49.64	55.50	60.80	65.67	70.21	74.47	78.50
3.4	25.85	36.55	44.77	51.69	57.79	63.31	68.38	73.10	77.54	81.73
3.6	26.85	37.97	46.51	53.70	60.04	65.77	71.04	75.94	80.55	84.91
3.8	27.84	39.37	48.21	55.67	62.24	68.18	73.65	78.73	83.51	88.02
4.0	28.80	40.73	49.89	57.61	64.41	70.55	76.21	81.47	86.41	91.09
4.2	29.76	42.08	51.54	59.51	66.54	72.89	78.73	84.16	89.27	94.10
4.4	30.69	43.41	53.16	61.39	68.63	75.18	81.21	86.81	92.08	97.06
4.6	31.62	44.71	54.76	63.23	70.70	77.44	83.65	89.43	94.85	99.98
4.8	32.53	46.00	56.34	65.05	72.73	79.67	86.06	92.00	97.58	102.86
5.0	33.42	47.27	57.89	66.85	74.74	81.87	88.43	94.54	100.27	105.70
5.2	34.31	48.52	59.43	68.62	76.72	84.04	90.77	97.04	102.93	108.50
5.4	35.18	49.76	60.94	70.37	78.67	86.18	93.09	99.51	105.55	111.26
5.6	36.05	50.98	62.43	72.09	80.60	88.30	95.37	101.96	108.14	113.99
5.8	36.90	52.18	63.91	73.80	82.51	90.39	97.63	104.37	110.70	116.69
6.0	37.74	53.38	65.37	75.49	84.40	92.45	99.86	106.75	113.23	119.36
6.2	38.58	54.56	66.82	77.16	86.26	94.50	102.07	109.11	115.73	121.99
6.4	39.40	55.72	68.25	78.81	88.11	96.52	104.25	111.45	118.21	124.60
6.6	40.22	56.88	69.66	80.44	89.93	98.52	106.41	113.76	120.66	127.19
6.8	41.03	58.02	71.06	82.06	91.74	100.50	108.55	116.04	123.08	129.74
7.0	41.83	59.15	72.45	83.66	93.53	102.46	110.67	118.31	125.49	132.27
7.5	43.80	61.94	75.86	87.60	97.93	107.28	115.88	123.88	131.39	138.50
8.0	45.72	64.66	79.19	91.45	102.24	112.00	120.97	129.32	137.17	144.59
8.5	47.61	67.33	82.46	95.22	106.46	116.62	125.96	134.66	142.83	150.55
9.0	49.46	69.94	85.66	98.92	110.59	121.15	130.85	139.89	148.37	156.40
9.5	51.27	72.51	88.81	102.55	114.65	125.59	135.66	145.02	153.82	162.14
10	53.06	75.03	91.90	106.11	118.64	129.96	140.38	150.07	159.17	167.78
11	56.54	79.96	97.93	113.08	126.42	138.49	149.58	159.91	169.61	178.79
12	59.91	84.73	103.77	119.83	133.97	146.76	158.52	169.46	179.74	189.47
13	63.20	89.38	109.46	126.40	141.32	154.80	167.21	178.75	189.59	199.85
14	66.40	93.90	115.01	132.80	148.47	162.64	175.67	187.80	199.20	209.97
15	69.52	98.32	120.42	139.05	155.46	170.30	183.94	196.64	208.57	219.85
16	72.58	102.64	125.71	145.16	162.30	177.79	192.03	205.29	217.74	229.52
17	75.57	106.88	130.90	151.15	168.99	185.12	199.95	213.76	226.72	238.99
18	78.51	111.03	135.98	157.02	175.55	192.31	207.72	222.06	235.53	248.27
19	81.39	115.10	140.97	162.78	182.00	199.37	215.34	230.21	244.17	257.38
20	84.22	119.11	145.88	168.45	188.33	206.30	222.83	238.22	252.67	266.34

Table 5.—*Velocity of water, v, in feet per second, based on Manning's formula* $v = (1.486/n)r^{2/3}s^{1/2}$, **n = .014**

s / r	.00005	.00010	.00015	.00020	.00025	.00030	.00035	.00040	.00045	.00050
0.2	0.26	0.36	0.44	0.51	0.57	0.63	0.68	0.73	0.77	0.81
0.4	.41	.58	.71	.81	.91	1.00	1.08	1.15	1.22	1.29
0.6	.53	.76	.92	1.07	1.19	1.31	1.41	1.51	1.60	1.69
0.8	.65	.91	1.12	1.29	1.45	1.58	1.71	1.83	1.94	2.05
1.0	.75	1.06	1.30	1.50	1.68	1.84	1.99	2.12	2.25	2.37
1.2	.85	1.20	1.47	1.70	1.90	2.08	2.24	2.40	2.54	2.68
1.4	.94	1.33	1.63	1.88	2.10	2.30	2.49	2.66	2.82	2.97
1.6	1.03	1.45	1.78	2.05	2.30	2.51	2.72	2.90	3.08	3.25
1.8	1.11	1.57	1.92	2.22	2.48	2.72	2.94	3.14	3.33	3.51
2.0	1.19	1.68	2.06	2.38	2.66	2.92	3.15	3.37	3.57	3.77
2.2	1.27	1.80	2.20	2.54	2.84	3.11	3.36	3.59	3.81	4.01
2.4	1.35	1.90	2.33	2.69	3.01	3.30	3.56	3.81	4.04	4.25
2.6	1.42	2.01	2.46	2.84	3.17	3.48	3.75	4.01	4.26	4.49
2.8	1.49	2.11	2.58	2.98	3.33	3.65	3.94	4.22	4.47	4.71
3.0	1.56	2.21	2.70	3.12	3.49	3.82	4.13	4.42	4.68	4.94
3.2	1.63	2.30	2.82	3.26	3.64	3.99	4.31	4.61	4.89	5.15
3.4	1.70	2.40	2.94	3.39	3.79	4.16	4.49	4.80	5.09	5.37
3.6	1.76	2.49	3.05	3.53	3.94	4.32	4.66	4.99	5.29	5.57
3.8	1.83	2.58	3.17	3.66	4.09	4.48	4.84	5.17	5.48	5.78
4.0	1.89	2.67	3.28	3.78	4.23	4.63	5.00	5.35	5.67	5.98
4.2	1.95	2.76	3.38	3.91	4.37	4.79	5.17	5.53	5.86	6.18
4.4	2.02	2.85	3.49	4.03	4.51	4.94	5.33	5.70	6.05	6.37
4.6	2.08	2.94	3.60	4.15	4.64	5.08	5.49	5.87	6.23	6.56
4.8	2.14	3.02	3.70	4.27	4.78	5.23	5.65	6.04	6.41	6.75
5.0	2.19	3.10	3.80	4.39	4.91	5.38	5.81	6.21	6.58	6.94
5.2	2.25	3.19	3.90	4.51	5.04	5.52	5.96	6.37	6.76	7.12
5.4	2.31	3.27	4.00	4.62	5.17	5.66	6.11	6.53	6.93	7.31
5.6	2.37	3.35	4.10	4.73	5.29	5.80	6.26	6.69	7.10	7.48
5.8	2.42	3.43	4.20	4.85	5.42	5.93	6.41	6.85	7.27	7.66
6.0	2.48	3.50	4.29	4.96	5.54	6.07	6.56	7.01	7.43	7.84
6.2	2.53	3.58	4.39	5.07	5.66	6.20	6.70	7.16	7.60	8.01
6.4	2.59	3.66	4.48	5.17	5.79	6.34	6.85	7.32	7.76	8.18
6.6	2.64	3.73	4.57	5.28	5.91	6.47	6.99	7.47	7.92	8.35
6.8	2.69	3.81	4.67	5.39	6.02	6.60	7.13	7.62	8.08	8.52
7.0	2.75	3.88	4.76	5.49	6.14	6.73	7.27	7.77	8.24	8.69
7.5	2.88	4.07	4.98	5.75	6.43	7.04	7.61	8.13	8.63	9.09
8.0	3.00	4.25	5.20	6.00	6.71	7.35	7.94	8.49	9.01	9.49
8.5	3.13	4.42	5.41	6.25	6.99	7.66	8.27	8.84	9.38	9.89
9.0	3.25	4.59	5.62	6.49	7.26	7.95	8.59	9.19	9.74	10.27
9.5	3.37	4.76	5.83	6.73	7.53	8.25	8.91	9.52	10.10	10.65
10	3.48	4.93	6.03	6.97	7.79	8.53	9.22	9.85	10.45	11.02
11	3.71	5.25	6.43	7.42	8.30	9.09	9.82	10.50	11.14	11.74
12	3.93	5.56	6.81	7.87	8.80	9.64	10.41	11.13	11.80	12.44
13	4.15	5.87	7.19	8.30	9.28	10.16	10.98	11.74	12.45	13.12
14	4.36	6.17	7.55	8.72	9.75	10.68	11.53	12.33	13.08	13.79
15	4.56	6.46	7.91	9.13	10.21	11.18	12.08	12.91	13.69	14.44
16	4.77	6.74	8.25	9.53	10.66	11.67	12.61	13.48	14.30	15.07
17	4.96	7.02	8.59	9.92	11.10	12.15	13.13	14.04	14.89	15.69
18	5.15	7.29	8.93	10.31	11.53	12.63	13.64	14.58	15.46	16.30
19	5.34	7.56	9.26	10.69	11.95	13.09	14.14	15.12	16.03	16.90
20	5.53	7.82	9.58	11.06	12.37	13.55	14.63	15.64	16.59	17.49

Table 5.—*Velocity of water, v, in feet per second, based on Manning's formula* $v = (1.486/n)r^{2/3}s^{1/2}$, **n = .014**—Continued

s r	.00055	.00060	.00065	.00070	.00075	.00080	.00085	.00090	.00095	.00100
0.2	0.85	0.89	0.93	0.96	0.99	1.03	1.06	1.09	1.12	1.15
0.4	1.35	1.41	1.47	1.52	1.58	1.63	1.68	1.73	1.78	1.82
0.6	1.77	1.85	1.93	2.00	2.07	2.14	2.20	2.27	2.33	2.39
0.8	2.15	2.24	2.33	2.42	2.51	2.59	2.67	2.74	2.82	2.89
1.0	2.49	2.60	2.71	2.81	2.91	3.00	3.09	3.18	3.27	3.36
1.2	2.81	2.94	3.06	3.17	3.28	3.39	3.49	3.60	3.69	3.79
1.4	3.12	3.25	3.39	3.51	3.64	3.76	3.87	3.99	4.09	4.20
1.6	3.41	3.56	3.70	3.84	3.98	4.11	4.23	4.36	4.48	4.59
1.8	3.68	3.85	4.00	4.16	4.30	4.44	4.58	4.71	4.84	4.97
2.0	3.95	4.13	4.30	4.46	4.61	4.77	4.91	5.05	5.19	5.33
2.2	4.21	4.40	4.58	4.75	4.92	5.08	5.23	5.39	5.53	5.68
2.4	4.46	4.66	4.85	5.03	5.21	5.38	5.55	5.71	5.86	6.02
2.6	4.71	4.92	5.12	5.31	5.50	5.68	5.85	6.02	6.19	6.35
2.8	4.95	5.17	5.38	5.58	5.77	5.96	6.15	6.33	6.50	6.67
3.0	5.18	5.41	5.63	5.84	6.05	6.24	6.44	6.62	6.81	6.98
3.2	5.41	5.65	5.88	6.10	6.31	6.52	6.72	6.91	7.10	7.29
3.4	5.63	5.88	6.12	6.35	6.57	6.79	7.00	7.20	7.40	7.59
3.6	5.85	6.11	6.36	6.60	6.83	7.05	7.27	7.48	7.68	7.88
3.8	6.06	6.33	6.59	6.84	7.08	7.31	7.54	7.75	7.97	8.17
4.0	6.27	6.55	6.82	7.08	7.32	7.57	7.80	8.02	8.24	8.46
4.2	6.48	6.77	7.04	7.31	7.57	7.82	8.06	8.29	8.52	8.74
4.4	6.68	6.98	7.27	7.54	7.81	8.06	8.31	8.55	8.78	9.01
4.6	6.89	7.19	7.48	7.77	8.04	8.30	8.56	8.81	9.05	9.28
4.8	7.08	7.40	7.70	7.99	8.27	8.54	8.81	9.06	9.31	9.55
5.0	7.28	7.60	7.91	8.21	8.50	8.78	9.05	9.31	9.57	9.81
5.2	7.47	7.80	8.12	8.43	8.72	9.01	9.29	9.56	9.82	10.07
5.4	7.66	8.00	8.33	8.64	8.95	9.24	9.52	9.80	10.07	10.33
5.6	7.85	8.20	8.53	8.86	9.17	9.47	9.76	10.04	10.32	10.58
5.8	8.04	8.39	8.74	9.07	9.38	9.69	9.99	10.28	10.56	10.84
6.0	8.22	8.58	8.94	9.27	9.60	9.91	10.22	10.51	10.80	11.08
6.2	8.40	8.77	9.13	9.48	9.81	10.13	10.44	10.75	11.04	11.33
6.4	8.58	8.96	9.33	9.68	10.02	10.35	10.67	10.98	11.28	11.57
6.6	8.76	9.15	9.52	9.88	10.23	10.56	10.89	11.20	11.51	11.81
6.8	8.93	9.33	9.71	10.08	10.43	10.78	11.11	11.43	11.74	12.05
7.0	9.11	9.51	9.90	10.28	10.64	10.99	11.32	11.65	11.97	12.28
7.5	9.54	9.96	10.37	10.76	11.14	11.50	11.86	12.20	12.54	12.86
8.0	9.96	10.40	10.82	11.23	11.63	12.01	12.38	12.74	13.09	13.43
8.5	10.37	10.83	11.27	11.70	12.11	12.50	12.89	13.26	13.63	13.98
9.0	10.77	11.25	11.71	12.15	12.58	12.99	13.39	13.78	14.16	14.52
9.5	11.17	11.66	12.14	12.60	13.04	13.47	13.88	14.28	14.67	15.06
10	11.55	12.07	12.56	13.03	13.49	13.93	14.36	14.78	15.19	15.58
11	12.31	12.86	13.38	13.89	14.38	14.85	15.31	15.75	16.18	16.60
12	13.05	13.63	14.18	14.72	15.24	15.74	16.22	16.69	17.15	17.59
13	13.76	14.37	14.96	15.53	16.07	16.60	17.11	17.61	18.09	18.56
14	14.46	15.10	15.72	16.31	16.89	17.44	17.98	18.50	19.00	19.50
15	15.14	15.81	16.46	17.08	17.68	18.26	18.82	19.37	19.90	20.42
16	15.81	16.51	17.18	17.83	18.46	19.06	19.65	20.22	20.77	21.31
17	16.46	17.19	17.89	18.57	19.22	19.85	20.46	21.05	21.63	22.19
18	17.10	17.86	18.59	19.29	19.97	20.62	21.25	21.87	22.47	23.05
19	17.72	18.51	19.27	20.00	20.70	21.38	22.03	22.67	23.29	23.90
20	18.34	19.16	19.94	20.69	21.42	22.12	22.80	23.46	24.10	24.73

Table 5.—*Velocity of water, v, in feet per second, based on Manning's formula* $v = (1.486/n)r^{2/3}s^{1/2}$, **n = .014**—Continued

r \ s	.0010	.0015	.0020	.0025	.0030	.0035	.0040	.0045	.0050
0.2	1.15	1.41	1.62	1.82	1.99	2.15	2.30	2.44	2.57
0.4	1.82	2.23	2.58	2.88	3.16	3.41	3.64	3.87	4.07
0.6	2.39	2.92	3.38	3.78	4.14	4.47	4.78	5.07	5.34
0.8	2.89	3.54	4.09	4.57	5.01	5.41	5.79	6.14	6.47
1.0	3.36	4.11	4.75	5.31	5.81	6.28	6.71	7.12	7.51
1.2	3.79	4.64	5.36	5.99	6.57	7.09	7.58	8.04	8.48
1.4	4.20	5.14	5.94	6.64	7.28	7.86	8.40	8.91	9.39
1.6	4.59	5.62	6.49	7.26	7.95	8.59	9.18	9.74	10.27
1.8	4.97	6.08	7.02	7.85	8.60	9.29	9.93	10.54	11.11
2.0	5.33	6.53	7.54	8.42	9.23	9.97	10.66	11.30	11.91
2.2	5.68	6.95	8.03	8.98	9.83	10.62	11.36	12.04	12.70
2.4	6.02	7.37	8.51	9.51	10.42	11.26	12.03	12.76	13.45
2.6	6.35	7.77	8.98	10.03	10.99	11.87	12.69	13.46	14.19
2.8	6.67	8.17	9.43	10.54	11.55	12.47	13.34	14.14	14.91
3.0	6.98	8.55	9.87	11.04	12.09	13.06	13.96	14.81	15.61
3.2	7.29	8.93	10.31	11.52	12.62	13.64	14.58	15.46	16.30
3.4	7.59	9.30	10.73	12.00	13.15	14.20	15.18	16.10	16.97
3.6	7.88	9.66	11.15	12.47	13.66	14.75	15.77	16.72	17.63
3.8	8.17	10.01	11.56	12.92	14.16	15.29	16.35	17.34	18.28
4.0	8.46	10.36	11.96	13.37	14.65	15.82	16.92	17.94	18.91
4.2	8.74	10.70	12.36	13.82	15.13	16.35	17.48	18.54	19.54
4.4	9.01	11.03	12.75	14.25	15.61	16.86	18.03	19.12	20.15
4.6	9.28	11.37	13.13	14.68	16.08	17.37	18.57	19.69	20.76
4.8	9.55	11.70	13.51	15.10	16.54	17.87	19.10	20.26	21.36
5.0	9.81	12.02	13.88	15.52	17.00	18.36	19.63	20.82	21.95
5.2	10.07	12.34	14.25	15.93	17.45	18.85	20.15	21.37	22.53
5.4	10.33	12.65	14.61	16.34	17.89	19.33	20.66	21.92	23.10
5.6	10.58	12.96	14.97	16.74	18.33	19.80	21.17	22.45	23.67
5.8	10.84	13.27	15.32	17.13	18.77	20.27	21.67	22.99	24.23
6.0	11.08	13.57	15.67	17.52	19.20	20.73	22.17	23.51	24.78
6.2	11.33	13.87	16.02	17.91	19.62	21.19	22.66	24.03	25.33
6.4	11.57	14.17	16.36	18.29	20.04	21.65	23.14	24.54	25.87
6.6	11.81	14.46	16.70	18.67	20.46	22.09	23.62	25.05	26.41
6.8	12.05	14.76	17.04	19.05	20.87	22.54	24.09	25.56	26.94
7.0	12.28	15.04	17.37	19.42	21.27	22.98	24.57	26.06	27.46
7.5	12.86	15.75	18.19	20.33	22.28	24.06	25.72	27.28	28.76
8.0	13.43	16.44	18.99	21.23	23.25	25.12	26.85	28.48	30.02
8.5	13.98	17.13	19.77	22.10	24.21	26.15	27.96	29.66	31.26
9.0	14.52	17.79	20.54	22.96	25.15	27.17	29.05	30.81	32.47
9.5	15.06	18.44	21.29	23.81	26.08	28.17	30.11	31.94	33.67
10	15.58	19.08	22.03	24.63	26.98	29.15	31.16	33.05	34.84
11	16.60	20.33	23.48	26.25	28.75	31.06	33.20	35.22	37.12
12	17.59	21.55	24.88	27.82	30.47	32.91	35.19	37.32	39.34
13	18.56	22.73	26.24	29.34	32.14	34.72	37.12	39.37	41.50
14	19.50	23.88	27.57	30.83	33.77	36.48	38.99	41.36	43.60
15	20.42	25.00	28.87	32.28	35.36	38.19	40.83	43.31	45.65
16	21.31	26.10	30.14	33.70	36.91	39.87	42.63	45.21	47.66
17	22.19	27.18	31.38	35.09	38.44	41.52	44.38	47.08	49.62
18	23.05	28.23	32.60	36.45	39.93	43.13	46.11	48.90	51.55
19	23.90	29.27	33.80	37.79	41.40	44.71	47.80	50.70	53.44
20	24.73	30.29	34.98	39.10	42.84	46.27	49.46	52.46	55.30

Table 5.—*Velocity of water, v, in feet per second, based on Manning's formula* $v = (1.486/n) r^{2/3} s^{1/2}$, **n** = **.014**—Continued

s / r	.0055	.0060	.0065	.0070	.0075	.0080	.0085	.0090	.0095	.0100
0.2	2.69	2.81	2.93	3.04	3.14	3.25	3.35	3.44	3.54	3.63
0.4	4.27	4.46	4.65	4.82	4.99	5.15	5.31	5.47	5.62	5.76
0.6	5.60	5.85	6.09	6.32	6.54	6.75	6.96	7.16	7.36	7.55
0.8	6.78	7.09	7.37	7.65	7.92	8.18	8.43	8.68	8.92	9.15
1.0	7.87	8.22	8.56	8.88	9.19	9.49	9.79	10.07	10.35	10.61
1.2	8.89	9.28	9.66	10.03	10.38	10.72	11.05	11.37	11.68	11.99
1.4	9.85	10.29	10.71	11.11	11.50	11.88	12.25	12.60	12.95	13.28
1.6	10.77	11.25	11.71	12.15	12.57	12.99	13.39	13.78	14.15	14.52
1.8	11.65	12.17	12.66	13.14	13.60	14.05	14.48	14.90	15.31	15.71
2.0	12.50	13.05	13.58	14.10	14.59	15.07	15.53	15.98	16.42	16.85
2.2	13.32	13.91	14.48	15.02	15.55	16.06	16.55	17.03	17.50	17.95
2.4	14.11	14.74	15.34	15.92	16.48	17.02	17.54	18.05	18.54	19.03
2.6	14.88	15.55	16.18	16.79	17.38	17.95	18.50	19.04	19.56	20.07
2.8	15.64	16.33	17.00	17.64	18.26	18.86	19.44	20.00	20.55	21.09
3.0	16.37	17.10	17.80	18.47	19.12	19.75	20.36	20.95	21.52	22.08
3.2	17.09	17.85	18.58	19.28	19.96	20.62	21.25	21.87	22.47	23.05
3.4	17.80	18.59	19.35	20.08	20.78	21.47	22.13	22.77	23.39	24.00
3.6	18.49	19.31	20.10	20.86	21.59	22.30	22.99	23.65	24.30	24.93
3.8	19.17	20.02	20.84	21.63	22.38	23.12	23.83	24.52	25.19	25.85
4.0	19.84	20.72	21.56	22.38	23.16	23.92	24.66	25.37	26.00	26.75
4.2	20.49	21.40	22.28	23.12	23.93	24.71	25.47	26.21	26.93	27.63
4.4	21.14	22.08	22.98	23.85	24.68	25.49	26.28	27.04	27.78	28.50
4.6	21.77	22.74	23.67	24.56	25.42	26.26	27.07	27.85	28.61	29.36
4.8	22.40	23.40	24.35	25.27	26.16	27.01	27.85	28.65	29.44	30.20
5.0	23.02	24.04	25.02	25.97	26.88	27.76	28.61	29.44	30.25	31.04
5.2	23.63	24.68	25.69	26.65	27.59	28.50	29.37	30.22	31.05	31.86
5.4	24.23	25.31	26.34	27.33	28.29	29.22	30.12	30.99	31.84	32.67
5.6	24.82	25.93	26.99	28.00	28.99	29.94	30.86	31.75	32.62	33.47
5.8	25.41	26.54	27.62	28.67	29.67	30.65	31.59	32.51	33.40	34.26
6.0	25.99	27.15	28.26	29.32	30.35	31.35	32.31	33.25	34.16	35.05
6.2	26.57	27.75	28.88	29.97	31.02	32.04	33.03	33.98	34.92	35.82
6.4	27.13	28.34	29.50	30.61	31.69	32.73	33.73	34.71	35.66	36.59
6.6	27.70	28.93	30.11	31.25	32.34	33.40	34.43	35.43	36.40	37.35
6.8	28.25	29.51	30.72	31.87	32.99	34.08	35.12	36.14	37.13	38.10
7.0	28.81	30.09	31.31	32.50	33.64	34.74	35.81	36.85	37.85	38.84
7.5	30.16	31.50	32.79	34.03	35.22	36.38	37.50	38.58	39.64	40.67
8.0	31.49	32.89	34.23	35.52	36.77	37.97	39.14	40.28	41.38	42.46
8.5	32.79	34.24	35.64	36.99	38.29	39.54	40.76	41.94	43.09	44.21
9.0	34.06	35.57	37.03	38.42	39.77	41.08	42.34	43.57	44.76	45.93
9.5	35.31	36.88	38.39	39.83	41.23	42.58	43.90	45.17	46.41	47.61
10	36.54	38.16	39.72	41.22	42.67	44.07	45.42	46.74	48.02	49.27
11	38.93	40.67	42.33	43.92	45.47	46.96	48.40	49.81	51.17	52.50
12	41.26	43.09	44.85	46.55	48.18	49.76	51.29	52.78	54.23	55.63
13	43.52	45.46	47.31	49.10	50.82	52.49	54.10	55.67	57.20	58.68
14	45.73	47.76	49.71	51.59	53.40	55.15	56.84	58.49	60.09	61.66
15	47.88	50.01	52.05	54.01	55.91	57.74	59.52	61.25	62.92	64.56
16	49.98	52.21	54.34	56.39	58.37	60.28	62.14	63.94	65.69	67.40
17	52.04	54.36	56.58	58.71	60.77	62.77	64.70	66.58	68.40	70.18
18	54.07	56.47	58.78	60.99	63.13	65.21	67.21	69.16	71.06	72.90
19	56.05	58.54	60.93	63.23	65.45	67.60	69.68	71.70	73.66	75.58
20	58.00	60.58	63.05	65.43	67.73	69.95	72.10	74.19	76.23	78.21

Table 5.—*Velocity of water, v, in feet per second, based on Manning's formula* $v = (1.486/n)r^{2/3}s^{1/2}$, **n = .014**—Continued

r	.01	.02	.03	.04	.05	.06	.07	.08	.09	.10
0.2	3.63	5.13	6.29	7.26	8.12	8.89	9.60	10.27	10.89	11.48
0.4	5.76	8.15	9.98	11.52	12.88	14.11	15.25	16.30	17.29	18.22
0.6	7.55	10.68	13.08	15.10	16.88	18.50	19.98	21.36	22.65	23.88
0.8	9.15	12.94	15.84	18.29	20.45	22.41	24.20	25.87	27.44	28.93
1.0	10.61	15.01	18.38	21.23	23.73	26.00	28.08	30.02	31.84	33.57
1.2	11.99	16.95	20.76	23.97	26.80	29.36	31.71	33.90	35.96	37.90
1.4	13.28	18.79	23.01	26.57	29.70	32.54	35.14	37.57	39.85	42.01
1.6	14.52	20.53	25.15	29.04	32.47	35.57	38.42	41.07	43.56	45.92
1.8	15.71	22.21	27.20	31.41	35.12	38.47	41.55	44.42	47.12	49.67
2.0	16.85	23.83	29.18	33.70	37.68	41.27	44.58	47.66	50.55	53.28
2.2	17.95	25.39	31.10	35.91	40.15	43.98	47.50	50.78	53.86	56.78
2.4	19.03	26.91	32.96	38.05	42.55	46.61	50.34	53.82	57.08	60.17
2.6	20.07	28.38	34.76	40.14	44.88	49.16	53.10	56.77	60.21	63.47
2.8	21.09	29.82	36.52	42.17	47.15	51.65	55.79	59.64	63.26	66.68
3.0	22.08	31.22	38.24	44.16	49.37	54.08	58.41	62.45	66.24	69.82
3.2	23.05	32.60	39.92	46.10	51.54	56.46	60.98	65.19	69.15	72.89
3.4	24.00	33.94	41.57	48.00	53.67	58.79	63.50	67.88	72.00	75.89
3.6	24.93	35.26	43.18	49.86	55.75	61.07	65.96	70.52	74.80	78.84
3.8	25.85	36.55	44.77	51.69	57.80	63.31	68.39	73.11	77.54	81.74
4.0	26.75	37.83	46.33	53.49	59.81	65.51	70.76	75.65	80.24	84.58
4.2	27.63	39.08	47.86	55.26	61.78	67.68	73.10	78.15	82.89	87.38
4.4	28.50	40.31	49.37	57.00	63.73	69.81	75.41	80.61	85.50	90.13
4.6	29.36	41.52	50.85	58.72	65.65	71.91	77.67	83.04	88.07	92.84
4.8	30.20	42.71	52.31	60.41	67.54	73.98	79.91	85.43	90.61	95.51
5.0	31.04	43.89	53.76	62.07	69.40	76.02	82.11	87.78	93.11	98.15
5.2	31.86	45.05	55.18	63.72	71.24	78.04	84.29	90.11	95.58	100.75
5.4	32.67	46.20	56.59	65.34	73.05	80.03	86.44	92.41	98.01	103.31
5.6	33.47	47.34	57.98	66.94	74.85	81.99	88.56	94.67	100.42	105.85
5.8	34.26	48.46	59.35	68.53	76.62	83.93	90.66	96.91	102.79	108.35
6.0	35.05	49.56	60.70	70.10	78.37	85.85	92.73	99.13	105.14	110.83
6.2	35.82	50.66	62.05	71.64	80.10	87.75	94.78	101.32	107.47	113.28
6.4	36.59	51.74	63.37	73.18	81.81	89.62	96.80	103.49	109.77	115.70
6.6	37.35	52.82	64.69	74.69	83.51	91.48	98.81	105.63	112.04	118.10
6.8	38.10	53.88	65.99	76.20	85.19	93.32	100.80	107.76	114.29	120.47
7.0	38.84	54.93	67.27	77.68	86.85	95.14	102.76	109.86	116.52	122.83
7.5	40.67	57.51	70.44	81.34	90.94	99.62	107.60	115.03	122.01	128.61
8.0	42.46	60.04	73.54	84.91	94.94	104.00	112.33	120.09	127.37	134.26
8.5	44.21	62.52	76.57	88.42	98.85	108.29	116.96	125.04	132.62	139.80
9.0	45.93	64.95	79.55	91.85	102.69	112.49	121.51	129.90	137.78	145.23
9.5	47.61	67.33	82.46	95.22	106.46	116.62	125.97	134.66	142.83	150.56
10	49.27	69.67	85.33	98.53	110.16	120.68	130.35	139.35	147.80	155.80
11	52.50	74.25	90.93	105.00	117.39	128.60	138.90	148.49	157.50	166.02
12	55.63	78.68	96.36	111.27	124.40	136.28	147.20	157.36	166.90	175.93
13	58.68	82.99	101.64	117.37	131.22	143.75	155.26	165.98	176.05	185.58
14	61.66	87.19	106.79	123.31	137.87	151.03	163.13	174.39	184.97	194.97
15	64.56	91.30	111.82	129.12	144.36	158.13	170.81	182.60	193.67	204.15
16	67.40	95.31	116.73	134.79	150.70	165.09	178.31	190.63	202.19	213.13
17	70.18	99.24	121.55	140.35	156.92	171.90	185.67	198.49	210.53	221.92
18	72.90	103.10	126.27	145.80	163.01	178.57	192.88	206.20	218.71	230.54
19	75.58	106.88	130.90	151.16	169.00	185.13	199.96	213.77	226.73	239.00
20	78.21	110.60	135.46	156.41	174.88	191.57	206.92	221.20	234.62	247.31

Table 6.—*Velocity of water, v, in feet per second, based on Manning's formula* $v = (1.486/n)r^{2/3}s^{1/2}$, **n = .015**

r \ s	.00005	.00010	.00015	.00020	.00025	.00030	.00035	.00040	.00045	.00050
0.2	0.24	0.34	0.41	0.48	0.54	0.59	0.63	0.68	0.72	0.76
0.4	.38	.54	.66	.76	.85	.93	1.01	1.08	1.14	1.20
0.6	.50	.70	.86	1.00	1.11	1.22	1.32	1.41	1.49	1.58
0.8	.60	.85	1.05	1.21	1.35	1.48	1.60	1.71	1.81	1.91
1.0	.70	.99	1.21	1.40	1.57	1.72	1.85	1.98	2.10	2.22
1.2	.79	1.12	1.37	1.58	1.77	1.94	2.09	2.24	2.37	2.50
1.4	.88	1.24	1.52	1.75	1.96	2.15	2.32	2.48	2.63	2.77
1.6	.96	1.36	1.66	1.92	2.14	2.35	2.54	2.71	2.87	3.03
1.8	1.04	1.47	1.80	2.07	2.32	2.54	2.74	2.93	3.11	3.28
2.0	1.11	1.57	1.93	2.22	2.49	2.72	2.94	3.15	3.34	3.52
2.2	1.18	1.68	2.05	2.37	2.65	2.90	3.14	3.35	3.55	3.75
2.4	1.26	1.78	2.17	2.51	2.81	3.08	3.32	3.55	3.77	3.97
2.6	1.32	1.87	2.29	2.65	2.96	3.24	3.50	3.75	3.97	4.19
2.8	1.39	1.97	2.41	2.78	3.11	3.41	3.68	3.94	4.17	4.40
3.0	1.46	2.06	2.52	2.91	3.26	3.57	3.86	4.12	4.37	4.61
3.2	1.52	2.15	2.63	3.04	3.40	3.73	4.02	4.30	4.56	4.81
3.4	1.58	2.24	2.74	3.17	3.54	3.88	4.19	4.48	4.75	5.01
3.6	1.65	2.33	2.85	3.29	3.68	4.03	4.35	4.65	4.94	5.20
3.8	1.71	2.41	2.95	3.41	3.81	4.18	4.51	4.82	5.12	5.39
4.0	1.77	2.50	3.06	3.53	3.95	4.32	4.67	4.99	5.30	5.58
4.2	1.82	2.58	3.16	3.65	4.08	4.47	4.82	5.16	5.47	5.77
4.4	1.88	2.66	3.26	3.76	4.21	4.61	4.98	5.32	5.64	5.95
4.6	1.94	2.74	3.36	3.88	4.33	4.75	5.13	5.48	5.81	6.13
4.8	1.99	2.82	3.45	3.99	4.46	4.88	5.27	5.64	5.98	6.30
5.0	2.05	2.90	3.55	4.10	4.58	5.02	5.42	5.79	6.14	6.48
5.2	2.10	2.97	3.64	4.21	4.70	5.15	5.56	5.95	6.31	6.65
5.4	2.16	3.05	3.73	4.31	4.82	5.28	5.70	6.10	6.47	6.82
5.6	2.21	3.12	3.83	4.42	4.94	5.41	5.84	6.25	6.63	6.99
5.8	2.26	3.20	3.92	4.52	5.06	5.54	5.98	6.40	6.78	7.15
6.0	2.31	3.27	4.01	4.63	5.17	5.67	6.12	6.54	6.94	7.31
6.2	2.36	3.34	4.09	4.73	5.29	5.79	6.25	6.69	7.09	7.48
6.4	2.41	3.41	4.18	4.83	5.40	5.91	6.39	6.83	7.24	7.64
6.6	2.46	3.49	4.27	4.93	5.51	6.04	6.52	6.97	7.39	7.79
6.8	2.51	3.56	4.35	5.03	5.62	6.16	6.65	7.11	7.54	7.95
7.0	2.56	3.63	4.44	5.13	5.73	6.28	6.78	7.25	7.69	8.11
7.5	2.68	3.80	4.65	5.37	6.00	6.57	7.10	7.59	8.05	8.49
8.0	2.80	3.96	4.85	5.60	6.27	6.86	7.41	7.93	8.41	8.86
8.5	2.92	4.13	5.05	5.84	6.52	7.15	7.72	8.25	8.75	9.23
9.0	3.03	4.29	5.25	6.06	6.78	7.42	8.02	8.57	9.09	9.58
9.5	3.14	4.44	5.44	6.28	7.03	7.70	8.31	8.89	9.43	9.94
10	3.25	4.60	5.63	6.50	7.27	7.96	8.60	9.20	9.75	10.28
11	3.46	4.90	6.00	6.93	7.75	8.49	9.17	9.80	10.39	10.96
12	3.67	5.19	6.36	7.34	8.21	8.99	9.71	10.39	11.02	11.61
13	3.87	5.48	6.71	7.75	8.66	9.49	10.25	10.95	11.62	12.25
14	4.07	5.75	7.05	8.14	9.10	9.97	10.77	11.51	12.21	12.87
15	4.26	6.03	7.38	8.52	9.53	10.44	11.27	12.05	12.78	13.47
16	4.45	6.29	7.70	8.90	9.95	10.90	11.77	12.58	13.34	14.07
17	4.63	6.55	8.02	9.26	10.36	11.34	12.25	13.10	13.89	14.65
18	4.81	6.80	8.33	9.62	10.76	11.79	12.73	13.61	14.43	15.21
19	4.99	7.05	8.64	9.98	11.15	12.22	13.20	14.11	14.96	15.77
20	5.16	7.30	8.94	10.32	11.54	12.64	13.66	14.60	15.48	16.32

Table 6.—*Velocity of water, v, in feet per second, based on Manning's formula* $v = (1.486/n)r^{2/3}s^{1/2}$, **n = .015**—Continued

r / s	.00055	.00060	.00065	.00070	.00075	.00080	.00085	.00090	.00095	.00100
0.2	0.79	0.83	0.86	0.90	0.93	0.96	0.99	1.02	1.04	1.07
0.4	1.26	1.32	1.37	1.42	1.47	1.52	1.57	1.61	1.66	1.70
0.6	1.65	1.73	1.80	1.86	1.93	1.99	2.05	2.11	2.17	2.23
0.8	2.00	2.09	2.18	2.26	2.34	2.41	2.49	2.56	2.63	2.70
1.0	2.32	2.43	2.53	2.62	2.71	2.80	2.89	2.97	3.05	3.13
1.2	2.62	2.74	2.85	2.96	3.06	3.16	3.26	3.36	3.45	3.54
1.4	2.91	3.04	3.16	3.28	3.40	3.51	3.61	3.72	3.82	3.92
1.6	3.18	3.32	3.46	3.59	3.71	3.83	3.95	4.07	4.18	4.29
1.8	3.44	3.59	3.74	3.88	4.01	4.15	4.27	4.40	4.52	4.64
2.0	3.69	3.85	4.01	4.16	4.31	4.45	4.58	4.72	4.85	4.97
2.2	3.93	4.10	4.27	4.43	4.59	4.74	4.89	5.03	5.17	5.30
2.4	4.16	4.35	4.53	4.70	4.86	5.02	5.18	5.33	5.47	5.62
2.6	4.39	4.59	4.78	4.96	5.13	5.30	5.46	5.62	5.77	5.92
2.8	4.62	4.82	5.02	5.21	5.39	5.57	5.74	5.90	6.07	6.22
3.0	4.83	5.05	5.25	5.45	5.64	5.83	6.01	6.18	6.35	6.52
3.2	5.05	5.27	5.48	5.69	5.89	6.08	6.27	6.45	6.63	6.80
3.4	5.25	5.49	5.71	5.93	6.13	6.34	6.53	6.72	6.90	7.08
3.6	5.46	5.70	5.93	6.16	6.37	6.58	6.78	6.98	7.17	7.36
3.8	5.66	5.91	6.15	6.38	6.61	6.82	7.03	7.24	7.44	7.63
4.0	5.85	6.11	6.36	6.60	6.84	7.06	7.28	7.49	7.69	7.89
4.2	6.05	6.32	6.57	6.82	7.06	7.29	7.52	7.74	7.95	8.16
4.4	6.24	6.52	6.78	7.04	7.28	7.52	7.76	7.98	8.20	8.41
4.6	6.43	6.71	6.99	7.25	7.50	7.75	7.99	8.22	8.45	8.66
4.8	6.61	6.91	7.19	7.46	7.72	7.97	8.22	8.46	8.69	8.91
5.0	6.79	7.10	7.39	7.66	7.93	8.19	8.45	8.69	8.93	9.16
5.2	6.97	7.28	7.58	7.87	8.14	8.41	8.67	8.92	9.16	9.40
5.4	7.15	7.47	7.77	8.07	8.35	8.62	8.89	9.15	9.40	9.64
5.6	7.33	7.65	7.96	8.27	8.56	8.84	9.11	9.37	9.63	9.88
5.8	7.50	7.83	8.15	8.46	8.76	9.05	9.32	9.59	9.86	10.11
6.0	7.67	8.01	8.34	8.65	8.96	9.25	9.54	9.81	10.08	10.34
6.2	7.84	8.19	8.52	8.85	9.16	9.46	9.75	10.03	10.31	10.57
6.4	8.01	8.36	8.71	9.04	9.35	9.66	9.96	10.24	10.53	10.80
6.6	8.17	8.54	8.89	9.22	9.55	9.86	10.16	10.46	10.74	11.02
6.8	8.34	8.71	9.07	9.41	9.74	10.06	10.37	10.67	10.96	11.24
7.0	8.50	8.88	9.24	9.59	9.93	10.25	10.57	10.88	11.17	11.46
7.5	8.90	9.30	9.68	10.04	10.40	10.74	11.07	11.39	11.70	12.00
8.0	9.29	9.71	10.10	10.48	10.85	11.21	11.55	11.89	12.21	12.53
8.5	9.68	10.11	10.52	10.92	11.30	11.67	12.03	12.38	12.72	13.05
9.0	10.05	10.50	10.93	11.34	11.74	12.12	12.50	12.86	13.21	13.55
9.5	10.42	10.88	11.33	11.76	12.17	12.57	12.96	13.33	13.70	14.05
10	10.78	11.26	11.72	12.17	12.59	13.01	13.41	13.79	14.17	14.54
11	11.49	12.00	12.49	12.96	13.42	13.86	14.29	14.70	15.10	15.49
12	12.18	12.72	13.24	13.74	14.22	14.69	15.14	15.58	16.00	16.42
13	12.85	13.42	13.96	14.49	15.00	15.49	15.97	16.43	16.88	17.32
14	13.50	14.10	14.67	15.23	15.76	16.28	16.78	17.26	17.74	18.20
15	14.13	14.76	15.36	15.94	16.50	17.04	17.57	18.08	18.57	19.05
16	14.75	15.41	16.04	16.64	17.23	17.79	18.34	18.87	19.39	19.89
17	15.36	16.04	16.70	17.33	17.94	18.53	19.10	19.65	20.19	20.71
18	15.96	16.67	17.35	18.00	18.63	19.25	19.84	20.41	20.97	21.52
19	16.54	17.28	17.98	18.66	19.32	19.95	20.57	21.16	21.74	22.31
20	17.12	17.88	18.61	19.31	19.99	20.65	21.28	21.90	22.50	23.08

Table 6.—*Velocity of water, v, in feet per second, based on Manning's formula* $v = (1.486/n)r^{2/3}s^{1/2}$, **n = .015**—Continued

r	.0010	.0015	.0020	.0025	.0030	.0035	.0040	.0045	.0050
0.2	1.07	1.31	1.52	1.69	1.86	2.00	2.14	2.27	2.40
0.4	1.70	2.08	2.41	2.69	2.95	3.18	3.40	3.61	3.80
0.6	2.23	2.73	3.15	3.52	3.86	4.17	4.46	4.73	4.98
0.8	2.70	3.31	3.82	4.27	4.68	5.05	5.40	5.73	6.04
1.0	3.13	3.84	4.43	4.95	5.43	5.86	6.27	6.65	7.01
1.2	3.54	4.33	5.00	5.59	6.13	6.62	7.08	7.50	7.91
1.4	3.92	4.80	5.54	6.20	6.79	7.33	7.84	8.32	8.77
1.6	4.29	5.25	6.06	6.78	7.42	8.02	8.57	9.09	9.58
1.8	4.64	5.68	6.56	7.33	8.03	8.67	9.27	9.83	10.37
2.0	4.97	6.09	7.03	7.86	8.61	9.30	9.95	10.55	11.12
2.2	5.30	6.49	7.49	8.38	9.18	9.91	10.60	11.24	11.85
2.4	5.62	6.88	7.94	8.88	9.73	10.51	11.23	11.91	12.56
2.6	5.92	7.25	8.38	9.37	10.26	11.08	11.85	12.57	13.25
2.8	6.22	7.62	8.80	9.84	10.78	11.64	12.45	13.20	13.92
3.0	6.52	7.98	9.22	10.30	11.29	12.19	13.03	13.82	14.57
3.2	6.80	8.33	9.62	10.76	11.78	12.73	13.61	14.43	15.21
3.4	7.08	8.68	10.02	11.20	12.27	13.25	14.17	15.03	15.84
3.6	7.36	9.01	10.41	11.63	12.75	13.77	14.72	15.61	16.45
3.8	7.63	9.34	10.79	12.06	13.21	14.27	15.26	16.18	17.06
4.0	7.89	9.67	11.16	12.48	13.67	14.77	15.79	16.75	17.65
4.2	8.16	9.99	11.53	12.89	14.12	15.26	16.31	17.30	18.24
4.4	8.41	10.30	11.90	13.30	14.57	15.74	16.82	17.84	18.81
4.6	8.66	10.61	12.25	13.70	15.01	16.21	17.33	18.38	19.38
4.8	8.91	10.92	12.61	14.09	15.44	16.68	17.83	18.91	19.93
5.0	9.16	11.22	12.95	14.48	15.87	17.14	18.32	19.43	20.48
5.2	9.40	11.52	13.30	14.87	16.29	17.59	18.81	19.95	21.03
5.4	9.64	11.81	13.64	15.25	16.70	18.04	19.29	20.45	21.56
5.6	9.88	12.10	13.97	15.62	17.11	18.48	19.76	20.96	22.09
5.8	10.11	12.39	14.30	15.99	17.52	18.92	20.23	21.45	22.61
6.0	10.34	12.67	14.63	16.36	17.92	19.35	20.69	21.94	23.13
6.2	10.57	12.95	14.95	16.72	18.31	19.78	21.15	22.43	23.64
6.4	10.80	13.23	15.27	17.07	18.70	20.20	21.60	22.91	24.15
6.6	11.02	13.50	15.59	17.43	19.09	20.62	22.05	23.38	24.65
6.8	11.24	13.77	15.90	17.78	19.48	21.04	22.49	23.85	25.14
7.0	11.46	14.04	16.21	18.13	19.86	21.45	22.93	24.32	25.63
7.5	12.00	14.70	16.98	18.98	20.79	22.46	24.01	25.46	26.84
8.0	12.53	15.35	17.72	19.81	21.70	23.44	25.06	26.58	28.02
8.5	13.05	15.98	18.45	20.63	22.60	24.41	26.10	27.68	29.18
9.0	13.55	16.60	19.17	21.43	23.48	25.36	27.11	28.75	30.31
9.5	14.05	17.21	19.87	22.22	24.34	26.29	28.10	29.81	31.42
10	14.54	17.81	20.56	22.99	25.19	27.20	29.08	30.85	32.51
11	15.49	18.98	21.91	24.50	26.84	28.99	30.99	32.87	34.65
12	16.42	20.11	23.22	25.96	28.44	30.72	32.84	34.83	36.72
13	17.32	21.21	24.49	27.39	30.00	32.40	34.64	36.74	38.73
14	18.20	22.29	25.74	28.77	31.52	34.04	36.40	38.60	40.69
15	19.05	23.34	26.95	30.13	33.00	35.65	38.11	40.42	42.61
16	19.89	24.36	28.13	31.45	34.45	37.21	39.78	42.20	44.48
17	20.71	25.37	29.29	32.75	35.87	38.75	41.42	43.94	46.31
18	21.52	26.35	30.43	34.02	37.27	40.25	43.03	45.64	48.11
19	22.31	27.32	31.55	35.27	38.64	41.73	44.61	47.32	49.88
20	23.08	28.27	32.64	36.50	39.98	43.18	46.16	48.97	51.61

Table 6.—*Velocity of water, v, in feet per second, based on Manning's formula* $v = (1.486/n)r^{2/3}s^{1/2}$, **n = .015**—Continued

r \ s	.0055	.0060	.0065	.0070	.0075	.0080	.0085	.0090	.0095	.0100
0.2	2.51	2.62	2.73	2.83	2.93	3.03	3.12	3.21	3.30	3.39
0.4	3.99	4.17	4.34	4.50	4.66	4.81	4.96	5.10	5.24	5.38
0.6	5.23	5.46	5.68	5.90	6.10	6.30	6.50	6.69	6.87	7.05
0.8	6.33	6.61	6.88	7.14	7.39	7.64	7.87	8.10	8.32	8.54
1.0	7.35	7.67	7.99	8.29	8.58	8.86	9.13	9.40	9.66	9.91
1.2	8.30	8.67	9.02	9.36	9.69	10.01	10.31	10.61	10.90	11.19
1.4	9.19	9.60	10.00	10.37	10.74	11.09	11.43	11.76	12.08	12.40
1.6	10.05	10.50	10.93	11.34	11.74	12.12	12.49	12.86	13.21	13.55
1.8	10.87	11.35	11.82	12.26	12.70	13.11	13.52	13.91	14.29	14.66
2.0	11.66	12.18	12.68	13.16	13.62	14.07	14.50	14.92	15.33	15.73
2.2	12.43	12.98	13.51	14.02	14.51	14.99	15.45	15.90	16.33	16.76
2.4	13.17	13.76	14.32	14.86	15.38	15.88	16.37	16.85	17.31	17.76
2.6	13.89	14.51	15.10	15.67	16.22	16.75	17.27	17.77	18.26	18.73
2.8	14.60	15.24	15.87	16.47	17.04	17.60	18.14	18.67	19.18	19.68
3.0	15.28	15.96	16.61	17.24	17.85	18.43	19.00	19.55	20.08	20.61
3.2	15.95	16.66	17.34	18.00	18.63	19.24	19.83	20.41	20.97	21.51
3.4	16.61	17.35	18.06	18.74	19.40	20.04	20.65	21.25	21.83	22.40
3.6	17.26	18.02	18.76	19.47	20.15	20.81	21.45	22.08	22.68	23.27
3.8	17.89	18.69	19.45	20.18	20.89	21.58	22.24	22.89	23.51	24.12
4.0	18.51	19.34	20.13	20.89	21.62	22.33	23.01	23.68	24.33	24.96
4.2	19.13	19.98	20.79	21.58	22.33	23.07	23.78	24.47	25.14	25.79
4.4	19.73	20.60	21.45	22.26	23.04	23.79	24.52	25.24	25.93	26.60
4.6	20.32	21.22	22.09	22.93	23.73	24.51	25.26	25.99	26.71	27.40
4.8	20.91	21.84	22.73	23.59	24.41	25.21	25.99	26.74	27.48	28.19
5.0	21.48	22.44	23.35	24.24	25.09	25.91	26.71	27.48	28.23	28.97
5.2	22.05	23.03	23.97	24.88	25.75	26.60	27.41	28.21	28.98	29.73
5.4	22.61	23.62	24.58	25.51	26.41	27.27	28.11	28.93	29.72	30.49
5.6	23.17	24.20	25.19	26.14	27.06	27.94	28.80	29.64	30.45	31.24
5.8	23.72	24.77	25.78	26.76	27.70	28.60	29.48	30.34	31.17	31.98
6.0	24.26	25.34	26.37	27.37	28.33	29.26	30.16	31.03	31.88	32.71
6.2	24.80	25.90	26.96	27.97	28.95	29.90	30.82	31.72	32.59	33.43
6.4	25.33	26.45	27.53	28.57	29.57	30.54	31.48	32.40	33.28	34.15
6.6	25.85	27.00	28.10	29.16	30.19	31.18	32.14	33.07	33.97	34.86
6.8	26.37	27.54	28.67	29.75	30.79	31.80	32.78	33.73	34.66	35.56
7.0	26.88	28.08	29.23	30.33	31.39	32.42	33.42	34.39	35.33	36.25
7.5	28.15	29.40	30.60	31.76	32.87	33.95	35.00	36.01	37.00	37.96
8.0	29.39	30.69	31.95	33.15	34.32	35.44	36.53	37.59	38.62	39.63
8.5	30.60	31.96	33.27	34.52	35.73	36.90	38.04	39.14	40.22	41.26
9.0	31.79	33.20	34.56	35.86	37.12	38.34	39.52	40.66	41.78	42.86
9.5	32.96	34.42	35.83	37.18	38.48	39.75	40.97	42.16	43.31	44.44
10	34.10	35.62	37.07	38.47	39.82	41.13	42.39	43.62	44.82	45.98
11	36.34	37.95	39.50	41.00	42.43	43.83	45.18	46.48	47.76	49.00
12	38.51	40.22	41.86	43.44	44.97	46.44	47.87	49.26	50.61	51.93
13	40.62	42.43	44.16	45.83	47.43	48.99	50.50	51.96	53.38	54.77
14	42.68	44.57	46.39	48.15	49.84	51.47	53.05	54.59	56.09	57.55
15	44.69	46.67	48.58	50.41	52.18	53.89	55.55	57.16	58.73	60.25
16	46.65	48.71	50.71	52.63	54.48	56.26	57.99	59.68	61.31	62.90
17	48.57	50.73	52.81	54.80	56.72	58.58	60.39	62.14	63.84	65.50
18	50.46	52.70	54.86	56.93	58.93	60.86	62.73	64.55	66.32	68.04
19	52.31	54.64	56.87	59.02	61.09	63.09	65.03	66.92	68.75	70.54
20	54.13	56.54	58.85	61.07	63.21	65.29	67.30	69.25	71.14	72.99

Table 6.—*Velocity of water, v, in feet per second, based on Manning's formula* $v=(1.486/n)r^{2/3}s^{1/2}$. **n = .015**—Continued

r	.01	.02	.03	.04	.05	.06	.07	.08	.09	.10
0.2	3.39	4.79	5.87	6.78	7.58	8.30	8.96	9.58	10.16	10.71
0.4	5.38	7.61	9.32	10.76	12.03	13.17	14.23	15.21	16.13	17.01
0.6	7.05	9.97	12.21	14.09	15.76	17.26	18.65	19.93	21.14	22.29
0.8	8.54	12.07	14.80	17.07	19.09	20.91	22.59	24.15	25.61	27.00
1.0	9.91	14.01	17.16	19.81	22.15	24.27	26.21	28.02	29.72	31.33
1.2	11.19	15.82	19.38	22.37	25.01	27.40	29.60	31.64	33.56	35.38
1.4	12.40	17.53	21.47	24.80	27.72	30.37	32.80	35.07	37.19	39.21
1.6	13.55	19.17	23.47	27.10	30.30	33.20	35.86	38.33	40.66	42.86
1.8	14.66	20.73	25.39	29.32	32.78	35.91	38.78	41.46	43.98	46.36
2.0	15.73	22.24	27.24	31.45	35.16	38.52	41.61	44.48	47.18	49.73
2.2	16.76	23.70	29.02	33.52	37.47	41.05	44.34	47.40	50.27	52.99
2.4	17.76	25.11	30.76	35.52	39.71	43.50	46.98	50.23	53.27	56.16
2.6	18.73	26.49	32.44	37.46	41.89	45.88	49.56	52.98	56.20	59.23
2.8	19.68	27.83	34.09	39.36	44.01	48.21	52.07	55.66	59.04	62.23
3.0	20.61	29.14	35.69	41.21	46.08	50.48	54.52	58.28	61.82	65.16
3.2	21.51	30.42	37.26	43.03	48.10	52.70	56.92	60.85	64.54	68.03
3.4	22.40	31.68	38.80	44.80	50.09	54.87	59.26	63.36	67.20	70.83
3.6	23.27	32.91	40.30	46.54	52.03	57.00	61.57	65.82	69.81	73.59
3.8	24.12	34.12	41.78	48.25	53.94	59.09	63.83	68.23	72.37	76.29
4.0	24.96	35.30	43.24	49.93	55.82	61.15	66.05	70.61	74.89	78.94
4.2	25.79	36.47	44.67	51.58	57.66	63.17	68.23	72.94	77.37	81.55
4.4	26.60	37.62	46.07	53.20	59.48	65.16	70.38	75.24	79.80	84.12
4.6	27.40	38.75	47.46	54.80	61.27	67.12	72.50	77.50	82.20	86.65
4.8	28.19	39.87	48.83	56.38	63.03	69.05	74.58	79.73	84.57	89.14
5.0	28.97	40.97	50.17	57.93	64.77	70.96	76.64	81.93	86.90	91.60
5.2	29.73	42.05	51.50	59.47	66.49	72.83	78.67	84.10	89.20	94.03
5.4	30.49	43.12	52.81	60.98	68.18	74.69	80.68	86.25	91.48	96.43
5.6	31.24	44.18	54.11	62.48	69.86	76.52	82.65	88.36	93.72	98.79
5.8	31.98	45.23	55.39	63.96	71.51	78.33	84.61	90.45	95.94	101.13
6.0	32.71	46.26	56.66	65.42	73.14	80.13	86.55	92.52	98.13	103.44
6.2	33.43	47.28	57.91	66.87	74.76	81.90	88.46	94.57	100.30	105.73
6.4	34.15	48.29	59.15	68.30	76.36	83.65	90.35	96.59	102.45	107.99
6.6	34.86	49.30	60.37	69.71	77.94	85.38	92.22	98.59	104.57	110.23
6.8	35.56	50.29	61.59	71.12	79.51	87.10	94.08	100.57	106.67	112.44
7.0	36.25	51.27	62.79	72.50	81.06	88.80	95.91	102.53	108.75	114.64
7.5	37.96	53.68	65.74	75.92	84.88	92.98	100.43	107.36	113.87	120.03
8.0	39.63	56.04	68.64	79.25	88.61	97.07	104.84	112.08	118.88	125.31
8.5	41.26	58.35	71.47	82.52	92.26	101.07	109.17	116.70	123.78	130.48
9.0	42.86	60.62	74.24	85.73	95.85	104.99	113.41	121.24	128.59	135.55
9.5	44.44	62.84	76.97	88.87	99.36	108.85	117.57	125.69	133.31	140.52
10	45.98	65.03	79.64	91.97	102.82	112.63	121.66	130.06	137.95	145.41
11	49.00	69.30	84.87	98.00	109.57	120.02	129.64	138.59	147.00	154.95
12	51.93	73.43	89.94	103.85	116.11	127.19	137.38	146.87	155.78	164.20
13	54.77	77.46	94.87	109.54	122.47	134.16	144.91	154.92	164.32	173.20
14	57.55	81.38	99.67	115.09	128.68	140.96	152.25	162.76	172.64	181.98
15	60.25	85.21	104.36	120.51	134.73	147.59	159.42	170.43	180.76	190.54
16	62.90	88.96	108.95	125.81	140.66	154.08	166.43	177.92	188.71	198.92
17	65.50	92.63	113.45	131.00	146.46	160.44	173.29	185.26	196.49	207.12
18	68.04	96.23	117.85	136.08	152.15	166.67	180.02	192.45	204.13	215.17
19	70.54	99.76	122.18	141.08	157.73	172.78	186.63	199.51	211.62	223.06
20	72.99	103.23	126.43	145.99	163.22	178.80	193.12	206.46	218.98	230.82

Table 7.—*Velocity of water, v, in feet per second, based on Manning's formula* $v=(1.486/n)r^{2/3}s^{1/2}$, **n=.0175**

r \ s	.00005	.00010	.00015	.00020	.00025	.00030	.00035	.00040	.00045	.00050
0.2	0.21	0.29	0.36	0.41	0.46	0.50	0.54	0.58	0.62	0.65
0.4	.33	.46	.56	.65	.73	.80	.86	.92	.98	1.03
0.6	.43	.60	.74	.85	.96	1.05	1.13	1.21	1.28	1.35
0.8	.52	.73	.90	1.03	1.16	1.27	1.37	1.46	1.55	1.64
1.0	.60	.85	1.04	1.20	1.34	1.47	1.59	1.70	1.80	1.90
1.2	.68	.96	1.17	1.36	1.52	1.66	1.79	1.92	2.03	2.14
1.4	.75	1.06	1.30	1.50	1.68	1.84	1.99	2.13	2.25	2.38
1.6	.82	1.16	1.42	1.64	1.84	2.01	2.17	2.32	2.46	2.60
1.8	.89	1.26	1.54	1.78	1.99	2.18	2.35	2.51	2.67	2.81
2.0	.95	1.35	1.65	1.91	2.13	2.33	2.52	2.70	2.86	3.01
2.2	1.02	1.44	1.76	2.03	2.27	2.49	2.69	2.87	3.05	3.21
2.4	1.08	1.52	1.86	2.15	2.41	2.64	2.85	3.04	3.23	3.40
2.6	1.14	1.61	1.97	2.27	2.54	2.78	3.00	3.21	3.41	3.59
2.8	1.19	1.69	2.07	2.39	2.67	2.92	3.16	3.37	3.58	3.77
3.0	1.25	1.77	2.16	2.50	2.79	3.06	3.30	3.53	3.75	3.95
3.2	1.30	1.84	2.26	2.61	2.92	3.19	3.45	3.69	3.91	4.12
3.4	1.36	1.92	2.35	2.72	3.04	3.33	3.59	3.84	4.07	4.29
3.6	1.41	1.99	2.44	2.82	3.15	3.45	3.73	3.99	4.23	4.46
3.8	1.46	2.07	2.53	2.92	3.27	3.58	3.87	4.14	4.39	4.62
4.0	1.51	2.14	2.62	3.03	3.38	3.71	4.00	4.28	4.54	4.78
4.2	1.56	2.21	2.71	3.13	3.50	3.83	4.14	4.42	4.69	4.94
4.4	1.61	2.28	2.79	3.22	3.61	3.95	4.27	4.56	4.84	5.10
4.6	1.66	2.35	2.88	3.32	3.71	4.07	4.39	4.70	4.98	5.25
4.8	1.71	2.42	2.96	3.42	3.82	4.19	4.52	4.83	5.13	5.40
5.0	1.76	2.48	3.04	3.51	3.93	4.30	4.65	4.97	5.27	5.55
5.2	1.80	2.55	3.12	3.60	4.03	4.41	4.77	5.10	5.41	5.70
5.4	1.85	2.61	3.20	3.70	4.13	4.53	4.89	5.23	5.54	5.84
5.6	1.89	2.68	3.28	3.79	4.23	4.64	5.01	5.36	5.68	5.99
5.8	1.94	2.74	3.36	3.88	4.33	4.75	5.13	5.48	5.81	6.13
6.0	1.98	2.80	3.43	3.97	4.43	4.86	5.25	5.61	5.95	6.27
6.2	2.03	2.87	3.51	4.05	4.53	4.96	5.36	5.73	6.08	6.41
6.4	2.07	2.93	3.58	4.14	4.63	5.07	5.48	5.85	6.21	6.55
6.6	2.11	2.99	3.66	4.23	4.72	5.17	5.59	5.98	6.34	6.68
6.8	2.16	3.05	3.73	4.31	4.82	5.28	5.70	6.10	6.47	6.82
7.0	2.20	3.11	3.81	4.39	4.91	5.38	5.81	6.21	6.59	6.95
7.5	2.30	3.25	3.98	4.60	5.14	5.64	6.09	6.51	6.90	7.28
8.0	2.40	3.40	4.16	4.80	5.37	5.88	6.35	6.79	7.21	7.59
8.5	2.50	3.54	4.33	5.00	5.59	6.13	6.62	7.07	7.50	7.91
9.0	2.60	3.67	4.50	5.20	5.81	6.36	6.87	7.35	7.79	8.22
9.5	2.69	3.81	4.66	5.39	6.02	6.60	7.13	7.62	8.08	8.52
10	2.79	3.94	4.83	5.57	6.23	6.83	7.37	7.88	8.36	8.81
11	2.97	4.20	5.14	5.94	6.64	7.27	7.86	8.40	8.91	9.39
12	3.15	4.45	5.45	6.29	7.04	7.71	8.33	8.90	9.44	9.95
13	3.32	4.69	5.75	6.64	7.42	8.13	8.78	9.39	9.96	10.50
14	3.49	4.93	6.04	6.98	7.80	8.54	9.23	9.86	10.46	11.03
15	3.65	5.16	6.33	7.30	8.17	8.95	9.66	10.33	10.96	11.55
16	3.81	5.39	6.60	7.63	8.53	9.34	10.09	10.78	11.44	12.06
17	3.97	5.61	6.88	7.94	8.88	9.72	10.50	11.23	11.91	12.55
18	4.12	5.83	7.14	8.25	9.22	10.10	10.91	11.66	12.37	13.04
19	4.28	6.05	7.41	8.55	9.56	10.47	11.31	12.09	12.83	13.52
20	4.42	6.26	7.66	8.85	9.89	10.84	11.70	12.51	13.27	13.99

Table 7.—*Velocity of water, v, in feet per second, based on Manning's formula* $v = (1.486/n)r^{2/3}s^{1/2}$, **n = .0175**—Continued

r	.00055	.00060	.00065	.00070	.00075	.00080	.00085	.00090	.00095	.00100
0.2	0.68	0.71	0.74	0.77	0.80	0.82	0.85	0.87	0.90	0.92
0.4	1.08	1.13	1.18	1.22	1.26	1.30	1.34	1.38	1.42	1.46
0.6	1.42	1.48	1.54	1.60	1.65	1.71	1.76	1.81	1.86	1.91
0.8	1.72	1.79	1.87	1.94	2.00	2.07	2.13	2.20	2.26	2.31
1.0	1.99	2.08	2.16	2.25	2.33	2.40	2.48	2.55	2.62	2.69
1.2	2.25	2.35	2.44	2.54	2.63	2.71	2.80	2.88	2.96	3.03
1.4	2.49	2.60	2.71	2.81	2.91	3.01	3.10	3.19	3.28	3.36
1.6	2.72	2.85	2.96	3.07	3.18	3.29	3.39	3.48	3.58	3.67
1.8	2.95	3.08	3.20	3.32	3.44	3.55	3.66	3.77	3.87	3.97
2.0	3.16	3.30	3.44	3.57	3.69	3.81	3.93	4.04	4.15	4.26
2.2	3.37	3.52	3.66	3.80	3.93	4.06	4.19	4.31	4.43	4.54
2.4	3.57	3.73	3.88	4.03	4.17	4.31	4.44	4.57	4.69	4.81
2.6	3.77	3.93	4.09	4.25	4.40	4.54	4.68	4.82	4.95	5.08
2.8	3.96	4.13	4.30	4.46	4.62	4.77	4.92	5.06	5.20	5.33
3.0	4.14	4.33	4.50	4.67	4.84	5.00	5.15	5.30	5.44	5.59
3.2	4.32	4.52	4.70	4.88	5.05	5.22	5.38	5.53	5.68	5.83
3.4	4.50	4.70	4.90	5.08	5.26	5.43	5.60	5.76	5.92	6.07
3.6	4.68	4.89	5.09	5.28	5.46	5.64	5.82	5.98	6.15	6.31
3.8	4.85	5.06	5.27	5.47	5.66	5.85	6.03	6.20	6.37	6.54
4.0	5.02	5.24	5.46	5.66	5.86	6.05	6.24	6.42	6.60	6.77
4.2	5.18	5.41	5.64	5.85	6.05	6.25	6.44	6.63	6.81	6.99
4.4	5.35	5.59	5.81	6.03	6.24	6.45	6.65	6.84	7.03	7.21
4.6	5.51	5.75	5.99	6.21	6.43	6.64	6.85	7.05	7.24	7.43
4.8	5.67	5.92	6.16	6.39	6.62	6.83	7.04	7.25	7.45	7.64
5.0	5.82	6.08	6.33	6.57	6.80	7.02	7.24	7.45	7.65	7.85
5.2	5.98	6.24	6.50	6.74	6.98	7.21	7.43	7.65	7.86	8.06
5.4	6.13	6.40	6.66	6.92	7.16	7.39	7.62	7.84	8.06	8.27
5.6	6.28	6.56	6.83	7.08	7.33	7.57	7.81	8.03	8.25	8.47
5.8	6.43	6.71	6.99	7.25	7.51	7.75	7.99	8.22	8.45	8.67
6.0	6.58	6.87	7.15	7.42	7.68	7.93	8.17	8.41	8.64	8.87
6.2	6.72	7.02	7.31	7.58	7.85	8.11	8.36	8.60	8.83	9.06
6.4	6.86	7.17	7.46	7.74	8.02	8.28	8.53	8.78	9.02	9.26
6.6	7.01	7.32	7.62	7.90	8.18	8.45	8.71	8.96	9.21	9.45
6.8	7.15	7.47	7.77	8.06	8.35	8.62	8.89	9.14	9.39	9.64
7.0	7.29	7.61	7.92	8.22	8.51	8.79	9.06	9.32	9.58	9.85
7.5	7.63	7.97	8.29	8.61	8.91	9.20	9.49	9.76	10.03	10.29
8.0	7.97	8.32	8.66	8.99	9.30	9.61	9.90	10.19	10.47	10.74
8.5	8.29	8.66	9.02	9.36	9.69	10.00	10.31	10.61	10.90	11.18
9.0	8.62	9.00	9.37	9.72	10.06	10.39	10.71	11.02	11.32	11.62
9.5	8.93	9.33	9.71	10.08	10.43	10.77	11.10	11.43	11.74	12.04
10	9.24	9.65	10.05	10.43	10.79	11.15	11.49	11.82	12.15	12.46
11	9.85	10.29	10.71	11.11	11.50	11.88	12.24	12.60	12.95	13.28
12	10.44	10.90	11.35	11.78	12.19	12.59	12.98	13.35	13.72	14.07
13	11.01	11.50	11.97	12.42	12.86	13.28	13.69	14.08	14.47	14.85
14	11.57	12.08	12.58	13.05	13.51	13.95	14.38	14.80	15.20	15.60
15	12.11	12.65	13.17	13.66	14.14	14.61	15.06	15.49	15.92	16.33
16	12.64	13.21	13.75	14.27	14.77	15.25	15.72	16.18	16.62	17.05
17	13.17	13.75	14.31	14.85	15.37	15.88	16.37	16.84	17.30	17.75
18	13.68	14.29	14.87	15.43	15.97	16.50	17.00	17.50	17.98	18.44
19	14.18	14.81	15.41	16.00	16.56	17.10	17.63	18.14	18.64	19.12
20	14.67	15.33	15.95	16.55	17.13	17.70	18.24	18.77	19.28	19.78

Table 7.—*Velocity of water, v, in feet per second, based on Manning's formula* $v = (1.486/n)r^{2/3}s^{1/2}$, **n = .0175**—Continued

r	.0010	.0015	.0020	.0025	.0030	.0035	.0040	.0045	.0050
0.2	0.92	1.12	1.30	1.45	1.59	1.72	1.84	1.95	2.05
0.4	1.46	1.79	2.06	2.30	2.52	2.73	2.92	3.09	3.26
0.6	1.91	2.34	2.70	3.02	3.31	3.57	3.82	4.05	4.27
0.8	2.31	2.83	3.27	3.66	4.01	4.33	4.63	4.91	5.17
1.0	2.69	3.29	3.80	4.25	4.65	5.02	5.37	5.70	6.00
1.2	3.03	3.71	4.29	4.79	5.25	5.67	6.06	6.43	6.78
1.4	3.36	4.12	4.75	5.31	5.82	6.29	6.72	7.13	7.51
1.6	3.67	4.50	5.19	5.81	6.36	6.87	7.35	7.79	8.21
1.8	3.97	4.87	5.62	6.28	6.88	7.43	7.95	8.43	8.88
2.0	4.26	5.22	6.03	6.74	7.38	7.97	8.53	9.04	9.53
2.2	4.54	5.56	6.42	7.18	7.87	8.50	9.08	9.64	10.16
2.4	4.81	5.90	6.81	7.61	8.34	9.01	9.63	10.21	10.76
2.6	5.08	6.22	7.18	8.03	8.79	9.50	10.15	10.77	11.35
2.8	5.33	6.53	7.54	8.43	9.24	9.98	10.67	11.32	11.93
3.0	5.59	6.84	7.90	8.83	9.67	10.45	11.17	11.85	12.49
3.2	5.83	7.14	8.25	9.22	10.10	10.91	11.66	12.37	13.04
3.4	6.07	7.44	8.59	9.60	10.52	11.36	12.14	12.88	13.58
3.6	6.31	7.72	8.92	9.97	10.92	11.80	12.61	13.38	14.10
3.8	6.54	8.01	9.25	10.34	11.33	12.23	13.08	13.87	14.62
4.0	6.77	8.29	9.57	10.70	11.72	12.66	13.53	14.35	15.13
4.2	6.99	8.56	9.89	11.05	12.11	13.08	13.98	14.83	15.63
4.4	7.21	8.83	10.20	11.40	12.49	13.49	14.42	15.30	16.12
4.6	7.43	9.10	10.50	11.74	12.86	13.89	14.85	15.76	16.61
4.8	7.64	9.36	10.81	12.08	13.23	14.29	15.28	16.21	17.09
5.0	7.85	9.62	11.10	12.41	13.60	14.69	15.70	16.66	17.56
5.2	8.06	9.87	11.40	12.74	13.96	15.08	16.12	17.10	18.02
5.4	8.27	10.12	11.69	13.07	14.32	15.46	16.53	17.53	18.48
5.6	8.47	10.37	11.98	13.39	14.67	15.84	16.94	17.96	18.93
5.8	8.67	10.62	12.26	13.71	15.01	16.22	17.34	18.39	19.38
6.0	8.87	10.86	12.54	14.02	15.36	16.59	17.73	18.81	19.83
6.2	9.06	11.10	12.82	14.33	15.70	16.95	18.12	19.22	20.26
6.4	9.26	11.34	13.09	14.64	16.03	17.32	18.51	19.64	20.70
6.6	9.45	11.57	13.36	14.94	16.36	17.68	18.90	20.04	21.13
6.8	9.64	11.80	13.63	15.24	16.69	18.03	19.28	20.45	21.55
7.0	9.83	12.03	13.90	15.54	17.02	18.38	19.65	20.84	21.97
7.5	10.29	12.60	14.55	16.27	17.82	19.25	20.58	21.83	23.01
8.0	10.74	13.15	15.19	16.98	18.60	20.09	21.48	22.78	24.02
8.5	11.18	13.70	15.82	17.68	19.37	20.92	22.37	23.72	25.01
9.0	11.62	14.23	16.43	18.37	20.12	21.74	23.24	24.65	25.98
9.5	12.04	14.75	17.03	19.04	20.86	22.53	24.09	25.55	26.93
10	12.46	15.26	17.63	19.71	21.59	23.32	24.93	26.44	27.87
11	13.28	16.27	18.78	21.00	23.00	24.85	26.56	28.17	29.70
12	14.07	17.24	19.90	22.25	24.38	26.33	28.15	29.86	31.47
13	14.85	18.18	21.00	23.47	25.71	27.77	29.69	31.49	33.20
14	15.60	19.10	22.06	24.66	27.02	29.18	31.20	33.09	34.88
15	16.33	20.00	23.10	25.82	28.29	30.55	32.66	34.65	36.52
16	17.05	20.88	24.11	26.96	29.53	31.90	34.10	36.17	38.13
17	17.75	21.74	25.11	28.07	30.75	33.21	35.51	37.66	39.70
18	18.44	22.59	26.08	29.16	31.94	34.50	36.89	39.12	41.24
19	19.12	23.42	27.04	30.23	33.12	35.77	38.24	40.56	42.75
20	19.78	24.23	27.98	31.28	34.27	37.01	39.57	41.97	44.24

Table 7.—*Velocity of water, v, in feet per second, based on Manning's formula* $v=(1.486/n)r^{2/3}s^{1/2}$, **n=.0175**—Continued

r \ s	.0055	.0060	.0065	.0070	.0075	.0080	.0085	.0090	.0095	.0100
0.2	2.15	2.25	2.34	2.43	2.51	2.60	2.68	2.76	2.83	2.90
0.4	3.42	3.57	3.72	3.86	3.99	4.12	4.25	4.37	4.49	4.61
0.6	4.48	4.68	4.87	5.05	5.23	5.40	5.57	5.73	5.89	6.04
0.8	5.43	5.67	5.90	6.12	6.34	6.55	6.75	6.94	7.13	7.32
1.0	6.30	6.58	6.85	7.10	7.35	7.59	7.83	8.06	8.28	8.49
1.2	7.11	7.43	7.73	8.02	8.30	8.58	8.84	9.10	9.35	9.59
1.4	7.88	8.23	8.57	8.89	9.20	9.50	9.80	10.08	10.36	10.63
1.6	8.61	9.00	9.37	9.72	10.06	10.39	10.71	11.02	11.32	11.62
1.8	9.32	9.73	10.13	10.51	10.88	11.24	11.58	11.92	12.25	12.56
2.0	10.00	10.44	10.87	11.28	11.67	12.06	12.43	12.79	13.14	13.48
2.2	10.65	11.13	11.58	12.02	12.44	12.85	13.24	13.63	14.00	14.36
2.4	11.29	11.79	12.27	12.74	13.18	13.61	14.03	14.44	14.84	15.22
2.6	11.91	12.44	12.94	13.43	13.90	14.36	14.80	15.23	15.65	16.06
2.8	12.51	13.07	13.60	14.11	14.61	15.09	15.55	16.00	16.44	16.87
3.0	13.10	13.68	14.24	14.78	15.30	15.80	16.28	16.76	17.22	17.66
3.2	13.68	14.28	14.87	15.43	15.97	16.49	17.00	17.49	17.97	18.44
3.4	14.24	14.87	15.48	16.06	16.63	17.17	17.70	18.21	18.71	19.20
3.6	14.79	15.45	16.08	16.69	17.27	17.84	18.39	18.92	19.44	19.95
3.8	15.34	16.02	16.67	17.30	17.91	18.49	19.06	19.62	20.15	20.68
4.0	15.87	16.57	17.25	17.90	18.53	19.14	19.73	20.30	20.86	21.40
4.2	16.39	17.12	17.82	18.49	19.14	19.77	20.38	20.97	21.54	22.10
4.4	16.91	17.66	18.38	19.08	19.75	20.39	21.02	21.63	22.22	22.80
4.6	17.42	18.19	18.94	19.65	20.34	21.01	21.65	22.28	22.89	23.49
4.8	17.92	18.72	19.48	20.22	20.93	21.61	22.28	22.92	23.55	24.16
5.0	18.41	19.23	20.02	20.77	21.50	22.21	22.89	23.55	24.20	24.83
5.2	18.90	19.74	20.55	21.32	22.07	22.80	23.50	24.18	24.84	25.49
5.4	19.38	20.25	21.07	21.87	22.63	23.38	24.10	24.80	25.47	26.14
5.6	19.86	20.74	21.59	22.40	23.19	23.95	24.69	25.40	26.10	26.78
5.8	20.33	21.23	22.10	22.93	23.74	24.52	25.27	26.00	26.72	27.41
6.0	20.79	21.72	22.61	23.46	24.28	25.08	25.85	26.60	27.33	28.04
6.2	21.25	22.20	23.10	23.98	24.82	25.63	26.42	27.19	27.93	28.66
6.4	21.71	22.67	23.60	24.49	25.35	26.18	26.99	27.77	28.53	29.27
6.6	22.16	23.14	24.09	25.00	25.87	26.72	27.55	28.34	29.12	29.88
6.8	22.60	23.61	24.57	25.50	26.39	27.26	28.10	28.91	29.71	30.48
7.0	23.04	24.07	25.05	26.00	26.91	27.79	28.65	29.48	30.29	31.07
7.5	24.13	25.20	26.23	27.22	28.18	29.10	30.00	30.87	31.71	32.54
8.0	25.19	26.31	27.38	28.42	29.42	30.38	31.31	32.22	33.11	33.97
8.5	26.23	27.39	28.51	29.59	30.63	31.63	32.61	33.55	34.47	35.37
9.0	27.25	28.46	29.62	30.74	31.82	32.86	33.87	34.85	35.81	36.74
9.5	28.25	29.50	30.71	31.87	32.99	34.07	35.12	36.13	37.12	38.09
10	29.23	30.53	31.78	32.98	34.13	35.25	36.34	37.39	38.42	39.41
11	31.15	32.53	33.86	35.14	36.37	37.57	38.72	39.84	40.94	42.00
12	33.01	34.48	35.88	37.24	38.54	39.81	41.03	42.22	43.38	44.51
13	34.82	36.37	37.85	39.28	40.66	41.99	43.28	44.54	45.76	46.95
14	36.58	38.21	39.77	41.27	42.72	44.12	45.48	46.79	48.08	49.32
15	38.30	40.01	41.64	43.21	44.73	46.19	47.62	49.00	50.34	51.65
16	39.99	41.76	43.47	45.11	46.69	48.23	49.71	51.15	52.55	53.92
17	41.64	43.49	45.26	46.97	48.62	50.21	51.76	53.26	54.72	56.14
18	43.25	45.18	47.02	48.80	50.51	52.16	53.77	55.33	56.84	58.32
19	44.84	46.83	48.75	50.59	52.36	54.08	55.74	57.36	58.93	60.46
20	46.40	48.46	50.44	52.35	54.18	55.96	57.68	59.35	60.98	62.57

Table 7.—*Velocity of water, v, in feet per second, based on Manning's formula* $v=(1.486/n)r^{2/3}s^{1/2}$, **n = .0175**—Continued

r \ s	.01	.02	.03	.04	.05	.06	.07	.08	.09	.10
0.2	2.90	4.11	5.03	5.81	6.49	7.11	7.68	8.21	8.71	9.18
0.4	4.61	6.52	7.98	9.22	10.31	11.29	12.20	13.04	13.83	14.58
0.6	6.04	8.54	10.46	12.08	13.51	14.80	15.98	17.09	18.12	19.10
0.8	7.32	10.35	12.67	14.64	16.36	17.92	19.36	20.70	21.95	23.14
1.0	8.49	12.01	14.71	16.98	18.99	20.80	22.47	24.02	25.47	26.85
1.2	9.59	13.56	16.61	19.18	21.44	23.49	25.37	27.12	28.77	30.32
1.4	10.63	15.03	18.41	21.25	23.76	26.03	28.12	30.06	31.88	33.60
1.6	11.62	16.43	20.12	23.23	25.97	28.45	30.73	32.86	34.85	36.73
1.8	12.56	17.77	21.76	25.13	28.10	30.78	33.24	35.54	37.69	39.73
2.0	13.48	19.06	23.35	26.96	30.14	33.02	35.66	38.13	40.44	42.63
2.2	14.36	20.31	24.88	28.73	32.12	35.18	38.00	40.63	43.09	45.42
2.4	15.22	21.53	26.36	30.44	34.04	37.28	40.27	43.05	45.66	48.13
2.6	16.06	22.71	27.81	32.11	35.90	39.33	42.48	45.41	48.17	50.77
2.8	16.87	23.86	29.22	33.74	37.72	41.32	44.63	47.71	50.61	53.34
3.0	17.66	24.98	30.59	35.33	39.50	43.27	46.73	49.96	52.99	55.85
3.2	18.44	26.08	31.94	36.88	41.23	45.17	48.79	52.15	55.32	58.31
3.4	19.20	27.15	33.26	38.40	42.93	47.03	50.80	54.31	57.60	60.72
3.6	19.95	28.21	34.55	39.89	44.60	48.86	52.77	56.41	59.84	63.07
3.8	20.68	29.24	35.81	41.36	46.24	50.65	54.71	58.49	62.03	65.39
4.0	21.40	30.26	37.06	42.79	47.85	52.41	56.61	60.52	64.19	67.66
4.2	22.10	31.26	38.29	44.21	49.43	54.14	58.48	62.52	66.31	69.90
4.4	22.80	32.25	39.49	45.60	50.98	55.85	60.33	64.49	68.40	72.10
4.6	23.49	33.22	40.68	46.97	52.52	57.53	62.14	66.43	70.46	74.27
4.8	24.16	34.17	41.85	48.32	54.03	59.19	63.93	68.34	72.49	76.41
5.0	24.83	35.11	43.01	49.66	55.52	60.82	65.69	70.23	74.49	78.52
5.2	25.49	36.04	44.14	50.97	56.99	62.43	67.43	72.09	76.46	80.60
5.4	26.14	36.96	45.27	52.27	58.44	64.02	69.15	73.92	78.41	82.65
5.6	26.78	37.87	46.38	53.56	59.88	65.59	70.85	75.74	80.33	84.68
5.8	27.41	38.77	47.48	54.82	61.29	67.14	72.52	77.53	82.23	86.68
6.0	28.04	39.65	48.56	56.08	62.70	68.68	74.18	79.30	84.11	88.66
6.2	28.66	40.53	49.64	57.32	64.08	70.20	75.82	81.06	85.97	90.62
6.4	29.27	41.40	50.70	58.54	65.45	71.70	77.44	82.79	87.81	92.56
6.6	29.88	42.25	51.75	59.75	66.81	73.18	79.05	84.51	89.63	94.48
6.8	30.48	43.10	52.79	60.96	68.15	74.64	80.64	86.20	91.43	96.38
7.0	31.07	43.94	53.82	62.15	69.48	76.11	82.21	87.89	93.22	98.26
7.5	32.54	46.01	56.35	65.07	72.75	79.69	86.08	92.02	97.61	102.89
8.0	33.97	48.03	58.83	67.93	75.95	83.20	89.86	96.07	101.90	107.41
8.5	35.37	50.02	61.26	70.73	79.08	86.63	93.57	100.03	106.10	111.84
9.0	36.74	51.96	63.64	73.48	82.15	89.99	97.21	103.92	110.22	116.18
9.5	38.09	53.87	65.97	76.18	85.17	93.30	100.77	107.73	114.27	120.45
10	39.41	55.74	68.27	78.83	88.13	96.54	104.28	111.48	118.24	124.64
11	42.00	59.40	72.75	84.00	93.91	102.88	111.12	118.79	126.00	132.81
12	44.51	62.94	77.09	89.02	99.52	109.02	117.76	125.89	133.52	140.75
13	46.95	66.39	81.31	93.89	104.98	115.00	124.21	132.79	140.84	148.46
14	49.32	69.76	85.43	98.65	110.29	120.82	130.50	139.51	147.97	155.98
15	51.65	73.04	89.45	103.29	115.49	126.51	136.64	146.08	154.94	163.32
16	53.92	76.25	93.39	107.83	120.56	132.07	142.64	152.50	161.75	170.50
17	56.14	79.40	97.24	112.28	125.54	137.52	148.54	158.79	168.42	177.53
18	58.32	82.48	101.02	116.64	130.41	142.86	154.30	164.96	174.96	184.43
19	60.46	85.51	104.72	120.92	135.20	148.10	159.97	171.01	181.39	191.20
20	62.57	88.48	108.37	125.13	139.90	153.25	165.53	176.96	187.70	197.85

Table 8.—*Velocity of water, v, in feet per second, based on Manning's formula* $v=(1.486/n)r^{2/3}s^{1/2}$, n=.020

s / r	.00005	.00010	.00015	.00020	.00025	.00030	.00035	.00040	.00045	.00050
0.2	0.18	0.25	0.31	0.36	0.40	0.44	0.48	0.51	0.54	0.57
0.4	.29	.40	.49	.57	.64	.70	.75	.81	.86	.90
0.6	.37	.53	.65	.75	.84	.92	.99	1.06	1.12	1.18
0.8	.45	.64	.78	.91	1.01	1.11	1.20	1.28	1.36	1.43
1.0	.53	.74	.91	1.05	1.17	1.29	1.39	1.49	1.58	1.66
1.2	.59	.84	1.03	1.19	1.33	1.45	1.57	1.68	1.78	1.88
1.4	.66	.93	1.14	1.31	1.47	1.61	1.74	1.86	1.97	2.08
1.6	.72	1.02	1.24	1.44	1.61	1.76	1.90	2.03	2.16	2.27
1.8	.78	1.10	1.35	1.55	1.74	1.90	2.06	2.20	2.33	2.46
2.0	.83	1.18	1.44	1.67	1.86	2.04	2.21	2.36	2.50	2.64
2.2	.89	1.26	1.54	1.78	1.99	2.18	2.35	2.51	2.67	2.81
2.4	.94	1.33	1.63	1.88	2.11	2.31	2.49	2.66	2.83	2.98
2.6	.99	1.40	1.72	1.99	2.22	2.43	2.63	2.81	2.98	3.14
2.8	1.04	1.48	1.81	2.09	2.33	2.56	2.76	2.95	3.13	3.30
3.0	1.09	1.55	1.89	2.19	2.44	2.68	2.89	3.09	3.28	3.46
3.2	1.14	1.61	1.98	2.28	2.55	2.79	3.02	3.23	3.42	3.61
3.4	1.19	1.68	2.06	2.38	2.66	2.91	3.14	3.36	3.56	3.76
3.6	1.23	1.75	2.14	2.47	2.76	3.02	3.27	3.49	3.70	3.90
3.8	1.28	1.81	2.22	2.56	2.86	3.13	3.38	3.62	3.84	4.05
4.0	1.32	1.87	2.29	2.65	2.96	3.24	3.50	3.74	3.97	4.19
4.2	1.37	1.93	2.37	2.74	3.06	3.35	3.62	3.87	4.10	4.32
4.4	1.41	2.00	2.44	2.82	3.15	3.46	3.73	3.99	4.23	4.46
4.6	1.45	2.06	2.52	2.91	3.25	3.56	3.84	4.11	4.36	4.60
4.8	1.49	2.11	2.59	2.99	3.34	3.66	3.96	4.23	4.48	4.73
5.0	1.54	2.17	2.66	3.07	3.44	3.76	4.06	4.35	4.61	4.86
5.2	1.58	2.23	2.73	3.15	3.53	3.86	4.17	4.46	4.73	4.99
5.4	1.62	2.29	2.80	3.23	3.62	3.96	4.28	4.57	4.85	5.11
5.6	1.66	2.34	2.87	3.31	3.70	4.06	4.38	4.69	4.97	5.24
5.8	1.70	2.40	2.94	3.39	3.79	4.15	4.49	4.80	5.09	5.36
6.0	1.73	2.45	3.00	3.47	3.88	4.25	4.59	4.91	5.20	5.49
6.2	1.77	2.51	3.07	3.55	3.96	4.34	4.69	5.02	5.32	5.61
6.4	1.81	2.56	3.14	3.62	4.05	4.44	4.79	5.12	5.43	5.73
6.6	1.85	2.61	3.20	3.70	4.13	4.53	4.89	5.23	5.55	5.85
6.8	1.89	2.67	3.27	3.77	4.22	4.62	4.99	5.33	5.66	5.96
7.0	1.92	2.72	3.33	3.85	4.30	4.71	5.09	5.44	5.77	6.08
7.5	2.01	2.85	3.49	4.03	4.50	4.93	5.33	5.69	6.04	6.37
8.0	2.10	2.97	3.64	4.20	4.70	5.15	5.56	5.94	6.30	6.65
8.5	2.19	3.09	3.79	4.38	4.89	5.36	5.79	6.19	6.56	6.92
9.0	2.27	3.21	3.94	4.55	5.08	5.57	6.01	6.43	6.82	7.19
9.5	2.36	3.33	4.08	4.71	5.27	5.77	6.24	6.67	7.07	7.45
10	2.44	3.45	4.22	4.88	5.45	5.97	6.45	6.90	7.32	7.71
11	2.60	3.67	4.50	5.20	5.81	6.37	6.88	7.35	7.80	8.22
12	2.75	3.89	4.77	5.51	6.16	6.75	7.29	7.79	8.26	8.71
13	2.90	4.11	5.03	5.81	6.50	7.12	7.69	8.22	8.71	9.19
14	3.05	4.32	5.29	6.10	6.82	7.48	8.07	8.63	9.16	9.65
15	3.20	4.52	5.53	6.39	7.15	7.83	8.45	9.04	9.59	10.10
16	3.34	4.72	5.78	6.67	7.46	8.17	8.83	9.44	10.01	10.55
17	3.47	4.91	6.02	6.95	7.77	8.51	9.19	9.82	10.42	10.98
18	3.61	5.10	6.25	7.22	8.07	8.84	9.55	10.21	10.83	11.41
19	3.74	5.29	6.48	7.48	8.36	9.16	9.90	10.58	11.22	11.83
20	3.87	5.47	6.70	7.74	8.66	9.48	10.24	10.95	11.61	12.24

Table 8.—*Velocity of water, v, in feet per second, based on Manning's formula* $v = (1.486/n)r^{2/3}s^{1/2}$, **n** = **.020**—Continued

r \ s	.00055	.00060	.00065	.00070	.00075	.00080	.00085	.00090	.00095	.00100
0.2	0.60	0.62	0.65	0.67	0.70	0.72	0.74	0.76	0.78	0.80
0.4	.95	.99	1.03	1.07	1.10	1.14	1.18	1.21	1.24	1.28
0.6	1.24	1.29	1.35	1.40	1.45	1.49	1.54	1.59	1.63	1.67
0.8	1.50	1.57	1.63	1.69	1.75	1.81	1.87	1.92	1.97	2.02
1.0	1.74	1.82	1.89	1.97	2.03	2.10	2.17	2.23	2.29	2.35
1.2	1.97	2.06	2.14	2.22	2.30	2.37	2.45	2.52	2.59	2.65
1.4	2.18	2.28	2.37	2.46	2.55	2.63	2.71	2.79	2.87	2.94
1.6	2.38	2.49	2.59	2.69	2.78	2.87	2.96	3.05	3.13	3.21
1.8	2.58	2.69	2.80	2.91	3.01	3.11	3.21	3.30	3.39	3.48
2.0	2.77	2.89	3.01	3.12	3.23	3.34	3.44	3.54	3.64	3.73
2.2	2.95	3.08	3.20	3.33	3.44	3.55	3.66	3.77	3.87	3.97
2.4	3.12	3.26	3.40	3.52	3.65	3.77	3.88	4.00	4.11	4.21
2.6	3.29	3.44	3.58	3.72	3.85	3.97	4.10	4.21	4.33	4.44
2.8	3.46	3.62	3.76	3.91	4.04	4.17	4.30	4.43	4.55	4.67
3.0	3.62	3.79	3.94	4.09	4.23	4.37	4.51	4.64	4.76	4.89
3.2	3.78	3.95	4.11	4.27	4.42	4.56	4.70	4.84	4.97	5.10
3.4	3.94	4.12	4.28	4.44	4.60	4.75	4.90	5.04	5.18	5.31
3.6	4.09	4.27	4.45	4.62	4.78	4.94	5.09	5.24	5.38	5.52
3.8	4.24	4.43	4.61	4.79	4.95	5.12	5.27	5.43	5.58	5.72
4.0	4.39	4.59	4.77	4.95	5.13	5.30	5.46	5.62	5.77	5.92
4.2	4.54	4.74	4.93	5.12	5.30	5.47	5.64	5.80	5.96	6.12
4.4	4.68	4.89	5.09	5.28	5.46	5.64	5.82	5.99	6.15	6.31
4.6	4.82	5.03	5.24	5.44	5.63	5.81	5.99	6.17	6.33	6.50
4.8	4.96	5.18	5.39	5.59	5.79	5.98	6.16	6.34	6.52	6.69
5.0	5.10	5.32	5.54	5.75	5.95	6.14	6.33	6.52	6.70	6.87
5.2	5.23	5.46	5.69	5.90	6.11	6.31	6.50	6.69	6.87	7.05
5.4	5.36	5.60	5.83	6.05	6.26	6.47	6.67	6.86	7.05	7.23
5.6	5.49	5.74	5.97	6.20	6.42	6.63	6.83	7.03	7.22	7.41
5.8	5.62	5.88	6.12	6.35	6.57	6.78	6.99	7.20	7.39	7.58
6.0	5.75	6.01	6.25	6.49	6.72	6.94	7.15	7.36	7.56	7.76
6.2	5.88	6.14	6.39	6.63	6.87	7.09	7.31	7.52	7.73	7.93
6.4	6.01	6.27	6.53	6.78	7.01	7.24	7.47	7.68	7.89	8.10
6.6	6.13	6.40	6.67	6.92	7.16	7.39	7.62	7.84	8.06	8.27
6.8	6.25	6.53	6.80	7.06	7.30	7.54	7.78	8.00	8.22	8.43
7.0	6.38	6.66	6.93	7.19	7.45	7.69	7.93	8.16	8.38	8.60
7.5	6.68	6.97	7.26	7.53	7.80	8.05	8.30	8.54	8.77	9.00
8.0	6.97	7.28	7.58	7.86	8.14	8.41	8.66	8.92	9.16	9.40
8.5	7.26	7.58	7.89	8.19	8.47	8.75	9.02	9.28	9.54	9.79
9.0	7.54	7.87	8.20	8.51	8.80	9.09	9.37	9.64	9.91	10.17
9.5	7.82	8.16	8.50	8.82	9.13	9.43	9.72	10.00	10.27	10.54
10	8.09	8.45	8.79	9.12	9.44	9.75	10.05	10.35	10.63	10.91
11	8.62	9.00	9.37	9.72	10.06	10.39	10.71	11.02	11.33	11.62
12	9.13	9.54	9.93	10.30	10.67	11.02	11.35	11.68	12.00	12.32
13	9.63	10.06	10.47	10.87	11.25	11.62	11.98	12.32	12.66	12.99
14	10.12	10.57	11.00	11.42	11.82	12.21	12.58	12.95	13.30	13.65
15	10.60	11.07	11.52	11.96	12.38	12.78	13.18	13.56	13.93	14.29
16	11.06	11.56	12.03	12.48	12.92	13.34	13.75	14.15	14.54	14.92
17	11.52	12.03	12.52	13.00	13.45	13.89	14.32	14.74	15.14	15.53
18	11.97	12.50	13.01	13.50	13.98	14.43	14.88	15.31	15.73	16.14
19	12.41	12.96	13.49	14.00	14.49	14.96	15.42	15.87	16.31	16.73
20	12.84	13.41	13.96	14.48	14.99	15.48	15.96	16.42	16.87	17.31

Table 8.—*Velocity of water, v, in feet per second, based on Manning's formula* $v = (1.486/n)r^{2/3}s^{1/2}$, **n = .020**—Continued

r	.0010	.0015	.0020	.0025	.0030	.0035	.0040	.0045	.0050
0.2	0.80	0.98	1.14	1.27	1.39	1.50	1.61	1.70	1.80
0.4	1.28	1.56	1.80	2.02	2.21	2.39	2.55	2.71	2.85
0.6	1.67	2.05	2.36	2.64	2.90	3.13	3.34	3.55	3.74
0.8	2.02	2.48	2.86	3.20	3.51	3.79	4.05	4.30	4.53
1.0	2.35	2.88	3.32	3.72	4.07	4.40	4.70	4.98	5.25
1.2	2.65	3.25	3.75	4.20	4.60	4.96	5.31	5.63	5.93
1.4	2.94	3.60	4.16	4.65	5.09	5.50	5.88	6.24	6.57
1.6	3.21	3.94	4.55	5.08	5.57	6.01	6.43	6.82	7.19
1.8	3.48	4.26	4.92	5.50	6.02	6.50	6.95	7.38	7.77
2.0	3.73	4.57	5.27	5.90	6.46	6.98	7.46	7.91	8.34
2.2	3.97	4.87	5.62	6.28	6.88	7.44	7.95	8.43	8.89
2.4	4.21	5.16	5.96	6.66	7.29	7.88	8.42	8.93	9.42
2.6	4.44	5.44	6.28	7.02	7.69	8.31	8.89	9.42	9.93
2.8	4.67	5.72	6.60	7.38	8.08	8.73	9.34	9.90	10.44
3.0	4.89	5.99	6.91	7.73	8.47	9.14	9.77	10.37	10.93
3.2	5.10	6.25	7.22	8.07	8.84	9.55	10.20	10.82	11.41
3.4	5.31	6.51	7.51	8.40	9.20	9.94	10.63	11.27	11.88
3.6	5.52	6.76	7.80	8.73	9.56	10.33	11.04	11.71	12.34
3.8	5.72	7.01	8.09	9.05	9.91	10.70	11.44	12.14	12.79
4.0	5.92	7.25	8.37	9.36	10.25	11.08	11.84	12.56	13.24
4.2	6.12	7.49	8.65	9.67	10.59	11.44	12.23	12.97	13.68
4.4	6.31	7.73	8.92	9.98	10.93	11.80	12.62	13.38	14.11
4.6	6.50	7.96	9.19	10.28	11.26	12.16	13.00	13.79	14.53
4.8	6.69	8.19	9.46	10.57	11.58	12.51	13.37	14.18	14.95
5.0	6.87	8.41	9.72	10.86	11.90	12.85	13.74	14.57	15.36
5.2	7.05	8.64	9.97	11.15	12.21	13.19	14.10	14.96	15.77
5.4	7.23	8.86	10.23	11.43	12.53	13.53	14.46	15.34	16.17
5.6	7.41	9.07	10.48	11.72	12.83	13.86	14.82	15.72	16.57
5.8	7.58	9.29	10.73	11.99	13.14	14.19	15.17	16.09	16.96
6.0	7.76	9.50	10.97	12.27	13.44	14.51	15.52	16.46	17.35
6.2	7.93	9.71	11.21	12.54	13.73	14.83	15.86	16.82	17.73
6.4	8.10	9.92	11.45	12.81	14.03	15.15	16.20	17.18	18.11
6.6	8.27	10.13	11.69	13.07	14.32	15.47	16.53	17.54	18.49
6.8	8.43	10.33	11.93	13.33	14.61	15.78	16.87	17.89	18.86
7.0	8.60	10.53	12.16	13.59	14.89	16.09	17.20	18.24	19.23
7.5	9.00	11.03	12.73	14.23	15.59	16.84	18.00	19.10	20.13
8.0	9.40	11.51	13.29	14.86	16.28	17.58	18.80	19.94	21.02
8.5	9.79	11.99	13.84	15.47	16.95	18.31	19.57	20.76	21.88
9.0	10.17	12.45	14.38	16.07	17.61	19.02	20.33	21.57	22.73
9.5	10.54	12.91	14.90	16.66	18.25	19.72	21.08	22.36	23.57
10	10.91	13.36	15.42	17.24	18.89	20.40	21.81	23.13	24.39
11	11.62	14.23	16.43	18.37	20.13	21.74	23.24	24.65	25.99
12	12.32	15.08	17.42	19.47	21.33	23.04	24.63	26.12	27.54
13	12.99	15.91	18.37	20.54	22.50	24.30	25.98	27.56	29.05
14	13.65	16.72	19.30	21.58	23.64	25.53	27.30	28.95	30.52
15	14.29	17.50	20.21	22.60	24.75	26.74	28.58	30.31	31.95
16	14.92	18.27	21.10	23.59	25.84	27.91	29.84	31.65	33.36
17	15.53	19.03	21.97	24.56	26.91	29.06	31.07	32.95	34.74
18	16.14	19.76	22.82	25.52	27.95	30.19	32.28	34.23	36.08
19	16.73	20.49	23.66	26.45	28.98	31.30	33.46	35.49	37.41
20	17.31	21.20	24.48	27.37	29.98	32.39	34.62	36.72	38.71

Table 8.—*Velocity of water, v, in feet per second, based on Manning's formula* $v = (1.486/n)r^{2/3}s^{1/2}$, **n** = **.020**—Continued

s / r	.0055	.0060	.0065	.0070	.0075	.0080	.0085	.0090	.0095	.0100
0.2	1.88	1.97	2.05	2.13	2.20	2.27	2.34	2.41	2.48	2.54
0.4	2.99	3.12	3.25	3.37	3.49	3.61	3.72	3.83	3.93	4.03
0.6	3.92	4.09	4.26	4.42	4.58	4.73	4.87	5.01	5.15	5.29
0.8	4.75	4.96	5.16	5.36	5.55	5.73	5.90	6.07	6.24	6.40
1.0	5.51	5.76	5.99	6.22	6.43	6.65	6.85	7.05	7.24	7.43
1.2	6.22	6.50	6.76	7.02	7.27	7.50	7.74	7.96	8.18	8.39
1.4	6.90	7.20	7.50	7.78	8.05	8.32	8.57	8.82	9.06	9.30
1.6	7.54	7.87	8.19	8.50	8.80	9.09	9.37	9.64	9.91	10.16
1.8	8.15	8.52	8.86	9.20	9.52	9.83	10.14	10.43	10.72	10.99
2.0	8.75	9.14	9.51	9.87	10.21	10.55	10.87	11.19	11.50	11.79
2.2	9.32	9.74	10.13	10.52	10.88	11.24	11.59	11.92	12.25	12.57
2.4	9.88	10.32	10.74	11.14	11.53	11.91	12.28	12.64	12.98	13.32
2.6	10.42	10.88	11.33	11.75	12.17	12.57	12.95	13.33	13.69	14.05
2.8	10.95	11.43	11.90	12.35	12.78	13.20	13.61	14.00	14.39	14.76
3.0	11.46	11.97	12.46	12.93	13.38	13.82	14.25	14.66	15.06	15.46
3.2	11.97	12.50	13.01	13.50	13.97	14.43	14.88	15.31	15.73	16.13
3.4	12.46	13.01	13.54	14.06	14.55	15.03	15.49	15.94	16.37	16.80
3.6	12.94	13.52	14.07	14.60	15.11	15.61	16.09	16.56	17.01	17.45
3.8	13.42	14.01	14.59	15.14	15.67	16.18	16.68	17.16	17.63	18.09
4.0	13.88	14.50	15.09	15.66	16.21	16.75	17.26	17.76	18.25	18.72
4.2	14.34	14.98	15.59	16.18	16.75	17.30	17.83	18.35	18.85	19.34
4.4	14.80	15.45	16.08	16.69	17.28	17.84	18.39	18.93	19.45	19.95
4.6	15.24	15.92	16.57	17.19	17.80	18.38	18.95	19.50	20.03	20.55
4.8	15.68	16.38	17.05	17.69	18.31	18.91	19.49	20.06	20.61	21.14
5.0	16.11	16.83	17.52	18.18	18.81	19.43	20.03	20.61	21.18	21.73
5.2	16.54	17.27	17.98	18.66	19.31	19.95	20.56	21.16	21.74	22.30
5.4	16.96	17.71	18.44	19.13	19.81	20.45	21.08	21.70	22.29	22.87
5.6	17.38	18.15	18.89	19.60	20.29	20.96	21.60	22.23	22.84	23.43
5.8	17.79	18.58	19.34	20.07	20.77	21.45	22.11	22.75	23.38	23.99
6.0	18.19	19.00	19.78	20.53	21.25	21.94	22.62	23.27	23.91	24.53
6.2	18.60	19.42	20.22	20.98	21.72	22.43	23.12	23.79	24.44	25.08
6.4	18.99	19.84	20.65	21.43	22.18	22.91	23.61	24.30	24.96	25.61
6.6	19.39	20.25	21.08	21.87	22.64	23.38	24.10	24.80	25.48	26.14
6.8	19.78	20.66	21.50	22.31	23.10	23.85	24.59	25.30	25.99	26.67
7.0	20.16	21.06	21.92	22.75	23.55	24.32	25.07	25.79	26.50	27.19
7.5	21.11	22.05	22.95	23.82	24.65	25.46	26.25	27.01	27.75	28.47
8.0	22.04	23.02	23.96	24.87	25.74	26.58	27.40	28.19	28.97	29.72
8.5	22.95	23.97	24.95	25.89	26.80	27.68	28.53	29.36	30.16	30.95
9.0	23.84	24.90	25.92	26.90	27.84	28.75	29.64	30.50	31.33	32.15
9.5	24.72	25.82	26.87	27.88	28.86	29.81	30.73	31.62	32.48	33.33
10	25.58	26.71	27.80	28.85	29.87	30.85	31.80	32.72	33.61	34.49
11	27.25	28.47	29.63	30.75	31.83	32.87	33.88	34.86	35.82	36.75
12	28.88	30.17	31.40	32.58	33.73	34.83	35.90	36.95	37.96	38.94
13	30.46	31.82	33.12	34.37	35.58	36.74	37.87	38.97	40.04	41.08
14	32.01	33.43	34.80	36.11	37.38	38.60	39.79	40.94	42.07	43.16
15	33.51	35.00	36.43	37.81	39.14	40.42	41.66	42.87	44.05	45.19
16	34.99	36.54	38.04	39.47	40.86	42.20	43.50	44.76	45.98	47.18
17	36.43	38.05	39.60	41.10	42.54	43.94	45.29	46.60	47.88	49.12
18	37.85	39.53	41.14	42.70	44.19	45.64	47.05	48.41	49.74	51.03
19	39.23	40.98	42.65	44.26	45.82	47.32	48.78	50.19	51.56	52.90
20	40.60	42.41	44.14	45.80	47.41	48.97	50.47	51.94	53.36	54.74

Table 8.—*Velocity of water, v, in feet per second, based on Manning's formula* $v = (1.486/n)r^{2/3}s^{1/2}$, **n** = **.020**—Continued

r \ s	.01	.02	.03	.04	.05	.06	.07	.08	.09	.10
0.2	2.54	3.59	4.40	5.08	5.68	6.22	6.72	7.19	7.62	8.04
0.4	4.03	5.70	6.99	8.07	9.02	9.88	10.67	11.41	12.10	12.76
0.6	5.29	7.47	9.15	10.57	11.82	12.95	13.98	14.95	15.86	16.71
0.8	6.40	9.06	11.09	12.81	14.32	15.68	16.94	18.11	19.21	20.25
1.0	7.43	10.51	12.87	14.86	16.61	18.20	19.66	21.02	22.29	23.50
1.2	8.39	11.87	14.53	16.78	18.76	20.55	22.20	23.73	25.17	26.53
1.4	9.30	13.15	16.11	18.60	20.79	22.78	24.60	26.30	27.90	29.40
1.6	10.16	14.37	17.60	20.33	22.73	24.90	26.89	28.75	30.49	32.14
1.8	10.99	15.55	19.04	21.99	24.58	26.93	29.09	31.10	32.98	34.77
2.0	11.79	16.68	20.43	23.59	26.37	28.89	31.21	33.36	35.38	37.30
2.2	12.57	17.77	21.77	25.14	28.10	30.79	33.25	35.55	37.70	39.74
2.4	13.32	18.84	23.07	26.64	29.78	32.62	35 24.	37.67	39.96	42.12
2.6	14.05	19.87	24.33	28.10	31.41	34.41	37.17	39.74	42.15	44.43
2.8	14.76	20.87	25.57	29.52	33.00	36.16	39.05	41.75	44.28	46.68
3.0	15.46	21.86	26.77	30.91	34.56	37.86	40.89	43.71	46.37	48.87
3.2	16.13	22.82	27.95	32.27	36.08	39.52	42.69	45.64	48.40	51.02
3.4	16.80	23.76	29.10	33.60	37.57	41.15	44.45	47.52	50.40	53.13
3.6	17.45	24.68	30.23	34.90	39.02	42.75	46.17	49.36	52.36	55.19
3.8	18.09	25.59	31.34	36.19	40.46	44.32	47.87	51.17	54.28	57.22
4.0	18.72	26.48	32.43	37.44	41.86	45.86	49.53	52.96	56.17	59.21
4.2	19.34	27.35	33.50	38.68	43.25	47.38	51.17	54.71	58.02	61.16
4.4	19.95	28.21	34.56	39.90	44.61	48.87	52.78	56.43	59.85	63.09
4.6	20.55	29.06	35.59	41.10	45.95	50.34	54.37	58.13	61.65	64.99
4.8	21.14	29.90	36.62	42.28	47.28	51.79	55.94	59.80	63.43	66.86
5.0	21.73	30.72	37.63	43.45	48.58	53.22	57.48	61.45	65.18	68.70
5.2	22.30	31.54	38.63	44.60	49.87	54.63	59.00	63.08	66.90	70.52
5.4	22.87	32.34	39.61	45.74	51.14	56.02	60.51	64.68	68.61	72.32
5.6	23.43	33.14	40.58	46.86	52.39	57.39	61.99	66.27	70.29	74.09
5.8	23.99	33.92	41.54	47.97	53.63	58.75	63.46	67.84	71.96	75.85
6.0	24.53	34.70	42.49	49.07	54.86	60.09	64.91	69.39	73.60	77.58
6.2	25.08	35.46	43.43	50.15	56.07	61.42	66.34	70.92	75.23	79.30
6.4	25.61	36.22	44.36	51.22	57.27	62.74	67.76	72.44	76.84	80.99
6.6	26.14	36.97	45.28	52.29	58.46	64.04	69.17	73.94	78.43	82.67
6.8	26.67	37.71	46.19	53.34	59.63	65.32	70.56	75.43	80.00	84.33
7.0	27.19	38.45	47.09	54.38	60.80	66.60	71.93	76.90	81.57	85.98
7.5	28.47	40.26	49.31	56.94	63.66	69.73	75.32	80.52	85.41	90.02
8.0	29.72	42.03	51.48	59.44	66.46	72.80	78.63	84.06	89.16	93.98
8.5	30.95	43.76	53.60	61.89	69.20	75.80	81.87	87.53	92.84	97.86
9.0	32.15	45.46	55.68	64.30	71.88	78.75	85.05	90.93	96.44	101.66
9.5	33.33	47.13	57.73	66.66	74.52	81.64	88.18	94.26	99.98	105.39
10	34.49	48.77	59.73	68.97	77.12	84.48	91.24	97.54	103.46	109.06
11	36.75	51.97	63.65	73.50	82.17	90.02	97.23	103.94	110.25	116.21
12	38.94	55.08	67.45	77.89	87.08	95.39	103.04	110.15	116.83	123.15
13	41.08	58.09	71.15	82.16	91.85	100.62	108.68	116.19	123.24	129.90
14	43.16	61.04	74.75	86.32	96.51	105.72	114.19	122.07	129.48	136.48
15	45.19	63.91	78.27	90.38	101.05	110.69	119.56	127.82	135.57	142.91
16	47.18	66.72	81.71	94.36	105.49	115.56	124.82	133.44	141.53	149.19
17	49.12	69.47	85.08	98.25	109.84	120.33	129.97	138.94	147.37	155.34
18	51.03	72.17	88.39	102.06	114.11	125.00	135.02	144.34	153.09	161.38
19	52.90	74.82	91.63	105.81	118.30	129.59	139.97	149.64	158.71	167.30
20	54.74	77.42	94.82	109.49	122.41	134.10	144.84	154.84	164.23	173.12

Table 9.—*Velocity of water, v, in feet per second, based on Manning's formula $v = (1.486/n) r^{2/3} s^{1/2}$, n = .0225*

r \ s	.00005	.00010	.00015	.00020	.00025	.00030	.00035	.00040	.00045	.00050
0.2	0.16	0.23	0.28	0.32	0.36	0.39	0.42	0.45	0.48	0.51
0.4	.25	.36	.44	.51	.57	.62	.67	.72	.76	.80
0.6	.33	.47	.58	.66	.74	.81	.88	.94	1.00	1.05
0.8	.40	.57	.70	.80	.90	.99	1.06	1.14	1.21	1.27
1.0	.47	.66	.81	.93	1.04	1.14	1.24	1.32	1.40	1.48
1.2	.53	.75	.91	1.05	1.18	1.29	1.40	1.49	1.58	1.67
1.4	.58	.83	1.01	1.17	1.31	1.43	1.55	1.65	1.75	1.85
1.6	.64	.90	1.11	1.28	1.43	1.56	1.69	1.81	1.92	2.02
1.8	.69	.98	1.20	1.38	1.55	1.69	1.83	1.95	2.07	2.19
2.0	.74	1.05	1.28	1.48	1.66	1.82	1.96	2.10	2.22	2.34
2.2	.79	1.12	1.37	1.58	1.77	1.93	2.09	2.23	2.37	2.50
2.4	.84	1.18	1.45	1.67	1.87	2.05	2.21	2.37	2.51	2.65
2.6	.88	1.25	1.53	1.77	1.97	2.16	2.34	2.50	2.65	2.79
2.8	.93	1.31	1.61	1.86	2.07	2.27	2.45	2.62	2.78	2.93
3.0	.97	1.37	1.68	1.94	2.17	2.38	2.57	2.75	2.91	3.07
3.2	1.01	1.43	1.76	2.03	2.27	2.48	2.68	2.87	3.04	3.21
3.4	1.06	1.49	1.83	2.11	2.36	2.59	2.79	2.99	3.17	3.34
3.6	1.10	1.55	1.90	2.19	2.45	2.69	2.90	3.10	3.29	3.47
3.8	1.14	1.61	1.97	2.27	2.54	2.79	3.01	3.22	3.41	3.60
4.0	1.18	1.66	2.04	2.35	2.63	2.88	3.11	3.33	3.53	3.72
4.2	1.22	1.72	2.11	2.43	2.72	2.98	3.22	3.44	3.65	3.84
4.4	1.25	1.77	2.17	2.51	2.80	3.07	3.32	3.55	3.76	3.97
4.6	1.29	1.83	2.24	2.58	2.89	3.16	3.42	3.65	3.88	4.08
4.8	1.33	1.88	2.30	2.66	2.97	3.26	3.52	3.76	3.99	4.20
5.0	1.37	1.93	2.37	2.73	3.05	3.34	3.61	3.86	4.10	4.32
5.2	1.40	1.98	2.43	2.80	3.13	3.43	3.71	3.96	4.21	4.43
5.4	1.44	2.03	2.49	2.87	3.21	3.52	3.80	4.07	4.31	4.55
5.6	1.47	2.08	2.55	2.95	3.29	3.61	3.90	4.17	4.42	4.66
5.8	1.51	2.13	2.61	3.02	3.37	3.69	3.99	4.26	4.52	4.77
6.0	1.54	2.18	2.67	3.08	3.45	3.78	4.08	4.36	4.63	4.88
6.2	1.58	2.23	2.73	3.15	3.52	3.86	4.17	4.46	4.73	4.98
6.4	1.61	2.28	2.79	3.22	3.60	3.94	4.26	4.55	4.83	5.09
6.6	1.64	2.32	2.85	3.29	3.67	4.02	4.35	4.65	4.93	5.20
6.8	1.68	2.37	2.90	3.35	3.75	4.11	4.43	4.74	5.03	5.30
7.0	1.71	2.42	2.96	3.42	3.82	4.19	4.52	4.83	5.13	5.40
7.5	1.79	2.53	3.10	3.58	4.00	4.38	4.73	5.06	5.37	5.66
8.0	1.87	2.64	3.24	3.74	4.18	4.58	4.94	5.28	5.60	5.91
8.5	1.95	2.75	3.37	3.89	4.35	4.76	5.15	5.50	5.84	6.15
9.0	2.02	2.86	3.50	4.04	4.52	4.95	5.35	5.72	6.06	6.39
9.5	2.09	2.96	3.63	4.19	4.68	5.13	5.54	5.92	6.28	6.62
10	2.17	3.07	3.75	4.34	4.85	5.31	5.74	6.13	6.50	6.85
11	2.31	3.27	4.00	4.62	5.16	5.66	6.11	6.53	6.93	7.30
12	2.45	3.46	4.24	4.90	5.47	6.00	6.48	6.92	7.34	7.74
13	2.58	3.65	4.47	5.16	5.77	6.32	6.83	7.30	7.75	8.16
14	2.71	3.84	4.70	5.43	6.07	6.64	7.18	7.67	8.14	8.58
15	2.84	4.02	4.92	5.68	6.35	6.96	7.52	8.03	8.52	8.98
16	2.97	4.19	5.14	5.93	6.63	7.26	7.85	8.39	8.90	9.38
17	3.09	4.37	5.35	6.18	6.90	7.56	8.17	8.73	9.26	9.76
18	3.21	4.54	5.56	6.42	7.17	7.86	8.49	9.07	9.62	10.14
19	3.33	4.70	5.76	6.65	7.44	8.15	8.80	9.41	9.98	10.52
20	3.44	4.87	5.96	6.88	7.69	8.43	9.10	9.73	10.32	10.88

Table 9.—*Velocity of water, v, in feet per second, based on Manning's formula* $v = (1.486/n)r^{2/3}s^{1/2}$, **n = .0225**—Continued

r \ s	.00055	.00060	.00065	.00070	.00075	.00080	.00085	.00090	.00095	.00100
0.2	0.53	0.55	0.58	0.60	0.62	0.64	0.66	0.68	0.70	0.71
0.4	.84	.88	.91	.95	.98	1.01	1.05	1.08	1.11	1.13
0.6	1.10	1.15	1.20	1.24	1.29	1.33	1.37	1.41	1.45	1.49
0.8	1.33	1.39	1.45	1.51	1.56	1.61	1.66	1.71	1.75	1.80
1.0	1.55	1.62	1.68	1.75	1.81	1.87	1.93	1.98	2.04	2.09
1.2	1.75	1.83	1.90	1.97	2.04	2.11	2.17	2.24	2.30	2.36
1.4	1.94	2.02	2.11	2.19	2.26	2.34	2.41	2.48	2.55	2.61
1.6	2.12	2.21	2.30	2.39	2.47	2.56	2.63	2.71	2.78	2.86
1.8	2.29	2.39	2.49	2.59	2.68	2.76	2.85	2.93	3.01	3.09
2.0	2.46	2.57	2.67	2.77	2.87	2.97	3.06	3.15	3.23	3.32
2.2	2.62	2.74	2.85	2.96	3.06	3.16	3.26	3.35	3.44	3.53
2.4	2.78	2.90	3.02	3.13	3.24	3.35	3.45	3.55	3.65	3.74
2.6	2.93	3.06	3.18	3.30	3.42	3.53	3.64	3.75	3.85	3.95
2.8	3.08	3.21	3.35	3.47	3.59	3.71	3.83	3.94	4.04	4.15
3.0	3.22	3.37	3.50	3.63	3.76	3.89	4.01	4.12	4.23	4.34
3.2	3.36	3.51	3.66	3.79	3.93	4.06	4.18	4.30	4.42	4.54
3.4	3.50	3.66	3.81	3.95	4.09	4.22	4.35	4.48	4.60	4.72
3.6	3.64	3.80	3.96	4.10	4.25	4.39	4.52	4.65	4.78	4.91
3.8	3.77	3.94	4.10	4.26	4.40	4.55	4.69	4.82	4.96	5.09
4.0	3.90	4.08	4.24	4.40	4.56	4.71	4.85	4.99	5.13	5.26
4.2	4.03	4.21	4.38	4.55	4.71	4.86	5.01	5.16	5.30	5.44
4.4	4.16	4.34	4.52	4.69	4.86	5.02	5.17	5.32	5.47	5.61
4.6	4.28	4.47	4.66	4.83	5.00	5.17	5.33	5.48	5.63	5.78
4.8	4.41	4.60	4.79	4.97	5.15	5.32	5.48	5.64	5.79	5.94
5.0	4.53	4.73	4.92	5.11	5.29	5.46	5.63	5.79	5.95	6.11
5.2	4.65	4.86	5.05	5.24	5.43	5.61	5.78	5.95	6.11	6.27
5.4	4.77	4.98	5.18	5.38	5.57	5.75	5.93	6.10	6.27	6.43
5.6	4.88	5.10	5.31	5.51	5.71	5.89	6.07	6.25	6.42	6.59
5.8	5.00	5.22	5.44	5.64	5.84	6.03	6.22	6.40	6.57	6.74
6.0	5.11	5.34	5.56	5.77	5.97	6.17	6.36	6.54	6.72	6.90
6.2	5.23	5.46	5.68	5.90	6.10	6.30	6.50	6.69	6.87	7.05
6.4	5.34	5.58	5.80	6.02	6.23	6.44	6.64	6.83	7.02	7.20
6.6	5.45	5.69	5.92	6.15	6.36	6.57	6.77	6.97	7.16	7.35
6.8	5.56	5.81	6.04	6.27	6.49	6.70	6.91	7.11	7.31	7.50
7.0	5.67	5.92	6.16	6.39	6.62	6.84	7.05	7.25	7.45	7.64
7.5	5.93	6.20	6.45	6.70	6.93	7.16	7.38	7.59	7.80	8.00
8.0	6.20	6.47	6.74	6.99	7.23	7.47	7.70	7.93	8.14	8.35
8.5	6.45	6.74	7.01	7.28	7.53	7.78	8.02	8.25	8.48	8.70
9.0	6.70	7.00	7.29	7.56	7.83	8.08	8.33	8.57	8.81	9.04
9.5	6.95	7.26	7.55	7.84	8.11	8.38	8.64	8.89	9.13	9.37
10	7.19	7.51	7.82	8.11	8.40	8.67	8.94	9.20	9.45	9.69
11	7.66	8.00	8.33	8.64	8.95	9.24	9.52	9.80	10.07	10.33
12	8.12	8.48	8.83	9.16	9.48	9.79	10.09	10.39	10.67	10.95
13	8.56	8.94	9.31	9.66	10.00	10.33	10.65	10.95	11.25	11.55
14	9.00	9.40	9.78	10.15	10.51	10.85	11.18	11.51	11.82	12.13
15	9.42	9.84	10.24	10.63	11.00	11.36	11.71	12.05	12.38	12.70
16	9.83	10.27	10.69	11.10	11.48	11.86	12.23	12.58	12.93	13.26
17	10.24	10.70	11.13	11.55	11.96	12.35	12.73	13.10	13.46	13.81
18	10.64	11.11	11.56	12.00	12.42	12.83	13.22	13.61	13.98	14.34
19	11.03	11.52	11.99	12.44	12.88	13.30	13.71	14.11	14.49	14.87
20	11.41	11.92	12.41	12.87	13.33	13.76	14.19	14.60	15.00	15.39

Table 9.—*Velocity of water, v, in feet per second, based on Manning's formula* $v = (1.486/n)r^{2/3}s^{1/2}$, **n = .0225**—Continued

r \ s	.0010	.0015	.0020	.0025	.0030	.0035	.0040	.0045	.0050
0.2	0.71	0.87	1.01	1.13	1.24	1.34	1.43	1.52	1.60
0.4	1.13	1.39	1.60	1.79	1.96	2.12	2.27	2.41	2.54
0.6	1.49	1.82	2.10	2.35	2.57	2.78	2.97	3.15	3.32
0.8	1.80	2.20	2.55	2.85	3.12	3.37	3.60	3.82	4.02
1.0	2.09	2.56	2.95	3.30	3.62	3.91	4.18	4.43	4.67
1.2	2.36	2.89	3.34	3.73	4.08	4.41	4.72	5.00	5.27
1.4	2.61	3.20	3.70	4.13	4.53	4.89	5.23	5.54	5.84
1.6	2.86	3.50	4.04	4.52	4.95	5.35	5.71	6.06	6.39
1.8	3.09	3.78	4.37	4.89	5.35	5.78	6.18	6.56	6.91
2.0	3.32	4.06	4.69	5.24	5.74	6.20	6.63	7.03	7.41
2.2	3.53	4.33	5.00	5.59	6.12	6.61	7.07	7.49	7.90
2.4	3.74	4.59	5.29	5.92	6.48	7.00	7.49	7.94	8.37
2.6	3.95	4.84	5.58	6.24	6.84	7.39	7.90	8.38	8.83
2.8	4.15	5.08	5.87	6.56	7.19	7.76	8.30	8.80	9.28
3.0	4.34	5.32	6.14	6.87	7.52	8.13	8.69	9.22	9.71
3.2	4.54	5.55	6.41	7.17	7.86	8.48	9.07	9.62	10.14
3.4	4.72	5.78	6.68	7.47	8.18	8.83	9.44	10.02	10.56
3.6	4.91	6.01	6.94	7.76	8.50	9.18	9.81	10.41	10.97
3.8	5.09	6.23	7.19	8.04	8.81	9.51	10.17	10.79	11.37
4.0	5.26	6.45	7.44	8.32	9.12	9.85	10.53	11.16	11.77
4.2	5.44	6.66	7.69	8.60	9.42	10.17	10.87	11.53	12.16
4.4	5.61	6.87	7.93	8.87	9.71	10.49	11.22	11.90	12.54
4.6	5.78	7.07	8.17	9.13	10.01	10.81	11.55	12.25	12.92
4.8	5.94	7.28	8.40	9.40	10.29	11.12	11.89	12.61	13.29
5.0	6.11	7.48	8.64	9.66	10.58	11.42	12.21	12.95	13.66
5.2	6.27	7.68	8.87	9.91	10.86	11.73	12.54	13.30	14.02
5.4	6.43	7.87	9.09	10.16	11.13	12.03	12.86	13.64	14.37
5.6	6.59	8.07	9.31	10.41	11.41	12.32	13.17	13.97	14.73
5.8	6.74	8.26	9.53	10.66	11.68	12.61	13.48	14.30	15.08
6.0	6.90	8.45	9.75	10.90	11.94	12.90	13.79	14.63	15.42
6.2	7.05	8.63	9.97	11.14	12.21	13.19	14.10	14.95	15.76
6.4	7.20	8.82	10.18	11.38	12.47	13.47	14.40	15.27	16.10
6.6	7.35	9.00	10.39	11.62	12.73	13.75	14.70	15.59	16.43
6.8	7.50	9.18	10.60	11.85	12.98	14.02	14.99	15.90	16.76
7.0	7.64	9.36	10.81	12.08	13.24	14.30	15.28	16.21	17.09
7.5	8.00	9.80	11.32	12.65	13.86	14.97	16.00	16.98	17.89
8.0	8.35	10.23	11.81	13.21	14.47	15.63	16.71	17.72	18.68
8.5	8.70	10.65	12.30	13.75	15.07	16.27	17.40	18.45	19.45
9.0	9.04	11.07	12.78	14.29	15.65	16.91	18.07	19.17	20.21
9.5	9.37	11.47	13.25	14.81	16.23	17.53	18.74	19.87	20.95
10	9.69	11.87	13.71	15.33	16.79	18.14	19.39	20.56	21.68
11	10.33	12.65	14.61	16.33	17.89	19.33	20.66	21.91	23.10
12	10.95	13.41	15.48	17.31	18.96	20.48	21.89	23.22	24.48
13	11.55	14.14	16.33	18.26	20.00	21.60	23.09	24.49	25.82
14	12.13	14.86	17.16	19.18	21.01	22.70	24.26	25.74	27.13
15	12.70	15.56	17.96	20.08	22.00	23.76	25.41	26.95	28.40
16	13.26	16.24	18.75	20.97	22.97	24.81	26.52	28.13	29.65
17	13.81	16.91	19.53	21.83	23.92	25.83	27.62	29.29	30.88
18	14.34	17.57	20.29	22.68	24.85	26.84	28.69	30.43	32.08
19	14.87	18.21	21.03	23.51	25.76	27.82	29.74	31.55	33.25
20	15.39	18.85	21.76	24.33	26.65	28.79	30.78	32.64	34.41

Table 9.—*Velocity of water, v, in feet per second, based on Manning's formula* $v = (1.486/n)r^{2/3}s^{1/2}$, **n** = .0225—Continued

r / s	.0055	.0060	.0065	.0070	.0075	.0080	.0085	.0090	.0095	.0100
0.2	1.68	1.75	1.82	1.89	1.96	2.02	2.08	2.14	2.20	2.26
0.4	2.66	2.78	2.89	3.00	3.11	3.21	3.31	3.40	3.49	3.59
0.6	3.48	3.64	3.79	3.93	4.07	4.20	4.33	4.46	4.58	4.70
0.8	4.22	4.41	4.59	4.76	4.93	5.09	5.25	5.40	5.55	5.69
1.0	4.90	5.12	5.32	5.53	5.72	5.91	6.09	6.27	6.44	6.60
1.2	5.53	5.78	6.01	6.24	6.46	6.67	6.88	7.08	7.27	7.46
1.4	6.13	6.40	6.66	6.92	7.16	7.39	7.62	7.84	8.06	8.27
1.6	6.70	7.00	7.28	7.56	7.82	8.08	8.33	8.57	8.81	9.03
1.8	7.25	7.57	7.88	8.18	8.46	8.74	9.01	9.27	9.53	9.77
2.0	7.78	8.12	8.45	8.77	9.08	9.38	9.67	9.95	10.22	10.48
2.2	8.29	8.65	9.01	9.35	9.67	9.99	10.30	10.60	10.89	11.17
2.4	8.78	9.17	9.54	9.91	10.25	10.59	10.91	11.23	11.54	11.84
2.6	9.26	9.67	10.07	10.45	10.81	11.17	11.51	11.85	12.17	12.49
2.8	9.73	10.16	10.58	10.98	11.36	11.74	12.10	12.45	12.79	13.12
3.0	10.19	10.64	11.08	11.49	11.90	12.29	12.67	13.03	13.39	13.74
3.2	10.64	11.11	11.56	12.00	12.42	12.83	13.22	13.61	13.98	14.34
3.4	11.07	11.57	12.04	12.49	12.93	13.36	13.77	14.17	14.56	14.93
3.6	11.50	12.02	12.51	12.98	13.43	13.88	14.30	14.72	15.12	15.51
3.8	11.93	12.46	12.97	13.46	13.93	14.38	14.83	15.26	15.68	16.08
4.0	12.34	12.89	13.42	13.92	14.41	14.89	15.34	15.79	16.22	16.64
4.2	12.75	13.32	13.86	14.38	14.89	15.38	15.85	16.31	16.76	17.19
4.4	13.15	13.74	14.30	14.84	15.36	15.86	16.35	16.82	17.28	17.73
4.6	13.55	14.15	14.73	15.28	15.82	16.34	16.84	17.33	17.80	18.27
4.8	13.94	14.56	15.15	15.72	16.28	16.81	17.33	17.83	18.32	18.79
5.0	14.32	14.96	15.57	16.16	16.72	17.27	17.80	18.32	18.82	19.31
5.2	14.70	15.35	15.98	16.59	17.17	17.73	18.28	18.81	19.32	19.82
5.4	15.08	15.75	16.39	17.01	17.60	18.18	18.74	19.29	19.81	20.33
5.6	15.45	16.13	16.79	17.43	18.04	18.63	19.20	19.76	20.30	20.83
5.8	15.81	16.51	17.19	17.84	18.46	19.07	19.66	20.23	20.78	21.32
6.0	16.17	16.89	17.58	18.25	18.89	19.51	20.11	20.69	21.26	21.81
6.2	16.53	17.27	17.97	18.65	19.30	19.94	20.55	21.15	21.72	22.29
6.4	16.88	17.63	18.35	19.05	19.72	20.36	20.99	21.60	22.19	22.77
6.6	17.23	18.00	18.74	19.44	20.12	20.78	21.42	22.05	22.65	23.24
6.8	17.58	18.36	19.11	19.83	20.53	21.20	21.86	22.49	23.10	23.71
7.0	17.92	18.72	19.48	20.22	20.93	21.62	22.28	22.93	23.56	24.17
7.5	18.77	19.60	20.40	21.17	21.91	22.63	23.33	24.01	24.66	25.31
8.0	19.59	20.46	21.30	22.10	22.88	23.63	24.36	25.06	25.75	26.42
8.5	20.40	21.31	22.18	23.01	23.82	24.60	25.36	26.10	26.81	27.51
9.0	21.19	22.13	23.04	23.91	24.75	25.56	26.35	27.11	27.85	28.58
9.5	21.97	22.95	23.88	24.79	25.66	26.50	27.31	28.10	28.87	29.62
10	22.73	23.75	24.71	25.65	26.55	27.42	28.26	29.08	29.88	30.66
11	24.23	25.30	26.34	27.33	28.29	29.22	30.12	30.99	31.84	32.67
12	25.67	26.81	27.91	28.96	29.98	30.96	31.92	32.84	33.74	34.62
13	27.08	28.28	29.44	30.55	31.62	32.66	33.66	34.64	35.59	36.51
14	28.45	29.72	30.93	32.10	33.22	34.31	35.37	36.40	37.39	38.36
15	29.79	31.12	32.39	33.61	34.79	35.93	37.03	38.11	39.15	40.17
16	31.10	32.48	33.81	35.09	36.32	37.51	38.66	39.78	40.87	41.94
17	32.38	33.82	35.20	36.53	37.82	39.06	40.26	41.42	42.56	43.67
18	33.64	35.14	36.57	37.95	39.28	40.57	41.82	43.03	44.21	45.36
19	34.88	36.43	37.91	39.34	40.73	42.06	43.36	44.61	45.84	47.03
20	36.09	37.69	39.23	40.71	42.14	43.52	44.86	46.16	47.43	48.66

Table 9.—*Velocity of water, v, in feet per second, based on Manning's formula* $v = (1.486/n)r^{2/3}s^{1/2}$, **n = .0225**—Continued

r \ s	.01	.02	.03	.04	.05	.06	.07	.08	.09	.10
0.2	2.26	3.19	3.91	4.52	5.05	5.53	5.98	6.39	6.78	7.14
0.4	3.59	5.07	6.21	7.17	8.02	8.78	9.49	10.14	10.76	11.34
0.6	4.70	6.64	8.14	9.40	10.51	11.51	12.43	13.29	14.09	14.86
0.8	5.69	8.05	9.86	11.38	12.73	13.94	15.06	16.10	17.07	18.00
1.0	6.60	9.34	11.44	13.21	14.77	16.18	17.47	18.68	19.81	20.89
1.2	7.46	10.55	12.92	14.92	16.68	18.27	19.73	21.09	22.37	23.58
1.4	8.27	11.69	14.32	16.53	18.48	20.25	21.87	23.38	24.80	26.14
1.6	9.03	12.78	15.65	18.07	20.20	22.13	23.90	25.55	27.10	28.57
1.8	9.77	13.82	16.93	19.55	21.85	23.94	25.86	27.64	29.32	30.90
2.0	10.48	14.83	18.16	20.97	23.44	25.68	27.74	29.65	31.45	33.15
2.2	11.17	15.80	19.35	22.34	24.98	27.36	29.56	31.60	33.52	35.33
2.4	11.84	16.74	20.51	23.68	26.47	29.00	31.32	33.49	35.52	37.44
2.6	12.49	17.66	21.63	24.98	27.92	30.59	33.04	35.32	37.46	39.49
2.8	13.12	18.55	22.72	26.24	29.34	32.14	34.71	37.11	39.36	41.49
3.0	13.74	19.43	23.79	27.48	30.72	33.65	36.35	38.86	41.21	43.44
3.2	14.34	20.28	24.84	28.68	32.07	35.13	37.94	40.56	43.03	45.35
3.4	14.93	21.12	25.87	29.87	33.39	36.58	39.51	42.24	44.80	47.22
3.6	15.51	21.94	26.87	31.03	34.69	38.00	41.04	43.88	46.54	49.06
3.8	16.08	22.74	27.86	32.17	35.96	39.39	42.55	45.49	48.25	50.86
4.0	16.64	23.54	28.83	33.28	37.21	40.76	44.03	47.07	49.93	52.63
4.2	17.19	24.31	29.78	34.38	38.44	42.11	45.49	48.63	51.58	54.37
4.4	17.73	25.08	30.72	35.47	39.65	43.44	46.92	50.16	53.20	56.08
4.6	18.27	25.83	31.64	36.53	40.85	44.75	48.33	51.67	54.80	57.77
4.8	18.79	26.58	32.55	37.59	42.02	46.03	49.72	53.15	56.38	59.43
5.0	19.31	27.31	33.45	38.62	43.18	47.30	51.09	54.62	57.93	61.07
5.2	19.82	28.03	34.33	39.65	44.33	48.56	52.45	56.07	59.47	62.69
5.4	20.33	28.75	35.21	40.66	45.46	49.79	53.78	57.50	60.98	64.28
5.6	20.83	29.45	36.07	41.65	46.57	51.02	55.10	58.91	62.48	65.86
5.8	21.32	30.15	36.93	42.64	47.67	52.22	56.41	60.30	63.96	67.42
6.0	21.81	30.84	37.77	43.61	48.76	53.42	57.70	61.68	65.42	68.96
6.2	22.29	31.52	38.61	44.58	49.84	54.60	58.97	63.04	66.87	70.49
6.4	22.77	32.20	39.43	45.53	50.91	55.77	60.23	64.39	68.30	71.99
6.6	23.24	32.86	40.25	46.48	51.96	56.92	61.48	65.73	69.71	73.49
6.8	23.71	33.52	41.06	47.41	53.01	58.07	62.72	67.05	71.12	74.96
7.0	24.17	34.18	41.86	48.34	54.04	59.20	63.94	68.36	72.50	76.42
7.5	25.31	35.79	43.83	50.61	56.58	61.98	66.95	71.57	75.92	80.02
8.0	26.42	37.36	45.76	52.84	59.07	64.71	69.89	74.72	79.25	83.54
8.5	27.51	38.90	47.64	55.01	61.51	67.38	72.78	77.80	82.52	86.99
9.0	28.58	40.41	49.49	57.15	63.90	70.00	75.60	80.82	85.73	90.36
9.5	29.62	41.90	51.31	59.25	66.24	72.57	78.38	83.79	88.87	93.68
10	30.66	43.35	53.10	61.31	68.55	75.09	81.11	86.71	91.97	96.94
11	32.67	46.20	56.58	65.33	73.04	80.02	86.43	92.39	98.00	103.30
12	34.62	48.96	59.96	69.23	77.41	84.79	91.59	97.91	103.85	109.47
13	36.51	51.64	63.24	73.00	81.65	89.44	96.61	103.28	109.54	115.47
14	38.36	54.25	66.45	76.73	85.78	93.97	101.50	108.51	115.09	121.32
15	40.17	56.81	69.58	80.34	89.82	98.39	106.28	113.62	120.51	127.03
16	41.94	59.31	72.63	83.87	93.77	102.72	110.95	118.61	125.81	132.61
17	43.67	61.75	75.63	87.33	97.64	106.96	115.53	123.50	131.00	138.08
18	45.36	64.15	78.57	90.72	101.43	111.11	120.01	128.30	136.08	143.44
19	47.03	66.50	81.45	94.05	105.15	115.19	124.42	133.01	141.08	148.71
20	48.66	68.82	84.28	97.32	108.81	119.20	128.75	137.64	145.99	153.88

Table 10.—*Velocity of water, v, in feet per second, based on Manning's formula* $v = (1.486/n)r^{2/3}s^{1/2}$, **n = .025**

r / s	.00005	.00010	.00015	.00020	.00025	.00030	.00035	.00040	.00045	.00050
0.2	0.14	0.20	0.25	0.29	0.32	0.35	0.38	0.41	0.43	0.45
0.4	.23	.32	.40	.46	.51	.56	.60	.65	.68	.72
0.6	.30	.42	.52	.60	.67	.73	.79	.85	.90	.95
0.8	.36	.51	.63	.72	.81	.89	.96	1.02	1.09	1.15
1.0	.42	.59	.73	.84	.94	1.03	1.11	1.19	1.26	1.33
1.2	.47	.67	.82	.95	1.06	1.16	1.26	1.34	1.42	1.50
1.4	.53	.74	.91	1.05	1.18	1.29	1.39	1.49	1.58	1.66
1.6	.57	.81	1.00	1.15	1.29	1.41	1.52	1.63	1.72	1.82
1.8	.62	.88	1.08	1.24	1.39	1.52	1.65	1.76	1.87	1.97
2.0	.67	.94	1.16	1.33	1.49	1.63	1.77	1.89	2.00	2.11
2.2	.71	1.01	1.23	1.42	1.59	1.74	1.88	2.01	2.13	2.25
2.4	.75	1.07	1.30	1.51	1.68	1.85	1.99	2.13	2.26	2.38
2.6	.79	1.12	1.38	1.59	1.78	1.95	2.10	2.25	2.38	2.51
2.8	.83	1.18	1.45	1.67	1.87	2.05	2.21	2.36	2.50	2.64
3.0	.87	1.24	1.51	1.75	1.95	2.14	2.31	2.47	2.62	2.76
3.2	.91	1.29	1.58	1.83	2.04	2.24	2.41	2.58	2.74	2.89
3.4	.95	1.34	1.65	1.90	2.13	2.33	2.51	2.69	2.85	3.01
3.6	.99	1.40	1.71	1.97	2.21	2.42	2.61	2.79	2.96	3.12
3.8	1.02	1.45	1.77	2.05	2.29	2.51	2.71	2.89	3.07	3.24
4.0	1.06	1.50	1.83	2.12	2.37	2.59	2.80	3.00	3.18	3.35
4.2	1.09	1.55	1.90	2.19	2.45	2.68	2.89	3.09	3.28	3.46
4.4	1.13	1.60	1.95	2.26	2.52	2.76	2.99	3.19	3.39	3.57
4.6	1.16	1.64	2.01	2.33	2.60	2.85	3.08	3.29	3.49	3.68
4.8	1.20	1.69	2.07	2.39	2.67	2.93	3.16	3.38	3.59	3.78
5.0	1.23	1.74	2.13	2.46	2.75	3.01	3.25	3.48	3.69	3.89
5.2	1.26	1.78	2.19	2.52	2.82	3.09	3.34	3.57	3.78	3.99
5.4	1.29	1.83	2.24	2.59	2.89	3.17	3.42	3.66	3.88	4.09
5.6	1.33	1.87	2.30	2.65	2.96	3.25	3.51	3.75	3.98	4.19
5.8	1.36	1.92	2.35	2.71	3.03	3.32	3.59	3.84	4.07	4.29
6.0	1.39	1.96	2.40	2.78	3.10	3.40	3.67	3.93	4.16	4.39
6.2	1.42	2.01	2.46	2.84	3.17	3.47	3.75	4.01	4.26	4.49
6.4	1.45	2.05	2.51	2.90	3.24	3.55	3.83	4.10	4.35	4.58
6.6	1.48	2.09	2.56	2.96	3.31	3.62	3.91	4.18	4.44	4.68
6.8	1.51	2.13	2.61	3.02	3.37	3.70	3.99	4.27	4.53	4.77
7.0	1.54	2.18	2.66	3.08	3.44	3.77	4.07	4.35	4.61	4.86
7.5	1.61	2.28	2.79	3.22	3.60	3.94	4.26	4.55	4.83	5.09
8.0	1.68	2.38	2.91	3.36	3.76	4.12	4.45	4.76	5.04	5.32
8.5	1.75	2.48	3.03	3.50	3.91	4.29	4.63	4.95	5.25	5.54
9.0	1.82	2.57	3.15	3.64	4.07	4.45	4.81	5.14	5.46	5.75
9.5	1.89	2.67	3.27	3.77	4.22	4.62	4.99	5.33	5.66	5.96
10	1.95	2.76	3.38	3.90	4.36	4.78	5.16	5.52	5.85	6.17
11	2.08	2.94	3.60	4.16	4.65	5.09	5.50	5.88	6.24	6.57
12	2.20	3.12	3.82	4.41	4.93	5.40	5.83	6.23	6.61	6.97
13	2.32	3.29	4.02	4.65	5.20	5.69	6.15	6.57	6.97	7.35
14	2.44	3.45	4.23	4.88	5.46	5.98	6.46	6.91	7.32	7.72
15	2.56	3.62	4.43	5.11	5.72	6.26	6.76	7.23	7.67	8.08
16	2.67	3.77	4.62	5.34	5.97	6.54	7.06	7.55	8.01	8.44
17	2.78	3.93	4.81	5.56	6.21	6.81	7.35	7.86	8.34	8.79
18	2.89	4.08	5.00	5.77	6.46	7.07	7.64	8.17	8.66	9.13
19	2.99	4.23	5.18	5.99	6.69	7.33	7.92	8.46	8.98	9.46
20	3.10	4.38	5.36	6.19	6.92	7.59	8.19	8.76	9.29	9.79

Table 10.—*Velocity of water, v, in feet per second, based on Manning's formula* $v = (1.486/n)r^{2/3}s^{1/2}$, **n** = **.025**—Continued

s / r	.00055	.00060	.00065	.00070	.00075	.00080	.00085	.00090	.00095	.00100
0.2	0.48	0.50	0.52	0.54	0.56	0.57	0.59	0.61	0.63	0.64
0.4	.76	.79	.82	.85	.88	.91	.94	.97	.99	1.02
0.6	.99	1.04	1.08	1.12	1.16	1.20	1.23	1.27	1.30	1.34
0.8	1.20	1.25	1.31	1.36	1.40	1.45	1.49	1.54	1.58	1.62
1.0	1.39	1.46	1.52	1.57	1.63	1.68	1.73	1.78	1.83	1.88
1.2	1.57	1.64	1.71	1.78	1.84	1.90	1.96	2.01	2.07	2.12
1.4	1.74	1.82	1.90	1.97	2.04	2.10	2.17	2.23	2.29	2.35
1.6	1.91	1.99	2.07	2.15	2.23	2.30	2.37	2.44	2.51	2.57
1.8	2.06	2.15	2.24	2.33	2.41	2.49	2.56	2.64	2.71	2.78
2.0	2.21	2.31	2.41	2.50	2.58	2.67	2.75	2.83	2.91	2.98
2.2	2.36	2.46	2.56	2.66	2.75	2.84	2.93	3.02	3.10	3.18
2.4	2.50	2.61	2.72	2.82	2.92	3.01	3.11	3.20	3.28	3.37
2.6	2.64	2.75	2.87	2.97	3.08	3.18	3.28	3.37	3.46	3.55
2.8	2.77	2.89	3.01	3.12	3.23	3.34	3.44	3.54	3.64	3.73
3.0	2.90	3.03	3.15	3.27	3.39	3.50	3.60	3.71	3.81	3.91
3.2	3.03	3.16	3.29	3.42	3.53	3.65	3.76	3.87	3.98	4.08
3.4	3.15	3.29	3.43	3.56	3.68	3.80	3.92	4.03	4.14	4.25
3.6	3.27	3.42	3.56	3.69	3.82	3.95	4.07	4.19	4.30	4.42
3.8	3.39	3.55	3.69	3.83	3.96	4.09	4.22	4.34	4.46	4.58
4.0	3.51	3.67	3.82	3.96	4.10	4.24	4.37	4.49	4.62	4.74
4.2	3.63	3.79	3.94	4.09	4.24	4.38	4.51	4.64	4.77	4.89
4.4	3.74	3.91	4.07	4.22	4.37	4.51	4.65	4.79	4.92	5.05
4.6	3.86	4.03	4.19	4.35	4.50	4.65	4.79	4.93	5.07	5.20
4.8	3.97	4.14	4.31	4.47	4.63	4.78	4.93	5.07	5.21	5.35
5.0	4.08	4.26	4.43	4.60	4.76	4.92	5.07	5.21	5.36	5.50
5.2	4.18	4.37	4.55	4.72	4.89	5.05	5.20	5.35	5.50	5.64
5.4	4.29	4.48	4.66	4.84	5.01	5.17	5.33	5.49	5.64	5.79
5.6	4.40	4.59	4.78	4.96	5.13	5.30	5.46	5.62	5.78	5.93
5.8	4.50	4.70	4.89	5.08	5.25	5.43	5.59	5.76	5.91	6.07
6.0	4.60	4.81	5.00	5.19	5.37	5.55	5.72	5.89	6.05	6.21
6.2	4.70	4.91	5.11	5.31	5.49	5.67	5.85	6.02	6.18	6.34
6.4	4.81	5.02	5.22	5.42	5.61	5.80	5.97	6.15	6.32	6.48
6.6	4.90	5.12	5.33	5.53	5.73	5.92	6.10	6.27	6.45	6.61
6.8	5.00	5.23	5.44	5.64	5.84	6.03	6.22	6.40	6.58	6.75
7.0	5.10	5.33	5.55	5.75	5.96	6.15	6.34	6.53	6.70	6.88
7.5	5.34	5.58	5.81	6.03	6.24	6.44	6.64	6.83	7.02	7.20
8.0	5.58	5.82	6.06	6.29	6.51	6.72	6.93	7.13	7.33	7.52
8.5	5.81	6.06	6.31	6.55	6.78	7.00	7.22	7.43	7.63	7.83
9.0	6.03	6.30	6.56	6.80	7.04	7.27	7.50	7.72	7.93	8.13
9.5	6.25	6.53	6.80	7.05	7.30	7.54	7.77	8.00	8.22	8.43
10	6.47	6.76	7.03	7.30	7.56	7.80	8.04	8.28	8.50	8.72
11	6.89	7.20	7.50	7.78	8.05	8.32	8.57	8.82	9.06	9.30
12	7.31	7.63	7.94	8.24	8.53	8.81	9.08	9.35	9.60	9.85
13	7.71	8.05	8.38	8.69	9.00	9.30	9.58	9.86	10.13	10.39
14	8.10	8.46	8.80	9.14	9.46	9.77	10.07	10.36	10.64	10.92
15	8.48	8.86	9.22	9.57	9.90	10.23	10.54	10.85	11.14	11.43
16	8.85	9.24	9.62	9.99	10.34	10.68	11.00	11.32	11.63	11.94
17	9.22	9.63	10.02	10.40	10.76	11.12	11.46	11.79	12.11	12.43
18	9.57	10.00	10.41	10.80	11.18	11.55	11.90	12.25	12.58	12.91
19	9.93	10.37	10.79	11.20	11.59	11.97	12.34	12.70	13.04	13.38
20	10.27	10.73	11.17	11.59	11.99	12.39	12.77	13.14	13.50	13.85

Table 10.—*Velocity of water, v, in feet per second, based on Manning's formula* $v = (1.486/n)r^{2/3}s^{1/2}$, **n** $= .025$—Continued

r	.0010	.0015	.0020	.0025	.0030	.0035	.0040	.0045	.0050
0.2	0.64	0.79	0.91	1.02	1.11	1.20	1.29	1.36	1.44
0.4	1.02	1.25	1.44	1.61	1.77	1.91	2.04	2.16	2.28
0.6	1.34	1.64	1.89	2.11	2.32	2.50	2.67	2.84	2.99
0.8	1.62	1.98	2.29	2.56	2.81	3.03	3.24	3.44	3.62
1.0	1.88	2.30	2.66	2.97	3.26	3.52	3.76	3.99	4.20
1.2	2.12	2.60	3.00	3.36	3.68	3.97	4.25	4.50	4.75
1.4	2.35	2.88	3.33	3.72	4.07	4.40	4.70	4.99	5.26
1.6	2.57	3.15	3.64	4.07	4.45	4.81	5.14	5.45	5.75
1.8	2.78	3.41	3.93	4.40	4.82	5.20	5.56	5.90	6.22
2.0	2.98	3.65	4.22	4.72	5.17	5.58	5.97	6.33	6.67
2.2	3.18	3.89	4.50	5.03	5.51	5.95	6.36	6.74	7.11
2.4	3.37	4.13	4.77	5.33	5.84	6.30	6.74	7.15	7.53
2.6	3.55	4.35	5.03	5.62	6.16	6.65	7.11	7.54	7.95
2.8	3.73	4.57	5.28	5.90	6.47	6.99	7.47	7.92	8.35
3.0	3.91	4.79	5.53	6.18	6.77	7.31	7.82	8.29	8.74
3.2	4.08	5.00	5.77	6.45	7.07	7.64	8.16	8.66	9.13
3.4	4.25	5.21	6.01	6.72	7.36	7.95	8.50	9.02	9.50
3.6	4.42	5.41	6.24	6.98	7.65	8.26	8.83	9.37	9.87
3.8	4.58	5.61	6.47	7.24	7.93	8.56	9.15	9.71	10.23
4.0	4.74	5.80	6.70	7.49	8.20	8.86	9.47	10.05	10.59
4.2	4.89	5.99	6.92	7.74	8.47	9.15	9.79	10.38	10.94
4.4	5.05	6.18	7.14	7.98	8.74	9.44	10.09	10.71	11.29
4.6	5.20	6.37	7.35	8.22	9.00	9.73	10.40	11.03	11.63
4.8	5.35	6.55	7.56	8.46	9.26	10.01	10.70	11.35	11.96
5.0	5.50	6.73	7.77	8.69	9.52	10.28	10.99	11.66	12.29
5.2	5.64	6.91	7.98	8.92	9.77	10.55	11.28	11.97	12.62
5.4	5.79	7.09	8.18	9.15	10.02	10.82	11.57	12.27	12.94
5.6	5.93	7.26	8.38	9.37	10.27	11.09	11.85	12.57	13.25
5.8	6.07	7.43	8.58	9.59	10.51	11.35	12.14	12.87	13.57
6.0	6.21	7.60	8.78	9.81	10.75	11.61	12.41	13.17	13.88
6.2	6.34	7.77	8.97	10.03	10.99	11.87	12.69	13.46	14.18
6.4	6.48	7.94	9.16	10.24	11.22	12.12	12.96	13.74	14.49
6.6	6.61	8.10	9.35	10.46	11.46	12.37	13.23	14.03	14.79
6.8	6.75	8.26	9.54	10.67	11.69	12.62	13.49	14.31	15.09
7.0	6.88	8.42	9.73	10.88	11.91	12.87	13.76	14.59	15.38
7.5	7.20	8.82	10.19	11.39	12.47	13.47	14.40	15.28	16.10
8.0	7.52	9.21	10.63	11.89	13.02	14.07	15.04	15.95	16.81
8.5	7.83	9.59	11.07	12.38	13.56	14.65	15.66	16.61	17.51
9.0	8.13	9.96	11.50	12.86	14.09	15.22	16.27	17.25	18.19
9.5	8.43	10.33	11.92	13.33	14.60	15.77	16.86	17.89	18.85
10	8.72	10.69	12.34	13.79	15.11	16.32	17.45	18.51	19.51
11	9.30	11.39	13.15	14.70	16.10	17.39	18.59	19.72	20.79
12	9.85	12.07	13.93	15.58	17.06	18.43	19.70	20.90	22.03
13	10.39	12.73	14.70	16.43	18.00	19.44	20.78	22.05	23.24
14	10.92	13.37	15.44	17.26	18.91	20.43	21.84	23.16	24.41
15	11.43	14.00	16.17	18.08	19.80	21.39	22.86	24.25	25.56
16	11.94	14.62	16.88	18.87	20.67	22.33	23.87	25.32	26.69
17	12.43	15.22	17.57	19.65	21.52	23.25	24.85	26.36	27.79
18	12.91	15.81	18.26	20.41	22.36	24.15	25.82	27.39	28.87
19	13.38	16.39	18.93	21.16	23.18	25.04	26.77	28.39	29.93
20	13.85	16.96	19.59	21.90	23.99	25.91	27.70	29.38	30.97

Table 10.—*Velocity of water, v, in feet per second, based on Manning's formula* $v = (1.486/n)r^{2/3}s^{1/2}$, **n = .025**—Continued

r \ s	.0055	.0060	.0065	.0070	.0075	.0080	.0085	.0090	.0095	.0100
0.2	1.51	1.57	1.64	1.70	1.76	1.82	1.87	1.93	1.98	2.03
0.4	2.39	2.50	2.60	2.70	2.79	2.89	2.98	3.06	3.15	3.23
0.6	3.14	3.28	3.41	3.54	3.66	3.78	3.90	4.01	4.12	4.23
0.8	3.80	3.97	4.13	4.29	4.44	4.58	4.72	4.86	4.99	5.12
1.0	4.41	4.60	4.79	4.97	5.15	5.32	5.48	5.64	5.79	5.94
1.2	4.98	5.20	5.41	5.62	5.81	6.00	6.19	6.37	6.54	6.71
1.4	5.52	5.76	6.00	6.22	6.44	6.65	6.86	7.06	7.25	7.44
1.6	6.03	6.30	6.56	6.80	7.04	7.27	7.50	7.71	7.93	8.13
1.8	6.52	6.81	7.09	7.36	7.62	7.87	8.11	8.34	8.57	8.80
2.0	7.00	7.31	7.61	7.89	8.17	8.44	8.70	8.95	9.20	9.44
2.2	7.46	7.79	8.11	8.41	8.71	8.99	9.27	9.54	9.80	10.05
2.4	7.90	8.25	8.59	8.91	9.23	9.53	9.82	10.11	10.39	10.65
2.6	8.34	8.71	9.06	9.40	9.73	10.05	10.36	10.66	10.95	11.24
2.8	8.76	9.15	9.52	9.88	10.23	10.56	10.89	11.20	11.51	11.81
3.0	9.17	9.58	9.97	10.34	10.71	11.06	11.40	11.73	12.05	12.36
3.2	9.57	10.00	10.41	10.80	11.18	11.54	11.90	12.25	12.58	12.91
3.4	9.97	10.41	10.84	11.24	11.64	12.02	12.39	12.75	13.10	13.44
3.6	10.35	10.81	11.26	11.68	12.09	12.49	12.87	13.25	13.61	13.96
3.8	10.73	11.21	11.67	12.11	12.54	12.95	13.34	13.73	14.11	14.47
4.0	11.11	11.60	12.08	12.53	12.97	13.40	13.81	14.21	14.60	14.98
4.2	11.48	11.99	12.47	12.95	13.40	13.84	14.27	14.68	15.08	15.47
4.4	11.84	12.36	12.87	13.35	13.82	14.28	14.71	15.14	15.56	15.96
4.6	12.19	12.73	13.25	13.76	14.24	14.70	15.16	15.60	16.02	16.44
4.8	12.54	13.10	13.64	14.15	14.65	15.13	15.59	16.05	16.49	16.91
5.0	12.89	13.46	14.01	14.54	15.05	15.55	16.02	16.49	16.94	17.38
5.2	13.23	13.82	14.38	14.93	15.45	15.96	16.45	16.93	17.39	17.84
5.4	13.57	14.17	14.75	15.31	15.84	16.36	16.87	17.36	17.83	18.30
5.6	13.90	14.52	15.11	15.68	16.23	16.77	17.28	17.78	18.27	18.74
5.8	14.23	14.86	15.47	16.05	16.62	17.16	17.69	18.20	18.70	19.19
6.0	14.56	15.20	15.82	16.42	17.00	17.55	18.09	18.62	19.13	19.63
6.2	14.88	15.54	16.17	16.78	17.37	17.94	18.49	19.03	19.55	20.06
6.4	15.20	15.87	16.52	17.14	17.74	18.33	18.89	19.44	19.97	20.49
6.6	15.51	16.20	16.86	17.50	18.11	18.71	19.28	19.84	20.38	20.91
6.8	15.82	16.53	17.20	17.85	18.48	19.08	19.67	20.24	20.79	21.33
7.0	16.13	16.85	17.54	18.20	18.84	19.45	20.05	20.63	21.20	21.75
7.5	16.89	17.64	18.36	19.05	19.72	20.37	21.00	21.61	22.20	22.77
8.0	17.63	18.42	19.17	19.89	20.59	21.27	21.92	22.56	23.17	23.78
8.5	18.36	19.18	19.96	20.71	21.44	22.14	22.82	23.49	24.13	24.76
9.0	19.07	19.92	20.73	21.52	22.27	23.00	23.71	24.40	25.07	25.72
9.5	19.77	20.65	21.50	22.31	23.09	23.85	24.58	25.29	25.99	26.66
10	20.46	21.37	22.24	23.08	23.89	24.68	25.44	26.17	26.89	27.59
11	21.80	22.77	23.70	24.60	25.46	26.30	27.11	27.89	28.66	29.40
12	23.11	24.13	25.12	26.07	26.98	27.87	28.72	29.56	30.37	31.16
13	24.37	25.46	26.50	27.50	28.46	29.39	30.30	31.18	32.03	32.86
14	25.61	26.74	27.84	28.89	29.90	30.88	31.83	32.76	33.65	34.53
15	26.81	28.00	29.15	30.25	31.31	32.34	33.33	34.30	35.24	36.15
16	27.99	29.23	30.43	31.58	32.69	33.76	34.80	35.81	36.79	37.74
17	29.14	30.44	31.68	32.88	34.03	35.15	36.23	37.28	38.30	39.30
18	30.28	31.62	32.91	34.16	35.36	36.52	37.64	38.73	39.79	40.83
19	31.39	32.78	34.12	35.41	36.65	37.86	39.02	40.15	41.25	42.32
20	32.48	33.92	35.31	36.64	37.93	39.17	40.38	41.55	42.69	43.80

Table 10.—*Velocity of water, v, in feet per second, based on Manning's formula* $v = (1.486/n) r^{2/3} s^{1/2}$, **n = .025**—Continued

r \ s	.01	.02	.03	.04	.05	.06	.07	.08	.09	.10
0.2	2.03	2.87	3.52	4.07	4.55	4.98	5.38	5.75	6.10	6.43
0.4	3.23	4.56	5.59	6.45	7.22	7.90	8.54	9.13	9.68	10.20
0.6	4.23	5.98	7.32	8.46	9.46	10.36	11.19	11.96	12.69	13.37
0.8	5.12	7.24	8.87	10.24	11.45	12.55	13.55	14.49	15.37	16.20
1.0	5.94	8.41	10.30	11.89	13.29	14.56	15.73	16.81	17.83	18.80
1.2	6.71	9.49	11.63	13.42	15.01	16.44	17.76	18.99	20.14	21.23
1.4	7.44	10.52	12.88	14.88	16.63	18.22	19.68	21.04	22.32	23.52
1.6	8.13	11.50	14.08	16.26	18.18	19.92	21.51	23.00	24.39	25.71
1.8	8.80	12.44	15.23	17.59	19.67	21.54	23.27	24.88	26.39	27.81
2.0	9.44	13.34	16.34	18.87	21.10	23.11	24.96	26.69	28.31	29.84
2.2	10.05	14.22	17.41	20.11	22.48	24.63	26.60	28.44	30.16	31.80
2.4	10.65	15.07	18.45	21.31	23.83	26.10	28.19	30.14	31.96	33.69
2.6	11.24	15.89	19.47	22.48	25.13	27.53	29.74	31.79	33.72	35.54
2.8	11.81	16.70	20.45	23.62	26.40	28.92	31.24	33.40	35.42	37.34
3.0	12.36	17.49	21.42	24.73	27.65	30.29	32.71	34.97	37.09	39.10
3.2	12.91	18.25	22.36	25.82	28.86	31.62	34.15	36.51	38.72	40.82
3.4	13.44	19.01	23.28	26.88	30.05	32.92	35.56	38.01	40.32	42.50
3.6	13.96	19.75	24.18	27.92	31.22	34.20	36.94	39.49	41.89	44.15
3.8	14.47	20.47	25.07	28.95	32.37	35.45	38.30	40.94	43.42	45.77
4.0	14.98	21.18	25.94	29.96	33.49	36.69	39.63	42.36	44.93	47.36
4.2	15.47	21.88	26.80	30.95	34.60	37.90	40.94	43.76	46.42	48.93
4.4	15.96	22.57	27.64	31.92	35.69	39.10	42.23	45.14	47.88	50.47
4.6	16.44	23.25	28.48	32.88	36.76	40.27	43.50	46.50	49.32	51.99
4.8	16.91	23.92	29.30	33.83	37.82	41.43	44.75	47.84	50.74	53.49
5.0	17.38	24.58	30.10	34.76	38.86	42.57	45.98	49.16	52.14	54.96
5.2	17.84	25.23	30.90	35.68	39.89	43.70	47.20	50.46	53.52	56.42
5.4	18.30	25.87	31.69	36.59	40.91	44.81	48.41	51.75	54.89	57.86
5.6	18.74	26.51	32.47	37.49	41.91	45.91	49.59	53.02	56.23	59.27
5.8	19.19	27.14	33.23	38.38	42.91	47.00	50.77	54.27	57.56	60.68
6.0	19.63	27.76	33.99	39.25	43.89	48.08	51.93	55.51	58.88	62.06
6.2	20.06	28.37	34.75	40.12	44.86	49.14	53.07	56.74	60.18	63.44
6.4	20.49	28.98	35.49	40.98	45.82	50.19	54.21	57.95	61.47	64.79
6.6	20.91	29.58	36.22	41.83	46.77	51.23	55.33	59.15	62.74	66.14
6.8	21.33	30.17	36.95	42.67	47.71	52.26	56.45	60.34	64.00	67.47
7.0	21.75	30.76	37.67	43.50	48.64	53.28	57.55	61.52	65.25	68.78
7.5	22.77	32.21	39.45	45.55	50.93	55.79	60.26	64.42	68.32	72.02
8.0	23.78	33.62	41.18	47.55	53.16	58.24	62.91	67.25	71.33	75.19
8.5	24.76	35.01	42.88	49.51	55.36	60.64	65.50	70.02	74.27	78.29
9.0	25.72	36.37	44.55	51.44	57.51	63.00	68.04	72.74	77.15	81.33
9.5	26.66	37.71	46.18	53.32	59.62	65.31	70.54	75.41	79.99	84.31
10	27.59	39.02	47.79	55.18	61.69	67.58	73.00	78.04	82.77	87.25
11	29.40	41.58	50.92	58.80	65.74	72.01	77.78	83.15	88.20	92.97
12	31.16	44.06	53.96	62.31	69.67	76.31	82.43	88.12	93.47	98.52
13	32.86	46.48	56.92	65.73	73.48	80.50	86.95	92.95	98.59	103.92
14	34.53	48.83	59.80	69.05	77.21	84.57	91.35	97.66	103.58	109.19
15	36.15	51.13	62.62	72.31	80.84	88.56	95.65	102.26	108.46	114.32
16	37.74	53.38	65.37	75.48	84.39	92.45	99.86	106.75	113.23	119.35
17	39.30	55.58	68.07	78.60	87.87	96.26	103.97	111.15	117.90	124.27
18	40.83	57.74	70.71	81.65	91.29	100.00	108.01	115.47	122.48	129.10
19	42.32	59.85	73.31	84.65	94.64	103.67	111.98	119.71	126.97	133.84
20	43.80	61.94	75.86	87.59	97.93	107.28	115.87	123.87	131.39	138.49

Table 11.—*Velocity of water, v, in feet per second, based on Manning's formula* $v = (1.486/n)r^{2/3}s^{1/2}$, n = .0275

r \ s	.00005	.00010	.00015	.00020	.00025	.00030	.00035	.00040	.00045	.00050
0.2	0.13	0.18	0.23	0.26	0.29	0.32	0.35	0.37	0.39	0.41
0.4	.21	.29	.36	.41	.46	.51	.55	.59	.62	.66
0.6	.27	.38	.47	.54	.61	.67	.72	.77	.82	.86
0.8	.33	.47	.57	.66	.74	.81	.87	.93	.99	1.04
1.0	.38	.54	.66	.76	.85	.94	1.01	1.08	1.15	1.21
1.2	.43	.61	.75	.86	.96	1.06	1.14	1.22	1.29	1.36
1.4	.48	.68	.83	.96	1.07	1.17	1.27	1.35	1.43	1.51
1.6	.52	.74	.91	1.05	1.17	1.28	1.38	1.48	1.57	1.65
1.8	.57	.80	.98	1.13	1.26	1.38	1.50	1.60	1.70	1.79
2.0	.61	.86	1.05	1.21	1.36	1.49	1.60	1.72	1.82	1.92
2.2	.65	.91	1.12	1.29	1.45	1.58	1.71	1.83	1.94	2.04
2.4	.68	.97	1.19	1.37	1.53	1.68	1.81	1.94	2.05	2.17
2.6	.72	1.02	1.25	1.44	1.62	1.77	1.91	2.04	2.17	2.28
2.8	.76	1.07	1.31	1.52	1.70	1.86	2.01	2.15	2.28	2.40
3.0	.79	1.12	1.38	1.59	1.78	1.95	2.10	2.25	2.38	2.51
3.2	.83	1.17	1.44	1.66	1.86	2.03	2.20	2.35	2.49	2.62
3.4	86	1.22	1.50	1.73	1.93	2.12	2.29	2.44	2.59	2.73
3.6	.90	1.27	1.55	1.80	2.01	2.20	2.37	2.54	2.69	2.84
3.8	.93	1.32	1.61	1.86	2.08	2.28	2.46	2.63	2.79	2.94
4.0	.96	1.36	1.67	1.93	2.15	2.36	2.55	2.72	2.89	3.04
4.2	.99	1.41	1.72	1.99	2.22	2.44	2.63	2.81	2.98	3.15
4.4	1.03	1.45	1.78	2.05	2.29	2.51	2.71	2.90	3.08	3.24
4.6	1.06	1.49	1.83	2.11	2.36	2.59	2.80	2.99	3.17	3.34
4.8	1.09	1.54	1.88	2.17	2.43	2.66	2.88	3.08	3.26	3.44
5.0	1.12	1.58	1.94	2.23	2.50	2.74	2.96	3.16	3.35	3.53
5.2	1.15	1.62	1.99	2.29	2.56	2.81	3.03	3.24	3.44	3.63
5.4	1.18	1.66	2.04	2.35	2.63	2.88	3.11	3.33	3.53	3.72
5.6	1.20	1.70	2.09	2.41	2.69	2.95	3.19	3.41	3.61	3.81
5.8	1.23	1.74	2.14	2.47	2.76	3.02	3.26	3.49	3.70	3.90
6.0	1.26	1.78	2.19	2.52	2.82	3.09	3.34	3.57	3.78	3.99
6.2	1.29	1.82	2.23	2.58	2.88	3.16	3.41	3.65	3.87	4.08
6.4	1.32	1.86	2.28	2.63	2.95	3.23	3.48	3.73	3.95	4.17
6.6	1.34	1.90	2.33	2.69	3.01	3.29	3.56	3.80	4.03	4.25
6.8	1.37	1.94	2.38	2.74	3.07	3.36	3.63	3.88	4.11	4.34
7.0	1.40	1.98	2.42	2.80	3.13	3.42	3.70	3.95	4.19	4.42
7.5	1.46	2.07	2.54	2.93	3.27	3.59	3.87	4.14	4.39	4.63
8.0	1.53	2.16	2.65	3.06	3.42	3.74	4.04	4.32	4.59	4.83
8.5	1.59	2.25	2.76	3.18	3.56	3.90	4.21	4.50	4.77	5.03
9.0	1.65	2.34	2.86	3.31	3.70	4.05	4.37	4.68	4.96	5.23
9.5	1.71	2.42	2.97	3.43	3.83	4.20	4.53	4.85	5.14	5.42
10	1.77	2.51	3.07	3.55	3.97	4.34	4.69	5.02	5.32	5.61
11	1.89	2.67	3.27	3.78	4.23	4.63	5.00	5.35	5.67	5.98
12	2.00	2.83	3.47	4.01	4.48	4.91	5.30	5.66	6.01	6.33
13	2.11	2.99	3.66	4.23	4.72	5.17	5.59	5.98	6.34	6.68
14	2.22	3.14	3.84	4.44	4.96	5.44	5.87	6.28	6.66	7.02
15	2.32	3.29	4.03	4.65	5.20	5.69	6.15	6.57	6.97	7.35
16	2.43	3.43	4.20	4.85	5.43	5.94	6.42	6.86	7.28	7.67
17	2.53	3.57	4.38	5.05	5.65	6.19	6.68	7.15	7.58	7.99
18	2.62	3.71	4.55	5.25	5.87	6.43	6.94	7.42	7.87	8.30
19	2.72	3.85	4.71	5.44	6.08	6.66	7.20	7.70	8.16	8.60
20	2.82	3.98	4.88	5.63	6.30	6.90	7.45	7.96	8.45	8.90

Table 11.—*Velocity of water, v, in feet per second, based on Manning's formula* $v=(1.486/n)r^{2/3}s^{1/2}$, **n=.0275**—Continued

s / r	.00055	.00060	.00065	.00070	.00075	.00080	.00085	.00090	.00095	.00100
0.2	0.43	0.45	0.47	0.49	0.51	0.52	0.54	0.55	0.57	0.58
0.4	.69	.72	.75	.78	.80	.83	.86	.88	.90	.93
0.6	.90	.94	.98	1.02	1.05	1.09	1.12	1.15	1.18	1.22
0.8	1.09	1.14	1.19	1.23	1.28	1.32	1.36	1.40	1.44	1.47
1.0	1.27	1.32	1.38	1.43	1.48	1.53	1.58	1.62	1.67	1.71
1.2	1.43	1.49	1.56	1.61	1.67	1.73	1.78	1.83	1.88	1.93
1.4	1.59	1.66	1.72	1.79	1.85	1.91	1.97	2.03	2.08	2.14
1.6	1.73	1.81	1.88	1.96	2.02	2.09	2.16	2.22	2.28	2.34
1.8	1.88	1.96	2.04	2.12	2.19	2.26	2.33	2.40	2.46	2.53
2.0	2.01	2.10	2.19	2.27	2.35	2.43	2.50	2.57	2.64	2.71
2.2	2.14	2.24	2.33	2.42	2.50	2.59	2.66	2.74	2.82	2.89
2.4	2.27	2.37	2.47	2.56	2.65	2.74	2.82	2.91	2.99	3.06
2.6	2.40	2.50	2.60	2.70	2.80	2.89	2.98	3.07	3.15	3.23
2.8	2.52	2.63	2.74	2.84	2.94	3.04	3.13	3.22	3.31	3.39
3.0	2.64	2.75	2.87	2.97	3.08	3.18	3.28	3.37	3.46	3.55
3.2	2.75	2.87	2.99	3.10	3.21	3.32	3.42	3.52	3.62	3.71
3.4	2.87	2.99	3.12	3.23	3.35	3.46	3.56	3.67	3.77	3.86
3.6	2.98	3.11	3.24	3.36	3.48	3.59	3.70	3.81	3.91	4.01
3.8	3.09	3.22	3.35	3.48	3.60	3.72	3.84	3.95	4.06	4.16
4.0	3.19	3.34	3.47	3.60	3.73	3.85	3.97	4.08	4.20	4.31
4.2	3.30	3.45	3.59	3.72	3.85	3.98	4.10	4.22	4.34	4.45
4.4	3.40	3.55	3.70	3.84	3.97	4.10	4.23	4.35	4.47	4.59
4.6	3.51	3.66	3.81	3.95	4.09	4.23	4.36	4.48	4.61	4.73
4.8	3.61	3.77	3.92	4.07	4.21	4.35	4.48	4.61	4.74	4.86
5.0	3.71	3.87	4.03	4.18	4.33	4.47	4.61	4.74	4.87	5.00
5.2	3.80	3.97	4.14	4.29	4.44	4.59	4.73	4.87	5.00	5.13
5.4	3.90	4.07	4.24	4.40	4.55	4.70	4.85	4.99	5.13	5.26
5.6	4.00	4.17	4.34	4.51	4.67	4.82	4.97	5.11	5.25	5.39
5.8	4.09	4.27	4.45	4.62	4.78	4.93	5.09	5.23	5.38	5.52
6.0	4.18	4.37	4.55	4.72	4.89	5.05	5.20	5.35	5.50	5.64
6.2	4.28	4.47	4.65	4.82	4.99	5.16	5.32	5.47	5.62	5.77
6.4	4.37	4.56	4.75	4.93	5.10	5.27	5.43	5.59	5.74	5.89
6.6	4.46	4.66	4.85	5.03	5.21	5.38	5.54	5.70	5.86	6.01
6.8	4.55	4.75	4.94	5.13	5.31	5.49	5.65	5.82	5.98	6.13
7.0	4.64	4.84	5.04	5.23	5.42	5.59	5.76	5.93	6.09	6.25
7.5	4.86	5.07	5.28	5.48	5.67	5.86	6.04	6.21	6.38	6.55
8.0	5.07	5.29	5.51	5.72	5.92	6.11	6.30	6.48	6.66	6.84
8.5	5.28	5.51	5.74	5.95	6.16	6.37	6.56	6.75	6.94	7.12
9.0	5.48	5.73	5.96	6.19	6.40	6.61	6.82	7.01	7.21	7.39
9.5	5.68	5.94	6.18	6.41	6.64	6.86	7.07	7.27	7.47	7.66
10	5.88	6.14	6.39	6.64	6.87	7.09	7.31	7.52	7.73	7.93
11	6.27	6.55	6.81	7.07	7.32	7.56	7.79	8.02	8.24	8.45
12	6.64	6.94	7.22	7.49	7.76	8.01	8.26	8.50	8.73	8.96
13	7.01	7.32	7.62	7.90	8.18	8.45	8.71	8.96	9.21	9.45
14	7.36	7.69	8.00	8.30	8.60	8.88	9.15	9.42	9.67	9.93
15	7.71	8.05	8.38	8.70	9.00	9.30	9.58	9.86	10.13	10.39
16	8.05	8.40	8.75	9.08	9.40	9.70	10.00	10.29	10.58	10.85
17	8.38	8.75	9.11	9.45	9.78	10.10	10.42	10.72	11.01	11.30
18	8.70	9.09	9.46	9.82	10.16	10.50	10.82	11.13	11.44	11.74
19	9.02	9.42	9.81	10.18	10.54	10.88	11.22	11.54	11.86	12.17
20	9.34	9.75	10.15	10.53	10.90	11.26	11.61	11.94	12.27	12.59

Table 11.—*Velocity of water, v, in feet per second, based on Manning's formula* $v=(1.486/n)r^{2/3}s^{1/2}$, **n=.0275**—Continued

r \ s	.0010	.0015	.0020	.0025	.0030	.0035	.0040	.0045	.0050
0.2	0.58	0.72	0.83	0.92	1.01	1.09	1.17	1.24	1.31
0.4	.93	1.14	1.31	1.47	1.61	1.74	1.86	1.97	2.07
0.6	1.22	1.49	1.72	1.92	2.11	2.27	2.43	2.58	2.72
0.8	1.47	1.80	2.08	2.33	2.55	2.75	2.95	3.12	3.29
1.0	1.71	2.09	2.42	2.70	2.96	3.20	3.42	3.62	3.82
1.2	1.93	2.36	2.73	3.05	3.34	3.61	3.86	4.09	4.31
1.4	2.14	2.62	3.02	3.38	3.70	4.00	4.28	4.54	4.78
1.6	2.34	2.86	3.31	3.70	4.05	4.37	4.68	4.96	5.23
1.8	2.53	3.10	3.58	4.00	4.38	4.73	5.06	5.36	5.65
2.0	2.71	3.32	3.84	4.29	4.70	5.07	5.43	5.75	6.07
2.2	2.89	3.54	4.09	4.57	5.01	5.41	5.78	6.13	6.46
2.4	3.06	3.75	4.33	4.84	5.31	5.73	6.13	6.50	6.85
2.6	3.23	3.96	4.57	5.11	5.60	6.04	6.46	6.85	7.22
2.8	3.39	4.16	4.80	5.37	5.88	6.35	6.79	7.20	7.59
3.0	3.55	4.35	5.03	5.62	6.16	6.65	7.11	7.54	7.95
3.2	3.71	4.54	5.25	5.87	6.43	6.94	7.42	7.87	8.30
3.4	3.86	4.73	5.46	6.11	6.69	7.23	7.73	8.20	8.64
3.6	4.01	4.92	5.68	6.35	6.95	7.51	8.03	8.51	8.98
3.8	4.16	5.10	5.88	6.58	7.21	7.78	8.32	8.83	9.30
4.0	4.31	5.27	6.09	6.81	7.46	8.06	8.61	9.13	9.63
4.2	4.45	5.45	6.29	7.03	7.70	8.32	8.90	9.44	9.95
4.4	4.59	5.62	6.49	7.25	7.95	8.58	9.18	9.73	10.26
4.6	4.73	5.79	6.68	7.47	8.19	8.84	9.45	10.03	10.57
4.8	4.86	5.96	6.88	7.69	8.42	9.10	9.72	10.31	10.87
5.0	5.00	6.12	7.07	7.90	8.65	9.35	9.99	10.60	11.17
5.2	5.13	6.28	7.25	8.11	8.88	9.60	10.26	10.88	11.47
5.4	5.26	6.44	7.44	8.32	9.11	9.84	10.52	11.16	11.76
5.6	5.39	6.60	7.62	8.52	9.33	10.08	10.78	11.43	12.05
5.8	5.52	6.76	7.80	8.72	9.55	10.32	11.03	11.70	12.33
6.0	5.64	6.91	7.98	8.92	9.77	10.56	11.28	11.97	12.62
6.2	5.77	7.06	8.16	9.12	9.99	10.79	11.53	12.23	12.90
6.4	5.89	7.21	8.33	9.31	10.20	11.02	11.78	12.50	13.17
6.6	6.01	7.36	8.50	9.51	10.41	11.25	12.02	12.75	13.44
6.8	6.13	7.51	8.67	9.70	10.62	11.47	12.27	13.01	13.71
7.0	6.25	7.66	8.84	9.89	10.83	11.70	12.51	13.26	13.98
7.5	6.55	8.02	9.26	10.35	11.34	12.25	13.09	13.89	14.64
8.0	6.84	8.37	9.67	10.81	11.84	12.79	13.67	14.50	15.28
8.5	7.12	8.72	10.06	11.25	12.33	13.31	14.23	15.10	15.91
9.0	7.39	9.06	10.46	11.69	12.81	13.83	14.79	15.68	16.53
9.5	7.66	9.39	10.84	12.12	13.28	14.34	15.33	16.26	17.14
10	7.93	9.71	11.22	12.54	13.74	14.84	15.86	16.83	17.74
11	8.45	10.35	11.95	13.36	14.64	15.81	16.90	17.93	18.90
12	8.96	10.97	12.67	14.16	15.51	16.76	17.91	19.00	20.03
13	9.45	11.57	13.36	14.94	16.36	17.67	18.89	20.04	21.13
14	9.93	12.16	14.04	15.69	17.19	18.57	19.85	21.06	22.20
15	10.39	12.73	14.70	16.43	18.00	19.44	20.79	22.05	23.24
16	10.85	13.29	15.34	17.16	18.79	20.30	21.70	23.02	24.26
17	11.30	13.84	15.98	17.86	19.57	21.14	22.60	23.97	25.26
18	11.74	14.37	16.60	18.56	20.33	21.96	23.47	24.90	26.24
19	12.17	14.90	17.21	19.24	21.07	22.76	24.33	25.81	27.21
20	12.59	15.42	17.81	19.91	21.81	23.55	25.18	26.71	28.15

Table 11.—*Velocity of water, v, in feet per second, based on Manning's formula $v = (1.486/n) r^{2/3} s^{1/2}$, n = .0275*—Continued

s r	.0055	.0060	.0065	.0070	.0075	.0080	.0085	.0090	.0095	.0100
0.2	1.37	1.43	1.49	1.55	1.60	1.65	1.70	1.75	1.80	1.85
0.4	2.18	2.27	2.37	2.45	2.54	2.62	2.70	2.78	2.86	2.93
0.6	2.85	2.98	3.10	3.22	3.33	3.44	3.54	3.65	3.75	3.84
0.8	3.45	3.61	3.75	3.90	4.03	4.17	4.29	4.42	4.54	4.66
1.0	4.01	4.19	4.36	4.52	4.68	4.83	4.98	5.13	5.27	5.40
1.2	4.53	4.73	4.92	5.11	5.28	5.46	5.63	5.79	5.95	6.10
1.4	5.02	5.24	5.45	5.66	5.86	6.05	6.23	6.42	6.59	6.76
1.6	5.48	5.73	5.96	6.18	6.40	6.61	6.82	7.01	7.20	7.39
1.8	5.93	6.19	6.45	6.69	6.92	7.15	7.37	7.59	7.79	8.00
2.0	6.36	6.64	6.92	7.18	7.43	7.67	7.91	8.14	8.36	8.58
2.2	6.78	7.08	7.37	7.65	7.92	8.18	8.43	8.67	8.91	9.14
2.4	7.18	7.50	7.81	8.10	8.39	8.66	8.93	9.19	9.44	9.69
2.6	7.58	7.91	8.24	8.55	8.85	9.14	9.42	9.69	9.96	10.22
2.8	7.96	8.32	8.65	8.98	9.30	9.60	9.90	10.18	10.46	10.73
3.0	8.34	8.71	9.06	9.40	9.73	10.05	10.36	10.66	10.96	11.24
3.2	8.70	9.09	9.46	9.82	10.16	10.50	10.82	11.13	11.44	11.73
3.4	9.06	9.46	9.85	10.22	10.58	10.93	11.26	11.59	11.91	12.22
3.6	9.41	9.83	10.23	10.62	10.99	11.35	11.70	12.04	12.37	12.69
3.8	9.76	10.19	10.61	11.01	11.40	11.77	12.13	12.48	12.83	13.16
4.0	10.10	10.55	10.98	11.39	11.79	12.18	12.55	12.92	13.27	13.62
4.2	10.43	10.90	11.34	11.77	12.18	12.58	12.97	13.34	13.71	14.07
4.4	10.76	11.24	11.70	12.14	12.57	12.98	13.38	13.76	14.14	14.51
4.6	11.08	11.58	12.05	12.50	12.94	13.37	13.78	14.18	14.57	14.95
4.8	11.40	11.91	12.40	12.86	13.32	13.75	14.18	14.59	14.99	15.38
5.0	11.72	12.24	12.74	13.22	13.68	14.13	14.57	14.99	15.40	15.80
5.2	12.03	12.56	13.08	13.57	14.05	14.51	14.95	15.39	15.81	16.22
5.4	12.33	12.88	13.41	13.92	14.40	14.88	15.33	15.78	16.21	16.63
5.6	12.64	13.20	13.74	14.26	14.76	15.24	15.71	16.17	16.61	17.04
5.8	12.94	13.51	14.06	14.59	15.11	15.60	16.08	16.55	17.00	17.44
6.0	13.23	13.82	14.39	14.93	15.45	15.96	16.45	16.93	17.39	17.84
6.2	13.52	14.13	14.70	15.26	15.79	16.31	16.81	17.30	17.77	18.24
6.4	13.81	14.43	15.02	15.58	16.13	16.66	17.17	17.67	18.16	18.63
6.6	14.10	14.73	15.33	15.91	16.47	17.01	17.53	18.04	18.53	19.01
6.8	14.38	15.02	15.64	16.23	16.80	17.35	17.88	18.40	18.90	19.40
7.0	14.66	15.32	15.94	16.54	17.12	17.69	18.23	18.76	19.27	19.77
7.5	15.35	16.04	16.69	17.32	17.93	18.52	19.09	19.64	20.18	20.70
8.0	16.03	16.74	17.43	18.08	18.72	19.33	19.93	20.51	21.07	21.61
8.5	16.69	17.43	18.14	18.83	19.49	20.13	20.75	21.35	21.94	22.51
9.0	17.34	18.11	18.85	19.56	20.25	20.91	21.56	22.18	22.79	23.38
9.5	17.98	18.77	19.54	20.28	20.99	21.68	22.35	22.99	23.62	24.24
10	18.60	19.43	20.22	20.98	21.72	22.43	23.12	23.79	24.45	25.08
11	19.82	20.70	21.55	22.36	23.15	23.91	24.64	25.36	26.05	26.73
12	21.00	21.94	22.83	23.70	24.53	25.33	26.11	26.87	27.61	28.32
13	22.16	23.14	24.09	25.00	25.87	26.72	27.54	28.34	29.12	29.88
14	23.28	24.31	25.31	26.26	27.18	28.08	28.94	29.78	30.59	31.39
15	24.37	25.46	26.50	27.50	28.46	29.40	30.30	31.18	32.03	32.87
16	25.45	26.58	27.66	28.71	29.71	30.69	31.63	32.55	33.44	34.31
17	26.50	27.67	28.80	29.89	30.94	31.95	32.94	33.89	34.82	35.73
18	27.52	28.75	29.92	31.05	32.14	33.20	34.22	35.21	36.17	37.11
19	28.53	29.80	31.02	32.19	33.32	34.41	35.47	36.50	37.50	38.48
20	29.53	30.84	32.10	33.31	34.48	35.61	36.71	37.77	38.81	39.81

Table 11.—*Velocity of water, v, in feet per second, based on Manning's formula* $v=(1.486/n)r^{2/3}s^{1/2}$, **n**=.0275—Continued

r \ s	.01	.02	.03	.04	.05	.06	.07	.08	.09	.10
0.2	1.85	2.61	3.20	3.70	4.13	4.53	4.89	5.23	5.54	5.84
0.4	2.93	4.15	5.08	5.87	6.56	7.19	7.76	8.30	8.80	9.28
0.6	3.84	5.44	6.66	7.69	8.60	9.42	10.17	10.87	11.53	12.16
0.8	4.66	6.59	8.07	9.31	10.41	11.41	12.32	13.17	13.97	14.73
1.0	5.40	7.64	9.36	10.81	12.08	13.24	14.30	15.28	16.21	17.09
1.2	6.10	8.63	10.57	12.20	13.64	14.95	16.14	17.26	18.31	19.30
1.4	6.76	9.56	11.71	13.52	15.12	16.56	17.89	19.13	20.29	21.38
1.6	7.39	10.45	12.80	14.78	16.53	18.11	19.56	20.91	22.18	23.38
1.8	8.00	11.31	13.85	15.99	17.88	19.59	21.16	22.62	23.99	25.29
2.0	8.58	12.13	14.86	17.16	19.18	21.01	22.69	24.26	25.73	27.13
2.2	9.14	12.93	15.83	18.28	20.44	22.39	24.18	25.85	27.42	28.90
2.4	9.69	13.70	16.78	19.37	21.66	23.73	25.63	27.40	29.06	30.63
2.6	10.22	14.45	17.70	20.43	22.85	25.03	27.03	28.90	30.65	32.31
2.8	10.73	15.18	18.59	21.47	24.00	26.29	28.40	30.36	32.20	33.95
3.0	11.24	15.90	19.47	22.48	25.13	27.53	29.74	31.79	33.72	35.54
3.2	11.73	16.59	20.32	23.47	26.24	28.74	31.05	33.19	35.20	37.11
3.4	12.22	17.28	21.16	24.44	27.32	29.93	32.33	34.56	36.65	38.64
3.6	12.69	17.95	21.98	25.39	28.38	31.09	33.58	35.90	38.08	40.14
3.8	13.16	18.61	22.79	26.32	29.42	32.23	34.81	37.22	39.48	41.61
4.0	13.62	19.26	23.58	27.23	30.45	33.35	36.03	38.51	40.85	43.06
4.2	14.07	19.89	24.36	28.13	31.45	34.46	37.22	39.79	42.20	44.48
4.4	14.51	20.52	25.13	29.02	32.44	35.54	38.39	41.04	43.53	45.88
4.6	14.95	21.14	25.89	29.89	33.42	36.61	39.54	42.27	44.84	47.26
4.8	15.38	21.75	26.63	30.75	34.38	37.66	40.68	43.49	46.13	48.62
5.0	15.80	22.35	27.37	31.60	35.33	38.70	41.80	44.69	47.40	49.97
5.2	16.22	22.94	28.09	32.44	36.27	39.73	42.91	45.87	48.66	51.29
5.4	16.63	23.52	28.81	33.26	37.19	40.74	44.00	47.04	49.90	52.60
5.6	17.04	24.10	29.51	34.08	38.10	41.74	45.08	48.20	51.12	53.89
5.8	17.44	24.67	30.21	34.89	39.01	42.73	46.15	49.34	52.33	55.16
6.0	17.84	25.23	30.90	35.68	39.90	43.70	47.21	50.47	53.53	56.42
6.2	18.24	25.79	31.59	36.47	40.78	44.67	48.25	51.58	54.71	57.67
6.4	18.63	26.34	32.26	37.25	41.65	45.63	49.28	52.68	55.88	58.90
6.6	19.01	26.89	32.93	38.03	42.51	46.57	50.30	53.78	57.04	60.12
6.8	19.40	27.43	33.59	38.79	43.37	47.51	51.31	54.86	58.19	61.33
7.0	19.77	27.96	34.25	39.55	44.22	48.44	52.32	55.93	59.32	62.53
7.5	20.70	29.28	35.86	41.41	46.30	50.71	54.78	58.56	62.11	65.47
8.0	21.61	30.57	37.44	43.23	48.33	52.94	57.19	61.14	64.84	68.35
8.5	22.51	31.83	38.98	45.01	50.32	55.13	59.55	63.66	67.52	71.17
9.0	23.38	33.06	40.50	46.76	52.28	57.27	61.86	66.13	70.14	73.93
9.5	24.24	34.28	41.98	48.48	54.20	59.37	64.13	68.56	72.71	76.65
10	25.08	35.47	43.44	50.16	56.08	61.44	66.36	70.94	75.24	79.31
11	26.73	37.80	46.29	53.45	59.76	65.47	70.71	75.59	80.18	84.52
12	28.32	40.05	49.06	56.65	63.33	69.38	74.94	80.11	84.97	89.57
13	29.88	42.25	51.75	59.75	66.80	73.18	79.04	84.50	89.63	94.47
14	31.39	44.39	54.37	62.78	70.19	76.89	83.05	88.78	94.17	99.26
15	32.87	46.48	56.93	65.73	73.49	80.50	86.96	92.96	98.60	103.93
16	34.31	48.52	59.43	68.62	76.72	84.04	90.78	97.05	102.93	108.50
17	35.73	50.52	61.88	71.45	79.89	87.51	94.52	101.05	107.18	112.98
18	37.11	52.49	64.28	74.23	82.99	90.91	98.19	104.97	111.34	117.36
19	38.48	54.41	66.64	76.95	86.03	94.25	101.80	108.83	115.43	121.67
20	39.81	56.31	68.96	79.63	89.03	97.52	105.34	112.61	119.44	125.90

Table 12.—*Velocity of water, v, in feet per second, based on Manning's formula* $v = (1.486/n)r^{2/3}s^{1/2}$, **n = .030**

s\r	.00005	.00010	.00015	.00020	.00025	.00030	.00035	.00040	.00045	.00050
0.2	0.12	0.17	0.21	0.24	0.27	0.29	0.32	0.34	0.36	0.38
0.4	.19	.27	.33	.38	.43	.47	.50	.54	.57	.60
0.6	.25	.35	.43	.50	.56	.61	.66	.70	.75	.79
0.8	.30	.43	.52	.60	.67	.74	.80	.85	.91	.95
1.0	.35	.50	.61	.70	.78	.86	.93	.99	1.05	1.11
1.2	.40	.56	.69	.79	.88	.97	1.05	1.12	1.19	1.25
1.4	.44	.62	.76	.88	.98	1.07	1.16	1.24	1.31	1.39
1.6	.48	.68	.83	.96	1.07	1.17	1.27	1.36	1.44	1.52
1.8	.52	.73	.90	1.04	1.16	1.27	1.37	1.47	1.55	1.64
2.0	.56	.79	.96	1.11	1.24	1.36	1.47	1.57	1.67	1.76
2.2	.59	.84	1.03	1.18	1.32	1.45	1.57	1.68	1.78	1.87
2.4	.63	.89	1.09	1.26	1.40	1.54	1.66	1.78	1.88	1.99
2.6	.66	.94	1.15	1.32	1.48	1.62	1.75	1.87	1.99	2.09
2.8	.70	.98	1.21	1.39	1.56	1.70	1.84	1.97	2.09	2.20
3.0	.73	1.03	1.26	1.46	1.63	1.78	1.93	2.06	2.19	2.30
3.2	.76	1.08	1.32	1.52	1.70	1.86	2.01	2.15	2.28	2.41
3.4	.79	1.12	1.37	1.58	1.77	1.94	2.10	2.24	2.38	2.50
3.6	.82	1.16	1.42	1.65	1.84	2.02	2.18	2.33	2.47	2.60
3.8	.85	1.21	1.48	1.71	1.91	2.09	2.26	2.41	2.56	2.70
4.0	.88	1.25	1.53	1.77	1.97	2.16	2.34	2.50	2.65	2.79
4.2	.91	1.29	1.58	1.82	2.04	2.23	2.41	2.58	2.74	2.88
4.4	.94	1.33	1.63	1.88	2.10	2.30	2.49	2.66	2.82	2.97
4.6	.97	1.37	1.68	1.94	2.17	2.37	2.56	2.74	2.91	3.06
4.8	1.00	1.41	1.73	1.99	2.23	2.44	2.64	2.82	2.99	3.15
5.0	1.02	1.45	1.77	2.05	2.29	2.51	2.71	2.90	3.07	3.24
5.2	1.05	1.49	1.82	2.10	2.35	2.58	2.78	2.97	3.15	3.32
5.4	1.08	1.52	1.87	2.16	2.41	2.64	2.85	3.05	3.23	3.41
5.6	1.10	1.56	1.91	2.21	2.47	2.71	2.92	3.12	3.31	3.49
5.8	1.13	1.60	1.96	2.26	2.53	2.77	2.99	3.20	3.39	3.58
6.0	1.16	1.64	2.00	2.31	2.59	2.83	3.06	3.27	3.47	3.66
6.2	1.18	1.67	2.05	2.36	2.64	2.90	3.13	3.34	3.55	3.74
6.4	1.21	1.71	2.09	2.41	2.70	2.96	3.19	3.41	3.62	3.82
6.6	1.23	1.74	2.13	2.46	2.76	3.02	3.26	3.49	3.70	3.90
6.8	1.26	1.78	2.18	2.51	2.81	3.08	3.33	3.56	3.77	3.98
7.0	1.28	1.81	2.22	2.56	2.87	3.14	3.39	3.63	3.85	4.05
7.5	1.34	1.90	2.32	2.68	3.00	3.29	3.55	3.80	4.03	4.24
8.0	1.40	1.98	2.43	2.80	3.13	3.43	3.71	3.96	4.20	4.43
8.5	1.46	2.06	2.53	2.92	3.26	3.57	3.86	4.13	4.38	4.61
9.0	1.52	2.14	2.62	3.03	3.39	3.71	4.01	4.29	4.55	4.79
9.5	1.57	2.22	2.72	3.14	3.51	3.85	4.16	4.44	4.71	4.97
10	1.63	2.30	2.82	3.25	3.64	3.98	4.30	4.60	4.88	5.14
11	1.73	2.45	3.00	3.46	3.87	4.24	4.58	4.90	5.20	5.48
12	1.84	2.60	3.18	3.67	4.11	4.50	4.86	5.19	5.51	5.81
13	1.94	2.74	3.35	3.87	4.33	4.74	5.12	5.48	5.81	6.12
14	2.03	2.88	3.52	4.07	4.55	4.98	5.38	5.75	6.10	6.43
15	2.13	3.01	3.69	4.26	4.76	5.22	5.64	6.03	6.39	6.74
16	2.22	3.15	3.85	4.45	4.97	5.45	5.88	6.29	6.67	7.03
17	2.32	3.27	4.01	4.63	5.18	5.67	6.13	6.55	6.95	7.32
18	2.41	3.40	4.17	4.81	5.38	5.89	6.36	6.80	7.22	7.61
19	2.49	3.53	4.32	4.99	5.58	6.11	6.60	7.05	7.48	7.89
20	2.58	3.65	4.47	5.16	5.77	6.32	6.83	7.30	7.74	8.16

Table 12.—*Velocity of water, v, in feet per second, based on Manning's formula* $v = (1.486/n)r^{2/3}s^{1/2}$, **n = .030**—Continued

r \ s	.00055	.00060	.00065	.00070	.00075	.00080	.00085	.00090	.00095	.00100
0.2	0.40	0.41	0.43	0.45	0.46	0.48	0.49	0.51	0.52	0.54
0.4	.63	.66	.69	.71	.74	.76	.78	.81	.83	.85
0.6	.83	..86	.90	.93	.97	1.00	1.03	1.06	1.09	1.11
0.8	1.00	1.05	1.09	1.13	1.17	1.21	1.24	1.28	1.32	1.35
1.0	1.16	1.21	1.26	1.31	1.36	1.40	1.44	1.49	1.53	1.57
1.2	1.31	1.37	1.43	1.48	1.53	1.58	1.63	1.68	1.72	1.77
1.4	1.45	1.52	1.58	1.64	1.70	1.75	1.81	1.86	1.91	1.96
1.6	1.59	1.66	1.73	1.79	1.86	1.92	1.98	2.03	2.09	2.14
1.8	1.72	1.80	1.87	1.94	2.01	2.07	2.14	2.20	2.26	2.32
2.0	1.84	1.93	2.00	2.08	2.15	2.22	2.29	2.36	2.42	2.49
2.2	1.96	2.05	2.14	2.22	2.29	2.37	2.44	2.51	2.58	2.65
2.4	2.08	2.17	2.26	2.35	2.43	2.51	2.59	2.66	2.74	2.81
2.6	2.20	2.29	2.39	2.48	2.56	2.65	2.73	2.81	2.89	2.96
2.8	2.31	2.41	2.51	2.60	2.69	2.78	2.87	2.95	3.03	3.11
3.0	2.42	2.52	2.63	2.73	2.82	2.91	3.00	3.09	3.18	3.26
3.2	2.52	2.63	2.74	2.85	2.95	3.04	3.14	3.23	3.32	3.40
3.4	2.63	2.74	2.86	2.96	3.07	3.17	3.27	3.36	3.45	3.54
3.6	2.73	2.85	2.97	3.08	3.19	3.29	3.39	3.49	3.59	3.68
3.8	2.83	2.95	3.08	3.19	3.30	3.41	3.52	3.62	3.72	3.81
4.0	2.93	3.06	3.18	3.30	3.42	3.53	3.64	3.74	3.85	3.95
4.2	3.02	3.16	3.29	3.41	3.53	3.65	3.76	3.87	3.97	4.08
4.4	3.12	3.26	3.39	3.52	3.64	3.76	3.88	3.99	4.10	4.21
4.6	3.21	3.36	3.49	3.62	3.75	3.88	3.99	4.11	4.22	4.33
4.8	3.31	3.45	3.59	3.73	3.86	3.99	4.11	4.23	4.34	4.46
5.0	3.40	3.55	3.69	3.83	3.97	4.10	4.22	4.35	4.46	4.58
5.2	3.49	3.64	3.79	3.93	4.07	4.21	4.33	4.46	4.58	4.70
5.4	3.58	3.73	3.89	4.03	4.18	4.31	4.44	4.57	4.70	4.82
5.6	3.66	3.83	3.98	4.13	4.28	4.42	4.55	4.69	4.81	4.94
5.8	3.75	3.92	4.08	4.23	4.38	4.52	4.66	4.80	4.93	5.06
6.0	3.84	4.01	4.17	4.33	4.48	4.63	4.77	4.91	5.04	5.17
6.2	3.92	4.09	4.26	4.42	4.58	4.73	4.87	5.02	5.15	5.29
6.4	4.00	4.18	4.35	4.52	4.68	4.83	4.98	5.12	5.26	5.40
6.6	4.09	4.27	4.44	4.61	4.77	4.93	5.08	5.23	5.37	5.51
6.8	4.17	4.35	4.53	4.70	4.87	5.03	5.18	5.33	5.48	5.62
7.0	4.25	4.44	4.62	4.80	4.96	5.13	5.28	5.44	5.59	5.73
7.5	4.45	4.65	4.84	5.02	5.20	5.37	5.53	5.69	5.85	6.00
8.0	4.65	4.85	5.05	5.24	5.43	5.60	5.78	5.94	6.11	6.27
8.5	4.84	5.05	5.26	5.46	5.65	5.84	6.01	6.19	6.36	6.52
9.0	5.03	5.25	5.46	5.67	5.87	6.06	6.25	6.43	6.61	6.78
9.5	5.21	5.44	5.66	5.88	6.08	6.28	6.48	6.67	6.85	7.03
10	5.39	5.63	5.86	6.08	6.30	6.50	6.70	6.90	7.09	7.27
11	5.75	6.00	6.25	6.48	6.71	6.93	7.14	7.35	7.55	7.75
12	6.09	6.36	6.62	6.87	7.11	7.34	7.57	7.79	8.00	8.21
13	6.42	6.71	6.98	7.25	7.50	7.75	7.98	8.22	8.44	8.66
14	6.75	7.05	7.34	7.61	7.88	8.14	8.39	8.63	8.87	9.10
15	7.07	7.38	7.68	7.97	8.25	8.52	8.78	9.04	9.29	9.53
16	7.38	7.70	8.02	8.32	8.61	8.90	9.17	9.44	9.69	9.95
17	7.68	8.02	8.35	8.66	8.97	9.26	9.55	9.82	10.09	10.36
18	7.98	8.33	8.67	9.00	9.32	9.62	9.92	10.21	10.49	10.76
19	8.27	8.64	8.99	9.33	9.66	9.98	10.28	10.58	10.87	11.15
20	8.56	8.94	9.30	9.66	9.99	10.32	10.64	10.95	11.25	11.54

Table 12.—*Velocity of water, v, in feet per second, based on Manning's formula* $v = (1.486/n)r^{2/3}s^{1/2}$, **n = .030**—Continued

s \ r	.0010	.0015	.0020	.0025	.0030	.0035	.0040	.0045	.0050
0.2	0.54	0.66	0.76	0.85	0.93	1.00	1.07	1.14	1.20
0.4	.85	1.04	1.20	1.34	1.47	1.59	1.70	1.80	1.90
0.6	1.11	1.36	1.58	1.76	1.93	2.08	2.23	2.36	2.49
0.8	1.35	1.65	1.91	2.13	2.34	2.53	2.70	2.86	3.02
1.0	1.57	1.92	2.22	2.48	2.71	2.93	3.13	3.32	3.50
1.2	1.77	2.17	2.50	2.80	3.06	3.31	3.54	3.75	3.96
1.4	1.96	2.40	2.77	3.10	3.40	3.67	3.92	4.16	4.38
1.6	2.14	2.62	3.03	3.39	3.71	4.01	4.29	4.55	4.79
1.8	2.32	2.84	3.28	3.66	4.01	4.34	4.64	4.92	5.18
2.0	2.49	3.05	3.52	3.93	4.31	4.65	4.97	5.27	5.56
2.2	2.65	3.25	3.75	4.19	4.59	4.96	5.30	5.62	5.92
2.4	2.81	3.44	3.97	4.44	4.86	5.25	5.62	5.96	6.28
2.6	2.96	3.63	4.19	4.68	5.13	5.54	5.92	6.28	6.62
2.8	3.11	3.81	4.40	4.92	5.39	5.82	6.22	6.60	6.96
3.0	3.26	3.99	4.61	5.15	5.64	6.10	6.52	6.91	7.29
3.2	3.40	4.17	4.81	5.38	5.89	6.36	6.80	7.22	7.61
3.4	3.54	4.34	5.01	5.60	6.13	6.63	7.08	7.51	7.92
3.6	3.68	4.51	5.20	5.82	6.37	6.88	7.36	7.80	8.23
3.8	3.81	4.67	5.39	6.03	6.61	7.14	7.63	8.09	8.53
4.0	3.95	4.83	5.58	6.24	6.84	7.38	7.89	8.37	8.83
4.2	4.08	4.99	5.77	6.45	7.06	7.63	8.16	8.65	9.12
4.4	4.21	5.15	5.95	6.65	7.28	7.87	8.41	8.92	9.40
4.6	4.33	5.31	6.13	6.85	7.50	8.11	8.66	9.19	9.69
4.8	4.46	5.46	6.30	7.05	7.72	8.34	8.91	9.46	9.97
5.0	4.58	5.61	6.48	7.24	7.93	8.57	9.16	9.72	10.24
5.2	4.70	5.76	6.65	7.43	8.14	8.80	9.40	9.97	10.51
5.4	4.82	5.90	6.82	7.62	8.35	9.02	9.64	10.23	10.78
5.6	4.94	6.05	6.99	7.81	8.56	9.24	9.88	10.48	11.05
5.8	5.06	6.19	7.15	8.00	8.76	9.46	10.11	10.73	11.31
6.0	5.17	6.33	7.31	8.18	8.96	9.68	10.34	10.97	11.57
6.2	5.29	6.47	7.48	8.36	9.16	9.89	10.57	11.21	11.82
6.4	5.40	6.61	7.64	8.54	9.35	10.10	10.80	11.45	12.07
6.6	5.51	6.75	7.79	8.71	9.55	10.31	11.02	11.69	12.32
6.8	5.62	6.89	7.95	8.89	9.74	10.52	11.24	11.93	12.57
7.0	5.73	7.02	8.11	9.06	9.93	10.72	11.46	12.16	12.82
7.5	6.00	7.35	8.49	9.49	10.40	11.23	12.00	12.73	13.42
8.0	6.27	7.67	8.86	9.91	10.85	11.72	12.53	13.29	14.01
8.5	6.52	7.99	9.23	10.32	11.30	12.21	13.05	13.84	14.59
9.0	6.78	8.30	9.58	10.72	11.74	12.68	13.55	14.38	15.15
9.5	7.03	8.61	9.94	11.11	12.17	13.14	14.05	14.90	15.71
10	7.27	8.90	10.28	11.50	12.59	13.60	14.54	15.42	16.26
11	7.75	9.49	10.96	12.25	13.42	14.49	15.49	16.43	17.32
12	8.21	10.06	11.61	12.98	14.22	15.36	16.42	17.42	18.36
13	8.66	10.61	12.25	13.69	15.00	16.20	17.32	18.37	19.36
14	9.10	11.14	12.87	14.39	15.76	17.02	18.20	19.30	20.35
15	9.53	11.67	13.47	15.06	16.50	17.82	19.05	20.21	21.30
16	9.95	12.18	14.07	15.73	17.23	18.61	19.89	21.10	22.24
17	10.36	12.68	14.65	16.37	17.94	19.37	20.71	21.97	23.16
18	10.76	13.18	15.21	17.01	18.63	20.13	21.52	22.82	24.06
19	11.15	13.66	15.77	17.63	19.32	20.87	22.31	23.66	24.94
20	11.54	14.14	16.32	18.25	19.99	21.59	23.08	24.48	25.81

Table 12.—*Velocity of water, v, in feet per second, based on Manning's formula* $v=(1.486/n)r^{2/3}s^{1/2}$, **n=.030**—Continued

r \ s	.0055	.0060	.0065	.0070	.0075	.0080	.0085	.0090	.0095	.0100
0.2	1.26	1.31	1.37	1.42	1.47	1.52	1.56	1.61	1.65	1.69
0.4	1.99	2.08	2.17	2.25	2.33	2.41	2.48	2.55	2.62	2.69
0.6	2.61	2.73	2.84	2.95	3.05	3.15	3.25	3.34	3.43	3.52
0.8	3.17	3.31	3.44	3.57	3.70	3.82	3.94	4.05	4.16	4.27
1.0	3.67	3.84	3.99	4.14	4.29	4.43	4.57	4.70	4.83	4.95
1.2	4.15	4.33	4.51	4.68	4.84	5.00	5.16	5.31	5.45	5.59
1.4	4.60	4.80	5.00	5.19	5.37	5.54	5.72	5.88	6.04	6.20
1.6	5.03	5.25	5.46	5.67	5.87	6.06	6.25	6.43	6.60	6.78
1.8	5.44	5.68	5.91	6.13	6.35	6.56	6.76	6.95	7.14	7.33
2.0	5.83	6.09	6.34	6.58	6.81	7.03	7.25	7.46	7.66	7.86
2.2	6.21	6.49	6.76	7.01	7.26	7.49	7.72	7.95	8.17	8.38
2.4	6.58	6.88	7.16	7.43	7.69	7.94	8.19	8.42	8.65	8.88
2.6	6.95	7.25	7.55	7.84	8.11	8.38	8.63	8.89	9.13	9.37
2.8	7.30	7.62	7.93	8.23	8.52	8.80	9.07	9.34	9.59	9.84
3.0	7.64	7.98	8.31	8.62	8.92	9.22	9.50	9.77	10.04	10.30
3.2	7.98	8.33	8.67	9.00	9.32	9.62	9.92	10.20	10.48	10.76
3.4	8.31	8.68	9.03	9.37	9.70	10.02	10.33	10.63	10.92	11.20
3.6	8.63	9.01	9.38	9.73	10.08	10.41	10.73	11.04	11.34	11.63
3.8	8.95	9.34	9.72	10.09	10.45	10.79	11.12	11.44	11.76	12.06
4.0	9.26	9.67	10.06	10.44	10.81	11.16	11.51	11.84	12.17	12.48
4.2	9.56	9.99	10.40	10.79	11.17	11.53	11.89	12.23	12.57	12.89
4.4	9.86	10.30	10.72	11.13	11.52	11.90	12.26	12.62	12.96	13.30
4.6	10.16	10.61	11.05	11.46	11.86	12.25	12.63	13.00	13.35	13.70
4.8	10.45	10.92	11.36	11.79	12.21	12.61	12.99	13.37	13.74	14.09
5.0	10.74	11.22	11.68	12.12	12.54	12.95	13.35	13.74	14.12	14.48
5.2	11.03	11.52	11.99	12.44	12.88	13.30	13.71	14.10	14.49	14.87
5.4	11.31	11.81	12.29	12.76	13.20	13.64	14.06	14.46	14.86	15.25
5.6	11.58	12.10	12.59	13.07	13.53	13.97	14.40	14.82	15.22	15.62
5.8	11.86	12.39	12.89	13.38	13.85	14.30	14.74	15.17	15.59	15.99
6.0	12.13	12.67	13.19	13.68	14.16	14.63	15.08	15.52	15.94	16.36
6.2	12.40	12.95	13.48	13.99	14.48	14.95	15.41	15.86	16.29	16.72
6.4	12.66	13.23	13.77	14.29	14.79	15.27	15.74	16.20	16.64	17.07
6.6	12.93	13.50	14.05	14.58	15.09	15.59	16.07	16.53	16.99	17.43
6.8	13.19	13.77	14.33	14.87	15.40	15.90	16.39	16.87	17.33	17.78
7.0	13.44	14.04	14.61	15.17	15.70	16.21	16.71	17.20	17.67	18.13
7.5	14.08	14.70	15.30	15.88	16.44	16.98	17.50	18.00	18.50	18.98
8.0	14.69	15.35	15.97	16.58	17.16	17.72	18.27	18.80	19.31	19.81
8.5	15.30	15.98	16.63	17.26	17.87	18.45	19.02	19.57	20.11	20.63
9.0	15.89	16.60	17.28	17.93	18.56	19.17	19.76	20.33	20.89	21.43
9.5	16.48	17.21	17.91	18.59	19.24	19.87	20.48	21.08	21.66	22.22
10	17.05	17.81	18.54	19.24	19.91	20.56	21.20	21.81	22.41	22.99
11	18.17	18.98	19.75	20.50	21.22	21.91	22.59	23.24	23.88	24.50
12	19.25	20.11	20.93	21.72	22.48	23.22	23.94	24.63	25.31	25.96
13	20.31	21.21	22.08	22.91	23.72	24.49	25.25	25.98	26.69	27.39
14	21.34	22.29	23.20	24.07	24.92	25.74	26.53	27.30	28.04	28.77
15	22.34	23.34	24.29	25.21	26.09	26.95	27.78	28.58	29.36	30.13
16	23.33	24.36	25.36	26.31	27.24	28.13	29.00	29.84	30.66	31.45
17	24.29	25.37	26.40	27.40	28.36	29.29	30.19	31.07	31.92	32.75
18	25.23	26.35	27.43	28.46	29.46	30.43	31.37	32.28	33.16	34.02
19	26.16	27.32	28.44	29.51	30.54	31.55	32.52	33.46	34.38	35.27
20	27.07	28.27	29.42	30.54	31.61	32.64	33.65	34.62	35.57	36.50

Table 12.—*Velocity of water, v, in feet per second, based on Manning's formula* $v=(1.486/n)r^{2/3}s^{1/2}$, **n = .030**—Continued

r \ s	.01	.02	.03	.04	.05	.06	.07	.08	.09	.10
0.2	1.69	2.40	2.93	3.39	3.79	4.15	4.48	4.79	5.08	5.36
0.4	2.69	3.80	4.66	5.38	6.01	6.59	7.11	7.61	8.07	8.50
0.6	3.52	4.98	6.10	7.05	7.88	8.63	9.32	9.97	10.57	11.14
0.8	4.27	6.04	7.39	8.54	9.54	10.46	11.29	12.07	12.81	13.50
1.0	4.95	7.01	8.58	9.91	11.08	12.13	13.11	14.01	14.86	15.66
1.2	5.59	7.91	9.69	11.19	12.51	13.70	14.80	15.82	16.79	17.69
1.4	6.20	8.77	10.74	12.40	13.86	15.18	16.40	17.53	18.60	19.60
1.6	6.78	9.58	11.74	13.55	15.15	16.60	17.93	19.17	20.33	21.43
1.8	7.33	10.37	12.70	14.66	16.39	17.95	19.39	20.73	21.99	23.18
2.0	7.86	11.12	13.62	15.73	17.58	19.26	20.80	22.24	23.59	24.86
2.2	8.38	11.85	14.51	16.76	18.74	20.52	22.17	23.70	25.14	26.50
2.4	8.88	12.56	15.38	17.76	19.85	21.75	23.49	25.11	26.64	28.08
2.6	9.37	13.25	16.22	18.73	20.94	22.94	24.78	26.49	28.10	29.62
2.8	9.84	13.92	17.04	19.68	22.00	24.10	26.03	27.83	29.52	31.12
3.0	10.30	14.57	17.85	20.61	23.04	25.24	27.26	29.14	30.91	32.58
3.2	10.76	15.21	18.63	21.51	24.05	26.35	28.46	30.42	32.27	34.01
3.4	11.20	15.84	19.40	22.40	25.04	27.43	29.63	31.68	33.60	35.42
3.6	11.63	16.45	20.15	23.27	26.02	28.50	30.78	32.91	34.90	36.79
3.8	12.06	17.06	20.89	24.12	26.97	29.55	31.91	34.12	36.19	38.14
4.0	12.48	17.65	21.62	24.96	27.91	30.57	33.02	35.30	37.44	39.47
4.2	12.89	18.24	22.33	25.79	28.83	31.58	34.12	36.47	38.68	40.78
4.4	13.30	18.81	23.04	26.60	29.74	32.58	35.19	37.62	39.90	42.06
4.6	13.70	19.38	23.73	27.40	30.64	33.56	36.25	38.75	41.10	43.32
4.8	14.09	19.93	24.41	28.19	31.52	34.53	37.29	39.87	42.28	44.57
5.0	14.48	20.48	25.09	28.97	32.39	35.48	38.32	40.97	43.45	45.80
5.2	14.87	21.03	25.75	29.73	33.24	36.42	39.34	42.05	44.60	47.01
5.4	15.25	21.56	26.41	30.49	34.09	37.35	40.34	43.12	45.74	48.21
5.6	15.62	22.09	27.06	31.24	34.93	38.26	41.33	44.18	46.86	49.40
5.8	15.99	22.61	27.70	31.98	35.75	39.17	42.31	45.23	47.97	50.56
6.0	16.36	23.13	28.33	32.71	36.57	40.06	43.27	46.26	49.07	51.72
6.2	16.72	23.64	28.95	33.43	37.38	40.95	44.23	47.28	50.15	52.86
6.4	17.07	24.15	29.57	34.15	38.18	41.82	45.18	48.29	51.22	53.99
6.6	17.43	24.65	30.19	34.86	38.97	42.69	46.11	49.30	52.29	55.11
6.8	17.78	25.14	30.79	35.56	39.75	43.55	47.04	50.29	53.34	56.22
7.0	18.13	25.63	31.39	36.25	40.53	44.40	47.96	51.27	54.38	57.32
7.5	18.98	26.84	32.87	37.96	42.44	46.49	50.21	53.68	56.94	60.02
8.0	19.81	28.02	34.32	39.63	44.30	48.53	52.42	56.04	59.44	62.66
8.5	20.63	29.18	35.73	41.26	46.13	50.53	54.58	58.35	61.89	65.24
9.0	21.43	30.31	37.12	42.86	47.92	52.50	56.70	60.62	64.30	67.77
9.5	22.22	31.42	38.48	44.44	49.68	54.42	58.78	62.84	66.66	70.26
10	22.99	32.51	39.82	45.98	51.41	56.32	60.83	65.03	68.97	72.70
11	24.50	34.65	42.43	49.00	54.78	60.01	64.82	69.30	73.50	77.47
12	25.96	36.72	44.97	51.93	58.05	63.60	68.69	73.43	77.89	82.10
13	27.39	38.73	47.43	54.77	61.24	67.08	72.46	77.46	82.16	86.60
14	28.77	40.69	49.84	57.55	64.34	70.48	76.13	81.38	86.32	90.99
15	30.13	42.61	52.18	60.25	67.37	73.80	79.71	85.21	90.38	95.27
16	31.45	44.48	54.48	62.90	70.33	77.04	83.21	88.96	94.36	99.46
17	32.75	46.31	56.72	65.50	73.23	80.22	86.65	92.63	98.25	103.56
18	34.02	48.11	58.93	68.04	76.07	83.33	90.01	96.23	102.06	107.58
19	35.27	49.88	61.09	70.54	78.87	86.39	93.31	99.76	105.81	111.53
20	36.50	51.61	63.21	72.99	81.61	89.40	96.56	103.23	109.49	115.41

Table 13.—*Velocity of water, v, in feet per second, based on Manning's formula* $v=(1.486/n)r^{2/3}s^{1/2}$, n=.035

r \ s'	.00005	.00010	.00015	.00020	.00025	.00030	.00035	.00040	.00045	.00050
0.2	0.10	0.15	0.18	0.21	0.23	0.25	0.27	0.29	0.31	0.32
0.4	.16	.23	.28	.33	.36	.40	.43	.46	.49	.52
0.6	.21	.30	.37	.43	.48	.52	.57	.60	.64	.68
0.8	.26	.37	.45	.52	.58	.63	.68	.73	.78	.82
1.0	.30	.42	.52	.60	.67	.74	.79	.85	.90	.95
1.2	.34	.48	.59	.68	.76	.83	.90	.96	1.02	1.07
1.4	.38	.53	.65	.75	.84	.92	.99	1.06	1.13	1.19
1.6	.41	.58	.71	.82	.92	1.01	1.09	1.16	1.23	1.30
1.8	.44	.63	.77	.89	.99	1.09	1.18	1.26	1.33	1.40
2.0	.48	.67	.83	.95	1.07	1.17	1.26	1.35	1.43	1.51
2.2	.51	.72	.88	1.02	1.14	1.24	1.34	1.44	1.52	1.61
2.4	.54	.76	.93	1.08	1.20	1.32	1.42	1.52	1.61	1.70
2.6	.57	.80	.98	1.14	1.27	1.39	1.50	1.61	1.70	1.80
2.8	.60	.84	1.03	1.19	1.33	1.46	1.58	1.69	1.79	1.89
3.0	.62	.88	1.08	1.25	1.40	1.53	1.65	1.77	1.87	1.97
3.2	.65	.92	1.13	1.30	1.46	1.60	1.72	1.84	1.96	2.06
3.4	.68	.96	1.18	1.36	1.52	1.66	1.80	1.92	2.04	2.15
3.6	.71	1.00	1.22	1.41	1.58	1.73	1.87	1.99	2.12	2.23
3.8	.73	1.03	1.27	1.46	1.63	1.79	1.93	2.07	2.19	2.31
4.0	.76	1.07	1.31	1.51	1.69	1.85	2.00	2.14	2.27	2.39
4.2	.78	1.11	1.35	1.56	1.75	1.91	2.07	2.21	2.34	2.47
4.4	.81	1.14	1.40	1.61	1.80	1.97	2.13	2.28	2.42	2.55
4.6	.83	1.17	1.44	1.66	1.86	2.03	2.20	2.35	2.49	2.63
4.8	.85	1.21	1.48	1.71	1.91	2.09	2.26	2.42	2.56	2.70
5.0	.88	1.24	1.52	1.76	1.96	2.15	2.32	2.48	2.63	2.78
5.2	.90	1.27	1.56	1.80	2.01	2.21	2.38	2.55	2.70	2.85
5.4	.92	1.31	1.60	1.85	2.07	2.26	2.44	2.61	2.77	2.92
5.6	.95	1.34	1.64	1.89	2.12	2.32	2.50	2.68	2.84	2.99
5.8	.97	1.37	1.68	1.94	2.17	2.37	2.56	2.74	2.91	3.06
6.0	.99	1.40	1.72	1.98	2.22	2.43	2.62	2.80	2.97	3.13
6.2	1.01	1.43	1.75	2.03	2.27	2.48	2.68	2.87	3.04	3.20
6.4	1.03	1.46	1.79	2.07	2.31	2.53	2.74	2.93	3.10	3.27
6.6	1.06	1.49	1.83	2.11	2.36	2.59	2.79	2.99	3.17	3.34
6.8	1.08	1.52	1.87	2.16	2.41	2.64	2.85	3.05	3.23	3.41
7.0	1.10	1.55	1.90	2.20	2.46	2.69	2.91	3.11	3.30	3.47
7.5	1.15	1.63	1.99	2.30	2.57	2.82	3.04	3.25	3.45	3.64
8.0	1.20	1.70	2.08	2.40	2.69	2.94	3.18	3.40	3.60	3.80
8.5	1.25	1.77	2.17	2.50	2.80	3.06	3.31	3.54	3.75	3.95
9.0	1.30	1.84	2.25	2.60	2.90	3.18	3.44	3.67	3.90	4.11
9.5	1.35	1.90	2.33	2.69	3.01	3.30	3.56	3.81	4.04	4.26
10	1.39	1.97	2.41	2.79	3.12	3.41	3.69	3.94	4.18	4.41
11	1.48	2.10	2.57	2.97	3.32	3.64	3.93	4.20	4.45	4.70
12	1.57	2.23	2.73	3.15	3.52	3.85	4.16	4.45	4.72	4.98
13	1.66	2.35	2.87	3.32	3.71	4.07	4.39	4.69	4.98	5.25
14	1.74	2.47	3.02	3.49	3.90	4.27	4.61	4.93	5.23	5.51
15	1.83	2.58	3.16	3.65	4.08	4.47	4.83	5.16	5.48	5.77
16	1.91	2.70	3.30	3.81	4.26	4.67	5.04	5.39	5.72	6.03
17	1.98	2.81	3.44	3.97	4.44	4.86	5.25	5.61	5.95	6.28
18	2.06	2.92	3.57	4.12	4.61	5.05	5.46	5.83	6.19	6.52
19	2.14	3.02	3.70	4.28	4.78	5.24	5.66	6.05	6.41	6.76
20	2.21	3.13	3.83	4.42	4.95	5.42	5.85	6.26	6.64	7.00

Table 13.—*Velocity of water, v, in feet per second, based on Manning's formula* $v = (1.486/n)r^{2/3}s^{1/2}$, **n = .035**—Continued

$\frac{s}{r}$	.00055	.00060	.00065	.00070	.00075	.00080	.00085	.00090	.00095	.00100
0.2	0.34	0.36	0.37	0.38	0.40	0.41	0.42	0.44	0.45	0.46
0.4	0.54	0.56	0.59	0.61	0.63	0.65	0.67	0.69	0.71	0.73
0.6	0.71	0.74	0.77	0.80	0.83	0.85	0.88	0.91	0.93	0.96
0.8	0.86	0.90	0.93	0.97	1.00	1.03	1.07	1.10	1.13	1.16
1.0	1.00	1.04	1.08	1.12	1.16	1.20	1.24	1.27	1.31	1.34
1.2	1.12	1.17	1.22	1.27	1.31	1.36	1.40	1.44	1.48	1.52
1.4	1.25	1.30	1.35	1.41	1.46	1.50	1.55	1.59	1.64	1.68
1.6	1.36	1.42	1.48	1.54	1.59	1.64	1.69	1.74	1.79	1.84
1.8	1.47	1.54	1.60	1.66	1.72	1.78	1.83	1.88	1.94	1.99
2.0	1.58	1.65	1.72	1.78	1.85	1.91	1.96	2.02	2.08	2.13
2.2	1.68	1.76	1.83	1.90	1.97	2.03	2.09	2.15	2.21	2.27
2.4	1.78	1.86	1.94	2.01	2.08	2.15	2.22	2.28	2.35	2.41
2.6	1.88	1.97	2.05	2.12	2.20	2.27	2.34	2.41	2.47	2.54
2.8	1.98	2.07	2.15	2.23	2.31	2.39	2.46	2.53	2.60	2.67
3.0	2.07	2.16	2.25	2.34	2.42	2.50	2.57	2.65	2.72	2.79
3.2	2.16	2.26	2.35	2.44	2.52	2.61	2.69	2.77	2.84	2.92
3.4	2.25	2.35	2.45	2.54	2.63	2.72	2.80	2.88	2.96	3.04
3.6	2.34	2.44	2.54	2.64	2.73	2.82	2.91	2.99	3.07	3.15
3.8	2.42	2.53	2.64	2.74	2.83	2.92	3.01	3.10	3.19	3.27
4.0	2.51	2.62	2.73	2.83	2.93	3.03	3.12	3.21	3.30	3.38
4.2	2.59	2.71	2.82	2.92	3.03	3.13	3.22	3.32	3.41	3.50
4.4	2.67	2.79	2.91	3.02	3.12	3.22	3.32	3.42	3.51	3.61
4.6	2.75	2.88	2.99	3.11	3.22	3.32	3.42	3.52	3.62	3.71
4.8	2.83	2.96	3.08	3.20	3.31	3.42	3.52	3.62	3.72	3.82
5.0	2.91	3.04	3.17	3.28	3.40	3.51	3.62	3.72	3.83	3.93
5.2	2.99	3.12	3.25	3.37	3.49	3.60	3.72	3.82	3.93	4.03
5.4	3.06	3.20	3.33	3.46	3.58	3.70	3.81	3.92	4.03	4.13
5.6	3.14	3.28	3.41	3.54	3.67	3.79	3.90	4.02	4.13	4.23
5.8	3.21	3.36	3.49	3.63	3.75	3.88	4.00	4.11	4.22	4.33
6.0	3.29	3.43	3.57	3.71	3.84	3.97	4.09	4.21	4.32	4.43
6.2	3.36	3.51	3.65	3.79	3.92	4.05	4.18	4.30	4.42	4.53
6.4	3.43	3.58	3.73	3.87	4.01	4.14	4.27	4.39	4.51	4.63
6.6	3.50	3.66	3.81	3.95	4.09	4.23	4.36	4.48	4.60	4.72
6.8	3.57	3.73	3.89	4.03	4.17	4.31	4.44	4.57	4.70	4.82
7.0	3.64	3.81	3.96	4.11	4.25	4.39	4.53	4.66	4.79	4.91
7.5	3.82	3.98	4.15	4.30	4.46	4.60	4.74	4.88	5.01	5.14
8.0	3.98	4.16	4.33	4.49	4.65	4.80	4.95	5.09	5.23	5.37
8.5	4.15	4.33	4.51	4.68	4.84	5.00	5.16	5.30	5.45	5.59
9.0	4.31	4.50	4.68	4.86	5.03	5.20	5.36	5.51	5.66	5.81
9.5	4.47	4.66	4.86	5.04	5.22	5.39	5.55	5.71	5.87	6.02
10	4.62	4.83	5.02	5.21	5.40	5.57	5.75	5.91	6.07	6.23
11	4.92	5.14	5.35	5.56	5.75	5.94	6.12	6.30	6.47	6.64
12	5.22	5.45	5.67	5.89	6.09	6.29	6.49	6.68	6.86	7.04
13	5.51	5.75	5.98	6.21	6.43	6.64	6.84	7.04	7.24	7.42
14	5.78	6.04	6.29	6.53	6.75	6.98	7.19	7.40	7.60	7.80
15	6.06	6.33	6.58	6.83	7.07	7.30	7.53	7.75	7.96	8.17
16	6.32	6.60	6.87	7.13	7.38	7.63	7.86	8.09	8.31	8.53
17	6.58	6.88	7.16	7.43	7.69	7.94	8.18	8.42	8.65	8.88
18	6.84	7.14	7.43	7.72	7.99	8.25	8.50	8.75	8.99	9.22
19	7.09	7.41	7.71	8.00	8.28	8.55	8.81	9.07	9.32	9.56
20	7.34	7.66	7.98	8.28	8.57	8.85	9.12	9.38	9.64	9.89

Table 13.—*Velocity of water, v, in feet per second, based on Manning's formula* $v = (1.486/n) r^{2/3} \delta^{1/2}$, **n = .035**—Continued

r	.0010	.0015	.0020	.0025	.0030	.0035	.0040	.0045	.0050
0.2	0.46	0.56	0.65	0.73	0.80	0.86	0.92	0.97	1.03
0.4	.73	.89	1.03	1.15	1.26	1.36	1.46	1.55	1.63
0.6	.96	1.17	1.35	1.51	1.65	1.79	1.91	2.03	2.14
0.8	1.16	1.42	1.64	1.83	2.00	2.16	2.31	2.45	2.59
1.0	1.34	1.64	1.90	2.12	2.33	2.51	2.69	2.85	3.00
1.2	1.52	1.86	2.14	2.40	2.63	2.84	3.03	3.22	3.39
1.4	1.68	2.06	2.38	2.66	2.91	3.14	3.36	3.56	3.76
1.6	1.84	2.25	2.60	2.90	3.18	3.44	3.67	3.90	4.11
1.8	1.99	2.43	2.81	3.14	3.44	3.72	3.97	4.21	4.44
2.0	2.13	2.61	3.01	3.37	3.69	3.99	4.26	4.52	4.77
2.2	2.27	2.78	3.21	3.59	3.93	4.25	4.54	4.82	5.08
2.4	2.41	2.95	3.40	3.81	4.17	4.50	4.81	5.11	5.38
2.6	2.54	3.11	3.59	4.01	4.40	4.75	5.08	5.39	5.68
2.8	2.67	3.27	3.77	4.22	4.62	4.99	5.33	5.66	5.96
3.0	2.79	3.42	3.95	4.42	4.84	5.22	5.59	5.92	6.24
3.2	2.92	3.57	4.12	4.61	5.05	5.45	5.83	6.18	6.52
3.4	3.04	3.72	4.29	4.80	5.26	5.68	6.07	6.44	6.79
3.6	3.15	3.86	4.46	4.99	5.46	5.90	6.31	6.69	7.05
3.8	3.27	4.00	4.62	5.17	5.66	6.12	6.54	6.94	7.31
4.0	3.38	4.14	4.78	5.35	5.86	6.33	6.77	7.18	7.57
4.2	3.50	4.28	4.94	5.53	6.05	6.54	6.99	7.41	7.82
4.4	3.61	4.42	5.10	5.70	6.24	6.74	7.21	7.65	8.06
4.6	3.71	4.55	5.25	5.87	6.43	6.95	7.43	7.88	8.30
4.8	3.82	4.68	5.40	6.04	6.62	7.15	7.64	8.10	8.54
5.0	3.93	4.81	5.55	6.21	6.80	7.34	7.85	8.33	8.78
5.2	4.03	4.94	5.70	6.37	6.98	7.54	8.06	8.55	9.01
5.4	4.13	5.06	5.84	6.53	7.16	7.73	8.27	8.77	9.24
5.6	4.23	5.19	5.99	6.69	7.33	7.92	8.47	8.98	9.47
5.8	4.33	5.31	6.13	6.85	7.51	8.11	8.67	9.19	9.69
6.0	4.43	5.43	6.27	7.01	7.68	8.29	8.87	9.40	9.91
6.2	4.53	5.55	6.41	7.16	7.85	8.48	9.06	9.61	10.13
6.4	4.63	5.67	6.55	7.32	8.02	8.66	9.26	9.82	10.35
6.6	4.72	5.79	6.68	7.47	8.18	8.84	9.45	10.02	10.56
6.8	4.82	5.90	6.82	7.62	8.35	9.02	9.64	10.22	10.78
7.0	4.91	6.02	6.95	7.77	8.51	9.19	9.83	10.42	10.99
7.5	5.14	6.30	7.28	8.13	8.91	9.62	10.29	10.91	11.50
8.0	5.37	6.58	7.59	8.49	9.30	10.05	10.74	11.39	12.01
8.5	5.59	6.85	7.91	8.84	9.69	10.46	11.18	11.86	12.50
9.0	5.81	7.11	8.22	9.19	10.06	10.87	11.62	12.32	12.99
9.5	6.02	7.38	8.52	9.52	10.43	11.27	12.04	12.78	13.47
10	6.23	7.63	8.81	9.85	10.79	11.66	12.46	13.22	13.93
11	6.64	8.13	9.39	10.50	11.50	12.42	13.28	14.09	14.85
12	7.04	8.62	9.95	11.13	12.19	13.17	14.07	14.93	15.74
13	7.42	9.09	10.50	11.74	12.86	13.89	14.85	15.75	16.60
14	7.80	9.55	11.03	12.33	13.51	14.59	15.60	16.54	17.44
15	8.17	10.00	11.55	12.91	14.14	15.28	16.33	17.32	18.26
16	8.53	10.44	12.06	13.48	14.77	15.95	17.05	18.08	19.06
17	8.88	10.87	12.55	14.04	15.37	16.61	17.75	18.83	19.85
18	9.22	11.29	13.04	14.58	15.97	17.25	18.44	19.56	20.62
19	9.56	11.71	13.52	15.12	16.56	17.88	19.12	20.28	21.38
20	9.89	12.12	13.99	15.64	17.13	18.51	19.78	20.99	22.12

Table 13.—*Velocity of water, v, in feet per second, based on Manning's formula* $v = (1.486/n)r^{2/3}s^{1/2}$, **n = .035**—Continued

s r	.0055	.0060	.0065	.0070	.0075	.0080	.0085	.0090	.0095	.0100
0.2	1.08	1.12	1.17	1.21	1.26	1.30	1.34	1.38	1.42	1.45
0.4	1.71	1.79	1.86	1.93	2.00	2.06	2.13	2.19	2.25	2.30
0.6	2.24	2.34	2.44	2.53	2.62	2.70	2.78	2.87	2.94	3.02
0.8	2.71	2.83	2.95	3.06	3.17	3.27	3.37	3.47	3.57	3.66
1.0	3.15	3.29	3.42	3.55	3.68	3.80	3.91	4.03	4.14	4.25
1.2	3.56	3.71	3.87	4.01	4.15	4.29	4.42	4.55	4.67	4.79
1.4	3.94	4.12	4.28	4.45	4.60	4.75	4.90	5.04	5.18	5.31
1.6	4.31	4.50	4.68	4.86	5.03	5.19	5.35	5.51	5.66	5.81
1.8	4.66	4.87	5.07	5.26	5.44	5.62	5.79	5.96	6.12	6.28
2.0	5.00	5.22	5.43	5.64	5.84	6.03	6.21	6.39	6.57	6.74
2.2	5.33	5.56	5.79	6.01	6.22	6.42	6.62	6.81	7.00	7.18
2.4	5.64	5.90	6.14	6.37	6.59	6.81	7.02	7.22	7.42	7.61
2.6	5.95	6.22	6.47	6.72	6.95	7.18	7.40	7.62	7.82	8.03
2.8	6.26	6.53	6.80	7.06	7.30	7.54	7.78	8.00	8.22	8.43
3.0	6.55	6.84	7.12	7.39	7.65	7.90	8.14	8.38	8.61	8.83
3.2	6.84	7.14	7.43	7.71	7.98	8.25	8.50	8.75	8.99	9.22
3.4	7.12	7.44	7.74	8.03	8.31	8.59	8.85	9.11	9.36	9.60
3.6	7.40	7.72	8.04	8.34	8.64	8.92	9.19	9.46	9.72	9.97
3.8	7.67	8.01	8.34	8.65	8.95	9.25	9.53	9.81	10.08	10.34
4.0	7.93	8.29	8.63	8.95	9.27	9.57	9.86	10.15	10.43	10.70
4.2	8.20	8.56	8.91	9.25	9.57	9.89	10.19	10.49	10.77	11.05
4.4	8.45	8.83	9.19	9.54	9.87	10.20	10.51	10.82	11.11	11.40
4.6	8.71	9.10	9.47	9.83	10.17	10.50	10.83	11.14	11.45	11.74
4.8	8.96	9.36	9.74	10.11	10.46	10.81	11.14	11.46	11.78	12.08
5.0	9.21	9.62	10.01	10.39	10.75	11.10	11.45	11.78	12.10	12.41
5.2	9.45	9.87	10.27	10.66	11.04	11.40	11.75	12.09	12.42	12.74
5.4	9.69	10.12	10.54	10.93	11.32	11.69	12.05	12.40	12.74	13.07
5.6	9.93	10.37	10.79	11.20	11.60	11.98	12.34	12.70	13.05	13.39
5.8	10.16	10.62	11.05	11.47	11.87	12.26	12.64	13.00	13.36	13.71
6.0	10.40	10.86	11.30	11.73	12.14	12.54	12.92	13.30	13.66	14.02
6.2	10.63	11.10	11.55	11.99	12.41	12.82	13.21	13.59	13.97	14.33
6.4	10.85	11.34	11.80	12.24	12.67	13.09	13.49	13.88	14.26	14.64
6.6	11.08	11.57	12.04	12.50	12.94	13.36	13.77	14.17	14.56	14.94
6.8	11.30	11.80	12.29	12.75	13.20	13.63	14.05	14.46	14.85	15.24
7.0	11.52	12.03	12.53	13.00	13.45	13.90	14.32	14.74	15.14	15.54
7.5	12.06	12.60	13.12	13.61	14.09	14.55	15.00	15.43	15.86	16.27
8.0	12.59	13.15	13.69	14.21	14.71	15.19	15.66	16.11	16.55	16.98
8.5	13.11	13.70	14.26	14.79	15.31	15.82	16.30	16.78	17.24	17.68
9.0	13.62	14.23	14.81	15.37	15.91	16.43	16.94	17.43	17.90	18.37
9.5	14.12	14.75	15.35	15.93	16.49	17.03	17.56	18.07	18.56	19.04
10	14.61	15.26	15.89	16.49	17.07	17.63	18.17	18.70	19.21	19.71
11	15.57	16.27	16.93	17.57	18.18	18.78	19.36	19.92	20.47	21.00
12	16.50	17.24	17.94	18.62	19.27	19.90	20.52	21.11	21.69	22.25
13	17.41	18.18	18.93	19.64	20.33	21.00	21.64	22.27	22.88	23.47
14	18.29	19.10	19.88	20.63	21.36	22.06	22.74	23.40	24.04	24.66
15	19.15	20.00	20.82	21.61	22.36	23.10	23.81	24.50	25.17	25.82
16	19.99	20.88	21.73	22.56	23.35	24.11	24.85	25.58	26.28	26.96
17	20.82	21.74	22.63	23.49	24.31	25.11	25.88	26.63	27.36	28.07
18	21.63	22.59	23.51	24.40	25.25	26.08	26.88	27.66	28.42	29.16
19	22.42	23.42	24.37	25.29	26.18	27.04	27.87	28.68	29.47	30.23
20	23.20	24.23	25.22	26.17	27.09	27.98	28.84	29.68	30.49	31.28

Table 13.—*Velocity of water, v, in feet per second, based on Manning's formula* $v=(1.486/n)r^{2/3}s^{1/2}$, **n = .035**—Continued

r \ s	.01	.02	.03	.04	.05	.06	.07	.08	.09	.10
0.2	1.45	2.05	2.51	2.90	3.25	3.56	3.84	4.11	4.36	4.59
0.4	2.30	3.26	3.99	4.61	5.15	5.65	6.10	6.52	6.91	7.29
0.6	3.02	4.27	5.23	6.04	6.75	7.40	7.99	8.54	9.06	9.55
0.8	3.66	5.17	6.34	7.32	8.18	8.96	9.68	10.35	10.98	11.57
1.0	4.25	6.00	7.35	8.49	9.49	10.40	11.23	12.01	12.74	13.43
1.2	4.79	6.78	8.30	9.59	10.72	11.74	12.68	13.56	14.38	15.16
1.4	5.31	7.51	9.20	10.63	11.88	13.02	14.06	15.03	15.94	16.80
1.6	5.81	8.21	10.06	11.62	12.99	14.23	15.37	16.43	17.42	18.37
1.8	6.28	8.88	10.88	12.56	14.05	15.39	16.62	17.77	18.85	19.87
2.0	6.74	9.53	11.67	13.48	15.07	16.51	17.83	19.06	20.22	21.31
2.2	7.18	10.16	12.44	14.36	16.06	17.59	19.00	20.31	21.55	22.71
2.4	7.61	10.76	13.18	15.22	17.02	18.64	20.14	21.53	22.83	24.07
2.6	8.03	11.35	13.90	16.06	17.95	19.66	21.24	22.71	24.08	25.39
2.8	8.43	11.93	14.61	16.87	18.86	20.66	22.32	23.86	25.30	26.67
3.0	8.83	12.49	15.30	17.66	19.75	21.63	23.37	24.98	26.49	27.93
3.2	9.22	13.04	15.97	18.44	20.62	22.58	24.39	26.08	27.66	29.16
3.4	9.60	13.58	16.63	19.20	21.47	23.52	25.40	27.15	28.80	30.36
3.6	9.97	14.10	17.27	19.95	22.30	24.43	26.39	28.21	29.92	31.54
3.8	10.34	14.62	17.91	20.68	23.12	25.32	27.35	29.24	31.02	32.69
4.0	10.70	15.13	18.53	21.40	23.92	26.21	28.31	30.26	32.10	33.83
4.2	11.05	15.63	19.14	22.10	24.71	27.07	29.24	31.26	33.16	34.95
4.4	11.40	16.12	19.75	22.80	25.49	27.93	30.16	32.25	34.20	36.05
4.6	11.74	16.61	20.34	23.49	26.26	28.77	31.07	33.22	35.23	37.14
4.8	12.08	17.09	20.93	24.16	27.01	29.59	31.96	34.17	36.24	38.20
5.0	12.41	17.56	21.50	24.83	27.76	30.41	32.85	35.11	37.24	39.26
5.2	12.74	18.02	22.07	25.49	28.50	31.21	33.72	36.04	38.23	40.30
5.4	13.07	18.48	22.63	26.14	29.22	32.01	34.58	36.96	39.20	41.33
5.6	13.39	18.93	23.19	26.78	29.94	32.80	35.42	37.87	40.17	42.34
5.8	13.71	19.38	23.74	27.41	30.65	33.57	36.26	38.77	41.12	43.34
6.0	14.02	19.83	24.28	28.04	31.35	34.34	37.09	39.65	42.06	44.33
6.2	14.33	20.26	24.82	28.66	32.04	35.10	37.91	40.53	42.99	45.31
6.4	14.64	20.70	25.35	29.27	32.73	35.85	38.72	41.40	43.91	46.28
6.6	14.94	21.13	25.87	29.88	33.40	36.59	39.52	42.25	44.82	47.24
6.8	15.24	21.55	26.39	30.48	34.08	37.33	40.32	43.10	45.72	48.19
7.0	15.54	21.97	26.91	31.07	34.74	38.06	41.11	43.94	46.61	49.13
7.5	16.27	23.01	28.18	32.54	36.38	39.85	43.04	46.01	48.80	51.44
8.0	16.98	24.02	29.42	33.97	37.97	41.60	44.93	48.03	50.95	53.70
8.5	17.68	25.01	30.63	35.37	39.54	43.32	46.79	50.02	53.05	55.92
9.0	18.37	25.98	31.82	36.74	41.08	45.00	48.60	51.96	55.11	58.09
9.5	19.04	26.93	32.99	38.09	42.58	46.65	50.39	53.87	57.13	60.22
10	19.71	27.87	34.13	39.41	44.07	48.27	52.14	55.74	59.12	62.32
11	21.00	29.70	36.37	42.00	46.96	51.44	55.56	59.40	63.00	66.41
12	22.25	31.47	38.54	44.51	49.76	54.51	58.88	62.94	66.76	70.37
13	23.47	33.20	40.66	46.95	52.49	57.50	62.11	66.39	70.42	74.23
14	24.66	34.88	42.72	49.32	55.15	60.41	65.25	69.76	73.99	77.99
15	25.82	36.52	44.73	51.65	57.74	63.25	68.32	73.04	77.47	81.66
16	26.96	38.13	46.69	53.92	60.28	66.03	71.33	76.25	80.88	85.25
17	28.07	39.70	48.62	56.14	62.77	68.76	74.27	79.40	84.21	88.77
18	29.16	41.24	50.51	58.32	65.21	71.43	77.15	82.48	87.48	92.21
19	30.23	42.75	52.36	60.46	67.60	74.05	79.98	85.51	90.69	95.60
20	31.28	44.24	54.18	62.57	69.95	76.63	82.77	88.48	93.85	98.92

Table 14.—*Velocity of water, v, in feet per second, based on Manning's formula* $v = (1.486/n)r^{2/3}s^{1/2}$, **n = .040**

r	.00005	.00010	.00015	.00020	.00025	.00030	.00035	.00040	.00045	.00050
0.2	0.09	0.13	0.16	0.18	0.20	0.22	0.24	0.25	0.27	0.28
0.4	.14	.20	.25	.29	.32	.35	.38	.40	.43	.45
0.6	.19	.26	.32	.37	.42	.46	.49	.53	.56	.59
0.8	.23	.32	.39	.45	.51	.55	.60	.64	.68	.72
1.0	.26	.37	.45	.53	.59	.64	.70	.74	.79	.83
1.2	.30	.42	.51	.59	.66	.73	.78	.84	.89	.94
1.4	.33	.46	.57	.66	.74	.81	.87	.93	.99	1.04
1.6	.36	.51	.62	.72	.80	.88	.95	1.02	1.08	1.14
1.8	.39	.55	.67	.78	.87	.95	1.03	1.10	1.17	1.23
2.0	.42	.59	.72	.83	.93	1.02	1.10	1.18	1.25	1.32
2.2	.44	.63	.77	.89	.99	1.09	1.18	1.26	1.33	1.41
2.4	.47	.67	.82	.94	1.05	1.15	1.25	1.33	1.41	1.49
2.6	.50	.70	.86	.99	1.11	1.22	1.31	1.40	1.49	1.57
2.8	.52	.74	.90	1.04	1.17	1.28	1.38	1.48	1.57	1.65
3.0	.55	.77	.95	1.09	1.22	1.34	1.45	1.55	1.64	1.73
3.2	.57	.81	.99	1.14	1.28	1.40	1.51	1.61	1.71	1.80
3.4	.59	.84	1.03	1.19	1.33	1.45	1.57	1.68	1.78	1.88
3.6	.62	.87	1.07	1.23	1.38	1.51	1.63	1.75	1.85	1.95
3.8	.64	.90	1.11	1.28	1.43	1.57	1.69	1.81	1.92	2.02
4.0	.66	.94	1.15	1.32	1.48	1.62	1.75	1.87	1.99	2.09
4.2	.68	.97	1.18	1.37	1.53	1.68	1.81	1.93	2.05	2.16
4.4	.71	1.00	1.22	1.41	1.58	1.73	1.87	2.00	2.12	2.23
4.6	.73	1.03	1.26	1.45	1.62	1.78	1.92	2.06	2.18	2.30
4.8	.75	1.06	1.29	1.49	1.67	1.83	1.98	2.11	2.24	2.36
5.0	.77	1.09	1.33	1.54	1.72	1.88	2.03	2.17	2.30	2.43
5.2	.79	1.12	1.37	1.58	1.76	1.93	2.09	2.23	2.37	2.49
5.4	.81	1.14	1.40	1.62	1.81	1.98	2.14	2.29	2.43	2.56
5.6	.83	1.17	1.43	1.66	1.85	2.03	2.19	2.34	2.49	2.62
5.8	.85	1.20	1.47	1.70	1.90	2.08	2.24	2.40	2.54	2.68
6.0	.87	1.23	1.50	1.73	1.94	2.12	2.29	2.45	2.60	2.74
6.2	.89	1.25	1.54	1.77	1.98	2.17	2.35	2.51	2.66	2.80
6.4	.91	1.28	1.57	1.81	2.02	2.22	2.40	2.56	2.72	2.86
6.6	.92	1.31	1.60	1.85	2.07	2.26	2.45	2.61	2.77	2.92
6.8	.94	1.33	1.63	1.89	2.11	2.31	2.49	2.67	2.83	2.98
7.0	.96	1.36	1.66	1.92	2.15	2.35	2.54	2.72	2.88	3.04
7.5	1.01	1.42	1.74	2.01	2.25	2.47	2.66	2.85	3.02	3.18
8.0	1.05	1.49	1.82	2.10	2.35	2.57	2.78	2.97	3.15	3.32
8.5	1.09	1.55	1.90	2.19	2.45	2.68	2.89	3.09	3.28	3.46
9.0	1.14	1.61	1.97	2.27	2.54	2.78	3.01	3.21	3.41	3.59
9.5	1.18	1.67	2.04	2.36	2.63	2.89	3.12	3.33	3.53	3.73
10	1.22	1.72	2.11	2.44	2.73	2.99	3.23	3.45	3.66	3.86
11	1.30	1.84	2.25	2.60	2.91	3.18	3.44	3.67	3.90	4.11
12	1.38	1.95	2.38	2.75	3.08	3.37	3.64	3.89	4.13	4.35
13	1.45	2.05	2.52	2.90	3.25	3.56	3.84	4.11	4.36	4.59
14	1.53	2.16	2.64	3.05	3.41	3.74	4.04	4.32	4.58	4.83
15	1.60	2.26	2.77	3.20	3.57	3.91	4.23	4.52	4.79	5.05
16	1.67	2.36	2.89	3.34	3.73	4.09	4.41	4.72	5.00	5.27
17	1.74	2.46	3.01	3.47	3.88	4.25	4.60	4.91	5.21	5.49
18	1.80	2.55	3.13	3.61	4.03	4.42	4.77	5.10	5.41	5.71
19	1.87	2.65	3.24	3.74	4.18	4.58	4.95	5.29	5.61	5.91
20	1.94	2.74	3.35	3.87	4.33	4.74	5.12	5.47	5.81	6.12

Table 14.—*Velocity of water, v, in feet per second, based on Manning's formula* $v = (1.486/n)r^{2/3}s^{1/2}$, **n = .040**—Continued

r \ s	.00055	.00060	.00065	.00070	.00075	.00080	.00085	.00090	.00095	.00100
0.2	0.30	0.31	0.32	0.34	0.35	0.36	0.37	0.38	0.39	0.40
0.4	.47	.49	.51	.53	.55	.57	.59	.61	.62	.64
0.6	.62	.65	.67	.70	.72	.75	.77	.79	.81	.84
0.8	.75	.78	.82	.85	.88	.91	.93	.96	.99	1.01
1.0	.87	.91	.95	.98	1.02	1.05	1.08	1.11	1.15	1.17
1.2	.98	1.03	1.07	1.11	1.15	1.19	1.22	1.26	1.29	1.33
1.4	1.09	1.14	1.19	1.23	1.27	1.31	1.36	1.39	1.43	1.47
1.6	1.19	1.24	1.30	1.34	1.39	1.44	1.48	1.52	1.57	1.61
1.8	1.29	1.35	1.40	1.45	1.51	1.55	1.60	1.65	1.69	1.74
2.0	1.38	1.44	1.50	1.56	1.62	1.67	1.72	1.77	1.82	1.86
2.2	1.47	1.54	1.60	1.66	1.72	1.78	1.83	1.89	1.94	1.99
2.4	1.56	1.63	1.70	1.76	1.82	1.88	1.94	2.00	2.05	2.11
2.6	1.65	1.72	1.79	1.86	1.92	1.99	2.05	2.11	2.17	2.22
2.8	1.73	1.81	1.88	1.95	2.02	2.09	2.15	2.21	2.27	2.33
3.0	1.81	1.89	1.97	2.04	2.12	2.19	2.25	2.32	2.38	2.44
3.2	1.89	1.98	2.06	2.13	2.21	2.28	2.35	2.42	2.49	2.55
3.4	1.97	2.06	2.14	2.22	2.30	2.38	2.45	2.52	2.59	2.66
3.6	2.05	2.14	2.22	2.31	2.39	2.47	2.54	2.62	2.69	2.76
3.8	2.12	2.22	2.31	2.39	2.48	2.56	2.64	2.71	2.79	2.86
4.0	2.20	2.29	2.39	2.48	2.56	2.65	2.73	2.81	2.89	2.96
4.2	2.27	2.37	2.47	2.56	2.65	2.74	2.82	2.90	2.98	3.06
4.4	2.34	2.44	2.54	2.64	2.73	2.82	2.91	2.99	3.07	3.15
4.6	2.41	2.52	2.62	2.72	2.81	2.91	3.00	3.08	3.17	3.25
4.8	2.48	2.59	2.70	2.80	2.90	2.99	3.08	3.17	3.26	3.34
5.0	2.55	2.66	2.77	2.87	2.97	3.07	3.17	3.26	3.35	3.44
5.2	2.62	2.73	2.84	2.95	3.05	3.15	3.25	3.35	3.44	3.53
5.4	2.68	2.80	2.92	3.03	3.13	3.23	3.33	3.43	3.52	3.62
5.6	2.75	2.87	2.99	3.10	3.21	3.31	3.42	3.51	3.61	3.70
5.8	2.81	2.94	3.06	3.17	3.28	3.39	3.50	3.60	3.70	3.79
6.0	2.88	3.00	3.13	3.25	3.36	3.47	3.58	3.68	3.78	3.88
6.2	2.94	3.07	3.20	3.32	3.43	3.55	3.66	3.76	3.86	3.96
6.4	3.00	3.14	3.26	3.39	3.51	3.62	3.73	3.84	3.95	4.05
6.6	3.07	3.20	3.33	3.46	3.58	3.70	3.81	3.92	4.03	4.13
6.8	3.13	3.27	3.40	3.53	3.65	3.77	3.89	4.00	4.11	4.22
7.0	3.19	3.33	3.47	3.60	3.72	3.85	3.96	4.08	4.19	4.30
7.5	3.34	3.49	3.63	3.77	3.90	4.03	4.15	4.27	4.39	4.50
8.0	3.48	3.64	3.79	3.93	4.07	4.20	4.33	4.46	4.58	4.70
8.5	3.63	3.79	3.94	4.09	4.24	4.38	4.51	4.64	4.77	4.89
9.0	3.77	3.94	4.10	4.25	4.40	4.55	4.69	4.82	4.95	5.08
9.5	3.91	4.08	4.25	4.41	4.56	4.71	4.86	5.00	5.14	5.27
10	4.04	4.22	4.40	4.56	4.72	4.88	5.03	5.17	5.31	5.45
11	4.31	4.50	4.68	4.86	5.03	5.20	5.36	5.51	5.66	5.81
12	4.57	4.77	4.96	5.15	5.33	5.51	5.68	5.84	6.00	6.16
13	4.82	5.03	5.24	5.43	5.62	5.81	5.99	6.16	6.33	6.50
14	5.06	5.29	5.50	5.71	5.91	6.10	6.29	6.47	6.65	6.82
15	5.30	5.53	5.76	5.98	6.19	6.39	6.59	6.78	6.96	7.15
16	5.53	5.78	6.01	6.24	6.46	6.67	6.88	7.08	7.27	7.46
17	5.76	6.02	6.26	6.50	6.73	6.95	7.16	7.37	7.57	7.77
18	5.98	6.25	6.51	6.75	6.99	7.22	7.44	7.65	7.86	8.07
19	6.20	6.48	6.74	7.00	7.24	7.48	7.71	7.94	8.15	8.36
20	6.42	6.70	6.98	7.24	7.50	7.74	7.98	8.21	8.44	8.66

Table 14.—*Velocity of water, v, in feet per second, based on Manning's formula* $v = (1.486/n)r^{2/3}s^{1/2}$, **n = .040**—Continued

r / s	.0010	.0015	.0020	.0025	.0030	.0035	.0040	.0045	.0050
0.2	0.40	049	0.57	0.64	0.70	0.75	0.80	0.85	0.90
0.4	.64	.78	.90	1.01	1.10	1.19	1.28	1.35	1.43
0.6	.84	1.02	1.18	1.32	1.45	1.56	1.67	1.77	1.87
0.8	1.01	1.24	1.43	1.60	1.75	1.89	2.02	2.15	2.26
1.0	1.17	1.44	1.66	1.86	2.03	2.20	2.35	2.49	2.63
1.2	1.33	1.62	1.88	2.10	2.30	2.48	2.65	2.81	2.97
1.4	1.47	1.80	2.08	2.32	2.55	2.75	2.94	3.12	3.29
1.6	1.61	1.97	2.27	2.54	2.78	3.01	3.21	3.41	3.59
1.8	1.74	2.13	2.46	2.75	3.01	3.25	3.48	3.69	3.89
2.0	1.86	2.28	2.64	2.95	3.23	3.49	3.73	3.96	4.17
2.2	1.99	2.43	2.81	3.14	3.44	3.72	3.97	4.22	4.44
2.4	2.11	2.58	2.98	3.33	3.65	3.94	4.21	4.47	4.71
2.6	2.22	2.72	3.14	3.51	3.85	4.16	4.44	4.71	4.97
2.8	2.33	2.86	3.30	3.69	4.04	4.37	4.67	4.95	5.22
3.0	2.44	2.99	3.46	3.86	4.23	4.57	4.89	5.18	5.46
3.2	2.55	3.12	3.61	4.03	4.42	4.77	5.10	5.41	5.70
3.4	2.66	3.25	3.76	4.20	4.60	4.97	5.31	5.63	5.94
3.6	2.76	3.38	3.90	4.36	4.78	5.16	5.52	5.85	6.17
3.8	2.86	3.50	4.05	4.52	4.95	5.35	5.72	6.07	6.40
4.0	2.96	3.63	4.19	4.68	5.13	5.54	5.92	6.28	6.62
4.2	3.06	3.75	4.32	4.84	5.30	5.72	6.12	6.49	6.84
4.4	3.15	3.86	4.46	4.99	5.46	5.90	6.31	6.69	7.05
4.6	3.25	3.98	4.60	5.14	5.63	6.08	6.50	6.89	7.27
4.8	3.34	4.09	4.73	5.29	5.79	6.25	6.69	7.09	7.47
5.0	3.44	4.21	4.86	5.43	5.95	6.43	6.87	7.29	7.68
5.2	3.53	4.32	4.99	5.58	6.11	6.60	7.05	7.48	7.88
5.4	3.62	4.43	6.11	5.72	6.26	6.76	7.23	7.67	8.09
5.6	3.70	4.54	5.24	5.86	6.42	6.93	7.41	7.86	8.28
5.8	3.79	4.64	5.36	6.00	6.57	7.09	7.58	8.04	8.48
6.0	3.88	4.75	5.49	6.13	6.72	7.26	7.76	8.23	8.67
6.2	3.96	4.86	5.61	6.27	6.87	7.42	7.93	8.41	8.87
6.4	4.05	4.96	5.73	6.40	7.01	7.58	8.10	8.59	9.06
6.6	4.13	5.06	5.85	6.54	7.16	7.73	8.27	8.77	9.24
6.8	4.22	5.16	5.96	6.67	7.30	7.89	8.43	8.94	9.43
7.0	4.30	5.27	6.08	6.80	7.45	8.04	8.60	9.12	9.61
7.5	4.50	5.51	6.37	7.12	7.80	8.42	9.00	9.55	10.07
8.0	4.70	5.76	6.65	7.43	8.14	8.79	9.40	9.97	10.51
8.5	4.89	5.99	6.92	7.74	8.47	9.15	9.79	10.38	10.94
9.0	5.08	6.23	7.19	8.04	8.80	9.51	10.17	10.78	11.37
9.5	5.27	6.45	7.45	8.33	9.13	9.86	10.54	11.18	11.78
10	5.45	6.68	7.71	8.62	9.44	10.20	10.91	11.57	12.19
11	5.81	7.12	8.22	9.19	10.06	10.87	11.62	12.33	12.99
12	6.16	7.54	8.71	9.74	10.67	11.52	12.32	13.06	13.77
13	6.50	7.95	9.19	10.27	11.25	12.15	12.99	13.78	14.52
14	6.82	8.36	9.65	10.79	11.82	12.77	13.65	14.48	15.26
15	7.15	8.75	10.10	11.30	12.38	13.37	14.29	15.16	15.98
16	7.46	9.14	10.55	11.79	12.92	13.96	14.92	15.82	16.68
17	7.77	9.51	10.98	12.28	13.45	14.53	15.53	16.48	17.37
18	8.07	9.88	11.41	12.76	13.98	15.10	16.14	17.12	18.04
19	8.36	10.24	11.83	13.23	14.49	15.65	16.73	17.74	18.70
20	8.66	10.60	12.24	13.69	14.99	16.19	17.31	18.36	19.36

Table 14.—*Velocity of water, v, in feet per second, based on Manning's formula* $v = (1.486/n)r^{2/3}s^{1/2}$, **n = .040**—Continued

r \ s	.0055	.0060	.0065	.0070	.0075	.0080	.0085	.0090	.0095	.0100
0.2	0.94	0.98	1.02	1.06	1.10	1.14	1.17	1.21	1.24	1.27
0.4	1.50	1.56	1.63	1.69	1.75	1.80	1.86	1.91	1.97	2.02
0.6	1.96	2.05	2.13	2.21	2.29	2.36	2.44	2.51	2.58	2.64
0.8	2.37	2.48	2.58	2.68	2.77	2.86	2.95	3.04	3.12	3.20
1.0	2.76	2.88	3.00	3.11	3.22	3.32	3.43	3.52	3.62	3.72
1.2	3.11	3.25	3.38	3.51	3.63	3.75	3.87	3.98	4.09	4.20
1.4	3.45	3.60	3.75	3.89	4.03	4.16	4.29	4.41	4.53	4.65
1.6	3.77	3.94	4.10	4.25	4.40	4.55	4.69	4.82	4.95	5.08
1.8	4.08	4.26	4.43	4.60	4.76	4.92	5.07	5.22	5.36	5.50
2.0	4.37	4.57	4.75	4.93	5.11	5.27	5.44	5.59	5.75	5.90
2.2	4.66	4.87	5.07	5.26	5.44	5.62	5.79	5.96	6.12	6.28
2.4	4.94	5.16	5.37	5.57	5.77	5.96	6.14	6.32	6.49	6.66
2.6	5.21	5.44	5.66	5.88	6.08	6.28	6.48	6.66	6.85	7.02
2.8	5.47	5.72	5.95	6.17	6.39	6.60	6.80	7.00	7.19	7.38
3.0	5.73	5.99	6.23	6.47	6.69	6.91	7.12	7.33	7.53	7.73
3.2	5.98	6.25	6.50	6.75	6.99	7.22	7.44	7.65	7.86	8.07
3.4	6.23	6.51	6.77	7.03	7.27	7.51	7.74	7.97	8.19	8.40
3.6	6.47	6.76	7.04	7.30	7.56	7.80	8.05	8.28	8.51	8.73
3.8	6.71	7.01	7.29	7.57	7.83	8.09	8.34	8.58	8.82	9.05
4.0	6.94	7.25	7.55	7.83	8.11	8.37	8.63	8.88	9.12	9.36
4.2	7.17	7.49	7.80	8.09	8.38	8.65	8.92	9.17	9.43	9.67
4.4	7.40	7.73	8.04	8.35	8.64	8.92	9.20	9.46	9.72	9.98
4.6	7.62	7.96	8.28	8.60	8.90	9.19	9.47	9.75	10.02	10.28
4.8	7.84	8.19	8.52	8.84	9.15	9.46	9.75	10.03	10.30	10.57
5.0	8.06	8.41	8.76	9.09	9.41	9.72	10.01	10.31	10.59	10.86
5.2	8.27	8.64	8.99	9.33	9.66	9.97	10.28	10.58	10.87	11.15
5.4	8.48	8.86	9.22	9.57	9.90	10.23	10.54	10.85	11.15	11.43
5.6	8.69	9.07	9.45	9.80	10.15	10.48	10.80	11.11	11.42	11.72
5.8	8.89	9.29	9.67	10.03	10.39	10.73	11.06	11.38	11.69	11.99
6.0	9.10	9.50	9.89	10.26	10.62	10.97	11.31	11.64	11.96	12.27
6.2	9.30	9.71	10.11	10.49	10.86	11.21	11.56	11.89	12.22	12.54
6.4	9.50	9.92	10.32	10.71	11.09	11.45	11.81	12.15	12.48	12.81
6.6	9.69	10.13	10.54	10.94	11.32	11.69	12.05	12.40	12.74	13.07
6.8	9.89	10.33	10.75	11.16	11.55	11.93	12.29	12.65	13.00	13.33
7.0	10.08	10.53	10.96	11.37	11.77	12.16	12.53	12.90	13.25	13.59
7.5	10.56	11.03	11.48	11.91	12.33	12.73	13.12	13.50	13.87	14.23
8.0	11.02	11.51	11.98	12.43	12.87	13.29	13.70	14.10	14.48	14.86
8.5	11.48	11.99	12.47	12.95	13.40	13.84	14.27	14.68	15.08	15.47
9.0	11.92	12.45	12.96	13.45	13.92	14.38	14.82	15.25	15.67	16.07
9.5	12.36	12.91	13.43	13.94	14.43	14.90	15.36	15.81	16.24	16.66
10	12.79	13.36	13.90	14.43	14.93	15.42	15.90	16.36	16.81	17.24
11	13.63	14.23	14.81	15.37	15.91	16.43	16.94	17.43	17.91	18.37
12	14.44	15.08	15.70	16.29	16.86	17.42	17.95	18.47	18.98	19.47
13	15.23	15.91	16.56	17.18	17.79	18.37	18.94	19.49	20.02	20.54
14	16.00	16.72	17.40	18.05	18.69	19.30	19.90	20.47	21.03	21.58
15	16.76	17.50	18.22	18.90	19.57	20.21	20.83	21.44	22.02	22.60
16	17.49	18.27	19.02	19.74	20.43	21.10	21.75	22.38	22.99	23.59
17	18.22	19.03	19.80	20.55	21.27	21.97	22.64	23.30	23.94	24.56
18	18.92	19.76	20.57	21.35	22.10	22.82	23.52	24.21	24.87	25.52
19	19.62	20.49	21.33	22.13	22.91	23.66	24.39	25.09	25.78	26.45
20	20.30	21.20	22.07	22.90	23.71	24.48	25.24	25.97	26.68	27.37

Table 14.—*Velocity of water, v, in feet per second, based on Manning's formula* $v = (1.486/n)r^{2/3}s^{1/2}$, **n = .040**—Continued

s	.01	.02	.03	.04	.05	.06	.07	.08	.09	.10
0.2	1.27	1.80	2.20	2.54	2.84	3.11	3.36	3.59	3.81	4.02
0.4	2.02	2.85	3.49	4.03	4.51	4.94	5.34	5.70	6.05	6.38
0.6	2.64	3.74	4.58	5.29	5.91	6.47	6.99	7.47	7.93	8.36
0.8	3.20	4.53	5.55	6.40	7.16	7.84	8.47	9.06	9.60	10.12
1.0	3.72	5.25	6.43	7.43	8.31	9.10	9.83	10.51	11.15	11.75
1.2	4.20	5.93	7.27	8.39	9.38	10.28	11.10	11.87	12.59	13.27
1.4	4.65	6.57	8.05	9.30	10.40	11.39	12.30	13.15	13.95	14.70
1.6	5.08	7.19	8.80	10.16	11.36	12.45	13.45	14.37	15.25	16.07
1.8	5.50	7.77	9.52	10.99	12.29	13.47	14.54	15.55	16.49	17.38
2.0	5.90	8.34	10.21	11.79	13.19	14.45	15.60	16.68	17.69	18.65
2.2	6.28	8.89	10.88	12.57	14.05	15.39	16.63	17.77	18.85	19.87
2.4	6.66	9.42	11.53	13.32	14.89	16.31	17.62	18.84	19.98	21.06
2.6	7.02	9.93	12.17	14.05	15.71	17.21	18.58	19.87	21.07	22.21
2.8	7.38	10.44	12.78	14.76	16.50	18.08	19.53	20.87	22.14	23.34
3.0	7.73	10.93	13.38	15.46	17.28	18.93	20.45	21.86	23.18	24.44
3.2	8.07	11.41	13.97	16.13	18.04	19.76	21.34	22.82	24.20	25.51
3.4	8.40	11.88	14.55	16.80	18.78	20.58	22.22	23.76	25.20	26.56
3.6	8.73	12.34	15.11	17.45	19.51	21.37	23.09	24.68	26.18	27.59
3.8	9.05	12.79	15.67	18.09	20.23	22.16	23.93	25.59	27.14	28.61
4.0	9.36	13.24	16.21	18.72	20.93	22.93	24.77	26.48	28.08	29.60
4.2	9.67	13.68	16.75	19.34	21.62	23.69	25.59	27.35	29.01	30.58
4.4	9.98	14.11	17.28	19.95	22.31	24.43	26.39	28.21	29.93	31.54
4.6	10.28	14.53	17.80	20.55	22.98	25.17	27.19	29.06	30.83	32.49
4.8	10.57	14.95	18.31	21.14	23.64	25.89	27.97	29.90	31.71	33.43
5.0	10.86	15.36	18.81	21.73	24.29	26.61	28.74	30.72	32.59	34.35
5.2	11.15	15.77	19.31	22.30	24.93	27.31	29.50	31.54	33.45	35.26
5.4	11.43	16.17	19.81	22.87	25.57	28.01	30.25	32.34	34.30	36.16
5.6	11.72	16.57	20.29	23.43	26.20	28.70	31.00	33.14	35.15	37.05
5.8	11.99	16.96	20.77	23.99	26.82	29.38	31.73	33.92	35.98	37.92
6.0	12.27	17.35	21.25	24.53	27.43	30.05	32.45	34.70	36.80	38.79
6.2	12.54	17.73	21.72	25.08	28.04	30.71	33.17	35.46	37.61	39.65
6.4	12.81	18.11	22.18	25.61	28.63	31.37	33.88	36.22	38.42	40.50
6.6	13.07	18.49	22.64	26.14	29.23	32.02	34.58	36.97	39.21	41.34
6.8	13.33	18.86	23.10	26.67	29.82	32.66	35.28	37.71	40.00	42.17
7.0	13.59	19.23	23.55	27.19	30.40	33.30	35.97	38.45	40.78	42.99
7.5	14.23	20.13	24.65	28.47	31.83	34.87	37.66	40.26	42.70	45.01
8.0	14.86	21.02	25.74	29.72	33.23	36.40	39.32	42.03	44.58	46.99
8.5	15.47	21.88	26.80	30.95	34.60	37.90	40.94	43.76	46.42	48.93
9.0	16.07	22.73	27.84	32.15	35.94	39.37	42.53	45.46	48.22	50.83
9.5	16.66	23.57	28.86	33.33	37.26	40.82	44.09	47.13	49.99	52.70
10	17.24	24.39	29.87	34.49	38.56	42.24	45.62	48.77	51.73	54.53
11	18.37	25.99	31.83	36.75	41.09	45.01	48.61	51.97	55.12	58.11
12	19.47	27.54	33.73	38.94	43.54	47.70	51.52	55.08	58.42	61.58
13	20.54	29.05	35.58	41.08	45.93	50.31	54.34	58.09	61.62	64.95
14	21.58	30.52	37.38	43.16	48.25	52.86	57.09	61.04	64.74	68.24
15	22.60	31.95	39.14	45.19	50.52	55.35	59.78	63.91	67.79	71.45
16	23.59	33.36	40.86	47.18	52.75	57.78	62.41	66.72	70.77	74.59
17	24.56	34.74	42.54	49.12	54.92	60.16	64.98	69.47	73.69	77.67
18	25.52	36.08	44.19	51.03	57.05	62.50	67.51	72.17	76.55	80.69
19	26.45	37.41	45.82	52.90	59.15	64.79	69.99	74.82	79.36	83.65
20	27.37	38.71	47.41	54.74	61.21	67.05	72.42	77.42	82.12	86.56

Table 15.—*Values of nv corresponding to different values of r and s in Manning's formula,* $v = (1.486/n)r^{2/3}s^{1/2}$

(To determine the mean velocity of water, v, divide the tabulated value by the coefficient of roughness, n)

$s=$ slope	\multicolumn{10}{c}{$r=$ hydraulic radius in feet}									
	0.1	0.2	0.3	0.4	0.5	0.6	0.7	0.8	0.9	1.0
.00005	.0023	.0036	.0047	.0057	.0066	.0075	.0083	.0091	.0098	.0105
10	.0032	.0051	.0067	.0081	.0094	.0106	.0117	.0128	.0139	.0149
15	.0039	.0062	.0082	.0099	.0115	.0130	.0144	.0157	.0170	.0182
20	.0045	.0072	.0094	.0114	.0132	.0150	.0166	.0181	.0196	.0210
25	.0051	.0080	.0105	.0128	.0148	.0167	.0185	.0203	.0219	.0235
.00030	.0056	.0088	.0115	.0140	.0162	.0183	.0203	.0222	.0240	.0257
35	.0060	.0095	.0125	.0151	.0175	.0198	.0219	.0240	.0259	.0278
40	.0064	.0102	.0133	.0161	.0187	.0211	.0234	.0256	.0277	.0297
45	.0068	.0108	.0141	.0171	.0199	.0224	.0249	.0272	.0294	.0316
50	.0072	.0114	.0149	.0180	.0209	.0236	.0262	.0286	.0310	.0332
.00055	.0075	.0119	.0156	.0189	.0220	.0248	.0275	.0300	.0325	.0349
60	.0078	.0125	.0163	.0198	.0229	.0259	.0287	.0314	.0339	.0364
65	.0082	.0130	.0170	.0206	.0239	.0270	.0299	.0327	.0353	.0379
70	.0085	.0135	.0176	.0213	.0248	.0280	.0310	.0339	.0367	.0393
75	.0088	.0139	.0182	.0221	.0256	.0290	.0321	.0351	.0379	.0407
.00080	.0091	.0144	.0188	.0228	.0265	.0299	.0331	.0362	.0392	.0420
85	.0093	.0148	.0194	.0235	.0273	.0308	.0342	.0374	.0404	.0433
90	.0096	.0153	.0200	.0242	.0281	.0317	.0351	.0384	.0416	.0446
95	.0099	.0157	.0205	.0249	.0289	.0326	.0361	.0395	.0427	.0458
100	.0101	.0161	.0211	.0255	.0296	.0334	.0370	.0405	.0439	.0470
.0011	.0106	.0169	.0221	.0268	.0311	.0351	.0389	.0425	.0459	.0493
12	.0111	.0176	.0231	.0280	.0324	.0366	.0406	.0444	.0480	.0515
13	.0115	.0183	.0240	.0291	.0338	.0381	.0422	.0462	.0500	.0536
14	.0119	.0190	.0249	.0302	.0350	.0395	.0438	.0479	.0518	.0556
15	.0124	.0197	.0258	.0312	.0363	.0410	.0454	.0496	.0537	.0576
.0016	.0128	.0203	.0266	.0323	.0375	.0423	.0469	.0512	.0554	.0594
17	.0132	.0210	.0275	.0333	.0387	.0436	.0483	.0528	.0571	.0613
18	.0136	.0216	.0283	.0342	.0397	.0449	.0497	.0543	.0587	.0630
19	.0140	.0222	.0290	.0352	.0409	.0461	.0511	.0558	.0604	.0648
20	.0143	.0227	.0298	.0361	.0419	.0473	.0524	.0573	.0620	.0665
.0025	.0160	.0254	.0333	.0403	.0468	.0529	.0586	.0641	.0693	.0743
30	.0175	.0278	.0365	.0442	.0513	.0579	.0642	.0702	.0759	.0814
35	.0189	.0301	.0394	.0477	.0554	.0625	.0693	.0758	.0820	.0879
40	.0202	.0321	.0421	.0510	.0592	.0669	.0741	.0810	.0876	.0940
45	.0215	.0341	.0447	.0541	.0628	.0709	.0786	.0859	.0929	.0997
.0050	.0226	.0359	.0471	.0570	.0662	.0748	.0828	.0906	.0980	.1051
55	.0237	.0377	.0494	.0598	.0694	.0784	.0869	.0950	.1027	.1102
60	.0248	.0394	.0516	.0625	.0725	.0819	.0908	.0992	.1073	.1151
65	.0258	.0410	.0537	.0650	.0755	.0852	.0945	.1033	.1117	.1198
70	.0268	.0425	.0557	.0675	.0783	.0884	.0980	.1071	.1159	.1243
.0075	.0277	.0440	.0577	.0699	.0811	.0916	.1015	.1109	.1200	.1287
80	.0286	.0455	.0596	.0722	.0837	.0946	.1048	.1145	.1239	.1329
85	.0295	.0469	.0614	.0744	.0863	.0975	.1080	.1181	.1277	.1370
90	.0304	.0482	.0632	.0765	.0888	.1003	.1111	.1215	.1314	.1410
95	.0312	.0495	.0649	.0786	.0912	.1030	.1142	.1248	.1350	.1448
.0100	.0320	.0508	.0666	.0807	.0936	.1057	.1172	.1281	.1385	.1486

Table 15.—*Values of nv corresponding to different values of r and s in Manning's formula,* $v = (1.486/n)r^{2/3}s^{1/2}$—Continued

(To determine the mean velocity of water, v, divide the tabulated value by the coefficient of roughness, n)

s=slope	\multicolumn{10}{c}{r=hydraulic radius in feet}									
	1.1	1.2	1.3	1.4	1.5	1.6	1.7	1.8	1.9	2.0
.00005	.0112	.0119	.0125	.0132	.0138	.0144	.0150	.0156	.0161	.0167
10	.0158	.0168	.0177	.0186	.0195	.0203	.0212	.0220	.0228	.0236
15	.0194	.0206	.0217	.0228	.0239	.0249	.0259	.0269	.0279	.0289
20	.0224	.0237	.0250	.0263	.0275	.0288	.0299	.0311	.0322	.0334
25	.0250	.0265	.0280	.0294	.0308	.0321	.0335	.0348	.0360	.0373
.00030	.0274	.0291	.0307	.0322	.0337	.0352	.0367	.0381	.0395	.0409
35	.0296	.0314	.0331	.0348	.0364	.0380	.0396	.0411	.0427	.0441
40	.0317	.0336	.0354	.0372	.0389	.0407	.0423	.0440	.0456	.0472
45	.0336	.0356	.0376	.0395	.0413	.0431	.0449	.0467	.0484	.0500
50	.0354	.0375	.0396	.0416	.0435	.0454	.0473	.0492	.0510	.0528
.00055	.0371	.0394	.0415	.0436	.0457	.0477	.0496	.0516	.0535	.0553
60	.0388	.0411	.0434	.0456	.0477	.0498	.0519	.0539	.0558	.0578
65	.0405	.0428	.0451	.0474	.0497	.0518	.0540	.0561	.0581	.0601
70	.0419	.0444	.0468	.0492	.0515	.0538	.0560	.0582	.0603	.0624
75	.0434	.0460	.0485	.0509	.0533	.0556	.0580	.0602	.0624	.0646
.00080	.0448	.0475	.0501	.0526	.0551	.0575	.0599	.0622	.0645	.0667
85	.0462	.0489	.0516	.0542	.0568	.0593	.0617	.0641	.0665	.0688
90	.0475	.0503	.0531	.0558	.0584	.0610	.0635	.0660	.0684	.0708
95	.0488	.0517	.0546	.0573	.0600	.0627	.0653	.0678	.0703	.0727
100	.0501	.0531	.0560	.0588	.0616	.0643	.0669	.0695	.0721	.0746
.0011	.0525	.0557	.0587	.0617	.0646	.0674	.0702	.0729	.0756	.0782
12	.0549	.0581	.0613	.0644	.0675	.0704	.0733	.0762	.0790	.0817
13	.0571	.0605	.0638	.0671	.0702	.0733	.0763	.0793	.0822	.0851
14	.0593	.0628	.0662	.0696	.0729	.0761	.0792	.0823	.0853	.0883
15	.0613	.0650	.0686	.0720	.0754	.0787	.0820	.0852	.0883	.0914
.0016	.0633	.0671	.0708	.0744	.0779	.0813	.0847	.0880	.0912	.0944
17	.0653	.0692	.0730	.0767	.0803	.0838	.0873	.0907	.0940	.0973
18	.0672	.0712	.0751	.0789	.0826	.0862	.0898	.0933	.0967	.1001
19	.0690	.0732	.0772	.0811	.0849	.0886	.0923	.0959	.0994	.1028
20	.0708	.0751	.0792	.0832	.0871	.0909	.0947	.0984	.1020	.1055
.0025	.0792	.0839	.0885	.0930	.0974	.1016	.1058	.1099	.1140	.1180
30	.0867	.0919	.0970	.1019	.1067	.1113	.1159	.1204	.1249	.1292
35	.0937	.0993	.1047	.1100	.1152	.1203	.1252	.1301	.1349	.1396
40	.1001	.1061	.1119	.1176	.1231	.1286	.1339	.1391	.1442	.1492
45	.1062	.1126	.1188	.1248	.1306	.1364	.1420	.1475	.1529	.1582
.0050	.1120	.1187	.1252	.1315	.1377	.1438	.1497	.1555	.1612	.1668
55	.1174	.1245	.1313	.1379	.1444	.1508	.1570	.1631	.1691	.1749
60	.1227	.1300	.1371	.1441	.1509	.1575	.1640	.1703	.1766	.1827
65	.1277	.1353	.1427	.1499	.1570	.1639	.1707	.1773	.1838	.1902
70	.1325	.1404	.1481	.1556	.1629	.1701	.1771	.1840	.1908	.1974
.0075	.1371	.1453	.1533	.1611	.1687	.1761	.1833	.1904	.1974	.2043
80	.1416	.1501	.1583	.1664	.1742	.1818	.1893	.1967	.2039	.2110
85	.1460	.1547	.1632	.1715	.1795	.1874	.1952	.2027	.2102	.2175
90	.1502	.1592	.1679	.1764	.1847	.1929	.2008	.2086	.2163	.2238
95	.1544	.1636	.1726	.1813	.1898	.1981	.2063	.2143	.2222	.2299
.0100	.1584	.1678	.1770	.1860	.1947	.2033	.2117	.2199	.2280	.2359

Table 15.—*Values of nv corresponding to different values of r and s in Manning's formula,* $v = (1.486/n)r^{2/3}s^{1/2}$—Continued

(To determine the mean velocity of water, v, divide the tabulated value by the coefficient of roughness, n)

| s=slope | \multicolumn{10}{c}{r=hydraulic radius in feet} |||||||||| |
	2.1	2.2	2.3	2.4	2.5	2.6	2.7	2.8	2.9	3.0
.00005	.0172	.0178	.0183	.0188	.0194	.0199	.0204	.0209	.0214	.0219
10	.0244	.0251	.0259	.0266	.0274	.0281	.0288	.0295	.0302	.0309
15	.0299	.0308	.0317	.0326	.0335	.0344	.0353	.0362	.0370	.0379
20	.0345	.0356	.0366	.0377	.0387	.0397	.0407	.0417	.0427	.0437
25	.0385	.0397	.0409	.0421	.0433	.0444	.0456	.0467	.0478	.0489
.00030	.0422	.0435	.0449	.0462	.0474	.0487	.0499	.0511	.0523	.0535
35	.0456	.0470	.0484	.0498	.0512	.0526	.0539	.0552	.0565	.0578
40	.0487	.0503	.0518	.0533	.0548	.0562	.0576	.0590	.0604	.0618
45	.0517	.0533	.0549	.0565	.0581	.0596	.0611	.0626	.0641	.0656
50	.0545	.0562	.0579	.0596	.0612	.0628	.0644	.0660	.0676	.0691
.00055	.0572	.0590	.0607	.0625	.0642	.0659	.0676	.0692	.0709	.0725
60	.0596	.0616	.0634	.0653	.0671	.0688	.0706	.0723	.0740	.0757
65	.0621	.0641	.0660	.0679	.0698	.0716	.0735	.0753	.0771	.0788
70	.0645	.0665	.0685	.0705	.0724	.0743	.0762	.0781	.0800	.0818
75	.0667	.0688	.0709	.0730	.0750	.0770	.0789	.0809	.0828	.0847
.00080	.0689	.0711	.0732	.0753	.0774	.0795	.0815	.0835	.0855	.0874
85	.0711	.0733	.0755	.0777	.0798	.0819	.0840	.0861	.0881	.0901
90	.0731	.0754	.0777	.0799	.0821	.0843	.0864	.0886	.0907	.0927
95	.0751	.0775	.0798	.0821	.0844	.0866	.0888	.0910	.0931	.0953
100	.0771	.0795	.0819	.0842	.0866	.0889	.0911	.0934	.0956	.0978
.0011	.0808	.0834	0859	.0884	.0908	.0932	.0956	.0979	.1002	.1025
12	.0844	.0871	.0897	.0923	.0948	.0973	.0998	.1023	.1047	.1071
13	.0879	.0906	.0934	.0960	.0987	.1013	.1039	.1064	.1090	.1115
14	.0912	.0941	.0969	.0997	.1024	.1051	.1078	.1105	.1131	.1157
15	.0944	.0974	.1003	.1032	.1060	.1088	.1116	.1143	.1170	.1197
.0016	.0975	.1006	.1036	.1066	.1095	.1124	.1153	.1181	.1209	.1236
17	.1005	.1036	.1068	.1098	.1129	.1159	.1188	.1218	.1246	.1274
18	.1034	.1066	.1099	.1130	.1161	.1192	.1222	.1252	.1282	.1311
19	.1062	.1096	.1129	.1161	.1193	.1225	.1256	.1287	.1317	.1347
20	.1090	.1124	.1158	.1191	.1224	.1257	.1289	.1320	.1352	.1382
.0025	.1218	.1257	.1295	.1332	.1369	.1405	.1441	.1476	.1511	.1546
30	.1335	.1377	.1418	.1459	.1499	.1539	.1578	.1617	.1655	.1693
35	.1442	.1487	.1532	.1576	.1619	.1662	.1705	.1747	.1788	.1829
40	.1541	.1590	.1638	.1685	.1731	.1777	.1822	.1867	.1911	.1955
45	.1635	.1686	.1737	.1787	.1836	.1885	.1933	.1980	.2027	.2074
.0050	.1723	.1777	.1831	.1884	.1936	.1987	.2037	.2087	.2137	.2186
55	.1807	.1864	.1920	.1976	.2030	.2084	.2137	.2189	.2241	.2292
60	.1888	.1947	.2006	.2063	.2120	.2177	.2232	.2287	.2341	.2394
65	.1965	.2027	.2088	.2148	.2207	.2265	.2323	.2380	.2436	.2492
70	.2039	.2103	.2166	.2229	.2290	.2351	.2411	.2470	.2528	.2586
.0075	.2110	.2177	.2242	.2307	.2371	.2433	.2495	.2557	.2617	.2677
80	.2180	.2248	.2316	.2383	.2448	.2513	.2577	.2640	.2703	.2765
85	.2247	.2317	.2387	.2456	.2524	.2591	.2657	.2722	.2786	.2850
90	.2312	.2385	.2456	.2527	.2597	.2666	.2734	.2801	.2867	.2932
95	.2375	.2450	.2524	.2596	.2668	.2739	.2808	.2877	.2946	.3013
.0100	.2437	.2514	.2589	.2664	.2737	.2810	.2881	.2952	.3022	.3091

Table 15.—*Values of nv corresponding to different values of r and s in Manning's formula,* $v = (1.486/n) r^{2/3} s^{1/2}$—Continued

(To determine the mean velocity of water, *v*, divide the tabulated value by the coefficient of roughness, *n*)

s=slope	r = hydraulic radius in feet									
	3.1	3.2	3.3	3.4	3.5	3.6	3.7	3.8	3.9	4.0
.00005	.0223	.0228	.0233	.0238	.0242	.0247	.0251	.0256	.0260	.0265
10	.0316	.0323	.0329	.0336	.0343	.0349	.0356	.0362	.0368	.0374
15	.0387	.0395	.0403	.0412	.0420	.0428	.0435	.0443	.0451	.0459
20	.0447	.0456	.0466	.0475	.0484	.0494	.0503	.0512	.0521	.0530
25	.0500	.0510	.0521	.0531	.0542	.0552	.0562	.0572	.0582	.0592
.00030	.0547	.0559	.0571	.0582	.0593	.0605	.0616	.0627	.0638	.0649
35	.0591	.0604	.0616	.0629	.0641	.0653	.0665	.0677	.0689	.0701
40	.0632	.0645	.0659	.0672	.0685	.0698	.0711	.0724	.0736	.0749
45	.0670	.0685	.0699	.0713	.0727	.0741	.0754	.0768	.0781	.0794
50	.0706	.0722	.0737	.0751	.0766	.0781	.0795	.0809	.0823	.0837
.00055	.0741	.0757	.0773	.0788	.0803	.0819	.0834	.0849	.0864	.0878
60	.0774	.0791	.0807	.0823	.0839	.0855	.0871	.0886	.0902	.0917
65	.0806	.0823	.0840	.0857	.0873	.0890	.0906	.0923	.0939	.0955
70	.0836	.0854	.0872	.0889	.0906	.0924	.0941	.0957	.0974	.0991
75	.0865	.0884	.0902	.0920	.0938	.0956	.0974	.0991	.1008	.1026
.00080	.0894	.0913	.0932	.0950	.0969	.0987	.1005	.1024	.1041	.1059
85	.0921	.0941	.0960	.0980	.0999	.1018	.1036	.1055	.1073	.1092
90	.0948	.0968	.0988	.1008	.1028	.1047	.1066	.1086	.1105	.1123
95	.0974	.0995	.1015	.1036	.1056	.1076	.1096	.1115	.1135	.1154
100	.0999	.1021	.1042	.1063	.1083	.1104	.1124	.1144	.1164	.1184
.0011	.1048	.1070	.1093	.1114	.1136	.1158	.1179	.1200	.1221	.1242
12	.1094	.1118	.1141	.1164	.1187	.1209	.1231	.1254	.1275	.1297
13	.1139	.1164	.1188	.1212	.1235	.1259	.1282	.1305	.1328	.1350
14	.1182	.1207	.1232	.1257	.1282	.1306	.1330	.1354	.1378	.1401
15	.1224	.1250	.1276	.1301	.1327	.1352	.1377	.1401	.1426	.1450
.0016	.1264	.1291	.1318	.1344	.1370	.1396	.1422	.1447	.1473	.1498
17	.1303	.1331	.1358	.1385	.1412	.1439	.1466	.1492	.1518	.1544
18	.1340	.1369	.1397	.1426	.1453	.1481	.1508	.1535	.1562	.1589
19	.1377	.1407	.1436	.1465	.1493	.1522	.1550	.1577	.1605	.1633
20	.1413	.1443	.1473	.1503	.1532	.1561	.1590	.1618	.1647	.1675
.0025	.1580	.1614	.1647	.1680	.1713	.1745	.1777	.1809	.1841	.1872
30	.1730	.1768	.1804	.1840	.1876	.1912	.1947	.1982	.2017	.2051
35	.1869	.1909	.1949	.1988	.2027	.2065	.2103	.2141	.2178	.2215
40	.1998	.2041	.2083	.2125	.2167	.2208	.2248	.2289	.2329	.2368
45	.2119	.2165	.2210	.2254	.2298	.2342	.2385	.2427	.2470	.2512
.0050	.2234	.2282	.2329	.2376	.2422	.2468	.2514	.2559	.2603	.2648
55	.2343	.2393	.2443	.2492	.2541	.2589	.2636	.2684	.2731	.2777
60	.2447	.2500	.2552	.2603	.2654	.2704	.2754	.2803	.2852	.2901
65	.2548	.2602	.2656	.2709	.2762	.2814	.2866	.2917	.2968	.3019
70	.2643	.2700	.2756	.2811	.2866	.2920	.2974	.3028	.3080	.3133
.0075	.2736	.2795	.2852	.2910	.2967	.3023	.3079	.3134	.3189	.3242
80	.2826	.2886	.2946	.3005	.3064	.3122	.3180	.3237	.3293	.3349
85	.2913	.2975	.3037	.3098	.3158	.3218	.3277	.3336	.3395	.3452
90	.2997	.3061	.3125	.3188	.3250	.3311	.3372	.3433	.3493	.3552
95	.3079	.3145	.3210	.3275	.3339	.3402	.3465	.3527	.3589	.3650
.0100	.3159	.3227	.3294	.3360	.3426	.3491	.3555	.3619	.3682	.3745

Table 15.—*Values of nv corresponding to different values of r and s in Manning's formula,* $v = (1.486/n)r^{2/3}s^{1/2}$—Continued

(To determine the mean velocity of water, v, divide the tabulated value by the coefficient of roughness, n)

s = slope	r = hydraulic radius in feet									
	4.2	4.4	4.6	4.8	5.0	5.2	5.4	5.6	5.8	6.0
.00005	.0274	.0282	.0291	.0299	.0307	.0315	.0323	.0331	.0339	.0347
10	.0387	.0399	.0411	.0423	.0435	.0446	.0457	.0469	.0480	.0491
15	.0474	.0489	.0503	.0518	.0532	.0546	.0560	.0574	.0588	.0601
20	.0547	.0564	.0581	.0598	.0615	.0631	.0647	.0663	.9678	.0694
25	.0612	.0631	.0650	.0669	.0687	.0705	.0723	.0741	.0759	.0776
.00030	.0670	.0691	.0712	.0732	.0753	.0773	.0792	.0812	.0831	.0850
35	.0724	.0747	.0769	.0791	.0813	.0834	.0856	.0877	.0897	.0918
40	.0774	.0798	.0822	.0846	.0869	.0892	.0915	.0937	.0959	.0981
45	.0821	.0846	.0872	.0897	.0922	.0946	.0970	.0994	.1018	.1041
50	.0865	.0892	.0919	.0946	.0972	.0997	.1023	.1048	.1073	.1097
.00055	.0907	.0936	.0964	.0992	.1019	.1046	.1073	.1099	.1125	.1152
60	.0948	.0977	.1007	.1036	.1064	.1092	.1120	.1148	.1175	.1202
65	.0986	.1017	.1048	.1078	.1108	.1137	.1166	.1195	.1223	.1251
70	.1024	.1056	.1087	.1119	.1150	.1180	.1210	.1240	.1269	.1298
75	.1059	.1093	.1125	.1158	.1190	.1222	.1253	.1284	.1314	.1344
.00080	.1094	.1129	.1163	.1196	.1229	.1262	.1294	.1325	.1357	.1388
85	.1128	.1163	.1198	.1233	.1267	.1300	.1334	.1366	.1399	.1431
90	.1161	.1197	.1233	.1269	.1304	.1338	.1372	.1406	.1439	.1472
95	.1192	.1230	.1267	.1303	.1339	.1375	.1410	.1444	.1479	.1512
100	.1223	.1262	.1300	.1337	.1374	.1410	.1446	.1482	.1517	.1553
.0011	.1283	.1323	.1363	.1402	.1441	.1479	.1517	.1554	.1591	.1627
12	.1340	.1382	.1424	.1465	.1505	.1545	.1584	.1623	.1662	.1700
13	.1395	.1439	.1482	.1525	.1567	.1608	.1649	.1690	.1730	.1769
14	.1447	.1493	.1538	.1582	.1626	.1669	.1711	.1753	.1795	.1836
15	.1498	.1545	.1592	.1638	.1683	.1727	.1771	.1815	.1858	.1900
.0016	.1547	.1596	.1644	.1691	.1738	.1784	.1830	.1874	.1919	.1963
17	.1595	.1645	.1695	.1743	.1792	.1839	.1886	.1932	.1978	.2023
18	.1641	.1693	.1744	.1794	.1844	.1892	.1941	.1988	.2035	.2082
19	.1686	.1739	.1792	.1843	.1894	.1944	.1994	.2043	.2091	.2139
20	.1730	.1784	.1838	.1891	.1943	.1995	.2046	.2096	.2145	.2194
.0025	.1934	.1995	.2055	.2114	.2173	.2230	.2287	.2343	.2399	.2454
30	.2119	.2186	.2251	.2316	.2380	.2443	.2505	.2567	.2628	.2688
35	.2289	.2361	.2432	.2502	.2571	.2639	.2706	.2772	.2838	.2903
40	.2447	.2524	.2600	.2674	.2748	.2821	.2893	.2964	.3034	.3103
45	.2595	.2677	.2757	.2837	.2915	.2992	.3068	.3144	.3218	.3292
.0050	.2735	.2821	.2906	.2990	.3072	.3154	.3234	.3314	.3392	.3470
55	.2869	.2959	.3048	.3136	.3222	.3308	.3392	.3475	.3558	.3639
60	.2996	.3091	.3184	.3275	.3366	.3455	.3543	.3630	.3716	.3801
65	.3119	.3217	.3314	.3409	.3503	.3596	.3688	.3778	.3867	.3956
70	.3237	.3338	.3439	.3538	.3635	.3732	.3827	.3921	.4014	.4105
.0075	.3350	.3456	.3560	.3662	.3763	.3863	.3961	.4058	.4154	.4249
80	.3460	.3569	.3676	.3782	.3886	.3989	.4091	.4191	.4291	.4389
85	.3566	.3679	.3790	.3898	.4006	.4112	.4217	.4320	.4423	.4524
90	.3670	.3785	.3899	.4012	.4122	.4231	.4339	.4446	.4551	.4655
95	.3770	.3889	.4006	.4121	.4235	.4347	.4458	.4567	.4676	.4782
.0100	.3868	.3990	.4110	.4228	.4345	.4460	.4574	.4686	.4797	.4907

Table 15.—*Values of nv corresponding to different values of r and s in Manning's formula,* $v = (1.486/n)\,r^{2/3}s^{1/2}$—Continued

(To determine the mean velocity of water, v, divide the tabulated value by the coefficient of roughness, n)

s = slope	r=hydraulic radius in feet									
	6.2	6.4	6.6	6.8	7.0	7.2	7.4	7.6	7.8	8.0
.00005	.0355	.0362	.0370	.0377	.0385	.0392	.0399	.0406	.0413	.0420
10	.0502	.0512	.0523	.0533	.0544	.0554	.0564	.0574	.0584	.0594
15	.0614	.0627	.0640	.0553	.0666	.0679	.0691	.0704	.0716	.0728
20	.0709	.0724	.0739	.0754	.0769	.0784	.0798	.0812	.0826	.0841
25	.0793	.0810	.0827	.0843	.0860	.0876	.0892	.0908	.0924	.0940
.00030	.0869	.0887	.0906	.0924	.0942	.0960	.0977	.0995	.1012	.1030
35	.0938	.0958	.0978	.0998	.1017	.1037	.1056	.1075	.1093	.1112
40	.1003	.1025	.1046	.1067	.1088	.1108	.1129	.1149	.1169	.1189
45	.1064	.1087	.1109	.1132	.1154	.1175	.1197	.1219	.1240	.1261
50	.1121	.1145	.1169	.1193	.1216	.1239	.1262	.1284	.1307	.1329
.00055	.1176	.1201	.1226	.1251	.1275	.1299	.1323	.1347	.1371	.1394
60	.1229	.1255	.1281	.1307	.1332	.1357	.1382	.1407	.1432	.1456
65	.1279	.1306	.1333	.1360	.1386	.1413	.1439	.1464	.1490	.1515
70	.1327	.1355	.1383	.1411	.1439	.1466	.1493	.1520	.1546	.1573
75	.1374	.1403	.1432	.1461	.1489	.1517	.1545	.1573	.1601	.1628
.00080	.1419	.1449	.1479	.1509	.1538	.1567	.1596	.1625	.1653	.1681
85	.1462	.1493	.1524	.1555	.1585	.1615	.1645	.1675	.1704	.1733
90	.1505	.1537	.1569	.1600	.1631	.1662	.1693	.1723	.1753	.1783
95	.1546	.1579	.1612	.1644	.1676	.1708	.1739	.1771	.1801	.1832
100	.1586	.1620	.1653	.1687	.1720	.1752	.1785	.1817	.1848	.1880
.0011	.1663	.1699	.1734	.1769	.1804	.1838	.1872	.1905	.1938	.1971
12	.1737	.1774	.1811	.1848	.1884	.1919	.1955	.1990	.2025	.2059
13	.1808	.1847	.1885	.1923	.1961	.1998	.2035	.2071	.2107	.2143
14	.1877	.1917	.1956	.1996	.2035	.2073	.2111	.2149	.2187	.2224
15	.1942	.1984	.2025	.2066	.2106	.2146	.2186	.2225	.2264	.2302
.0016	.2006	.2049	.2091	.2134	.2175	.2216	.2257	.2298	.2338	.2378
17	.2068	.2112	.2156	.2199	.2242	.2285	.2327	.2368	.2410	.2451
18	.2128	.2173	.2218	.2263	.2307	.2351	.2394	.2437	.2480	.2522
19	.2186	.2233	.2279	.2325	.2370	.2415	.2460	.2504	.2548	.2591
20	.2243	.2291	.2338	.2385	.2432	.2478	.2524	.2569	.2614	.2658
.0025	.2508	.2561	.2614	.2667	.2719	.2770	.2822	.2872	.2922	.2972
30	.2747	.2806	.2864	.2921	.2978	.3035	.3091	.3146	.3201	.3256
35	.2967	.3030	.3093	.3156	.3217	.3278	.3338	.3398	.3458	.3517
40	.3172	.3240	.3307	.3373	.3439	.3504	.3569	.3633	.3696	.3759
45	.3364	.3436	.3507	.3578	.3648	.3717	.3785	.3853	.3921	.3987
.0050	.3546	.3622	.3697	.3771	.3845	.3918	.3990	.4062	.4133	.4203
55	.3719	.3799	.3878	.3956	.4033	.4109	.4185	.4260	.4334	.4408
60	.3885	.3968	.4050	.4132	.4212	.4292	.4371	.4449	.4527	.4604
65	.4043	.4130	.4215	.4300	.4384	.4467	.4549	.4631	.4712	.4792
70	.4196	.4286	.4375	.4462	.4550	.4636	.4721	.4806	.4890	.4973
.0075	.4343	.4436	.4528	.4619	.4709	.4798	.4887	.4975	.5061	.5148
80	.4486	.4582	.4677	.4771	.4864	.4956	.5047	.5138	.5227	.5316
85	.4624	.4723	.4820	.4917	.5013	.5108	.5203	.5296	.5388	.5480
90	.4758	.4860	.4960	.5060	.5159	.5256	.5353	.5449	.5545	.5639
95	.4888	.4993	.5096	.5199	.5300	.5401	.5500	.5599	.5697	.5793
.0100	.5015	.5122	.5229	.5334	.5438	.5541	.5643	.5744	.5845	.5944

Table 15.—*Values of nv corresponding to different values of r and s in Manning's formula,* $v = (1.486/n)r^{2/3}s^{1/2}$—Continued

(To determine the mean velocity of water, v, divide the tabulated value by the coefficient of roughness, n)

s=slope	\multicolumn{10}{c}{r=hydraulic radius in feet}									
	8.2	8.4	8.6	8.8	9.0	9.2	9.4	9.6	9.8	10.0
.00005	.0427	.0434	.0441	.0448	.0455	.0461	.0468	.0475	.0481	.0488
10	.0604	.0614	.0624	.0633	.0643	.0653	.0662	.0671	.0681	.0690
15	.0740	.0752	.0764	.0776	.0787	.0799	.0811	.0822	.0833	.0845
20	.0855	.0868	.0882	.0896	.0909	.0923	.0936	.0949	.0962	.0975
25	.0955	.0971	.0986	.1002	.1017	.1032	.1047	.1061	.1076	.1091
.00030	.1047	.1064	.1080	.1097	.1114	.1130	.1146	.1163	.1179	.1195
35	.1131	.1149	.1167	.1185	.1203	.1221	.1238	.1256	.1273	.1290
40	.1209	.1228	.1248	.1267	.1286	.1305	.1324	.1343	.1361	.1379
45	.1282	.1303	.1323	.1344	.1364	.1384	.1404	.1424	.1444	.1463
50	.1351	.1373	.1395	.1416	.1438	.1459	.1480	.1501	.1522	.1542
.00055	.1417	.1440	.1463	.1485	.1508	.1530	.1552	.1574	.1596	.1618
60	.1480	.1504	.1528	.1552	.1575	.1598	.1621	.1644	.1667	.1690
65	.1541	.1566	.1590	.1615	.1639	.1664	.1687	.1711	.1735	.1758
70	.1599	.1625	.1650	.1676	.1701	.1726	.1751	.1776	.1801	.1825
75	.1655	.1682	.1708	.1735	.1761	.1787	.1813	.1838	.1864	.1889
.00080	.1709	.1737	.1764	.1792	.1819	.1845	.1872	.1899	.1925	.1951
85	.1762	.1790	.1819	.1847	.1875	.1902	.1930	.1957	.1984	.2011
90	.1813	.1842	.1871	.1900	.1929	.1957	.1986	.2014	.2042	.2069
95	.1863	.1893	.1923	.1952	.1982	.2011	.2040	.2069	.2098	.2126
100	.1911	.1942	.1973	.2003	.2033	.2063	.2093	.2123	.2152	.2181
.0011	.2004	.2037	.2069	.2101	.2133	.2164	.2195	.2226	.2257	.2288
12	.2093	.2127	.2161	.2194	.2227	.2260	.2293	.2325	.2357	.2389
13	.2179	.2214	.2249	.2284	.2318	.2352	.2386	.2420	.2454	.2487
14	.2261	.2298	.2334	.2370	.2406	.2441	.2477	.2512	.2546	.2581
15	.2340	.2378	.2416	.2453	.2490	.2527	.2563	.2600	.2636	.2671
.0016	.2417	.2456	.2495	.2534	.2572	.2610	.2648	.2685	.2722	.2759
17	.2491	.2532	.2572	.2612	.2651	.2690	.2729	.2768	.2806	.2844
18	.2564	.2605	.2646	.2687	.2728	.2768	.2808	.2848	.2887	.2926
19	.2634	.2677	.2719	.2761	.2803	.2844	.2885	.2926	.2966	.3007
20	.2702	.2746	.2790	.2833	.2875	.2918	.2960	.3002	.3044	.3085
.0025	.3021	.3070	.3119	.3167	.3215	.3262	.3309	.3356	.3403	.3449
30	.3310	.3363	.3417	.3469	.3522	.3574	.3625	.3676	.3727	.3778
35	.3575	.3633	.3690	.3747	.3804	.3860	.3916	.3971	.4026	.4081
40	.3822	.3884	.3945	.4006	.4066	.4127	.4186	.4245	.4304	.4362
45	.4054	.4119	.4184	.4249	.4313	.4377	.4440	.4503	.4565	.4627
.0050	.4273	.4342	.4411	.4479	.4546	.4613	.4680	.4746	.4812	.4877
55	.4481	.4554	.4626	.4697	.4768	.4839	.4909	.4978	.5047	.5115
60	.4681	.4756	.4832	.4906	.4980	.5054	.5127	.5199	.5271	.5343
65	.4872	.4951	.5029	.5107	.5184	.5260	.5336	.5412	.5486	.5561
70	.5056	.5138	.5219	.5299	.5379	.5459	.5538	.5616	.5694	.5771
.0075	.5233	.5318	.5402	.5485	.5568	.5650	.5732	.5813	.5893	.5973
80	.5405	.5492	.5579	.5665	.5751	.5836	.5920	.6004	.6087	.6169
85	.5571	.5661	.5751	.5840	.5928	.6015	.6102	.6188	.6274	.6359
90	.5733	.5825	.5918	.6009	.6100	.6190	.6279	.6368	.6456	.6543
95	.5890	.5985	.6080	.6174	.6267	.6359	.6451	.6542	.6633	.6723
.0100	.6043	.6141	.6238	.6334	.6430	.6524	.6619	.6712	.6805	.6897

Table 15.—*Values of nv corresponding to different values of r and s in Manning's formula,* $v=(1.486/n)r^{2/3}s^{1/2}$—Continued

(To determine the mean velocity of water, v, divide the tabulated value by the coefficient of roughness, n)

| s = slope | \multicolumn{10}{c}{r = hydraulic radius in feet} | | | | | | | | | |
	10.5	11.0	11.5	12.0	12.5	13.0	13.5	14.0	14.5	15.0
.00005	.0504	.0520	.0535	.0551	.0566	.0581	.0596	.0610	.0625	.0639
10	.0713	.0735	.0757	.0779	.0800	.0822	.0843	.0863	.0884	.0904
15	.0873	.0900	.0927	.0954	.0980	.1006	.1032	.1057	.1082	.1107
20	.1008	.1039	.1071	.1102	.1132	.1162	.1191	.1221	.1250	.1278
25	.1127	.1162	.1197	.1232	.1266	.1299	.1332	.1365	.1397	.1429
.00030	.1234	.1273	.1311	.1349	.1386	.1423	.1459	.1495	.1530	.1565
35	.1333	.1375	.1416	.1457	.1497	.1537	.1576	.1615	.1653	.1691
40	.1425	.1470	.1514	.1558	.1601	.1643	.1685	.1726	.1767	.1808
45	.1512	.1559	.1606	.1652	.1698	.1743	.1787	.1831	.1874	.1917
50	.1593	.1643	.1693	.1742	.1790	.1837	.1884	.1930	.1976	.2021
.00055	.1671	.1724	.1776	.1827	.1877	.1927	.1976	.2024	.2072	.2120
60	.1745	.1800	.1854	.1908	.1961	.2012	.2064	.2114	.2164	.2214
65	.1817	.1874	.1930	.1986	.2041	.2095	.2148	.2201	.2253	.2304
70	.1885	.1945	.2003	.2061	.2118	.2174	.2229	.2284	.2338	.2391
75	.1951	.2013	.2073	.2133	.2192	.2250	.2307	.2364	.2420	.2475
.00080	.2015	.2079	.2141	.2203	.2264	.2324	.2383	.2441	.2499	.2556
85	.2077	.2143	.2207	.2271	.2333	.2395	.2456	.2517	.2576	.2635
90	.2138	.2205	.2271	.2337	.2401	.2465	.2528	.2590	.2651	.2711
95	.2196	.2265	.2334	.2401	.2467	.2532	.2597	.2661	.2723	.2786
100	.2253	.2324	.2394	.2463	.2531	.2598	.2664	.2730	.2794	.2858
.0011	.2363	.2438	.2511	.2583	.2655	.2725	.2794	.2863	.2931	.2998
12	.2468	.2546	.2623	.2698	.2773	.2846	.2919	.2990	.3061	.3131
13	.2569	.2650	.2730	.2808	.2886	.2962	.3038	.3112	.3186	.3259
14	.2666	.2750	.2833	.2914	.2995	.3074	.3152	.3230	.3306	.3382
15	.2760	.2847	.2932	.3017	.3100	.3182	.3263	.3343	.3422	.3500
.0016	.2850	.2940	.3028	.3116	.3201	.3286	.3370	.3453	.3534	.3615
17	.2938	.3030	.3122	.3211	.3300	.3387	.3474	.3559	.3643	.3727
18	.3023	.3118	.3212	.3305	.3396	.3486	.3574	.3662	.3749	.3835
19	.3106	.3204	.3300	.3395	.3489	.3581	.3672	.3763	.3852	.3940
20	.3187	.3287	.3386	.3483	.3579	.3674	.3768	.3860	.3952	.4042
.0025	.3563	.3675	.3785	.3894	.4002	.4108	.4213	.4316	.4418	.4519
30	.3903	.4026	.4147	.4266	.4384	.4500	.4615	.4728	.4840	.4950
35	.4215	.4348	.4479	.4608	.4735	.4861	.4984	.5107	.5228	.5347
40	.4507	.4648	.4788	.4926	.5062	.5196	.5328	.5459	.5588	.5716
45	.4780	.4930	.5079	.5225	.5369	.5511	.5652	.5790	.5927	.6063
.0050	.5038	.5197	.5353	.5508	.5659	.5809	.5957	.6104	.6248	.6391
55	.5284	.5451	.5615	.5776	.5936	.6093	.6248	.6402	.6553	.6703
60	.5519	.5693	.5864	.6033	.6200	.6364	.6526	.6686	.6844	.7001
65	.5745	.5926	.6104	.6280	.6453	.6624	.6793	.6959	.7124	.7287
70	.5962	.6149	.6334	.6517	.6696	.6874	.7049	.7222	.7393	.7562
.0075	.6171	.6365	.6557	.6745	.6931	.7115	.7296	.7475	.7652	.7827
80	.6373	.6574	.6772	.6967	.7159	.7348	.7536	.7721	.7903	.8084
85	.6569	.6776	.6980	.7181	.7379	.7575	.7768	.7958	.8147	.8333
90	.6760	.6973	.7182	.7389	.7593	.7794	.7993	.8189	.8383	.8574
95	.6945	.7164	.7379	.7592	.7801	.8008	.8212	.8413	.8612	.8809
.0100	.7125	.7350	.7571	.7789	.8004	.8216	.8425	.8632	.8836	.9038

Table 15.—*Values of nv corresponding to different values of r and s in Manning's formula,* $v = (1.486/n)r^{2/3}s^{1/2}$—Continued

(To determine the mean velocity of water, v, divide the tabulated value by the coefficient of roughness, n)

s=slope	r=hydraulic radius in feet									
	15.5	16.0	16.5	17.0	17.5	18.0	18.5	19.0	19.5	20
.00005	.0653	.0667	.0681	.0695	.0708	.0722	.0735	.0748	.0761	.0774
10	.0924	.0944	.0963	.0982	.1002	.1021	.1039	.1058	.1077	.1095
15	.1131	.1156	.1180	.1203	.1227	.1250	.1273	.1296	.1319	.1341
20	.1306	.1334	.1362	.1389	.1417	.1443	.1470	.1496	.1522	.1548
25	.1461	.1492	.1523	.1553	.1584	.1614	.1644	.1673	.1702	.1731
.00030	.1600	.1634	.1668	.1702	.1735	.1768	.1800	.1833	.1865	.1896
35	.1728	.1765	.1802	.1838	.1874	.1909	.1945	.1979	.2014	.2048
40	.1848	.1887	.1926	.1965	.2003	.2041	.2079	.2116	.2153	.2190
45	.1960	.2002	.2043	.2084	.2125	.2165	.2205	.2245	.2284	.2323
50	.2066	.2110	.2154	.2197	.2240	.2282	.2324	.2366	.2407	.2448
.00055	.2166	.2213	.2259	.2304	.2349	.2394	.2438	.2481	.2525	.2568
60	.2263	.2311	.2359	.2407	.2453	.2500	.2546	.2592	.2637	.2682
65	.2355	.2406	.2455	.2505	.2554	.2602	.2650	.2698	.2745	.2791
70	.2444	.2496	.2548	.2599	.2650	.2700	.2750	.2799	.2848	.2897
75	.2530	.2584	.2638	.2691	.2743	.2795	.2847	.2898	.2948	.2998
.00080	.2613	.2669	.2724	.2779	.2833	.2887	.2940	.2993	.3045	.3097
85	.2693	.2751	.2808	.2864	.2920	.2976	.3030	.3085	.3139	.3192
90	.2771	.2831	.2889	.2947	.3005	.3062	.3118	.3174	.3230	.3285
95	.2847	.2908	.2968	.3028	.3087	.3146	.3204	.3261	.3318	.3375
100	.2921	.2984	.3046	.3107	.3167	.3228	.3287	.3346	.3404	.3462
.0011	.3064	.3129	.3194	.3258	.3322	.3385	.3447	.3509	.3571	.3631
12	.3200	.3269	.3336	.3403	.3470	.3536	.3601	.3665	.3729	.3793
13	.3331	.3402	.3473	.3542	.3611	.3680	.3748	.3815	.3882	.3948
14	.3457	.3530	.3604	.3676	.3748	.3819	.3889	.3959	.4028	.4097
15	.3578	.3654	.3730	.3805	.3879	.3953	.4026	.4098	.4170	.4241
.0016	.3695	.3774	.3852	.3930	.4007	.4083	.4158	.4232	.4306	.4380
17	.3809	.3890	.3971	.4051	.4130	.4208	.4286	.4363	.4439	.4514
18	.3919	.4003	.4086	.4168	.4250	.4330	.4410	.4489	.4567	.4645
19	.4027	.4113	.4198	.4282	.4366	.4449	.4531	.4612	.4693	.4773
20	.4131	.4220	.4307	.4394	.4479	.4564	.4649	.4732	.4815	.4897
.0025	.4619	.4718	.4816	.4912	.5008	.5103	.5197	.5290	.5383	.5474
30	.5060	.5168	.5275	.5381	.5486	.5590	.5693	.5795	.5897	.5997
35	.5465	.5582	.5698	.5812	.5926	.6038	.6149	.6260	.6369	.6477
40	.5843	.5968	.6091	.6214	.6335	.6455	.6574	.6692	.6809	.6925
45	.6197	.6330	.6461	.6591	.6719	.6847	.6973	.7098	.7222	.7345
.0050	.6532	.6672	.6810	.6947	.7083	.7217	.7350	.7482	.7612	.7742
55	.6851	.6998	.7143	.7286	.7428	.7569	.7709	.7847	.7984	.8120
60	.7156	.7309	.7460	.7610	.7759	.7906	.8051	.8196	.8339	.8481
65	.7448	.7607	.7765	.7921	.8075	.8229	.8380	.8531	.8680	.8827
70	.7729	.7894	.8058	.8220	.8380	.8539	.8697	.8853	.9007	.9161
.0075	.8000	.8171	.8341	.8508	.8674	.8839	.9002	.9163	.9323	.9482
80	.8263	.8439	.8614	.8787	.8959	.9129	.9297	.9464	.9629	.9793
85	.8517	.8699	.8879	.9058	.9235	.9410	.9583	.9755	.9925	1.0094
90	.8764	.8951	.9137	.9321	.9502	.9683	.9861	1.0038	1.0213	1.0387
95	.9004	.9197	.9387	.9576	.9763	.9948	1.0131	1.0313	1.0493	1.0672
.0100	.9238	.9436	.9631	.9825	1.0016	1.0206	1.0394	1.0581	1.0766	1.0949

Table 16.—*Trigonometric functions for various slopes*

Slope-cotangent	Angle	Sine	Cosine	Tangent	Cosecant	2 x cosecant (for wetted perimeter)
.25000	75°57'50''	.97014	.24254	4.00000	1.03078	2.06156
.50000	63°26'06''	.89443	.44721	2.00000	1.11803	2.23606
.75000	53°07'48''	.80000	.60000	1.33333	1.25000	2.50000
1.00000	45°00'00''	.70711	.70711	1.00000	1.41421	2.82842
1.25000	38°39'35''	.62469	.78087	.80000	1.60078	3.20156
1.50000	33°41'24''	.55470	.83205	.66667	1.80278	3.60556
1.75000	29°44'42''	.49614	.86824	.57143	2.01556	4.03112
2.00000	26°33'54''	.44721	.89443	.50000	2.23607	4.47214
2.25000	23°57'45''	.40614	.91381	.44444	2.46221	4.92442
2.50000	21°48'05''	.37139	.92848	.40000	2.69258	5.38516
2.75000	19°58'59''	.34174	.93979	.36364	2.92617	5.85234
3.00000	18°26'06''	.31623	.94868	.33333	3.16228	6.32456
3.25000	17°06'10''	.29409	.95578	.30769	3.40037	6.80074
3.50000	15°56'43''	.27472	.96152	.28571	3.64005	7.28010
3.75000	14°55'53''	.25766	.96624	.26667	3.88104	7.76208
4.00000	14°02'10''	.24254	.97014	.25000	4.12311	8.24622
4.50000	12°31'44''	.21693	.97619	.22222	4.60977	9.21954
5.00000	11°18'36''	.19612	.98058	.20000	5.09902	10.19804
5.50000	10°18'17''	.17889	.98387	.18182	5.59017	11.18034
6.00000	9°27'44''	.16440	.98639	.16667	6.08276	12.16552

Table 17.—*Critical depth in rectangular sections*

$$d_c = \frac{q^{2/3}}{g^{1/3}} = 0.31433 \, q^{2/3}; \; q = \text{cubic feet per second per foot of width}; \; g = 32.2$$

$$d_c = \text{critical depth in feet}$$

q	d_c	q	d_c	q	d_c	q	d_c	q	d_c
1	.3143	57	4.655	230	11.80	790	26.86	4500	85.68
2	.4988	58	4.710	240	12.14	800	27.09	4600	86.94
3	.6538	59	4.764	250	12.47	810	27.31	4700	88.19
4	.7921	60	4.817	260	12.80	820	27.54	4800	89.44
5	.9191	61	4.871	270	13.13	830	27.76	4900	90.67
6	1.038	62	4.924	280	13.45	840	27.98	5000	91.91
7	1.150	63	4.977	290	13.77	850	28.21	5100	93.13
8	1.257	64	5.029	300	14.09	860	28.43	5200	94.34
9	1.360	65	5.082	310	14.40	870	28.65	5300	95.54
10	1.459	66	5.134	320	14.70	880	28.86	5400	96.74
11	1.555	67	5.186	330	15.01	890	29.08	5500	97.93
12	1.647	68	5.237	340	15.31	900	29.30	5600	99.12
13	1.738	69	5.288	350	15.61	910	29.52	5700	100.29
14	1.826	70	5.339	360	15.91	920	29.73	5800	101.46
15	1.912	71	5.390	370	16.20	930	29.95	5900	102.63
16	1.996	72	5.440	380	16.49	940	30.16	6000	103.79
17	2.078	73	5.490	390	16.78	950	30.38	6100	104.94
18	2.159	74	5.540	400	17.06	960	30.59	6200	106.08
19	2.238	75	5.590	410	17.35	970	30.80	6300	107.22
20	2.316	76	5.640	420	17.63	980	31.01	6400	108.35
21	2.393	77	5.689	430	17.91	990	31.22	6500	109.48
22	2.468	78	5.738	440	18.18	1000	31.43	6600	110.60
23	2.542	79	5.787	450	18.46	1100	33.49	6700	111.72
24	2.615	80	5.836	460	18.73	1200	35.49	6800	112.83
25	2.688	81	5.884	470	19.00	1300	37.44	6900	113.92
26	2.759	82	5.933	480	19.27	1400	39.34	7000	115.02
27	2.829	83	5.981	490	19.54	1500	41.19	7100	116.11
28	2.898	84	6.029	500	19.80	1600	43.00	7200	117.20
29	2.967	85	6.077	510	20.06	1700	44.77	7300	118.28
30	3.035	86	6.124	520	20.33	1800	46.51	7400	119.36
31	3.102	87	6.172	530	20.58	1900	48.22	7500	120.43
32	3.168	88	6.219	540	20.84	2000	49.90	7600	121.50
33	3.234	89	6.266	550	21.10	2100	51.55	7700	122.56
34	3.299	90	6.313	560	21.35	2200	53.17	7800	123.62
35	3.363	91	6.359	570	21.61	2300	54.77	7900	124.68
36	3.427	92	6.406	580	21.86	2400	56.34	8000	125.73
37	3.490	93	6.452	590	22.11	2500	57.90	8100	126.77
38	3.553	94	6.498	600	22.36	2600	59.43	8200	127.81
39	3.615	95	6.544	610	22.61	2700	60.95	8300	128.85
40	3.676	96	6.590	620	22.85	2800	62.44	8400	129.88
41	3.738	97	6.636	630	23.10	2900	63.92	8500	130.91
42	3.798	98	6.681	640	23.34	3000	65.38	8600	131.94
43	3.858	99	6.727	650	23.59	3100	66.82	8700	132.96
44	3.918	100	6.772	660	23.83	3200	68.25	8800	133.98
45	3.977	110	7.216	670	24.07	3300	69.66	8900	134.99
46	4.036	120	7.647	680	24.31	3400	71.06	9000	136.00
47	4.094	130	8.067	690	24.54	3500	72.45	9100	137.01
48	4.152	140	8.475	700	24.78	3600	73.82	9200	138.01
49	4.209	150	8.874	710	25.02	3700	75.19	9300	139.01
50	4.266	160	9.265	720	25.25	3800	76.54	9400	140.00
51	4.323	170	9.645	730	25.48	3900	77.87	9500	140.99
52	4.379	180	10.200	740	25.71	4000	79.20	9600	141.98
53	4.435	190	10.39	750	25.95	4100	80.52	9700	142.96
54	4.490	200	10.75	760	26.18	4200	81.82	9800	143.94
55	4.546	210	11.11	770	26.41	4300	83.12	9900	144.92
56	4.601	220	11.46	780	26.63	4400	84.40	10000	145.89

Table 18.—*Uniform flow in trapezoidal channels by Manning's formula*

Values of $\dfrac{Qn}{b^{8/3}S^{1/2}}$

D/b[1]	$z=0$	$z=\frac{1}{4}$	$z=\frac{1}{2}$	$z=\frac{3}{4}$	$z=1$	$z=1\frac{1}{4}$	$z=1\frac{1}{2}$	$z=1\frac{3}{4}$	$z=2$	$z=2\frac{1}{2}$	$z=3$	$z=4$
.02	.00213	.00215	.00216	.00217	.00218	.00219	.00220	.00220	.00221	.00222	.00223	.00225
.03	.00414	.00419	.00423	.00426	.00429	.00431	.00433	.00434	.00437	.00440	.00443	.00449
.04	.00661	.00670	.00679	.00685	.00690	.00696	.00700	.00704	.00707	.00715	.00722	.00735
.05	.00947	.00964	.00980	.00991	.0100	.0101	.0102	.0103	.0103	.0104	.0106	.0109
.06	.0127	.0130	.0132	.0134	.0136	.0137	.0138	.0140	.0141	.0143	.0145	.0149
.07	.0162	.0166	.0170	.0173	.0176	.0177	.0180	.0182	.0183	.0186	.0190	.0196
.08	.0200	.0206	.0211	.0215	.0219	.0222	.0225	.0228	.0231	.0235	.0240	.0250
.09	.0240	.0249	.0256	.0262	.0267	.0271	.0275	.0279	.0282	.0289	.0296	.0310
.10	.0283	.0294	.0305	.0311	.0318	.0324	.0329	.0334	.0339	.0348	.0358	.0375
.11	.0329	.0342	.0354	.0364	.0373	.0380	.0387	.0394	.0400	.0413	.0424	.0448
.12	.0376	.0393	.0408	.0420	.0431	.0441	.0450	.0458	.0466	.0482	.0497	.0527
.13	.0425	.0446	.0464	.0480	.0493	.0505	.0516	.0527	.0537	.0556	.0575	.0613
.14	.0476	.0501	.0524	.0542	.0559	.0573	.0587	.0599	.0612	.0636	.0659	.0705
.15	.0528	.0559	.0585	.0608	.0628	.0645	.0662	.0677	.0692	.0721	.0749	.0805
.16	.0582	.0619	.0650	.0676	.0699	.0720	.0740	.0759	.0776	.0811	.0845	.0912
.17	.0638	.0680	.0717	.0748	.0775	.0800	.0823	.0845	.0867	.0907	.0947	.103
.18	.0695	.0744	.0786	.0822	.0854	.0883	.0910	.0936	.0961	.101	.105	.115
.19	.0753	.0809	.0857	.0900	.0936	.0970	.100	.103	.106	.112	.117	.128
.20	.0813	.0875	.0932	.0979	.102	.106	.110	.113	.116	.123	.129	.141
.21	.0873	.0944	.101	.106	.111	.115	.120	.123	.127	.134	.142	.156
.22	.0935	.101	.109	.115	.120	.125	.130	.134	.139	.147	.155	.171
.23	.0997	.109	.117	.124	.130	.135	.141	.146	.151	.160	.169	.187
.24	.106	.116	.125	.133	.139	.146	.152	.157	.163	.173	.184	.204
.25	.113	.124	.133	.142	.150	.157	.163	.170	.176	.187	.199	.222
.26	.119	.131	.142	.152	.160	.168	.175	.182	.189	.202	.215	.241
.27	.126	.139	.151	.162	.171	.180	.188	.195	.203	.218	.232	.260
.28	.133	.147	.160	.172	.182	.192	.201	.209	.217	.234	.249	.281
.29	.139	.155	.170	.182	.193	.204	.214	.223	.232	.250	.267	.302
.30	.146	.163	.179	.193	.205	.217	.227	.238	.248	.267	.286	.324
.31	.153	.172	.189	.204	.217	.230	.242	.253	.264	.285	.306	.347
.32	.160	.180	.199	.215	.230	.243	.256	.269	.281	.304	.327	.371
.33	.167	.189	.209	.227	.243	.257	.271	.285	.298	.323	.348	.396
.34	.174	.198	.219	.238	.256	.272	.287	.301	.315	.343	.369	.422
.35	.181	.207	.230	.251	.270	.287	.303	.318	.334	.363	.392	.450
.36	.190	.216	.241	.263	.283	.302	.319	.336	.353	.384	.416	.477
.37	.196	.225	.251	.275	.297	.317	.336	.354	.372	.406	.440	.507
.38	.203	.234	.263	.289	.311	.333	.354	.373	.392	.429	.465	.536
.39	.210	.244	.274	.301	.326	.349	.371	.392	.412	.452	.491	.568
.40	.218	.254	.286	.314	.341	.366	.389	.412	.433	.476	.518	.600
.41	.225	.263	.297	.328	.357	.383	.408	.432	.455	.501	.545	.634
.42	.233	.272	.310	.342	.373	.401	.427	.453	.478	.526	.574	.668
.43	.241	.282	.321	.356	.389	.418	.447	.474	.501	.553	.604	.703
.44	.249	.292	.334	.371	.405	.437	.467	.496	.524	.579	.634	.739

[1] For D/b less than 0.04, use of the assumption $R=D$ is more convenient and more accurate than interpolation in the table.

Table 18.—*Uniform flow in trapezoidal channels by Manning's formula*—Continued

D/b^1	$z=0$	$z=\frac{1}{4}$	$z=\frac{1}{2}$	$z=\frac{3}{4}$	$z=1$	$z=1\frac{1}{4}$	$z=1\frac{1}{2}$	$z=1\frac{3}{4}$	$z=2$	$z=2\frac{1}{2}$	$z=3$	$z=4$
								Values of $\dfrac{Qn}{b^{8/3}S^{1/2}}$				
.45	.256	.303	.346	.385	.422	.455	.487	.519	.548	.607	.665	.778
.46	.263	.313	.359	.401	.439	.475	.509	.541	.574	.635	.696	.816
.47	.271	.323	.371	.417	.457	.494	.530	.565	.600	.665	.729	.856
.48	.279	.333	.384	.432	.475	.514	.552	.589	.626	.695	.763	.897
.49	.287	.345	.398	.448	.492	.534	.575	.614	.652	.725	.797	.939
.50	.295	.356	.411	.463	.512	.556	.599	.639	.679	.758	.833	.983
.52	.310	.377	.438	.496	.548	.599	.646	.692	.735	.820	.906	1.07
.54	.327	.398	.468	.530	.590	.644	.696	.746	.795	.891	.984	1.17
.56	.343	.421	.496	.567	.631	.690	.748	.803	.856	.963	1.07	1.27
.58	.359	.444	.526	.601	.671	.739	.802	.863	.922	1.04	1.15	1.37
.60	.375	.468	.556	.640	.717	.789	.858	.924	.988	1.12	1.24	1.49
.62	.391	.492	.590	.679	.763	.841	.917	.989	1.06	1.20	1.33	1.60
.64	.408	.516	.620	.718	.809	.894	.976	1.05	1.13	1.28	1.43	1.72
.66	.424	.541	.653	.759	.858	.951	1.04	1.13	1.21	1.37	1.53	1.85
.68	.441	.566	.687	.801	.908	1.01	1.10	1.20	1.29	1.47	1.64	1.98
.70	.457	.591	.722	.842	.958	1.07	1.17	1.27	1.37	1.56	1.75	2.12
.72	.474	.617	.757	.887	1.01	1.13	1.24	1.35	1.45	1.66	1.87	2.27
.74	.491	.644	.793	.932	1.07	1.19	1.31	1.43	1.55	1.77	1.98	2.41
.76	.508	.670	.830	.981	1.12	1.26	1.39	1.51	1.64	1.88	2.11	2.57
.78	.525	.698	.868	1.03	1.18	1.32	1.46	1.60	1.73	1.98	2.24	2.73
.80	.542	.725	.906	1.08	1.24	1.40	1.54	1.69	1.83	2.10	2.37	2.90
.82	.559	.753	.945	1.13	1.30	1.47	1.63	1.78	1.93	2.22	2.51	3.07
.84	.576	.782	.985	1.18	1.36	1.54	1.71	1.87	2.03	2.34	2.65	3.25
.86	.593	.810	1.03	1.23	1.43	1.61	1.79	1.97	2.14	2.47	2.80	3.44
.88	.610	.839	1.07	1.29	1.49	1.69	1.88	2.07	2.25	2.60	2.95	3.63
.90	.627	.871	1.11	1.34	1.56	1.77	1.98	2.17	2.36	2.74	3.11	3.83
.92	.645	.898	1.15	1.40	1.63	1.86	2.07	2.28	2.48	2.88	3.27	4.04
.94	.662	.928	1.20	1.46	1.70	1.94	2.16	2.38	2.60	3.03	3.43	4.25
.96	.680	.960	1.25	1.52	1.78	2.03	2.27	2.50	2.73	3.17	3.61	4.48
.98	.697	.991	1.29	1.58	1.85	2.11	2.37	2.61	2.85	3.33	3.79	4.70
1.00	.714	1.02	1.33	1.64	1.93	2.21	2.47	2.73	2.99	3.48	3.97	4.93
1.05	.759	1.10	1.46	1.80	2.13	2.44	2.75	3.04	3.33	3.90	4.45	5.55
1.10	.802	1.19	1.58	1.97	2.34	2.69	3.04	3.37	3.70	4.34	4.96	6.21
1.15	.846	1.27	1.71	2.14	2.56	2.96	3.34	3.72	4.09	4.82	5.52	6.91
1.20	.891	1.36	1.85	2.33	2.79	3.24	3.68	4.09	4.50	5.32	6.11	7.68
1.25	.936	1.45	1.99	2.52	3.04	3.54	4.03	4.49	4.95	5.86	6.73	8.48
1.30	.980	1.54	2.14	2.73	3.30	3.85	4.39	4.90	5.42	6.42	7.39	9.34
1.35	1.02	1.64	2.29	2.94	3.57	4.18	4.76	5.34	5.90	7.01	8.10	10.2
1.40	1.07	1.74	2.45	3.16	3.85	4.52	5.18	5.80	6.43	7.65	8.83	11.2
1.45	1.11 .	1.84	2.61	3.39	4.15	4.88	5.60	6.29	6.98	8.30	9.62	12.2

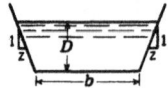

[1] For D/b less than 0.04, use of the assumption $R=D$ is more convenient and more accurate than interpolation in the table.

Table 18.—*Uniform flow in trapezoidal channels by Manning's formula*—Continued

D/b[1]	Values of $\dfrac{Qn}{b^{8/3}S^{1/2}}$											
	$z=0$	$z=\frac14$	$z=\frac12$	$z=\frac34$	$z=1$	$z=1\frac14$	$z=1\frac12$	$z=1\frac34$	$z=2$	$z=2\frac12$	$z=3$	$z=4$
1.50	1.16	1.94	2.78	3.63	4.46	5.26	6.04	6.81	7.55	9.02	10.4	13.3
1.55	1.20	2.05	2.96	3.88	4.78	5.65	6.50	7.33	8.14	9.74	11.3	14.4
1.60	1.25	2.15	3.14	4.14	5.12	6.06	6.99	7.89	8.79	10.5	12.2	15.6
1.65	1.30	2.27	3.33	4.41	5.47	6.49	7.50	8.47	9.42	11.3	13.2	16.8
1.70	1.34	2.38	3.52	4.69	5.83	6.94	8.02	9.08	10.1	12.2	14.2	18.1
1.75	1.39	2.50	3.73	4.98	6.21	7.41	8.57	9.72	10.9	13.0	15.2	19.5
1.80	1.43	2.62	3.93	5.28	6.60	7.89	9.13	10.4	11.6	14.0	16.3	20.9
1.85	1.48	2.74	4.15	5.59	7.01	8.40	9.75	11.1	12.4	15.0	17.4	22.4
1.90	1.52	2.86	4.36	5.91	7.43	8.91	10.4	11.8	13.2	15.9	18.7	24.0
1.95	1.57	2.99	4.59	6.24	7.87	9.46	11.0	12.5	14.0	17.0	19.9	25.6
2.00	1.61	3.12	4.83	6.58	8.32	10.0	11.7	13.3	14.9	18.0	21.1	27.2
2.10	1.71	3.39	5.31	7.30	9.27	11.2	13.1	15.0	16.8	20.3	23.9	30.8
2.20	1.79	3.67	5.82	8.06	10.3	12.5	14.6	16.7	18.7	22.8	26.8	34.7
2.30	1.89	3.96	6.36	8.86	11.3	13.8	16.2	18.6	20.9	25.4	30.0	38.8
2.40	1.98	4.26	6.93	9.72	12.5	15.3	17.9	20.6	23.1	28.3	33.4	43.3
2.50	2.07	4.58	7.52	10.6	13.7	16.8	19.8	22.7	25.6	31.3	37.0	48.0
2.60	2.16	4.90	8.14	11.6	15.0	18.4	21.7	25.0	28.2	34.5	40.8	53.0
2.70	2.26	5.24	8.80	12.6	16.3	20.1	23.8	27.4	31.0	37.9	44.8	58.4
2.80	2.35	5.59	9.49	13.6	17.8	21.9	25.9	29.9	33.8	41.6	49.1	64.0
2.90	2.44	5.95	10.2	14.7	19.3	23.8	28.2	32.6	36.9	45.3	53.7	70.1
3.00	2.53	6.33	11.0	15.9	20.9	25.8	30.6	35.4	40.1	49.4	58.4	76.4
3.20	2.72	7.12	12.5	18.3	24.2	30.1	35.8	41.5	47.1	58.0	68.9	90.3
3.40	2.90	7.97	14.2	21.0	27.9	34.8	41.5	48.2	54.6	67.7	80.2	105
3.60	3.09	8.86	16.1	24.0	32.0	39.9	47.8	55.5	63.0	78.2	92.8	122
3.80	3.28	9.81	18.1	27.1	36.3	45.5	54.6	63.5	72.4	89.6	107	141
4.00	3.46	10.8	20.2	30.5	41.1	51.6	61.9	72.1	82.2	102	122	160
4.50	3.92	13.5	26.2	40.1	54.5	68.8	82.9	96.9	111	136	164	217
5.00	4.39	16.7	33.1	51.5	70.3	89.2	108	126	145	181	216	287

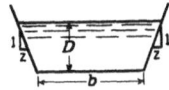

[1] For D/b less than 0.04, use of the assumption $R=D$ is more convenient and more accurate than interpolation in the table.

Table 19.—*Uniform flow in trapezoidal channels by Manning's formula*

D/b [1]	Values of $\dfrac{Qn}{D^{8/3}\,S^{1/2}}$											
	$z=0$	$z=\frac{1}{4}$	$z=\frac{1}{2}$	$z=\frac{3}{4}$	$z=1$	$z=1\frac{1}{4}$	$z=1\frac{1}{2}$	$z=1\frac{3}{4}$	$z=2$	$z=2\frac{1}{2}$	$z=3$	$z=4$
.01	146.7	147.2	147.6	148.0	148.3	148.6	148.8	148.9	149.2	149.5	149.9	150.5
.02	72.4	72.9	73.4	73.7	74.0	74.3	74.5	74.8	74.9	75.3	75 6	76 3
.03	47.6	48.2	48.6	49.0	49.3	49.5	49.8	50.0	50.2	50.6	50.9	51.6
.04	35.3	35.8	36.3	36.6	36.9	37.2	37.4	37.6	37.8	38.2	38.6	39.3
.05	27.9	28.4	28.9	29.2	29.5	29.8	30.0	30.2	30.5	30.8	31.2	32.0
.06	23.0	23.5	23.9	24.3	24.6	24.8	25.1	25.3	25.5	25.9	26.3	27.1
.07	19.45	19.97	20.4	20.8	21.1	21.3	21.6	21.8	22.0	22.4	22.8	23.6
.08	16.82	17.34	17.73	18.13	18.43	18.70	18.95	19.18	19.40	19.82	20.2	21.0
.09	14.78	15.29	15.72	16.08	16.39	16.66	16.91	17.14	17.36	17.79	18.21	19.04
.10	13.16	13.66	14.14	14.44	14.75	15.02	15.28	15.51	15.74	16.17	16.60	17.43
.11	11.83	12.33	12.76	13.11	13.42	13.69	13.94	14.18	14.41	14.85	15.28	16.13
.12	10.73	11.23	11.65	12.00	12.31	12.59	12.84	13.08	13.31	13.75	14.19	15.05
.13	9.80	10.29	10.71	11.06	11.37	11.65	11.90	12.14	12.38	12.83	13.26	14.13
.14	9.00	9.49	9.91	10.26	10.57	10.85	11.10	11.35	11.58	12.03	12.48	13.35
.15	8.32	8.80	9.21	9.57	9.88	10.16	10.42	10.67	10.89	11.35	11.80	12.68
.16	7.72	8.20	8.61	8.96	9.27	9.55	9.81	10.06	10.29	10.75	11.20	12.09
.17	7.19	7.67	8.08	8.43	8.74	9.02	9.28	9.53	9.77	10.23	10.68	11.57
.18	6.73	7.20	7.61	7.96	8.27	8.55	8.81	9.05	9.30	9.76	10.21	11.11
.19	6.31	6.78	7.18	7.54	7.85	8.13	8.39	8.64	8.88	9.35	9.80	10.70
.20	5.94	6.40	6.81	7.16	7.47	7.75	8.01	8.26	8.50	8.97	9.43	10.33
.21	5.60	6.06	6.47	6.82	7.13	7.41	7.67	7.92	8.16	8.63	9.09	10.00
.22	5.30	5.75	6.16	6.50	6.82	7.10	7.36	7.61	7.86	8.33	8.79	9.70
.23	5.02	5.47	5.87	6.22	6.53	6.81	7.08	7.33	7.58	8.05	8.51	9.43
.24	4.77	5.22	5.62	5.96	6.27	6.56	6.82	7.07	7.32	7.79	8.26	9.18
.25	4.54	4.98	5.38	5.73	6.04	6.32	6.58	6.84	7.08	7.56	8.03	8.95
.26	4.32	4.77	5.16	5.51	5.82	6.10	6.37	6.62	6.87	7.35	7.81	8.74
.27	4.13	4.57	4.96	5.31	5.62	5.90	6.16	6.42	6.67	7.15	7.62	8.54
.28	3.95	4.38	4.77	5.12	5.43	5.71	5.98	6.23	6.48	6.96	7.43	8.36
.29	3.78	4.21	4.60	4.95	5.25	5.54	5.81	6.06	6.31	6.79	7.26	8.19
.30	3.62	4.05	4.44	4.78	5.09	5.38	5.64	5.90	6.15	6.63	7.10	8.04
.31	3.48	3.90	4.29	4.63	4.94	5.23	5.49	5.75	6.00	6.48	6.96	7.89
.32	3.34	3.76	4.15	4.49	4.80	5.08	5.35	5.61	5.86	6.34	6.82	7.75
.33	3.21	3.64	4.02	4.36	4.67	4.95	5.22	5.48	5.73	6.21	6.69	7.62
.34	3.09	3.51	3.89	4.23	4.54	4.83	5.09	5.35	5.60	6.09	6.56	7.50
.35	2.98	3.40	3.78	4.12	4.43	4.71	4.98	5.23	5.49	5.97	6.45	7.39
.36	2.88	3.29	3.67	4.01	4.32	4.60	4.87	5.12	5.38	5.86	6.34	7.28
.37	2.78	3.19	3.56	3.90	4.21	4.49	4.76	5.02	5.27	5.76	6.24	7.18
.38	2.68	3.09	3.47	3.81	4.11	4.40	4.67	4.92	5.17	5.66	6.14	7.08
.39	2.59	3.00	3.37	3.71	4.02	4.30	4.57	4.83	5.08	5.57	6.05	6.99
.40	2.51	2.92	3.29	3.62	3.93	4.21	4.48	4.74	4.99	5.48	5.96	6.91
.41	2.43	2.84	3.20	3.54	3.85	4.13	4.40	4.66	4.91	5.40	5.88	6.83
.42	2.36	2.76	3.13	3.46	3.77	4.05	4.32	4.58	4.83	5.32	5.80	6.75
.43	2.29	2.68	3.05	3.38	3.69	3.97	4.24	4.50	4.76	5.25	5.73	6.67
.44	2.22	2.61	2.98	3.31	3.62	3.90	4.17	4.43	4.68	5.17	5.66	6.60

[1] For D/b less than 0.04, use of the assumption $R=D$ is more convenient and more accurate than interpolation in the table.

Table 19.—*Uniform flow of trapezoidal channels by Manning's formula*—Continued

Values of $\dfrac{Qn}{D^{8/3}\,S^{1/2}}$

D/b [1]	$z=0$	$z=\frac{1}{4}$	$z=\frac{1}{2}$	$z=\frac{3}{4}$	$z=1$	$z=1\frac{1}{4}$	$z=1\frac{1}{2}$	$z=1\frac{3}{4}$	$z=2$	$z=2\frac{1}{2}$	$z=3$	$z=4$
.45	2.15	2.55	2.91	3.24	3.55	3.83	4.10	4.36	4.61	5.11	5.59	6.54
.46	2.09	2.48	2.85	3.18	3.48	3.77	4.04	4.29	4.55	5.04	5.52	6.47
.47	2 03	2.42	2.78	3.12	3.42	3.70	3.97	4.23	4.49	4.98	5.46	6.41
.48	1.977	2.36	2.72	3.06	3.36	3.64	3.91	4.17	4.43	4.92	5.40	6.35
.49	1.923	2.31	2.67	3.00	3.30	3.58	3.85	4.12	4.37	4.86	5.34	6.29
.50	1.872	2.26	2.61	2.94	3.25	3.53	3.80	4.06	4.31	4.81	5.29	6.24
.52	1.777	2.16	2.51	2.84	3.14	3.43	3.70	3.96	4.21	4.70	5.19	6.14
.54	1.689	2.06	2.42	2.74	3.05	3.33	3.60	3.86	4.11	4.61	5.09	6.04
.56	1.608	1.977	2.33	2.66	2.96	3.24	3.51	3.77	4.02	4.52	5.00	5.96
.58	1.533	1.900	2.25	2.57	2.87	3.16	3.43	3.69	3.94	4.44	4.92	5.87
.60	1.464	1.827	2.17	2.50	2.80	3.08	3.35	3.61	3.86	4.36	4.85	5.80
.62	1.400	1.760	2.11	2.43	2.73	3.01	3.28	3.54	3.79	4.29	4.77	5.73
.64	1.340	1.697	2.04	2.36	2.66	2.94	3.21	3.47	3.72	4.22	4.71	5.66
.66	1.285	1.638	1.979	2.30	2.60	2.88	3.15	3.41	3.66	4.16	4.64	5.60
.68	1.234	1.583	1.922	2.24	2.54	2.82	3.09	3.35	3.60	4.10	4.59	5.54
.70	1.184	1.531	1.868	2.18	2.48	2.76	3.03	3.29	3.55	4.04	4.53	5.49
.72	1.139	1.482	1.818	2.13	2.43	2.71	2.98	3.24	3.49	3.99	4.48	5.44
.74	1.096	1.437	1.770	2.08	2.38	2.66	2.93	3.19	3.45	3.94	4.43	5.39
.76	1.056	1.393	1.725	2.04	2.33	2.61	2.88	3.15	3.40	3.90	4.38	5.34
.78	1.018	1.353	1.683	1.998	2.29	2.57	2.84	3.10	3.35	3.85	4.34	5.30
.80	.982	1.315	1.642	1.954	2.25	2.53	2.80	3.06	3.31	3.81	4.30	5.26
.82	.949	1.278	1.604	1.916	2.21	2.49	2.76	3.02	3.27	3.77	4.26	5.22
.84	.917	1.245	1.568	1.886	2.17	2.45	2.72	2.98	3.23	3.73	4.22	5.18
.86	.887	1.211	1.534	1.843	2.14	2.41	2.68	2.94	3.20	3.69	4.18	5.14
.88	.858	1.180	1.501	1.810	2.10	2.38	2.65	2.91	3.16	3.66	4.15	5.11
.90	.831	1.153	1.470	1.777	2.07	2.35	2.62	2.87	3.13	3.63	4.12	5.07
.92	.805	1.122	1.441	1.747	2.04	2.32	2.58	2.84	3.10	3.60	4.08	5.04
.94	.781	1.095	1.413	1.718	2.01	2.29	2.55	2.81	3.07	3.57	4.05	5.01
.96	.758	1.070	1.396	1.690	1.981	2.26	2.53	2.78	3.04	3.54	4.03	4.99
.98	.736	1.046	1.360	1.663	1.954	2.23	2.50	2.76	3.01	3.51	4.00	4.96
1.00	.714	1.022	1.335	1.638	1.928	2.21	2.47	2.73	2.99	3.48	3.97	4.93
1.05	.666	.969	1.278	1.579	1.871	2.14	2.41	2.67	2.92	3.42	3.91	4.87
1.10	.622	.920	1.226	1.525	1.813	2.09	2.36	2.61	2.87	3.37	3.85	4.82
1.15	.583	.876	1.178	1.477	1.763	2.04	2.30	2.56	2.82	3.32	3.80	4.76
1.20	.548	.836	1.136	1.432	1.717	1.993	2.26	2.51	2.77	3.27	3.76	4.72
1.25	.516	.800	1.098	1.392	1.676	1.950	2.22	2.47	2.73	3.23	3.71	4.68
1.30	.487	.767	1.062	1.354	1.638	1.912	2.18	2.43	2.69	3.19	3.67	4.64
1.35	.460	.736	1.028	1.320	1.603	1.876	2.14	2.40	2.65	3.15	3.64	4.60
1.40	.436	.708	.998	1.288	1.570	1.843	2.11	2.37	2.62	3.12	3.60	4.57
1.45	.414	.682	.970	1.259	1.540	1.812	2.08	2.34	2.59	3.08	3.57	4.53
1.50	.393	.658	.944	1.231	1.512	1.784	2.05	2.31	2.56	3.06	3.54	4.51
1.55	.374	.636	.920	1.206	1.486	1.757	2.02	2.28	2.53	3.03	3.52	4.48
1.60	.357	.615	.897	1.182	1.461	1.731	1.995	2.25	2.51	3.00	3.49	4.45
1.65	.341	.596	.876	1.160	1.438	1.708	1.972	2.23	2.48	2.98	3.47	4.43
1.70	.325	.578	.856	1.139	1.416	1.686	1.949	2.21	2.46	2.96	3.44	4.40

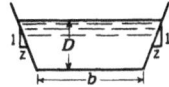

Table 19.—*Uniform flow of trapezoidal channels by Manning's formula*—Continued

Values of $\dfrac{Qn}{D^{8/3}\,S^{1/2}}$

D/b [1]	$z=0$	$z=\frac14$	$z=\frac12$	$z=\frac34$	$z=1$	$z=1\frac14$	$z=1\frac12$	$z=1\frac34$	$z=2$	$z=2\frac12$	$z=3$	$z=4$
1.75	.312	.561	.838	1.119	1.396	1.666	1.928	2.19	2.44	2.93	3.42	4.38
1.80	.298	.546	.820	1.101	1.377	1.646	1.905	2.17	2.42	2.91	3.40	4.36
1.85	.286	.531	.804	1.083	1.359	1.628	1.890	2.15	2.40	2.90	3.38	4.34
1.90	.275	.517	.788	1.067	1.342	1.610	1.872	2.13	2.38	2.88	3.37	4.33
1.95	.264	.504	.773	1.051	1.326	1.594	1.856	2.11	2.36	2.86	3.35	4.31
2.00	.254	.491	.760	1.036	1.310	1.578	1.840	2.10	2.35	2.84	3.33	4.29
2.10	.236	.469	.734	1.009	1.282	1.549	1.811	2.07	2.32	2.81	3.30	4.26
2.20	.219	.448	.711	.984	1.256	1.523	1.784	2.04	2.29	2.79	3.27	4.24
2.30	.205	.430	.690	.962	1.233	1.499	1.760	2.02	2.27	2.76	3.25	4.21
2.40	.1919	.413	.671	.941	1.212	1.477	1.737	1.993	2.24	2.74	3.23	4.19
2.50	.1800	.398	.653	.922	1.192	1.457	1.717	1.972	2.22	2.72	3.21	4.17
2.60	.1693	.383	.637	.905	1.174	1.438	1.698	1.954	2.21	2.70	3.19	4.15
2.70	.1597	.371	.623	.889	1.157	1.422	1.681	1.937	2.19	2.68	3.17	4.13
2.80	.1508	.359	.609	.874	1.142	1.406	1.665	1.920	2.17	2.67	3.15	4.11
2.90	.1427	.348	.596	.861	1.128	1.391	1.650	1.905	2.16	2.65	3.14	4.10
3.00	.1354	.338	.585	.848	1.114	1.377	1.636	1.891	2.14	2.64	3.12	4.08
3.20	.1223	.320	.563	.825	1.090	1.353	1.611	1.865	2.12	2.61	3.10	4.06
3.40	.1111	.305	.545	.805	1.069	1.331	1.589	1.843	2.09	2.59	3.07	4.03
3.60	.1015	.291	.529	.787	1.050	1.312	1.569	1.823	2.07	2.57	3.05	4.01
3.80	.0932	.279	.514	.771	1.033	1.294	1.552	1.805	2.06	2.55	3.04	4.00
4.00	.0859	.268	.501	.757	1.019	1.279	1.536	1.790	2.04	2.53	3.02	3.98
4.50	.0711	.245	.474	.727	.987	1.246	1.502	1.755	2.01	2.50	2.98	3.94
5.00	.0601	.228	.453	.704	.962	1.220	1.476	1.729	1.979	2.47	2.96	3.92

Table 20.—*Values of c_n for use in the formula for normal depth in channels of infinite width,* $D_n = c_n q^{3/5}$

(q=discharge per foot of width)

n s	.010	.011	.012	.013	.014	.015	.0175	.020
.00001	1.573	1.666	1.755	1.841	1.925	2.007	2.201	2.385
.00002	1.278	1.353	1.426	1.496	1.564	1.630	1.788	1.937
.00003	1.131	1.198	1.262	1.324	1.385	1.443	1.583	1.715
.00004	1.038	1.099	1.158	1.215	1.270	1.324	1.452	1.573
.00005	.9707	1.028	1.083	1.136	1.188	1.238	1.358	1.471
.00006	.9191	.9731	1.025	1.076	1.125	1.172	1.286	1.393
.00007	.8775	.9292	.9790	1.027	1.074	1.119	1.228	1.330
.00008	.8431	.8927	.9405	.9868	1.032	1.075	1.179	1.278
.00009	.8138	.8617	.9079	.9525	.9958	1.038.	1.139	1.233
.0001	.7885	.8349	.8796	.9229	.9649	1.006	1.103	1.195
.0002	.6404	.6781	.7145	.7496	.7837	.8168	.8960	.9707
.0003	.5671	.6005	.6326	.6638	.6940	.7233	.7934	.8595
.0004	.5202	.5508	.5803	.6089	.6366	.6635	.7278	.7885
.0005	.4865	.5152	.5428	.5695	.5954	.6205	.6807	.7374
.0006	.4606	.4877	.5139	.5391	.5637	.5875	.6444	.6982
.0007	.4398	.4657	.4906	.5148	.5382	.5609	.6153	.6666
.0008	.4225	.4474	.4714	.4946	.5171	.5389	.5911	.6404
.0009	.4079	.4319	.4550	.4774	.4991	.5202	.5706	.6182
.001	.3952	.4184	.4409	.4625	.4836	.5040	.5529	.5990
.002	.3210	.3399	.3581	.3757	.3928	.4094	.4491	.4865
.003	.2842	.3009	.3171	.3327	.3478	.3625	.3976	.4308
.004	.2607	.2761	.2909	.3052	.3190	.3325	.3647	.3952
.005	.2438	.2582	.2720	.2854	.2984	.3110	.3411	.3696
.006	.2309	.2444	.2575	.2702	.2825	.2944	.3230	.3499
.007	.2204	.2334	.2459	.2580	.2697	.2811	.3084	.3341
.008	.2118	.2242	.2362	.2479	.2591	.2701	.2963	.3210
.009	.2044	.2164	.2280	.2393	.2501	.2607	.2860	.3098
.01	.1981	.2097	.2210	.2318	.2424	.2526	.2771	.3002
.02	.1609	.1703	.1795	.1883	.1969	.2052	.2251	.2438
.03	.1424	.1508	.1589	.1667	.1743	.1817	.1993	.2159
.04	.1307	.1384	.1458	.1529	.1599	.1667	.1828	.1981
.05	.1222	.1294	.1363	.1430	.1495	.1559	.1710	.1852
.06	.1157	.1225	.1291	.1354	.1416	.1476	.1619	.1754
.07	.1105	.1170	.1232	.1293	.1352	.1409	.1546	.1674
.08	.1061	.1124	.1184	.1242	.1299	.1354	.1485	.1609
.09	.1025	.1085	.1143	.1199	.1254	.1307	.1433	.1553
.1	.0993	.1051	.1107	.1162	.1215	.1266	.1389	.1505
.2	.0806	.0854	.0899	.0944	.0987	.1028	.1128	.1222
.3	.0714	.0756	.0796	.0836	.0874	.0911	.0999	.1082
.4	.0655	.0693	.0731	.0767	.0801	.0835	.0916	.0993
.5	.0612	.0649	.0683	.0717	.0750	.0781	.0857	.0928
.6	.0580	.0614	.0647	.0679	.0710	.0740	.0811	.0879
.7	.0554	.0586	.0618	.0648	.0678	.0706	.0775	.0839
.8	.0532	.0563	.0593	.0623	.0651	.0678	.0744	.0806
.9	.0513	.0544	.0573	.0601	.0628	.0655	.0718	.0778

Table 20.—*Values of c_n for use in the formula for normal depths in channels of infinite width, $D_n = c_n q^{3/5}$*—Continued

n s	.0225	.025	.0275	.030	.035	.040	.045	.050
.00001	2.559	2.726	2.887	3.041	3.336	3.614	3.879	4.132
.00002	2.079	2.214	2.345	2.470	2.710	2.936	3.151	3.356
.00003	1.841	1.961	2.076	2.187	2.399	2.599	2.790	2.972
.00004	1.688	1.799	1.904	2.007	2.201	2.385	2.559	2.726
.00005	1.579	1.682	1.781	1.877	2.058	2.230	2.393	2.550
.00006	1.495	1.593	1.686	1.777	1.949	2.111	2.266	2.414
.00007	1.427	1.521	1.610	1.696	1.861	2.016	2.164	2.305
.00008	1.371	1.461	1.547	1.630	1.788	1.937	2.079	2.214
.00009	1.324	1.410	1.493	1.573	1.726	1.870	2.007	2.137
.0001	1.283	1.366	1.447	1.524	1.672	1.811	1.944	2.071
.0002	1.042	1.110	1.175	1.238	1.358	1.471	1.579	1.682
.0003	.9225	.9827	1.041	1.096	1.203	1.303	1.398	1.489
.0004	.8462	.9014	.9545	1.006	1.103	1.195	1.283	1.366
.0005	.7914	.8431	.8928	.9405	1.032	1.118	1.200	1.278
.0006	.7493	.7982	.8452	.8905	.9768	1.058	1.136	1.210
.0007	.7154	.7621	.8070	.8502	.9326	1.010	1.084	1.155
.0008	.6873	.7322	.7753	.8168	.8960	.9707	1.042	1.110
.0009	.6635	.7068	.7484	.7885	.8649	.9370	1.006	1.071
.001	.6428	.6848	.7251	.7639	.8380	.9079	.9744	1.038
.002	.5221	.5562	.5890	.6205	.6806	.7374	.7914	.8431
.003	.4623	.4925	.5215	.5494	.6027	.6530	.7008	.7465
.004	.4241	.4518	.4784	.5040	.5529	.5990	.6428	.6848
.005	.3967	.4225	.4474	.4714	.5171	.5602	.6012	.6404
.006	.3755	.4000	.4236	.4463	.4895	.5304	.5692	.6064
.007	.3586	.3820	.4044	.4261	.4674	.5064	.5435	.5790
.008	.3445	.3670	.3886	.4094	.4491	.4865	.5221	.5562
.009	.3325	.3542	.3751	.3952	.4335	.4696	.5040	.5369
.01	.3222	.3432	.3634	.3829	.4200	.4550	.4883	.5202
.02	.2788	.2760	.2952	.3110	.3411	.3696	.3967	.4225
.03	.2317	.2468	.2614	.2754	.3021	.3273	.3512	.3741
.04	.2126	.2264	.2398	.2526	.2771	.3002	.3222	.3432
.05	.1988	.2118	.2242	.2362	.2591	.2808	.3013	.3210
.06	.1882	.2005	.2123	.2237	.2453	.2658	.2853	.3039
.07	.1797	.1914	.2027	.2136	.2343	.2538	.2724	.2902
.08	.1727	.1839	.1947	.2052	.2251	.2438	.2616	.2788
.09	.1667	.1775	.1880	.1981	.2172	.2354	.2526	.2691
.1	.1615	.1720	.1821	.1919	.2105	.2280	.2447	.2607
.2	.1312	.1397	.1479	.1559	.1710	.1852	.1988	.2118
.3	.1161	.1237	.1310	.1380	.1514	.1640	.1760	.1875
.4	.1065	.1135	.1202	.1266	.1388	.1505	.1615	.1720
.5	.0996	.1061	.1124	.1184	.1298	.1407	.1510	.1609
.6	.0943	.1005	.1064	.1121	.1230	.1332	.1430	.1523
.7	.0901	.0959	.1016	.1070	.1174	.1272	.1365	.1454
.8	.0865	.0922	.0976	.1028	.1127	.1222	.1312	.1397
.9	.0835	.0890	.0942	.0993	.1088	.1180	.1266	.1349

Table 20.—*Values of c_n for use in the formula for normal depth in channels of infinite width, $D_n = c_n q^{3/5}$*—Continued

s \ n	.055	.060	.065	.070	.075	.100	.125	.150	.175
.00001	4.375	4.610	4.837	5.056	5.270	6.263	7.160	7.988	8.762
.00002	3.554	3.744	3.929	4.107	4.281	5.087	5.816	6.488	7.117
.00003	3.147	3.315	3.479	3.637	3.790	4.505	5.150	5.745	6.302
.00004	2.887	3.041	3.191	3.336	3.477	4.132	4.724	5.270	5.781
.00005	2.700	2.844	2.984	3.120	3.252	3.865	4.418	4.929	5.407
.00006	2.556	2.693	2.825	2.954	3.079	3.659	4.183	4.667	5.119
.00007	2.440	2.571	2.698	2.820	2.940	3.493	3.994	4.456	4.887
.00008	2.345	2.470	2.592	2.710	2.824	3.356	3.837	4.281	4.696
.00009	2.263	2.385	2.502	2.616	2.726	3.240	3.704	4.132	4.533
.0001	2.193	2.310	2.424	2.534	2.641	3.139	3.589	4.004	4.391
.0002	1.781	1.877	1.969	2.058	2.145	2.550	2.915	3.252	3.567
.0003	1.577	1.662	1.743	1.823	1.900	2.258	2.581	2.879	3.158
.0004	1.447	1.524	1.599	1.672	1.743	2.071	2.368	2.641	2.897
.0005	1.353	1.426	1.496	1.564	1.630	1.937	2.214	2.470	2.710
.0006	1.281	1.350	1.416	1.480	1.543	1.834	2.096	2.339	2.565
.0007	1.223	1.289	1.352	1.414	1.473	1.751	2.002	2.233	2.450
.0008	1.175	1.238	1.299	1.358	1.415	1.682	1.923	2.145	2.353
.0009	1.134	1.195	1.254	1.311	1.366	1.624	1.856	2.071	2.272
.001	1.099	1.158	1.215	1.270	1.324	1.573	1.799	2.007	2.201
.002	.8927	.9405	.9868	1.032	1.075	1.278	1.461	1.630	1.788
.003	.7904	.8328	.8738	.9135	.9521	1.131	1.294	1.443	1.583
.004	.7251	.7639	.8015	.8380	.8734	1.038	1.187	1.324	1.452
.005	.6781	.7145	.7496	.7837	.8168	.9707	1.110	1.238	1.358
.006	.6420	.6764	.7097	.7420	.7734	.9191	1.051	1.172	1.286
.007	.6130	.6459	.6777	.7085	.7384	.8775	1.003	1.119	1.228
.008	.5890	.6205	.6510	.6806	.7094	.8431	.9638	1.075	1.179
.009	.5685	.5990	.6284	.6570	.6848	.8138	.9304	1.038	1.139
.01	.5508	.5803	.6089	.6366	.6635	.7885	.9014	1.006	1.103
.02	.4475	.4714	.4946	.5171	.5389	.6404	.7322	.8168	.8960
.03	.3962	.4174	.4379	.4578	.4772	.5671	.6483	.7233	.7934
.04	.3634	.3829	.4017	.4200	.4377	.5202	.5947	.6635	.7278
.05	.3399	.3581	.3757	.3928	.4094	.4865	.5562	.6205	.6806
.06	.3218	.3390	.3557	.3719	.3876	.4606	.5266	.5875	.6444
.07	.3072	.3237	.3396	.3551	.3701	.4398	.5028	.5609	.6153
.08	.2952	.3110	.3263	.3411	.3555	.4225	.4831	.5389	.5911
.09	.2849	.3002	.3150	.3293	.3432	.4079	.4663	.5202	.5706
.1	.2761	.2909	.3052	.3190	.3325	.3952	.4518	.5040	.5529
.2	.2242	.2362	.2479	.2591	.2701	.3210	.3670	.4094	.4491
.3	.1986	.2092	.2195	.2295	.2392	.2842	.3249	.3625	.3976
.4	.1821	.1919	.2013	.2105	.2194	.2607	.2981	.3325	.3647
.5	.1703	.1795	.1883	.1969	.2052	.2438	.2788	.3110	.3411
.6	.1613	.1699	.1783	.1864	.1943	.2309	.2639	.2944	.3230
.7	.1540	.1622	.1702	.1780	.1855	.2204	.2520	.2811	.3084
.8	.1479	.1559	.1635	.1710	.1782	.2118	.2421	.2701	.2963
.9	.1428	.1505	.1579	.1650	.1720	.2044	.2337	.2607	.2860

Table 21.—*Uniform flow in circular sections flowing partly full*

d = Depth of flow
D = Diameter of pipe
A = Area of flow
R = Hydraulic radius

Q = Discharge in second-feet by Manning's formula
n = Manning's coefficient
S = Slope of the channel bottom and of the water surface

$\dfrac{d}{D}$	$\dfrac{A}{D^2}$	$\dfrac{R}{D}$	$\dfrac{Qn}{D^{8/3}S^{1/2}}$	$\dfrac{Qn}{d^{8/3}S^{1/2}}$	$\dfrac{d}{D}$	$\dfrac{A}{D^2}$	$\dfrac{R}{D}$	$\dfrac{Qn}{D^{8/3}S^{1/2}}$	$\dfrac{Qn}{d^{8/3}S^{1/2}}$
0.01	0.0013	0.0066	0.00007	15.04	0.51	0.4027	0.2531	0.239	1.442
0.02	0.0037	0.0132	0.00031	10.57	0.52	0.4127	0.2562	0.247	1.415
0.03	0.0069	0.0197	0.00074	8.56	0.53	0.4227	0.2592	0.255	1.388
0.04	0.0105	0.0262	0.00138	7.38	0.54	0.4327	0.2621	0.263	1.362
0.05	0.0147	0.0325	0.00222	6.55	0.55	0.4426	0.2649	0.271	1.336
0.06	0.0192	0.0389	0.00328	5.95	0.56	0.4526	0.2676	0.279	1.311
0.07	0.0242	0.0451	0.00455	5.47	0.57	0.4625	0.2703	0.287	1.286
0.08	0.0294	0.0513	0.00604	5.09	0.58	0.4724	0.2728	0.295	1.262
0.09	0.0350	0.0575	0.00775	4.76	0.59	0.4822	0.2753	0.303	1.238
0.10	0.0409	0.0635	0.00967	4.49	0.60	0.4920	0.2776	0.311	1.215
0.11	0.0470	0.0695	0.01181	4.25	0.61	0.5018	0.2799	0.319	1.192
0.12	0.0534	0.0755	0.01417	4.04	0.62	0.5115	0.2821	0.327	1.170
0.13	0.0600	0.0813	0.01674	3.86	0.63	0.5212	0.2842	0.335	1.148
0.14	0.0668	0.0871	0.01952	3.69	0.64	0.5308	0.2862	0.343	1.126
0.15	0.0739	0.0929	0.0225	3.54	0.65	0.5404	0.2882	0.350	1.105
0.16	0.0811	0.0985	0.0257	3.41	0.66	0.5499	0.2900	0.358	1.084
0.17	0.0885	0.1042	0.0291	3.28	0.67	0.5594	0.2917	0.366	1.064
0.18	0.0961	0.1097	0.0327	3.17	0.68	0.5687	0.2933	0.373	1.044
0.19	0.1039	0.1152	0.0365	3.06	0.69	0.5780	0.2948	0.380	1.024
0.20	0.1118	0.1206	0.0406	2.96	0.70	0.5872	0.2962	0.388	1.004
0.21	0.1199	0.1259	0.0448	2.87	0.71	0.5964	0.2975	0.395	0.985
0.22	0.1281	0.1312	0.0492	2.79	0.72	0.6054	0.2987	0.402	0.965
0.23	0.1365	0.1364	0.0537	2.71	0.73	0.6143	0.2998	0.409	0.947
0.24	0.1449	0.1416	0.0585	2.63	0.74	0.6231	0.3008	0.416	0.928
0.25	0.1535	0.1466	0.0634	2.56	0.75	0.6319	0.3017	0.422	0.910
0.26	0.1623	0.1516	0.0686	2.49	0.76	0.6405	0.3024	0.429	0.891
0.27	0.1711	0.1566	0.0739	2.42	0.77	0.6489	0.3031	0.435	0.873
0.28	0.1800	0.1614	0.0793	2.36	0.78	0.6573	0.3036	0.441	0.856
0.29	0.1890	0.1662	0.0849	2.30	0.79	0.6655	0.3039	0.447	0.838
0.30	0.1982	0.1709	0.0907	2.25	0.80	0.6736	0.3042	0.453	0.821
0.31	0.2074	0.1756	0.0966	2.20	0.81	0.6815	0.3043	0.458	0.804
0.32	0.2167	0.1802	0.1027	2.14	0.82	0.6893	0.3043	0.463	0.787
0.33	0.2260	0.1847	0.1089	2.09	0.83	0.6969	0.3041	0.468	0.770
0.34	0.2355	0.1891	0.1153	2.05	0.84	0.7043	0.3038	0.473	0.753
0.35	0.2450	0.1935	0.1218	2.00	0.85	0.7115	0.3033	0.477	0.736
0.36	0.2546	0.1978	0.1284	1.958	0.86	0.7186	0.3026	0.481	0.720
0.37	0.2642	0.2020	0.1351	1.915	0.87	0.7254	0.3018	0.485	0.703
0.38	0.2739	0.2062	0.1420	1.875	0.88	0.7320	0.3007	0.488	0.687
0.39	0.2836	0.2102	0.1490	1.835	0.89	0.7384	0.2995	0.491	0.670
0.40	0.2934	0.2142	0.1561	1.797	0.90	0.7445	0.2980	0.494	0.654
0.41	0.3032	0.2182	0.1633	1.760	0.91	0.7504	0.2963	0.496	0.637
0.42	0.3130	0.2220	0.1705	1.724	0.92	0.7560	0.2944	0.497	0.621
0.43	0.3229	0.2258	0.1779	1.689	0.93	0.7612	0.2921	0.498	0.604
0.44	0.3328	0.2295	0.1854	1.655	0.94	0.7662	0.2895	0.498	0.588
0.45	0.3428	0.2331	0.1929	1.622	0.95	0.7707	0.2865	0.498	0.571
0.46	0.3527	0.2366	0.201	1.590	0.96	0.7749	0.2829	0.496	0.553
0.47	0.3627	0.2401	0.208	1.559	0.97	0.7785	0.2787	0.494	0.535
0.48	0.3727	0.2435	0.216	1.530	0.98	0.7817	0.2735	0.489	0.517
0.49	0.3827	0.2468	0.224	1.500	0.99	0.7841	0.2666	0.483	0.496
0.50	0.3927	0.2500	0.232	1.471	1.00	0.7854	0.2500	0.463	0.463

Table 22.—*Velocity head and discharge at critical depths and static pressures in circular conduits partly full*

D=Diameter of circle.
d=Depth of water.
h_v=Velocity head for a critical depth of d.
Q=Discharge when the critical depth is d.
P=Pressure on cross section of water prism in cubic units of water. To get P in pounds, when d and D are in feet, multiply by 62.5.

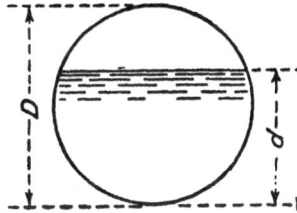

$\dfrac{d}{D}$	$\dfrac{h_v}{D}$	$\dfrac{Q}{D^{\frac{5}{2}}}$	$\dfrac{P}{D^3}$	$\dfrac{d}{D}$	$\dfrac{h_v}{D}$	$\dfrac{Q}{D^{\frac{5}{2}}}$	$\dfrac{P}{D^3}$	$\dfrac{d}{D}$	$\dfrac{h_v}{D}$	$\dfrac{Q}{D^{\frac{5}{2}}}$	$\dfrac{P}{D^3}$
1	2	3	4	1	2	3	4	1	2	3	4
.01	.0033	.0006	.0000	.34	.1243	.6657	.0332	.67	.2974	2.4464	.1644
.02	.0067	.0025	.0000	.35	.1284	.7040	.0356	.68	.3048	2.5182	.1700
.03	.0101	.0055	.0001	.36	.1326	.7433	.0381	.69	.3125	2.5912	.1758
.04	.0134	.0098	.0002	.37	.1368	.7836	.0407	.70	.3204	2.6656	.1816
.05	.0168	.0153	.0003	.38	.1411	.8249	.0434	.71	.3286	2.7414	.1875
.06	.0203	.0220	.0005	.39	.1454	.8671	.0462	.72	.3371	2.8188	.1935
.07	.0237	.0298	.0007	.40	.1497	.9103	.0491	.73	.3459	2.8977	.1996
.08	.0271	.0389	.0010	.41	.1541	.9545	.0520	.74	.3552	2.9783	.2058
.09	.0306	.0491	.0013	.42	.1586	.9996	.0551	.75	.3648	3.0607	.2121
.10	.0341	.0605	.0017	.43	.1631	1.0458	.0583	.76	.3749	3.1450	.2185
.11	.0376	.0731	.0021	.44	.1676	1.0929	.0616	.77	.3855	3.2314	.2249
.12	.0411	.0868	.0026	.45	.1723	1.1410	.0650	.78	.3967	3.3200	.2314
.13	.0446	.1016	.0032	.46	.1769	1.1899	.0684	.79	.4085	3.4112	.2380
.14	.0482	.1176	.0038	.47	.1817	1.2399	.0720	.80	.4210	3.5050	.2447
.15	.0517	.1347	.0045	.48	.1865	1.2908	.0757	.81	.4343	3.6019	.2515
.16	.0553	.1530	.0053	.49	.1914	1.3427	.0795	.82	.4485	3.7021	.2584
.17	.0589	.1724	.0061	.50	.1964	1.3955	.0833	.83	.4638	3.8061	.2653
.18	.0626	.1928	.0070	.51	.2014	1.4493	.0873	.84	.4803	3.9144	.2723
.19	.0662	.2144	.0080	.52	.2065	1.5041	.0914	.85	.4982	4.0276	.2794
.20	.0699	.2371	.0091	.53	.2117	1.5598	.0956	.86	.5177	4.1465	.2865
.21	.0736	.2609	.0103	.54	.2170	1.6164	.0998	.87	.5392	4.2721	.2938
.22	.0773	.2857	.0115	.55	.2224	1.6735	.1042	.88	.5632	4.4056	.3011
.23	.0811	.3116	.0128	.56	.2279	1.7327	.1087	.89	.5900	4.5486	.3084
.24	.0848	.3386	.0143	.57	.2335	1.7923	.1133	.90	.6204	4.7033	.3158
.25	.0887	.3667	.0157	.58	.2393	1.8530	.1179	.91	.6555	4.8725	.3233
.26	.0925	.3957	.0173	.59	.2451	1.9146	.1227	.92	.6966	5.0603	.3308
.27	.0963	.4259	.0190	.60	.2511	1.9773	.1276	.93	.7459	5.2726	.3384
.28	.1002	.4571	.0207	.61	.2572	2.0409	.1326	.94	.8065	5.5183	.3460
.29	.1042	.4893	.0226	.62	.2635	2.1057	.1376	.95	.8841	5.8118	.3537
.30	.1081	.5225	.0255	.63	.2699	2.1716	.1428	.96	.9885	6.1787	.3615
.31	.1121	.5568	.0266	.64	.2765	2.2386	.1481	.97	1.1410	6.6692	.3692
.32	.1161	.5921	.0287	.65	.2833	2.3067	.1534	.98	1.3958	7.4063	.3770
.33	.1202	.6284	.0309	.66	.2902	2.3760	.1589	.99	1.9700	8.8263	.3848
								1.00	------	------	.3927

Table 23.—*Uniform flow in horseshoe sections flowing partly full*

d = Depth of flow
D = Diameter
A = Area of flow
R = Hydraulic radius

Q = Discharge in second-feet by Manning's formula
n = Manning's coefficient.
S = Slope of the channel bottom and of the water surface

$\dfrac{d}{D}$	$\dfrac{A}{D^2}$	$\dfrac{R}{D}$	$\dfrac{Qn}{D^{8/3}S^{1/2}}$	$\dfrac{Qn}{d^{8/3}S^{1/2}}$	$\dfrac{d}{D}$	$\dfrac{A}{D^2}$	$\dfrac{R}{D}$	$\dfrac{Qn}{D^{8/3}S^{1/2}}$	$\dfrac{Qn}{d^{8/3}S^{1/2}}$
0.01	0.0019	0.0066	0.00010	21.40	0.51	0.4466	0.2602	0.2705	1.629
0.02	0.0053	0.0132	0.00044	14.93	0.52	0.4566	0.2630	0.2785	1.593
0.03	0.0097	0.0198	0.00105	12.14	0.53	0.4666	0.2657	0.2866	1.558
0.04	0.0150	0.0264	0.00198	10.56	0.54	0.4766	0.2683	0.2946	1.524
0.05	0.0209	0.0329	0.00319	9.40	0.55	0.4865	0.2707	0.303	1.490
0.06	0.0275	0.0394	0.00473	8.58	0.56	0.4965	0.2733	0.311	1.458
0.07	0.0346	0.0459	0.00659	7.92	0.57	0.5064	0.2757	0.319	1.427
0.08	0.0421	0.0524	0.00876	7.37	0.58	0.5163	0.2781	0.327	1.397
0.09	0.0502	0.0590	0.01131	6.95	0.59	0.5261	0.2804	0.335	1.368
0.10	0.0585	0.0670	0.01434	6.66	0.60	0.5359	0.2824	0.343	1.339
0.11	0.0670	0.0748	0.01768	6.36	0.61	0.5457	0.2844	0.351	1.310
0.12	0.0753	0.0823	0.02117	6.04	0.62	0.5555	0.2864	0.359	1.283
0.13	0.0839	0.0895	0.02495	5.75	0.63	0.5651	0.2884	0.367	1.257
0.14	0.0925	0.0964	0.02890	5.47	0.64	0.5748	0.2902	0.374	1.231
0.15	0.1012	0.1031	0.0331	5.21	0.65	0.5843	0.2920	0.382	1.206
0.16	0.1100	0.1097	0.0375	4.96	0.66	0.5938	0.2937	0.390	1.181
0.17	0.1188	0.1161	0.0420	4.74	0.67	0.6033	0.2953	0.398	1.157
0.18	0.1277	0.1222	0.0467	4.52	0.68	0.6126	0.2967	0.405	1.133
0.19	0.1367	0.1282	0.0516	4.33	0.69	0.6219	0.2981	0.412	1.109
0.20	0.1457	0.1341	0.0567	4.15	0.70	0.6312	0.2994	0.420	1.087
0.21	0.1549	0.1398	0.0620	3.98	0.71	0.6403	0.3006	0.427	1.064
0.22	0.1640	0.1454	0.0674	3.82	0.72	0.6493	0.3018	0.434	1.042
0.23	0.1733	0.1508	0.0730	3.68	0.73	0.6582	0.3028	0.441	1.021
0.24	0.1825	0.1560	0.0786	3.53	0.74	0.6671	0.3036	0.448	1.000
0.25	0.1919	0.1611	0.0844	3.40	0.75	0.6758	0.3044	0.454	0.979
0.26	0.2013	0.1662	0.0904	3.28	0.76	0.6844	0.3050	0.461	0.958
0.27	0.2107	0.1710	0.0965	3.17	0.77	0.6929	0.3055	0.467	0.938
0.28	0.2202	0.1758	0.1027	3.06	0.78	0.7012	0.3060	0.473	0.918
0.29	0.2297	0.1804	0.1090	2.96	0.79	0.7094	0.3064	0.479	0.898
0.30	0.2393	0.1850	0.1155	2.86	0.80	0.7175	0.3067	0.485	0.879
0.31	0.2489	0.1895	0.1220	2.77	0.81	0.7254	0.3067	0.490	0.860
0.32	0.2586	0.1938	0.1287	2.69	0.82	0.7332	0.3066	0.495	0.841
0.33	0.2683	0.1981	0.1355	2.61	0.83	0.7408	0.3064	0.500	0.822
0.34	0.2780	0.2023	0.1424	2.53	0.84	0.7482	0.3061	0.505	0.804
0.35	0.2878	0.2063	0.1493	2.45	0.85	0.7554	0.3056	0.509	0.786
0.36	0.2975	0.2103	0.1563	2.38	0.86	0.7625	0.3050	0.513	0.768
0.37	0.3074	0.2142	0.1635	2.32	0.87	0.7693	0.3042	0.517	0.750
0.38	0.3172	0.2181	0.1708	2.25	0.88	0.7759	0.3032	0.520	0.732
0.39	0.3271	0.2217	0.1781	2.19	0.89	0.7823	0.3020	0.523	0.714
0.40	0.3370	0.2252	0.1854	2.13	0.90	0.7884	0.3005	0.526	0.696
0.41	0.3469	0.2287	0.1928	2.08	0.91	0.7943	0.2988	0.528	0.678
0.42	0.3568	0.2322	0.2003	2.02	0.92	0.7999	0.2969	0.529	0.661
0.43	0.3667	0.2356	0.2079	1.973	0.93	0.8052	0.2947	0.530	0.643
0.44	0.3767	0.2390	0.2156	1.925	0.94	0.8101	0.2922	0.530	0.625
0.45	0.3867	0.2422	0.2233	1.878	0.95	0.8146	0.2893	0.529	0.607
0.46	0.3966	0.2454	0.2310	1.832	0.96	0.8188	0.2858	0.528	0.589
0.47	0.4066	0.2484	0.2388	1.788	0.97	0.8224	0.2816	0.525	0.569
0.48	0.4166	0.2514	0.2466	1.746	0.98	0.8256	0.2766	0.521	0.550
0.49	0.4266	0.2544	0.2545	1.705	0.99	0.8280	0.2696	0.513	0.527
0.50	0.4366	0.2574	0.2625	1.667	1.00	0.8293	0.2538	0.494	0.494

Table 24.—*Velocity head and discharge at critical depths and static pressures in horseshoe conduits partly full*

D=Diameter of horseshoe.
d=Depth of water.
h_v=Velocity head for a critical depth of d.
Q=Discharge when the critical depth is d.
P=Pressure on cross section of water prism in cubic units of water. To get P in pounds, when d and D are in feet, multiply by 62.5.

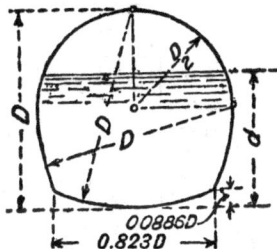

0.0886D
0.823D

$\dfrac{d}{D}$	$\dfrac{h_v}{D}$	$\dfrac{Q}{D^{5/2}}$	$\dfrac{P}{D^3}$	$\dfrac{d}{D}$	$\dfrac{h_v}{D}$	$\dfrac{Q}{D^{5/2}}$	$\dfrac{P}{D^3}$	$\dfrac{d}{D}$	$\dfrac{h_v}{D}$	$\dfrac{Q}{D^{5/2}}$	$\dfrac{P}{D^3}$
1	2	3	4	1	2	3	4	1	2	3	4
.01	.0033	.0009	.0000	.35	.1472	.8854	.0449	.69	.3362	2.8922	.1999
.02	.0067	.0035	.0000	.36	.1518	.9296	.0478	.70	.3443	2.9702	.2062
.03	.0100	.0079	.0001	.37	.1563	.9746	.0508	.71	.3528	3.0499	.2125
.04	.0134	.0139	.0002	.38	.1609	1.0205	.0540	.72	.3615	3.1311	.2190
.05	.0168	.0217	.0004	.39	.1655	1.0673	.0572	.73	.3707	3.2140	.2255
.06	.0201	.0312	.0007	.40	.1702	1.1148	.0605	.74	.3802	3.2987	.2321
.07	.0235	.0425	.0010	.41	.1749	1.1633	.0639	.75	.3902	3.3853	.2385
.08	.0269	.0554	.0014	.42	.1795	1.2125	.0675	.76	.4006	3.4740	.2457
.09	.0305	.0703	.0018	.43	.1843	1.2626	.0711	.77	.4116	3.5650	.2525
.10	.0351	.0879	.0024	.44	.1890	1.3135	.0748	.78	.4232	3.6584	.2595
.11	.0397	.1069	.0030	.45	.1938	1.3652	.0786	.79	.4354	3.7544	.2666
.12	.0443	.1272	.0037	.46	.1986	1.4178	.0825	.80	.4484	3.8534	.2737
.13	.0489	.1487	.0045	.47	.2035	1.4712	.0865	.81	.4623	3.9557	.2809
.14	.0534	.1714	.0054	.48	.2084	1.5253	.0907	.82	.4771	4.0616	.2882
.15	.0579	.1953	.0063	.49	.2133	1.5803	.0949	.83	.4930	4.1716	.2956
.16	.0624	.2203	.0074	.50	.2183	1.6361	.0992	.84	.5102	4.2863	.3030
.17	.0669	.2465	.0085	.51	.2234	1.6923	.1036	.85	.5289	4.4063	.3105
.18	.0714	.2736	.0098	.52	.2285	1.7505	.1081	.86	.5494	4.5325	.3181
.19	.0758	.3019	.0111	.53	.2337	1.8092	.1127	.87	.5719	4.6660	.3258
.20	.0803	.3312	.0125	.54	.2391	1.8688	.1174	.88	.5969	4.8080	.3335
.21	.0847	.3615	.0140	.55	.2445	1.9294	.1223	.89	.6251	4.9605	.3413
.22	.0891	.3923	.0156	.56	.2500	1.9911	.1272	.90	.6570	5.1256	.3492
.23	.0936	.4251	.0173	.57	.2557	2.0537	.1322	.91	.6939	5.3065	.3572
.24	.0980	.4583	.0191	.58	.2615	2.1174	.1373	.92	.7371	5.5077	.3653
.25	.1024	.4926	.0210	.59	.2674	2.1821	.1425	.93	.7889	5.7354	.3733
.26	.1069	.5277	.0229	.60	.2735	2.2479	.1478	.94	.8528	5.9996	.3813
.27	.1113	.5638	.0250	.61	.2797	2.3148	.1532	.95	.9345	6.3157	.3894
.28	.1158	.6009	.0271	.62	.2861	2.3823	.1587	.96	1.0446	6.7114	.3976
.29	.1202	.6389	.0294	.63	.2926	2.4519	.1643	.97	1.2053	7.2417	.4058
.30	.1247	.6775	.0317	.64	.2994	2.5221	.1700	.98	1.4742	8.0392	.4140
.31	.1292	.7175	.0342	.65	.3063	2.5936	.1758	.99	2.0304	9.5780	.4223
.32	.1337	.7582	.0367	.66	.3134	2.6663	.1817	1.00	------	------	.4306
.33	.1382	.7997	.0393	.67	.3203	2.7402	.1877				
.34	.1427	.8421	.0421	.68	.3283	2.8155	.1937				

Table 25.—*Area in square feet, A, and hydraulic radius in feet, r, of semicircular flumes for various values of freeboard in feet, F*

Flume No.	Diameter in feet	F=0.0		F=0.1		F=0.2		F=0.3		F=0.4	
		A	r	A	r	A	r	A	r	A	r
24	1.273	0.64	0.32	0.51	0.28	0.39	0.24	0.27	0.20	.....	.....
30	1.592	1.00	0.40	0.84	0.36	0.68	0.32	0.53	0.28	0.38	0.23
36	1.910	1.43	0.48	1.24	0.44	1.05	0.41	0.87	0.36	0.69	0.32
42	2.228	1.95	0.55	1.72	0.53	1.50	0.48	1.29	0.45	1.08	0.40
48	2.546	2.55	0.64	2.29	0.60	2.04	0.57	1.79	0.53	1.54	0.49
60	3.183	3.98	0.80	3.66	0.76	3.34	0.73	3.03	0.69	2.72	0.65
72	3.820	5.73	0.96	5.35	0.92	4.97	0.89	4.59	0.85	4.21	0.81
84	4.456	7.80	1.11	7.35	1.08	6.91	1.05	6.47	1.01	6.03	0.98
96	5.093	10.2	1.27	9.68	1.24	9.17	1.21	8.66	1.17	8.16	1.13
108	5.730	12.9	1.43	12.3	1.40	11.8	1.37	11.2	1.33	10.6	1.29
120	6.366	15.9	1.59	15.3	1.56	14.6	1.53	14.0	1.49	13.4	1.45
132	7.003	19.3	1.75	18.6	1.72	17.9	1.69	17.2	1.65	16.5	1.61
144	7.639	22.9	1.91	22.2	1.88	21.4	1.84	20.6	1.81	19.9	1.77
156	8.276	26.9	2.07	26.1	2.04	25.2	2.00	24.4	1.97	23.6	1.93
168	8.913	31.2	2.23	30.3	2.18	29.4	2.16	28.5	2.13	27.6	2.09
180	9.549	35.8	2.39	34.9	2.35	33.9	2.32	32.9	2.29	32.0	2.25
192	10.186	40.7	2.54	39.7	2.51	38.7	2.48	37.7	2.45	36.7	2.42
204	10.823	46.0	2.70	44.9	2.68	43.8	2.64	42.8	2.61	41.7	2.57
216	11.459	51.6	2.86	50.4	2.81	49.3	2.80	48.1	2.76	47.0	2.73
228	12.096	57.5	3.02	56.2	2.99	55.0	2.96	53.8	2.93	52.6	2.89
240	12.732	63.7	3.18	62.4	3.15	61.1	3.12	59.8	3.08	58.6	3.05
252	13.369	70.2	3.34	68.9	3.31	67.5	3.26	66.2	3.24	64.8	3.21

Flume No.	Diameter in feet	F=0.5		F=0.6		F=0.7		F=0.8		F=0.9	
		A	r	A	r	A	r	A	r	A	r
30	1.592	0.25	0.18	.....	.....	.....	.....	.....	.....	.....	.....
36	1.910	0.52	0.27	.....	.....	.....	.....	.....	.....	.....	.....
42	2.228	0.87	0.36	0.68	0.31	.....	.....	.....	.....	.....	.....
48	2.546	1.31	0.44	1.08	0.39	0.85	0.34	.....	.....	.....	.....
60	3.183	2.41	0.61	2.11	0.56	1.82	0.51	1.54	0.46	.....	.....
72	3.820	3.84	0.77	3.48	0.73	3.12	0.68	2.76	0.64	2.42	0.59
84	4.456	5.59	0.93	5.16	0.89	4.73	0.85	4.31	0.80	3.90	0.76
96	5.093	7.66	1.09	7.16	1.06	6.67	1.01	6.18	0.97	5.70	0.93
108	5.730	10.0	1.26	9.48	1.22	8.92	1.18	8.37	1.14	7.82	1.09
120	6.366	12.7	1.42	12.1	1.38	11.5	1.34	10.9	1.30	10.3	1.25
132	7.003	15.8	1.58	15.1	1.54	14.4	1.50	13.7	1.46	13.0	1.42
144	7.639	19.1	1.74	18.4	1.70	17.6	1.66	16.9	1.62	16.1	1.58
156	8.276	22.8	1.90	21.9	1.86	21.1	1.82	20.3	1.79	19.5	1.74
168	8.913	26.7	2.06	25.9	2.02	25.0	1.98	24.1	1.95	23.2	1.91
180	9.549	31.0	2.22	30.1	2.18	29.1	2.14	28.2	2.11	27.3	2.07
192	10.186	35.7	2.38	34.6	2.34	33.6	2.31	32.6	2.27	31.6	2.23
204	10.823	40.6	2.54	39.5	2.50	38.4	2.46	37.4	2.43	36.3	2.39
216	11.459	45.8	2.70	44.7	2.66	43.6	2.62	42.4	2.59	41.3	2.55
228	12.096	51.4	2.86	50.2	2.82	49.0	2.78	47.8	2.75	46.6	2.71
240	12.732	57.3	3.01	56.0	2.98	54.8	2.94	53.5	2.91	52.2	2.87
252	13.369	63.5	3.18	62.2	3.14	60.8	3.10	59.4	3.07	58.2	3.03

Table 25.—*Area in square feet, A, and hydraulic radius in feet, r, of semicircular flumes for various values of freeboard in feet, F*—Con.

Flume No.	Diameter in feet	F=1.0 A	r	F=1.1 A	r	F=1.2 A	r	F=1.3 A	r	F=1.4 A	r	F=1.5 A	r
72	3.820	2.08	0.54	....	....	....	....	....	....	....	....	....	....
84	4.456	3.49	0.71	3.10	0.66	2.71	0.61	....	....	....	....	....	....
96	5.093	5.22	0.88	4.76	0.83	4.30	0.78	3.85	0.73	3.41	0.68	....	....
108	5.730	7.28	1.05	6.74	1.00	6.22	0.96	5.70	0.91	5.19	0.86	4.69	0.81
120	6.366	9.65	1.21	9.05	1.16	8.46	1.12	7.87	1.08	7.29	1.03	6.72	0.98
132	7.003	12.4	1.38	11.7	1.33	11.0	1.29	10.4	1.24	9.72	1.20	9.07	1.15
144	7.639	15.4	1.54	14.6	1.50	13.9	1.46	13.2	1.41	12.5	1.37	11.8	1.32
156	8.276	18.7	1.70	17.9	1.66	17.1	1.62	16.3	1.58	15.5	1.53	14.8	1.49
168	8.913	22.4	1.87	21.5	1.82	20.6	1.78	19.8	1.74	18.9	1.70	18.1	1.65
180	9.549	26.3	2.03	25.4	1.99	24.5	1.95	23.5	1.90	22.6	1.86	21.7	1.82
192	10.186	30.6	2.19	29.6	2.15	28.6	2.11	27.6	2.07	26.7	2.03	25.7	1.98
204	10.823	35.2	2.35	34.2	2.31	33.1	2.27	32.1	2.23	31.0	2.19	30.0	2.14
216	11.459	40.2	2.51	39.0	2.47	37.9	2.43	36.8	2.39	35.7	2.35	34.6	2.31
228	12.096	45.4	2.67	44.2	2.64	43.0	2.60	41.9	2.56	40.7	2.52	39.5	2.47
240	12.732	51.0	2.83	49.7	2.80	48.5	2.76	47.2	2.72	46.0	2.68	44.7	2.63
252	13.369	56.9	2.99	55.5	2.96	54.2	2.92	52.9	2.88	51.6	2.84	50.3	2.80

Table 26.—*Area in square feet, A, and hydraulic radius in feet, r, of rectangular channels*

Depth, feet	Bottom width 4 feet A	$r=\frac{area}{wet\ per.}$	Bottom width 6 feet A	$r=\frac{area}{wet\ per.}$	Bottom width 8 feet A	$r=\frac{area}{wet\ per.}$	Bottom width 10 feet A	$r=\frac{area}{wet\ per.}$	Bottom width 12 feet A	$r=\frac{area}{wet\ per.}$	Bottom width 14 feet A	$r=\frac{area}{wet\ per.}$
1.0	4	.67	6	.75	8	.80	10	.83	12	.86	14	.88
1.5	6	.86	9	1.00	12	1.09	15	1.15	18	1.20	21	1.24
2.0	8	1.00	12	1.20	16	1.33	20	1.43	24	1.50	28	1.56
2.5	10	1.11	15	1.36	20	1.54	25	1.67	30	1.76	35	1.84
3.0	12	1.20	18	1.50	24	1.71	30	1.88	36	2.00	42	2.10
3.5	14	1.27	21	1.62	28	1.87	35	2.06	42	2.21	49	2.33
4.0	16	1.33	24	1.71	32	2.00	40	2.22	48	2.40	56	2.55
4.5	18	1.38	27	1.80	36	2.12	45	2.37	54	2.57	63	2.74
5.0	20	1.43	30	1.88	40	2.22	50	2.50	60	2.73	70	2.92
5.5	22	1.47	33	1.94	44	2.32	55	2.62	66	2.88	77	3.08
6.0	24	1.50	36	2.00	48	2.40	60	2.73	72	3.00	84	3.23
6.5	26	1.53	39	2.05	52	2.48	65	2.83	78	3.12	91	3.37
7.0	28	1.56	42	2.10	56	2.55	70	2.92	84	3.23	98	3.50
7.5	30	1.58	45	2.14	60	2.61	75	3.00	90	3.33	105	3.62
8.0	32	1.60	48	2.18	64	2.67	80	3.08	96	3.43	112	3.73
8.5	34	1.62	51	2.22	68	2.72	85	3.15	102	3.52	119	3.84
9.0	36	1.64	54	2.25	72	2.77	90	3.21	108	3.60	126	3.94
9.5	38	1.65	57	2.28	76	2.82	95	3.28	114	3.68	133	4.03
10.0	40	1.67	60	2.31	80	2.86	100	3.33	120	3.75	140	4.12

Table 26.—*Area in square feet, A, and hydraulic radius in feet, r, of rectangular channels*—Continued

Depth, feet	Bottom width 16 feet		Bottom width 18 feet		Bottom width 20 feet		Bottom width 25 feet		Bottom width 30 feet		Bottom width 40 feet	
	A	$r=\frac{area}{wet\ per.}$	A	$r=\frac{area}{wet\ per.}$	A	$r=\frac{area}{wet\ per.}$	A	$r=\frac{area}{wet\ per.}$	A	$r=\frac{area}{wet\ per.}$	A	$r=\frac{area}{wet\ per.}$
1.0	16	.89	18	.90	20	.91	25	.93	30	.94	40	.95
1.5	24	1.26	27	1.29	30	1.30	38	1.34	45	1.36	60	1.40
2.0	32	1.60	36	1.64	40	1.67	50	1.72	60	1.77	80	1.82
2.5	40	1.90	45	1.95	50	2.00	62	2.08	75	2.14	100	2.22
3.0	48	2.18	54	2.25	60	2.31	75	2.42	90	2.50	120	2.61
3.5	56	2.43	63	2.52	70	2.59	88	2.73	105	2.84	140	2.98
4.0	64	2.67	72	2.77	80	2.86	100	3.03	120	3.16	160	3.33
4.5	72	2.88	81	3.00	90	3.10	112	3.31	135	3.46	180	3.67
5.0	80	3.08	90	3.21	100	3.33	125	3.57	150	3.75	200	4.00
5.5	88	3.26	99	3.42	110	3.55	138	3.82	165	4.03	220	4.31
6.0	96	3.43	108	3.60	120	3.75	150	4.05	180	4.29	240	4.61
6.5	104	3.59	117	3.78	130	3.94	162	4.27	195	4.54	260	4.91
7.0	112	3.73	126	3.94	140	4.12	175	4.48	210	4.77	280	5.18
7.5	120	3.87	135	4.09	150	4.29	188	4.69	225	5.00	300	5.46
8.0	128	4.00	144	4.24	160	4.44	200	4.88	240	5.22	320	5.71
8.5	136	4.12	153	4.37	170	4.59	212	5.06	255	5.43	340	5.96
9.0	144	4.24	162	4.50	180	4.74	225	5.23	270	5.63	360	6.21
9.5	152	4.34	171	4.62	190	4.87	238	5.40	285	5.82	380	6.44
10.0	160	4.44	180	4.74	200	5.00	250	5.56	300	6.00	400	6.67

Depth, feet	Bottom width 50 feet		Bottom width 60 feet		Bottom width 70 feet		Bottom width 80 feet		Bottom width 90 feet		Bottom width 100 feet	
	A	$r=\frac{area}{wet\ per.}$	A	$r=\frac{area}{wet\ per.}$	A	$r=\frac{area}{wet\ per.}$	A	$r=\frac{area}{wet\ per.}$	A	$r=\frac{area}{wet\ per.}$	A	$r=\frac{area}{wet\ per.}$
1.0	50	.96	60	.97	70	.97	80	.98	90	.98	100	.98
1.5	75	1.42	90	1.43	105	1.44	120	1.45	135	1.45	150	1.46
2.0	100	1.85	120	1.88	140	1.89	160	1.91	180	1.92	200	1.92
2.5	125	2.27	150	2.31	175	2.33	200	2.35	225	2.37	250	2.38
3.0	150	2.68	180	2.73	210	2.76	240	2.79	270	2.81	300	2.83
3.5	175	3.07	210	3.13	245	3.18	280	3.22	315	3.25	350	3.27
4.0	200	3.45	240	3.53	280	3.59	320	3.64	360	3.67	400	3.70
4.5	225	3.81	270	3.91	315	3.99	360	4.04	405	4.09	450	4.13
5.0	250	4.17	300	4.29	350	4.38	400	4.44	450	4.50	500	4.55
5.5	275	4.51	330	4.65	385	4.75	440	4.83	495	4.90	550	4.95
6.0	300	4.84	360	5.00	420	5.12	480	5.22	540	5.29	600	5.36
6.5	325	5.16	390	5.34	455	5.48	520	5.59	585	5.68	650	5.75
7.0	350	5.47	420	5.68	490	5.83	560	5.96	630	6.06	700	6.14
7.5	375	5.77	450	6.00	525	6.18	600	6.32	675	6.43	750	6.52
8.0	400	6.06	480	6.32	560	6.51	640	6.67	720	6.79	800	6.90
8.5	425	6.34	510	6.62	595	6.84	680	7.01	765	7.15	850	7.26
9.0	450	6.62	540	6.92	630	7.16	720	7.35	810	7.50	900	7.63
9.5	475	6.88	570	7.22	665	7.47	760	7.68	855	7.84	950	7.98
10.0	500	7.14	600	7.50	700	7.78	800	8.00	900	8.18	1000	8.33

Table 27.—*Area in square feet, A, top width in feet, T, and hydraulic radius in feet, r, of trapezoidal channels,*
side slopes ½ to 1

Depth	Bottom width 2 feet			Bottom width 3 feet			Bottom width 4 feet			Bottom width 5 feet		
	T	A	r	T	A	r	T	A	r	T	A	r
0.4	2.4	0.88	.30	3.4	1.28	.33	4.4	1.68	.34	5.4	2.08	.35
0.6	2.6	1.38	.41	3.6	1.98	.46	4.6	2.58	.48	5.6	3.18	.50
0.8	2.8	1.92	.51	3.8	2.72	.57	4.8	3.52	.61	5.8	4.32	.64
1.0	3.0	2.50	.59	4.0	3.50	.67	5.0	4.50	.72	6.0	5.50	.76
1.2	3.2	3.12	.67	4.2	4.32	.76	5.2	5.52	.83	6.2	6.72	.87
1.4	3.4	3.78	.74	4.4	5.18	.85	5.4	6.58	.92	6.4	7.98	.98
1.6	3.6	4.48	.80	4.6	6.08	.92	5.6	7.68	1.01	6.6	9.28	1.08
1.8	3.8	5.22	.87	4.8	7.02	1.00	5.8	8.82	1.10	6.8	10.62	1.17
2.0	4.0	6.00	.93	5.0	8.00	1.07	6.0	10.00	1.18	7.0	12.00	1.27
2.2	4.2	6.82	.99	5.2	9.02	1.14	6.2	11.22	1.25	7.2	13.42	1.35
2.4	4.4	7.68	1.04	5.4	10.08	1.21	6.4	12.48	1.33	7.4	14.88	1.44
2.6	4.6	8.58	1.10	5.6	11.18	1.27	6.6	13.78	1.41	7.6	16.38	1.52
2.8	4.8	9.52	1.15	5.8	12.32	1.33	6.8	15.12	1.47	7.8	17.92	1.59
3.0	5.0	10.50	1.21	6.0	13.50	1.39	7.0	16.50	1.54	8.0	19.50	1.67
3.2	5.2	11.52	1.26	6.2	14.72	1.45	7.2	17.92	1.60	8.2	21.12	1.74
3.4	5.4	12.58	1.31	6.4	15.98	1.51	7.4	19.38	1.67	8.4	22.78	1.81
3.6	5.6	13.68	1.36	6.6	17.28	1.57	7.6	20.88	1.73	8.6	24.48	1.88
3.8	5.8	14.82	1.41	6.8	18.62	1.62	7.8	22.42	1.79	8.8	26.22	1.94
4.0	6.0	16.00	1.46	7.0	20.00	1.67	8.0	24.00	1.85	9.0	28.00	2.01
4.2	6.2	17.22	1.51	7.2	21.42	1.73	8.2	25.62	1.91	9.2	29.82	2.07
4.4	6.4	18.48	1.56	7.4	22.88	1.78	8.4	27.28	1.97	9.4	31.68	2.14
4.6	6.6	19.78	1.61	7.6	24.38	1.84	8.6	28.98	2.03	9.6	33.58	2.20
4.8	6.8	21.12	1.66	7.8	25.92	1.89	8.8	30.72	2.08	9.8	35.52	2.26
5.0	7.0	22.50	1.71	8.0	27.50	1.94	9.0	32.50	2.14	10.0	37.50	2.32
5.2	7.2	23.92	1.75	8.2	29.12	1.99	9.2	34.32	2.19	10.2	39.52	2.38
5.4	7.4	25.38	1.80	8.4	30.78	2.04	9.4	36.18	2.25	10.4	41.58	2.43
5.6	7.6	26.88	1.85	8.6	32.48	2.09	9.6	38.08	2.30	10.6	43.68	2.49
5.8	7.8	28.42	1.90	8.8	34.22	2.14	9.8	40.02	2.36	10.8	45.82	2.55
6.0	8.0	30.00	1.95	9.0	36.00	2.19	10.0	42.00	2.41	11.0	48.00	2.61
6.2	8.2	31.62	1.99	9.2	37.82	2.24	10.2	44.02	2.46	11.2	50.22	2.66
6.4	8.4	33.28	2.04	9.4	39.68	2.29	10.4	46.08	2.52	11.4	52.48	2.72
6.6	8.6	34.98	2.09	9.6	41.58	2.34	10.6	48.18	2.57	11.6	54.78	2.77
6.8	8.8	36.72	2.13	9.8	43.52	2.39	10.8	50.32	2.62	11.8	57.12	2.82
7.0	9.0	38.50	2.18	10.0	45.50	2.44	11.0	52.50	2.67	12.0	59.50	2.88
7.2	9.2	40.32	2.23	10.2	47.52	2.49	11.2	54.72	2.72	12.2	61.92	2.93
7.4	9.4	42.18	2.27	10.4	49.58	2.54	11.4	56.98	2.77	12.4	64.38	2.99
7.6	9.6	44.08	2.32	10.6	51.68	2.58	11.6	59.28	2.82	12.6	66.88	3.04
7.8	9.8	46.02	2.37	10.8	53.82	2.63	11.8	61.62	2.87	12.8	69.42	3.09
8.0	10.0	48.00	2.41	11.0	56.00	2.68	12.0	64.00	2.92	13.0	72.00	3.15
8.2	--------	--------	----	11.2	58.22	2.73	12.2	66.42	2.97	13.2	74.62	3.20
8.4	--------	--------	----	11.4	60.48	2.78	12.4	68.88	3.02	13.4	77.28	3.25
8.6	--------	--------	----	11.6	62.78	2.82	12.6	71.38	3.07	13.6	79.98	3.30
8.8	--------	--------	----	11.8	65.12	2.87	12.8	73.92	3.12	13.8	82.72	3.35
9.0	--------	--------	----	12.0	67.50	2.92	13.0	76.50	3.17	14.0	85.50	3.40
9.2	--------	--------	----	12.2	69.92	2.97	13.2	79.12	3.22	14.2	88.32	3.45
9.4	--------	--------	----	12.4	72.38	3.01	13.4	81.78	3.27	14.4	91.18	3.50
9.6	--------	--------	----	12.6	74.88	3.06	13.6	84.48	3.32	14.6	94.08	3.55
9.8	--------	--------	----	12.8	77.42	3.11	13.8	87.22	3.37	14.8	97.02	3.60

Table 27.—*Area in square feet, A, top width in feet, T, and hydraulic radius in feet, r, of trapezoidal channels,* **side slopes ½ to 1**—Continued

Depth	Bottom width 2 feet			Bottom width 3 feet			Bottom width 4 feet			Bottom width 5 feet		
	T	A	r	T	A	r	T	A	r	T	A	r
10.0				13.0	80.00	3.15	14.0	90.00	3.41	15.0	100.00	3.65
10.5				13.5	86.62	3.27	14.5	97.12	3.53	15.5	107.62	3.78
11.0				14.0	93.5	3.39	15.0	104.5	3.65	16.0	115.5	3.90
11.5				14.5	100.6	3.50	15.5	112.1	3.77	16.5	123.6	4.02
12.0				15.0	108.0	3.62	16.0	120.0	3.89	17.0	132.0	4.15
12.5				15.5	115.6	3.74	16.5	128.1	4.01	17.5	140.6	4.27
13.0							17.0	136.5	4.13	18.0	149.5	4.39
13.5							17.5	145.1	4.24	18.5	158.6	4.51
14.0							18.0	154.0	4.36	19.0	168.0	4.63
14.5							18.5	163.1	4.48	19.5	177.6	4.75
15.0							19.0	172.5	4.59	20.0	187.5	4.86
15.5							19.5	182.1	4.71	20.5	197.6	4.98
16.0							20.0	192.0	4.83	21.0	208.0	5.10
16.5										21.5	218.6	5.22
17.0										22.0	229.5	5.34
17.5										22.5	240.6	5.45
18.0										23.0	252.0	5.57
18.5										23.5	263.6	5.69
19.0										24.0	275.5	5.80
19.5										24.5	287.6	5.92
20.0										25.0	300.0	6.03

Depth	Bottom width 6 feet			Bottom width 7 feet			Bottom width 8 feet			Bottom width 9 feet		
	T	A	r	T	A	r	T	A	r	T	A	r
0.4	6.4	2.48	.36	7.4	2.88	.36	8.4	3.28	.37	9.4	3.68	.37
0.6	6.6	3.78	.52	7.6	4.38	.52	8.6	4.98	.53	9.6	5.58	.54
0.8	6.8	5.12	.66	7.8	5.92	.67	8.8	6.72	.69	9.8	7.52	.70
1.0	7.0	6.50	.79	8.0	7.50	.81	9.0	8.50	.83	10.0	9.50	.85
1.2	7.2	7.92	.91	8.2	9.12	.94	9.2	10.32	.96	10.2	11.52	.98
1.4	7.4	9.38	1.03	8.4	10.78	1.07	9.4	12.18	1.10	10.4	13.58	1.12
1.6	7.6	10.88	1.14	8.6	12.48	1.18	9.6	14.08	1.22	10.6	15.68	1.25
1.8	7.8	12.42	1.24	8.8	14.22	1.29	9.8	16.02	1.33	10.8	17.82	1.37
2.0	8.0	14.00	1.34	9.0	16.00	1.39	10.0	18.00	1.44	11.0	20.00	1.48
2.2	8.2	15.62	1.43	9.2	17.82	1.49	10.2	20.02	1.55	11.2	22.22	1.59
2.4	8.4	17.28	1.52	9.4	19.68	1.59	10.4	22.08	1.65	11.4	24.48	1.71
2.6	8.6	18.98	1.61	9.6	21.58	1.69	10.6	24.18	1.75	11.6	26.78	1.81
2.8	8.8	20.72	1.69	9.8	23.52	1.77	10.8	26.32	1.84	11.8	29.12	1.91
3.0	9.0	22.50	1.78	10.0	25.50	1.86	11.0	28.50	1.94	12.0	31.50	2.01
3.2	9.2	24.32	1.85	10.2	27.52	1.94	11.2	30.72	2.03	12.2	33.92	2.10
3.4	9.4	26.18	1.93	10.4	29.58	2.03	11.4	32.98	2.12	12.4	36.38	2.19
3.6	9.6	28.08	2.00	10.6	31.68	2.11	11.6	35.28	2.20	12.6	38.88	2.28
3.8	9.8	30.02	2.07	10.8	33.82	2.18	11.8	37.62	2.28	12.8	41.42	2.37
4.0	10.0	32.00	2.14	11.0	36.00	2.26	12.0	40.00	2.36	13.0	44.00	2.45
4.2	10.2	34.02	2.21	11.2	38.22	2.33	12.2	42.42	2.44	13.2	46.62	2.53
4.4	10.4	36.08	2.28	11.4	40.48	2.41	12.4	44.88	2.52	13.4	49.28	2.62
4.6	10.6	38.18	2.35	11.6	42.78	2.48	12.6	47.38	2.59	13.6	51.98	2.70
4.8	10.8	40.32	2.41	11.8	45.12	2.54	12.8	49.92	2.66	13.8	54.72	2.77

Table 27.—*Area in square feet, A, top width in feet, T, and hydraulic radius in feet, r, of trapezoidal channels,* **side slopes ½ to 1**—Continued

Depth	Bottom width 6 feet			Bottom width 7 feet			Bottom width 8 feet			Bottom width 9 feet		
	T	A	r	T	A	r	T	A	r	T	A	r
5.0	11.0	42.50	2.47	12.0	47.50	2.61	13.0	52.50	2.74	14.0	57.50	2.85
5.2	11.2	44.72	2.54	12.2	49.92	2.68	13.2	55.12	2.81	14.2	60.32	2.92
5.4	11.4	46.98	2.60	12.4	52.38	2.74	13.4	57.78	2.88	14.4	63.18	3.00
5.6	11.6	49.28	2.66	12.6	54.88	2.81	13.6	60.48	2.95	14.6	66.08	3.07
5.8	11.8	51.62	2.72	12.8	57.42	2.87	13.8	63.22	3.01	14.8	69.02	3.14
6.0	12.0	54.00	2.78	13.0	60.00	2.94	14.0	66.00	3.08	15.0	72.00	3.21
6.2	12.2	56.42	2.84	13.2	62.62	3.00	14.2	68.82	3.15	15.2	75.02	3.28
6.4	12.4	58.88	2.90	13.4	65.28	3.06	14.4	71.68	3.21	15.4	78.08	3.35
6.6	12.6	61.38	2.96	13.6	67.98	3.12	14.6	74.58	3.28	15.6	81.18	3.42
6.8	12.8	63.92	3.01	13.8	70.72	3.18	14.8	77.52	3.34	15.8	84.32	3.48
7.0	13.0	66.50	3.07	14.0	73.50	3.24	15.0	80.50	3.40	16.0	87.50	3.55
7.2	13.2	69.12	3.13	14.2	76.32	3.30	15.2	83.52	3.47	16.2	90.72	3.61
7.4	13.4	71.78	3.18	14.4	79.18	3.36	15.4	86.58	3.53	16.4	93.98	3.68
7.6	13.6	74.48	3.24	14.6	82.08	3.42	15.6	89.68	3.59	16.6	97.28	3.74
7.8	13.8	77.22	3.29	14.8	85.02	3.48	15.8	92.82	3.65	16.8	100.62	3.81
8.0	14.0	80.00	3.35	15.0	88.00	3.54	16.0	96.00	3.71	17.0	104.00	3.87
8.2	14.2	82.82	3.40	15.2	91.02	3.59	16.2	99.22	3.77	17.2	107.42	3.93
8.4	14.4	85.68	3.46	15.4	94.08	3.65	16.4	102.48	3.83	17.4	110.88	3.99
8.6	14.6	88.58	3.51	15.6	97.18	3.70	16.6	105.78	3.88	17.6	114.38	4.05
8.8	14.8	91.52	3.56	15.8	100.32	3.76	16.8	109.12	3.94	17.8	117.92	4.11
9.0	15.0	94.50	3.62	16.0	103.50	3.82	17.0	112.50	4.00	18.0	121.50	4.17
9.2	15.2	97.52	3.67	16.2	106.72	3.87	17.2	115.92	4.06	18.2	125.12	4.23
9.4	15.4	100.58	3.72	16.4	109.98	3.93	17.4	119.38	4.11	18.4	128.78	4.29
9.6	15.6	103.68	3.77	16.6	113.28	3.98	17.6	122.88	4.17	18.6	132.48	4.35
9.8	15.8	106.82	3.83	16.8	116.62	4.03	17.8	126.42	4.23	18.8	136.22	4.41
10.0	16.0	110.00	3.88	17.0	120.00	4.09	18.0	130.00	4.28	19.0	140.00	4.46
10.5	16.5	118.12	4.01	17.5	128.62	4.22	18.5	139.12	4.42	19.5	149.62	4.61
11.0	17.0	126.5	4.13	18.0	137.5	4.35	19.0	148.5	4.56	20.0	159.5	4.75
11.5	17.5	135.1	4.26	18.5	146.6	4.48	19.5	158.1	4.69	20.5	169.6	4.89
12.0	18.0	144.0	4.39	19.0	156.0	4.61	20.0	168.0	4.82	21.0	180.0	5.02
12.5	18.5	153.1	4.51	19.5	165.6	4.74	20.5	178.1	4.95	21.5	190.6	5.16
13.0	19.0	162.5	4.63	20.0	175.5	4.87	21.0	188.5	5.09	22.0	201.5	5.29
13.5	19.5	172.1	4.76	20.5	185.6	4.99	21.5	199.1	5.21	22.5	212.6	5.43
14.0	20.0	182.0	4.88	21.0	196.0	5.12	22.0	210.0	5.34	23.0	224.0	5.56
14.5	20.5	192.1	5.00	21.5	206.6	5.24	22.5	221.1	5.47	23.5	235.6	5.69
15.0	21.0	202.5	5.12	22.0	217.5	5.36	23.0	232.5	5.60	24.0	247.5	5.82
15.5	21.5	213.1	5.24	22.5	228.6	5.49	23.5	244.1	5.72	24.5	259.6	5.95
16.0	22.0	224.0	5.36	23.0	240.0	5.61	24.0	256.0	5.85	25.0	272.0	6.07
16.5	22.5	235.1	5.48	23.5	251.6	5.73	24.5	268.1	5.97	25.5	284.6	6.20
17.0	23.0	246.5	5.60	24.0	263.5	5.85	25.0	280.5	6.10	26.0	297.5	6.33
17.5	23.5	258.1	5.72	24.5	275.6	5.97	25.5	293.1	6.22	26.5	310.6	6.45
18.0	24.0	270.0	5.84	25.0	288.0	6.10	26.0	306.0	6.34	27.0	324.0	6.58
18.5	24.5	282.1	5.96	25.5	300.6	6.22	26.5	319.1	6.46	27.5	337.6	6.70
19.0	25.0	294.5	6.07	26.0	313.5	6.34	27.0	332.5	6.59	28.0	351.5	6.83
19.5	25.5	307.1	6.19	26.5	326.6	6.45	27.5	346.1	6.71	28.5	365.6	6.95
20.0	26.0	320.0	6.31	27.0	340.0	6.57	28.0	360.0	6.83	29.0	380.0	7.07

Table 27.—*Area in square feet, A, top width in feet, T, and hydraulic radius in feet, r, of trapezoidal channels,* **side slopes ½ to 1**—Continued

Depth	Bottom width 10 feet			Bottom width 12 feet			Bottom width 14 feet			Bottom width 16 feet		
	T	A	r	T	A	r	T	A	r	T	A	r
0.4	10.4	4.08	0.37	12.4	4.88	0.38	14.4	5.68	0.38	16.4	6.48	0.38
0.6	10.6	6.18	.54	12.6	7.38	.55	14.6	8.58	.56	16.6	9.78	.56
0.8	10.8	8.32	.71	12.8	9.92	.72	14.8	11.52	.73	16.8	13.12	.74
1.0	11.0	10.50	.86	13.0	12.50	.88	15.0	14.50	.89	17.0	16.50	.90
1.2	11.2	12.72	1.00	13.2	15.12	1.03	15.2	17.52	1.05	17.2	19.92	1.07
1.4	11.4	14.98	1.14	13.4	17.78	1.18	15.4	20.58	1.20	17.4	23.38	1.22
1.6	11.6	17.28	1.27	13.6	20.48	1.31	15.6	23.68	1.35	17.6	26.88	1.37
1.8	11.8	19.62	1.40	13.8	23.22	1.45	15.8	26.82	1.49	17.8	30.42	1.52
2.0	12.0	22.00	1.52	14.0	26.00	1.58	16.0	30.00	1.62	18.0	34.00	1.66
2.2	12.2	24.42	1.64	14.2	28.82	1.70	16.2	33.22	1.76	18.2	37.62	1.80
2.4	12.4	26.88	1.75	14.4	31.68	1.82	16.4	36.48	1.88	18.4	41.28	1.93
2.6	12.6	29.38	1.86	14.6	34.58	1.94	16.6	39.78	2.01	18.6	44.98	2.06
2.8	12.8	31.92	1.96	14.8	37.52	2.05	16.8	43.12	2.13	18.8	48.72	2.19
3.0	13.0	34.50	2.06	15.0	40.50	2.16	17.0	46.50	2.25	19.0	52.50	2.31
3.2	13.2	37.12	2.16	15.2	43.52	2.27	17.2	49.92	2.36	19.2	56.32	2.43
3.4	13.4	39.78	2.26	15.4	46.58	2.38	17.4	53.38	2.47	19.4	60.18	2.55
3.6	13.6	42.48	2.35	15.6	49.68	2.48	17.6	56.88	2.58	19.6	64.08	2.66
3.8	13.8	45.22	2.44	15.8	52.82	2.58	17.8	60.42	2.69	19.8	68.02	2.78
4.0	14.0	48.00	2.53	16.0	56.00	2.67	18.0	64.00	2.79	20.0	72.00	2.89
4.2	14.2	50.82	2.62	16.2	59.22	2.77	18.2	67.62	2.89	20.2	76.02	2.99
4.4	14.4	53.68	2.71	16.4	62.48	2.86	18.4	71.28	2.99	20.4	80.08	3.10
4.6	14.6	56.58	2.79	16.6	65.78	2.95	18.6	74.98	3.09	20.6	84.18	3.20
4.8	14.8	59.52	2.87	16.8	69.12	3.04	18.8	78.72	3.18	20.8	88.32	3.30
5.0	15.0	62.50	2.95	17.0	72.50	3.13	19.0	82.50	3.28	21.0	92.50	3.40
5.2	15.2	65.52	3.03	17.2	75.92	3.21	19.2	86.32	3.37	21.2	96.72	3.50
5.4	15.4	68.58	3.11	17.4	79.38	3.30	19.4	90.18	3.46	21.4	100.98	3.60
5.6	15.6	71.68	3.18	17.6	82.88	3.38	19.6	94.08	3.55	21.6	105.28	3.69
5.8	15.8	74.82	3.26	17.8	86.42	3.46	19.8	98.02	3.63	21.8	109.62	3.78
6.0	16.0	78.00	3.33	18.0	90.00	3.54	20.0	102.00	3.72	22.0	114.00	3.88
6.2	16.2	81.22	3.40	18.2	93.62	3.62	20.2	106.02	3.80	22.2	118.42	3.97
6.4	16.4	84.48	3.47	18.4	97.28	3.70	20.4	110.08	3.89	22.4	122.88	4.05
6.6	16.6	87.78	3.55	18.6	100.98	3.77	20.6	114.18	3.97	22.6	127.38	4.14
6.8	16.8	91.12	3.62	18.8	104.72	3.85	20.8	118.32	4.05	22.8	131.92	4.23
7.0	17.0	94.50	3.68	19.0	108.50	3.92	21.0	122.50	4.13	23.0	136.50	4.31
7.2	17.2	97.92	3.75	19.2	112.32	4.00	21.2	126.72	4.21	23.2	141.12	4.40
7.4	17.4	101.38	3.82	19.4	116.18	4.07	21.4	130.98	4.29	23.4	145.78	4.48
7.6	17.6	104.88	3.89	19.6	120.08	4.14	21.6	135.28	4.36	23.6	150.48	4.56
7.8	17.8	108.42	3.95	19.8	124.02	4.21	21.8	139.62	4.44	23.8	155.22	4.64
8.0	18.0	112.00	4.02	20.0	128.00	4.28	22.0	144.00	4.52	24.0	160.00	4.72
8.2	18.2	115.62	4.08	20.2	132.02	4.35	22.2	148.42	4.59	24.2	164.82	4.80
8.4	18.4	119.28	4.14	20.4	136.08	4.42	22.4	152.88	4.66	24.4	169.68	4.88
8.6	18.6	122.98	4.21	20.6	140.18	4.49	22.6	157.38	4.74	24.6	174.58	4.96
8.8	18.8	126.72	4.27	20.8	144.32	4.56	22.8	161.92	4.81	24.8	179.52	5.03
9.0	19.0	130.50	4.33	21.0	148.50	4.62	23.0	166.50	4.88	25.0	184.50	5.11
9.2	19.2	134.32	4.39	21.2	152.72	4.69	23.2	171.12	4.95	25.2	189.52	5.18
9.4	19.4	138.18	4.45	21.4	156.98	4.75	23.4	175.78	5.02	25.4	194.58	5.26
9.6	19.6	142.08	4.52	21.6	161.28	4.82	23.6	180.48	5.09	25.6	199.68	5.33
9.8	19.8	146.02	4.58	21.8	165.62	4.88	23.8	185.22	5.16	25.8	204.82	5.40

Table 27.—*Area in square feet, A, top width in feet, T, and hydraulic radius in feet, r, of trapezoidal channels,*
side slopes ½ to 1—Continued

Depth	Bottom width 10 feet			Bottom width 12 feet			Bottom width 14 feet			Bottom width 16 feet		
	T	A	r	T	A	r	T	A	r	T	A	r
10.0	20.0	150.00	4.64	22.0	170.00	4.95	24.0	190.00	5.23	26.0	210.00	5.47
10.5	20.5	160.12	4.78	22.5	181.12	5.11	24.5	202.12	5.39	26.5	223.12	5.65
11.0	21.0	170.5	4.93	23.0	192.5	5.26	25.0	214.5	5.56	27.0	236.5	5.83
11.5	21.5	181.1	5.07	23.5	204.1	5.41	25.5	227.1	5.72	27.5	250.1	6.00
12.0	22.0	192.0	5.21	24.0	216.0	5.56	26.0	240.0	5.88	28.0	264.0	6.16
12.5	22.5	203.1	5.35	24.5	228.1	5.71	26.5	253.1	6.03	28.5	278.1	6.33
13.0	23.0	214.5	5.49	25.0	240.5	5.86	27.0	266.5	6.19	29.0	292.5	6.49
13.5	23.5	226.1	5.63	25.5	253.1	6.00	27.5	280.1	6.34	29.5	307.1	6.65
14.0	24.0	238.0	5.76	26.0	266.0	6.14	28.0	294.0	6.49	30.0	322.0	6.81
14.5	24.5	250.1	5.90	26.5	279.1	6.28	28.5	308.1	6.64	30.5	337.1	6.96
15.0	25.0	262.5	6.03	27.0	292.5	6.42	29.0	322.5	6.78	31.0	352.5	7.12
15.5	25.5	275.1	6.16	27.5	306.1	6.56	29.5	337.1	6.93	31.5	368.1	7.27
16.0	26.0	288.0	6.29	28.0	320.0	6.70	30.0	352.0	7.07	32.0	384.0	7.42
16.5	26.5	301.1	6.42	28.5	334.1	6.83	30.5	367.1	7.21	32.5	400.1	7.56
17.0	27.0	314.5	6.55	29.0	348.5	6.97	31.0	382.5	7.35	33.0	416.5	7.71
17.5	27.5	328.1	6.68	29.5	363.1	7.10	31.5	398.1	7.49	33.5	433.1	7.86
18.0	28.0	342.0	6.81	30.0	378.0	7.23	32.0	414.0	7.63	34.0	450.0	8.00
18.5	28.5	356.1	6.93	30.5	393.1	7.37	32.5	430.1	7.77	34.5	467.1	8.14
19.0	29.0	370.5	7.06	31.0	408.5	7.50	33.0	446.5	7.90	35.0	484.5	8.28
19.5	29.5	385.1	7.18	31.5	424.1	7.63	33.5	463.1	8.04	35.5	502.1	8.42
20.0	30.0	400.0	7.31	32.0	440.0	7.76	34.0	480.0	8.17	36.0	520.0	8.56

Depth	Bottom width 18 feet			Bottom width 20 feet			Bottom width 22 feet			Bottom width 24 feet		
	T	A	r	T	A	r	T	A	r	T	A	r
0.4	18.4	7.28	.39	20.4	8.08	.39	22.4	8.88	.39	24.4	9.68	.39
0.6	18.6	10.98	.57	20.6	12.18	.57	22.6	13.38	.57	24.6	14.58	.58
0.8	18.8	14.72	.74	20.8	16.32	.75	22.8	17.92	.75	24.8	19.52	.76
1.0	19.0	18.50	.91	21.0	20.50	.92	23.0	22.50	.93	25.0	24.50	.93
1.2	19.2	22.32	1.08	21.2	24.72	1.09	23.2	27.12	1.10	25.2	29.52	1.11
1.4	19.4	26.18	1.24	21.4	28.98	1.25	23.4	31.78	1.26	25.4	34.58	1.27
1.6	19.6	30.08	1.39	21.6	33.28	1.41	23.6	36.48	1.43	25.6	39.68	1.44
1.8	19.8	34.02	1.54	21.8	37.62	1.57	23.8	41.22	1.58	25.8	44.82	1.60
2.0	20.0	38.00	1.69	22.0	42.00	1.72	24.0	46.00	1.74	26.0	50.00	1.76
2.2	20.2	42.02	1.83	22.2	46.42	1.86	24.2	50.82	1.89	26.2	55.22	1.91
2.4	20.4	46.08	1.97	22.4	50.88	2.01	24.4	55.68	2.03	26.4	60.48	2.06
2.6	20.6	50.18	2.11	22.6	55.38	2.15	24.6	60.58	2.18	26.6	65.78	2.21
2.8	20.8	54.32	2.24	22.8	59.92	2.28	24.8	65.52	2.32	26.8	71.12	2.35
3.0	21.0	58.50	2.37	23.0	64.50	2.41	25.0	70.50	2.46	27.0	76.50	2.49
3.2	21.2	62.72	2.49	23.2	69.12	2.55	25.2	75.52	2.59	27.2	81.92	2.63
3.4	21.4	66.98	2.62	23.4	73.78	2.67	25.4	80.58	2.72	27.4	87.38	2.76
3.6	21.6	71.28	2.74	23.6	78.48	2.80	25.6	85.68	2.85	27.6	92.88	2.90
3.8	21.8	75.62	2.85	23.8	83.22	2.92	25.8	90.82	2.98	27.8	98.42	3.03
4.0	22.0	80.00	2.97	24.0	88.00	3.04	26.0	96.00	3.10	28.0	104.00	3.16
4.2	22.2	84.42	3.08	24.2	92.82	3.16	26.2	101.22	3.22	28.2	109.62	3.28
4.4	22.4	88.88	3.19	24.4	97.68	3.27	26.4	106.48	3.34	28.4	115.28	3.41
4.6	22.6	93.38	3.30	24.6	102.58	3.39	26.6	111.78	3.46	28.6	120.98	3.53
4.8	22.8	97.92	3.41	24.8	107.52	3.50	26.8	117.12	3.58	28.8	126.72	3.65

Table 27.—*Area in square feet, A, top width in feet, T, and hydraulic radius in feet, r, of trapezoidal channels,* **side slopes ½ to 1**—Continued

Depth	Bottom width 18 feet			Bottom width 20 feet			Bottom width 22 feet			Bottom width 24 feet		
	T	A	r	T	A	r	T	A	r	T	A	r
5.0	23.0	102.50	3.51	25.0	112.50	3.61	27.0	122.50	3.69	29.0	132.50	3.77
5.2	23.2	107.12	3.62	25.2	117.52	3.72	27.2	127.92	3.80	29.2	138.32	3.88
5.4	23.4	111.78	3.72	25.4	122.58	3.82	27.4	133.38	3.91	29.4	144.18	4.00
5.6	23.6	116.48	3.82	25.6	127.68	3.93	27.6	138.88	4.02	29.6	150.08	4.11
5.8	23.8	121.22	3.91	25.8	132.82	4.03	27.8	144.42	4.13	29.8	156.02	4.22
6.0	24.0	126.00	4.01	26.0	138.00	4.13	28.0	150.00	4.24	30.0	162.00	4.33
6.2	24.2	130.82	4.11	26.2	143.22	4.23	28.2	155.62	4.34	30.2	168.02	4.44
6.4	24.4	135.68	4.20	26.4	148.48	4.33	28.4	161.28	4.44	30.4	174.08	4.54
6.6	24.6	140.58	4.29	26.6	153.78	4.42	28.6	166.98	4.54	30.6	180.18	4.65
6.8	24.8	145.52	4.38	26.8	159.12	4.52	28.8	172.72	4.64	30.8	186.32	4.75
7.0	25.0	150.50	4.47	27.0	164.50	4.61	29.0	178.50	4.74	31.0	192.50	4.85
7.2	25.2	155.52	4.56	27.2	169.92	4.71	29.2	184.32	4.84	31.2	198.72	4.96
7.4	25.4	160.58	4.65	27.4	175.38	4.80	29.4	190.18	4.93	31.4	204.98	5.06
7.6	25.6	165.68	4.73	27.6	180.88	4.89	29.6	196.08	5.03	31.6	211.28	5.15
7.8	25.8	170.82	4.82	27.8	186.42	4.98	29.8	202.02	5.12	31.8	217.62	5.25
8.0	26.0	176.00	4.90	28.0	192.00	5.07	30.0	208.00	5.21	32.0	224.00	5.35
8.2	26.2	181.22	4.99	28.2	197.62	5.15	30.2	214.02	5.31	32.2	230.42	5.44
8.4	26.4	186.48	5.07	28.4	203.28	5.24	30.4	220.08	5.40	32.4	236.88	5.54
8.6	26.6	191.78	5.15	28.6	208.98	5.33	30.6	226.18	5.49	32.6	243.38	5.63
8.8	26.8	197.12	5.23	28.8	214.72	5.41	30.8	232.32	5.57	32.8	249.92	5.72
9.0	27.0	202.50	5.31	29.0	220.50	5.50	31.0	238.50	5.66	33.0	256.50	5.81
9.2	27.2	207.92	5.39	29.2	226.32	5.58	31.2	244.72	5.75	33.2	263.12	5.90
9.4	27.4	213.38	5.47	29.4	232.18	5.66	31.4	250.98	5.83	33.4	269.78	5.99
9.6	27.6	218.88	5.55	29.6	238.08	5.74	31.6	257.28	5.92	33.6	276.48	6.08
9.8	27.8	224.42	5.62	29.8	244.02	5.82	31.8	263.62	6.00	33.8	283.22	6.17
10.0	28.0	230.00	5.70	30.0	250.00	5.90	32.0	270.00	6.09	34.0	290.00	6.26
10.5	28.5	244.12	5.89	30.5	265.12	6.10	32.5	286.12	6.29	34.5	307.12	6.47
11.0	29.0	258.5	6.07	31.0	280.5	6.29	33.0	302.5	6.49	35.0	324.5	6.68
11.5	29.5	273.1	6.25	31.5	296.1	6.48	33.5	319.1	6.69	35.5	342.1	6.88
12.0	30.0	288.0	6.42	32.0	312.0	6.66	34.0	336.0	6.88	36.0	360.0	7.08
12.5	30.5	303.1	6.60	32.5	328.1	6.84	34.5	353.1	7.07	36.5	378.1	7.28
13.0	31.0	318.5	6.77	33.0	344.5	7.02	35.0	370.5	7.25	37.0	396.5	7.47
13.5	31.5	334.1	6.93	33.5	361.1	7.20	35.5	388.1	7.44	37.5	415.1	7.66
14.0	32.0	350.0	7.10	34.0	378.0	7.37	36.0	406.0	7.62	38.0	434.0	7.85
14.5	32.5	366.1	7.26	34.5	395.1	7.54	36.5	424.1	7.79	38.5	453.1	8.03
15.0	33.0	382.5	7.42	35.0	412.5	7.70	37.0	442.5	7.97	39.0	472.5	8.21
15.5	33.5	399.1	7.58	35.5	430.1	7.87	37.5	461.1	8.14	39.5	492.1	8.39
16.0	34.0	416.0	7.74	36.0	448.0	8.03	38.0	480.0	8.31	40.0	512.0	8.57
16.5	34.5	433.1	7.89	36.5	466.1	8.19	38.5	499.1	8.47	40.5	532.1	8.74
17.0	35.0	450.5	8.04	37.0	484.5	8.35	39.0	518.5	8.64	41.0	552.5	8.91
17.5	35.5	468.1	8.19	37.5	503.1	8.51	39.5	538.1	8.80	41.5	573.1	9.08
18.0	36.0	486.0	8.34	38.0	522.0	8.66	40.0	558.0	8.96	42.0	594.0	9.25
18.5	36.5	504.1	8.49	38.5	541.1	8.82	40.5	578.1	9.12	42.5	615.1	9.41
19.0	37.0	522.5	8.64	39.0	560.5	8.97	41.0	598.5	9.28	43.0	636.5	9.57
19.5	37.5	541.1	8.78	39.5	580.1	9.12	41.5	619.1	9.44	43.5	658.1	9.73
20.0	38.0	560.0	8.93	40.0	600.0	9.27	42.0	640.0	9.59	44.0	680.0	9.90

Table 27.—*Area in square feet, A, top width in feet, T, and hydraulic radius in feet, r, of trapezoidal channels,* side slopes ½ to 1—Continued

Depth	Bottom width 26 feet			Bottom width 28 feet			Bottom width 30 feet			Bottom width 32 feet		
	T	A	r	T	A	r	T	A	r	T	A	r
0.4	26.4	10.48	.39	28.4	11.28	.39	30.4	12.08	.39	32.4	12.88	.39
0.6	26.6	15.78	.58	28.6	16.98	.58	30.6	18.18	.58	32.6	19.38	.58
0.8	26.8	21.12	.76	28.8	22.72	.76	30.8	24.32	.77	32.8	25.92	.77
1.0	27.0	26.50	.94	29.0	28.50	.94	31.0	30.50	.95	33.0	32.50	.95
1.2	27.2	31.92	1.11	29.2	34.32	1.12	31.2	36.72	1.12	33.2	39.12	1.13
1.4	27.4	37.38	1.28	29.4	40.18	1.29	31.4	42.98	1.30	33.4	45.78	1.30
1.6	27.6	42.88	1.45	29.6	46.08	1.46	31.6	49.28	1.47	33.6	52.48	1.48
1.8	27.8	48.42	1.61	29.8	52.02	1.62	31.8	55.62	1.63	33.8	59.22	1.64
2.0	28.0	54.00	1.77	30.0	58.00	1.79	32.0	62.00	1.80	34.0	66.00	1.81
2.2	28.2	59.62	1.93	30.2	64.02	1.94	32.2	68.42	1.96	34.2	72.82	1.97
2.4	28.4	65.28	2.08	30.4	70.08	2.10	32.4	74.88	2.12	34.4	79.68	2.13
2.6	28.6	70.98	2.23	30.6	76.18	2.25	32.6	81.38	2.27	34.6	86.58	2.29
2.8	28.8	76.72	2.38	30.8	82.32	2.40	32.8	87.92	2.42	34.8	93.52	2.44
3.0	29.0	82.50	2.52	31.0	88.50	2.55	33.0	94.50	2.57	35.0	100.50	2.60
3.2	29.2	88.32	2.66	31.2	94.72	2.69	33.2	101.12	2.72	35.2	107.52	2.75
3.4	29.4	94.18	2.80	31.4	100.98	2.84	33.4	107.78	2.87	35.4	114.58	2.89
3.6	29.6	100.08	2.94	31.6	107.28	2.98	33.6	114.48	3.01	35.6	121.68	3.04
3.8	29.8	106.02	3.07	31.8	113.62	3.11	33.8	121.22	3.15	35.8	128.82	3.18
4.0	30.0	112.00	3.21	32.0	120.00	3.25	34.0	128.00	3.29	36.0	136.00	3.32
4.2	30.2	118.02	3.33	32.2	126.42	3.38	34.2	134.82	3.42	36.2	143.22	3.46
4.4	30.4	124.08	3.46	32.4	132.88	3.51	34.4	141.68	3.56	36.4	150.48	3.60
4.6	30.6	130.18	3.59	32.6	139.38	3.64	34.6	148.58	3.69	36.6	157.78	3.73
4.8	30.8	136.32	3.71	32.8	145.92	3.77	34.8	155.52	3.82	36.8	165.12	3.86
5.0	31.0	142.50	3.83	33.0	152.50	3.89	35.0	162.50	3.95	37.0	172.50	3.99
5.2	31.2	148.72	3.95	33.2	159.12	4.02	35.2	169.52	4.07	37.2	179.92	4.12
5.4	31.4	154.98	4.07	33.4	165.78	4.14	35.4	176.58	4.20	37.4	187.38	4.25
5.6	31.6	161.28	4.19	33.6	172.48	4.26	35.6	183.68	4.32	37.6	194.88	4.38
5.8	31.8	167.62	4.30	33.8	179.22	4.37	35.8	190.82	4.44	37.8	202.42	4.50
6.0	32.0	174.00	4.41	34.0	186.00	4.49	36.0	198.00	4.56	38.0	210.00	4.62
6.2	32.2	180.42	4.53	34.2	192.82	4.61	36.2	205.22	4.68	38.2	217.62	4.74
6.4	32.4	186.88	4.64	34.4	199.68	4.72	36.4	212.48	4.80	38.4	225.28	4.86
6.6	32.6	193.38	4.74	34.6	206.58	4.83	36.6	219.78	4.91	38.6	232.98	4.98
6.8	32.8	199.92	4.85	34.8	213.52	4.94	36.8	227.12	5.02	38.8	240.72	5.10
7.0	33.0	206.50	4.96	35.0	220.50	5.05	37.0	234.50	5.14	39.0	248.50	5.21
7.2	33.2	213.12	5.06	35.2	227.52	5.16	37.2	241.92	5.25	39.2	256.32	5.33
7.4	33.4	219.78	5.17	35.4	234.58	5.27	37.4	249.38	5.36	39.4	264.18	5.44
7.6	33.6	226.48	5.27	35.6	241.68	5.37	37.6	256.88	5.47	39.6	272.08	5.55
7.8	33.8	233.22	5.37	35.8	248.82	5.48	37.8	264.42	5.57	39.8	280.02	5.66
8.0	34.0	240.00	5.47	36.0	256.00	5.58	38.0	272.00	5.68	40.0	288.00	5.77
8.2	34.2	246.82	5.57	36.2	263.22	5.68	38.2	279.62	5.78	40.2	296.02	5.88
8.4	34.4	253.68	5.66	36.4	270.48	5.78	38.4	287.28	5.89	40.4	304.08	5.99
8.6	34.6	260.58	5.76	36.6	277.78	5.88	38.6	294.98	5.99	40.6	312.18	6.09
8.8	34.8	267.52	5.86	36.8	285.12	5.98	38.8	302.72	6.09	40.8	320.32	6.20
9.0	35.0	274.50	5.95	37.0	292.50	6.08	39.0	310.50	6.19	41.0	328.50	6.30
9.2	35.2	281.52	6.04	37.2	299.92	6.17	39.2	318.32	6.29	41.2	336.72	6.40
9.4	35.4	288.58	6.14	37.4	307.38	6.27	39.4	326.18	6.39	41.4	344.98	6.51
9.6	35.6	295.68	6.23	37.6	314.88	6.37	39.6	334.08	6.49	41.6	353.28	6.61
9.8	35.8	302.82	6.32	37.8	322.42	6.46	39.8	342.02	6.59	41.8	361.62	6.71

Table 27.—*Area in square feet, A, top width in feet, T, and hydraulic radius in feet, r, of trapezoidal channels,* **side slopes ½ to 1**—Continued

Depth	Bottom width 26 feet			Bottom width 28 feet			Bottom width 30 feet			Bottom width 32 feet		
	T	A	r	T	A	r	T	A	r	T	A	r
10.0	36.0	310.00	6.41	38.0	330.00	6.55	40.0	350.00	6.68	42.0	370.00	6.81
10.5	36.5	328.12	6.63	38.5	349.12	6.78	40.5	370.12	6.92	42.5	391.12	7.05
11.0	37.0	346.5	6.85	39.0	368.5	7.01	41.0	390.5	7.15	43.0	412.5	7.29
11.5	37.5	365.1	7.06	39.5	388.1	7.23	41.5	411.1	7.38	43.5	434.1	7.52
12.0	38.0	384.0	7.27	40.0	408.0	7.44	42.0	432.0	7.60	44.0	456.0	7.75
12.5	38.5	403.1	7.47	40.5	428.1	7.65	42.5	453.1	7.82	44.5	478.1	7.97
13.0	39.0	422.5	7.67	41.0	448.5	7.86	43.0	474.5	8.03	45.0	500.5	8.20
13.5	39.5	442.1	7.87	41.5	469.1	8.06	43.5	496.1	8.24	45.5	523.1	8.41
14.0	40.0	462.0	8.06	42.0	490.0	8.26	44.0	518.0	8.45	46.0	546.0	8.62
14.5	40.5	482.1	8.25	42.5	511.1	8.46	44.5	540.1	8.65	46.5	569.1	8.83
15.0	41.0	502.5	8.44	43.0	532.5	8.65	45.0	562.5	8.85	47.0	592.5	9.04
15.5	41.5	523.1	8.62	43.5	554.1	8.84	45.5	585.1	9.05	47.5	616.1	9.24
16.0	42.0	544.0	8.81	44.0	576.0	9.03	46.0	608.0	9.24	48.0	640.0	9.44
16.5	42.5	565.1	8.98	44.5	598.1	9.22	46.5	631.1	9.43	48.5	664.1	9.64
17.0	43.0	586.5	9.16	45.0	620.5	9.40	47.0	654.5	9.62	49.0	688.5	9.83
17.5	43.5	608.1	9.34	45.5	643.1	9.58	47.5	678.1	9.81	49.5	713.1	10.03
18.0	44.0	630.0	9.51	46.0	666.0	9.76	48.0	702.0	9.99	50.0	738.0	10.21
18.5	44.5	652.1	9.68	46.5	689.1	9.93	48.5	726.1	10.17	50.5	763.1	10.40
19.0	45.0	674.5	9.85	47.0	712.5	10.11	49.0	750.5	10.35	51.0	788.5	10.59
19.5	45.5	697.1	10.02	47.5	736.1	10.28	49.5	775.1	10.53	51.5	814.1	10.77
20.0	46.0	720.0	10.18	48.0	760.0	10.45	50.0	800.0	10.71	52.0	840.0	10.95

Depth	Bottom width 35 feet			Bottom width 40 feet			Bottom width 45 feet			Bottom width 50 feet		
	T	A	r	T	A	r	T	A	r	T	A	r
0.4	35.4	14.08	.39	40.4	16.08	.39	45.4	18.08	.39	50.4	20.08	.39
0.6	35.6	21.18	.58	40.6	24.18	.58	45.6	27.18	.59	50.6	30.18	.59
0.8	35.8	28.32	.77	40.8	32.32	.77	45.8	36.32	.78	50.8	40.32	.78
1.0	36.0	35.50	.95	40.0	40.50	.96	45.0	45.50	.96	51.0	50.50	.97
1.2	36.2	42.72	1.13	41.2	48.72	1.14	46.2	54.72	1.15	51.2	60.72	1.15
1.4	36.4	49.98	1.31	41.4	56.98	1.32	46.4	63.98	1.33	51.4	70.98	1.34
1.6	36.6	57.28	1.48	41.6	65.28	1.50	46.6	73.28	1.51	51.6	81.28	1.52
1.8	36.8	64.62	1.66	41.8	73.62	1.67	46.8	82.62	1.69	51.8	91.62	1.70
2.0	37.0	72.00	1.82	42.0	82.00	1.84	47.0	92.00	1.86	52.0	102.00	1.87
2.2	37.2	79.42	1.99	42.2	90.42	2.01	47.2	101.42	2.03	52.2	112.42	2.05
2.4	37.4	86.88	2.15	42.4	98.88	2.18	47.4	110.88	2.20	52.4	122.88	2.22
2.6	37.6	94.38	2.31	42.6	107.38	2.34	47.6	120.38	2.37	52.6	133.38	2.39
2.8	37.8	101.92	2.47	42.8	115.92	2.51	47.8	129.92	2.53	52.8	143.92	2.56
3.0	38.0	109.50	2.63	43.0	124.50	2.67	48.0	139.50	2.70	53.0	154.50	2.72
3.2	38.2	117.12	2.78	43.2	133.12	2.82	48.2	149.12	2.86	53.2	165.12	2.89
3.4	38.4	124.78	2.93	43.4	141.78	2.98	48.4	158.78	3.02	53.4	175.78	3.05
3.6	38.6	132.48	3.08	43.6	150.48	3.13	48.6	168.48	3.18	53.6	186.48	3.21
3.8	38.8	140.22	3.22	43.8	159.22	3.28	48.8	178.22	3.33	53.8	197.22	3.37
4.0	39.0	148.00	3.37	44.0	168.00	3.43	49.0	188.00	3.49	54.0	208.00	3.53
4.2	39.2	155.82	3.51	44.2	176.82	3.58	49.2	197.82	3.64	54.2	218.82	3.68
4.4	39.4	163.68	3.65	44.4	185.68	3.73	49.4	207.68	3.79	54.4	229.68	3.84
4.6	39.6	171.58	3.79	44.6	194.58	3.87	49.6	217.58	3.94	54.6	240.58	3.99
4.8	39.8	179.52	3.93	44.8	203.52	4.01	49.8	227.52	4.08	54.8	251.52	4.14

Table 27.—*Area in square feet, A, top width in feet, T, and hydraulic radius in feet, r, of trapezoidal channels,* **side slopes ½ to 1**—Continued

Depth	Bottom width 35 feet			Bottom width 40 feet			Bottom width 45 feet			Bottom width 50 feet		
	T	A	r	T	A	r	T	A	r	T	A	r
5.0	40.0	187.50	4.06	45.0	212.50	4.15	50.0	237.50	4.23	55.0	262.50	4.29
5.2	40.2	195.52	4.19	45.2	221.52	4.29	50.2	247.52	4.37	55.2	273.52	4.44
5.4	40.4	203.58	4.32	45.4	230.58	4.43	50.4	257.58	4.51	55.4	284.58	4.58
5.6	40.6	211.68	4.45	45.6	239.68	4.56	50.6	267.68	4.65	55.6	295.68	4.73
5.8	40.8	219.82	4.58	45.8	248.82	4.70	50.8	277.82	4.79	55.8	306.82	4.87
6.0	41.0	228.00	4.71	46.0	258.00	4.83	51.0	288.00	4.93	56.0	318.00	5.01
6.2	41.2	236.22	4.83	46.2	267.22	4.96	51.2	298.22	5.07	56.2	329.22	5.16
6.4	41.4	244.48	4.96	46.4	276.48	5.09	51.4	308.48	5.20	56.4	340.48	5.29
6.6	41.6	252.78	5.08	46.6	285.78	5.22	51.6	318.78	5.33	56.6	351.78	5.43
6.8	41.8	261.12	5.20	46.8	295.12	5.35	51.8	329.12	5.47	56.8	363.12	5.57
7.0	42.0	269.50	5.32	47.0	304.50	5.47	52.0	339.50	5.60	57.0	374.50	5.70
7.2	42.2	277.92	5.44	47.2	313.92	5.60	52.2	349.92	5.73	57.2	385.92	5.84
7.4	42.4	286.38	5.56	47.4	323.38	5.72	52.4	360.38	5.86	57.4	397.38	5.97
7.6	42.6	294.88	5.67	47.6	332.88	5.84	52.6	370.88	5.98	57.6	408.88	6.10
7.8	42.8	303.42	5.79	47.8	342.42	5.96	52.8	381.42	6.11	57.8	420.42	6.23
8.0	43.0	312.00	5.90	48.0	352.00	6.08	53.0	392.00	6.23	58.0	432.00	6.36
8.2	43.2	320.62	6.01	48.2	361.62	6.20	53.2	402.62	6.36	58.2	443.62	6.49
8.4	43.4	329.28	6.12	48.4	371.28	6.32	53.4	413.28	6.48	58.4	455.28	6.62
8.6	43.6	337.98	6.23	48.6	380.98	6.43	53.6	423.98	6.60	58.6	466.98	6.75
8.8	43.8	346.72	6.34	48.8	390.72	6.55	53.8	434.72	6.72	58.8	478.72	6.87
9.0	44.0	355.50	6.45	49.0	400.50	6.66	54.0	445.50	6.84	59.0	490.50	6.99
9.2	44.2	364.32	6.56	49.2	410.32	6.77	54.2	456.32	6.96	59.2	502.32	7.12
9.4	44.4	373.18	6.66	49.4	420.18	6.89	54.4	467.18	7.08	59.4	514.18	7.24
9.6	44.6	382.08	6.77	49.6	430.08	7.00	54.6	478.08	7.19	59.6	526.08	7.36
9.8	44.8	391.02	6.87	49.8	440.02	7.11	54.8	489.02	7.31	59.8	538.02	7.48
10.0	45.0	400.00	6.97	50.0	450.00	7.22	55.0	500.00	7.42	60.0	550.00	7.60
10.5	45.5	422.62	7.23	50.5	475.12	7.48	55.5	527.62	7.70	60.5	580.12	7.90
11.0	46.0	445.5	7.48	51.0	500.5	7.75	56.0	555.5	7.98	61.0	610.5	8.18
11.5	46.5	468.6	7.72	51.5	526.1	8.01	56.5	583.6	8.25	61.5	641.1	8.47
12.0	47.0	492.0	7.96	52.0	552.0	8.26	57.0	612.0	8.52	62.0	672.0	8.75
12.5	47.5	515.6	8.19	52.5	578.1	8.51	57.5	640.6	8.78	62.5	703.1	9.02
13.0	48.0	539.5	8.42	53.0	604.5	8.75	58.0	669.5	9.04	63.0	734.5	9.29
13.5	48.5	563.6	8.65	53.5	631.1	8.99	58.5	698.6	9.29	63.5	766.1	9.55
14.0	49.0	588.0	8.87	54.0	658.0	9.23	59.0	728.0	9.54	64.0	798.0	9.81
14.5	49.5	612.6	9.09	54.5	685.1	9.46	59.5	757.6	9.79	64.5	830.1	10.07
15.0	50.0	637.5	9.30	55.0	712.5	9.69	60.0	787.5	10.03	65.0	862.5	10.32
15.5	50.5	662.6	9.51	55.5	740.1	9.91	60.5	817.6	10.26	65.5	895.1	10.57
16.0	51.0	688.0	9.72	56.0	768.0	10.13	61.0	848.0	10.50	66.0	928.0	10.82
16.5	51.5	713.6	9.93	56.5	796.1	10.35	61.5	878.6	10.73	66.5	961.1	11.06
17.0	52.0	739.5	10.13	57.0	824.5	10.57	62.0	909.5	10.96	67.0	994.5	11.30
17.5	52.5	765.6	10.33	57.5	853.1	10.78	62.5	940.6	11.18	67.5	1,028.1	11.53
18.0	53.0	792.0	10.53	58.0	882.0	10.99	63.0	972.0	11.40	68.0	1,062.0	11.77
18.5	53.5	818.6	10.72	58.5	911.1	11.20	63.5	1,003.6	11.62	68.5	1,096.1	12.00
19.0	54.0	845.5	10.91	59.0	940.5	11.40	64.0	1,035.5	11.84	69.0	1,130.5	12.22
19.5	54.5	872.6	11.10	59.5	970.1	11.60	64.5	1,067.6	12.05	69.5	1,165.1	12.45
20.0	55.0	900.0	11.29	60.0	1,000.0	11.80	65.0	1,100.0	12.26	70.0	1,200.0	12.67

Table 27.—*Area in square feet, A, top width in feet, T, and hydraulic radius in feet, r, of trapezoidal channels,*
side slopes ½ to 1—Continued

Depth	Bottom width 60 feet			Bottom width 70 feet			Bottom width 80 feet			Bottom width 90 feet		
	T	A	r	T	A	r	T	A	r	T	A	r
0.4	60.4	24.08	.40	70.4	28.08	.40	80.4	32.08	.40	90.4	36.08	.40
0.6	60.6	36.18	.59	70.6	42.18	.59	80.6	48.18	.59	90.6	54.18	.59
0.8	60.8	48.32	.78	70.8	56.32	.78	80.8	64.32	.79	90.8	72.32	.79
1.0	61.0	60.50	.97	71.0	70.50	.98	81.0	80.50	.98	91.0	90.50	.98
1.2	61.2	72.72	1.16	71.2	84.72	1.17	81.2	96.72	1.17	91.2	108.72	1.17
1.4	61.4	84.98	1.35	71.4	98.98	1.35	81.4	112.98	1.36	91.4	126.98	1.36
1.6	61.6	97.28	1.53	71.6	113.28	1.54	81.6	129.28	1.55	91.6	145.28	1.55
1.8	61.8	109.62	1.71	71.8	127.62	1.72	81.8	145.62	1.73	91.8	163.62	1.74
2.0	62.0	122.00	1.89	72.0	142.00	1.91	82.0	162.00	1.92	92.0	182.00	1.93
2.2	62.2	134.42	2.07	72.2	156.42	2.09	82.2	178.42	2.10	92.2	200.42	2.11
2.4	62.4	146.88	2.25	72.4	170.88	2.27	82.4	194.88	2.28	92.4	218.88	2.30
2.6	62.6	159.38	2.42	72.6	185.38	2.45	82.6	211.38	2.46	92.6	237.38	2.48
2.8	62.8	171.92	2.59	72.8	199.92	2.62	82.8	227.92	2.64	92.8	255.92	2.66
3.0	63.0	184.50	2.77	73.0	214.50	2.80	83.0	244.50	2.82	93.0	274.50	2.84
3.2	63.2	197.12	2.94	73.2	229.12	2.97	83.2	261.12	3.00	93.2	293.12	3.02
3.4	63.4	209.78	3.10	73.4	243.78	3.14	83.4	277.78	3.17	93.4	311.78	3.19
3.6	63.6	222.48	3.27	73.6	258.48	3.31	83.6	294.48	3.34	93.6	330.48	3.37
3.8	63.8	235.22	3.43	73.8	273.22	3.48	83.8	311.22	3.52	93.8	349.22	3.55
4.0	64.0	248.00	3.60	74.0	288.00	3.65	84.0	328.00	3.69	94.0	368.00	3.72
4.2	64.2	260.82	3.76	74.2	302.82	3.81	84.2	344.82	3.86	94.2	386.82	3.89
4.4	64.4	273.68	3.92	74.4	317.68	3.98	84.4	361.68	4.03	94.4	405.68	4.06
4.6	64.6	286.58	4.08	74.6	332.58	4.14	84.6	378.58	4.19	94.6	424.58	4.23
4.8	64.8	299.52	4.23	74.8	347.52	4.30	84.8	395.52	4.36	94.8	443.52	4.40
5.0	65.0	312.50	4.39	75.0	362.50	4.47	85.0	412.50	4.52	95.0	462.50	4.57
5.2	65.2	325.52	4.54	75.2	377.52	4.62	85.2	429.52	4.69	95.2	481.52	4.74
5.4	65.4	338.58	4.70	75.4	392.58	4.78	85.4	446.58	4.85	95.4	500.58	4.90
5.6	65.6	351.68	4.85	75.6	407.68	4.94	85.6	463.68	5.01	95.6	519.68	5.07
5.8	65.8	364.82	5.00	75.8	422.82	5.10	85.8	480.82	5.17	95.8	538.82	5.23
6.0	66.0	378.00	5.15	76.0	438.00	5.25	86.0	498.00	5.33	96.0	558.00	5.40
6.2	66.2	391.22	5.30	76.2	453.22	5.40	86.2	515.22	5.49	96.2	577.22	5.56
6.4	66.4	404.48	5.44	76.4	468.48	5.56	86.4	532.48	5.65	96.4	596.48	5.72
6.6	66.6	417.78	5.59	76.6	483.78	5.71	86.6	549.78	5.80	96.6	615.78	5.88
6.8	66.8	431.12	5.73	76.8	499.12	5.86	86.8	567.12	5.96	96.8	635.12	6.04
7.0	67.0	444.50	5.88	77.0	514.50	6.01	87.0	584.50	6.11	97.0	654.50	6.19
7.2	67.2	457.92	6.02	77.2	529.92	6.15	87.2	601.92	6.26	97.2	673.92	6.35
7.4	67.4	471.38	6.16	77.4	545.38	6.30	87.4	619.38	6.42	97.4	693.38	6.51
7.6	67.6	484.88	6.30	77.6	560.88	6.45	87.6	636.88	6.57	97.6	712.88	6.66
7.8	67.8	498.42	6.44	77.8	576.42	6.59	87.8	654.42	6.72	97.8	732.42	6.82
8.0	68.0	512.00	6.57	78.0	592.00	6.74	88.0	672.00	6.86	98.0	752.00	6.97
8.2	68.2	525.62	6.71	78.2	607.62	6.88	88.2	689.62	7.01	98.2	771.62	7.12
8.4	68.4	539.28	6.85	78.4	623.28	7.02	88.4	707.28	7.16	98.4	791.28	7.27
8.6	68.6	552.98	6.98	78.6	638.98	7.16	88.6	724.98	7.31	98.6	810.98	7.42
8.8	68.8	566.72	7.11	78.8	654.72	7.30	88.8	742.72	7.45	98.8	830.72	7.57
9.0	69.0	580.50	7.24	79.0	670.50	7.44	89.0	760.50	7.60	99.0	850.50	7.72
9.2	69.2	594.32	7.38	79.2	686.32	7.58	89.2	778.32	7.74	99.2	870.32	7.87
9.4	69.4	608.18	7.51	79.4	702.18	7.71	89.4	796.18	7.88	99.4	890.18	8.02
9.6	69.6	622.08	7.64	79.6	718.08	7.85	89.6	814.08	8.02	99.6	910.08	8.16
9.8	69.8	636.02	7.76	79.8	734.02	7.99	89.8	832.02	8.16	99.8	930.02	8.31

Table 27.—*Area in square feet, A, top width in feet, T, and hydraulic radius in feet, r, of trapezoidal channels,* **side slopes ½ to 1**—Continued

Depth	Bottom width 60 feet			Bottom width 70 feet			Bottom width 80 feet			Bottom width 90 feet		
	T	A	r	T	A	r	T	A	r	T	A	r
10.0	70.0	650.00	7.89	80.0	750.00	8.12	90.0	850.00	8.30	100.0	950.00	8.45
10.5	70.5	685.12	8.21	80.5	790.12	8.45	90.5	895.12	8.65	100.5	1,000.12	8.81
11.0	71.0	720.5	8.52	81.0	830.5	8.78	91.0	940.5	8.99	101.0	1,050.5	9.17
11.5	71.5	756.1	8.82	81.5	871.1	9.10	91.5	986.1	9.33	101.5	1,101.1	9.52
12.0	72.0	792.0	9.12	82.0	912.0	9.42	92.0	1,032.0	9.66	102.0	1,152.0	9.86
12.5	72.5	828.1	9.42	82.5	953.1	9.73	92.5	1,078.1	9.99	102.5	1,203.1	10.20
13.0	73.0	864.5	9.71	83.0	994.5	10.04	93.0	1,124.5	10.31	103.0	1,254.5	10.54
13.5	73.5	901.1	9.99	83.5	1,036.1	10.34	93.5	1,171.1	10.63	103.5	1,306.1	10.87
14.0	74.0	938.0	10.27	84.0	1,078.0	10.64	94.0	1,218.0	10.94	104.0	1,358.0	11.19
14.5	74.5	975.1	10.55	84.5	1,120.1	10.94	94.5	1,265.1	11.25	104.5	1,410.1	11.52
15.0	75.0	1,012.5	10.82	85.0	1,162.5	11.23	95.0	1,312.5	11.56	105.0	1,462.5	11.84
15.5	75.5	1,050.1	11.09	85.5	1,205.1	11.51	95.5	1,360.1	11.86	105.5	1,515.1	12.15
16.0	76.0	1,088.0	11.36	86.0	1,248.0	11.80	96.0	1,408.0	12.16	106.0	1,568.0	12.47
16.5	76.5	1,126.1	11.62	86.5	1,291.1	12.08	96.5	1,456.1	12.46	106.5	1,621.1	12.78
17.0	77.0	1,164.5	11.88	87.0	1,334.5	12.35	97.0	1,504.5	12.75	107.0	1,674.5	13.08
17.5	77.5	1,203.1	12.14	87.5	1,378.1	12.63	97.5	1,553.1	13.04	107.5	1,728.1	13.38
18.0	78.0	1,242.0	12.39	88.0	1,422.0	12.90	98.0	1,602.0	13.32	108.0	1,782.0	13.68
18.5	78.5	1,281.1	12.64	88.5	1,466.1	13.16	98.5	1,651.1	13.60	108.5	1,836.1	13.98
19.0	79.0	1,320.5	12.88	89.0	1,510.5	13.43	99.0	1,700.5	13.88	109.0	1,890.5	14.27
19.5	79.5	1,360.1	13.13	89.5	1,555.1	13.69	99.5	1,750.1	14.16	109.5	1,945.1	14.56
20.0	80.0	1,400.0	13.37	90.0	1,600.0	13.95	100.0	1,800.0	14.43	110.0	2,000.0	14.85

Table 28.—*Area in square feet, A, top width in feet, T, and hydraulic radius in feet, r, of trapezoidal channels,*
side slopes ¾ to 1

Depth	Bottom width 2 feet			Bottom width 3 feet			Bottom width 4 feet			Bottom width 5 feet		
	T	A	r	T	A	r	T	A	r	T	A	r
0.4	2.6	0.92	0.31	3.6	1.32	0.33	4.6	1.72	0.34	5.6	2.12	0.35
0.6	2.9	1.47	.42	3.9	2.07	.46	4.9	2.67	.49	5.9	3.27	.50
0.8	3.2	2.08	.52	4.2	2.88	.58	5.2	3.68	.61	6.2	4.48	.64
1.0	3.5	2.75	.61	4.5	3.75	.68	5.5	4.75	.73	6.5	5.75	.77
1.2	3.8	3.48	.70	4.8	4.68	.78	5.8	5.88	.84	6.8	7.08	.89
1.4	4.1	4.27	.78	5.1	5.67	.87	6.1	7.07	.94	7.1	8.47	1.00
1.6	4.4	5.12	.85	5.4	6.72	.96	6.4	8.32	1.04	7.4	9.92	1.10
1.8	4.7	6.03	.93	5.7	7.83	1.04	6.7	9.63	1.13	7.7	11.43	1.20
2.0	5.0	7.00	1.00	6.0	9.00	1.13	7.0	11.00	1.22	8.0	13.00	1.30
2.2	5.3	8.03	1.07	6.3	10.23	1.20	7.3	12.43	1.31	8.3	14.63	1.39
2.4	5.6	9.12	1.14	6.6	11.52	1.28	7.6	13.92	1.39	8.6	16.32	1.48
2.6	5.9	10.27	1.21	6.9	12.87	1.35	7.9	15.47	1.47	8.9	18.07	1.57
2.8	6.2	11.48	1.28	7.2	14.28	1.43	8.2	17.08	1.55	9.2	19.88	1.66
3.0	6.5	12.75	1.34	7.5	15.75	1.50	8.5	18.75	1.63	9.5	21.75	1.74
3.2	6.8	14.08	1.41	7.8	17.28	1.57	8.8	20.48	1.71	9.8	23.68	1.82
3.4	7.1	15.47	1.47	8.1	18.87	1.64	9.1	22.27	1.78	10.1	25.67	1.90
3.6	7.4	16.92	1.54	8.4	20.52	1.71	9.4	24.12	1.86	10.4	27.72	1.98
3.8	7.7	18.43	1.60	8.7	22.23	1.78	9.7	26.03	1.93	10.7	29.83	2.06
4.0	8.0	20.00	1.67	9.0	24.00	1.85	10.0	28.00	2.00	11.0	32.00	2.13
4.2	8.3	21.63	1.73	9.3	25.83	1.91	10.3	30.03	2.07	11.3	34.23	2.21
4.4	8.6	23.32	1.79	9.6	27.72	1.98	10.6	32.12	2.14	11.6	36.52	2.28
4.6	8.9	25.07	1.86	9.9	29.67	2.05	10.9	34.27	2.21	11.9	38.87	2.36
4.8	9.2	26.88	1.92	10.2	31.68	2.11	11.2	36.48	2.28	12.2	41.28	2.43
5.0	9.5	28.75	1.98	10.5	33.75	2.18	11.5	38.75	2.35	12.5	43.75	2.50
5.2	9.8	30.68	2.05	10.8	35.88	2.24	11.8	41.08	2.42	12.8	46.28	2.57
5.4	10.1	32.67	2.11	11.1	38.07	2.31	12.1	43.47	2.48	13.1	48.87	2.64
5.6	10.4	34.72	2.17	11.4	40.32	2.37	12.4	45.92	2.55	13.4	51.52	2.71
5.8	10.7	36.83	2.23	11.7	42.63	2.44	12.7	48.43	2.62	13.7	54.23	2.78
6.0	11.0	39.00	2.29	12.0	45.00	2.50	13.0	51.00	2.68	14.0	57.00	2.85
6.2	11.3	41.23	2.36	12.3	47.43	2.56	13.3	53.63	2.75	14.3	59.83	2.92
6.4	11.6	43.52	2.42	12.6	49.92	2.63	13.6	56.32	2.82	14.6	62.72	2.99
6.6	11.9	45.87	2.48	12.9	52.47	2.69	13.9	59.07	2.88	14.9	65.67	3.05
6.8	12.2	48.28	2.54	13.2	55.08	2.75	14.2	61.88	2.95	15.2	68.68	3.12
7.0	12.5	50.75	2.60	13.5	57.75	2.82	14.5	64.75	3.01	15.5	71.75	3.19
7.2	12.8	53.28	2.66	13.8	60.48	2.88	14.8	67.68	3.08	15.8	74.88	3.26
7.4	13.1	55.87	2.73	14.1	63.27	2.94	15.1	70.67	3.14	16.1	78.07	3.32
7.6	13.4	58.52	2.79	14.4	66.12	3.01	15.4	73.72	3.21	16.4	81.32	3.39
7.8	13.7	61.23	2.85	14.7	69.03	3.07	15.7	76.83	3.27	16.7	84.63	3.45
8.0	14.0	64.00	2.91	15.0	72.00	3.13	16.0	80.00	3.33	17.0	88.00	3.52
8.2	-----	-----	-----	15.3	75.03	3.19	16.3	83.23	3.40	17.3	91.43	3.59
8.4				15.6	78.12	3.26	16.6	86.52	3.46	17.6	94.92	3.65
8.6				15.9	81.27	3.32	16.9	89.87	3.52	17.9	98.47	3.72
8.8				16.2	84.48	3.38	17.2	93.28	3.59	18.2	102.08	3.78
9.0				16.5	87.75	3.44	17.5	96.75	3.65	18.5	105.75	3.85
9.2				16.8	91.08	3.50	17.8	100.28	3.71	18.8	109.48	3.91
9.4				17.1	94.47	3.56	18.1	103.87	3.78	19.1	113.27	3.97
9.6				17.4	97.92	3.63	18.4	107.52	3.84	19.4	117.12	4.04
9.8				17.7	101.43	3.69	18.7	111.23	3.90	19.7	121.03	4.10

Table 28.—*Area in square feet, A, top width in feet, T, and hydraulic radius in feet, r, of trapezoidal channels,* **side slopes ¾ to 1**—Continued

Depth	Bottom width 2 feet			Bottom width 3 feet			Bottom width 4 feet			Bottom width 5 feet		
	T	A	r	T	A	r	T	A	r	T	A	r
10.0	----	----	----	18.0	105.00	3.75	19.0	115.00	3.97	20.0	125.00	4.17
10.5	----	----	----	18.75	114.19	3.90	19.75	124.69	4.12	20.75	135.19	4.33
11.0	----	----	----	19.5	123.8	4.06	20.5	134.8	4.28	21.5	145.8	4.48
11.5	----	----	----	20.25	133.7	4.21	21.25	145.2	4.43	22.25	156.7	4.64
12.0	----	----	----	21.0	144.0	4.36	22.0	156.0	4.59	23.0	168.0	4.80
12.5	----	----	----	21.75	154.7	4.52	22.75	167.2	4.74	23.75	179.7	4.96
13.0				----	----	----	23.5	178.8	4.90	24.5	191.8	5.11
13.5				----	----	----	24.25	190.7	5.05	25.25	204.2	5.27
14.0				----	----	----	25.0	203.0	5.21	26.0	217.0	5.43
14.5				----	----	----	25.75	215.7	5.36	26.75	230.2	5.58
15.0				----	----	----	26.5	228.8	5.51	27.5	243.8	5.74
15.5				----	----	----	27.25	242.2	5.67	28.25	257.7	5.89
16.0							28.0	256.0	5.82	29.0	272.0	6.04
16.5							----	----	----	29.75	286.7	6.20
17.0							----	----	----	30.5	301.8	6.35
17.5							----	----	----	31.25	317.2	6.51
18.0							----	----	----	32.0	333.0	6.66
18.5							----	----	----	32.75	349.2	6.81
19.0							----	----	----	33.5	365.8	6.97
19.5							----	----	----	34.25	382.7	7.12
20.0							----	----	----	35.0	400.0	7.27

Depth	Bottom width 6 feet			Bottom width 7 feet			Bottom width 8 feet			Bottom width 9 feet		
	T	A	r	T	A	r	T	A	r	T	A	r
0.4	6.6	2.52	0.36	7.6	2.92	0.37	8.6	3.32	0.37	9.6	3.72	0.37
0.6	6.9	3.87	.52	7.9	4.47	.53	8.9	5.07	.53	9.9	5.67	.54
0.8	7.2	5.28	.66	8.2	6.08	.68	9.2	6.88	.69	10.2	7.68	.70
1.0	7.5	6.75	.79	8.5	7.75	.82	9.5	8.75	.83	10.5	9.75	.85
1.2	7.8	8.28	.92	8.8	9.48	.95	9.8	10.68	.97	10.8	11.88	.99
1.4	8.1	9.87	1.04	9.1	11.27	1.07	10.1	12.67	1.10	11.1	14.07	1.13
1.6	8.4	11.52	1.15	9.4	13.12	1.19	10.4	14.72	1.23	11.4	16.32	1.26
1.8	8.7	13.23	1.26	9.7	15.03	1.31	10.7	16.83	1.35	11.7	18.63	1.38
2.0	9.0	15.00	1.36	10.0	17.00	1.42	11.0	19.00	1.46	12.0	21.00	1.50
2.2	9.3	16.83	1.46	10.3	19.03	1.52	11.3	21.23	1.57	12.3	23.43	1.62
2.4	9.6	18.72	1.56	10.6	21.12	1.62	11.6	23.52	1.68	12.6	25.92	1.73
2.6	9.9	20.67	1.65	10.9	23.27	1.72	11.9	25.87	1.78	12.9	28.47	1.84
2.8	10.2	22.68	1.74	11.2	25.48	1.82	12.2	28.28	1.89	13.2	31.08	1.94
3.0	10.5	24.75	1.83	11.5	27.75	1.91	12.5	30.75	1.98	13.5	33.75	2.05
3.2	10.8	26.88	1.92	11.8	30.08	2.01	12.8	33.28	2.08	13.8	36.48	2.15
3.4	11.1	29.07	2.00	12.1	32.47	2.09	13.1	35.87	2.17	14.1	39.27	2.24
3.6	11.4	31.32	2.09	12.4	34.92	2.18	13.4	38.52	2.27	14.4	42.12	2.34
3.8	11.7	33.63	2.17	12.7	37.43	2.27	13.7	41.23	2.36	14.7	45.03	2.43
4.0	12.0	36.00	2.25	13.0	40.00	2.35	14.0	44.00	2.44	15.0	48.00	2.53
4.2	12.3	38.43	2.33	13.3	42.63	2.44	14.3	46.83	2.53	15.3	51.03	2.62
4.4	12.6	40.92	2.41	13.6	45.32	2.52	14.6	49.72	2.62	15.6	54.12	2.71
4.6	12.9	43.47	2.48	13.9	48.07	2.60	14.9	52.67	2.70	15.9	57.27	2.79
4.8	13.2	46.08	2.56	14.2	50.88	2.68	15.2	55.68	2.78	16.2	60.48	2.88

Table 28.—*Area in square feet, A, top width in feet, T, and hydraulic radius in feet, r, of trapezoidal channels,* side slopes ¾ to 1—Continued

Depth	Bottom width 6 feet			Bottom width 7 feet			Bottom width 8 feet			Bottom width 9 feet		
	T	A	r	T	A	r	T	A	r	T	A	r
5.0	13.5	48.75	2.64	14.5	53.75	2.76	15.5	58.75	2.87	16.5	63.75	2.97
5.2	13.8	51.48	2.71	14.8	56.68	2.83	15.8	61.88	2.95	16.8	67.08	3.05
5.4	14.1	54.27	2.78	15.1	59.67	2.91	16.1	65.07	3.03	17.1	70.47	3.13
5.6	14.4	57.12	2.86	15.4	62.72	2.99	16.4	68.32	3.11	17.4	73.92	3.21
5.8	14.7	60.03	2.93	15.7	65.83	3.06	16.7	71.63	3.18	17.7	77.43	3.29
6.0	15.0	63.00	3.00	16.0	69.00	3.14	17.0	75.00	3.26	18.0	81.00	3.38
6.2	15.3	66.03	3.07	16.3	72.23	3.21	17.3	78.43	3.34	18.3	84.63	3.45
6.4	15.6	69.12	3.14	16.6	75.52	3.28	17.6	81.92	3.41	18.6	88.32	3.53
6.6	15.9	72.27	3.21	16.9	78.87	3.36	17.9	85.47	3.49	18.9	92.07	3.61
6.8	16.2	75.48	3.28	17.2	82.28	3.43	18.2	89.08	3.56	19.2	95.88	3.69
7.0	16.5	78.75	3.35	17.5	85.75	3.50	18.5	92.75	3.64	19.5	99.75	3.76
7.2	16.8	82.08	3.42	17.8	89.28	3.57	18.8	96.48	3.71	19.8	103.68	3.84
7.4	17.1	85.47	3.49	18.1	92.87	3.64	19.1	100.27	3.78	20.1	107.67	3.92
7.6	17.4	88.92	3.56	18.4	96.52	3.71	19.4	104.12	3.86	20.4	111.72	3.99
7.8	17.7	92.43	3.62	18.7	100.23	3.78	19.7	108.03	3.93	20.7	115.83	4.06
8.0	18.0	96.00	3.69	19.0	104.00	3.85	20.0	112.00	4.00	21.0	120.00	4.14
8.2	18.3	99.63	3.76	19.3	107.83	3.92	20.3	116.03	4.07	21.3	124.23	4.21
8.4	18.6	103.32	3.83	19.6	111.72	3.99	20.6	120.12	4.14	21.6	128.52	4.28
8.6	18.9	107.07	3.89	19.9	115.67	4.06	20.9	124.27	4.21	21.9	132.87	4.36
8.8	19.2	110.88	3.96	20.2	119.68	4.13	21.2	128.48	4.28	22.2	137.28	4.43
9.0	19.5	114.75	4.03	20.5	123.75	4.19	21.5	132.75	4.35	22.5	141.75	4.50
9.2	19.8	118.68	4.09	20.8	127.88	4.26	21.8	137.08	4.42	22.8	146.28	4.57
9.4	20.1	122.67	4.16	21.1	132.07	4.33	22.1	141.47	4.49	23.1	150.87	4.64
9.6	20.4	126.72	4.22	21.4	136.32	4.40	22.4	145.92	4.56	23.4	155.52	4.71
9.8	20.7	130.83	4.29	21.7	140.63	4.46	22.7	150.43	4.63	23.7	160.23	4.78
10.0	21.0	135.00	4.35	22.0	145.00	4.53	23.0	155.00	4.70	24.0	165.00	4.85
10.5	21.75	145.69	4.52	22.75	156.19	4.70	23.75	166.69	4.87	24.75	177.19	5.03
11.0	22.5	156.8	4.68	23.5	167.8	4.86	24.5	178.8	5.04	25.5	189.8	5.20
11.5	23.25	168.2	4.84	24.25	179.7	5.03	25.25	191.2	5.20	26.25	202.7	5.37
12.0	24.0	180.0	5.00	25.0	192.0	5.19	26.0	204.0	5.37	27.0	216.0	5.54
12.5	24.75	192.2	5.16	25.75	204.7	5.35	26.75	217.2	5.53	27.75	229.7	5.71
13.0	25.5	204.8	5.32	26.5	217.8	5.51	27.5	230.8	5.70	28.5	243.8	5.87
13.5	26.25	217.7	5.48	27.25	231.2	5.67	28.25	244.7	5.86	29.25	258.2	6.04
14.0	27.0	231.0	5.63	28.0	245.0	5.83	29.0	259.0	6.02	30.0	273.0	6.20
14.5	27.75	244.7	5.79	28.75	259.2	5.99	29.75	273.7	6.19	30.75	288.2	6.37
15.0	28.5	258.8	5.95	29.5	273.8	6.15	30.5	288.8	6.35	31.5	303.8	6.53
15.5	29.25	273.2	6.10	30.25	288.7	6.31	31.25	304.2	6.51	32.25	319.7	6.70
16.0	30.0	288.0	6.26	31.0	304.0	6.47	32.0	320.0	6.67	33.0	336.0	6.86
16.5	30.75	303.2	6.42	31.75	319.7	6.63	32.75	336.2	6.83	33.75	352.7	7.02
17.0	31.5	318.8	6.57	32.5	335.8	6.78	33.5	352.8	6.99	34.5	369.8	7.18
17.5	32.25	334.7	6.73	33.25	352.2	6.94	34.25	369.7	7.14	35.25	387.2	7.34
18.0	33.0	351.0	6.88	34.0	369.0	7.10	35.0	387.0	7.30	36.0	405.0	7.50
18.5	33.75	367.7	7.04	34.75	386.2	7.25	35.75	404.7	7.46	36.75	423.2	7.66
19.0	34.5	384.8	7.19	35.5	403.8	7.41	36.5	422.8	7.62	37.5	441.8	7.82
19.5	35.25	402.2	7.35	36.25	421.7	7.56	37.25	441.2	7.77	38.25	460.7	7.98
20.0	36.0	420.0	7.50	37.0	440.0	7.72	38.0	460.0	7.93	39.0	480.0	8.14

Table 28.—*Area in square feet, A, top width in feet, T, and hydraulic radius in feet, r, of trapezoidal channels,*
side slopes ¾ to 1—Continued

Depth	Bottom width 10 feet			Bottom width 12 feet			Bottom width 14 feet			Bottom width 16 feet		
	T	A	r	T	A	r	T	A	r	T	A	r
0.4	10.6	4.12	.37	12.6	4.92	.38	14.6	5.72	.38	16.6	6.52	.38
0.6	10.9	6.27	.55	12.9	7.47	.55	14.9	8.67	.56	16.9	9.87	.56
0.8	11.2	8.48	.71	13.2	10.08	.72	15.2	11.68	.73	17.2	13.28	.74
1.0	11.5	10.75	.86	13.5	12.75	.88	15.5	14.75	.89	17.5	16.75	.91
1.2	11.8	13.08	1.01	13.8	15.48	1.03	15.8	17.88	1.05	17.8	20.28	1.07
1.4	12.1	15.47	1.15	14.1	18.27	1.18	16.1	21.07	1.20	18.1	23.87	1.22
1.6	12.4	17.92	1.28	14.4	21.12	1.32	16.4	24.32	1.35	18.4	27.52	1.38
1.8	12.7	20.43	1.41	14.7	24.03	1.46	16.7	27.63	1.49	18.7	31.23	1.52
2.0	13.0	23.00	1.53	15.0	27.00	1.59	17.0	31.00	1.63	19.0	35.00	1.67
2.2	13.3	25.63	1.65	15.3	30.03	1.72	17.3	34.43	1.77	19.3	38.83	1.81
2.4	13.6	28.32	1.77	15.6	33.12	1.84	17.6	37.92	1.90	19.6	42.72	1.94
2.6	13.9	31.07	1.88	15.9	36.27	1.96	17.9	41.47	2.02	19.9	46.67	2.07
2.8	14.2	33.88	1.99	16.2	39.48	2.08	18.2	45.08	2.15	20.2	50.68	2.20
3.0	14.5	36.75	2.10	16.5	42.75	2.19	18.5	48.75	2.27	20.5	54.75	2.33
3.2	14.8	39.68	2.20	16.8	46.08	2.30	18.8	52.48	2.39	20.8	58.88	2.45
3.4	15.1	42.67	2.31	17.1	49.47	2.41	19.1	56.27	2.50	21.1	63.07	2.57
3.6	15.4	45.72	2.41	17.4	52.92	2.52	19.4	60.12	2.61	21.4	67.32	2.69
3.8	15.7	48.83	2.50	17.7	56.43	2.62	19.7	64.03	2.72	21.7	71.63	2.81
4.0	16.0	52.00	2.60	18.0	60.00	2.73	20.0	68.00	2.83	22.0	76.00	2.92
4.2	16.3	55.23	2.69	18.3	63.63	2.83	20.3	72.03	2.94	22.3	80.43	3.04
4.4	16.6	58.52	2.79	18.6	67.32	2.93	20.6	76.12	3.04	22.6	84.92	3.15
4.6	16.9	61.87	2.88	18.9	71.07	3.02	20.9	80.27	3.15	22.9	89.47	3.25
4.8	17.2	65.28	2.97	19.2	74.88	3.12	21.2	84.48	3.25	23.2	94.08	3.36
5.0	17.5	68.75	3.06	19.5	78.75	3.21	21.5	88.75	3.35	23.5	98.75	3.46
5.2	17.8	72.28	3.14	19.8	82.68	3.31	21.8	93.08	3.45	23.8	103.48	3.57
5.4	18.1	75.87	3.23	20.1	86.67	3.40	22.1	97.47	3.54	24.1	108.27	3.67
5.6	18.4	79.52	3.31	20.4	90.72	3.49	22.4	101.92	3.64	24.4	113.12	3.77
5.8	18.7	83.23	3.40	20.7	94.83	3.58	22.7	106.43	3.73	24.7	118.03	3.87
6.0	19.0	87.00	3.48	21.0	99.00	3.67	23.0	111.00	3.83	25.0	123.00	3.97
6.2	19.3	90.83	3.56	21.3	103.23	3.75	23.3	115.63	3.92	25.3	128.03	4.06
6.4	19.6	94.72	3.64	21.6	107.52	3.84	23.6	120.32	4.01	25.6	133.12	4.16
6.6	19.9	98.67	3.72	21.9	111.87	3.93	23.9	125.07	4.10	25.9	138.27	4.25
6.8	20.2	102.68	3.80	22.2	116.28	4.01	24.2	129.88	4.19	26.2	143.48	4.35
7.0	20.5	106.75	3.88	22.5	120.75	4.09	24.5	134.75	4.28	26.5	148.75	4.44
7.2	20.8	110.88	3.96	22.8	125.28	4.18	24.8	139.68	4.37	26.8	154.08	4.53
7.4	21.1	115.07	4.04	23.1	129.87	4.26	25.1	144.67	4.45	27.1	159.47	4.62
7.6	21.4	119.32	4.11	23.4	134.52	4.34	25.4	149.72	4.54	27.4	164.92	4.71
7.8	21.7	123.63	4.19	23.7	139.23	4.42	25.7	154.83	4.62	27.7	170.43	4.80
8.0	22.0	128.00	4.27	24.0	144.00	4.50	26.0	160.00	4.71	28.0	176.00	4.89
8.2	22.3	132.43	4.34	24.3	148.83	4.58	26.3	165.23	4.79	28.3	181.63	4.98
8.4	22.6	136.92	4.42	24.6	153.72	4.66	26.6	170.52	4.87	28.6	187.32	5.06
8.6	22.9	141.47	4.49	24.9	158.67	4.74	26.9	175.87	4.95	28.9	193.07	5.15
8.8	23.2	146.08	4.57	25.2	163.68	4.81	27.2	181.28	5.04	29.2	198.88	5.23
9.0	23.5	150.75	4.64	25.5	168.75	4.89	27.5	186.75	5.12	29.5	204.75	5.32
9.2	23.8	155.48	4.71	25.8	173.88	4.97	27.8	192.32	5.20	29.8	210.68	5.40
9.4	24.1	160.27	4.78	26.1	179.07	5.04	28.1	197.87	5.28	30.1	216.67	5.49
9.6	24.4	165.12	4.86	26.4	184.32	5.12	28.4	203.52	5.36	30.4	222.72	5.57
9.8	24.7	170.03	4.93	26.7	189.63	5.20	28.7	209.23	5.43	30.7	228.83	5.65

Table 28.—*Area in square feet, A, top width in feet, T, and hydraulic radius in feet, r, of trapezoidal channels,*
side slopes ¾ to 1—Continued

Depth	Bottom width 10 feet			Bottom width 12 feet			Bottom width 14 feet			Bottom width 16 feet		
	T	A	r	T	A	r	T	A	r	T	A	r
10.0	25.0	175.00	5.00	27.0	195.00	5.27	29.0	215.00	5.51	31.0	235.00	5.73
10.5	25.75	187.69	5.18	27.75	208.69	5.46	29.75	229.69	5.71	31.75	250.69	5.93
11.0	26.5	200.8	5.35	28.5	222.8	5.64	30.5	244.8	5.90	32.5	266.8	6.13
11.5	27.25	214.2	5.53	29.25	237.2	5.82	31.25	260.2	6.09	33.25	283.2	6.33
12.0	28.0	228.0	5.70	30.0	252.0	6.00	32.0	276.0	6.27	34.0	300.0	6.52
12.5	28.75	242.2	5.87	30.75	267.2	6.18	32.75	292.2	6.46	34.75	317.2	6.71
13.0	29.5	256.8	6.04	31.5	282.8	6.35	33.5	308.8	6.64	35.5	334.8	6.90
13.5	30.25	271.7	6.21	32.25	298.7	6.53	34.25	325.7	6.82	36.25	352.7	7.09
14.0	31.0	287.0	6.38	33.0	315.0	6.70	35.0	343.0	7.00	37.0	371.0	7.27
14.5	31.75	302.7	6.54	33.75	331.7	6.87	35.75	360.7	7.18	37.75	389.7	7.46
15.0	32.5	318.8	6.71	34.5	348.8	7.05	36.5	378.8	7.35	38.5	408.8	7.64
15.5	33.25	335.2	6.88	35.25	366.2	7.22	37.25	397.2	7.53	39.25	428.2	7.82
16.0	34.0	352.0	7.04	36.0	384.0	7.38	38.0	416.0	7.70	40.0	448.0	8.00
16.5	34.75	369.2	7.20	36.75	402.2	7.55	38.75	435.2	7.88	40.75	468.2	8.18
17.0	35.5	386.8	7.37	37.5	420.8	7.72	39.5	454.8	8.05	41.5	488.8	8.35
17.5	36.25	404.7	7.53	38.25	439.7	7.89	40.25	474.7	8.22	42.25	509.7	8.53
18.0	37.0	423.0	7.69	39.0	459.0	8.05	41.0	495.0	8.39	43.0	531.0	8.70
18.5	37.75	441.7	7.85	39.75	478.7	8.22	41.75	515.7	8.56	43.75	552.7	8.88
19.0	38.5	460.8	8.01	40.5	498.8	8.38	42.5	536.8	8.73	44.5	574.8	9.05
19.5	39.25	480.2	8.17	41.25	519.2	8.55	43.25	558.2	8.90	45.25	597.2	9.22
20.0	40.0	500.0	8.33	42.0	540.0	8.71	44.0	580.0	9.06	46.0	620.0	9.39

Depth	Bottom width 18 feet			Bottom width 20 feet			Bottom width 22 feet			Bottom width 24 feet		
	T	A	r	T	A	r	T	A	r	T	A	r
0.4	18.6	7.32	.39	20.6	8.12	.39	22.6	8.92	.39	24.6	9.72	.39
0.6	18.9	11.07	.57	20.9	12.27	.57	22.9	13.47	.57	24.9	14.67	.58
0.8	19.2	14.88	.74	21.2	16.48	.75	23.2	18.08	.75	25.2	19.68	.76
1.0	19.5	18.75	.91	21.5	20.75	.92	23.5	22.75	.93	25.5	24.75	.93
1.2	19.8	22.68	1.08	21.8	25.08	1.09	23.8	27.48	1.10	25.8	29.88	1.11
1.4	20.1	26.67	1.24	22.1	29.47	1.25	24.1	32.27	1.27	26.1	35.07	1.28
1.6	20.4	30.72	1.40	22.4	33.92	1.41	24.4	37.12	1.43	26.4	40.32	1.44
1.8	20.7	34.83	1.55	22.7	38.43	1.57	24.7	42.03	1.59	26.7	45.63	1.60
2.0	21.0	39.00	1.70	23.0	43.00	1.72	25.0	47.00	1.74	27.0	51.00	1.76
2.2	21.3	43.23	1.84	23.3	47.63	1.87	25.3	52.03	1.89	27.3	56.43	1.91
2.4	21.6	47.52	1.98	23.6	52.32	2.01	25.6	57.12	2.04	27.6	61.92	2.06
2.6	21.9	51.87	2.12	23.9	57.07	2.15	25.9	62.27	2.18	27.9	67.47	2.21
2.8	22.2	56.28	2.25	24.2	61.88	2.29	26.2	67.48	2.33	28.2	73.08	2.36
3.0	22.5	60.75	2.38	24.5	66.75	2.43	26.5	72.75	2.47	28.5	78.75	2.50
3.2	22.8	65.28	2.51	24.8	71.68	2.56	26.8	78.08	2.60	28.8	84.48	2.64
3.4	23.1	69.87	2.64	25.1	76.67	2.69	27.1	83.47	2.74	29.1	90.27	2.78
3.6	23.4	74.52	2.76	25.4	81.72	2.82	27.4	88.92	2.87	29.4	96.12	2.91
3.8	23.7	79.23	2.88	25.7	86.83	2.94	27.7	94.43	3.00	29.7	102.03	3.05
4.0	24.0	84.00	3.00	26.0	92.00	3.07	28.0	100.00	3.13	30.0	108.00	3.18
4.2	24.3	88.83	3.12	26.3	97.23	3.19	28.3	105.63	3.25	30.3	114.03	3.31
4.4	24.6	93.72	3.23	26.6	102.52	3.31	28.6	111.32	3.37	30.6	120.12	3.43
4.6	24.9	98.67	3.34	26.9	107.87	3.42	28.9	117.07	3.49	30.9	126.27	3.56
4.8	25.2	103.68	3.46	27.2	113.28	3.54	29.2	122.88	3.61	31.2	132.48	3.68

Table 28.—*Area in square feet, A, top width in feet, T, and hydraulic radius in feet, r, of trapezoidal channels,*
side slopes ¾ to 1—Continued

Depth	Bottom width 18 feet			Bottom width 20 feet			Bottom width 22 feet			Bottom width 24 feet		
	T	A	r	T	A	r	T	A	r	T	A	r
5.0	25.5	108.75	3.57	27.5	118.75	3.65	29.5	128.75	3.73	31.5	138.75	3.80
5.2	25.8	113.88	3.67	27.8	124.28	3.77	29.8	134.68	3.85	31.8	145.08	3.92
5.4	26.1	119.07	3.78	28.1	129.87	3.88	30.1	140.67	3.96	32.1	151.47	4.04
5.6	26.4	124.32	3.89	28.4	135.52	3.99	30.4	146.72	4.08	32.4	157.92	4.16
5.8	26.7	129.63	3.99	28.7	141.23	4.09	30.7	152.83	4.19	32.7	164.43	4.27
6.0	27.0	135.00	4.09	29.0	147.00	4.20	31.0	159.00	4.30	33.0	171.00	4.38
6.2	27.3	140.43	4.19	29.3	152.83	4.31	31.3	165.23	4.41	33.3	177.63	4.50
6.4	27.6	145.92	4.29	29.6	158.72	4.41	31.6	171.52	4.51	33.6	184.32	4.61
6.6	27.9	151.47	4.39	29.9	164.67	4.51	31.9	177.87	4.62	33.9	191.07	4.72
6.8	28.2	157.08	4.49	30.2	170.68	4.61	32.2	184.28	4.73	34.2	197.88	4.83
7.0	28.5	162.75	4.58	30.5	176.75	4.71	32.5	190.75	4.83	34.5	204.75	4.93
7.2	28.8	168.48	4.68	30.8	182.88	4.81	32.8	197.28	4.93	34.8	211.68	5.04
7.4	29.1	174.27	4.77	31.1	189.07	4.91	33.1	203.87	5.03	35.1	218.67	5.15
7.6	29.4	180.12	4.87	31.4	195.32	5.01	33.4	210.52	5.13	35.4	225.72	5.25
7.8	29.7	186.03	4.96	31.7	201.63	5.10	33.7	217.23	5.23	35.7	232.83	5.35
8.0	30.0	192.00	5.05	32.0	208.00	5.20	34.0	224.00	5.33	36.0	240.00	5.45
8.2	30.3	198.03	5.14	32.3	214.43	5.29	34.3	230.83	5.43	36.3	247.23	5.56
8.4	30.6	204.12	5.23	32.6	220.92	5.39	34.6	237.72	5.53	36.6	254.52	5.66
8.6	30.9	210.27	5.32	32.9	227.47	5.48	34.9	244.67	5.62	36.9	261.87	5.76
8.8	31.2	216.48	5.41	33.2	234.08	5.57	35.2	251.68	5.72	37.2	269.28	5.85
9.0	31.5	222.75	5.50	33.5	240.75	5.66	35.5	258.75	5.81	37.5	276.75	5.95
9.2	31.8	229.08	5.59	33.8	247.48	5.76	35.8	265.88	5.91	37.8	284.28	6.05
9.4	32.1	235.47	5.67	34.1	254.27	5.85	36.1	273.07	6.00	38.1	291.87	6.14
9.6	32.4	241.92	5.76	34.4	261.12	5.93	36.4	280.32	6.09	38.4	299.52	6.24
9.8	32.7	248.43	5.85	34.7	268.03	6.02	36.7	287.63	6.19	38.7	307.23	6.33
10.0	33.0	255.00	5.93	35.0	275.00	6.11	37.0	295.00	6.28	39.0	315.00	6.43
10.5	33.75	271.69	6.14	35.75	292.69	6.33	37.75	313.69	6.50	39.75	334.69	6.66
11.0	34.5	288.8	6.35	36.5	310.8	6.54	38.5	332.8	6.72	40.5	354.8	6.89
11.5	35.25	306.2	6.55	37.25	329.2	6.75	39.25	352.2	6.94	41.25	375.2	7.11
12.0	36.0	324.0	6.75	38.0	348.0	6.96	40.0	372.0	7.15	42.0	396.0	7.33
12.5	36.75	342.2	6.95	38.75	367.2	7.16	40.75	392.2	7.37	42.75	417.2	7.55
13.0	37.5	360.8	7.14	39.5	386.8	7.37	41.5	412.8	7.57	43.5	438.8	7.77
13.5	38.25	379.7	7.34	40.25	406.7	7.57	42.25	433.7	7.78	44.25	460.7	7.98
14.0	39.0	399.0	7.53	41.0	427.0	7.76	43.0	455.0	7.98	45.0	483.0	8.19
14.5	39.75	418.7	7.72	41.75	447.7	7.96	43.75	476.7	8.18	45.75	505.7	8.39
15.0	40.5	438.8	7.91	42.5	468.8	8.15	44.5	498.8	8.38	46.5	528.8	8.60
15.5	41.25	459.2	8.09	43.25	490.2	8.34	45.25	521.2	8.58	47.25	552.2	8.80
16.0	42.0	480.0	8.28	44.0	512.0	8.53	46.0	544.0	8.77	48.0	576.0	9.00
16.5	42.75	501.2	8.46	44.75	534.2	8.72	46.75	567.2	8.97	48.75	600.2	9.20
17.0	43.5	522.8	8.64	45.5	556.8	8.91	47.5	590.8	9.16	49.5	624.8	9.39
17.5	44.25	544.7	8.82	46.25	579.7	9.09	48.25	614.7	9.35	50.25	649.7	9.59
18.0	45.0	567.0	9.00	47.0	603.0	9.28	49.0	639.0	9.54	51.0	675.0	9.78
18.5	45.75	589.7	9.18	47.75	626.7	9.46	49.75	663.7	9.72	51.75	700.7	9.97
19.0	46.5	612.8	9.35	48.5	650.8	9.64	50.5	688.8	9.91	52.5	726.8	10.16
19.5	47.25	636.2	9.53	49.25	675.2	9.82	51.25	714.2	10.09	53.25	753.2	10.35
20.0	48.0	660.0	9.71	50.0	700.0	10.00	52.0	740.0	10.28	54.0	780.0	10.54

130   HYDRAULIC AND EXCAVATION TABLES

Table 28.—*Area in square feet, A, top width in feet, T, and hydraulic radius in feet, r, of trapezoidal channels,*
side slopes ¾ to 1—Continued

Depth	Bottom width 26 feet			Bottom width 28 feet			Bottom width 30 feet			Bottom width 32 feet		
	T	A	r	T	A	r	T	A	r	T	A	r
0.4	26.6	10.52	.39	28.6	11.32	.39	30.6	12.12	.39	32.6	12.92	.39
0.6	26.9	15.87	.58	28.9	17.07	.58	30.9	18.27	.58	32.9	19.47	.58
0.8	27.2	21.28	.76	29.2	22.88	.76	31.2	24.48	.77	33.2	26.08	.77
1.0	27.5	26.75	.94	29.5	28.75	.94	31.5	30.75	.95	33.5	32.75	.95
1.2	27.8	32.28	1.11	29.8	34.68	1.12	31.8	37.08	1.12	33.8	39.48	1.13
1.4	28.1	37.87	1.28	30.1	40.67	1.29	32.1	43.47	1.30	34.1	46.27	1.30
1.6	28.4	43.52	1.45	30.4	46.72	1.46	32.4	49.92	1.47	34.4	53.12	1.48
1.8	28.7	49.23	1.61	30.7	52.83	1.63	32.7	56.43	1.64	34.7	60.03	1.64
2.0	29.0	55.00	1.77	31.0	59.00	1.79	33.0	63.00	1.80	35.0	67.00	1.81
2.2	29.3	60.83	1.93	31.3	65.23	1.95	33.3	69.63	1.96	35.3	74.03	1.97
2.4	29.6	66.72	2.09	31.6	71.52	2.10	33.6	76.32	2.12	35.6	81.12	2.13
2.6	29.9	72.67	2.24	31.9	77.87	2.26	33.9	83.07	2.28	35.9	88.27	2.29
2.8	30.2	78.68	2.38	32.2	84.28	2.41	34.2	89.88	2.43	36.2	95.48	2.45
3.0	30.5	84.75	2.53	32.5	90.75	2.56	34.5	96.75	2.58	36.5	102.75	2.60
3.2	30.8	90.88	2.67	32.8	97.28	2.70	34.8	103.68	2.73	36.8	110.08	2.75
3.4	31.1	97.07	2.81	33.1	103.87	2.85	35.1	110.67	2.87	37.1	117.47	2.90
3.6	31.4	103.32	2.95	33.4	110.52	2.99	35.4	117.72	3.02	37.4	124.92	3.05
3.8	31.7	109.63	3.09	33.7	117.23	3.13	35.7	124.83	3.16	37.7	132.43	3.19
4.0	32.0	116.00	3.22	34.0	124.00	3.26	36.0	132.00	3.30	38.0	140.00	3.33
4.2	32.3	122.43	3.35	34.3	130.83	3.40	36.3	139.23	3.44	38.3	147.63	3.47
4.4	32.6	128.92	3.48	34.6	137.72	3.53	36.6	146.52	3.57	38.6	155.32	3.61
4.6	32.9	135.47	3.61	34.9	144.67	3.66	36.9	153.87	3.71	38.9	163.07	3.75
4.8	33.2	142.08	3.74	35.2	151.68	3.79	37.2	161.28	3.84	39.2	170.88	3.88
5.0	33.5	148.75	3.86	35.5	158.75	3.92	37.5	168.75	3.97	39.5	178.75	4.02
5.2	33.8	155.48	3.99	35.8	165.88	4.05	37.8	176.28	4.10	39.8	186.68	4.15
5.4	34.1	162.27	4.11	36.1	173.07	4.17	38.1	183.87	4.23	40.1	194.67	4.28
5.6	34.4	169.12	4.23	36.4	180.32	4.29	38.4	191.52	4.35	40.4	202.72	4.41
5.8	34.7	176.03	4.35	36.7	187.63	4.41	38.7	199.23	4.48	40.7	210.83	4.53
6.0	35.0	183.00	4.46	37.0	195.00	4.53	39.0	207.00	4.60	41.0	219.00	4.66
6.2	35.3	190.03	4.58	37.3	202.43	4.65	39.3	214.83	4.72	41.3	227.23	4.78
6.4	35.6	197.12	4.69	37.6	209.92	4.77	39.6	222.72	4.84	41.6	235.52	4.91
6.6	35.9	204.27	4.81	37.9	217.47	4.89	39.9	230.67	4.96	41.9	243.87	5.03
6.8	36.2	211.48	4.92	38.2	225.08	5.00	40.2	238.68	5.08	42.2	252.28	5.15
7.0	36.5	218.75	5.03	38.5	232.75	5.12	40.5	246.75	5.19	42.5	260.75	5.27
7.2	36.8	226.08	5.14	38.8	240.48	5.23	40.8	254.88	5.31	42.8	269.28	5.39
7.4	37.1	233.47	5.25	39.1	248.27	5.34	41.1	263.07	5.42	43.1	277.87	5.50
7.6	37.4	240.92	5.35	39.4	256.12	5.45	41.4	271.32	5.54	43.4	286.52	5.62
7.8	37.7	248.43	5.46	39.7	264.03	5.56	41.7	279.63	5.65	43.7	295.23	5.73
8.0	38.0	256.00	5.57	40.0	272.00	5.67	42.0	288.00	5.76	44.0	304.00	5.85
8.2	38.3	263.63	5.67	40.3	280.03	5.77	42.3	296.43	5.87	44.3	312.83	5.96
8.4	38.6	271.32	5.77	40.6	288.12	5.88	42.6	304.92	5.98	44.6	321.72	6.07
8.6	38.9	279.07	5.88	40.9	296.27	5.99	42.9	313.47	6.09	44.9	330.67	6.18
8.8	39.2	286.88	5.98	41.2	304.48	6.09	43.2	322.08	6.19	45.2	339.68	6.29
9.0	39.5	294.75	6.08	41.5	312.75	6.19	43.5	330.75	6.30	45.5	348.75	6.40
9.2	39.8	302.68	6.18	41.8	321.08	6.30	43.8	339.48	6.41	45.8	357.88	6.51
9.4	40.1	310.67	6.28	42.1	329.47	6.40	44.1	348.27	6.51	46.1	367.07	6.61
9.6	40.4	318.72	6.37	42.4	337.92	6.50	44.4	357.12	6.61	46.4	376.32	6.72
9.8	40.7	326.83	6.47	42.7	346.43	6.60	44.7	366.03	6.72	46.7	385.63	6.83

Table 28.—*Area in square feet, A, top width in feet, T, and hydraulic radius in feet, r, of trapezoidal channels,* **side slopes ¾ to 1**—Continued

Depth	Bottom width 26 feet			Bottom width 28 feet			Bottom width 30 feet			Bottom width 32 feet		
	T	A	r	T	A	r	T	A	r	T	A	r
10.0	41.0	335.00	6.57	43.0	355.00	6.70	45.0	375.00	6.82	47.0	395.00	6.93
10.5	41.75	355.69	6.81	43.75	376.69	6.94	45.75	397.69	7.07	47.75	418.69	7.19
11.0	42.5	376.8	7.04	44.5	398.8	7.18	46.5	420.8	7.32	48.5	442.8	7.44
11.5	43.25	398.2	7.27	45.25	421.2	7.42	47.25	444.2	7.56	49.25	467.2	7.69
12.0	44.0	420.0	7.50	46.0	444.0	7.66	48.0	468.0	7.80	50.0	492.0	7.94
12.5	44.75	442.2	7.72	46.75	467.2	7.89	48.75	492.2	8.04	50.75	517.2	8.18
13.0	45.5	464.8	7.94	47.5	490.8	8.11	49.5	516.8	8.27	51.5	542.8	8.41
13.5	46.25	487.7	8.16	48.25	514.7	8.34	50.25	541.7	8.50	52.25	568.7	8.65
14.0	47.0	511.0	8.38	49.0	539.0	8.56	51.0	567.0	8.72	53.0	595.0	8.88
14.5	47.75	534.7	8.59	49.75	563.7	8.77	51.75	592.7	8.95	53.75	621.7	9.11
15.0	48.5	558.8	8.80	50.5	588.8	8.99	52.5	618.8	9.17	54.5	648.8	9.33
15.5	49.25	583.2	9.01	51.25	614.2	9.20	53.25	645.2	9.38	55.25	676.2	9.56
16.0	50.0	608.0	9.21	52.0	640.0	9.41	54.0	672.0	9.60	56.0	704.0	9.78
16.5	50.75	633.2	9.42	52.75	666.2	9.62	54.75	699.2	9.81	56.75	732.2	10.00
17.0	51.5	658.8	9.62	53.5	692.8	9.83	55.5	726.8	10.02	57.5	760.8	10.21
17.5	52.25	684.7	9.82	54.25	719.7	10.03	56.25	754.7	10.23	58.25	789.7	10.43
18.0	53.0	711.0	10.01	55.0	747.0	10.23	57.0	783.0	10.44	59.0	819.0	10.64
18.5	53.75	737.7	10.21	55.75	774.7	10.43	57.75	811.7	10.65	59.75	848.7	10.85
19.0	54.5	764.8	10.40	56.5	802.8	10.63	58.5	840.8	10.85	60.5	878.8	11.05
19.5	55.25	792.2	10.60	57.25	831.2	10.83	59.25	870.2	11.05	61.25	909.2	11.26
20.0	56.0	820.0	10.79	58.0	860.0	11.03	60.0	900.0	11.25	62.0	940.0	11.46

Depth	Bottom width 35 feet			Bottom width 40 feet			Bottom width 45 feet			Bottom width 50 feet		
	T	A	r	T	A	r	T	A	r	T	A	r
0.4	35.6	14.12	.39	40.6	16.12	.39	45.6	18.12	.39	50.6	20.12	.39
0.6	35.9	21.27	.58	40.9	24.27	.58	45.9	27.27	.59	50.9	30.27	.59
0.8	36.2	28.48	.77	41.2	32.48	.77	46.2	36.48	.78	51.2	40.48	.78
1.0	36.5	35.75	.95	41.5	40.75	.96	46.5	45.75	.96	51.5	50.75	.97
1.2	36.8	43.08	1.13	41.8	49.08	1.14	46.8	55.08	1.15	51.8	61.08	1.15
1.4	37.1	50.47	1.31	42.1	57.47	1.32	47.1	64.47	1.33	52.1	71.47	1.34
1.6	37.4	57.92	1.49	42.4	65.92	1.50	47.4	73.92	1.51	52.4	81.92	1.52
1.8	37.7	65.43	1.66	42.7	74.43	1.67	47.7	83.43	1.69	52.7	92.43	1.70
2.0	38.0	73.00	1.83	43.0	83.00	1.84	48.0	93.00	1.86	53.0	103.00	1.87
2.2	38.3	80.63	1.99	43.3	91.63	2.01	48.3	102.63	2.03	53.3	113.63	2.05
2.4	38.6	88.32	2.15	43.6	100.32	2.18	48.6	112.32	2.20	53.6	124.32	2.22
2.6	38.9	96.07	2.31	43.9	109.07	2.35	48.9	122.07	2.37	53.9	135.07	2.39
2.8	39.2	103.88	2.47	44.2	117.88	2.51	49.2	131.88	2.54	54.2	145.88	2.56
3.0	39.5	111.75	2.63	44.5	126.75	2.67	49.5	141.75	2.70	54.5	156.75	2.73
3.2	39.8	119.68	2.78	44.8	135.68	2.83	49.8	151.68	2.86	54.8	167.68	2.89
3.4	40.1	127.67	2.93	45.1	144.67	2.98	50.1	161.67	3.02	55.1	178.67	3.05
3.6	40.4	135.72	3.08	45.4	153.72	3.14	50.4	171.72	3.18	55.4	189.72	3.22
3.8	40.7	143.83	3.23	45.7	162.83	3.29	50.7	181.83	3.34	55.7	200.83	3.38
4.0	41.0	152.00	3.38	46.0	172.00	3.44	51.0	192.00	3.49	56.0	212.00	3.53
4.2	41.3	160.23	3.52	46.3	181.23	3.59	51.3	202.23	3.64	56.3	223.23	3.69
4.4	41.6	168.52	3.66	46.6	190.52	3.74	51.6	212.52	3.80	56.6	234.52	3.84
4.6	41.9	176.87	3.80	46.9	199.87	3.88	51.9	222.87	3.94	56.9	245.87	4.00
4.8	42.2	185.28	3.94	47.2	209.28	4.02	52.2	233.28	4.09	57.2	257.28	4.15

Table 28.—*Area in square feet, A, top width in feet, T, and hydraulic radius in feet, r, of trapezoidal channels,*
side slopes ¾ to 1—Continued

Depth	Bottom width 35 feet			Bottom width 40 feet			Bottom width 45 feet			Bottom width 50 feet		
	T	A	r	T	A	r	T	A	r	T	A	r
5.0	42.5	193.75	4.08	47.5	218.75	4.17	52.5	243.75	4.24	57.5	268.75	4.30
5.2	42.8	202.28	4.21	47.8	228.28	4.31	52.8	254.28	4.38	57.8	280.28	4.45
5.4	43.1	210.87	4.35	48.1	237.87	4.45	53.1	264.87	4.53	58.1	291.87	4.60
5.6	43.4	219.52	4.48	48.4	247.52	4.58	53.4	275.52	4.67	58.4	303.52	4.74
5.8	43.7	228.23	4.61	48.7	257.23	4.72	53.7	286.23	4.81	58.7	315.23	4.89
6.0	44.0	237.00	4.74	49.0	267.00	4.85	54.0	297.00	4.95	59.0	327.00	5.03
6.2	44.3	245.83	4.87	49.3	276.83	4.99	54.3	307.83	5.09	59.3	338.83	5.17
6.4	44.6	254.72	4.99	49.6	286.72	5.12	54.6	318.72	5.22	59.6	350.72	5.31
6.6	44.9	263.67	5.12	49.9	296.67	5.25	54.9	329.67	5.36	59.9	362.67	5.45
6.8	45.2	272.68	5.24	50.2	306.68	5.38	55.2	340.68	5.49	60.2	374.68	5.59
7.0	45.5	281.75	5.37	50.5	316.75	5.51	55.5	351.75	5.63	60.5	386.75	5.73
7.2	45.8	290.88	5.49	50.8	326.88	5.64	55.8	362.88	5.76	60.8	398.88	5.87
7.4	46.1	300.07	5.61	51.1	337.07	5.76	56.1	374.07	5.89	61.1	411.07	6.00
7.6	46.4	309.32	5.73	51.4	347.32	5.89	56.4	385.32	6.02	61.4	423.32	6.14
7.8	46.7	318.63	5.85	51.7	357.63	6.01	56.7	396.63	6.15	61.7	435.63	6.27
8.0	47.0	328.00	5.96	52.0	368.00	6.13	57.0	408.00	6.28	62.0	448.00	6.40
8.2	47.3	337.43	6.08	52.3	378.43	6.26	57.3	419.43	6.40	62.3	460.43	6.53
8.4	47.6	346.92	6.20	52.6	388.92	6.38	57.6	430.92	6.53	62.6	472.92	6.66
8.6	47.9	356.47	6.31	52.9	399.47	6.50	57.9	442.47	6.65	62.9	485.47	6.79
8.8	48.2	366.08	6.42	53.2	410.08	6.61	58.2	454.08	6.78	63.2	498.08	6.92
9.0	48.5	375.75	6.53	53.5	420.75	6.73	58.5	465.75	6.90	63.5	510.75	7.04
9.2	48.8	385.48	6.65	53.8	431.48	6.85	58.8	477.48	7.02	63.8	523.48	7.17
9.4	49.1	395.27	6.76	54.1	442.27	6.96	59.1	489.27	7.14	64.1	536.27	7.30
9.6	49.4	405.12	6.87	54.4	453.12	7.08	59.4	501.12	7.26	64.4	549.12	7.42
9.8	49.7	415.03	6.98	54.7	464.03	7.19	59.7	513.03	7.38	64.7	562.03	7.54
10.0	50.0	425.00	7.08	55.0	475.00	7.31	60.0	525.00	7.50	65.0	575.00	7.67
10.5	50.75	450.19	7.35	55.75	502.69	7.59	60.75	555.19	7.79	65.75	607.69	7.97
11.0	51.5	475.8	7.61	56.5	530.8	7.86	61.5	585.8	8.08	66.5	640.8	8.27
11.5	52.25	501.7	7.87	57.25	559.2	8.13	62.25	616.7	8.36	67.25	674.2	8.56
12.0	53.0	528.0	8.12	58.0	588.0	8.40	63.0	648.0	8.64	68.0	708.0	8.85
12.5	53.75	554.7	8.37	58.75	617.2	8.66	63.75	679.7	8.91	68.75	742.2	9.13
13.0	54.5	581.8	8.62	59.5	646.8	8.92	64.5	711.8	9.18	69.5	776.8	9.42
13.5	55.25	609.2	8.86	60.25	676.7	9.18	65.25	744.2	9.45	70.25	811.7	9.69
14.0	56.0	637.0	9.10	61.0	707.0	9.43	66.0	777.0	9.71	71.0	847.0	9.96
14.5	56.75	665.2	9.34	61.75	737.7	9.67	66.75	810.2	9.97	71.75	882.7	10.23
15.0	57.5	693.8	9.57	62.5	768.8	9.92	67.5	843.8	10.23	72.5	918.8	10.50
15.5	58.25	722.7	9.80	63.25	800.2	10.16	68.25	877.7	10.48	73.25	955.2	10.76
16.0	59.0	752.0	10.03	64.0	832.0	10.40	69.0	912.0	10.73	74.0	992.0	11.02
16.5	59.75	781.7	10.25	64.75	864.2	10.64	69.75	946.7	10.98	74.75	1,029.2	11.28
17.0	60.5	811.8	10.47	65.5	896.8	10.87	70.5	981.8	11.22	75.5	1,066.8	11.53
17.5	61.25	842.2	10.69	66.25	929.7	11.10	71.25	1,017.2	11.46	76.25	1,104.7	11.78
18.0	62.0	873.0	10.91	67.0	963.0	11.33	72.0	1,053.0	11.70	77.0	1,143.0	12.03
18.5	62.75	904.2	11.13	67.75	996.7	11.56	72.75	1,089.2	11.94	77.75	1,181.7	12.28
19.0	63.5	935.8	11.34	68.5	1,030.8	11.78	73.5	1,125.8	12.17	78.5	1,220.8	12.52
19.5	64.25	967.7	11.55	69.25	1,065.2	12.00	74.25	1,162.7	12.40	79.25	1,260.2	12.76
20.0	65.0	1,000.0	11.76	70.0	1,100.0	12.22	75.0	1,200.0	12.63	80.0	1,300.0	13.00

Table 28.—*Area in square feet, A; top width in feet, T, and hydraulic radius in feet, r, of trapezoidal channels,* **side slopes ¾ to 1**—Continued

Depth	Bottom width 60 feet			Bottom width 70 feet			Bottom width 80 feet			Bottom width 90 feet		
	T	A	r	T	A	r	T	A	r	T	A	r
0.4	60.6	24.12	0.40	70.6	28.12	0.40	80.6	32.12	0.40	90.6	36.12	0.40
0.6	60.9	36.27	.59	70.9	42.27	.59	80.9	48.27	.59	90.9	54.27	.59
0.8	61.2	48.48	.78	71.2	56.48	.78	81.2	64.48	.79	91.2	72.48	.79
1.0	61.5	60.75	.97	71.5	70.75	.98	81.5	80.75	.98	91.5	90.75	.98
1.2	61.8	73.08	1.16	71.8	85.08	1.17	81.8	97.08	1.17	91.8	109.08	1.17
1.4	62.1	85.47	1.35	72.1	99.47	1.35	82.1	113.47	1.36	92.1	127.47	1.36
1.6	62.4	97.92	1.53	72.4	113.92	1.54	82.4	129.92	1.55	92.4	145.92	1.55
1.8	62.7	110.43	1.71	72.7	128.43	1.72	82.7	146.43	1.73	92.7	164.43	1.74
2.0	63.0	123.00	1.89	73.0	143.00	1.91	83.0	163.00	1.92	93.0	183.00	1.93
2.2	63.3	135.63	2.07	73.3	157.63	2.09	83.3	179.63	2.10	93.3	201.63	2.11
2.4	63.6	148.32	2.25	73.6	172.32	2.27	83.6	196.32	2.28	93.6	220.32	2.30
2.6	63.9	161.07	2.42	73.9	187.07	2.45	83.9	213.07	2.46	93.9	239.07	2.48
2.8	64.2	173.88	2.60	74.2	201.88	2.62	84.2	229.88	2.64	94.2	257.88	2.66
3.0	64.5	186.75	2.77	74.5	216.75	2.80	84.5	246.75	2.82	94.5	276.75	2.84
3.2	64.8	199.68	2.94	74.8	231.68	2.97	84.8	263.68	3.00	94.8	295.68	3.02
3.4	65.1	212.67	3.10	75.1	246.67	3.14	85.1	280.67	3.17	95.1	314.67	3.19
3.6	65.4	225.72	3.27	75.4	261.72	3.31	85.4	297.72	3.35	95.4	333.72	3.37
3.8	65.7	238.83	3.44	75.7	276.83	3.48	85.7	314.83	3.52	95.7	352.83	3.55
4.0	66.0	252.00	3.60	76.0	292.00	3.65	86.0	332.00	3.69	96.0	372.00	3.72
4.2	66.3	265.23	3.76	76.3	307.23	3.82	86.3	349.23	3.86	96.3	391.23	3.89
4.4	66.6	278.52	3.92	76.6	322.52	3.98	86.6	366.52	4.03	96.6	410.52	4.06
4.6	66.9	291.87	4.08	76.9	337.87	4.15	86.9	383.87	4.20	96.9	429.87	4.24
4.8	67.2	305.28	4.24	77.2	353.28	4.31	87.2	401.28	4.36	97.2	449.28	4.40
5.0	67.5	318.75	4.40	77.5	368.75	4.47	87.5	418.75	4.53	97.5	468.75	4.57
5.2	67.8	332.28	4.55	77.8	384.28	4.63	87.8	436.28	4.69	97.8	488.28	4.74
5.4	68.1	345.87	4.71	78.1	399.87	4.79	88.1	453.87	4.85	98.1	507.87	4.91
5.6	68.4	359.52	4.86	78.4	415.52	4.95	88.4	471.52	5.02	98.4	527.52	5.07
5.8	68.7	373.23	5.01	78.7	431.23	5.10	88.7	489.23	5.18	98.7	547.23	5.24
6.0	69.0	387.00	5.16	79.0	447.00	5.26	89.0	507.00	5.34	99.0	567.00	5.40
6.2	69.3	400.83	5.31	79.3	462.83	5.41	89.3	524.83	5.50	99.3	586.83	5.56
6.4	69.6	414.72	5.46	79.6	478.72	5.57	89.6	542.72	5.65	99.6	606.72	5.72
6.6	69.9	428.67	5.60	79.9	494.67	5.72	89.9	560.67	5.81	99.9	626.67	5.88
6.8	70.2	442.68	5.75	80.2	510.68	5.87	90.2	578.68	5.97	100.2	646.68	6.04
7.0	70.5	456.75	5.89	80.5	526.75	6.02	90.5	596.75	6.12	100.5	666.75	6.20
7.2	70.8	470.88	6.04	80.8	542.88	6.17	90.8	614.88	6.27	100.8	686.88	6.36
7.4	71.1	485.07	6.18	81.1	559.07	6.32	91.1	633.07	6.43	101.1	707.07	6.52
7.6	71.4	499.32	6.32	81.4	575.32	6.46	91.4	651.32	6.58	101.4	727.32	6.67
7.8	71.7	513.63	6.46	81.7	591.63	6.61	91.7	669.63	6.73	101.7	747.63	6.83
8.0	72.0	528.00	6.60	82.0	608.00	6.76	92.0	688.00	6.88	102.0	768.00	6.98
8.2	72.3	542.43	6.74	82.3	624.43	6.90	92.3	706.43	7.03	102.3	788.43	7.14
8.4	72.6	556.92	6.88	82.6	640.92	7.04	92.6	724.92	7.18	102.6	808.92	7.29
8.6	72.9	571.47	7.01	82.9	657.47	7.19	92.9	743.47	7.32	102.9	829.47	7.44
8.8	73.2	586.08	7.15	83.2	674.08	7.33	93.2	762.08	7.47	103.2	850.08	7.59
9.0	73.5	600.75	7.28	83.5	690.75	7.47	93.5	780.75	7.62	103.5	870.75	7.74
9.2	73.8	615.48	7.42	83.8	707.48	7.61	93.8	799.48	7.76	103.8	891.48	7.89
9.4	74.1	630.27	7.55	84.1	724.27	7.75	94.1	818.27	7.91	104.1	912.27	8.04
9.6	74.4	645.12	7.68	84.4	741.12	7.88	94.4	837.12	8.05	104.4	933.12	8.19
9.8	74.7	660.03	7.81	84.7	758.03	8.02	94.7	856.03	8.19	104.7	954.03	8.33

Table 28.—*Area in square feet, A, top width in feet, T, and hydraulic radius in feet, r, of trapezoidal channels,* **side slopes ¾ to 1**—Continued

Depth	Bottom width 60 feet			Bottom width 70 feet			Bottom width 80 feet			Bottom width 90 feet		
	T	A	r	T	A	r	T	A	r	T	A	r
10.0	75.0	675.00	7.94	85.0	775.00	8.16	95.0	875.00	8.33	105.0	975.00	8.48
10.5	75.75	712.69	8.26	85.75	817.69	8.50	95.75	922.69	8.68	105.75	1,027.69	8.84
11.0	76.5	750.8	8.58	86.5	860.8	8.83	96.5	970.8	9.03	106.5	1,080.8	9.20
11.5	77.25	789.2	8.89	87.25	904.2	9.16	97.25	1,019.2	9.37	107.25	1,134.2	9.55
12.0	78.0	828.0	9.20	88.0	948.0	9.48	98.0	1,068.0	9.71	108.0	1,188.0	9.90
12.5	78.75	867.2	9.50	88.75	992.2	9.80	98.75	1,117.2	10.04	108.75	1,242.2	10.24
13.0	79.5	906.8	9.80	89.5	1,036.8	10.11	99.5	1,166.8	10.37	109.5	1,296.8	10.59
13.5	80.25	946.7	10.10	90.25	1,081.7	10.43	100.25	1,216.7	10.70	110.25	1,351.7	10.92
14.0	81.0	987.0	10.39	91.0	1,127.0	10.73	101.0	1,267.0	11.02	111.0	1,407.0	11.26
14.5	81.75	1,027.7	10.68	91.75	1,172.7	11.04	101.75	1,317.7	11.34	111.75	1,462.7	11.59
15.0	82.5	1,068.8	10.96	92.5	1,218.8	11.34	102.5	1,368.8	11.65	112.5	1,518.8	11.91
15.5	83.25	1,110.2	11.24	93.25	1,265.2	11.63	103.25	1,420.2	11.96	113.25	1,575.2	12.23
16.0	84.0	1,152.0	11.52	94.0	1,312.0	11.93	104.0	1,472.0	12.27	114.0	1,632.0	12.55
16.5	84.75	1,194.2	11.79	94.75	1,359.2	12.22	104.75	1,524.2	12.57	114.75	1,689.2	12.87
17.0	85.5	1,236.8	12.07	95.5	1,406.8	12.50	105.5	1,576.8	12.87	115.5	1,746.8	13.18
17.5	86.25	1,279.7	12.33	96.25	1,454.7	12.79	106.25	1,629.7	13.17	116.25	1,804.7	13.49
18.0	87.0	1,323.0	12.60	97.0	1,503.0	13.07	107.0	1,683.0	13.46	117.0	1,863.0	13.80
18.5	87.75	1,366.7	12.86	97.75	1,551.7	13.35	107.75	1,736.7	13.76	117.75	1,921.7	14.10
19.0	88.5	1,410.8	13.12	98.5	1,600.8	13.62	108.5	1,790.8	14.05	118.5	1,980.8	14.41
19.5	89.25	1,455.2	13.38	99.25	1,650.2	13.90	109.25	1,845.2	14.33	119.25	2,040.2	14.70
20.0	90.0	1,500.0	13.64	100.0	1,700.0	14.17	110.0	1,900.0	14.62	120.0	2,100.0	15.00

Table 29.—*Area in square feet, A, top width in feet, T, and hydraulic radius in feet, r, of trapezoidal channels,*
side slopes 1 to 1

Depth	Bottom width 2 feet			Bottom width 3 feet			Bottom width 4 feet			Bottom width 5 feet		
	T	A	r	T	A	r	T	A	r	T	A	r
0.4	2.8	.96	.31	3.8	1.36	.33	4.8	1.76	.34	5.8	2.16	0.35
0.6	3.2	1.56	.42	4.2	2.16	.46	5.2	2.76	.48	6.2	3.36	.50
0.8	3.6	2.24	.53	4.6	3.04	.58	5.6	3.84	.61	6.6	4.64	.64
1.0	4.0	3.00	.62	5.0	4.00	.69	6.0	5.00	.73	7.0	6.00	.77
1.2	4.4	3.84	.71	5.4	5.04	.79	6.4	6.24	.84	7.4	7.44	.89
1.4	4.8	4.76	.80	5.8	6.16	.89	6.8	7.56	.95	7.8	8.96	1.00
1.6	5.2	5.76	.88	6.2	7.36	.98	7.2	8.96	1.05	8.2	10.56	1.11
1.8	5.6	6.84	.96	6.6	8.64	1.07	7.6	10.44	1.15	8.6	12.24	1.21
2.0	6.0	8.00	1.04	7.0	10.00	1.16	8.0	12.00	1.24	9.0	14.00	1.31
2.2	6.4	9.24	1.12	7.4	11.44	1.24	8.4	13.64	1.33	9.4	15.84	1.41
2.4	6.8	10.56	1.20	7.8	12.96	1.32	8.8	15.36	1.42	9.8	17.76	1.51
2.6	7.2	11.96	1.28	8.2	14.56	1.41	9.2	17.16	1.51	10.2	19.76	1.60
2.8	7.6	13.44	1.35	8.6	16.24	1.49	9.6	19.04	1.60	10.6	21.84	1.69
3.0	8.0	15.00	1.43	9.0	18.00	1.57	10.0	21.00	1.68	11.0	24.00	1.78
3.2	8.4	16.64	1.51	9.4	19.84	1.65	10.4	23.04	1.77	11.4	26.24	1.87
3.4	8.8	18.36	1.58	9.8	21.76	1.72	10.8	25.16	1.85	11.8	28.56	1.95
3.6	9.2	20.16	1.65	10.2	23.76	1.80	11.2	27.36	1.93	12.2	30.96	2.04
3.8	9.6	22.04	1.73	10.6	25.84	1.88	11.6	29.64	2.01	12.6	33.44	2.12
4.0	10.0	24.00	1.80	11.0	28.00	1.96	12.0	32.00	2.09	13.0	36.00	2.21
4.2	10.4	26.04	1.88	11.4	30.24	2.03	12.4	34.44	2.17	13.4	38.64	2.29
4.4	10.8	28.16	1.95	11.8	32.56	2.11	12.8	36.96	2.25	13.8	41.36	2.37
4.6	11.2	30.36	2.02	12.2	34.96	2.18	13.2	39.56	2.33	14.2	44.16	2.45
4.8	11.6	32.64	2.10	12.6	37.44	2.26	13.6	42.24	2.40	14.6	47.04	2.53
5.0	12.0	35.00	2.17	13.0	40.00	2.33	14.0	45.00	2.48	15.0	50.00	2.61
5.2	12.4	37.44	2.24	13.4	42.64	2.41	14.4	47.84	2.56	15.4	53.04	2.69
5.4	12.8	39.96	2.31	13.8	45.36	2.48	14.8	50.76	2.63	15.8	56.16	2.77
5.6	13.2	42.56	2.39	14.2	48.16	2.56	15.2	53.76	2.71	16.2	59.36	2.85
5.8	13.6	45.24	2.46	14.6	51.04	2.63	15.6	56.84	2.79	16.6	62.64	2.93
6.0	14.0	48.00	2.53	15.0	54.00	2.70	16.0	60.00	2.86	17.0	66.00	3.00
6.2	14.4	50.84	2.60	15.4	57.04	2.78	16.4	63.24	2.94	17.4	69.44	3.08
6.4	14.8	53.76	2.67	15.8	60.16	2.85	16.8	66.56	3.01	17.8	72.96	3.16
6.6	15.2	56.76	2.75	16.2	63.36	2.92	17.2	69.96	3.09	18.2	76.56	3.23
6.8	15.6	59.84	2.82	16.6	66.64	3.00	17.6	73.44	3.16	18.6	80.24	3.31
7.0	16.0	63.00	2.89	17.0	70.00	3.07	18.0	77.00	3.24	19.0	84.00	3.39
7.2	16.4	66.24	2.96	17.4	73.44	3.14	18.4	80.64	3.31	19.4	87.84	3.46
7.4	16.8	69.56	3.03	17.8	76.96	3.22	18.8	84.36	3.38	19.8	91.76	3.54
7.6	17.2	72.96	3.11	18.2	80.56	3.29	19.2	88.16	3.46	20.2	95.76	3.61
7.8	17.6	76.44	3.18	18.6	84.24	3.36	19.6	92.04	3.53	20.6	99.84	3.69
8.0	18.0	80.00	3.25	19.0	88.00	3.43	20.0	96.00	3.61	21.0	104.00	3.76
8.2	-----	-------	----	19.4	91.84	3.51	20.4	100.04	3.68	21.4	108.24	3.84
8.4	-----	-------	----	19.8	95.76	3.58	20.8	104.16	3.75	21.8	112.56	3.91
8.6	-----	-------	----	20.2	99.76	3.65	21.2	108.36	3.83	22.2	116.96	3.99
8.8	-----	-------	----	20.6	103.84	3.72	21.6	112.64	3.90	22.6	121.44	4.06
9.0	-----	-------	----	21.0	108.00	3.80	22.0	117.00	3.97	23.0	126.00	4.14
9.2	-----	-------	----	21.4	112.24	3.87	22.4	121.44	4.05	23.4	130.64	4.21
9.4	-----	-------	----	21.8	116.56	3.94	22.8	125.96	4.12	23.8	135.36	4.29
9.6	-----	-------	----	22.2	120.96	4.01	23.2	130.56	4.19	24.2	140.16	4.36
9.8	-----	-------	----	22.6	125.44	4.08	23.6	135.24	4.26	24.6	145.04	4.43

Table 29.—*Area in square feet, A, top width in feet, T, and hydraulic radius in feet, r, of trapezoidal channels,*
side slopes 1 to 1—Continued

Depth	Bottom width 2 feet			Bottom width 3 feet			Bottom width 4 feet			Bottom width 5 feet		
	T	A	r	T	A	r	T	A	r	T	A	r
10.0				23.0	130.00	4.16	24.0	140.00	4.34	25.0	150.00	4.51
10.5				24.0	141.75	4.34	25.0	152.25	4.52	26.0	162.75	4.69
11.0				25.0	154.0	4.51	26.0	165.0	4.70	27.0	176.0	4.87
11.5				26.0	166.8	4.69	27.0	178.3	4.88	28.0	189.8	5.06
12.0				27.0	180.0	4.87	28.0	192.0	5.06	29.0	204.0	5.24
12.5				28.0	193.8	5.05	29.0	206.3	5.24	30.0	218.8	5.42
13.0							30.0	221.0	5.42	31.0	234.0	5.60
13.5							31.0	236.3	5.60	32.0	249.8	5.78
14.0							32.0	252.0	5.78	33.0	266.0	5.96
14.5							33.0	268.3	5.96	34.0	282.8	6.15
15.0							34.0	285.0	6.14	35.0	300.0	6.33
15.5							35.0	302.3	6.32	36.0	317.8	6.51
16.0							36.0	320.0	6.50	37.0	336.0	6.69
16.5										38.0	354.8	6.87
17.0										39.0	374.0	7.05
17.5										40.0	393.8	7.23
18.0										41.0	414.0	7.40
18.5										42.0	434.8	7.58
19.0										43.0	456.0	7.76
19.5										44.0	477.8	7.94
20.0										45.0	500.0	8.12

Depth	Bottom width 6 feet			Bottom width 7 feet			Bottom width 8 feet			Bottom width 9 feet		
	T	A	r	T	A	r.	T	A	r	T	A	r
0.4	6.8	2.56	.36	7.8	2.96	.36	8.8	3.36	.37	9.8	3.76	.37
0.6	7.2	3.96	.51	8.2	4.56	.52	9.2	5.16	.53	10.2	5.76	.54
0.8	7.6	5.44	.66	8.6	6.24	.67	9.6	7.04	.69	10.6	7.84	.70
1.0	8.0	7.00	.79	9.0	8.00	.81	10.0	9.00	.83	11.0	10.00	.85
1.2	8.4	8.64	.92	9.4	9.84	.95	10.4	11.04	.97	11.4	12.24	.99
1.4	8.8	10.36	1.04	9.8	11.76	1.07	10.8	13.16	1.10	11.8	14.56	1.12
1.6	9.2	12.16	1.16	10.2	13.76	1.19	11.2	15.36	1.23	12.2	16.96	1.25
1.8	9.6	14.04	1.27	10.6	15.84	1.31	11.6	17.64	1.35	12.6	19.44	1.38
2.0	10.0	16.00	1.37	11.0	18.00	1.42	12.0	20.00	1.46	13.0	22.00	1.50
2.2	10.4	18.04	1.48	11.4	20.24	1.53	12.4	22.44	1.58	13.4	24.64	1.62
2.4	10.8	20.16	1.58	11.8	22.56	1.64	12.8	24.96	1.69	13.8	27.36	1.73
2.6	11.2	22.36	1.67	12.2	24.96	1.74	13.2	27.56	1.79	14.2	30.16	1.84
2.8	11.6	24.64	1.77	12.6	27.44	1.84	13.6	30.24	1.90	14.6	33.04	1.95
3.0	12.0	27.00	1.86	13.0	30.00	1.94	14.0	33.00	2.00	15.0	36.00	2.06
3.2	12.4	29.44	1.96	13.4	32.64	2.03	14.4	35.84	2.10	15.4	39.04	2.16
3.4	12.8	31.96	2.05	13.8	35.36	2.13	14.8	38.76	2.20	15.8	42.16	2.26
3.6	13.2	34.56	2.14	14.2	38.16	2.22	15.2	41.76	2.30	16.2	45.36	2.36
3.8	13.6	37.24	2.22	14.6	41.04	2.31	15.6	44.84	2.39	16.6	48.64	2.46
4.0	14.0	40.00	2.31	15.0	44.00	2.40	16.0	48.00	2.49	17.0	52.00	2.56
4.2	14.4	42.84	2.40	15.4	47.04	2.49	16.4	51.24	2.58	17.4	55.44	2.66
4.4	14.8	45.76	2.48	15.8	50.16	2.58	16.8	54.56	2.67	17.8	58.96	2.75
4.6	15.2	48.76	2.56	16.2	53.36	2.67	17.2	57.96	2.76	18.2	62.56	2.84
4.8	15.6	51.84	2.65	16.6	56.64	2.75	17.6	61.44	2.85	18.6	66.24	2.93

Table 29.—*Area in square feet, A, top width in feet, T, and hydraulic radius in feet, r, of trapezoidal channels,*
side slopes 1 to 1—Continued

Depth	Bottom width 6 feet			Bottom width 7 feet			Bottom width 8 feet			Bottom width 9 feet		
	T	A	r	T	A	r	T	A	r	T	A	r
5.0	16.0	55.00	2.73	17.0	60.00	2.84	18.0	65.00	2.94	19.0	70.00	3.02
5.2	16.4	58.24	2.81	17.4	63.44	2.92	18.4	68.64	3.02	19.4	73.84	3.11
5.4	16.8	61.56	2.89	17.8	66.96	3.01	18.8	72.36	3.11	19.8	77.76	3.20
5.6	17.2	64.96	2.97	18.2	70.56	3.09	19.2	76.16	3.19	20.2	81.76	3.29
5.8	17.6	68.44	3.05	18.6	74.24	3.17	19.6	80.04	3.28	20.6	85.84	3.38
6.0	18.0	72.00	3.13	19.0	78.00	3.25	20.0	84.00	3.36	21.0	90.00	3.47
6.2	18.4	75.64	3.21	19.4	81.84	3.34	20.4	88.04	3.45	21.4	94.24	3.55
6.4	18.8	79.36	3.29	19.8	85.76	3.42	20.8	92.16	3.53	21.8	98.56	3.64
6.6	19.2	83.16	3.37	20.2	89.76	3.50	21.2	96.36	3.61	22.2	102.96	3.72
6.8	19.6	87.04	3.45	20.6	93.84	3.58	21.6	100.64	3.70	22.6	107.44	3.81
7.0	20.0	91.00	3.53	21.0	98.00	3.66	22.0	105.00	3.78	23.0	112.00	3.89
7.2	20.4	95.04	3.60	21.4	102.24	3.74	22.4	109.44	3.86	23.4	116.64	3.97
7.4	20.8	99.16	3.68	21.8	106.56	3.82	22.8	113.96	3.94	23.8	121.36	4.05
7.6	21.2	103.36	3.76	22.2	110.96	3.89	23.2	118.56	4.02	24.2	126.16	4.14
7.8	21.6	107.64	3.84	22.6	115.44	3.97	23.6	123.24	4.10	24.6	131.04	4.22
8.0	22.0	112.00	3.91	23.0	120.00	4.05	24.0	128.00	4.18	25.0	136.00	4.30
8.2	22.4	116.44	3.99	23.4	124.64	4.13	24.4	132.84	4.26	25.4	141.04	4.38
8.4	22.8	120.96	4.06	23.8	129.36	4.21	24.8	137.76	4.34	25.8	146.16	4.46
8.6	23.2	125.56	4.14	24.2	134.16	4.28	25.2	142.76	4.42	26.2	151.36	4.54
8.8	23.6	130.24	4.22	24.6	139.04	4.36	25.6	147.84	4.49	26.6	156.64	4.62
9.0	24.0	135.00	4.29	25.0	144.00	4.44	26.0	153.00	4.57	27.0	162.00	4.70
9.2	24.4	139.84	4.37	25.4	149.04	4.51	26.4	158.24	4.65	27.4	167.44	4.78
9.4	24.8	144.76	4.44	25.8	154.16	4.59	26.8	163.56	4.73	27.8	172.96	4.86
9.6	25.2	149.76	4.52	26.2	159.36	4.67	27.2	168.96	4.81	28.2	178.56	4.94
9.8	25.6	154.84	4.59	26.6	164.64	4.74	27.6	174.44	4.88	28.6	184.24	5.02
10.0	26.0	160.00	4.67	27.0	170.00	4.82	28.0	180.00	4.96	29.0	190.00	5.10
10.5	27.0	173.25	4.85	28.0	183.75	5.01	29.0	194.25	5.15	30.0	204.75	5.29
11.0	28.0	187.0	5.04	29.0	198.0	5.20	30.0	209.0	5.34	31.0	220.0	5.48
11.5	29.0	201.3	5.22	30.0	212.8	5.38	31.0	224.3	5.53	32.0	235.8	5.68
12.0	30.0	216.0	5.41	31.0	228.0	5.57	32.0	240.0	5.72	33.0	252.0	5.87
12.5	31.0	231.3	5.59	32.0	243.8	5.76	33.0	256.3	5.91	34.0	268.8	6.06
13.0	32.0	247.0	5.78	33.0	260.0	5.94	34.0	273.0	6.10	35.0	286.0	6.25
13.5	33.0	263.3	5.96	34.0	276.8	6.13	35.0	290.3	6.29	36.0	303.8	6.44
14.0	34.0	280.0	6.14	35.0	294.0	6.31	36.0	308.0	6.47	37.0	322.0	6.63
14.5	35.0	297.3	6.32	36.0	311.8	6.49	37.0	326.3	6.66	38.0	340.8	6.81
15.0	36.0	315.0	6.50	37.0	330.0	6.68	38.0	345.0	6.84	39.0	360.0	7.00
15.5	37.0	333.3	6.69	38.0	348.8	6.86	39.0	364.3	7.03	40.0	379.8	7.19
16.0	38.0	352.0	6.87	39.0	368.0	7.04	40.0	384.0	7.21	41.0	400.0	7.37
16.5	39.0	371.3	7.05	40.0	387.8	7.23	41.0	404.3	7.40	42.0	420.8	7.56
17.0	40.0	391.0	7.23	41.0	408.0	7.41	42.0	425.0	7.58	43.0	442.0	7.74
17.5	41.0	411.3	7.41	42.0	428.8	7.59	43.0	446.3	7.76	44.0	463.8	7.93
18.0	42.0	432.0	7.59	43.0	450.0	7.77	44.0	468.0	7.94	45.0	486.0	8.11
18.5	43.0	453.3	7.77	44.0	471.8	7.95	45.0	490.3	8.13	46.0	508.8	8.30
19.0	44.0	475.0	7.95	45.0	494.0	8.13	46.0	513.0	8.31	47.0	532.0	8.48
19.5	45.0	497.3	8.13	46.0	516.8	8.31	47.0	536.3	8.49	48.0	555.8	8.66
20.0	46.0	520.0	8.31	47.0	540.0	8.49	48.0	560.0	8.67	49.0	580.0	8.85

Table 29.—*Area in square feet, A, top width in feet, T, and hydraulic radius in feet, r, of trapezoidal channels,* side slopes 1 to 1—Continued

Depth	Bottom width 10 feet			Bottom width 12 feet			Bottom width 14 feet			Bottom width 16 feet		
	T	A	r	T	A	r	T	A	r	T	A	r
0.4	10.8	4.16	.37	12.8	4.96	.38	14.8	5.76	.38	16.8	6.56	.38
0.6	11.2	6.36	.54	13.2	7.56	.55	15.2	8.76	.56	17.2	9.96	.56
0.8	11.6	8.64	.70	13.6	10.24	.72	15.6	11.84	.73	17.6	13.44	.74
1.0	12.0	11.00	.86	14.0	13.00	.88	16.0	15.00	.89	18.0	17.00	.90
1.2	12.4	13.44	1.00	14.4	15.84	1.03	16.4	18.24	1.05	18.4	20.64	1.06
1.4	12.8	15.96	1.14	14.8	18.76	1.18	16.8	21.56	1.20	18.8	24.36	1.22
1.6	13.2	18.56	1.28	15.2	21.76	1.32	17.2	24.96	1.35	19.2	28.16	1.37
1.8	13.6	21.24	1.41	15.6	24.84	1.45	17.6	28.44	1.49	19.6	32.04	1.52
2.0	14.0	24.00	1.53	16.0	28.00	1.59	18.0	32.00	1.63	20.0	36.00	1.66
2.2	14.4	26.84	1.65	16.4	31.24	1.71	18.4	35.64	1.76	20.4	40.04	1.80
2.4	14.8	29.76	1.77	16.8	34.56	1.84	18.8	39.36	1.89	20.8	44.16	1.94
2.6	15.2	32.76	1.89	17.2	37.96	1.96	19.2	43.16	2.02	21.2	48.36	2.07
2.8	15.6	35.84	2.00	17.6	41.44	2.08	19.6	47.04	2.15	21.6	52.64	2.20
3.0	16.0	39.00	2.11	18.0	45.00	2.20	20.0	51.00	2.27	22.0	57.00	2.33
3.2	16.4	42.24	2.22	18.4	48.64	2.31	20.4	55.04	2.39	22.4	61.44	2.45
3.4	16.8	45.56	2.32	18.8	52.36	2.42	20.8	59.16	2.51	22.8	65.96	2.57
3.6	17.2	48.96	2.43	19.2	56.16	2.53	21.2	63.36	2.62	23.2	70.56	2.69
3.8	17.6	52.44	2.53	19.6	60.04	2.64	21.6	67.64	2.73	23.6	75.24	2.81
4.0	18.0	56.00	2.63	20.0	64.00	2.75	22.0	72.00	2.84	24.0	80.00	2.93
4.2	18.4	59.64	2.73	20.4	68.04	2.85	22.4	76.44	2.95	24.4	84.84	3.04
4.4	18.8	63.36	2.82	20.8	72.16	2.95	22.8	80.96	3.06	24.8	89.76	3.16
4.6	19.2	67.16	2.92	21.2	76.36	3.05	23.2	85.56	3.17	25.2	94.76	3.27
4.8	19.6	71.04	3.01	21.6	80.64	3.15	23.6	90.24	3.27	25.6	99.84	3.38
5.0	20.0	75.00	3.11	22.0	85.00	3.25	24.0	95.00	3.38	26.0	105.00	3.48
5.2	20.4	79.04	3.20	22.4	89.44	3.35	24.4	99.84	3.48	26.4	110.24	3.59
5.4	20.8	83.16	3.29	22.8	93.96	3.45	24.8	104.76	3.58	26.8	115.56	3.70
5.6	21.2	87.36	3.38	23.2	98.56	3.54	25.2	109.76	3.68	27.2	120.96	3.80
5.8	21.6	91.64	3.47	23.6	103.24	3.63	25.6	114.84	3.78	27.6	126.44	3.90
6.0	22.0	96.00	3.56	24.0	108.00	3.73	26.0	120.00	3.87	28.0	132.00	4.00
6.2	22.4	100.44	3.65	24.4	112.84	3.82	26.4	125.24	3.97	28.4	137.64	4.10
6.4	22.8	104.96	3.73	24.8	117.76	3.91	26.8	130.56	4.07	28.8	143.36	4.20
6.6	23.2	109.56	3.82	25.2	122.76	4.00	27.2	135.96	4.16	29.2	149.16	4.30
6.8	23.6	114.24	3.91	25.6	127.84	4.09	27.6	141.44	4.26	29.6	155.04	4.40
7.0	24.0	119.00	3.99	26.0	133.00	4.18	28.0	147.00	4.35	30.0	161.00	4.50
7.2	24.4	123.84	4.08	26.4	138.24	4.27	28.4	152.64	4.44	30.4	167.04	4.59
7.4	24.8	128.76	4.16	26.8	143.56	4.36	28.8	158.36	4.53	30.8	173.16	4.69
7.6	25.2	133.76	4.25	27.2	148.96	4.45	29.2	164.16	4.62	31.2	179.36	4.78
7.8	25.6	138.84	4.33	27.6	154.44	4.53	29.6	170.04	4.72	31.6	185.64	4.88
8.0	26.0	144.00	4.41	28.0	160.00	4.62	30.0	176.00	4.81	32.0	192.00	4.97
8.2	26.4	149.24	4.50	28.4	165.64	4.71	30.4	182.04	4.89	32.4	198.44	5.06
8.4	26.8	154.56	4.58	28.8	171.36	4.79	30.8	188.16	4.98	32.8	204.96	5.16
8.6	27.2	159.96	4.66	29.2	177.16	4.88	31.2	194.36	5.07	33.2	211.56	5.25
8.8	27.6	165.44	4.74	29.6	183.04	4.96	31.6	200.64	5.16	33.6	218.24	5.34
9.0	28.0	171.00	4.82	30.0	189.00	5.05	32.0	207.00	5.25	34.0	225.00	5.43
9.2	28.4	176.64	4.90	30.4	195.04	5.13	32.4	213.44	5.33	34.4	231.84	5.52
9.4	28.8	182.36	4.98	30.8	201.16	5.21	32.8	219.96	5.42	34.8	238.76	5.61
9.6	29.2	188.16	5.06	31.2	207.36	5.30	33.2	226.56	5.51	35.2	245.76	5.70
9.8	29.6	194.04	5.14	31.6	213.64	5.38	33.6	233.24	5.59	35.6	252.84	5.78

Table 29.—*Area in square feet, A, top width in feet, T, and hydraulic radius in feet, r, of trapezoidal channels,*
side slopes 1 to 1—Continued

Depth	Bottom width 10 feet			Bottom width 12 feet			Bottom width 14 feet			Bottom width 16 feet		
	T	A	r	T	A	r	T	A	r	T	A	r
10.0	30.0	200.00	5.22	32.0	220.00	5.46	34.0	240.00	5.68	36.0	260.00	5.87
10.5	31.0	215.25	5.42	33.0	236.25	5.67	35.0	257.25	5.89	37.0	278.25	6.09
11.0	32.0	231.0	5.62	34.0	253.0	5.87	36.0	275.0	6.10	38.0	297.0	6.30
11.5	33.0	247.3	5.82	35.0	270.3	6.07	37.0	293.3	6.30	39.0	316.3	6.52
12.0	34.0	264.0	6.01	36.0	288.0	6.27	38.0	312.0	6.51	40.0	336.0	6.73
12.5	35.0	281.3	6.20	37.0	306.3	6.47	39.0	331.3	6.71	41.0	356.3	6.94
13.0	36.0	299.0	6.39	38.0	325.0	6.66	40.0	351.0	6.91	42.0	377.0	7.14
13.5	37.0	317.3	6.59	39.0	344.3	6.86	41.0	371.3	7.12	43.0	398.3	7.35
14.0	38.0	336.0	6.77	40.0	364.0	7.05	42.0	392.0	7.31	44.0	420.0	7.55
14.5	39.0	355.3	6.97	41.0	384.3	7.25	43.0	413.3	7.51	45.0	442.3	7.76
15.0	40.0	375.0	7.15	42.0	405.0	7.44	44.0	435.0	7.71	46.0	465.0	7.96
15.5	41.0	395.3	7.34	43.0	426.3	7.63	45.0	457.3	7.91	47.0	488.3	8.16
16.0	42.0	416.0	7.53	44.0	448.0	7.82	46.0	480.0	8.10	48.0	512.0	8.36
16.5	43.0	437.3	7.72	45.0	470.3	8.02	47.0	503.3	8.30	49.0	536.3	8.56
17.0	44.0	459.0	7.90	46.0	493.0	8.21	48.0	527.0	8.49	50.0	561.0	8.75
17.5	45.0	481.3	8.09	47.0	516.3	8.40	49.0	551.3	8.68	51.0	586.3	8.95
18.0	46.0	504.0	8.27	48.0	540.0	8.58	50.0	576.0	8.87	52.0	612.0	9.15
18.5	47.0	527.3	8.46	49.0	564.3	8.77	51.0	601.3	9.07	53.0	638.3	9.34
19.0	48.0	551.0	8.64	50.0	589.0	8.96	52.0	627.0	9.26	54.0	665.0	9.54
19.5	49.0	575.3	8.83	51.0	614.3	9.15	53.0	653.3	9.45	55.0	692.3	9.73
20.0	50.0	600.0	9.01	52.0	640.0	9.33	54.0	680.0	9.64	56.0	720.0	9.92

Depth	Bottom width 18 feet			Bottom width 20 feet			Bottom width 22 feet			Bottom width 24 feet		
	T	A	r	T	A	r	T	A	r	T	A	r
0.4	18.8	7.36	.38	20.8	8.16	.39	22.8	8.96	.39	24.8	9.76	.39
0.6	19.2	11.16	.57	21.2	12.36	.57	23.2	13.56	.57	25.2	14.76	.57
0.8	19.6	15.04	.74	21.6	16.64	.75	23.6	18.24	.75	25.6	19.84	.76
1.0	20.0	19.00	.91	22.0	21.00	.92	24.0	23.00	.93	26.0	25.00	.93
1.2	20.4	23.04	1.08	22.4	25.44	1.09	24.4	27.84	1.10	26.4	30.24	1.10
1.4	20.8	27.16	1.24	22.8	29.96	1.25	24.8	32.76	1.26	26.8	35.56	1.27
1.6	21.2	31.36	1.39	23.2	34.56	1.41	25.2	37.76	1.42	27.2	40.96	1.44
1.8	21.6	35.64	1.54	23.6	39.24	1.56	25.6	42.84	1.58	27.6	46.44	1.60
2.0	22.0	40.00	1.69	24.0	44.00	1.71	26.0	48.00	1.74	28.0	52.00	1.75
2.2	22.4	44.44	1.83	24.4	48.84	1.86	26.4	53.24	1.89	28.4	57.64	1.91
2.4	22.8	48.96	1.98	24.8	53.76	2.01	26.8	58.56	2.03	28.8	63.36	2.06
2.6	23.2	53.56	2.11	25.2	58.76	2.15	27.2	63.96	2.18	29.2	69.16	2.21
2.8	23.6	58.24	2.25	25.6	63.84	2.29	27.6	69.44	2.32	29.6	75.04	2.35
3.0	24.0	63.00	2.38	26.0	69.00	2.42	28.0	75.00	2.46	30.0	81.00	2.49
3.2	24.4	67.84	2.51	26.4	74.24	2.56	28.4	80.64	2.60	30.4	87.04	2.63
3.4	24.8	72.76	2.63	26.8	79.56	2.69	28.8	86.36	2.73	30.8	93.16	2.77
3.6	25.2	77.76	2.76	27.2	84.96	2.81	29.2	92.16	2.86	31.2	99.36	2.91
3.8	25.6	82.84	2.88	27.6	90.44	2.94	29.6	98.04	2.99	31.6	105.64	3.04
4.0	26.0	88.00	3.00	28.0	96.00	3.07	30.0	104.00	3.12	32.0	112.00	3.17
4.2	26.4	93.24	3.12	28.4	101.64	3.19	30.4	110.04	3.25	32.4	118.44	3.30
4.4	26.8	98.56	3.24	28.8	107.36	3.31	30.8	116.16	3.37	32.8	124.96	3.43
4.6	27.2	103.96	3.35	29.2	113.16	3.43	31.2	122.36	3.49	33.2	131.56	3.55
4.8	27.6	109.44	3.47	29.6	119.04	3.55	31.6	128.64	3.62	33.6	138.24	3.68

Table 29.—*Area in square feet, A, top width in feet, T, and hydraulic radius in feet, r, of trapezoidal channels,*
side slopes 1 to 1—Continued

Depth	Bottom width 18 feet			Bottom width 20 feet			Bottom width 22 feet			Bottom width 24 feet		
	T	A	r	T	A	r	T	A	r	T	A	r
5.0	28.0	115.00	3.58	30.0	125.00	3.66	32.0	135.00	3.74	34.0	145.00	3.80
5.2	28.4	120.64	3.69	30.4	131.04	3.78	32.4	141.44	3.85	34.4	151.84	3.92
5.4	28.8	126.36	3.80	30.8	137.16	3.89	32.8	147.96	3.97	34.8	158.76	4.04
5.6	29.2	132.16	3.91	31.2	143.36	4.00	33.2	154.56	4.08	35.2	165.76	4.16
5.8	29.6	138.04	4.01	31.6	149.64	4.11	33.6	161.24	4.20	35.6	172.84	4.28
6.0	30.0	144.00	4.12	32.0	156.00	4.22	34.0	168.00	4.31	36.0	180.00	4.39
6.2	30.4	150.04	4.22	32.4	162.44	4.33	34.4	174.84	4.42	36.4	187.24	4.51
6.4	30.8	156.16	4.33	32.8	168.96	4.43	34.8	181.76	4.53	36.8	194.56	4.62
6.6	31.2	162.36	4.43	33.2	175.56	4.54	35.2	188.76	4.64	37.2	201.96	4.73
6.8	31.6	168.64	4.53	33.6	182.24	4.65	35.6	195.84	4.75	37.6	209.44	4.84
7.0	32.0	175.00	4.63	34.0	189.00	4.75	36.0	203.00	4.86	38.0	217.00	4.95
7.2	32.4	181.44	4.73	34.4	195.84	4.85	36.4	210.24	4.96	38.4	224.64	5.06
7.4	32.8	187.96	4.83	34.8	202.76	4.95	36.8	217.56	5.07	38.8	232.36	5.17
7.6	33.2	194.56	4.93	35.2	209.76	5.05	37.2	224.96	5.17	39.2	240.16	5.28
7.8	33.6	201.24	5.02	35.6	216.84	5.16	37.6	232.44	5.28	39.6	248.04	5.38
8.0	34.0	208.00	5.12	36.0	224.00	5.25	38.0	240.00	5.38	40.0	256.00	5.49
8.2	34.4	214.84	5.22	36.4	231.24	5.35	38.4	247.64	5.48	40.4	264.04	5.59
8.4	34.8	221.76	5.31	36.8	238.56	5.45	38.8	255.36	5.58	40.8	272.16	5.70
8.6	35.2	228.76	5.40	37.2	245.96	5.55	39.2	263.16	5.68	41.2	280.36	5.80
8.8	35.6	235.84	5.50	37.6	253.44	5.65	39.6	271.04	5.78	41.6	288.64	5.90
9.0	36.0	243.00	5.59	38.0	261.00	5.74	40.0	279.00	5.88	42.0	297.00	6.01
9.2	36.4	250.24	5.68	38.4	268.64	5.84	40.4	287.04	5.98	42.4	305.44	6.11
9.4	36.8	257.56	5.78	38.8	276.36	5.93	40.8	295.16	6.07	42.8	313.96	6.21
9.6	37.2	264.96	5.87	39.2	284.16	6.03	41.2	303.36	6.17	43.2	322.56	6.31
9.8	37.6	272.44	5.96	39.6	292.04	6.12	41.6	311.64	6.27	43.6	331.24	6.40
10.0	38.0	280.00	6.05	40.0	300.00	6.21	42.0	320.00	6.36	44.0	340.00	6.50
10.5	39.0	299.25	6.27	41.0	320.25	6.44	43.0	341.25	6.60	45.0	362.25	6.75
11.0	40.0	319.0	6.50	42.0	341.0	6.67	44.0	363.0	6.83	46.0	385.0	6.99
11.5	41.0	339.3	6.72	43.0	362.3	6.90	45.0	385.3	7.07	47.0	408.3	7.22
12.0	42.0	360.0	6.93	44.0	384.0	7.12	46.0	408.0	7.29	48.0	432.0	7.46
12.5	43.0	381.3	7.15	45.0	406.3	7.34	47.0	431.3	7.52	49.0	456.3	7.69
13.0	44.0	403.0	7.36	46.0	429.0	7.56	48.0	455.0	7.74	50.0	481.0	7.92
13.5	45.0	425.3	7.57	47.0	452.3	7.77	49.0	479.3	7.96	51.0	506.3	8.14
14.0	46.0	448.0	7.78	48.0	476.0	7.99	50.0	504.0	8.18	52.0	532.0	8.37
14.5	47.0	471.3	7.99	49.0	500.3	8.20	51.0	529.3	8.40	53.0	558.3	8.59
15.0	48.0	495.0	8.19	50.0	525.0	8.41	52.0	555.0	8.61	54.0	585.0	8.81
15.5	49.0	519.3	8.40	51.0	550.3	8.62	53.0	581.3	8.83	55.0	612.3	9.03
16.0	50.0	544.0	8.60	52.0	576.0	8.83	54.0	608.0	9.04	56.0	640.0	9.24
16.5	51.0	569.3	8.80	53.0	602.3	9.03	55.0	635.3	9.25	57.0	668.3	9.46
17.0	52.0	595.0	9.00	54.0	629.0	9.24	56.0	663.0	9.46	58.0	697.0	9.67
17.5	53.0	621.3	9.20	55.0	656.3	9.44	57.0	691.3	9.67	59.0	726.3	9.88
18.0	54.0	648.0	9.40	56.0	684.0	9.65	58.0	720.0	9.87	60.0	756.0	10.09
18.5	55.0	675.3	9.60	57.0	712.3	9.85	59.0	749.3	10.08	61.0	786.3	10.30
19.0	56.0	703.0	9.80	58.0	741.0	10.05	60.0	779.0	10.29	62.0	817.0	10.51
19.5	57.0	731.3	10.00	59.0	770.3	10.25	61.0	809.3	10.49	63.0	848.3	10.72
20.0	58.0	760.0	10.19	60.0	800.0	10.45	62.0	840.0	10.69	64.0	880.0	10.92

Table 29.—*Area in square feet, A, top width in feet, T, and hydraulic radius in feet, r, of trapezoidal channels,* **side slopes 1 to 1**—Continued

Depth	Bottom width 26 feet			Bottom width 28 feet			Bottom width 30 feet			Bottom width 32 feet		
	T	*A*	*r*	*T*	*A*	*r*	*T*	*A*	*r*	*T*	*A*	*r*
0.4	26.8	10.56	.39	28.8	11.36	.39	30.8	12.16	.39	32.8	12.96	.39
0.6	27.2	15.96	.58	29.2	17.16	.58	31.2	18.36	.58	33.2	19.56	.58
0.8	27.6	21.44	.76	29.6	23.04	.76	31.6	24.64	.76	33.6	26.24	.77
1.0	28.0	27.00	.94	30.0	29.00	.94	32.0	31.00	.94	34.0	33.00	.95
1.2	28.4	32.64	1.11	30.4	35.04	1.12	32.4	37.44	1.12	34.4	39.84	1.13
1.4	28.8	38.36	1.28	30.8	41.16	1.29	32.8	43.96	1.29	34.8	46.76	1.30
1.6	29.2	44.16	1.45	31.2	47.36	1.46	33.2	50.56	1.46	35.2	53.76	1.47
1.8	29.6	50.04	1.61	31.6	53.64	1.62	33.6	57.24	1.63	35.6	60.84	1.64
2.0	30.0	56.00	1.77	32.0	60.00	1.78	34.0	64.00	1.79	36.0	68.00	1.81
2.2	30.4	62.04	1.93	32.4	66.44	1.94	34.4	70.84	1.96	36.4	75.24	1.97
2.4	30.8	68.16	2.08	32.8	72.96	2.10	34.8	77.76	2.11	36.8	82.56	2.13
2.6	31.2	74.36	2.23	33.2	79.56	2.25	35.2	84.76	2.27	37.2	89.96	2.29
2.8	31.6	80.64	2.38	33.6	86.24	2.40	35.6	91.84	2.42	37.6	97.44	2.44
3.0	32.0	87.00	2.52	34.0	93.00	2.55	36.0	99.00	2.57	38.0	105.00	2.59
3.2	32.4	93.44	2.67	34.4	99.84	2.69	36.4	106.24	2.72	38.4	112.64	2.74
3.4	32.8	99.96	2.81	34.8	106.76	2.84	36.8	113.56	2.87	38.8	120.36	2.89
3.6	33.2	106.56	2.95	35.2	113.76	2.98	37.2	120.96	3.01	39.2	128.16	3.04
3.8	33.6	113.24	3.08	35.6	120.84	3.12	37.6	128.44	3.15	39.6	136.04	3.18
4.0	34.0	120.00	3.22	36.0	128.00	3.26	38.0	136.00	3.29	40.0	144.00	3.32
4.2	34.4	126.84	3.35	36.4	135.24	3.39	38.4	143.64	3.43	40.4	152.04	3.46
4.4	34.8	133.76	3.48	36.8	142.56	3.52	38.8	151.36	3.57	40.8	160.16	3.60
4.6	35.2	140.76	3.61	37.2	149.96	3.66	39.2	159.16	3.70	41.2	168.36	3.74
4.8	35.6	147.84	3.74	37.6	157.44	3.79	39.6	167.04	3.83	41.6	176.64	3.88
5.0	36.0	155.00	3.86	38.0	165.00	3.92	40.0	175.00	3.96	42.0	185.00	4.01
5.2	36.4	162.24	3.99	38.4	172.64	4.04	40.4	183.04	4.09	42.4	193.44	4.14
5.4	36.8	169.56	4.11	38.8	180.36	4.17	40.8	191.16	4.22	42.8	201.96	4.27
5.6	37.2	176.96	4.23	39.2	188.16	4.29	41.2	199.30	4.35	43.2	210.56	4.40
5.8	37.6	184.44	4.35	39.6	196.04	4.41	41.6	207.64	4.47	43.6	219.24	4.53
6.0	38.0	192.00	4.47	40.0	204.00	4.54	42.0	216.00	4.60	44.0	228.00	4.66
6.2	38.4	199.64	4.59	40.4	212.04	4.66	42.4	224.44	4.72	44.4	236.84	4.78
6.4	38.8	207.36	4.70	40.8	220.16	4.78	42.8	232.96	4.84	44.8	245.76	4.91
6.6	39.2	215.16	4.82	41.2	228.36	4.89	43.2	241.56	4.96	45.2	254.76	5.03
6.8	39.6	223.04	4.93	41.6	236.64	5.01	43.6	250.24	5.08	45.6	263.84	5.15
7.0	40.0	231.00	5.04	42.0	245.00	5.13	44.0	259.00	5.20	46.0	273.00	5.27
7.2	40.4	239.04	5.16	42.4	253.44	5.24	44.4	267.84	5.32	46.4	282.24	5.39
7.4	40.8	247.16	5.27	42.8	261.96	5.35	44.8	276.76	5.43	46.8	291.56	5.51
7.6	41.2	255.36	5.38	43.2	270.56	5.47	45.2	285.76	5.55	47.2	300.96	5.63
7.8	41.6	263.64	5.49	43.6	279.24	5.58	45.6	294.84	5.66	47.6	310.44	5.74
8.0	42.0	272.00	5.59	44.0	288.00	5.69	46.0	304.00	5.78	48.0	320.00	5.86
8.2	42.4	280.44	5.70	44.4	296.84	5.80	46.4	313.24	5.89	48.4	329.64	5.97
8.4	42.8	288.96	5.81	44.8	305.76	5.91	46.8	322.56	6.00	48.8	339.36	6.09
8.6	43.2	297.56	5.91	45.2	314.76	6.02	47.2	331.96	6.11	49.2	349.16	6.20
8.8	43.6	306.24	6.02	45.6	323.84	6.12	47.6	341.44	6.22	49.6	359.04	6.31
9.0	44.0	315.00	6.12	46.0	333.00	6.23	48.0	351.00	6.33	50.0	369.00	6.42
9.2	44.4	323.84	6.23	46.4	342.24	6.34	48.4	360.64	6.44	50.4	379.04	6.53
9.4	44.8	332.76	6.33	46.8	351.56	6.44	48.8	370.36	6.54	50.8	389.16	6.64
9.6	45.2	341.76	6.43	47.2	360.96	6.54	49.2	380.16	6.65	51.2	399.36	6.75
9.8	45.6	350.84	6.53	47.6	370.44	6.65	49.6	390.04	6.76	51.6	409.64	6.86

Table 29.—*Area in square feet, A, top width in feet, T, and hydraulic radius in feet, r, of trapezoidal channels,*
side slopes 1 to 1—Continued

Depth	Bottom width 26 feet			Bottom width 28 feet			Bottom width 30 feet			Bottom width 32 feet		
	T	A	r	T	A	r	T	A	r	T	A	r
10.0	46.0	360.00	6.63	48.0	380.00	6.75	50.0	400.00	6.86	52.0	420.00	6.97
10.5	47.0	383.25	6.88	49.0	404.25	7.01	51.0	425.25	7.12	53.0	446.25	7.23
11.0	48.0	407.0	7.13	50.0	429.0	7.26	52.0	451.0	7.38	54.0	473.0	7.49
11.5	49.0	431.3	7.37	51.0	454.3	7.51	53.0	477.3	7.63	55.0	500.3	7.75
12.0	50.0	456.0	7.61	52.0	480.0	7.75	54.0	504.0	7.88	56.0	528.0	8.01
12.5	51.0	481.3	7.84	53.0	506.3	7.99	55.0	531.3	8.13	57.0	556.3	8.26
13.0	52.0	507.0	8.08	54.0	533.0	8.23	56.0	559.0	8.37	58.0	585.0	8.51
13.5	53.0	533.3	8.31	55.0	560.3	8.47	57.0	587.3	8.61	59.0	614.3	8.75
14.0	54.0	560.0	8.54	56.0	588.0	8.70	58.0	616.0	8.85	60.0	644.0	8.99
14.5	55.0	587.3	8.76	57.0	616.3	8.93	59.0	645.3	9.09	61.0	674.3	9.24
15.0	56.0	615.0	8.99	58.0	645.0	9.16	60.0	675.0	9.32	62.0	705.0	9.47
15.5	57.0	643.3	9.21	59.0	674.3	9.39	61.0	705.3	9.55	63.0	736.3	9.71
16.0	58.0	672.0	9.43	60.0	704.0	9.61	62.0	736.0	9.78	64.0	768.0	9.94
16.5	59.0	701.3	9.65	61.0	734.3	9.83	63.0	767.3	10.01	65.0	800.3	10.17
17.0	60.0	731.0	9.87	62.0	765.0	10.05	64.0	799.0	10.23	66.0	833.0	10.40
17.5	61.0	761.3	10.08	63.0	796.3	10.28	65.0	831.3	10.46	67.0	866.3	10.63
18.0	62.0	792.0	10.30	64.0	828.0	10.49	66.0	864.0	10.68	68.0	900.0	10.85
18.5	63.0	823.3	10.51	65.0	860.3	10.71	67.0	897.3	10.90	69.0	934.3	11.08
19.0	64.0	855.0	10.72	66.0	893.0	10.92	68.0	931.0	11.12	70.0	969.0	11.30
19.5	65.0	887.3	10.93	67.0	926.3	11.14	69.0	965.3	11.34	71.0	1,004.3	11.52
20.0	66.0	920.0	11.14	68.0	960.0	11.35	70.0	1,000.0	11.55	72.0	1,040.0	11.74

Depth	Bottom width 35 feet			Bottom width 40 feet			Bottom width 45 feet			Bottom width 50 feet		
	T	A	r	T	A	r	T	A	r	T	A	r
0.4	35.8	14.16	.39	40.8	16.16	.39	45.8	18.16	.39	50.8	20.16	.39
0.6	36.2	21.36	.58	41.2	24.36	.58	46.2	27.36	.59	51.2	30.36	.59
0.8	36.6	28.64	.77	41.6	32.64	.77	46.6	36.64	.78	51.6	40.64	.78
1.0	37.0	36.00	.95	42.0	41.00	.96	47.0	46.00	.96	52.0	51.00	.97
1.2	37.4	43.44	1.13	42.4	49.44	1.14	47.4	55.44	1.15	52.4	61.44	1.15
1.4	37.8	50.96	1.31	42.8	57.96	1.32	47.8	64.96	1.33	52.8	71.96	1.33
1.6	38.2	58.56	1.48	43.2	66.56	1.49	48.2	74.56	1.51	53.2	82.56	1.51
1.8	38.6	66.24	1.65	43.6	75.24	1.67	48.6	84.24	1.68	53.6	93.24	1.69
2.0	39.0	74.00	1.82	44.0	84.00	1.84	49.0	94.00	1.86	54.0	104.00	1.87
2.2	39.4	81.84	1.99	44.4	92.84	2.01	49.4	103.84	2.03	54.4	114.84	2.04
2.4	39.8	89.76	2.15	44.8	101.76	2.17	49.8	113.76	2.20	54.8	125.76	2.21
2.6	40.2	97.76	2.31	45.2	110.76	2.34	50.2	123.76	2.36	55.2	136.76	2.38
2.8	40.6	105.84	2.47	45.6	119.84	2.50	50.6	133.84	2.53	55.6	147.84	2.55
3.0	41.0	114.00	2.62	46.0	129.00	2.66	51.0	144.00	2.69	56.0	159.00	2.72
3.2	41.4	122.24	2.77	46.4	138.24	2.82	51.4	154.24	2.85	56.4	170.24	2.88
3.4	41.8	130.56	2.93	46.8	147.56	2.97	51.8	164.56	3.01	56.8	181.56	3.05
3.6	42.2	138.96	3.08	47.2	156.96	3.13	52.2	174.96	3.17	57.2	192.96	3.21
3.8	42.6	147.44	3.22	47.6	166.44	3.28	52.6	185.44	3.33	57.6	204.44	3.37
4.0	43.0	156.00	3.37	48.0	176.00	3.43	53.0	196.00	3.48	58.0	216.00	3.52
4.2	43.4	164.64	3.51	48.4	185.64	3.58	53.4	206.64	3.63	58.4	227.64	3.68
4.4	43.8	173.36	3.65	48.8	195.36	3.73	53.8	217.36	3.78	58.8	239.36	3.83
4.6	44.2	182.16	3.79	49.2	205.16	3.87	54.2	228.16	3.93	59.2	251.16	3.99
4.8	44.6	191.04	3.93	49.6	215.04	4.01	54.6	239.04	4.08	59.6	263.04	4.14

Table 29.—*Area in square feet, A, top width in feet, T, and hydraulic radius in feet, r, of trapezoidal channels,* **side slopes 1 to 1**—Continued

Depth	Bottom width 35 feet			Bottom width 40 feet			Bottom width 45 feet			Bottom width 50 feet		
	T	A	r	T	A	r	T	A	r	T	A	r
5.0	45.0	200.00	4.07	50.0	225.00	4.16	55.0	250.00	4.23	60.0	275.00	4.29
5.2	45.4	209.04	4.21	50.4	235.04	4.30	55.4	261.04	4.37	60.4	287.04	4.44
5.4	45.8	218.16	4.34	50.8	245.16	4.44	55.8	272.16	4.52	60.8	299.16	4.58
5.6	46.2	227.36	4.47	51.2	255.36	4.57	56.2	283.36	4.66	61.2	311.36	4.73
5.8	46.6	236.64	4.60	51.6	265.64	4.71	56.6	294.64	4.80	61.6	323.64	4.87
6.0	47.0	246.00	4.73	52.0	276.00	4.84	57.0	306.00	4.94	62.0	336.00	5.02
6.2	47.4	255.44	4.86	52.4	286.44	4.98	57.4	317.44	5.08	62.4	348.44	5.16
6.4	47.8	264.96	4.99	52.8	296.96	5.11	57.8	328.96	5.21	62.8	360.96	5.30
6.6	48.2	274.56	5.12	53.2	307.56	5.24	58.2	340.56	5.35	63.2	373.56	5.44
6.8	48.6	284.24	5.24	53.6	318.24	5.37	58.6	352.24	5.48	63.6	386.24	5.58
7.0	49.0	294.00	5.37	54.0	329.00	5.50	59.0	364.00	5.62	64.0	399.00	5.72
7.2	49.4	303.84	5.49	54.4	339.84	5.63	59.4	375.84	5.75	64.4	411.84	5.85
7.4	49.8	313.76	5.61	54.8	350.76	5.76	59.8	387.76	5.88	64.8	424.76	5.99
7.6	50.2	323.76	5.73	55.2	361.76	5.88	60.2	399.76	6.01	65.2	437.76	6.12
7.8	50.6	333.84	5.85	55.6	372.84	6.01	60.6	411.84	6.14	65.6	450.84	6.26
8.0	51.0	344.00	5.97	56.0	384.00	6.13	61.0	424.00	6.27	66.0	464.00	6.39
8.2	51.4	354.24	6.09	56.4	395.24	6.25	61.4	436.24	6.40	66.4	477.24	6.52
8.4	51.8	364.56	6.20	56.8	406.56	6.38	61.8	448.56	6.52	66.8	490.56	6.65
8.6	52.2	374.96	6.32	57.2	417.96	6.50	62.2	460.96	6.65	67.2	503.96	6.78
8.8	52.6	385.44	6.44	57.6	429.44	6.62	62.6	473.44	6.77	67.6	517.44	6.91
9.0	53.0	396.00	6.55	58.0	441.00	6.74	63.0	486.00	6.90	68.0	531.00	7.04
9.2	53.4	406.64	6.66	58.4	452.64	6.86	63.4	498.64	7.02	68.4	544.64	7.16
9.4	53.8	417.36	6.78	58.8	464.36	6.97	63.8	511.36	7.14	68.8	558.36	7.29
9.6	54.2	428.16	6.89	59.2	476.16	7.09	64.2	524.16	7.26	69.2	572.16	7.42
9.8	54.6	439.04	7.00	59.6	488.04	7.21	64.6	537.04	7.39	69.6	586.04	7.54
10.0	55.0	450.00	7.11	60.0	500.00	7.32	65.0	550.00	7.51	70.0	600.00	7.66
10.5	56.0	477.75	7.38	61.0	530.25	7.61	66.0	582.75	7.80	71.0	635.25	7.97
11.0	57.0	506.0	7.65	62.0	561.0	7.89	67.0	616.0	8.09	72.0	671.0	8.27
11.5	58.0	534.8	7.92	63.0	592.3	8.17	68.0	649.8	8.38	73.0	707.3	8.57
12.0	59.0	564.0	8.18	64.0	624.0	8.44	69.0	684.0	8.66	74.0	744.0	8.86
12.5	60.0	593.8	8.44	65.0	656.3	8.71	70.0	718.8	8.95	75.0	781.3	9.15
13.0	61.0	624.0	8.69	66.0	689.0	8.97	71.0	754.0	9.22	76.0	819.0	9.44
13.5	62.0	654.8	8.95	67.0	722.3	9.24	72.0	789.8	9.49	77.0	857.3	9.72
14.0	63.0	686.0	9.20	68.0	756.0	9.50	73.0	826.0	9.76	78.0	896.0	10.00
14.5	64.0	717.8	9.44	69.0	790.3	9.76	74.0	862.8	10.03	79.0	935.3	10.28
15.0	65.0	750.0	9.69	70.0	825.0	10.01	75.0	900.0	10.29	80.0	975.0	10.55
15.5	66.0	782.8	9.93	71.0	860.3	10.26	76.0	937.8	10.56	81.0	1,015.3	10.82
16.0	67.0	816.0	10.17	72.0	896.0	10.51	77.0	976.0	10.81	82.0	1,056.0	11.09
16.5	68.0	849.8	10.41	73.0	932.3	10.76	78.0	1,014.8	11.07	83.0	1,097.3	11.35
17.0	69.0	884.0	10.64	74.0	969.0	11.00	79.0	1,054.0	11.32	84.0	1,139.0	11.61
17.5	70.0	918.8	10.87	75.0	1,006.3	11.24	80.0	1,093.8	11.57	85.0	1,181.3	11.87
18.0	71.0	954.0	11.10	76.0	1,044.0	11.48	81.0	1,134.0	11.82	86.0	1,224.0	12.13
18.5	72.0	989.8	11.33	77.0	1,082.3	11.72	82.0	1,174.8	12.07	87.0	1,267.3	12.38
19.0	73.0	1,026.0	11.56	78.0	1,121.0	11.96	83.0	1,216.0	12.32	88.0	1,311.0	12.64
19.5	74.0	1,062.8	11.79	79.0	1,160.3	12.19	84.0	1,257.8	12.56	89.0	1,355.3	12.89
20.0	75.0	1,100.0	12.01	80.0	1,200.0	12.43	85.0	1,300.0	12.80	90.0	1,400.0	13.14

Table 29.—*Area in square feet, A, top width in feet, T, and hydraulic radius in feet, r, of trapezoidal channels,*
side slopes 1 to 1—Continued

Depth	Bottom width 60 feet			Bottom width 70 feet			Bottom width 80 feet			Bottom width 90 feet		
	T	A	r	T	A	r	T	A	r	T	A	r
0.4	60.8	24.16	.40	70.8	28.16	.40	80.8	32.16	.40	90.8	36.16	.40
0.6	61.2	36.36	.59	71.2	42.36	.59	81.2	48.36	.59	91.2	54.36	.59
0.8	61.6	48.64	.78	71.6	56.64	.78	81.6	64.64	.79	91.6	72.64	.79
1.0	62.0	61.00	.97	72.0	71.00	.97	82.0	81.00	.98	92.0	91.00	.98
1.2	62.4	73.44	1.16	72.4	85.44	1.16	82.4	97.44	1.17	92.4	109.44	1.17
1.4	62.8	85.96	1.34	72.8	99.96	1.35	82.8	113.96	1.36	92.8	127.96	1.36
1.6	63.2	98.56	1.53	73.2	114.56	1.54	83.2	130.56	1.54	93.2	146.56	1.55
1.8	63.6	111.24	1.71	73.6	129.24	1.72	83.6	147.24	1.73	93.6	165.24	1.74
2.0	64.0	124.00	1.89	74.0	144.00	1.90	84.0	164.00	1.91	94.0	184.00	1.92
2.2	64.4	136.84	2.07	74.4	158.84	2.08	84.4	180.84	2.10	94.4	202.84	2.11
2.4	64.8	149.76	2.24	74.8	173.76	2.26	84.8	197.76	2.28	94.8	221.76	2.29
2.6	65.2	162.76	2.42	75.2	188.76	2.44	85.2	214.76	2.46	95.2	240.76	2.47
2.8	65.6	175.84	2.59	75.6	203.84	2.62	85.6	231.84	2.64	95.6	259.84	2.65
3.0	66.0	189.00	2.76	76.0	219.00	2.79	86.0	249.00	2.81	96.0	279.00	2.83
3.2	66.4	202.24	2.93	76.4	234.24	2.97	86.4	266.24	2.99	96.4	298.24	3.01
3.4	66.8	215.56	3.10	76.8	249.56	3.13	86.8	283.56	3.16	96.8	317.56	3.19
3.6	67.2	228.96	3.26	77.2	264.96	3.30	87.2	300.96	3.34	97.2	336.96	3.36
3.8	67.6	242.44	3.43	77.6	280.44	3.47	87.6	318.44	3.51	97.6	356.44	3.54
4.0	68.0	256.00	3.59	78.0	296.00	3.64	88.0	336.00	3.68	98.0	376.00	3.71
4.2	68.4	269.64	3.75	78.4	311.64	3.81	88.4	353.64	3.85	98.4	395.64	3.88
4.4	68.8	283.36	3.91	78.8	327.36	3.97	88.8	371.36	4.02	98.8	415.36	4.05
4.6	69.2	297.16	4.07	79.2	343.16	4.13	89.2	389.16	4.18	99.2	435.16	4.22
4.8	69.6	311.04	4.23	79.6	359.04	4.30	89.6	407.04	4.35	99.6	455.04	4.39
5.0	70.0	325.00	4.38	80.0	375.00	4.46	90.0	425.00	4.51	100.0	475.00	4.56
5.2	70.4	339.04	4.54	80.4	391.04	4.62	90.4	443.04	4.68	100.4	495.04	4.73
5.4	70.8	353.16	4.69	80.8	407.16	4.77	90.8	461.16	4.84	100.8	515.16	4.89
5.6	71.2	367.36	4.84	81.2	423.36	4.93	91.2	479.36	5.00	101.2	535.36	5.06
5.8	71.6	381.64	4.99	81.6	439.64	5.09	91.6	497.64	5.16	101.6	555.64	5.22
6.0	72.0	396.00	5.14	82.0	456.00	5.24	92.0	516.00	5.32	102.0	576.00	5.38
6.2	72.4	410.44	5.29	82.4	472.44	5.40	92.4	534.44	5.48	102.4	596.44	5.55
6.4	72.8	424.96	5.44	82.8	488.96	5.55	92.8	552.96	5.64	102.8	616.96	5.71
6.6	73.2	439.56	5.59	83.2	505.56	5.70	93.2	571.56	5.79	103.2	637.56	5.87
6.8	73.6	454.24	5.73	83.6	522.24	5.85	93.6	590.24	5.95	103.6	658.24	6.03
7.0	74.0	469.00	5.88	84.0	539.00	6.00	94.0	609.00	6.10	104.0	679.00	6.18
7.2	74.4	483.84	6.02	84.4	555.84	6.15	94.4	627.84	6.26	104.4	699.84	6.34
7.4	74.8	498.76	6.16	84.8	572.76	6.30	94.8	646.76	6.41	104.8	720.76	6.50
7.6	75.2	513.76	6.30	85.2	589.76	6.45	95.2	665.76	6.56	105.2	741.76	6.65
7.8	75.6	528.84	6.44	85.6	606.84	6.59	95.6	684.84	6.71	105.6	762.84	6.81
8.0	76.0	544.00	6.58	86.0	624.00	6.74	96.0	704.00	6.86	106.0	784.00	6.96
8.2	76.4	559.24	6.72	86.4	641.24	6.88	96.4	723.24	7.01	106.4	805.24	7.11
8.4	76.8	574.56	6.86	86.8	658.56	7.02	96.8	742.56	7.16	106.8	826.56	7.27
8.6	77.2	589.96	7.00	87.2	675.96	7.17	97.2	761.96	7.30	107.2	847.96	7.42
8.8	77.6	605.44	7.13	87.6	693.44	7.31	97.6	781.44	7.45	107.6	869.44	7.57
9.0	78.0	621.00	7.27	88.0	711.00	7.45	98.0	801.00	7.60	108.0	891.00	7.72
9.2	78.4	636.64	7.40	88.4	728.64	7.59	98.4	820.64	7.74	108.4	912.64	7.87
9.4	78.8	652.36	7.53	88.8	746.36	7.73	98.8	840.36	7.88	108.8	934.36	8.01
9.6	79.2	668.16	7.67	89.2	764.16	7.87	99.2	860.16	8.03	109.2	956.16	8.16
9.8	79.6	684.04	7.80	89.6	782.04	8.00	99.6	880.04	8.17	109.6	978.04	8.31

Table 29.—*Area in square feet, A, top width in feet, T, and hydraulic radius in feet, r, of trapezoidal channels,* **side slopes 1 to 1**—Continued

Depth	Bottom width 60 feet			Bottom width 70 feet			Bottom width 80 feet			Bottom width 90 feet		
	T	A	r	T	A	r	T	A	r	T	A	r
10.0	80.0	700.00	7.93	90.0	800.00	8.14	100.0	900.00	8.31	110.0	1,000.00	8.45
10.5	81.0	740.25	8.25	91.0	845.25	8.48	101.0	950.25	8.66	111.0	1,055.25	8.82
11.0	82.0	781.0	8.57	92.0	891.0	8.81	102.0	1,001.0	9.01	112.0	1,111.0	9.17
11.5	83.0	822.3	8.89	93.0	937.3	9.14	103.0	1,052.3	9.35	113.0	1,167.3	9.53
12.0	84.0	864.0	9.20	94.0	984.0	9.47	104.0	1,104.0	9.69	114.0	1,224.0	9.88
12.5	85.0	906.3	9.50	95.0	1,031.3	9.79	105.0	1,156.3	10.02	115.0	1,281.3	10.22
13.0	86.0	949.0	9.81	96.0	1,079.0	10.11	106.0	1,209.0	10.35	116.0	1,339.0	10.56
13.5	87.0	992.3	10.11	97.0	1,127.3	10.42	107.0	1,262.3	10.68	117.0	1,397.3	10.90
14.0	88.0	1,036.0	10.40	98.0	1,176.0	10.73	108.0	1,316.0	11.00	118.0	1,456.0	11.23
14.5	89.0	1,080.3	10.69	99.0	1,225.3	11.04	109.0	1,370.3	11.32	119.0	1,515.3	11.57
15.0	90.0	1,125.0	10.98	100.0	1,275.0	11.34	110.0	1,425.0	11.64	120.0	1,575.0	11.89
15.5	91.0	1,170.3	11.27	101.0	1,325.3	11.64	111.0	1,480.3	11.95	121.0	1,635.3	12.22
16.0	92.0	1,216.0	11.55	102.0	1,376.0	11.94	112.0	1,536.0	12.26	122.0	1,696.0	12.54
16.5	93.0	1,262.3	11.83	103.0	1,427.3	12.23	113.0	1,592.3	12.57	123.0	1,757.3	12.86
17.0	94.0	1,309.0	12.11	104.0	1,479.0	12.53	114.0	1,649.0	12.87	124.0	1,819.0	13.17
17.5	95.0	1,356.3	12.39	105.0	1,531.3	12.81	115.0	1,706.3	13.18	125.0	1,881.3	13.49
18.0	96.0	1,404.0	12.66	106.0	1,584.0	13.10	116.0	1,764.0	13.47	126.0	1,944.0	13.80
18.5	97.0	1,452.3	12.93	107.0	1,637.3	13.38	117.0	1,822.3	13.77	127.0	2,007.3	14.10
19.0	98.0	1,501.0	13.20	108.0	1,691.0	13.67	118.0	1,881.0	14.06	128.0	2,071.0	14.41
19.5	99.0	1,550.3	13.46	109.0	1,745.3	13.94	119.0	1,940.3	14.36	129.0	2,135.3	14.71
20.0	100.0	1,600.0	13.73	110.0	1,800.0	14.22	120.0	2,000.0	14.64	130.0	2,200.0	15.01

Table 30.—*Area in square feet, A, top width in feet, T, and hydraulic radius in feet, r, of trapezoidal channels,*
side slopes 1¼ to 1

Depth	Bottom width 2 feet			Bottom width 3 feet			Bottom width 4 feet			Bottom width 5 feet		
	T	A	r	T	A	r	T	A	r	T	A	r
0.4	3.0	1.00	0.30	4.0	1.40	0.33	5.0	1.80	0.34	6.0	2.20	0.35
0.6	3.5	1.65	.42	4.5	2.25	.46	5.5	2.85	.48	6.5	3.45	.50
0.8	4.0	2.40	.53	5.0	3.20	.58	6.0	4.00	.61	7.0	4.80	.63
1.0	4.5	3.25	.62	5.5	4.25	.69	6.5	5.25	.73	7.5	6.25	.76
1.2	5.0	4.20	.72	6.0	5.40	.79	7.0	6.60	.84	8.0	7.80	.88
1.4	5.5	5.25	.81	6.5	6.65	.89	7.5	8.05	.95	8.5	9.45	1.00
1.6	6.0	6.40	.90	7.0	8.00	.98	8.0	9.60	1.05	9.0	11.20	1.11
1.8	6.5	7.65	.99	7.5	9.45	1.08	8.5	11.25	1.15	9.5	13.05	1.21
2.0	7.0	9.00	1.07	8.0	11.00	1.17	9.0	13.00	1.25	10.0	15.00	1.32
2.2	7.5	10.45	1.16	8.5	12.65	1.26	9.5	14.85	1.34	10.5	17.05	1.42
2.4	8.0	12.00	1.24	9.0	14.40	1.35	10.0	16.80	1.44	11.0	19.20	1.51
2.6	8.5	13.65	1.32	9.5	16.25	1.43	10.5	18.85	1.53	11.5	21.45	1.61
2.8	9.0	15.40	1.40	10.0	18.20	1.52	11.0	21.00	1.62	12.0	23.80	1.70
3.0	9.5	17.25	1.49	10.5	20.25	1.61	11.5	23.25	1.71	12.5	26.25	1.80
3.2	10.0	19.20	1.57	11.0	22.40	1.69	12.0	25.60	1.80	13.0	28.80	1.89
3.4	10.5	21.25	1.65	11.5	24.65	1.78	12.5	28.05	1.88	13.5	31.45	1.98
3.6	11.0	23.40	1.73	12.0	27.00	1.86	13.0	30.60	1.97	14.0	34.20	2.07
3.8	11.5	25.65	1.81	12.5	29.45	1.94	13.5	33.25	2.06	14.5	37.05	2.16
4.0	12.0	28.00	1.89	13.0	32.00	2.02	14.0	36.00	2.14	15.0	40.00	2.25
4.2	12.5	30.45	1.97	13.5	34.65	2.11	14.5	38.85	2.23	15.5	43.05	2.33
4.4	13.0	33.00	2.05	14.0	37.40	2.19	15.0	41.80	2.31	16.0	46.20	2.42
4.6	13.5	35.65	2.13	14.5	40.25	2.27	15.5	44.85	2.39	16.5	49.45	2.51
4.8	14.0	38.40	2.21	15.0	43.20	2.35	16.0	48.00	2.48	17.0	52.80	2.59
5.0	14.5	41.25	2.29	15.5	46.25	2.43	16.5	51.25	2.56	17.5	56.25	2.68
5.2	15.0	44.20	2.37	16.0	49.40	2.51	17.0	54.60	2.64	18.0	59.80	2.76
5.4	15.5	47.25	2.45	16.5	52.65	2.60	17.5	58.05	2.73	18.5	63.45	2.85
5.6	16.0	50.40	2.53	17.0	56.00	2.68	18.0	61.60	2.81	19.0	67.20	2.93
5.8	16.5	53.65	2.61	17.5	59.45	2.76	18.5	65.25	2.89	19.5	71.05	3.01
6.0	17.0	57.00	2.69	18.0	63.00	2.84	19.0	69.00	2.97	20.0	75.00	3.10
6.2	17.5	60.45	2.77	18.5	66.65	2.92	19.5	72.85	3.05	20.5	79.05	3.18
6.4	18.0	64.00	2.85	19.0	70.40	3.00	20.0	76.80	3.14	21.0	83.20	3.26
6.6	18.5	67.65	2.92	19.5	74.25	3.08	20.5	80.85	3.22	21.5	87.45	3.35
6.8	19.0	71.40	3.00	20.0	78.20	3.16	21.0	85.00	3.30	22.0	91.80	3.43
7.0	19.5	75.25	3.08	20.5	82.25	3.24	21.5	89.25	3.38	22.5	96.25	3.51
7.2	20.0	79.20	3.16	21.0	86.40	3.32	22.0	93.60	3.46	23.0	100.80	3.59
7.4	20.5	83.25	3.24	21.5	90.65	3.40	22.5	98.05	3.54	23.5	105.45	3.68
7.6	21.0	87.40	3.32	22.0	95.00	3.48	23.0	102.60	3.62	24.0	110.20	3.76
7.8	21.5	91.65	3.40	22.5	99.45	3.56	23.5	107.25	3.70	24.5	115.05	3.84
8.0	22.0	96.00	3.48	23.0	104.00	3.63	24.0	112.00	3.78	25.0	120.00	3.92
8.2	-----	------	----	23.5	108.65	3.71	24.5	116.85	3.86	25.5	125.05	4.00
8.4	-----	------	----	24.0	113.40	3.79	25.0	121.80	3.94	26.0	130.20	4.08
8.6	-----	------	----	24.5	118.25	3.87	25.5	126.85	4.02	26.5	135.45	4.16
8.8	-----	------	----	25.0	123.20	3.95	26.0	132.00	4.10	27.0	140.80	4.24
9.0	-----	------	----	25.5	128.25	4.03	26.5	137.25	4.18	27.5	146.25	4.33
9.2	-----	------	----	26.0	133.40	4.11	27.0	142.60	4.26	28.0	151.80	4.41
9.4	-----	------	----	26.5	138.65	4.19	27.5	148.05	4.34	28.5	157.45	4.49
9.6	-----	------	----	27.0	144.00	4.27	28.0	153.60	4.42	29.0	163.20	4.57
9.8	-----	------	----	27.5	149.45	4.35	28.5	159.25	4.50	29.5	169.05	4.65

Table 30.—*Area in square feet, A, top width in feet, T, and hydraulic radius in feet, r, of trapezoidal channels,*
side slopes 1¼ to 1—Continued

Depth	Bottom width 2 feet			Bottom width 3 feet			Bottom width 4 feet			Bottom width 5 feet		
	T	A	r	T	A	r	T	A	r	T	A	r
10.0				28.00	155.00	4.43	29.00	165.00	4.58	30.0	175.00	4.73
10.5				29.25	169.31	4.62	30.25	179.81	4.78	31.25	190.31	4.93
11.0				30.50	184.3	4.82	31.50	195.3	4.98	32.50	206.3	5.13
11.5				31.75	199.8	5.02	32.75	211.3	5.18	33.75	222.8	5.33
12.0				33.00	216.0	5.22	34.00	228.0	5.37	35.00	240.0	5.53
12.5				34.25	232.8	5.41	35.25	245.3	5.57	36.25	257.8	5.73
13.0							36.50	263.3	5.77	37.50	276.3	5.93
13.5							37.75	281.8	5.97	38.75	295.3	6.12
14.0							39.00	301.0	6.17	40.00	315.0	6.32
14.5							40.25	320.8	6.36	41.25	335.3	6.52
15.0							41.50	341.3	6.56	42.50	356.3	6.72
15.5							42.75	362.3	6.76	43.75	377.8	6.92
16.0							44.00	384.0	6.95	45.00	400.0	7.11
16.5										46.25	422.8	7.31
17.0										47.50	446.3	7.51
17.5										48.75	470.3	7.71
18.0										50.00	495.0	7.90
18.5										51.25	520.3	8.10
19.0										52.50	546.3	8.30
19.5										53.75	572.8	8.49
20.0										55.00	600.0	8.69

Depth	Bottom width 6 feet			Bottom width 7 feet			Bottom width 8 feet			Bottom width 9 feet		
	T	A	r	T	A	r	T	A	r	T	A	r
0.4	7.0	2.60	0.36	8.0	3.00	0.36	9.0	3.40	0.37	10.0	3.80	0.37
0.6	7.5	4.05	.51	8.5	4.65	.52	9.5	5.25	.53	10.5	5.85	.54
0.8	8.0	5.60	.65	9.0	6.40	.67	10.0	7.20	.68	11.0	8.00	.69
1.0	8.5	7.25	.79	9.5	8.25	.81	10.5	9.25	.83	11.5	10.25	.84
1.2	9.0	9.00	.91	10.0	10.20	.94	11.0	11.40	.96	12.0	12.60	.98
1.4	9.5	10.85	1.04	10.5	12.25	1.07	11.5	13.65	1.09	12.5	15.05	1.12
1.6	10.0	12.80	1.15	11.0	14.40	1.19	12.0	16.00	1.22	13.0	17.60	1.25
1.8	10.5	14.85	1.26	11.5	16.65	1.30	12.5	18.45	1.34	13.5	20.25	1.37
2.0	11.0	17.00	1.37	12.0	19.00	1.42	13.0	21.00	1.46	14.0	23.00	1.49
2.2	11.5	19.25	1.48	12.5	21.45	1.53	13.5	23.65	1.57	14.5	25.85	1.61
2.4	12.0	21.60	1.58	13.0	24.00	1.63	14.0	26.40	1.68	15.0	28.80	1.73
2.6	12.5	24.05	1.68	13.5	26.65	1.74	14.5	29.25	1.79	15.5	31.85	1.84
2.8	13.0	26.60	1.78	14.0	29.40	1.84	15.0	32.20	1.90	16.0	35.00	1.95
3.0	13.5	29.25	1.87	14.5	32.25	1.94	15.5	35.25	2.00	16.5	38.25	2.06
3.2	14.0	32.00	1.97	15.0	35.20	2.04	16.0	38.40	2.10	17.0	41.60	2.16
3.4	14.5	34.85	2.06	15.5	38.25	2.14	16.5	41.65	2.21	17.5	45.05	2.27
3.6	15.0	37.80	2.16	16.0	41.40	2.23	17.0	45.00	2.30	18.0	48.60	2.37
3.8	15.5	40.85	2.25	16.5	44.65	2.33	17.5	48.45	2.40	18.5	52.25	2.47
4.0	16.0	44.00	2.34	17.0	48.00	2.42	18.0	52.00	2.50	19.0	56.00	2.57
4.2	16.5	47.25	2.43	17.5	51.45	2.52	18.5	55.65	2.59	19.5	59.85	2.67
4.4	17.0	50.60	2.52	18.0	55.00	2.61	19.0	59.40	2.69	20.0	63.80	2.76
4.6	17.5	54.05	2.61	18.5	58.65	2.70	19.5	63.25	2.78	20.5	67.85	2.86
4.8	18.0	57.60	2.70	19.0	62.40	2.79	20.0	67.20	2.88	21.0	72.00	2.95

Table 30.—*Area in square feet, A, top width in feet, T, and hydraulic radius in feet, r, of trapezoidal channels,* **side slopes 1¼ to 1**—Continued

Depth	Bottom width 6 feet			Bottom width 7 feet			Bottom width 8 feet			Bottom width 9 feet		
	T	A	r	T	A	r	T	A	r	T	A	r
5.0	18.5	61.25	2.78	19.5	66.25	2.88	20.5	71.25	2.97	21.5	76.25	3.05
5.2	19.0	65.00	2.87	20.0	70.20	2.97	21.0	75.40	3.06	22.0	80.60	3.14
5.4	19.5	68.85	2.96	20.5	74.25	3.06	21.5	79.65	3.15	22.5	85.05	3.24
5.6	20.0	72.80	3.04	21.0	78.40	3.14	22.0	84.00	3.24	23.0	89.60	3.33
5.8	20.5	76.85	3.13	21.5	82.65	3.23	22.5	88.45	3.33	23.5	94.25	3.42
6.0	21.0	81.00	3.21	22.0	87.00	3.32	23.0	93.00	3.42	24.0	99.00	3.51
6.2	21.5	85.25	3.30	22.5	91.45	3.41	23.5	97.65	3.51	24.5	103.85	3.60
6.4	22.0	89.60	3.38	23.0	96.00	3.49	24.0	102.40	3.59	25.0	108.80	3.69
6.6	22.5	94.05	3.47	23.5	100.65	3.58	24.5	107.25	3.68	25.5	113.85	3.78
6.8	23.0	98.60	3.55	24.0	105.40	3.66	25.0	112.20	3.77	26.0	119.00	3.87
7.0	23.5	103.25	3.63	24.5	110.25	3.75	25.5	117.25	3.86	26.5	124.25	3.96
7.2	24.0	108.00	3.72	25.0	115.20	3.83	26.0	122.40	3.94	27.0	129.60	4.04
7.4	24.5	112.85	3.80	25.5	120.25	3.92	26.5	127.65	4.03	27.5	135.05	4.13
7.6	25.0	117.80	3.88	26.0	125.40	4.00	27.0	133.00	4.11	28.0	140.60	4.22
7.8	25.5	122.85	3.97	26.5	130.65	4.09	27.5	138.45	4.20	28.5	146.25	4.30
8.0	26.0	128.00	4.05	27.0	136.00	4.17	28.0	144.00	4.28	29.0	152.00	4.39
8.2	26.5	133.25	4.13	27.5	141.45	4.25	28.5	149.65	4.37	29.5	157.85	4.48
8.4	27.0	138.60	4.27	28.0	147.00	4.34	29.0	155.40	4.45	30.0	163.80	4.56
8.6	27.5	144.05	4.30	28.5	152.65	4.42	29.5	161.25	4.54	30.5	169.85	4.65
8.8	28.0	149.60	4.38	29.0	158.40	4.50	30.0	167.20	4.62	31.0	176.00	4.73
9.0	28.5	155.25	4.46	29.5	164.25	4.59	30.5	173.25	4.71	31.5	182.25	4.82
9.2	29.0	161.00	4.54	30.0	170.20	4.67	31.0	179.40	4.79	32.0	188.60	4.90
9.4	29.5	166.85	4.62	30.5	176.25	4.75	31.5	185.65	4.87	32.5	195.05	4.99
9.6	30.0	172.80	4.70	31.0	182.40	4.83	32.0	192.00	4.96	33.0	201.60	5.07
9.8	30.5	178.85	4.79	31.5	188.65	4.92	32.5	198.45	5.04	33.5	208.25	5.16
10.0	31.00	185.00	4.87	32.00	195.00	5.00	33.00	205.00	5.12	34.00	215.00	5.24
10.5	32.25	200.81	5.07	33.25	211.31	5.20	34.25	221.81	5.33	35.25	232.31	5.45
11.0	33.50	217.3	5.27	34.50	228.3	5.41	35.50	239.3	5.54	36.50	250.3	5.66
11.5	34.75	234.3	5.47	35.75	245.8	5.61	36.75	257.3	5.74	37.75	268.8	5.87
12.0	36.00	252.0	5.67	37.00	264.0	5.81	38.00	276.0	5.95	39.00	288.0	6.07
12.5	37.25	270.3	5.87	38.25	282.8	6.01	39.25	295.3	6.15	40.25	307.8	6.28
13.0	38.50	289.3	6.07	39.50	302.3	6.22	40.50	315.3	6.35	41.50	328.3	6.48
13.5	39.75	308.8	6.27	40.75	322.3	6.42	41.75	335.8	6.56	42.75	349.3	6.69
14.0	41.00	329.0	6.47	42.00	343.0	6.62	43.00	357.0	6.76	44.00	371.0	6.89
14.5	42.25	349.8	6.67	43.25	364.3	6.82	44.25	378.8	6.96	45.25	393.3	7.10
15.0	43.50	371.3	6.87	44.50	386.3	7.02	45.50	401.3	7.16	46.50	416.3	7.30
15.5	44.75	393.3	7.07	45.75	408.8	7.22	46.75	424.3	7.36	47.75	439.8	7.50
16.0	46.00	416.0	7.27	47.00	432.0	7.42	48.00	448.0	7.56	49.00	464.0	7.70
16.5	47.25	439.3	7.47	48.25	455.8	7.62	49.25	472.3	7.77	50.25	488.8	7.91
17.0	48.50	463.3	7.67	49.50	480.3	7.82	50.50	497.3	7.97	51.50	514.3	8.11
17.5	49.75	487.8	7.86	50.75	505.3	8.02	51.75	522.8	8.17	52.75	540.3	8.31
18.0	51.00	513.0	8.06	52.00	531.0	8.22	53.00	549.0	8.37	54.00	567.0	8.51
18.5	52.25	538.8	8.26	53.25	557.3	8.41	54.25	575.8	8.56	55.25	594.3	8.71
19.0	53.50	565.3	8.46	54.50	584.3	8.61	55.50	603.3	8.76	56.50	622.3	8.91
19.5	54.75	592.3	8.66	55.75	611.8	8.81	56.75	631.3	8.96	57.75	650.8	9.11
20.0	56.00	620.0	8.85	57.00	640.0	9.01	58.00	660.0	9.16	59.00	680.0	9.31

Table 30.—*Area in square feet, A, top width in feet, T, and hydraulic radius in feet, r, of trapezoidal channels,*
side slopes 1¼ to 1—Continued

Depth	Bottom width 10 feet			Bottom width 12 feet			Bottom width 14 feet			Bottom width 16 feet		
	T	A	r	T	A	r	T	A	r	T	A	r
0.4	11.0	4.20	0.37	13.0	5.00	0.38	15.0	5.80	0.38	17.0	6.60	0.38
0.6	11.5	6.45	.54	13.5	7.65	.55	15.5	8.85	.56	17.5	10.05	.56
0.8	12.0	8.80	.70	14.0	10.40	.71	16.0	12.00	.72	18.0	13.60	.73
1.0	12.5	11.25	.85	14.5	13.25	.87	16.5	15.25	.89	18.5	17.25	.90
1.2	13.0	13.80	1.00	15.0	16.20	1.02	17.0	18.60	1.04	19.0	21.00	1.06
1.4	13.5	16.45	1.14	15.5	19.25	1.17	17.5	22.05	1.19	19.5	24.85	1.21
1.6	14.0	19.20	1.27	16.0	22.40	1.31	18.0	25.60	1.34	20.0	28.80	1.36
1.8	14.5	22.05	1.40	16.5	25.65	1.44	18.5	29.25	1.48	20.5	32.85	1.51
2.0	15.0	25.00	1.52	17.0	29.00	1.58	19.0	33.00	1.62	21.0	37.00	1.65
2.2	15.5	28.05	1.65	17.5	32.45	1.70	19.5	36.85	1.75	21.5	41.25	1.79
2.4	16.0	31.20	1.76	18.0	36.00	1.83	20.0	40.80	1.88	22.0	45.60	1.93
2.6	16.5	34.45	1.88	18.5	39.65	1.95	20.5	44.85	2.01	22.5	50.05	2.06
2.8	17.0	37.80	1.99	19.0	43.40	2.07	21.0	49.00	2.13	23.0	54.60	2.19
3.0	17.5	41.25	2.10	19.5	47.25	2.19	21.5	53.25	2.26	23.5	59.25	2.31
3.2	18.0	44.80	2.21	20.0	51.20	2.30	22.0	57.60	2.38	24.0	64.00	2.44
3.4	18.5	48.45	2.32	20.5	55.25	2.41	22.5	62.05	2.49	24.5	68.85	2.56
3.6	19.0	52.20	2.43	21.0	59.40	2.52	23.0	66.60	2.61	25.0	73.80	2.68
3.8	19.5	56.05	2.53	21.5	63.65	2.63	23.5	71.25	2.72	25.5	78.85	2.80
4.0	20.0	60.00	2.63	22.0	68.00	2.74	24.0	76.00	2.84	26.0	84.00	2.92
4.2	20.5	64.05	2.73	22.5	72.45	2.85	24.5	80.85	2.95	26.5	89.25	3.03
4.4	21.0	68.20	2.83	23.0	77.00	2.95	25.0	85.80	3.05	27.0	94.60	3.14
4.6	21.5	72.45	2.93	23.5	81.65	3.05	25.5	90.85	3.16	27.5	100.05	3.26
4.8	22.0	76.80	3.03	24.0	86.40	3.16	26.0	96.00	3.27	28.0	105.60	3.37
5.0	22.5	81.25	3.12	24.5	91.25	3.26	26.5	101.25	3.37	28.5	111.25	3.48
5.2	23.0	85.80	3.22	25.0	96.20	3.36	27.0	106.60	3.48	29.0	117.00	3.58
5.4	23.5	90.45	3.31	25.5	101.25	3.46	27.5	112.05	3.58	29.5	122.85	3.69
5.6	24.0	95.20	3.41	26.0	106.40	3.56	28.0	117.60	3.68	30.0	128.80	3.80
5.8	24.5	100.05	3.50	26.5	111.65	3.65	28.5	123.25	3.78	30.5	134.85	3.90
6.0	25.0	105.00	3.59	27.0	117.00	3.75	29.0	129.00	3.88	31.0	141.00	4.00
6.2	25.5	110.05	3.69	27.5	122.45	3.84	29.5	134.85	3.98	31.5	147.25	4.11
6.4	26.0	115.20	3.78	28.0	128.00	3.94	30.0	140.80	4.08	32.0	153.60	4.21
6.6	26.5	120.45	3.87	28.5	133.65	4.03	30.5	146.85	4.18	32.5	160.05	4.31
6.8	27.0	125.80	3.96	29.0	139.40	4.13	31.0	153.00	4.28	33.0	166.60	4.41
7.0	27.5	131.25	4.05	29.5	145.25	4.22	31.5	159.25	4.37	33.5	173.25	4.51
7.2	28.0	136.80	4.14	30.0	151.20	4.31	32.0	165.60	4.47	34.0	180.00	4.61
7.4	28.5	142.45	4.23	30.5	157.25	4.41	32.5	172.05	4.56	34.5	186.85	4.71
7.6	29.0	148.20	4.32	31.0	163.40	4.50	33.0	178.60	4.66	35.0	193.80	4.81
7.8	29.5	154.05	4.40	31.5	169.65	4.59	33.5	185.25	4.75	35.5	200.85	4.90
8.0	30.0	160.00	4.49	32.0	176.00	4.68	34.0	192.00	4.85	36.0	208.00	5.00
8.2	30.5	166.05	4.58	32.5	182.45	4.77	34.5	198.85	4.94	36.5	215.25	5.09
8.4	31.0	172.20	4.67	33.0	189.00	4.86	35.0	205.80	5.03	37.0	222.60	5.19
8.6	31.5	178.45	4.75	33.5	195.65	4.95	35.5	212.85	5.12	37.5	230.05	5.28
8.8	32.0	184.80	4.84	34.0	202.40	5.04	36.0	220.00	5.22	38.0	237.60	5.38
9.0	32.5	191.25	4.93	34.5	209.25	5.13	36.5	227.25	5.31	38.5	245.25	5.47
9.2	33.0	197.80	5.01	35.0	216.20	5.22	37.0	234.60	5.40	39.0	253.00	5.57
9.4	33.5	204.45	5.10	35.5	223.25	5.30	37.5	242.05	5.49	39.5	260.85	5.66
9.6	34.0	211.20	5.18	36.0	230.40	5.39	38.0	249.60	5.58	40.0	268.80	5.75
9.8	34.5	218.05	5.27	36.5	237.65	5.48	38.5	257.25	5.67	40.5	276.85	5.84

Table 30.—*Area in square feet, A, top width in feet, T, and hydraulic radius in feet, r, of trapezoidal channels,* side slopes 1¼ to 1—Continued

Depth	Bottom width 10 feet			Bottom width 12 feet			Bottom width 14 feet			Bottom width 16 feet		
	T	A	r	T	A	r	T	A	r	T	A	r
10.0	35.00	225.00	5.36	37.00	245.00	5.57	39.00	265.00	5.76	41.00	285.00	5.94
10.5	36.25	242.81	5.57	38.25	263.81	5.78	40.25	284.81	5.98	42.25	305.81	6.16
11.0	37.50	261.3	5.78	39.50	283.3	6.00	41.50	305.3	6.20	43.50	327.3	6.39
11.5	38.75	280.3	5.99	40.75	303.3	6.21	42.75	326.3	6.42	44.75	349.3	6.61
12.0	40.00	300.0	6.20	42.00	324.0	6.43	44.00	348.0	6.64	46.00	372.0	6.84
12.5	41.25	320.3	6.40	43.25	345.3	6.64	45.25	370.3	6.86	47.25	395.3	7.06
13.0	42.50	341.3	6.61	44.50	367.3	6.85	46.50	393.3	7.07	48.50	419.3	7.28
13.5	43.75	362.8	6.82	45.75	389.8	7.06	47.75	416.8	7.28	49.75	443.8	7.49
14.0	45.00	385.0	7.02	47.00	413.0	7.27	49.00	441.0	7.50	51.00	469.0	7.71
14.5	46.25	407.8	7.23	48.25	436.8	7.48	50.25	465.8	7.71	52.25	494.8	7.93
15.0	47.50	431.3	7.43	49.50	461.3	7.68	51.50	491.3	7.92	53.50	521.3	8.14
15.5	48.75	455.3	7.64	50.75	486.3	7.89	52.75	517.3	8.13	54.75	548.3	8.36
16.0	50.00	480.0	7.84	52.00	512.0	8.10	54.00	544.0	8.34	56.00	576.0	8.57
16.5	51.25	505.3	8.04	53.25	538.3	8.30	55.25	571.3	8.55	57.25	604.3	8.78
17.0	52.50	531.3	8.25	54.50	565.3	8.51	56.50	599.3	8.76	58.50	633.3	8.99
17.5	53.75	557.8	8.45	55.75	592.8	8.71	57.75	627.8	8.97	59.75	662.8	9.20
18.0	55.00	585.0	8.65	57.00	621.0	8.92	59.00	657.0	9.17	61.00	693.0	9.41
18.5	56.25	612.8	8.85	58.25	649.8	9.12	60.25	686.8	9.38	62.25	723.8	9.62
19.0	57.50	641.3	9.05	59.50	679.3	9.33	61.50	717.3	9.59	63.50	755.3	9.83
19.5	58.75	670.3	9.25	60.75	709.3	9.53	62.75	748.3	9.79	64.75	787.3	10.04
20.0	60.00	700.0	9.46	62.00	740.0	9.73	64.00	780.0	10.00	66.00	820.0	10.25

Depth	Bottom width 18 feet			Bottom width 20 feet			Bottom width 22 feet			Bottom width 24 feet		
	T	A	r	T	A	r	T	A	r	T	A	r
0.4	19.0	7.40	0.38	21.0	8.20	0.39	23.0	9.00	0.39	25.0	9.80	0.39
0.6	19.5	11.25	.56	21.5	12.45	.57	23.5	13.65	.57	25.5	14.85	.57
0.8	20.0	15.20	.74	22.0	16.80	.74	24.0	18.40	.75	26.0	20.00	.75
1.0	20.5	19.25	.91	22.5	21.25	.92	24.5	23.25	.92	26.5	25.25	.93
1.2	21.0	23.40	1.07	23.0	25.80	1.08	25.0	28.20	1.09	27.0	30.60	1.10
1.4	21.5	27.65	1.23	23.5	30.45	1.24	25.5	33.25	1.26	27.5	36.05	1.27
1.6	22.0	32.00	1.38	24.0	35.20	1.40	26.0	38.40	1.42	28.0	41.60	1.43
1.8	22.5	36.45	1.53	24.5	40.05	1.55	26.5	43.65	1.57	28.5	47.25	1.59
2.0	23.0	41.00	1.68	25.0	45.00	1.70	27.0	49.00	1.73	29.0	53.00	1.74
2.2	23.5	45.65	1.82	25.5	50.05	1.85	27.5	54.45	1.87	29.5	58.85	1.90
2.4	24.0	50.40	1.96	26.0	55.20	1.99	28.0	60.00	2.02	30.0	64.80	2.05
2.6	24.5	55.25	2.10	26.5	60.45	2.13	28.5	65.65	2.16	30.5	70.85	2.19
2.8	25.0	60.20	2.23	27.0	65.80	2.27	29.0	71.40	2.31	31.0	77.00	2.34
3.0	25.5	65.25	2.36	27.5	71.25	2.41	29.5	77.25	2.44	31.5	83.25	2.48
3.2	26.0	70.40	2.49	28.0	76.80	2.54	30.0	83.20	2.58	32.0	89.60	2.62
3.4	26.5	75.65	2.62	28.5	82.45	2.67	30.5	89.25	2.71	32.5	96.05	2.75
3.6	27.0	81.00	2.74	29.0	88.20	2.80	31.0	95.40	2.85	33.0	102.60	2.89
3.8	27.5	86.45	2.87	29.5	94.05	2.92	31.5	101.65	2.98	33.5	109.25	3.02
4.0	28.0	92.00	2.99	30.0	100.00	3.05	32.0	108.00	3.10	34.0	116.00	3.15
4.2	28.5	97.65	3.11	30.5	106.05	3.17	32.5	114.45	3.23	34.5	122.85	3.28
4.4	29.0	103.40	3.22	31.0	112.20	3.29	33.0	121.00	3.35	35.0	129.80	3.41
4.6	29.5	109.25	3.34	31.5	118.45	3.41	33.5	127.65	3.48	35.5	136.85	3.53
4.8	30.0	115.20	3.45	32.0	124.80	3.53	34.0	134.40	3.60	36.0	144.00	3.66

Table 30.—*Area in square feet, A, top width in feet, T, and hydraulic radius in feet, r, of trapezoidal channels,* side slopes 1¼ to 1—Continued

Depth	Bottom width 18 feet			Bottom width 20 feet			Bottom width 22 feet			Bottom width 24 feet		
	T	A	r	T	A	r	T	A	r	T	A	r
5.0	30.5	121.25	3.57	32.5	131.25	3.65	34.5	141.25	3.72	36.5	151.25	3.78
5.2	31.0	127.40	3.68	33.0	137.80	3.76	35.0	148.20	3.83	37.0	158.60	3.90
5.4	31.5	133.65	3.79	33.5	144.45	3.87	35.5	155.25	3.95	37.5	166.05	4.02
5.6	32.0	140.00	3.90	34.0	151.20	3.99	36.0	162.40	4.07	38.0	173.60	4.14
5.8	32.5	146.45	4.00	34.5	158.05	4.10	36.5	169.65	4.18	38.5	181.25	4.26
6.0	33.0	153.00	4.11	35.0	165.00	4.21	37.0	177.00	4.30	39.0	189.00	4.37
6.2	33.5	159.65	4.22	35.5	172.05	4.32	37.5	184.45	4.41	39.5	196.85	4.49
6.4	34.0	166.40	4.32	36.0	179.20	4.43	38.0	192.00	4.52	40.0	204.80	4.60
6.6	34.5	173.25	4.43	36.5	186.45	4.53	38.5	199.65	4.63	40.5	212.85	4.72
6.8	35.0	180.20	4.53	37.0	193.80	4.64	39.0	207.40	4.74	41.0	221.00	4.83
7.0	35.5	187.25	4.63	37.5	201.25	4.75	39.5	215.25	4.85	41.5	229.25	4.94
7.2	36.0	194.40	4.74	38.0	208.80	4.85	40.0	223.20	4.95	42.0	237.60	5.05
7.4	36.5	201.65	4.84	38.5	216.45	4.95	40.5	231.25	5.06	42.5	246.05	5.16
7.6	37.0	209.00	4.94	39.0	224.20	5.06	41.0	239.40	5.17	43.0	254.60	5.27
7.8	37.5	216.45	5.04	39.5	232.20	5.16	41.5	247.65	5.27	43.5	263.25	5.38
8.0	38.0	224.00	5.14	40.0	240.00	5.26	42.0	256.00	5.38	44.0	272.00	5.48
8.2	38.5	231.65	5.23	40.5	248.05	5.36	42.5	264.45	5.48	44.5	280.85	5.59
8.4	39.0	239.40	5.33	41.0	256.20	5.46	43.0	273.00	5.58	45.0	289.80	5.69
8.6	39.5	247.25	5.43	41.5	264.45	5.56	43.5	281.65	5.69	45.5	298.85	5.80
8.8	40.0	255.20	5.53	42.0	272.80	5.66	44.0	290.40	5.79	46.0	308.00	5.90
9.0	40.5	263.25	5.62	42.5	281.25	5.76	44.5	299.25	5.89	46.5	317.25	6.01
9.2	41.0	271.40	5.72	43.0	289.80	5.86	45.0	308.20	5.99	47.0	326.60	6.11
9.4	41.5	279.65	5.81	43.5	298.45	5.96	45.5	317.25	6.09	47.5	336.05	6.21
9.6	42.0	288.00	5.91	44.0	307.20	6.05	46.0	326.40	6.19	48.0	345.60	6.31
9.8	42.5	296.45	6.00	44.5	316.05	6.15	46.5	335.65	6.29	48.5	355.25	6.42
10.0	43.00	305.00	6.10	45.00	325.00	6.25	47.00	345.00	6.39	49.00	365.00	6.52
10.5	44.25	326.81	6.33	46.25	347.81	6.49	48.25	368.81	6.63	50.25	389.81	6.77
11.0	45.50	349.3	6.56	47.50	371.3	6.72	49.50	393.3	6.87	51.50	415.3	7.01
11.5	46.75	372.3	6.79	48.75	395.3	6.96	50.75	418.3	7.11	52.75	441.3	7.26
12.0	48.00	396.0	7.02	50.00	420.0	7.19	52.00	444.0	7.35	54.00	468.0	7.50
12.5	49.25	420.3	7.24	51.25	445.3	7.42	53.25	470.3	7.58	55.25	495.3	7.74
13.0	50.50	445.3	7.47	52.50	471.3	7.65	54.50	497.3	7.82	56.50	523.3	7.97
13.5	51.75	470.8	7.69	53.75	497.8	7.87	55.75	524.8	8.05	57.75	551.8	8.21
14.0	53.00	497.0	7.91	55.00	525.0	8.10	57.00	553.0	8.28	59.00	581.0	8.44
14.5	54.25	523.8	8.13	56.25	552.8	8.32	58.25	581.8	8.50	60.25	610.8	8.67
15.0	55.50	551.3	8.35	57.50	581.3	8.54	59.50	611.3	8.73	61.50	641.3	8.90
15.5	56.75	579.3	8.57	58.75	610.3	8.77	60.75	641.3	8.95	62.75	672.3	9.13
16.0	58.00	608.0	8.78	60.00	640.0	8.99	62.00	672.0	9.18	64.00	704.0	9.36
16.5	59.25	637.3	9.00	61.25	670.3	9.20	63.25	703.3	9.40	65.25	736.3	9.58
17.0	60.50	667.3	9.21	62.50	701.3	9.42	64.50	735.3	9.62	66.50	769.3	9.81
17.5	61.75	697.8	9.43	63.75	732.8	9.64	65.75	767.8	9.84	67.75	802.8	10.03
18.0	63.00	729.0	9.64	65.00	765.0	9.85	67.00	801.0	10.06	69.00	837.0	10.25
18.5	64.25	760.8	9.85	66.25	797.8	10.07	68.25	834.8	10.28	70.25	871.8	10.47
19.0	65.50	793.3	10.06	67.50	831.3	10.28	69.50	869.3	10.49	71.50	907.3	10.69
19.5	66.75	826.3	10.27	68.75	865.3	10.50	70.75	904.3	10.71	72.75	943.3	10.91
20.0	68.00	860.0	10.48	70.00	900.0	10.71	72.00	940.0	10.93	74.00	980.0	11.13

Table 30.—*Area in square feet, A, top width in feet, T, and hydraulic radius in feet, r, of trapezoidal channels,* **side slopes 1¼ to 1**—Continued

Depth	Bottom width 26 feet			Bottom width 28 feet			Bottom width 30 feet			Bottom width 32 feet		
	T	A	r	T	A	r	T	A	r	T	A	r
0.4	27.0	10.60	0.39	29.0	11.40	0.39	31.0	12.20	0.39	33.0	13.00	0.39
0.6	27.5	16.05	.57	29.5	17.25	.58	31.5	18.45	.58	33.5	19.65	.58
0.8	28.0	21.60	.76	30.0	23.20	.76	32.0	24.80	.76	34.0	26.40	.76
1.0	28.5	27.25	.93	30.5	29.25	.94	32.5	31.25	.94	34.5	33.25	.94
1.2	29.0	33.00	1.11	31.0	35.40	1.11	33.0	37.80	1.12	35.0	40.20	1.12
1.4	29.5	38.85	1.27	31.5	41.65	1.28	33.5	44.45	1.29	35.5	47.25	1.30
1.6	30.0	44.80	1.44	32.0	48.00	1.45	34.0	51.20	1.46	36.0	54.40	1.47
1.8	30.5	50.85	1.60	32.5	54.45	1.61	34.5	58.05	1.62	36.5	61.65	1.63
2.0	31.0	57.00	1.76	33.0	61.00	1.77	35.0	65.00	1.79	37.0	69.00	1.80
2.2	31.5	63.25	1.91	33.5	67.65	1.93	35.5	72.05	1.95	37.5	76.45	1.96
2.4	32.0	69.60	2.07	34.0	74.40	2.08	36.0	79.20	2.10	38.0	84.00	2.12
2.6	32.5	76.05	2.22	34.5	81.25	2.24	36.5	86.45	2.26	38.5	91.65	2.27
2.8	33.0	82.60	2.36	35.0	88.20	2.39	37.0	93.80	2.41	39.0	99.40	2.43
3.0	33.5	89.25	2.51	35.5	95.25	2.53	37.5	101.25	2.56	39.5	107.25	2.58
3.2	34.0	96.00	2.65	36.0	102.40	2.68	38.0	108.80	2.70	40.0	115.20	2.73
3.4	34.5	102.85	2.79	36.5	109.65	2.82	38.5	116.45	2.85	40.5	123.25	2.87
3.6	35.0	109.80	2.93	37.0	117.00	2.96	39.0	124.20	2.99	41.0	131.40	3.02
3.8	35.5	116.85	3.06	37.5	124.45	3.10	39.5	132.05	3.13	41.5	139.65	3.16
4.0	36.0	124.00	3.20	38.0	132.00	3.23	40.0	140.00	3.27	42.0	148.00	3.30
4.2	36.5	131.25	3.33	38.5	139.65	3.37	40.5	148.05	3.41	42.5	156.45	3.44
4.4	37.0	138.60	3.46	39.0	147.40	3.50	41.0	156.20	3.54	43.0	165.00	3.58
4.6	37.5	146.05	3.59	39.5	155.25	3.63	41.5	164.45	3.68	43.5	173.65	3.72
4.8	38.0	153.60	3.71	40.0	163.20	3.76	42.0	172.80	3.81	44.0	182.40	3.85
5.0	38.5	161.25	3.84	40.5	171.25	3.89	42.5	181.25	3.94	44.5	191.25	3.98
5.2	39.0	169.00	3.96	41.0	179.40	4.02	43.0	189.80	4.07	45.0	200.20	4.12
5.4	39.5	176.85	4.09	41.5	187.65	4.14	43.5	198.45	4.20	45.5	209.25	4.25
5.6	40.0	184.80	4.21	42.0	196.00	4.27	44.0	207.20	4.32	46.0	218.40	4.37
5.8	40.5	192.85	4.33	42.5	204.45	4.39	44.5	216.05	4.45	46.5	227.65	4.50
6.0	41.0	201.00	4.45	43.0	213.00	4.51	45.0	225.00	4.57	47.0	237.00	4.63
6.2	41.5	209.25	4.56	43.5	221.65	4.63	45.5	234.05	4.70	47.5	246.45	4.75
6.4	42.0	217.60	4.68	44.0	230.40	4.75	46.0	243.20	4.82	48.0	256.00	4.88
6.6	42.5	226.05	4.80	44.5	239.25	4.87	46.5	252.45	4.94	48.5	265.65	5.00
6.8	43.0	234.60	4.91	45.0	248.20	4.99	47.0	261.80	5.06	49.0	275.40	5.12
7.0	43.5	243.25	5.02	45.5	257.25	5.10	47.5	271.25	5.18	49.5	285.25	5.24
7.2	44.0	252.00	5.14	46.0	266.40	5.22	48.0	280.80	5.29	50.0	295.20	5.36
7.4	44.5	260.85	5.25	46.5	275.65	5.33	48.5	290.45	5.41	50.5	305.25	5.48
7.6	45.0	269.80	5.36	47.0	285.00	5.45	49.0	300.20	5.53	51.0	315.40	5.60
7.8	45.5	278.85	5.47	47.5	294.45	5.56	49.5	310.05	5.64	51.5	325.65	5.72
8.0	46.0	288.00	5.58	48.0	304.00	5.67	50.0	320.00	5.75	52.0	336.00	5.83
8.2	46.5	297.25	5.69	48.5	313.65	5.78	50.5	330.05	5.87	52.5	346.45	5.95
8.4	47.0	306.60	5.80	49.0	323.40	5.89	51.0	340.20	5.98	53.0	357.00	6.06
8.6	47.5	316.05	5.90	49.5	333.25	6.00	51.5	350.45	6.09	53.5	367.65	6.18
8.8	48.0	325.60	6.01	50.0	343.20	6.11	52.0	360.80	6.20	54.0	378.40	6.29
9.0	48.5	335.25	6.12	50.5	353.25	6.22	52.5	371.25	6.31	54.5	389.25	6.40
9.2	49.0	345.00	6.22	51.0	363.40	6.33	53.0	381.80	6.42	55.0	400.20	6.51
9.4	49.5	354.85	6.33	51.5	373.65	6.43	53.5	392.45	6.53	55.5	411.25	6.62
9.6	50.0	364.80	6.43	52.0	384.00	6.54	54.0	403.20	6.64	56.0	422.40	6.73
9.8	50.5	374.85	6.53	52.5	394.45	6.64	54.5	414.05	6.75	56.5	433.65	6.84

Table 30.—*Area in square feet, A, top width in feet, T, and hydraulic radius in feet, r, of trapezoidal channels,*
side slopes 1¼ to 1—Continued

Depth	Bottom width 26 feet			Bottom width 28 feet			Bottom width 30 feet			Bottom width 32 feet		
	T	A	r	T	A	r	T	A	r	T	A	r
10.0	51.00	385.00	6.64	53.00	405.00	6.75	55.00	425.00	6.85	57.00	445.00	6.95
10.5	52.25	410.81	6.89	54.25	431.81	7.01	56.25	452.81	7.12	58.25	473.81	7.22
11.0	53.50	437.3	7.14	55.50	459.3	7.26	57.50	481.3	7.38	59.50	503.3	7.49
11.5	54.75	464.3	7.39	56.75	487.3	7.52	58.75	510.3	7.64	60.75	533.3	7.75
12.0	56.00	492.0	7.64	58.00	516.0	7.77	60.00	540.0	7.89	62.00	564.0	8.01
12.5	57.25	520.3	7.88	59.25	545.3	8.02	61.25	570.3	8.15	63.25	595.3	8.27
13.0	58.50	549.3	8.12	60.50	575.3	8.26	62.50	601.3	8.39	64.50	627.3	8.52
13.5	59.75	578.8	8.36	61.75	605.8	8.51	63.75	632.8	8.64	65.75	659.8	8.77
14.0	61.00	609.0	8.60	63.00	637.0	8.75	65.00	665.0	8.89	67.00	693.0	9.02
14.5	62.25	639.8	8.83	64.25	668.8	8.99	66.25	697.8	9.13	68.25	726.8	9.27
15.0	63.50	671.3	9.07	65.50	701.3	9.22	67.50	731.3	9.37	69.50	761.3	9.51
15.5	64.75	703.3	9.30	66.75	734.3	9.46	68.75	765.3	9.61	70.75	796.3	9.76
16.0	66.00	736.0	9.53	68.00	768.0	9.69	70.00	800.0	9.85	72.00	832.0	10.00
16.5	67.25	769.3	9.76	69.25	802.3	9.93	71.25	835.3	10.09	73.25	868.3	10.24
17.0	68.50	803.3	9.99	70.50	837.3	10.16	72.50	871.3	10.32	74.50	905.3	10.47
17.5	69.75	837.8	10.21	71.75	872.8	10.39	73.75	907.8	10.55	75.75	942.8	10.71
18.0	71.00	873.0	10.44	73.00	909.0	10.62	75.00	945.0	10.78	77.00	981.0	10.95
18.5	72.25	908.8	10.66	74.25	945.8	10.84	76.25	982.8	11.01	78.25	1,019.8	11.18
19.0	73.50	945.3	10.89	75.50	983.3	11.07	77.50	1,021.3	11.24	79.50	1,059.3	11.41
19.5	74.75	982.3	11.11	76.75	1,021.3	11.29	78.75	1,060.3	11.47	80.75	1,099.3	11.64
20.0	76.00	1,020.0	11.33	78.00	1,060.0	11.52	80.00	1,100.0	11.70	82.00	1,140.0	11.87

Depth	Bottom width 35 feet			Bottom width 40 feet			Bottom width 45 feet			Bottom width 50 feet		
	T	A	r	T	A	r	T	A	r	T	A	r
0.4	36.0	14.20	0.39	41.0	16.20	0.39	46.0	18.20	0.39	51.0	20.20	0.39
0.6	36.5	21.45	.58	41.5	24.45	.58	46.5	27.45	.59	51.5	30.45	.59
0.8	37.0	28.80	.77	42.0	32.80	.77	47.0	36.80	.77	52.0	40.80	.78
1.0	37.5	36.25	.95	42.5	41.25	.95	47.5	46.25	.96	52.5	51.25	.96
1.2	38.0	43.80	1.13	43.0	49.80	1.14	48.0	55.80	1.14	53.0	61.80	1.15
1.4	38.5	51.45	1.30	43.5	58.45	1.31	48.5	65.45	1.32	53.5	72.45	1.33
1.6	39.0	59.20	1.48	44.0	67.20	1.49	49.0	75.20	1.50	54.0	83.20	1.51
1.8	39.5	67.05	1.64	44.5	76.05	1.66	49.5	85.05	1.68	54.5	94.05	1.69
2.0	40.0	75.00	1.81	45.0	85.00	1.83	50.0	95.00	1.85	55.0	105.00	1.86
2.2	40.5	83.05	1.98	45.5	94.05	2.00	50.5	105.05	2.02	55.5	116.05	2.03
2.4	41.0	91.20	2.14	46.0	103.20	2.16	51.0	115.20	2.19	56.0	127.20	2.21
2.6	41.5	99.45	2.30	46.5	112.45	2.33	51.5	125.45	2.35	56.5	138.45	2.37
2.8	42.0	107.80	2.45	47.0	121.80	2.49	52.0	135.80	2.52	57.0	149.80	2.54
3.0	42.5	116.25	2.61	47.5	131.25	2.65	52.5	146.25	2.68	57.5	161.25	2.71
3.2	43.0	124.80	2.76	48.0	140.80	2.80	53.0	156.80	2.84	58.0	172.80	2.87
3.4	43.5	133.45	2.91	48.5	150.45	2.96	53.5	167.45	3.00	58.5	184.45	3.03
3.6	44.0	142.20	3.06	49.0	160.20	3.11	54.0	178.20	3.15	59.0	196.20	3.19
3.8	44.5	151.05	3.20	49.5	170.05	3.26	54.5	189.05	3.31	59.5	208.05	3.35
4.0	45.0	160.00	3.35	50.0	180.00	3.41	55.0	200.00	3.46	60.0	220.00	3.50
4.2	45.5	169.05	3.49	50.5	190.05	3.56	55.5	211.05	3.61	60.5	232.05	3.66
4.4	46.0	178.20	3.63	51.0	200.20	3.70	56.0	222.20	3.76	61.0	244.20	3.81
4.6	46.5	187.45	3.77	51.5	210.45	3.85	56.5	233.45	3.91	61.5	256.45	3.96
4.8	47.0	196.80	3.91	52.0	220.80	3.99	57.0	244.80	4.06	62.0	268.80	4.11

Table 30.—*Area in square feet, A, top width in feet, T, and hydraulic radius in feet, r, of trapezoidal channels.*

side slopes 1¼ to 1—Continued

Depth	Bottom width 35 feet			Bottom width 40 feet			Bottom width 45 feet			Bottom width 50 feet		
	T	A	r	T	A	r	T	A	r	T	A	r
5.0	47.5	206.25	4.04	52.5	231.25	4.13	57.5	256.25	4.20	62.5	281.25	4.26
5.2	48.0	215.80	4.18	53.0	241.80	4.27	58.0	267.80	4.34	63.0	293.80	4.41
5.4	48.5	225.45	4.31	53.5	252.45	4.41	58.5	279.45	4.49	63.5	306.45	4.55
5.6	49.0	235.20	4.44	54.0	263.20	4.54	59.0	291.20	4.63	64.0	319.20	4.70
5.8	49.5	245.05	4.57	54.5	274.05	4.68	59.5	303.05	4.77	64.5	332.05	4.84
6.0	50.0	255.00	4.70	55.0	285.00	4.81	60.0	315.00	4.91	65.0	345.00	4.98
6.2	50.5	265.05	4.83	55.5	296.05	4.95	60.5	327.05	5.04	65.5	358.05	5.13
6.4	51.0	275.20	4.96	56.0	307.20	5.08	61.0	339.20	5.18	66.0	371.20	5.27
6.6	51.5	285.45	5.09	56.5	318.45	5.21	61.5	351.45	5.31	66.5	384.45	5.40
6.8	52.0	295.80	5.21	57.0	329.80	5.34	62.0	363.80	5.45	67.0	397.80	5.54
7.0	52.5	306.25	5.33	57.5	341.25	5.47	62.5	376.25	5.58	67.5	411.25	5.68
7.2	53.0	316.80	5.46	58.0	352.80	5.60	63.0	388.80	5.71	68.0	424.80	5.82
7.4	53.5	327.45	5.58	58.5	364.45	5.72	63.5	401.45	5.84	68.5	438.45	5.95
7.6	54.0	338.20	5.70	59.0	376.20	5.85	64.0	414.20	5.97	69.0	452.20	6.08
7.8	54.5	349.05	5.82	59.5	388.05	5.97	64.5	427.05	6.10	69.5	466.05	6.22
8.0	55.0	360.00	5.94	60.0	400.00	6.10	65.0	440.00	6.23	70.0	480.00	6.35
8.2	55.5	371.05	6.06	60.5	412.05	6.22	65.5	453.05	6.36	70.5	494.05	6.48
8.4	56.0	382.20	6.18	61.0	424.20	6.34	66.0	466.20	6.48	71.0	508.20	6.61
8.6	56.5	393.45	6.29	61.5	436.45	6.46	66.5	479.45	6.61	71.5	522.45	6.74
8.8	57.0	404.80	6.41	62.0	448.80	6.58	67.0	492.80	6.73	72.0	536.80	6.87
9.0	57.5	416.25	6.52	62.5	461.25	6.70	67.5	506.25	6.86	72.5	551.25	6.99
9.2	58.0	427.80	6.64	63.0	473.80	6.82	68.0	519.80	6.98	73.0	565.80	7.12
9.4	58.5	439.45	6.75	63.5	486.45	6.94	68.5	533.45	7.10	73.5	580.45	7.25
9.6	59.0	451.20	6.86	64.0	499.20	7.06	69.0	547.20	7.23	74.0	595.20	7.37
9.8	59.5	463.05	6.98	64.5	512.05	7.17	69.5	561.05	7.35	74.5	610.05	7.50
10.0	60.00	475.00	7.09	65.00	525.00	7.29	70.00	575.00	7.47	75.00	625.00	7.62
10.5	61.25	505.31	7.36	66.25	557.81	7.58	71.25	610.31	7.76	76.25	662.81	7.93
11.0	62.50	536.3	7.64	67.50	591.3	7.86	72.50	646.3	8.06	77.50	701.3	8.23
11.5	63.75	567.8	7.91	68.75	625.3	8.14	73.75	682.8	8.35	78.75	740.3	8.53
12.0	65.00	600.0	8.17	70.00	660.0	8.42	75.00	720.0	8.63	80.00	780.0	8.82
12.5	66.25	632.8	8.44	71.25	695.3	8.69	76.25	757.8	8.91	81.25	820.3	9.11
13.0	67.50	666.3	8.70	72.50	731.3	8.96	77.50	796.3	9.19	82.50	861.3	9.40
13.5	68.75	700.3	8.95	73.75	767.8	9.23	78.75	835.3	9.47	83.75	902.8	9.68
14.0	70.00	735.0	9.21	75.00	805.0	9.49	80.00	875.0	9.74	85.00	945.0	9.97
14.5	71.25	770.3	9.46	76.25	842.8	9.75	81.25	915.3	10.01	86.25	987.8	10.24
15.0	72.50	806.3	9.71	77.50	881.3	10.01	82.50	956.3	10.28	87.50	1,031.3	10.52
15.5	73.75	842.8	9.96	78.75	920.3	10.27	83.75	997.8	10.55	88.75	1,075.3	10.79
16.0	75.00	880.0	10.21	80.00	960.0	10.52	85.00	1,040.0	10.81	90.00	1,120.0	11.06
16.5	76.25	917.8	10.45	81.25	1,000.3	10.78	86.25	1,082.8	11.07	91.25	1,165.3	11.33
17.0	77.50	956.3	10.69	82.50	1,041.3	11.03	87.50	1,126.3	11.33	92.50	1,211.3	11.60
17.5	78.75	995.3	10.93	83.75	1,082.8	11.28	88.75	1,170.3	11.58	93.75	1,257.8	11.86
18.0	80.00	1,035.0	11.17	85.00	1,125.0	11.52	90.00	1,215.0	11.84	95.00	1,305.0	12.13
18.5	81.25	1,075.3	11.41	86.25	1,167.8	11.77	91.25	1,260.3	12.09	96.25	1,352.8	12.39
19.0	82.50	1,116.3	11.65	87.50	1,211.3	12.01	92.50	1,306.3	12.34	97.50	1,401.3	12.64
19.5	83.75	1,157.8	11.88	88.75	1,255.3	12.26	93.75	1,352.8	12.59	98.75	1,450.3	12.90
20.0	85.00	1,200.0	12.12	90.00	1,300.0	12.50	95.00	1,400.0	12.84	100.00	1,500.0	13.15

Table 30.—*Area in square feet, A, top width in feet, T, and hydraulic radius in feet, r, of trapezoidal channels,*
side slopes 1¼ to 1—Continued

Depth	Bottom width 60 feet			Bottom width 70 feet			Bottom width 80 feet			Bottom width 90 feet		
	T	A	r	T	A	r	T	A	r	T	A	r
0.4	61.0	24.20	0.39	71.0	28.20	0.40	81.0	32.20	0.40	91.0	36.20	0.40
0.6	61.5	36.45	.59	71.5	42.45	.59	81.5	48.45	.59	91.5	54.45	.59
0.8	62.0	48.80	.78	72.0	56.80	.78	82.0	64.80	.78	92.0	72.80	.79
1.0	62.5	61.25	.97	72.5	71.25	.97	82.5	81.25	.98	92.5	91.25	.98
1.2	63.0	73.80	1.16	73.0	85.80	1.16	83.0	97.80	1.17	93.0	109.80	1.17
1.4	63.5	86.45	1.34	73.5	100.45	1.35	83.5	114.45	1.35	93.5	128.45	1.36
1.6	64.0	99.20	1.52	74.0	115.20	1.53	84.0	131.20	1.54	94.0	147.20	1.55
1.8	64.5	112.05	1.70	74.5	130.05	1.72	84.5	148.05	1.73	94.5	166.05	1.73
2.0	65.0	125.00	1.88	75.0	145.00	1.90	85.0	165.00	1.91	95.0	185.00	1.92
2.2	65.5	138.05	2.06	75.5	160.05	2.08	85.5	182.05	2.09	95.5	204.05	2.10
2.4	66.0	151.20	2.23	76.0	175.20	2.26	86.0	199.20	2.27	96.0	223.20	2.28
2.6	66.5	164.45	2.41	76.5	190.45	2.43	86.5	216.45	2.45	96.5	242.45	2.47
2.8	67.0	177.80	2.58	77.0	205.80	2.61	87.0	233.80	2.63	97.0	261.80	2.65
3.0	67.5	191.25	2.75	77.5	221.25	2.78	87.5	251.25	2.80	97.5	281.25	2.82
3.2	68.0	204.80	2.92	78.0	236.80	2.95	88.0	268.80	2.98	98.0	300.80	3.00
3.4	68.5	218.45	3.08	78.5	252.45	3.12	88.5	286.45	3.15	98.5	320.45	3.18
3.6	69.0	232.20	3.25	79.0	268.20	3.29	89.0	304.20	3.32	99.0	340.20	3.35
3.8	69.5	246.05	3.41	79.5	284.05	3.46	89.5	322.05	3.49	99.5	360.05	3.52
4.0	70.0	260.00	3.57	80.0	300.00	3.62	90.0	340.00	3.66	100.0	380.00	3.70
4.2	70.5	274.05	3.73	80.5	316.05	3.79	90.5	358.05	3.83	100.5	400.05	3.87
4.4	71.0	288.20	3.89	81.0	332.20	3.95	91.0	376.20	4.00	101.0	420.20	4.04
4.6	71.5	302.45	4.05	81.5	348.45	4.11	91.5	394.45	4.16	101.5	440.45	4.21
4.8	72.0	316.80	4.20	82.0	364.80	4.27	92.0	412.80	4.33	102.0	460.80	4.37
5.0	72.5	331.25	4.36	82.5	381.25	4.43	92.5	431.25	4.49	102.5	481.25	4.54
5.2	73.0	345.80	4.51	83.0	397.80	4.59	93.0	449.80	4.65	103.0	501.80	4.71
5.4	73.5	360.45	4.66	83.5	414.45	4.75	93.5	468.45	4.82	103.5	522.45	4.87
5.6	74.0	375.20	4.81	84.0	431.20	4.90	94.0	487.20	4.98	104.0	543.20	5.03
5.8	74.5	390.05	4.96	84.5	448.05	5.06	94.5	506.05	5.13	104.5	564.05	5.20
6.0	75.0	405.00	5.11	85.0	465.00	5.21	95.0	525.00	5.29	105.0	585.00	5.36
6.2	75.5	420.05	5.26	85.5	482.05	5.37	95.5	544.05	5.45	105.5	606.05	5.52
6.4	76.0	435.20	5.41	86.0	499.20	5.52	96.0	563.20	5.60	106.0	627.20	5.68
6.6	76.5	450.45	5.55	86.5	516.45	5.67	96.5	582.45	5.76	106.5	648.45	5.84
6.8	77.0	465.80	5.70	87.0	533.80	5.82	97.0	601.80	5.91	107.0	669.80	5.99
7.0	77.5	481.25	5.84	87.5	551.25	5.97	97.5	621.25	6.07	107.5	691.25	6.15
7.2	78.0	496.80	5.98	88.0	568.80	6.11	98.0	640.80	6.22	108.0	712.80	6.31
7.4	78.5	512.45	6.12	88.5	586.45	6.26	98.5	660.45	6.37	108.5	734.45	6.46
7.6	79.0	528.20	6.26	89.0	604.20	6.41	99.0	680.20	6.52	109.0	756.20	6.61
7.8	79.5	544.05	6.40	89.5	622.05	6.55	99.5	700.05	6.67	109.5	778.05	6.77
8.0	80.0	560.00	6.54	90.0	640.00	6.69	100.0	720.00	6.82	110.0	800.00	6.92
8.2	80.5	576.05	6.68	90.5	658.05	6.84	100.5	740.05	6.96	110.5	822.05	7.07
8.4	81.0	592.20	6.82	91.0	676.20	6.98	101.0	760.20	7.11	111.0	844.20	7.22
8.6	81.5	608.45	6.95	91.5	694.45	7.12	101.5	780.45	7.26	111.5	866.45	7.37
8.8	82.0	624.80	7.09	92.0	712.80	7.26	102.0	800.80	7.40	112.0	888.80	7.52
9.0	82.5	641.25	7.22	92.5	731.25	7.40	102.5	821.25	7.55	112.5	911.25	7.67
9.2	83.0	657.80	7.35	93.0	749.80	7.54	103.0	841.80	7.69	113.0	933.80	7.82
9.4	83.5	674.45	7.49	93.5	768.45	7.68	103.5	862.45	7.83	113.5	956.45	7.96
9.6	84.0	691.20	7.62	94.0	787.20	7.81	104.0	883.20	7.98	114.0	979.20	8.11
9.8	84.5	708.05	7.75	94.5	806.05	7.95	104.5	904.05	8.12	114.5	1,002.05	8.26

Table 30.—*Area in square feet, A, top width in feet, T, and hydraulic radius in feet, r, of trapezoidal channels,*
side slopes 1¼ to 1—Continued

Depth	Bottom width 60 feet			Bottom width 70 feet			Bottom width 80 feet			Bottom width 90 feet		
	T	A	r	T	A	r	T	A	r	T	A	r
10.0	85.00	725.00	7.88	95.00	825.00	8.09	105.00	925.00	8.26	115.00	1,025.00	8.40
10.5	86.25	767.81	8.20	96.25	872.81	8.42	106.25	977.81	8.61	116.25	1,082.81	8.76
11.0	87.50	811.3	8.52	97.50	921.3	8.76	107.50	1,031.3	8.95	117.50	1,141.3	9.11
11.5	88.75	855.3	8.83	98.75	970.3	9.08	108.75	1,085.3	9.29	118.75	1,200.3	9.46
12.0	90.00	900.0	9.14	100.00	1,020.0	9.41	110.00	1,140.0	9.63	120.00	1,260.0	9.81
12.5	91.25	945.3	9.45	101.25	1,070.3	9.73	111.25	1,195.3	9.96	121.25	1,320.3	10.15
13.0	92.50	991.3	9.75	102.50	1,121.3	10.05	112.50	1,251.3	10.29	122.50	1,381.3	10.49
13.5	93.75	1,037.8	10.05	103.75	1,172.8	10.36	113.75	1,307.8	10.61	123.75	1,442.8	10.83
14.0	95.00	1,085.0	10.35	105.00	1,225.0	10.67	115.00	1,365.0	10.94	125.00	1,505.0	11.16
14.5	96.25	1,132.8	10.64	106.25	1,277.8	10.98	116.25	1,422.8	11.25	126.25	1,567.8	11.49
15.0	97.50	1,181.3	10.94	107.50	1,331.3	11.28	117.50	1,481.3	11.57	127.50	1,631.3	11.82
15.5	98.75	1,230.3	11.22	108.75	1,385.3	11.58	118.75	1,540.3	11.88	128.75	1,695.3	12.14
16.0	100.00	1,280.0	11.51	110.00	1,440.0	11.88	120.00	1,600.0	12.19	130.00	1,760.0	12.46
16.5	101.25	1,330.3	11.79	111.25	1,495.3	12.17	121.25	1,660.3	12.50	131.25	1,825.3	12.78
17.0	102.50	1,381.3	12.07	112.50	1,551.3	12.47	122.50	1,721.3	12.80	132.50	1,891.3	13.09
17.5	103.75	1,432.8	12.35	113.75	1,607.8	12.76	123.75	1,782.8	13.11	133.75	1,957.8	13.41
18.0	105.00	1,485.0	12.62	115.00	1,665.0	13.05	125.00	1,845.0	13.41	135.00	2,025.0	13.72
18.5	106.25	1,537.8	12.90	116.25	1,722.8	13.33	126.25	1,907.8	13.70	136.25	2,092.8	14.02
19.0	107.50	1,591.3	13.17	117.50	1,781.3	13.62	127.50	1,971.3	14.00	137.50	2,161.3	14.33
19.5	108.75	1,645.3	13.44	118.75	1,840.3	13.90	128.75	2,035.3	14.29	138.75	2,230.3	14.63
20.0	110.00	1,700.0	13.71	120.00	1,900.0	14.18	130.00	2,100.0	14.58	140.00	2,300.0	14.93

Table 31.—*Area in square feet, A, top width in feet, T, and hydraulic radius in feet, r, of trapezoidal channels,*
side slopes 1½ to 1

Depth	Bottom width 2 feet			Bottom width 3 feet			Bottom width 4 feet			Bottom width 5 feet		
	T	A	r	T	A	r	T	A	r	T	A	r
0.4	3.2	1.04	.30	4.2	1.44	.32	5.2	1.84	.34	6.2	2.24	.35
0.6	3.8	1.74	.42	4.8	2.34	.45	5.8	2.94	.48	6.8	3.54	.49
0.8	4.4	2.56	.52	5.4	3.36	.57	6.4	4.16	.60	7.4	4.96	.63
1.0	5.0	3.50	.62	6.0	4.50	.68	7.0	5.50	.72	8.0	6.50	.76
1.2	5.6	4.56	.72	6.6	5.76	.79	7.6	6.96	.84	8.6	8.16	.87
1.4	6.2	5.74	.81	7.2	7.14	.89	8.2	8.54	.94	9.2	9.94	.99
1.6	6.8	7.04	.91	7.8	8.64	.99	8.8	10.24	1.05	9.8	11.84	1.10
1.8	7.4	8.46	1.00	8.4	10.26	1.08	9.4	12.06	1.15	10.4	13.86	1.21
2.0	8.0	10.00	1.09	9.0	12.00	1.18	10.0	14.00	1.25	11.0	16.00	1.31
2.2	8.6	11.66	1.17	9.6	13.86	1.27	10.6	16.06	1.35	11.6	18.26	1.41
2.4	9.2	13.44	1.26	10.2	15.84	1.36	11.2	18.24	1.44	12.2	20.64	1.51
2.6	9.8	15.34	1.35	10.8	17.94	1.45	11.8	20.54	1.54	12.8	23.14	1.61
2.8	10.4	17.36	1.44	11.4	20.16	1.54	12.4	22.96	1.63	13.4	25.76	1.71
3.0	11.0	19.50	1.52	12.0	22.50	1.63	13.0	25.50	1.72	14.0	28.50	1.80
3.2	11.6	21.76	1.61	12.6	24.96	1.72	13.6	28.16	1.81	14.6	31.36	1.90
3.4	12.2	24.14	1.69	13.2	27.54	1.80	14.2	30.94	1.90	15.2	34.34	1.99
3.6	12.8	26.64	1.78	13.8	30.24	1.89	14.8	33.84	1.99	15.8	37.44	2.08
3.8	13.4	29.26	1.86	14.4	33.06	1.98	15.4	36.86	2.08	16.4	40.66	2.17
4.0	14.0	32.00	1.95	15.0	36.00	2.07	16.0	40.00	2.17	17.0	44.00	2.27
4.2	14.6	34.86	2.03	15.6	39.06	2.15	16.6	43.26	2.26	17.6	47.46	2.36
4.4	15.2	37.84	2.12	16.2	42.24	2.24	17.2	46.64	2.35	18.2	51.04	2.45
4.6	15.8	40.94	2.20	16.8	45.54	2.33	17.8	50.14	2.44	18.8	54.74	2.54
4.8	16.4	44.16	2.29	17.4	48.96	2.41	18.4	53.76	2.52	19.4	58.56	2.63
5.0	17.0	47.50	2.37	18.0	52.50	2.50	19.0	57.50	2.61	20.0	62.50	2.71
5.2	17.6	50.96	2.46	18.6	56.16	2.58	19.6	61.36	2.70	20.6	66.56	2.80
5.4	18.2	54.54	2.54	19.2	59.94	2.67	20.2	65.34	2.78	21.2	70.74	2.89
5.6	18.8	58.24	2.62	19.8	63.84	2.75	20.8	69.44	2.87	21.8	75.04	2.98
5.8	19.4	62.06	2.71	20.4	67.86	2.84	21.4	73.66	2.96	22.4	79.46	3.07
6.0	20.0	66.00	2.79	21.0	72.00	2.92	22.0	78.00	3.04	23.0	84.00	3.15
6.2	20.6	70.06	2.88	21.6	76.26	3.01	22.6	82.46	3.13	23.6	88.66	3.24
6.4	21.2	74.24	2.96	22.2	80.64	3.09	23.2	87.04	3.21	24.2	93.44	3.33
6.6	21.8	78.54	3.04	22.8	85.14	3.18	23.8	91.74	3.30	24.8	98.34	3.41
6.8	22.4	82.96	3.13	23.4	89.76	3.26	24.4	96.56	3.39	25.4	103.36	3.50
7.0	23.0	87.50	3.21	24.0	94.50	3.35	25.0	101.50	3.47	26.0	108.50	3.59
7.2	23.6	92.16	3.30	24.6	99.36	3.43	25.6	106.56	3.56	26.6	113.76	3.67
7.4	24.2	96.94	3.38	25.2	104.34	3.52	26.2	111.74	3.64	27.2	119.14	3.76
7.6	24.8	101.84	3.46	25.8	109.44	3.60	26.8	117.04	3.73	27.8	124.64	3.85
7.8	25.4	106.86	3.55	26.4	114.66	3.68	27.4	122.46	3.81	28.4	130.26	3.93
8.0	26.0	112.00	3.63	27.0	120.00	3.77	28.0	128.00	3.90	29.0	136.00	4.02
8.2	----	------	----	27.6	125.46	3.85	28.6	133.66	3.98	29.6	141.86	4.10
8.4	----	------	----	28.2	131.04	3.94	29.2	139.44	4.07	30.2	147.84	4.19
8.6	----	------	----	28.8	136.74	4.02	29.8	145.34	4.15	30.8	153.94	4.28
8.8	----	------	----	29.4	142.56	4.10	30.4	151.36	4.24	31.4	160.16	4.36
9.0	----	------	----	30.0	148.50	4.19	31.0	157.50	4.32	32.0	166.50	4.45
9.2	----	------	----	30.6	154.56	4.27	31.6	163.76	4.41	32.6	172.96	4.53
9.4	----	------	----	31.2	160.74	4.36	32.2	170.14	4.49	33.2	179.54	4.62
9.6	----	------	----	31.8	167.04	4.44	32.8	176.64	4.57	33.8	186.24	4.70
9.8	----	------	----	32.4	173.46	4.52	33.4	183.26	4.66	34.4	193.06	4.79

Table 31.—*Area in square feet, A, top width in feet, T, and hydraulic radius in feet, r, of trapezoidal channels,* **side slopes 1½ to 1**—Continued

Depth	Bottom width 2 feet			Bottom width 3 feet			Bottom width 4 feet			Bottom width 5 feet		
	T	A	r	T	A	r	T	A	r	T	A	r
10.0				33.0	180.00	4.61	34.0	190.00	4.74	35.0	200.00	4.87
10.5				34.5	196.87	4.82	35.5	207.37	4.95	36.5	217.87	5.08
11.0				36.0	214.5	5.03	37.0	225.5	5.16	38.0	236.5	5.30
11.5				37.5	232.9	5.24	38.5	244.4	5.38	39.5	255.9	5.51
12.0				39.0	252.0	5.45	40.0	264.0	5.59	41.0	276.0	5.72
12.5				40.5	271.9	5.66	41.5	284.4	5.80	42.5	296.9	5.93
13.0							43.0	305.5	6.01	44.0	318.5	6.14
13.5							44.5	327.4	6.22	45.5	340.9	6.35
14.0							46.0	350.0	6.42	47.0	364.0	6.56
14.5							47.5	373.4	6.63	48.5	387.9	6.77
15.0							49.0	397.5	6.84	50.0	412.5	6.98
15.5							50.5	422.4	7.05	51.5	437.9	7.19
16.0							52.0	448.0	7.26	53.0	464.0	7.40
16.5										54.5	490.9	7.61
17.0										56.0	518.5	7.82
17.5										57.5	546.9	8.03
18.0										59.0	576.0	8.24
18.5										60.5	605.9	8.45
19.0										62.0	636.5	8.66
19.5										63.5	667.9	8.87
20.0										65.0	700.0	9.08

Depth	Bottom width 6 feet			Bottom width 7 feet			Bottom width 8 feet			Bottom width 9 feet		
	T	A	r	T	A	r	T	A	r	T	A	r
0.4	7.2	2.64	.35	8.2	3.04	.36	9.2	3.44	.36	10.2	3.84	.37
0.6	7.8	4.14	.51	8.8	4.74	.52	9.8	5.34	.53	10.8	5.94	.53
0.8	8.4	5.76	.65	9.4	6.56	.66	10.4	7.36	.68	11.4	8.16	.69
1.0	9.0	7.50	.78	10.0	8.50	.80	11.0	9.50	.82	12.0	10.50	.83
1.2	9.6	9.36	.91	10.6	10.56	.93	11.6	11.76	.95	12.6	12.96	.97
1.4	10.2	11.34	1.03	11.2	12.74	1.06	12.2	14.14	1.08	13.2	15.54	1.11
1.6	10.8	13.44	1.14	11.8	15.04	1.18	12.8	16.64	1.21	13.8	18.24	1.24
1.8	11.4	15.66	1.25	12.4	17.46	1.29	13.4	19.26	1.33	14.4	21.06	1.36
2.0	12.0	18.00	1.36	13.0	20.00	1.41	14.0	22.00	1.45	15.0	24.00	1.48
2.2	12.6	20.46	1.47	13.6	22.66	1.52	14.6	24.86	1.56	15.6	27.06	1.60
2.4	13.2	23.04	1.57	14.2	25.44	1.63	15.2	27.84	1.67	16.2	30.24	1.71
2.6	13.8	25.74	1.67	14.8	28.34	1.73	15.8	30.94	1.78	16.8	33.54	1.83
2.8	14.4	28.56	1.77	15.4	31.36	1.83	16.4	34.16	1.89	17.4	36.96	1.94
3.0	15.0	31.50	1.87	16.0	34.50	1.94	17.0	37.50	1.99	18.0	40.50	2.04
3.2	15.6	34.56	1.97	16.6	37.76	2.04	17.6	40.96	2.10	18.6	44.16	2.15
3.4	16.2	37.74	2.07	17.2	41.14	2.14	18.2	44.54	2.20	19.2	47.94	2.26
3.6	16.8	41.04	2.16	17.8	44.64	2.23	18.8	48.24	2.30	19.8	51.84	2.36
3.8	17.4	44.46	2.26	18.4	48.26	2.33	19.4	52.06	2.40	20.4	55.86	2.46
4.0	18.0	48.00	2.35	19.0	52.00	2.43	20.0	56.00	2.50	21.0	60.00	2.56
4.2	18.6	51.66	2.44	19.6	55.86	2.52	20.6	60.06	2.60	21.6	64.26	2.66
4.4	19.2	55.44	2.54	20.2	59.84	2.62	21.2	64.24	2.69	22.2	68.64	2.76
4.6	19.8	59.34	2.63	20.8	63.94	2.71	21.8	68.54	2.79	22.8	73.14	2.86
4.8	20.4	63.36	2.72	21.4	68.16	2.80	22.4	72.96	2.88	23.4	77.76	2.96

Table 31.—*Area in square feet, A, top width in feet, T, and hydraulic radius in feet, r, of trapezoidal channels,* **side slopes 1½ to 1**—Continued

Depth	Bottom width 6 feet			Bottom width 7 feet			Bottom width 8 feet			Bottom width 9 feet		
	T	A	r	T	A	r	T	A	r	T	A	r
5.0	21.0	67.50	2.81	22.0	72.50	2.90	23.0	77.50	2.98	24.0	82.50	3.05
5.2	21.6	71.76	2.90	22.6	76.96	2.99	23.6	82.16	3.07	24.6	87.36	3.15
5.4	22.2	76.14	2.99	23.2	81.54	3.08	24.2	86.94	3.16	25.2	92.34	3.24
5.6	22.8	80.64	3.08	23.8	86.24	3.17	24.8	91.84	3.26	25.8	97.44	3.34
5.8	23.4	85.26	3.17	24.4	91.06	3.26	25.4	96.86	3.35	26.4	102.66	3.43
6.0	24.0	90.00	3.26	25.0	96.00	3.35	26.0	102.00	3.44	27.0	108.00	3.53
6.2	24.6	94.86	3.35	25.6	101.06	3.44	26.6	107.26	3.53	27.6	113.46	3.62
6.4	25.2	99.84	3.43	26.2	106.24	3.53	27.2	112.64	3.62	28.2	119.04	3.71
6.6	25.8	104.94	3.52	26.8	111.54	3.62	27.8	118.14	3.72	28.8	124.74	3.80
6.8	26.4	110.16	3.61	27.4	116.96	3.71	28.4	123.76	3.81	29.4	130.56	3.90
7.0	27.0	115.50	3.70	28.0	122.50	3.80	29.0	129.50	3.90	30.0	136.50	3.99
7.2	27.6	120.96	3.78	28.6	128.16	3.89	29.6	135.36	3.99	30.6	142.56	4.08
7.4	28.2	126.54	3.87	29.2	133.94	3.98	30.2	141.34	4.08	31.2	148.74	4.17
7.6	28.8	132.24	3.96	29.8	139.84	4.06	30.8	147.44	4.16	31.8	155.04	4.26
7.8	29.4	138.06	4.05	30.4	145.86	4.15	31.4	153.66	4.25	32.4	161.46	4.35
8.0	30.0	144.00	4.13	31.0	152.00	4.24	32.0	160.00	4.34	33.0	168.00	4.44
8.2	30.6	150.06	4.22	31.6	158.26	4.33	32.6	166.46	4.43	33.6	174.66	4.53
8.4	31.2	156.24	4.31	32.2	164.64	4.42	33.2	173.04	4.52	34.2	181.44	4.62
8.6	31.8	162.54	4.39	32.8	171.14	4.50	33.8	179.74	4.61	34.8	188.34	4.71
8.8	32.4	168.96	4.48	33.4	177.76	4.59	34.4	186.56	4.70	35.4	195.36	4.80
9.0	33.0	175.50	4.56	34.0	184.50	4.68	35.0	193.50	4.78	36.0	202.50	4.89
9.2	33.6	182.16	4.65	34.6	191.36	4.76	35.6	200.56	4.87	36.6	209.76	4.97
9.4	34.2	188.94	4.74	35.2	198.34	4.85	36.2	207.74	4.96	37.2	217.14	5.06
9.6	34.8	195.84	4.82	35.8	205.44	4.94	36.8	215.04	5.05	37.8	224.64	5.15
9.8	35.4	202.86	4.91	36.4	212.66	5.02	37.4	222.46	5.13	38.4	232.26	5.24
10.0	36.0	210.00	4.99	37.0	220.00	5.11	38.0	230.00	5.22	39.0	240.00	5.33
10.5	37.5	228.37	5.21	38.5	238.87	5.33	39.5	249.37	5.44	40.5	259.87	5.55
11.0	39.0	247.5	5.42	40.0	258.5	5.54	41.0	269.5	5.65	42.0	280.5	5.76
11.5	40.5	267.4	5.63	41.5	278.9	5.75	42.5	290.4	5.87	43.5	301.9	5.98
12.0	42.0	288.0	5.85	43.0	300.0	5.97	44.0	312.0	6.09	45.0	324.0	6.20
12.5	43.5	309.4	6.06	44.5	321.9	6.18	45.5	334.4	6.30	46.5	346.9	6.42
13.0	45.0	331.5	6.27	46.0	344.5	6.39	47.0	357.5	6.52	48.0	370.5	6.63
13.5	46.5	354.4	6.48	47.5	367.9	6.61	48.5	381.4	6.73	49.5	394.9	6.85
14.0	48.0	378.0	6.69	49.0	392.0	6.82	50.0	406.0	6.94	51.0	420.0	7.06
14.5	49.5	402.4	6.90	50.5	416.9	7.03	51.5	431.4	7.16	52.5	445.9	7.28
15.0	51.0	427.5	7.12	52.0	442.5	7.24	53.0	457.5	7.37	54.0	472.5	7.49
15.5	52.5	453.4	7.33	53.5	468.9	7.46	54.5	484.4	7.58	55.5	499.9	7.70
16.0	54.0	480.0	7.54	55.0	496.0	7.67	56.0	512.0	7.79	57.0	528.0	7.92
16.5	55.5	507.4	7.75	56.5	523.9	7.88	57.5	540.4	8.01	58.5	556.9	8.13
17.0	57.0	535.5	7.96	58.0	552.5	8.09	59.0	569.5	8.22	60.0	586.5	8.34
17.5	58.5	564.4	8.17	59.5	581.9	8.30	60.5	599.4	8.43	61.5	616.9	8.56
18.0	60.0	594.0	8.38	61.0	612.0	8.51	62.0	630.0	8.64	63.0	648.0	8.77
18.5	61.5	624.4	8.59	62.5	642.9	8.72	63.5	661.4	8.85	64.5	679.9	8.98
19.0	63.0	655.5	8.80	64.0	674.5	8.93	65.0	693.5	9.06	66.0	712.5	9.19
19.5	64.5	687.4	9.01	65.5	706.9	9.14	66.5	726.4	9.28	67.5	745.9	9.40
20.0	66.0	720.0	9.22	67.0	740.0	9.35	68.0	760.0	9.49	69.0	780.0	9.62

Table 31.—*Area in square feet, A, top width in feet, T, and hydraulic radius in feet, r, of trapezoidal channels,*
side slopes 1½ to 1—Continued

Depth	Bottom width 10 feet			Bottom width 12 feet			Bottom width 14 feet			Bottom width 16 feet		
	T	A	r	T	A	r	T	A	r	T	A	r
0.4	11.2	4.24	.37	13.2	5.04	.37	15.2	5.84	.38	17.2	6.64	.38
0.6	11.8	6.54	.54	13.8	7.74	.55	15.8	8.94	.55	17.8	10.14	.56
0.8	12.4	8.96	.70	14.4	10.56	.71	16.4	12.16	.72	18.4	13.76	.73
1.0	13.0	11.50	.85	15.0	13.50	.87	17.0	15.50	.88	19.0	17.50	.89
1.2	13.6	14.16	.99	15.6	16.56	1.01	17.6	18.96	1.03	19.6	21.36	1.05
1.4	14.2	16.94	1.13	16.2	19.74	1.16	18.2	22.54	1.18	20.2	25.34	1.20
1.6	14.8	19.84	1.26	16.8	23.04	1.30	18.8	26.24	1.33	20.8	29.44	1.35
1.8	15.4	22.86	1.39	17.4	26.46	1.43	19.4	30.06	1.47	21.4	33.66	1.50
2.0	16.0	26.00	1.51	18.0	30.00	1.56	20.0	34.00	1.60	22.0	38.00	1.64
2.2	16.6	29.26	1.63	18.6	33.66	1.69	20.6	38.06	1.74	22.6	42.46	1.77
2.4	17.2	32.64	1.75	19.2	37.44	1.81	21.2	42.24	1.86	23.2	47.04	1.91
2.6	17.8	36.14	1.87	19.8	41.34	1.93	21.8	46.54	1.99	23.8	51.74	2.04
2.8	18.4	39.76	1.98	20.4	45.36	2.05	22.4	50.96	2.11	24.4	56.56	2.17
3.0	19.0	43.50	2.09	21.0	49.50	2.17	23.0	55.50	2.24	25.0	61.50	2.29
3.2	19.6	47.36	2.20	21.6	53.76	2.28	23.6	60.16	2.36	25.6	66.56	2.42
3.4	20.2	51.34	2.31	22.2	58.14	2.40	24.2	64.94	2.47	26.2	71.74	2.54
3.6	20.8	55.44	2.41	22.8	62.64	2.51	24.8	69.84	2.59	26.8	77.04	2.66
3.8	21.4	59.66	2.52	23.4	67.26	2.62	25.4	74.86	2.70	27.4	82.46	2.78
4.0	22.0	64.00	2.62	24.0	72.00	2.72	26.0	80.00	2.81	28.0	88.00	2.89
4.2	22.6	68.46	2.72	24.6	76.86	2.83	26.6	85.26	2.93	28.6	93.66	3.01
4.4	23.2	73.04	2.82	25.2	81.84	2.94	27.2	90.64	3.04	29.2	99.44	3.12
4.6	23.8	77.74	2.92	25.8	86.94	3.04	27.8	96.14	3.14	29.8	105.34	3.23
4.8	24.4	82.56	3.02	26.4	92.16	3.14	28.4	101.76	3.25	30.4	111.36	3.34
5.0	25.0	87.50	3.12	27.0	97.50	3.25	29.0	107.50	3.36	31.0	117.50	3.45
5.2	25.6	92.56	3.22	27.6	102.96	3.35	29.6	113.36	3.46	31.6	123.76	3.56
5.4	26.2	97.74	3.32	28.2	108.54	3.45	30.2	119.34	3.57	32.2	130.14	3.67
5.6	26.8	103.04	3.41	28.8	114.24	3.55	30.8	125.44	3.67	32.8	136.64	3.78
5.8	27.4	108.46	3.51	29.4	120.06	3.65	31.4	131.66	3.77	33.4	143.26	3.88
6.0	28.0	114.00	3.60	30.0	126.00	3.75	32.0	138.00	3.87	34.0	150.00	3.99
6.2	28.6	119.66	3.70	30.6	132.06	3.84	32.6	144.46	3.97	34.6	156.86	4.09
6.4	29.2	125.44	3.79	31.2	138.24	3.94	33.2	151.04	4.07	35.2	163.84	4.19
6.6	29.8	131.34	3.89	31.8	144.54	4.04	33.8	157.74	4.17	35.8	170.94	4.30
6.8	30.4	137.36	3.98	32.4	150.96	4.13	34.4	164.56	4.27	36.4	178.16	4.40
7.0	31.0	143.50	4.07	33.0	157.50	4.23	35.0	171.50	4.37	37.0	185.50	4.50
7.2	31.6	149.76	4.16	33.6	164.16	4.32	35.6	178.56	4.47	37.6	192.96	4.60
7.4	32.2	156.14	4.26	34.2	170.94	4.42	36.2	185.74	4.57	38.2	200.54	4.70
7.6	32.8	162.64	4.35	34.8	177.84	4.51	36.8	193.04	4.66	38.8	208.24	4.80
7.8	33.4	169.26	4.44	35.4	184.86	4.61	37.4	200.46	4.76	39.4	216.06	4.90
8.0	34.0	176.00	4.53	36.0	192.00	4.70	38.0	208.00	4.85	40.0	224.00	5.00
8.2	34.6	182.86	4.62	36.6	199.26	4.79	38.6	215.66	4.95	40.6	232.06	5.09
8.4	35.2	189.84	4.71	37.2	206.64	4.89	39.2	223.44	5.05	41.2	240.24	5.19
8.6	35.8	196.94	4.80	37.8	214.14	4.98	39.8	231.34	5.14	41.8	248.54	5.29
8.8	36.4	204.16	4.89	38.4	221.76	5.07	40.4	239.36	5.23	42.4	256.96	5.38
9.0	37.0	211.50	4.98	39.0	229.50	5.16	41.0	247.50	5.33	43.0	265.50	5.48
9.2	37.6	218.96	5.07	39.6	237.36	5.25	41.6	255.76	5.42	43.6	274.16	5.58
9.4	38.2	226.54	5.16	40.2	245.34	5.35	42.2	264.14	5.52	44.2	282.94	5.67
9.6	38.8	234.24	5.25	40.8	253.44	5.44	42.8	272.64	5.61	44.8	291.84	5.77
9.8	39.4	242.06	5.34	41.4	261.66	5.53	43.4	281 26	5.70	45.4	300.86	5.86

Table 31.—*Area in square feet, A, top width in feet, T, and hydraulic radius in feet, r, of trapezoidal channels,* **side slopes 1½ to 1**—Continued

Depth	Bottom width 10 feet			Bottom width 12 feet			Bottom width 14 feet			Bottom width 16 feet		
	T	A	r	T	A	r	T	A	r	T	A	r
10.0	40.0	250.00	5.43	42.0	270.00	5.62	44.0	290.00	5.79	46.0	310.00	5.96
10.5	41.5	270.37	5.65	43.5	291.37	5.84	45.5	312.37	6.02	47.5	333.37	6.19
11.0	43.0	291.5	5.87	45.0	313.5	6.07	47.0	335.5	6.27	49.0	357.5	6.42
11.5	44.5	313.4	6.09	46.5	336.4	6.29	48.5	359.4	6.48	50.5	382.4	6.65
12.0	46.0	336.0	6.31	48.0	360.0	6.51	50.0	384.0	6.71	52.0	408.0	6.88
12.5	47.5	359.4	6.53	49.5	384.4	6.74	51.5	409.4	6.93	53.5	434.4	7.11
13.0	49.0	383.5	6.74	51.0	409.5	6.96	53.0	435.5	7.15	55.0	461.5	7.34
13.5	50.5	408.4	6.96	52.5	435.4	7.18	54.5	462.4	7.38	56.5	489.4	7.57
14.0	52.0	434.0	7.18	54.0	462.0	7.39	56.0	490.0	7.60	58.0	518.0	7.79
14.5	53.5	460.4	7.39	55.5	489.4	7.61	57.5	518.4	7.82	59.5	547.4	8.02
15.0	55.0	487.5	7.61	57.0	517.5	7.83	59.0	547.5	8.04	61.0	577.5	8.24
15.5	56.5	515.4	7.82	58.5	546.4	8.05	60.5	577.4	8.26	62.5	608.4	8.46
16.0	58.0	544.0	8.04	60.0	576.0	8.27	62.0	608.0	8.48	64.0	640.0	8.69
16.5	59.5	573.4	8.25	61.5	606.4	8.48	63.5	639.4	8.70	65.5	672.4	8.91
17.0	61.0	603.5	8.46	63.0	637.5	8.70	65.0	671.5	8.92	67.0	705.5	9.13
17.5	62.5	634.4	8.68	64.5	669.4	8.91	66.5	704.4	9.14	68.5	739.4	9.35
18.0	64.0	666.0	8.89	66.0	702.0	9.13	68.0	738.0	9.35	70.0	774.0	9.57
18.5	65.5	698.4	9.11	67.5	735.4	9.34	69.5	772.4	9.57	71.5	809.4	9.79
19.0	67.0	731.5	9.32	69.0	769.5	9.56	71.0	807.5	9.79	73.0	845.5	10.01
19.5	68.5	765.4	9.53	70.5	804.4	9.77	72.5	843.4	10.00	74.5	882.4	10.22
20.0	70.0	800.0	9.74	72.0	840.0	9.99	74.0	880.0	10.22	76.0	920.0	10.44

Depth	Bottom width 18 feet			Bottom width 20 feet			Bottom width 22 feet			Bottom width 24 feet		
	T	A	r	T	A	r	T	A	r	T	A	r
0.4	19.2	7.44	.38	21.2	8.24	.38	23.2	9.04	.39	25.2	9.84	.39
0.6	19.8	11.34	.56	21.8	12.54	.57	23.8	13.74	.57	25.8	14.94	.57
0.8	20.4	15.36	.74	22.4	16.96	.74	24.4	18.56	.75	26.4	20.16	.75
1.0	21.0	19.50	.90	23.0	21.50	.91	25.0	23.50	.92	27.0	25.50	.92
1.2	21.6	23.76	1.06	23.6	26.16	1.08	25.6	28.56	1.08	27.6	30.96	1.09
1.4	22.2	28.14	1.22	24.2	30.94	1.24	26.2	33.74	1.25	28.2	36.54	1.26
1.6	22.8	32.64	1.37	24.8	35.84	1.39	26.8	39.04	1.41	28.8	42.24	1.42
1.8	23.4	37.26	1.52	25.4	40.86	1.54	27.4	44.46	1.56	29.4	48.06	1.58
2.0	24.0	42.00	1.67	26.0	46.00	1.69	28.0	50.00	1.71	30.0	54.00	1.73
2.2	24.6	46.86	1.81	26.6	51.26	1.84	28.6	55.66	1.86	30.6	60.06	1.88
2.4	25.2	51.84	1.94	27.2	56.64	1.98	29.2	61.44	2.00	31.2	66.24	2.03
2.6	25.8	56.94	2.08	27.8	62.14	2.12	29.8	67.34	2.15	31.8	72.54	2.17
2.8	26.4	62.16	2.21	28.4	67.76	2.25	30.4	73.36	2.29	32.4	78.96	2.32
3.0	27.0	67.50	2.34	29.0	73.50	2.39	31.0	79.50	2.42	33.0	85.50	2.46
3.2	27.6	72.96	2.47	29.6	79.36	2.52	31.6	85.76	2.56	33.6	92.16	2.59
3.4	28.2	78.54	2.60	30.2	85.34	2.65	32.2	92.14	2.69	34.2	98.94	2.73
3.6	28.8	84.24	2.72	30.8	91.44	2.77	32.8	98.64	2.82	34.8	105.84	2.86
3.8	29.4	90.06	2.84	31.4	97.66	2.90	33.4	105.26	2.95	35.4	112.86	2.99
4.0	30.0	96.00	2.96	32.0	104.00	3.02	34.0	112.00	3.08	36.0	120.00	3.12
4.2	30.6	102.06	3.08	32.6	110.46	3.14	34.6	118.86	3.20	36.6	127.26	3.25
4.4	31.2	108.24	3.20	33.2	117.04	3.26	35.2	125.84	3.32	37.2	134.64	3.38
4.6	31.8	114.54	3.31	33.8	123.74	3.38	35.8	132.94	3.45	37.8	142.14	3.50
4.8	32.4	120.96	3.43	34.4	130.56	3.50	36.4	140.16	3.57	38.4	149.76	3.63

Table 31.—*Area in square feet, A, top width in feet, T, and hydraulic radius in feet, r, of trapezoidal channels,*
side slopes 1½ to 1—Continued

Depth	Bottom width 18 feet			Bottom width 20 feet			Bottom width 22 feet			Bottom width 24 feet		
	T	A	r	T	A	r	T	A	r	T	A	r
5.0	33.0	127.50	3.54	35.0	137.50	3.62	37.0	147.50	3.68	39.0	157.50	3.75
5.2	33.6	134.16	3.65	35.6	144.56	3.73	37.6	154.96	3.80	39.6	165.36	3.87
5.4	34.2	140.94	3.76	36.2	151.74	3.84	38.2	162.54	3.92	40.2	173.34	3.99
5.6	34.8	147.84	3.87	36.8	159.04	3.96	38.8	170.24	4.03	40.8	181.44	4.11
5.8	35.4	154.86	3.98	37.4	166.46	4.07	39.4	178.06	4.15	41.4	189.66	4.22
6.0	36.0	162.00	4.09	38.0	174.00	4.18	40.0	186.00	4.26	42.0	198.00	4.34
6.2	36.6	169.26	4.19	38.6	181.66	4.29	40.6	194.06	4.38	42.6	206.46	4.45
6.4	37.2	176.64	4.30	39.2	189.44	4.40	41.2	202.24	2.49	43.2	215.04	4.57
6.6	37.8	184.14	4.41	39.8	197.34	4.51	41.8	210.54	4.60	43.8	223.74	4.68
6.8	38.4	191.76	4.51	40.4	205.36	4.61	42.4	218.96	4.71	44.4	232.56	4.79
7.0	39.0	199.50	4.61	41.0	213.50	4.72	43.0	227.50	4.82	45.0	241.50	4.90
7.2	39.6	207.36	4.72	41.6	221.76	4.83	43.6	236.16	4.92	45.6	250.56	5.02
7.4	40.2	215.34	4.82	42.2	230.14	4.93	44.2	244.94	5.03	46.2	259.74	5.12
7.6	40.8	223.44	4.92	42.8	238.64	5.03	44.8	253.84	5.14	46.8	269.04	5.23
7.8	41.4	231.66	5.02	43.4	247.26	5.14	45.4	262.86	5.24	47.4	278.46	5.34
8.0	42.0	240.00	5.12	44.0	256.00	5.24	46.0	272.00	5.35	48.0	288.00	5.45
8.2	42.6	248.46	5.22	44.6	264.86	5.34	46.6	281.26	5.45	48.6	297.66	5.56
8.4	43.2	257.04	5.32	45.2	273.84	5.45	47.2	290.64	5.56	49.2	307.44	5.66
8.6	43.8	265.74	5.42	45.8	282.94	5.55	47.8	300.14	5.66	49.8	317.34	5.77
8.8	44.4	274.56	5.52	46.4	292.16	5.65	48.4	309.76	5.77	50.4	327.36	5.87
9.0	45.0	283.50	5.62	47.0	301.50	5.75	49.0	319.50	5.87	51.0	337.50	5.98
9.2	45.6	292.56	5.72	47.6	310.96	5.85	49.6	329.36	5.97	51.6	347.76	6.08
9.4	46.2	301.74	5.81	48.2	320.54	5.95	50.2	339.34	6.07	52.2	358.14	6.19
9.6	46.8	311.04	5.91	48.8	330.24	6.05	50.8	349.44	6.17	52.8	368.64	6.29
9.8	47.4	320.46	6.01	49.4	340.06	6.15	51.4	359.66	6.27	53.4	379.26	6.39
10.0	48.0	330.00	6.10	50.0	350.00	6.24	52.0	370.00	6.37	54.0	390.00	6.49
10.5	49.5	354.37	6.34	51.5	375.37	6.49	53.5	396.37	6.62	55.5	417.37	6.75
11.0	51.0	379.5	6.58	53.0	401.5	6.73	55.0	423.5	6.87	57.0	445.5	7.00
11.5	52.5	405.4	6.82	54.5	428.4	6.97	56.5	451.4	7.11	58.5	474.4	7.25
12.0	54.0	432.0	7.05	56.0	456.0	7.21	58.0	480.0	7.35	60.0	504.0	7.49
12.5	55.5	459.4	7.28	57.5	484.4	7.44	59.5	509.4	7.59	61.5	534.4	7.74
13.0	57.0	487.5	7.51	59.0	513.5	7.68	61.0	539.5	7.83	63.0	565.5	7.98
13.5	58.5	516.4	7.74	60.5	543.4	7.91	62.5	570.4	8.07	64.5	597.4	8.22
14.0	60.0	546.0	7.97	62.0	574.0	8.14	64.0	602.0	8.31	66.0	630.0	8.46
14.5	61.5	576.4	8.20	63.5	605.4	8.38	65.5	634.4	8.54	67.5	663.4	8.70
15.0	63.0	607.5	8.43	65.0	637.5	8.61	67.0	667.5	8.77	69.0	697.5	8.93
15.5	64.5	639.4	8.65	66.5	670.4	8.83	68.5	701.4	9.01	70.5	732.4	9.17
16.0	66.0	672.0	8.88	68.0	704.0	9.06	70.0	736.0	9.24	72.0	768.0	9.40
16.5	67.5	705.4	9.10	69.5	738.4	9.29	71.5	771.4	9.47	73.5	804.4	9.63
17.0	69.0	739.5	9.33	71.0	773.5	9.51	73.0	807.5	9.69	75.0	841.5	9.87
17.5	70.5	774.4	9.55	72.5	809.4	9.74	74.5	844.4	9.92	76.5	879.4	10.10
18.0	72.0	810.0	9.77	74.0	846.0	9.96	76.0	882.0	10.15	78.0	918.0	10.33
18.5	73.5	846.4	9.99	75.5	883.4	10.19	77.5	920.4	10.38	79.5	957.4	10.56
19.0	75.0	883.5	10.21	77.0	921.5	10.41	79.0	959.5	10.60	81.0	997.5	10.78
19.5	76.5	921.4	10.43	78.5	960.4	10.63	80.5	999.4	10.83	82.5	1,038.4	11.01
20.0	78.0	960.0	10.65	80.0	1,000.0	10.86	82.0	1,040.0	11.05	84.0	1,080.0	11.24

Table 31.—*Area in square feet, A, top width in feet, T, and hydraulic radius in feet, r, of trapezoidal channels,* **side slopes 1½ to 1**—Continued

Depth	Bottom width 26 feet			Bottom width 28 feet			Bottom width 30 feet			Bottom width 32 feet		
	T	A	r	T	A	r	T	A	r	T	A	r
0.4	27.2	10.64	.39	29.2	11.44	.39	31.2	12.24	.39	33.2	13.04	.39
0.6	27.8	16.14	.57	29.8	17.34	.57	31.8	18.54	.58	33.8	19.74	.58
0.8	28.4	21.76	.75	30.4	23.36	.76	32.4	24.96	.76	34.4	26.56	.76
1.0	29.0	27.50	.93	31.0	29.50	.93	33.0	31.50	.94	35.0	33.50	.94
1.2	29.6	33.36	1.10	31.6	35.76	1.11	33.6	38.16	1.11	35.6	40.56	1.12
1.4	30.2	39.34	1.27	32.2	42.14	1.28	34.2	44.94	1.28	36.2	47.74	1.29
1.6	30.8	45.44	1.43	32.8	48.64	1.44	34.8	51.84	1.45	36.8	55.04	1.46
1.8	31.4	51.66	1.59	33.4	55.26	1.60	35.4	58.86	1.61	37.4	62.46	1.62
2.0	32.0	58.00	1.75	34.0	62.00	1.76	36.0	66.00	1.77	38.0	70.00	1.79
2.2	32.6	64.46	1.90	34.6	68.86	1.92	36.6	73.26	1.93	38.6	77.66	1.94
2.4	33.2	71.04	2.05	35.2	75.84	2.07	37.2	80.64	2.09	39.2	85.44	2.10
2.6	33.8	77.74	2.20	35.8	82.94	2.22	37.8	88.14	2.24	39.8	93.34	2.26
2.8	34.4	84.56	2.34	36.4	90.16	2.37	38.4	95.76	2.39	40.4	101.36	2.41
3.0	35.0	91.50	2.49	37.0	97.50	2.51	39.0	103.50	2.54	41.0	109.50	2.56
3.2	35.6	98.56	2.63	37.6	104.96	2.65	39.6	111.36	2.68	41.6	117.76	2.70
3.4	36.2	105.74	2.76	38.2	112.54	2.80	40.2	119.34	2.82	42.2	126.14	2.85
3.6	36.8	113.04	2.90	38.8	120.24	2.93	40.8	127.44	2.97	42.8	134.64	2.99
3.8	37.4	120.46	3.03	39.4	128.06	3.07	41.4	135.66	3.10	43.4	143.26	3.13
4.0	38.0	128.00	3.17	40.0	136.00	3.21	42.0	144.00	3.24	44.0	152.00	3.27
4.2	38.6	135.66	3.30	40.6	144.06	3.34	42.6	152.46	3.38	44.6	160.86	3.41
4.4	39.2	143.44	3.43	41.2	152.24	3.47	43.2	161.04	3.51	45.2	169.84	3.55
4.6	39.8	151.34	3.55	41.8	160.54	3.60	43.8	169.74	3.64	45.8	178.94	3.68
4.8	40.4	159.36	3.68	42.4	168.96	3.73	44.4	178.56	3.77	46.4	188.16	3.82
5.0	41.0	167.50	3.80	43.0	177.50	3.86	45.0	187.50	3.90	47.0	197.50	3.95
5.2	41.6	175.76	3.93	43.6	186.16	3.98	45.6	196.56	4.03	47.6	206.96	4.08
5.4	42.2	184.14	4.05	44.2	194.94	4.11	46.2	205.74	4.16	48.2	216.54	4.21
5.6	42.8	192.64	4.17	44.8	203.84	4.23	46.8	215.04	4.28	48.8	226.24	4.33
5.8	43.4	201.26	4.29	45.4	212.86	4.35	47.4	224.46	4.41	49.4	236.06	4.46
6.0	44.0	210.00	4.41	46.0	222.00	4.47	48.0	234.00	4.53	50.0	246.00	4.59
6.2	44.6	218.86	4.53	46.6	231.26	4.59	48.6	243.66	4.65	50.6	256.06	4.71
6.4	45.2	227.84	4.64	47.2	240.64	4.71	49.2	253.44	4.78	51.2	266.24	4.83
6.6	45.8	236.94	4.76	47.8	250.14	4.83	49.8	263.34	4.90	51.8	276.54	4.96
6.8	46.4	246.16	4.87	48.4	259.76	4.95	50.4	273.36	5.01	52.4	286.96	5.08
7.0	47.0	255.50	4.99	49.0	269.50	5.06	51.0	283.50	5.13	53.0	297.50	5.20
7.2	47.6	264.96	5.10	49.6	279.36	5.18	51.6	293.76	5.25	53.6	308.16	5.32
7.4	48.2	274.54	5.21	50.2	289.34	5.29	52.2	304.14	5.37	54.2	318.94	5.44
7.6	48.8	284.24	5.32	50.8	299.44	5.40	52.8	314.64	5.48	54.8	329.84	5.55
7.8	49.4	294.06	5.43	51.4	309.66	5.52	53.4	325.26	5.60	55.4	340.86	5.67
8.0	50.0	304.00	5.54	52.0	320.00	5.63	54.0	336.00	5.71	56.0	352.00	5.79
8.2	50.6	314.06	5.65	52.6	330.46	5.74	54.6	346.86	5.82	56.6	363.26	5.90
8.4	51.2	324.24	5.76	53.2	341.04	5.85	55.2	357.84	5.94	57.2	374.64	6.01
8.6	51.8	334.54	5.87	53.8	351.74	5.96	55.8	368.94	6.05	57.8	386.14	6.13
8.8	52.4	344.96	5.98	54.4	362.56	6.07	56.4	380.16	6.16	58.4	397.76	6.24
9.0	53.0	355.50	6.08	55.0	373.50	6.18	57.0	391.50	6.27	59.0	409.50	6.35
9.2	53.6	366.16	6.19	55.6	384.56	6.29	57.6	402.96	6.38	59.6	421.36	6.47
9.4	54.2	376.94	6.29	56.2	395.74	6.39	58.2	414.54	6.49	60.2	433.34	6.58
9.6	54.8	387.84	6.40	56.8	407.04	6.50	58.8	426.24	6.60	60.8	445.44	6.69
9.8	55.4	398.86	6.50	57.4	418.46	6.61	59.4	438.06	6.70	61.4	457.66	6.80

Table 31.—*Area in square feet, A, top width in feet, T, and hydraulic radius in feet, r, of trapezoidal channels,* side slopes 1½ to 1—Continued

Depth	Bottom width 26 feet			Bottom width 28 feet			Bottom width 30 feet			Bottom width 32 feet		
	T	A	r	T	A	r	T	A	r	T	A	r
10.0	56.0	410.00	6.61	58.0	430.00	6.71	60.0	450.00	6.81	62.0	470.00	6.91
10.5	57.5	438.37	6.86	59.5	459.37	6.98	61.5	480.37	7.08	63.5	501.37	7.18
11.0	59.0	467.5	7.12	61.0	489.5	7.23	63.0	511.5	7.34	65.0	533.5	7.44
11.5	60.5	497.4	7.37	62.5	520.4	7.49	64.5	543.4	7.60	66.5	566.4	7.71
12.0	62.0	528.0	7.62	64.0	552.0	7.75	66.0	576.0	7.86	68.0	600.0	7.97
12.5	63.5	559.4	7.87	65.5	584.4	8.00	67.5	609.4	8.12	69.5	634.4	8.23
13.0	65.0	591.5	8.12	67.0	617.5	8.25	69.0	643.5	8.37	71.0	669.5	8.49
13.5	66.5	624.4	8.36	68.5	651.4	8.50	70.5	678.4	8.62	72.5	705.4	8.74
14.0	68.0	658.0	8.60	70.0	686.0	8.74	72.0	714.0	8.87	74.0	742.0	9.00
14.5	69.5	692.4	8.84	71.5	721.4	8.99	73.5	750.4	9.12	75.5	779.4	9.25
15.0	71.0	727.5	9.08	73.0	757.5	9.23	75.0	787.5	9.37	77.0	817.5	9.50
15.5	72.5	763.4	9.32	74.5	794.4	9.47	76.5	825.4	9.61	78.5	856.4	9.74
16.0	74.0	800.0	9.56	76.0	832.0	9.71	78.0	864.0	9.85	80.0	896.0	9.99
16.5	75.5	837.4	9.80	77.5	870.4	9.95	79.5	903.4	10.09	81.5	936.4	10.23
17.0	77.0	875.5	10.03	79.0	909.5	10.19	81.0	943.5	10.33	83.0	977.5	10.48
17.5	78.5	914.4	10.26	80.5	949.4	10.42	82.5	984.4	10.57	84.5	1,019.4	10.72
18.0	80.0	954.0	10.50	82.0	990.0	10.66	84.0	1,026.0	10.81	86.0	1,062.0	10.96
18.5	81.5	994.4	10.73	83.5	1,031.4	10.89	85.5	1,068.4	11.05	87.5	1,105.4	11.20
19.0	83.0	1,035.5	10.96	85.0	1,073.5	11.12	87.0	1,111.5	11.28	89.0	1,149.5	11.44
19.5	84.5	1,077.4	11.19	86.5	1,116.4	11.36	88.5	1,155.4	11.52	90.5	1,194.4	11.67
20.0	86.0	1,120.0	11.42	88.0	1,160.0	11.59	90.0	1,200.0	11.75	92.0	1,240.0	11.91

Depth	Bottom width 35 feet			Bottom width 40 feet			Bottom width 45 feet			Bottom width 50 feet		
	T	A	r	T	A	r	T	A	r	T	A	r
0.4	36.2	14.24	.39	41.2	16.24	.39	46.2	18.24	.39	51.2	20.24	.39
0.6	36.8	21.54	.58	41.8	24.54	.58	46.8	27.54	.58	51.8	30.54	.59
0.8	37.4	28.96	.76	42.4	32.96	.77	47.4	36.96	.77	52.4	40.96	.77
1.0	38.0	36.50	.95	43.0	41.50	.95	48.0	46.50	.96	53.0	51.50	.96
1.2	38.6	44.16	1.12	43.6	50.16	1.13	48.6	56.16	1.14	53.6	62.16	1.14
1.4	39.2	51.94	1.30	44.2	58.94	1.31	49.2	65.94	1.32	54.2	72.94	1.33
1.6	39.8	59.84	1.47	44.8	67.84	1.48	49.8	75.84	1.49	54.8	83.84	1.50
1.8	40.4	67.86	1.64	45.4	76.86	1.65	50.4	85.86	1.67	55.4	94.86	1.68
2.0	41.0	76.00	1.80	46.0	86.00	1.82	51.0	96.00	1.84	56.0	106.00	1.85
2.2	41.6	84.26	1.96	46.6	95.26	1.99	51.6	106.26	2.01	56.6	117.26	2.02
2.4	42.2	92.64	2.12	47.2	104.64	2.15	52.2	116.64	2.17	57.2	128.64	2.19
2.6	42.8	101.14	2.28	47.8	114.14	2.31	52.8	127.14	2.34	57.8	140.14	2.36
2.8	43.4	109.76	2.43	48.4	123.76	2.47	53.4	137.76	2.50	58.4	151.76	2.53
3.0	44.0	118.50	2.59	49.0	133.50	2.63	54.0	148.50	2.66	59.0	163.50	2.69
3.2	44.6	127.36	2.74	49.6	143.36	2.78	54.6	159.36	2.82	59.6	175.36	2.85
3.4	45.2	136.34	2.88	50.2	153.34	2.93	55.2	170.34	2.97	60.2	187.34	3.01
3.6	45.8	145.44	3.03	50.8	163.44	3.08	55.8	181.44	3.13	60.8	199.44	3.17
3.8	46.4	154.66	3.18	51.4	173.66	3.23	56.4	192.66	3.28	61.4	211.66	3.32
4.0	47.0	164.00	3.32	52.0	184.00	3.38	57.0	204.00	3.43	62.0	224.00	3.48
4.2	47.6	173.46	3.46	52.6	194.46	3.53	57.6	215.46	3.58	62.6	236.46	3.63
4.4	48.2	183.04	3.60	53.2	205.04	3.67	58.2	227.04	3.73	63.2	249.04	3.78
4.6	48.8	192.74	3.74	53.8	215.74	3.81	58.8	238.74	3.88	63.8	261.74	3.93
4.8	49.4	202.56	3.87	54.4	226.56	3.95	59.4	250.56	4.02	64.4	274.56	4.08

Table 31.—*Area in square feet, A, top width in feet, T, and hydraulic radius in feet, r, of trapezoidal channels,*
side slopes 1½ to 1—Continued

Depth	Bottom width 35 feet			Bottom width 40 feet			Bottom width 45 feet			Bottom width 50 feet		
	T	A	r	T	A	r	T	A	r	T	A	r
5.0	50.0	212.50	4.01	55.0	237.50	4.09	60.0	262.50	4.16	65.0	287.50	4.23
5.2	50.6	222.56	4.14	55.6	248.56	4.23	60.6	274.56	4.31	65.6	300.56	4.37
5.4	51.2	232.74	4.27	56.2	259.74	4.37	61.2	286.74	4.45	66.2	313.74	4.52
5.6	51.8	243.04	4.40	56.8	271.04	4.50	61.8	299.04	4.59	66.8	327.04	4.66
5.8	52.4	253.46	4.53	57.4	282.46	4.64	62.4	311.46	4.73	67.4	340.46	4.80
6.0	53.0	264.00	4.66	58.0	294.00	4.77	63.0	324.00	4.86	68.0	354.00	4.94
6.2	53.6	274.66	4.79	58.6	305.66	4.90	63.6	336.66	5.00	68.6	367.66	5.08
6.4	54.2	285.44	4.91	59.2	317.44	5.03	64.2	349.44	5.13	69.2	381.44	5.22
6.6	54.8	296.34	5.04	59.8	329.34	5.16	64.8	362.34	5.27	69.8	395.34	5.36
6.8	55.4	307.36	5.16	60.4	341.36	5.29	65.4	375.36	5.40	70.4	409.36	5.49
7.0	56.0	318.50	5.29	61.0	353.50	5.42	66.0	388.50	5.53	71.0	423.50	5.63
7.2	56.6	329.76	5.41	61.6	365.76	5.55	66.6	401.76	5.66	71.6	437.76	5.76
7.4	57.2	341.14	5.53	62.2	378.14	5.67	67.2	415.14	5.79	72.2	452.14	5.90
7.6	57.8	352.64	5.65	62.8	390.64	5.80	67.8	428.64	5.92	72.8	466.64	6.03
7.8	58.4	364.26	5.77	63.4	403.26	5.92	68.4	442.26	6.05	73.4	481.26	6.16
8.0	59.0	376.00	5.89	64.0	416.00	6.04	69.0	456.00	6.18	74.0	496.00	6.29
8.2	59.6	387.86	6.01	64.6	428.86	6.16	69.6	469.86	6.30	74.6	510.86	6.42
8.4	60.2	399.84	6.12	65.2	441.84	6.29	70.2	483.84	6.43	75.2	525.84	6.55
8.6	60.8	411.94	6.24	65.8	454.94	6.41	70.8	497.94	6.55	75.8	540.94	6.68
8.8	61.4	424.16	6.36	66.4	468.16	6.53	71.4	512.16	6.67	76.4	556.16	6.80
9.0	62.0	436.50	6.47	67.0	481.50	6.65	72.0	526.50	6.80	77.0	571.50	6.93
9.2	62.6	448.96	6.59	67.6	494.96	6.76	72.6	540.96	6.92	77.6	586.96	7.06
9.4	63.2	461.54	6.70	68.2	508.54	6.88	73.2	555.54	7.04	78.2	602.54	7.18
9.6	63.8	474.24	6.81	68.8	522.24	7.00	73.8	570.24	7.16	78.8	618.24	7.31
9.8	64.4	487.06	6.92	69.4	536.06	7.12	74.4	585.06	7.28	79.4	634.06	7.43
10.0	65.0	500.00	7.04	70.0	550.00	7.23	75.0	600.00	7.40	80.0	650.00	7.55
10.5	66.5	532.87	7.31	71.5	585.37	7.52	76.5	637.87	7.70	81.5	690.37	7.86
11.0	68.0	566.5	7.59	73.0	621.5	7.80	78.0	676.5	7.99	83.0	731.5	8.16
11.5	69.5	600.9	7.86	74.5	658.4	8.08	79.5	715.9	8.28	84.5	773.4	8.46
12.0	71.0	636.0	8.13	76.0	696.0	8.36	81.0	756.0	8.56	86.0	816.0	8.75
12.5	72.5	671.9	8.39	77.5	734.4	8.63	82.5	796.9	8.83	87.5	859.4	9.04
13.0	74.0	708.5	8.65	79.0	773.5	8.90	84.0	838.5	9.13	89.0	903.5	9.33
13.5	75.5	745.9	8.91	80.5	813.4	9.17	85.5	880.9	9.40	90.5	948.4	9.61
14.0	77.0	784.0	9.17	82.0	854.0	9.44	87.0	924.0	9.68	92.0	994.0	9.89
14.5	78.5	822.9	9.43	83.5	895.4	9.70	88.5	967.9	9.95	93.5	1,040.4	10.17
15.0	80.0	862.5	9.68	85.0	937.5	9.96	90.0	1,012.5	10.22	95.0	1,087.5	10.45
15.5	81.5	902.9	9.93	86.5	980.4	10.22	91.5	1,057.9	10.49	96.5	1,135.4	10.72
16.0	83.0	944.0	10.18	88.0	1,024.0	10.48	93.0	1,104.0	10.75	98.0	1,184.0	10.99
16.5	84.5	985.9	10.43	89.5	1,068.4	10.74	94.5	1,150.9	11.01	99.5	1,233.4	11.26
17.0	86.0	1,028.5	10.68	91.0	1,113.5	10.99	96.0	1,198.5	11.28	101.0	1,283.5	11.53
17.5	87.5	1,071.9	10.93	92.5	1,159.4	11.25	97.5	1,246.9	11.53	102.5	1,334.4	11.80
18.0	89.0	1,116.0	11.17	94.0	1,206.0	11.50	99.0	1,296.0	11.79	104.0	1,386.0	12.06
18.5	90.5	1,160.9	11.41	95.5	1,253.4	11.75	100.5	1,345.9	12.05	105.5	1,438.4	12.33
19.0	92.0	1,206.5	11.66	97.0	1,301.5	11.99	102.0	1,396.5	12.30	107.0	1,491.5	12.59
19.5	93.5	1,252.9	11.90	98.5	1,350.4	12.24	103.5	1,447.9	12.56	108.5	1,545.4	12.85
20.0	95.0	1,300.0	12.14	100.0	1,400.0	12.49	105.0	1,500.0	12.81	110.0	1,600.0	13.10

Table 31.—*Area in square feet, A, top width in feet, T, and hydraulic radius in feet, r, of trapezoidal channels,*
side slopes 1½ to 1—Continued

Depth	Bottom width 60 feet T	A	r	Bottom width 70 feet T	A	r	Bottom width 80 feet T	A	r	Bottom width 90 feet T	A	r
0.4	61.2	24.24	0.39	71.2	28.24	0.40	81.2	32.24	0.40	91.2	36.24	0.40
0.6	61.8	36.54	.59	71.8	42.54	.59	81.8	48.54	.59	91.8	54.54	.59
0.8	62.4	48.96	.78	72.4	56.96	.78	82.4	64.96	.78	92.4	72.96	.79
1.0	63.0	61.50	.97	73.0	71.50	.97	83.0	81.50	.97	93.0	91.50	.98
1.2	63.6	74.16	1.15	73.6	86.16	1.16	83.6	98.16	1.16	93.6	110.16	1.17
1.4	64.2	86.94	1.34	74.2	100.94	1.35	84.2	114.94	1.35	94.2	128.94	1.36
1.6	64.8	99.84	1.52	74.8	115.84	1.53	84.8	131.84	1.54	94.8	147.84	1.54
1.8	65.4	112.86	1.70	75.4	130.86	1.71	85.4	148.86	1.72	95.4	166.86	1.73
2.0	66.0	126.00	1.87	76.0	146.00	1.89	86.0	166.00	1.90	96.0	186.00	1.91
2.2	66.6	139.26	2.05	76.6	161.26	2.07	86.6	183.26	2.08	96.6	205.26	2.10
2.4	67.2	152.64	2.22	77.2	176.64	2.25	87.2	200.64	2.26	97.2	224.64	2.28
2.6	67.8	166.14	2.39	77.8	192.14	2.42	87.8	218.14	2.44	97.8	244.14	2.46
2.8	68.4	179.76	2.56	78.4	207.76	2.59	88.4	235.76	2.62	98.4	263.76	2.64
3.0	69.0	193.50	2.73	79.0	223.50	2.77	89.0	253.50	2.79	99.0	283.50	2.81
3.2	69.6	207.36	2.90	79.6	239.36	2.94	89.6	271.36	2.96	99.6	303.36	2.99
3.4	70.2	221.34	3.06	80.2	255.34	3.10	90.2	289.34	3.14	100.2	323.34	3.16
3.6	70.8	235.44	3.23	80.8	271.44	3.27	90.8	307.44	3.31	100.8	343.44	3.34
3.8	71.4	249.66	3.39	81.4	287.66	3.44	91.4	325.66	3.48	101.4	363.66	3.51
4.0	72.0	264.00	3.55	82.0	304.00	3.60	92.0	344.00	3.64	102.0	384.00	3.68
4.2	72.6	278.46	3.71	82.6	320.46	3.76	92.6	362.46	3.81	102.6	404.46	3.85
4.4	73.2	293.04	3.86	83.2	337.04	3.93	93.2	381.04	3.97	103.2	425.04	4.01
4.6	73.8	307.74	4.02	83.8	353.74	4.09	93.8	399.74	4.14	103.8	445.74	4.18
4.8	74.4	322.56	4.17	84.4	370.56	4.24	94.4	418.56	4.30	104.4	466.56	4.35
5.0	75.0	337.50	4.33	85.0	387.50	4.40	95.0	437.50	4.46	105.0	487.50	4.51
5.2	75.6	352.56	4.48	85.6	404.56	4.56	95.6	456.56	4.62	105.6	508.56	4.68
5.4	76.2	367.74	4.63	86.2	421.74	4.71	96.2	475.74	4.78	106.2	529.74	4.84
5.6	76.8	383.04	4.78	86.8	439.04	4.87	96.8	495.04	4.94	106.8	551.04	5.00
5.8	77.4	398.46	4.92	87.4	456.46	5.02	97.4	514.46	5.10	107.4	572.46	5.16
6.0	78.0	414.00	5.07	88.0	474.00	5.17	98.0	534.00	5.25	108.0	594.00	5.32
6.2	78.6	429.66	5.22	88.6	491.66	5.32	98.6	553.66	5.41	108.6	615.66	5.48
6.4	79.2	445.44	5.36	89.2	509.44	5.47	99.2	573.44	5.56	109.2	637.44	5.64
6.6	79.8	461.34	5.51	89.8	527.34	5.62	99.8	593.34	5.72	109.8	659.34	5.79
6.8	80.4	477.36	5.65	90.4	545.36	5.77	100.4	613.36	5.87	110.4	681.36	5.95
7.0	81.0	493.50	5.79	91.0	563.50	5.92	101.0	633.50	6.02	111.0	703.50	6.10
7.2	81.6	509.76	5.93	91.6	581.76	6.06	101.6	653.76	6.17	111.6	725.76	6.26
7.4	82.2	526.14	6.07	92.2	600.14	6.21	102.2	674.14	6.32	112.2	748.14	6.41
7.6	82.8	542.64	6.21	92.8	618.64	6.35	102.8	694.64	6.47	112.8	770.64	6.56
7.8	83.4	559.26	6.35	93.4	637.26	6.49	103.4	715.26	6.62	113.4	793.26	6.72
8.0	84.0	576.00	6.48	94.0	656.00	6.64	104.0	736.00	6.76	114.0	816.00	6.87
8.2	84.6	592.86	6.62	94.6	674.86	6.78	104.6	756.86	6.91	114.6	838.86	7.02
8.4	85.2	609.84	6.75	95.2	693.84	6.92	105.2	777.84	7.05	115.2	861.84	7.16
8.6	85.8	626.94	6.89	95.8	712.94	7.06	105.8	798.94	7.20	115.8	884.94	7.31
8.8	86.4	644.16	7.02	96.4	732.16	7.20	106.4	820.16	7.34	116.4	908.16	7.46
9.0	87.0	661.50	7.16	97.0	751.50	7.34	107.0	841.50	7.48	117.0	931.50	7.61
9.2	87.6	678.96	7.29	97.6	770.96	7.47	107.6	862.96	7.63	117.6	954.96	7.75
9.4	88.2	696.54	7.42	98.2	790.54	7.61	108.2	884.54	7.77	118.2	978.54	7.90
9.6	88.8	714.24	7.55	98.8	810.24	7.75	108.8	906.24	7.91	118.8	1,002.24	8.04
9.8	89.4	732.06	7.68	99.4	830.06	7.88	109.4	928.06	8.05	119.4	1,026.06	8.19

Table 31.—*Area in square feet, A, top width in feet, T, and hydraulic radius in feet, r, of trapezoidal channels,* **side slopes 1½ to 1**—Continued

Depth	Bottom width 60 feet			Bottom width 70 feet			Bottom width 80 feet			Bottom width 90 feet		
	T	A	r	T	A	r	T	A	r	T	A	r
10.0	90.0	750.00	7.81	100.0	850.00	8.01	110.0	950.00	8.19	120.0	1,050.00	8.33
10.5	91.5	795.37	8.13	101.5	900.37	8.35	111.5	1,005.37	8.53	121.5	1,110.37	8.68
11.0	93.0	841.5	8.44	103.0	951.5	8.68	113.0	1,061.5	8.87	123.0	1,171.5	9.04
11.5	94.5	888.4	8.76	104.5	1,003.4	9.00	114.5	1,118.4	9.21	124.5	1,233.4	9.38
12.0	96.0	936.0	9.06	106.0	1,056.0	9.32	116.0	1,176.0	9.54	126.0	1,296.0	9.72
12.5	97.5	984.4	9.37	107.5	1,109.4	9.64	117.5	1,234.4	9.87	127.5	1,359.4	10.06
13.0	99.0	1,033.5	9.67	109.0	1,163.5	9.96	119.0	1,293.5	10.20	129.0	1,423.5	10.40
13.5	100.5	1,083.4	9.97	110.5	1,218.4	10.27	120.5	1,353.4	10.52	130.5	1,488.4	10.73
14.0	102.0	1,134.0	10.26	112.0	1,274.0	10.57	122.0	1,414.0	10.84	132.0	1,554.0	11.06
14.5	103.5	1,185.4	10.56	113.5	1,330.4	10.88	123.5	1,475.4	11.15	133.5	1,620.4	11.39
15.0	105.0	1,237.5	10.85	115.0	1,387.5	11.18	125.0	1,537.5	11.47	135.0	1,687.5	11.71
15.5	106.5	1,290.4	11.14	116.5	1,445.4	11.48	126.5	1,600.4	11.78	136.5	1,755.4	12.03
16.0	108.0	1,344.0	11.42	118.0	1,504.0	11.78	128.0	1,664.0	12.09	138.0	1,824.0	12.35
16.5	109.5	1,398.4	11.70	119.5	1,563.4	12.07	129.5	1,728.4	12.39	139.5	1,893.4	12.67
17.0	111.0	1,453.5	11.98	121.0	1,623.5	12.37	131.0	1,793.5	12.69	141.0	1,963.5	12.98
17.5	112.5	1,509.4	12.26	122.5	1,684.4	12.66	132.5	1,859.4	12.99	142.5	2,034.4	13.29
18.0	114.0	1,566.0	12.54	124.0	1,746.0	12.94	134.0	1,926.0	13.29	144.0	2,106.0	13.60
18.5	115.5	1,623.4	12.81	125.5	1,808.4	13.23	135.5	1,993.4	13.59	145.5	2,178.4	13.90
19.0	117.0	1,681.5	13.09	127.0	1,871.5	13.51	137.0	2,061.5	13.88	147.0	2,251.5	14.20
19.5	118.5	1,740.4	13.36	128.5	1,935.4	13.79	138.5	2,130.4	14.17	148.5	2,325.4	14.51
20.0	120.0	1,800.0	13.62	130.0	2,000.0	14.07	140.0	2,200.0	14.46	150.0	2,400.0	14.80

168 HYDRAULIC AND EXCAVATION TABLES

Table 32.—*Area in square feet, A, top width in feet, T, and hydraulic radius in feet, r, of trapezoidal channels,*
side slopes 1¾ to 1

Depth	Bottom width 2 feet			Bottom width 3 feet			Bottom width 4 feet			Bottom width 5 feet		
	T	A	r	T	A	r	T	A	r	T	A	r
0.4	3.4	1.08	0.30	4.4	1.48	0.32	5.4	1.88	0.33	6.4	2.28	0.34
0.6	4.1	1.83	.41	5.1	2.43	.45	6.1	3.03	.47	7.1	3.63	.49
0.8	4.8	2.72	.52	5.8	3.52	.57	6.8	4.32	.60	7.8	5.12	.62
1.0	5.5	3.75	.62	6.5	4.75	.68	7.5	5.75	.72	8.5	6.75	.75
1.2	6.2	4.92	.72	7.2	6.12	.78	8.2	7.32	.83	9.2	8.52	.87
1.4	6.9	6.23	.82	7.9	7.63	.88	8.9	9.03	.94	9.9	10.43	.98
1.6	7.6	7.68	.91	8.6	9.28	.98	9.6	10.88	1.04	10.6	12.48	1.09
1.8	8.3	9.27	1.00	9.3	11.07	1.08	10.3	12.87	1.14	11.3	14.67	1.20
2.0	9.0	11.00	1.09	10.0	13.00	1.18	11.0	15.00	1.24	12.0	17.00	1.30
2.2	9.7	12.87	1.18	10.7	15.07	1.27	11.7	17.27	1.34	12.7	19.47	1.40
2.4	10.4	14.88	1.27	11.4	17.28	1.36	12.4	19.68	1.44	13.4	22.08	1.50
2.6	11.1	17.03	1.36	12.1	19.63	1.46	13.1	22.23	1.54	14.1	24.83	1.60
2.8	11.8	19.32	1.45	12.8	22.12	1.55	13.8	24.92	1.63	14.8	27.72	1.70
3.0	12.5	21.75	1.54	13.5	24.75	1.64	14.5	27.75	1.72	15.5	30.75	1.80
3.2	13.2	24.32	1.63	14.2	27.52	1.73	15.2	30.72	1.82	16.2	33.92	1.90
3.4	13.9	27.03	1.72	14.9	30.43	1.82	15.9	33.83	1.91	16.9	37.23	1.99
3.6	14.6	29.88	1.81	15.6	33.48	1.91	16.6	37.08	2.00	17.6	40.68	2.08
3.8	15.3	32.87	1.90	16.3	36.67	2.00	17.3	40.47	2.09	18.3	44.27	2.18
4.0	16.0	36.00	1.99	17.0	40.00	2.09	18.0	44.00	2.19	19.0	48.00	2.27
4.2	16.7	39.27	2.07	17.7	43.47	2.18	18.7	47.67	2.28	19.7	51.87	2.37
4.4	17.4	42.68	2.16	18.4	47.08	2.27	19.4	51.48	2.37	20.4	55.88	2.46
4.6	18.1	46.23	2.25	19.1	50.83	2.36	20.1	55.43	2.46	21.1	60.03	2.55
4.8	18.8	49.92	2.34	19.8	54.72	2.45	20.8	59.52	2.55	21.8	64.32	2.64
5.0	19.5	53.75	2.43	20.5	58.75	2.54	21.5	63.75	2.64	22.5	68.75	2.73
5.2	20.2	57.72	2.51	21.2	62.92	2.63	22.2	68.12	2.73	23.2	73.32	2.82
5.4	20.9	61.83	2.60	21.9	67.23	2.71	22.9	72.63	2.82	23.9	78.03	2.92
5.6	21.6	66.08	2.69	22.6	71.68	2.80	23.6	77.28	2.91	24.6	82.88	3.01
5.8	22.3	70.47	2.78	23.3	76.27	2.89	24.3	82.07	3.00	25.3	87.87	3.10
6.0	23.0	75.00	2.86	24.0	81.00	2.98	25.0	87.00	3.09	26.0	93.00	3.19
6.2	23.7	79.67	2.95	24.7	85.87	3.07	25.7	92.07	3.18	26.7	98.27	3.28
6.4	24.4	84.48	3.04	25.4	90.88	3.16	26.4	97.28	3.26	27.4	103.68	3.37
6.6	25.1	89.43	3.13	26.1	96.03	3.24	27.1	102.63	3.35	28.1	109.23	3.46
6.8	25.8	94.52	3.21	26.8	101.32	3.33	27.8	108.12	3.44	28.8	114.92	3.55
7.0	26.5	99.75	3.30	27.5	106.75	3.42	28.5	113.75	3.53	29.5	120.75	3.64
7.2	27.2	105.12	3.39	28.2	112.32	3.51	29.2	119.52	3.62	30.2	126.72	3.72
7.4	27.9	110.63	3.48	28.9	118.03	3.60	29.9	125.43	3.71	30.9	132.83	3.81
7.6	28.6	116.28	3.56	29.6	123.88	3.68	30.6	131.48	3.80	31.6	139.08	3.90
7.8	29.3	122.07	3.65	30.3	129.87	3.77	31.3	137.67	3.88	32.3	145.47	3.99
8.0	30.0	128.00	3.74	31.0	136.00	3.86	32.0	144.00	3.97	33.0	152.00	4.08
8.2	----	----	----	31.7	142.27	3.95	32.7	150.47	4.06	33.7	158.67	4.17
8.4	----	----	----	32.4	148.68	4.03	33.4	157.08	4.15	34.4	165.48	4.26
8.6	----	----	----	33.1	155.23	4.12	34.1	163.83	4.24	35.1	172.43	4.35
8.8	----	----	----	33.8	161.92	4.21	34.8	170.72	4.32	35.8	179.52	4.44
9.0	----	----	----	34.5	168.75	4.30	35.5	177.75	4.41	36.5	186.75	4.52
9.2	----	----	----	35.2	175.72	4.38	36.2	184.92	4.50	37.2	194.12	4.61
9.4	----	----	----	35.9	182.83	4.47	36.9	192.23	4.59	37.9	201.63	4.70
9.6	----	----	----	36.6	190.08	4.56	37.6	199.68	4.68	38.6	209.28	4.79
9.8	----	----	----	37.3	197.47	4.65	38.3	207.27	4.76	39.3	217.07	4.88

Table 32.—*Area in square feet, A, top width in feet, T, and hydraulic radius in feet, r, of trapezoidal channels,*
side slopes 1¾ to 1—Continued

Depth	Bottom width 2 feet T	A	r	Bottom width 3 feet T	A	r	Bottom width 4 feet T	A	r	Bottom width 5 feet T	A	r
10.0	----	----	----	38.00	205.00	4.73	39.00	215.00	4.85	40.00	225.00	4.97
10.5	----	----	----	39.75	224.44	4.95	40.75	234.94	5.07	41.75	245.44	5.19
11.0	----	----	----	41.50	244.8	5.17	42.50	255.8	5.29	43.50	266.8	5.41
11.5	----	----	----	43.25	265.9	5.39	44.25	277.4	5.51	45.25	288.9	5.63
12.0	----	----	----	45.00	288.0	5.61	46.00	300.0	5.73	47.00	312.0	5.85
12.5	----	----	----	46.75	310.9	5.82	47.75	323.4	5.95	48.75	335.9	6.07
13.0	----	----	----	----	----	----	49.50	347.8	6.17	50.50	360.8	6.28
13.5	----	----	----	----	----	----	51.25	372.9	6.38	52.25	386.4	6.50
14.0	----	----	----	----	----	----	53.00	399.0	6.60	54.00	413.0	6.72
14.5	----	----	----	----	----	----	54.75	425.9	6.82	55.75	440.4	6.94
15.0	----	----	----	----	----	----	56.50	453.8	7.04	57.50	468.8	7.16
15.5	----	----	----	----	----	----	58.25	482.4	7.26	59.25	497.9	7.38
16.0	----	----	----	----	----	----	60.00	512.0	7.47	61.00	528.0	7.60
16.5	----	----	----	----	----	----	----	----	----	62.75	558.9	7.82
17.0	----	----	----	----	----	----	----	----	----	64.50	590.8	8.03
17.5	----	----	----	----	----	----	----	----	----	66.25	623.4	8.25
18.0	----	----	----	----	----	----	----	----	----	68.00	657.0	8.47
18.5	----	----	----	----	----	----	----	----	----	69.75	691.4	8.69
19.0	----	----	----	----	----	----	----	----	----	71.50	726.8	8.91
19.5	----	----	----	----	----	----	----	----	----	73.25	762.9	9.13
20.0	----	----	----	----	----	----	----	----	----	75.00	800.0	9.34

Depth	Bottom width 6 feet T	A	r	Bottom width 7 feet T	A	r	Bottom width 8 feet T	A	r	Bottom width 9 feet T	A	r
0.4	7.4	2.68	0.35	8.4	3.08	0.36	9.4	3.48	0.36	10.4	3.88	0.37
0.6	8.1	4.23	.50	9.1	4.83	.51	10.1	5.43	.52	11.1	6.03	.53
0.8	8.8	5.92	.64	9.8	6.72	.66	10.8	7.52	.67	11.8	8.32	.68
1.0	9.5	7.75	.77	10.5	8.75	.79	11.5	9.75	.81	12.5	10.75	.82
1.2	10.2	9.72	.90	11.2	10.92	.92	12.2	12.12	.94	13.2	13.32	.96
1.4	10.9	11.83	1.02	11.9	13.23	1.05	12.9	14.63	1.07	13.9	16.03	1.09
1.6	11.6	14.08	1.13	12.6	15.68	1.17	13.6	17.28	1.20	14.6	18.88	1.22
1.8	12.3	16.47	1.24	13.3	18.27	1.28	14.3	20.07	1.32	15.3	21.87	1.35
2.0	13.0	19.00	1.35	14.0	21.00	1.39	15.0	23.00	1.43	16.0	25.00	1.47
2.2	13.7	21.67	1.46	14.7	23.87	1.50	15.7	26.07	1.55	16.7	28.27	1.58
2.4	14.4	24.48	1.56	15.4	26.88	1.61	16.4	29.28	1.66	17.4	31.68	1.70
2.6	15.1	27.43	1.66	16.1	30.03	1.72	17.1	32.63	1.77	18.1	35.23	1.81
2.8	15.8	30.52	1.77	16.8	33.32	1.82	17.8	36.12	1.87	18.8	38.92	1.92
3.0	16.5	33.75	1.87	17.5	36.75	1.92	18.5	39.75	1.98	19.5	42.75	2.03
3.2	17.2	37.12	1.96	18.2	40.32	2.03	19.2	43.52	2.08	20.2	46.72	2.13
3.4	17.9	40.63	2.06	18.9	44.03	2.13	19.9	47.43	2.19	20.9	50.83	2.24
3.6	18.6	44.28	2.16	19.6	47.88	2.23	20.6	51.48	2.29	21.6	55.08	2.34
3.8	19.3	48.07	2.25	20.3	51.87	2.32	21.3	55.67	2.39	22.3	59.47	2.45
4.0	20.0	52.00	2.35	21.0	56.00	2.42	22.0	60.00	2.49	23.0	64.00	2.55
4.2	20.7	56.07	2.45	21.7	60.27	2.52	22.7	64.47	2.59	23.7	68.67	2.65
4.4	21.4	60.28	2.54	22.4	64.68	2.61	23.4	69.08	2.68	24.4	73.48	2.75
4.6	22.1	64.63	2.63	23.1	69.23	2.71	24.1	73.83	2.78	25.1	78.43	2.85
4.8	22.8	69.12	2.73	23.8	73.92	2.81	24.8	78.72	2.88	25.8	83.52	2.95

Table 32.—*Area in square feet, A, top width in feet, T, and hydraulic radius in feet, r, of trapezoidal channels,* **side slopes 1¾ to 1**—Continued

Depth	Bottom width 6 feet			Bottom width 7 feet			Bottom width 8 feet			Bottom width 9 feet		
	T	A	r	T	A	r	T	A	r	T	A	r
5.0	23.5	73.75	2.82	24.5	78.75	2.90	25.5	83.75	2.97	26.5	88.75	3.04
5.2	24.2	78.52	2.91	25.2	83.72	2.99	26.2	88.92	3.07	27.2	94.12	3.14
5.4	24.9	83.43	3.00	25.9	88.83	3.09	26.9	94.23	3.17	27.9	99.63	3.24
5.6	25.6	88.48	3.10	26.6	94.08	3.18	27.6	99.68	3.26	28.6	105.28	3.33
5.8	26.3	93.67	3.19	27.3	99.47	3.27	28.3	105.27	3.35	29.3	111.07	3.43
6.0	27.0	99.00	3.28	28.0	105.00	3.37	29.0	111.00	3.45	30.0	117.00	3.53
6.2	27.7	104.47	3.37	28.7	110.67	3.46	29.7	116.87	3.54	30.7	123.07	3.62
6.4	28.4	110.08	3.46	29.4	116.48	3.55	30.4	122.88	3.64	31.4	129.28	3.72
6.6	29.1	115.83	3.55	30.1	122.43	3.64	31.1	129.03	3.73	32.1	135.63	3.81
6.8	29.8	121.72	3.64	30.8	128.52	3.73	31.8	135.32	3.82	32.8	142.12	3.90
7.0	30.5	127.75	3.73	31.5	134.75	3.83	32.5	141.75	3.91	33.5	148.75	4.00
7.2	31.2	133.92	3.82	32.2	141.12	3.92	33.2	148.32	4.01	34.2	155.52	4.09
7.4	31.9	140.23	3.91	32.9	147.63	4.01	33.9	155.03	4.10	34.9	162.43	4.18
7.6	32.6	146.68	4.00	33.6	154.28	4.10	34.6	161.88	4.19	35.6	169.48	4.28
7.8	33.3	153.27	4.09	34.3	161.07	4.19	35.3	168.87	4.28	36.3	176.67	4.37
8.0	34.0	160.00	4.18	35.0	168.00	4.28	36.0	176.00	4.37	37.0	184.00	4.46
8.2	34.7	166.87	4.27	35.7	175.07	4.37	36.7	183.27	4.46	37.7	191.47	4.55
8.4	35.4	173.88	4.36	36.4	182.28	4.46	37.4	190.68	4.56	38.4	199.08	4.64
8.6	36.1	181.03	4.45	37.1	189.63	4.55	38.1	198.23	4.65	39.1	206.83	4.74
8.8	36.8	188.32	4.54	37.8	197.12	4.64	38.8	205.92	4.74	39.8	214.72	4.83
9.0	37.5	195.75	4.63	38.5	204.75	4.73	39.5	213.75	4.83	40.5	222.75	4.92
9.2	38.2	203.32	4.72	39.2	212.52	4.82	40.2	221.72	4.92	41.2	230.92	5.01
9.4	38.9	211.03	4.81	39.9	220.43	4.91	40.9	229.83	5.01	41.9	239.23	5.10
9.6	39.6	218.88	4.90	40.6	228.48	5.00	41.6	238.08	5.10	42.6	247.68	5.19
9.8	40.3	226.87	4.99	41.3	236.67	5.09	42.3	246.47	5.19	43.3	256.27	5.28
10.0	41.00	235.00	5.07	42.00	245.00	5.18	43.00	255.00	5.28	44.00	265.00	5.37
10.5	42.75	255.94	5.30	43.75	266.44	5.40	44.75	276.94	5.50	45.75	287.44	5.60
11.0	44.50	277.8	5.52	45.50	288.8	5.62	46.50	299.8	5.73	47.50	310.8	5.83
11.5	46.25	300.4	5.74	47.25	311.9	5.85	48.25	323.4	5.95	49.25	334.9	6.05
12.0	48.00	324.0	5.96	49.00	336.0	6.07	50.00	348.0	6.17	51.00	360.0	6.27
12.5	49.75	348.4	6.18	50.75	360.9	6.29	51.75	373.4	6.40	52.75	385.9	6.50
13.0	51.50	373.8	6.40	52.50	386.8	6.51	53.50	399.8	6.62	54.50	412.8	6.72
13.5	53.25	399.9	6.62	54.25	413.4	6.73	55.25	426.9	6.84	56.25	440.4	6.94
14.0	55.00	427.0	6.84	56.00	441.0	6.95	57.00	455.0	7.06	58.00	469.0	7.17
14.5	56.75	454.9	7.06	57.75	469.4	7.17	58.75	483.9	7.28	59.75	498.4	7.39
15.0	58.50	483.8	7.28	59.50	498.8	7.39	60.50	513.8	7.50	61.50	528.8	7.61
15.5	60.25	513.4	7.50	61.25	528.9	7.61	62.25	544.4	7.72	63.25	559.9	7.83
16.0	62.00	544.0	7.72	63.00	560.0	7.83	64.00	576.0	7.95	65.00	592.0	8.05
16.5	63.75	575.4	7.94	64.75	591.9	8.05	65.75	608.4	8.17	66.75	624.9	8.28
17.0	65.50	607.8	8.15	66.50	624.8	8.27	67.50	641.8	8.39	68.50	658.8	8.50
17.5	67.25	640.9	8.37	68.25	658.4	8.49	69.25	675.9	8.61	70.25	693.4	8.72
18.0	69.00	675.0	8.59	70.00	693.0	8.71	71.00	711.0	8.83	72.00	729.0	8.94
18.5	70.75	709.9	8.81	71.75	728.4	8.93	72.75	746.9	9.05	73.75	765.4	9.16
19.0	72.50	745.8	9.03	73.50	764.8	9.15	74.50	783.8	9.27	75.50	802.8	9.38
19.5	74.25	782.4	9.25	75.25	801.9	9.37	76.25	821.4	9.48	77.25	840.9	9.60
20.0	76.00	820.0	9.47	77.00	840.0	9.59	78.00	860.0	9.70	79.00	880.0	9.82

Table 32.—*Area in square feet, A, top width in feet, T, and hydraulic radius in feet, r, of trapezoidal channels,* **side slopes 1¾ to 1**—Continued

Depth	Bottom width 10 feet			Bottom width 12 feet			Bottom width 14 feet			Bottom width 16 feet		
	T	A	r	T	A	r	T	A	r	T	A	r
0.4	11.4	4.28	0.37	13.4	5.08	0.37	15.4	5.88	0.38	17.4	6.68	0.38
0.6	12.1	6.63	.53	14.1	7.83	.54	16.1	9.03	.55	18.1	10.23	.56
0.8	12.8	9.12	.69	14.8	10.72	.70	16.8	12.32	.72	18.8	13.92	.72
1.0	13.5	11.75	.84	15.5	13.75	.86	17.5	15.75	.87	19.5	17.75	.89
1.2	14.2	14.52	.98	16.2	16.92	1.00	18.2	19.32	1.03	20.2	21.72	1.04
1.4	14.9	17.43	1.11	16.9	20.23	1.15	18.9	23.03	1.17	20.9	25.83	1.19
1.6	15.6	20.48	1.24	17.6	23.68	1.28	19.6	26.88	1.31	21.6	30.08	1.34
1.8	16.3	23.67	1.37	18.3	27.27	1.42	20.3	30.87	1.45	22.3	34.47	1.48
2.0	17.0	27.00	1.49	19.0	31.00	1.55	21.0	35.00	1.59	23.0	39.00	1.62
2.2	17.7	30.47	1.61	19.7	34.87	1.67	21.7	39.27	1.72	23.7	43.67	1.76
2.4	18.4	34.08	1.73	20.4	38.88	1.79	22.4	43.68	1.85	24.4	48.48	1.89
2.6	19.1	37.83	1.85	21.1	43.03	1.91	23.1	48.23	1.97	25.1	53.43	2.02
2.8	19.8	41.72	1.96	21.8	47.32	2.03	23.8	52.92	2.09	25.8	58.52	2.14
3.0	20.5	45.75	2.07	22.5	51.75	2.15	24.5	57.75	2.21	26.5	63.75	2.27
3.2	21.2	49.92	2.18	23.2	56.32	2.26	25.2	62.72	2.33	27.2	69.12	2.39
3.4	21.9	54.23	2.29	23.9	61.03	2.37	25.9	67.83	2.45	27.9	74.63	2.51
3.6	22.6	58.68	2.39	24.6	65.88	2.48	26.6	73.08	2.56	28.6	80.28	2.63
3.8	23.3	63.27	2.50	25.3	70.87	2.59	27.3	78.47	2.68	29.3	86.07	2.75
4.0	24.0	68.00	2.60	26.0	76.00	2.70	28.0	84.00	2.79	30.0	92.00	2.86
4.2	24.7	72.87	2.71	26.7	81.27	2.81	28.7	89.67	2.90	30.7	98.07	2.98
4.4	25.4	77.88	2.81	27.4	86.68	2.91	29.4	95.48	3.01	31.4	104.28	3.09
4.6	26.1	83.03	2.91	28.1	92.23	3.02	30.1	101.43	3.12	32.1	110.63	3.20
4.8	26.8	88.32	3.01	28.8	97.92	3.12	30.8	107.52	3.22	32.8	117.12	3.31
5.0	27.5	93.75	3.11	29.5	103.75	3.23	31.5	113.75	3.33	33.5	123.75	3.42
5.2	28.2	99.32	3.21	30.2	109.72	3.33	32.2	120.12	3.44	34.2	130.52	3.53
5.4	28.9	105.03	3.31	30.9	115.83	3.43	32.9	126.63	3.54	34.9	137.43	3.64
5.6	29.6	110.88	3.40	31.6	122.08	3.53	33.6	133.28	3.64	35.6	144.48	3.75
5.8	30.3	116.87	3.50	32.3	128.47	3.63	34.3	140.07	3.75	36.3	151.67	3.85
6.0	31.0	123.00	3.60	33.0	135.00	3.73	35.0	147.00	3.85	37.0	159.00	3.96
6.2	31.7	129.27	3.69	33.7	141.67	3.83	35.7	154.07	3.95	37.7	166.47	4.06
6.4	32.4	135.68	3.79	34.4	148.48	3.93	36.4	161.28	4.05	38.4	174.08	4.16
6.6	33.1	142.23	3.89	35.1	155.43	4.03	37.1	168.63	4.15	39.1	181.83	4.27
6.8	33.8	148.92	3.98	35.8	162.52	4.12	37.8	176.12	4.25	39.8	189.72	4.37
7.0	34.5	155.75	4.08	36.5	169.75	4.22	38.5	183.75	4.35	40.5	197.75	4.47
7.2	35.2	162.72	4.17	37.2	177.12	4.32	39.2	191.52	4.45	41.2	205.92	4.57
7.4	35.9	169.83	4.26	37.9	184.63	4.41	39.9	199.43	4.55	41.9	214.23	4.67
7.6	36.6	177.08	4.36	38.6	192.28	4.51	40.6	207.48	4.65	42.6	222.68	4.77
7.8	37.3	184.47	4.45	39.3	200.07	4.61	41.3	215.67	4.75	43.3	231.27	4.87
8.0	38.0	192.00	4.54	40.0	208.00	4.70	42.0	224.00	4.84	44.0	240.00	4.97
8.2	38.7	199.67	4.64	40.7	216.07	4.80	42.7	232.47	4.94	44.7	248.87	5.07
8.4	39.4	207.48	4.73	41.4	224.28	4.89	43.4	241.08	5.04	45.4	257.88	5.17
8.6	40.1	215.43	4.82	42.1	232.63	4.98	44.1	249.83	5.13	46.1	267.03	5.27
8.8	40.8	223.52	4.92	42.8	241.12	5.08	44.8	258.72	5.23	46.8	276.32	5.37
9.0	41.5	231.75	5.01	43.5	249.75	5.17	45.5	267.75	5.33	47.5	285.75	5.47
9.2	42.2	240.12	5.10	44.2	258.52	5.27	46.2	276.92	5.42	48.2	295.32	5.56
9.4	42.9	248.63	5.19	44.9	267.43	5.36	46.9	286.23	5.52	48.9	305.03	5.66
9.6	43.6	257.28	5.28	45.6	276.48	5.45	47.6	295.68	5.61	49.6	314.88	5.76
9.8	44.3	266.07	5.37	46.3	285.67	5.55	48.3	305.27	5.71	50.3	324.87	5.85

Table 32.—*Area in square feet, A, top width in feet, T, and hydraulic radius in feet, r, of trapezoidal channels,*
side slopes 1¾ to 1—Continued

Depth	Bottom width 10 feet			Bottom width 12 feet			Bottom width 14 feet			Bottom width 16 feet		
	T	A	r	T	A	r	T	A	r	T	A	r
10.0	45.00	275.00	5.47	47.00	295.00	5.64	49.00	315.00	5.80	51.00	335.00	5.95
10.5	46.75	297.94	5.69	48.75	318.94	5.87	50.75	339.94	6.04	52.75	360.94	6.19
11.0	48.50	321.8	5.92	50.50	343.8	6.10	52.50	365.8	6.27	54.50	387.8	6.43
11.5	50.25	346.4	6.15	52.25	369.4	6.33	54.25	392.4	6.50	56.25	415.4	6.66
12.0	52.00	372.0	6.37	54.00	396.0	6.56	56.00	420.0	6.73	58.00	444.0	6.90
12.5	53.75	398.4	6.60	55.75	423.4	6.79	57.75	448.4	6.96	59.75	473.4	7.13
13.0	55.50	425.8	6.82	57.50	451.8	7.01	59.50	477.8	7.19	61.50	503.8	7.36
13.5	57.25	453.9	7.05	59.25	480.9	7.24	61.25	507.9	7.42	63.25	534.9	7.60
14.0	59.00	483.0	7.27	61.00	511.0	7.47	63.00	539.0	7.65	65.00	567.0	7.83
14.5	60.75	512.9	7.49	62.75	541.9	7.69	64.75	570.9	7.88	66.75	599.9	8.06
15.0	62.50	543.8	7.72	64.50	573.8	7.92	66.50	603.8	8.11	68.50	633.8	8.29
15.5	64.25	575.4	7.94	66.25	606.4	8.14	68.25	637.4	8.33	70.25	668.4	8.52
16.0	66.00	608.0	8.16	68.00	640.0	8.37	70.00	672.0	8.56	72.00	704.0	8.75
16.5	67.75	641.4	8.38	69.75	674.4	8.59	71.75	707.4	8.79	73.75	740.4	8.97
17.0	69.50	675.8	8.61	71.50	709.8	8.81	73.50	743.8	9.01	75.50	777.8	9.20
17.5	71.25	710.9	8.83	73.25	745.9	9.04	75.25	780.9	9.24	77.25	815.9	9.43
18.0	73.00	747.0	9.05	75.00	783.0	9.26	77.00	819.0	9.46	79.00	855.0	9.65
18.5	74.75	783.9	9.27	76.75	820.9	9.48	78.75	857.9	9.69	80.75	894.9	9.88
19.0	76.50	821.8	9.49	78.50	859.8	9.70	80.50	897.8	9.91	82.50	935.8	10.11
19.5	78.25	860.4	9.71	80.25	899.4	9.93	82.25	938.4	10.13	84.25	977.4	10.33
20.0	80.00	900.0	9.93	82.00	940.0	10.15	84.00	980.0	10.36	86.00	1020.0	10.56

Depth	Bottom width 18 feet			Bottom width 20 feet			Bottom width 22 feet			Bottom width 24 feet		
	T	A	r	T	A	r	T	A	r	T	A	r
0.4	19.4	7.48	0.38	21.4	8.28	0.38	23.4	9.08	0.38	25.4	9.88	0.39
0.6	20.1	11.43	.56	22.1	12.63	.56	24.1	13.83	.57	26.1	15.03	.57
0.8	20.8	15.52	.73	22.8	17.12	.74	24.8	18.72	.74	26.8	20.32	.75
1.0	21.5	19.75	.90	23.5	21.75	.91	25.5	23.75	.91	27.5	25.75	.92
1.2	22.2	24.12	1.06	24.2	26.52	1.07	26.2	28.92	1.08	28.2	31.32	1.09
1.4	22.9	28.63	1.21	24.9	31.43	1.23	26.9	34.23	1.24	28.9	37.03	1.25
1.6	23.6	33.28	1.36	25.6	36.48	1.38	27.6	39.68	1.39	29.6	42.88	1.41
1.8	24.3	38.07	1.51	26.3	41.67	1.53	28.3	45.27	1.55	30.3	48.87	1.56
2.0	25.0	43.00	1.65	27.0	47.00	1.67	29.0	51.00	1.70	31.0	55.00	1.72
2.2	25.7	48.07	1.79	27.7	52.47	1.82	29.7	56.87	1.84	31.7	61.27	1.86
2.4	26.4	53.28	1.93	28.4	58.08	1.96	30.4	62.88	1.99	32.4	67.68	2.01
2.6	27.1	58.63	2.06	29.1	63.83	2.09	31.1	69.03	2.13	33.1	74.23	2.15
2.8	27.8	64.12	2.19	29.8	69.72	2.23	31.8	75.32	2.26	33.8	80.92	2.29
3.0	28.5	69.75	2.32	30.5	75.75	2.36	32.5	81.75	2.40	34.5	87.75	2.43
3.2	29.2	75.52	2.44	31.2	81.92	2.49	33.2	88.32	2.53	35.2	94.72	2.57
3.4	29.9	81.43	2.57	31.9	88.23	2.62	33.9	95.03	2.66	35.9	101.83	2.70
3.6	30.6	87.48	2.69	32.6	94.68	2.74	34.6	101.88	2.79	36.6	109.08	2.83
3.8	31.3	93.67	2.81	33.3	101.27	2.87	35.3	108.87	2.92	37.3	116.47	2.96
4.0	32.0	100.00	2.93	34.0	108.00	2.99	36.0	116.00	3.04	38.0	124.00	3.09
4.2	32.7	106.47	3.05	34.7	114.87	3.11	36.7	123.27	3.17	38.7	131.67	3.22
4.4	33.4	113.08	3.16	35.4	121.88	3.23	37.4	130.68	3.29	39.4	139.48	3.34
4.6	34.1	119.83	3.28	36.1	129.03	3.35	38.1	138.23	3.41	40.1	147.43	3.47
4.8	34.8	126.72	3.39	36.8	136.32	3.46	38.8	145.92	3.53	40.8	155.52	3.59

Table 32.—*Area in square feet, A, top width in feet, T, and hydraulic radius in feet, r, of trapezoidal channels,*
side slopes 1¾ to 1—Continued

Depth	Bottom width 18 feet			Bottom width 20 feet			Bottom width 22 feet			Bottom width 24 feet		
	T	A	r	T	A	r	T	A	r	T	A	r
5.0	35.5	133.75	3.51	37.5	143.75	3.58	39.5	153.75	3.65	41.5	163.75	3.71
5.2	36.2	140.92	3.62	38.2	151.32	3.69	40.2	161.72	3.76	42.2	172.12	3.83
5.4	36.9	148.23	3.73	38.9	159.03	3.81	40.9	169.83	3.88	42.9	180.63	3.95
5.6	37.6	155.68	3.84	39.6	166.88	3.92	41.6	178.08	4.00	43.6	189.28	4.06
5.8	38.3	163.27	3.95	40.3	174.87	4.03	42.3	186.47	4.11	44.3	198.07	4.18
6.0	39.0	171.00	4.05	41.0	183.00	4.14	43.0	195.00	4.22	45.0	207.00	4.30
6.2	39.7	178.87	4.16	41.7	191.27	4.25	43.7	203.67	4.33	45.7	216.07	4.41
6.4	40.4	186.88	4.27	42.4	199.68	4.36	44.4	212.48	4.45	46.4	225.28	4.52
6.6	41.1	195.03	4.37	43.1	208.23	4.47	45.1	221.43	4.56	47.1	234.63	4.64
6.8	41.8	203.32	4.48	43.8	216.92	4.58	45.8	230.52	4.67	47.8	244.12	4.75
7.0	42.5	211.75	4.58	44.5	225.75	4.68	46.5	239.75	4.77	48.5	253.75	4.86
7.2	43.2	220.32	4.69	45.2	234.72	4.79	47.2	249.12	4.88	49.2	263.52	4.97
7.4	43.9	229.03	4.79	45.9	243.83	4.89	47.9	258.63	4.99	49.9	273.43	5.08
7.6	44.6	237.88	4.89	46.6	253.08	5.00	48.6	268.28	5.10	50.6	283.48	5.19
7.8	45.3	246.87	4.99	47.3	262.47	5.10	49.3	278.07	5.20	51.3	293.67	5.30
8.0	46.0	256.00	5.09	48.0	272.00	5.21	50.0	288.00	5.31	52.0	304.00	5.40
8.2	46.7	265.27	5.20	48.7	281.67	5.31	50.7	298.07	5.41	52.7	314.47	5.51
8.4	47.4	274.68	5.30	49.4	291.48	5.41	51.4	308.28	5.52	53.4	325.08	5.62
8.6	48.1	284.23	5.40	50.1	301.43	5.51	52.1	318.63	5.62	54.1	335.83	5.72
8.8	48.8	293.92	5.50	50.8	311.52	5.62	52.8	329.12	5.73	54.8	346.72	5.83
9.0	49.5	303.75	5.60	51.5	321.75	5.72	53.5	339.75	5.83	55.5	357.75	5.93
9.2	50.2	313.72	5.70	52.2	332.12	5.82	54.2	350.52	5.93	56.2	368.92	6.04
9.4	50.9	323.83	5.79	52.9	342.63	5.92	54.9	361.43	6.03	56.9	380.23	6.14
9.6	51.6	334.08	5.89	53.6	353.28	6.02	55.6	372.48	6.14	57.6	391.68	6.25
9.8	52.3	344.47	5.99	54.3	364.07	6.12	56.3	383.67	6.24	58.3	403.27	6.35
10.0	53.00	355.00	6.09	55.00	375.00	6.22	57.00	395.00	6.34	59.00	415.00	6.45
10.5	54.75	381.94	6.33	56.75	402.94	6.46	58.75	423.94	6.59	60.75	444.94	6.71
11.0	56.50	409.8	6.57	58.50	431.8	6.71	60.50	453.8	6.84	62.50	475.8	6.96
11.5	58.25	438.4	6.81	60.25	461.4	6.95	62.25	484.4	7.09	64.25	507.4	7.21
12.0	60.00	468.0	7.05	62.00	492.0	7.20	64.00	516.0	7.33	66.00	540.0	7.46
12.5	61.75	498.4	7.29	63.75	523.4	7.44	65.75	548.4	7.58	67.75	573.4	7.71
13.0	63.50	529.8	7.52	65.50	555.8	7.68	67.50	581.8	7.82	69.50	607.8	7.95
13.5	65.25	561.9	7.76	67.25	588.9	7.91	69.25	615.9	8.06	71.25	642.9	8.20
14.0	67.00	595.0	7.99	69.00	623.0	8.15	71.00	651.0	8.30	73.00	679.0	8.44
14.5	68.75	628.9	8.23	70.75	657.9	8.39	72.75	686.9	8.54	74.75	715.9	8.68
15.0	70.50	663.8	8.46	72.50	693.8	8.62	74.50	723.8	8.78	76.50	753.8	8.92
15.5	72.25	699.4	8.69	74.25	730.4	8.86	76.25	761.4	9.01	78.25	792.4	9.16
16.0	74.00	736.0	8.92	76.00	768.0	9.09	78.00	800.0	9.25	80.00	832.0	9.40
16.5	75.75	773.4	9.15	77.75	806.4	9.32	79.75	839.4	9.48	81.75	872.4	9.64
17.0	77.50	811.8	9.38	79.50	845.8	9.55	81.50	879.8	9.72	83.50	913.8	9.88
17.5	79.25	850.9	9.61	81.25	885.9	9.78	83.25	920.9	9.95	85.25	955.9	10.11
18.0	81.00	891.0	9.84	83.00	927.0	10.02	85.00	963.0	10.18	87.00	999.0	10.35
18.5	82.75	931.9	10.07	84.75	968.9	10.25	86.75	1,005.9	10.42	88.75	1,042.9	10.58
19.0	84.50	973.8	10.29	86.50	1,011.8	10.47	88.50	1,049.8	10.65	90.50	1,087.8	10.81
19.5	86.25	1,016.4	10.52	88.25	1,055.4	10.70	90.25	1,094.4	10.88	92.25	1,133.4	11.05
20.0	88.00	1,060.0	10.75	90.00	1,100.0	10.93	92.00	1,140.0	11.11	94.00	1,180.0	11.28

Table 32.—*Area in square feet, A, top width in feet, T, and hydraulic radius in feet, r, of trapezoidal channels,*
side slopes 1¾ to 1—Continued

Depth	Bottom width 26 feet			Bottom width 28 feet			Bottom width 30 feet			Bottom width 32 feet		
	T	A	r	T	A	r	T	A	r	T	A	r
0.4	27.4	10.68	0.39	29.4	11.48	0.39	31.4	12.28	0.39	33.4	13.08	0.39
0.6	28.1	16.23	.57	30.1	17.43	.57	32.1	18.63	.57	34.1	19.83	.58
0.8	28.8	21.92	.75	30.8	23.52	.75	32.8	25.12	.76	34.8	26.72	.76
1.0	29.5	27.75	.92	31.5	29.75	.93	33.5	31.75	.93	35.5	33.75	.94
1.2	30.2	33.72	1.09	32.2	36.12	1.10	34.2	38.52	1.11	36.2	40.92	1.11
1.4	30.9	39.83	1.26	32.9	42.63	1.27	34.9	45.43	1.27	36.9	48.23	1.28
1.6	31.6	46.08	1.42	33.6	49.28	1.43	35.6	52.48	1.44	37.6	55.68	1.45
1.8	32.3	52.47	1.58	34.3	56.07	1.59	36.3	59.67	1.60	38.3	63.27	1.61
2.0	33.0	59.00	1.73	35.0	63.00	1.75	37.0	67.00	1.76	39.0	71.00	1.77
2.2	33.7	65.67	1.88	35.7	70.07	1.90	37.7	74.47	1.92	39.7	78.87	1.93
2.4	34.4	72.48	2.03	36.4	77.28	2.05	38.4	82.08	2.07	40.4	86.88	2.08
2.6	35.1	79.43	2.18	37.1	84.63	2.20	39.1	89.83	2.22	41.1	95.03	2.24
2.8	35.8	86.52	2.32	37.8	92.12	2.34	39.8	97.72	2.37	41.8	103.32	2.39
3.0	36.5	93.75	2.46	38.5	99.75	2.49	40.5	105.75	2.51	42.5	111.75	2.53
3.2	37.2	101.12	2.60	39.2	107.52	2.63	41.2	113.92	2.66	43.2	120.32	2.68
3.4	37.9	108.63	2.74	39.9	115.43	2.77	41.9	122.23	2.80	43.9	129.03	2.82
3.6	38.6	116.28	2.87	40.6	123.48	2.90	42.6	130.68	2.94	44.6	137.88	2.96
3.8	39.3	124.07	3.00	41.3	131.67	3.04	43.3	139.27	3.07	45.3	146.87	3.10
4.0	40.0	132.00	3.13	42.0	140.00	3.17	44.0	148.00	3.21	46.0	156.00	3.24
4.2	40.7	140.07	3.26	42.7	148.47	3.30	44.7	156.87	3.34	46.7	165.27	3.38
4.4	41.4	148.28	3.39	43.4	157.08	3.43	45.4	165.88	3.47	47.4	174.68	3.51
4.6	42.1	156.63	3.52	44.1	165.83	3.56	46.1	175.03	3.61	48.1	184.23	3.65
4.8	42.8	165.12	3.64	44.8	174.72	3.69	46.8	184.32	3.73	48.8	193.92	3.78
5.0	43.5	173.75	3.76	45.5	183.75	3.82	47.5	193.75	3.86	49.5	203.75	3.91
5.2	44.2	182.52	3.89	46.2	192.92	3.94	48.2	203.32	3.99	50.2	213.72	4.04
5.4	44.9	191.43	4.01	46.9	202.23	4.06	48.9	213.03	4.12	50.9	223.83	4.16
5.6	45.6	200.48	4.13	47.6	211.68	4.19	49.6	222.88	4.24	51.6	234.08	4.29
5.8	46.3	209.67	4.25	48.3	221.27	4.31	50.3	232.87	4.36	52.3	244.47	4.41
6.0	47.0	219.00	4.36	49.0	231.00	4.43	51.0	243.00	4.48	53.0	255.00	4.54
6.2	47.7	228.47	4.48	49.7	240.87	4.55	51.7	253.27	4.61	53.7	265.67	4.66
6.4	48.4	238.08	4.60	50.4	250.88	4.66	52.4	263.68	4.73	54.4	276.48	4.78
6.6	49.1	247.83	4.71	51.1	261.03	4.78	53.1	274.23	4.84	55.1	287.43	4.90
6.8	49.8	257.72	4.83	51.8	271.32	4.90	53.8	284.92	4.96	55.8	298.52	5.02
7.0	50.5	267.75	4.94	52.5	281.75	5.01	54.5	295.75	5.08	56.5	309.75	5.14
7.2	51.2	277.92	5.05	53.2	292.32	5.13	55.2	306.72	5.20	57.2	321.12	5.26
7.4	51.9	288.23	5.16	53.9	303.03	5.24	55.9	317.83	5.31	57.9	332.63	5.38
7.6	52.6	298.68	5.27	54.6	313.88	5.35	56.6	329.08	5.43	58.6	344.28	5.50
7.8	53.3	309.27	5.38	55.3	324.87	5.47	57.3	340.47	5.54	59.3	356.07	5.61
8.0	54.0	320.00	5.49	56.0	336.00	5.58	58.0	352.00	5.65	60.0	368.00	5.73
8.2	54.7	330.87	5.60	56.7	347.27	5.69	58.7	363.67	5.77	60.7	380.07	5.84
8.4	55.4	341.88	5.71	57.4	358.68	5.80	59.4	375.48	5.88	61.4	392.28	5.96
8.6	56.1	353.03	5.82	58.1	370.23	5.91	60.1	387.43	5.99	62.1	404.63	6.07
8.8	56.8	364.32	5.93	58.8	381.92	6.02	60.8	399.52	6.10	62.8	417.12	6.18
9.0	57.5	375.75	6.03	59.5	393.75	6.13	61.5	411.75	6.21	63.5	429.75	6.29
9.2	58.2	387.32	6.14	60.2	405.72	6.23	62.2	424.12	6.32	64.2	442.52	6.41
9.4	58.9	399.03	6.25	60.9	417.83	6.34	62.9	436.53	6.43	64.9	455.43	6.52
9.6	59.6	410.88	6.35	61.6	430.08	6.45	63.6	449.28	6.54	65.6	468.48	6.63
9.8	60.3	422.87	6.46	62.3	442.47	6.55	64.3	462.07	6.65	66.3	481.67	6.74

Table 32.—*Area in square feet, A, top width in feet, T, and hydraulic radius in feet, r, of trapezoidal channels,*
side slopes 1¾ to 1—Continued

Depth	Bottom width 26 feet			Bottom width 28 feet			Bottom width 30 feet			Bottom width 32 feet		
	T	A	r	T	A	r	T	A	r	T	A	r
10.0	61.00	435.00	6.56	63.00	455.00	6.66	65.00	475.00	6.76	67.00	495.00	6.85
10.5	62.75	465.94	6.82	64.75	486.94	6.92	66.75	507.94	7.02	68.75	528.94	7.12
11.0	64.50	497.8	7.08	66.50	519.8	7.18	68.50	541.8	7.29	70.50	563.8	7.38
11.5	66.25	530.4	7.33	68.25	553.4	7.44	70.25	576.4	7.55	72.25	599.4	7.65
12.0	68.00	564.0	7.58	70.00	588.0	7.70	72.00	612.0	7.81	74.00	636.0	7.91
12.5	69.75	598.4	7.83	71.75	623.4	7.95	73.75	648.4	8.07	75.75	673.4	8.17
13.0	71.50	633.8	8.08	73.50	659.8	8.21	75.50	685.8	8.32	77.50	711.8	8.43
13.5	73.25	669.9	8.33	75.25	696.9	8.46	77.25	723.9	8.58	79.25	750.9	8.69
14.0	75.00	707.0	8.58	77.00	735.0	8.70	79.00	763.0	8.83	81.00	791.0	8.94
14.5	76.75	744.9	8.82	78.75	773.9	8.95	80.75	802.9	9.08	82.75	831.9	9.20
15.0	78.50	783.8	9.06	80.50	813.8	9.20	82.50	843.8	9.33	84.50	873.8	9.45
15.5	80.25	823.4	9.31	82.25	854.4	9.44	84.25	885.4	9.57	86.25	916.4	9.70
16.0	82.00	864.0	9.55	84.00	896.0	9.69	86.00	928.0	9.82	88.00	960.0	9.95
16.5	83.75	905.4	9.79	85.75	938.4	9.93	87.75	971.4	10.07	89.75	1,004.4	10.20
17.0	85.50	947.8	10.03	87.50	981.8	10.17	89.50	1,015.8	10.31	91.50	1,049.8	10.44
17.5	87.25	990.0	10.26	89.25	1,025.9	10.41	91.25	1,060.9	10.55	93.25	1,095.9	10.69
18.0	89.00	1,035.0	10.50	91.00	1,071.0	10.65	93.00	1,107.0	10.79	95.00	1,143.0	10.93
18.5	90.75	1,079.9	10.74	92.75	1,116.9	10.89	94.75	1,153.9	11.03	96.75	1,190.9	11.17
19.0	92.50	1,125.8	10.97	94.50	1,163.8	11.13	96.50	1,201.8	11.27	98.50	1,239.8	11.42
19.5	94.25	1,172.4	11.21	96.25	1,211.4	11.36	98.25	1,250.4	11.51	100.25	1,289.4	11.66
20.0	96.00	1,220.0	11.44	98.00	1,260.0	11.60	100.00	1,300.0	11.75	102.00	1,340.0	11.90

Depth	Bottom width 35 feet			Bottom width 40 feet			Bottom width 45 feet			Bottom width 50 feet		
	T	A	r	T	A	r	T	A	r	T	A	r
0.4	36.4	14.28	0.39	41.4	16.28	0.39	46.4	18.28	0.39	51.4	20.28	0.39
0.6	37.1	21.63	.58	42.1	24.63	.58	47.1	27.63	.58	52.1	30.63	.58
0.8	37.8	29.12	.76	42.8	33.12	.77	47.8	37.12	.77	52.8	41.12	.77
1.0	38.5	36.75	.94	43.5	41.75	.95	48.5	46.75	.95	53.5	51.75	.96
1.2	39.2	44.52	1.12	44.2	50.52	1.13	49.2	56.52	1.13	54.2	62.52	1.14
1.4	39.9	52.43	1.29	44.9	59.43	1.30	49.9	66.43	1.31	54.9	73.43	1.32
1.6	40.6	60.48	1.46	45.6	68.48	1.47	50.6	76.48	1.49	55.6	84.48	1.50
1.8	41.3	68.67	1.63	46.3	77.67	1.64	51.3	86.67	1.66	56.3	95.67	1.67
2.0	42.0	77.00	1.79	47.0	87.00	1.81	52.0	97.00	1.83	57.0	107.00	1.84
2.2	42.7	85.47	1.95	47.7	96.47	1.97	52.7	107.47	2.00	57.7	118.47	2.01
2.4	43.4	94.08	2.11	48.4	106.08	2.14	53.4	118.08	2.16	58.4	130.08	2.18
2.6	44.1	102.83	2.26	49.1	115.83	2.29	54.1	128.83	2.32	59.1	141.83	2.35
2.8	44.8	111.72	2.41	49.8	125.72	2.45	54.8	139.72	2.48	59.8	153.72	2.51
3.0	45.5	120.75	2.56	50.5	135.75	2.61	55.5	150.75	2.64	60.5	165.75	2.67
3.2	46.2	129.92	2.71	51.2	145.92	2.76	56.2	161.92	2.80	61.2	177.92	2.83
3.4	46.9	139.23	2.86	51.9	156.23	2.91	56.9	173.23	2.95	61.9	190.23	2.99
3.6	47.6	148.68	3.00	52.6	166.68	3.06	57.6	184.68	3.10	62.6	202.68	3.14
3.8	48.3	158.27	3.15	53.3	177.27	3.20	58.3	196.27	3.25	63.3	215.27	3.30
4.0	49.0	168.00	3.29	54.0	188.00	3.35	59.0	208.00	3.40	64.0	228.00	3.45
4.2	49.7	177.87	3.43	54.7	198.87	3.49	59.7	219.87	3.55	64.7	240.87	3.60
4.4	50.4	187.88	3.56	55.4	209.88	3.64	60.4	231.88	3.70	65.4	253.88	3.75
4.6	51.1	198.03	3.70	56.1	221.03	3.78	61.1	244.03	3.84	66.1	267.03	3.90
4.8	51.8	208.32	3.83	56.8	232.32	3.91	61.8	256.32	3.98	66.8	280.32	4.04

Table 32.—*Area in square feet, A, top width in feet, T, and hydraulic radius in feet, r, of trapezoidal channels,* **side slopes 1¾ to 1**—Continued

Depth	Bottom width 35 feet			Bottom width 40 feet			Bottom width 45 feet			Bottom width 50 feet		
	T	A	r	T	A	r	T	A	r	T	A	r
5.0	52.5	218.75	3.97	57.5	243.75	4.05	62.5	268.75	4.12	67.5	293.75	4.19
5.2	53.2	229.32	4.10	58.2	255.32	4.19	63.2	281.32	4.26	68.2	307.32	4.33
5.4	53.9	240.03	4.23	58.9	267.03	4.32	63.9	294.03	4.40	68.9	321.03	4.47
5.6	54.6	250.88	4.36	59.6	278.88	4.46	64.6	306.88	4.54	69.6	334.88	4.61
5.8	55.3	261.87	4.49	60.3	290.87	4.59	65.3	319.87	4.68	70.3	348.87	4.75
6.0	56.0	273.00	4.61	61.0	303.00	4.72	66.0	333.00	4.81	71.0	363.00	4.89
6.2	56.7	284.27	4.74	61.7	315.27	4.85	66.7	346.27	4.95	71.7	377.27	5.03
6.4	57.4	295.68	4.86	62.4	327.68	4.98	67.4	359.68	5.08	72.4	391.68	5.17
6.6	58.1	307.23	4.99	63.1	340.23	5.11	68.1	373.23	5.21	73.1	406.23	5.30
6.8	58.8	318.92	5.11	63.8	352.92	5.24	68.8	386.92	5.34	73.8	420.92	5.44
7.0	59.5	330.75	5.23	64.5	365.75	5.36	69.5	400.75	5.47	74.5	435.75	5.57
7.2	60.2	342.72	5.35	65.2	378.72	5.49	70.2	414.72	5.60	75.2	450.72	5.70
7.4	60.9	354.83	5.47	65.9	391.83	5.61	70.9	428.83	5.73	75.9	465.83	5.84
7.6	61.6	367.08	5.59	66.6	405.08	5.73	71.6	443.08	5.86	76.6	481.08	5.97
7.8	62.3	379.47	5.71	67.3	418.47	5.86	72.3	457.47	5.98	77.3	496.47	6.10
8.0	63.0	392.00	5.83	68.0	432.00	5.98	73.0	472.00	6.11	78.0	512.00	6.22
8.2	63.7	404.67	5.95	68.7	445.67	6.10	73.7	486.67	6.23	78.7	527.67	6.35
8.4	64.4	417.48	6.06	69.4	459.48	6.22	74.4	501.48	6.36	79.4	543.48	6.48
8.6	65.1	430.43	6.18	70.1	473.43	6.34	75.1	516.43	6.48	80.1	559.43	6.61
8.8	65.8	443.52	6.29	70.8	487.52	6.46	75.8	531.52	6.60	80.8	575.52	6.73
9.0	66.5	456.75	6.41	71.5	501.75	6.58	76.5	546.75	6.73	81.5	591.75	6.86
9.2	67.2	470.12	6.52	72.2	516.12	6.70	77.2	562.12	6.85	82.2	608.12	6.98
9.4	67.9	483.63	6.63	72.9	530.63	6.81	77.9	577.63	6.97	82.9	624.63	7.11
9.6	68.6	497.28	6.75	73.6	545.28	6.93	78.6	593.28	7.09	83.6	641.28	7.23
9.8	69.3	511.07	6.86	74.3	560.07	7.04	79.3	609.07	7.21	84.3	658.07	7.35
10.0	70.00	525.00	6.97	75.00	575.00	7.16	80.00	625.00	7.33	85.00	675.00	7.47
10.5	71.75	560.44	7.25	76.75	612.94	7.45	81.75	665.44	7.62	86.75	717.94	7.78
11.0	73.50	596.8	7.52	78.50	651.8	7.73	83.50	706.8	7.91	88.50	761.8	8.07
11.5	75.25	633.9	7.79	80.25	691.4	8.01	85.25	748.9	8.20	90.25	806.4	8.37
12.0	77.00	672.0	8.06	82.00	732.0	8.28	87.00	792.0	8.48	92.00	852.0	8.66
12.5	78.75	710.9	8.33	83.75	773.4	8.56	88.75	835.9	8.76	93.75	898.4	8.95
13.0	80.50	750.8	8.59	85.50	815.8	8.83	90.50	880.8	9.04	95.50	945.8	9.24
13.5	82 25	791 4	8.85	87.25	858.9	9.10	92.25	926.4	9.32	97.25	993.9	9.52
14.0	84.00	833.0	9.11	89.00	903.0	9.36	94.00	973.0	9.59	99.00	1,043.0	9.80
14.5	85.75	875.4	9.37	90.75	947.9	9.63	95.75	1,020.4	9.86	100.75	1,092.9	10.08
15.0	87.50	918.8	9.62	92.50	993.8	9.89	97.50	1,068.8	10.13	102.50	1,143.8	10.35
15.5	89.25	962.9	9.88	94.25	1,040.4	10.15	99.25	1,117.9	10.40	104.25	1,195.4	10.63
16.0	91.00	1,008.0	10.13	96.00	1,088.0	10.41	101.00	1,168.0	10.67	106.00	1,248.0	10.90
16.5	92.75	1,053.9	10.38	97.75	1,136.4	10.67	102.75	1,218.9	10.93	107.75	1,301.4	11.17
17.0	94.50	1,100.8	10.63	99.50	1,185.8	10.93	104.50	1,270.8	11.19	109.50	1,355.8	11.44
17.5	96.25	1,148.4	10.88	101.25	1,235.9	11.18	106.25	1,323.4	11.45	111.25	1,410.9	11.70
18.0	98.00	1,197.0	11.13	103.00	1,287.0	11.43	108.00	1,377.0	11.71	113.00	1,467.0	11.97
18.5	99.75	1,246.4	11.38	104.75	1,338.9	11.69	109.75	1,431.4	11.97	114.75	1,523.9	12.23
19.0	101.50	1,296.8	11.62	106.50	1,391.8	11.94	111.50	1,486.8	12.23	116.50	1,581.8	12.49
19.5	103.25	1,347.9	11.86	108.25	1,445.4	12.19	113.25	1,542.9	12.48	118.25	1,640.4	12.76
20.0	105.00	1,400.0	12.11	110.00	1,500.0	12.44	115.00	1,600.0	12.74	120.00	1,700.0	13.01

Table 32.—*Area in square feet, A, top width in feet, T, and hydraulic radius in feet, r, of trapezoidal channels,* **side slopes 1¾ to 1**—Continued

Depth	Bottom width 60 feet			Bottom width 70 feet			Bottom width 80 feet			Bottom width 90 feet		
	T	A	r	T	A	r	T	A	r	T	A	r
0.4	61.4	24.28	0.39	71.4	28.28	0.39	81.4	32.28	0.40	91.4	36.28	0.40
0.6	62.1	36.63	.59	72.1	42.63	.59	82.1	48.63	.59	92.1	54.63	.59
0.8	62.8	49.12	.78	72.8	57.12	.78	82.8	65.12	.78	92.8	73.12	.78
1.0	63.5	61.75	.96	73.5	71.75	.97	83.5	81.75	.97	93.5	91.75	.98
1.2	64.2	74.52	1.15	74.2	86.52	1.16	84.2	98.52	1.16	94.2	110.52	1.17
1.4	64.9	87.43	1.33	74.9	101.43	1.34	84.9	115.43	1.35	94.9	129.43	1.35
1.6	65.6	100.48	1.51	75.6	116.48	1.52	85.6	132.48	1.53	95.6	148.48	1.54
1.8	66.3	113.67	1.69	76.3	131.67	1.70	86.3	149.67	1.72	96.3	167.67	1.72
2.0	67.0	127.00	1.87	77.0	147.00	1.88	87.0	167.00	1.90	97.0	187.00	1.91
2.2	67.7	140.47	2.04	77.7	162.47	2.06	87.7	184.47	2.08	97.7	206.47	2.09
2.4	68.4	154.08	2.21	78.4	178.08	2.24	88.4	202.08	2.25	98.4	226.08	2.27
2.6	69.1	167.83	2.38	79.1	193.83	2.41	89.1	219.83	2.43	99.1	245.83	2.45
2.8	69.8	181.72	2.55	79.8	209.72	2.58	89.8	237.72	2.60	99.8	265.72	2.62
3.0	70.5	195.75	2.72	80.5	225.75	2.75	90.5	255.75	2.78	100.5	285.75	2.80
3.2	71.2	209.92	2.88	81.2	241.92	2.92	91.2	273.92	2.95	101.2	305.92	2.97
3.4	71.9	224.23	3.04	81.9	258.23	3.08	91.9	292.23	3.12	101.9	326.23	3.15
3.6	72.6	238.68	3.20	82.6	274.68	3.25	92.6	310.68	3.29	102.6	346.68	3.32
3.8	73.3	253.27	3.36	83.3	291.27	3.41	93.3	329.27	3.45	103.3	367.27	3.49
4.0	74.0	268.00	3.52	84.0	308.00	3.58	94.0	348.00	3.62	104.0	388.00	3.66
4.2	74.7	282.87	3.68	84.7	324.87	3.74	94.7	366.87	3.78	104.7	408.87	3.82
4.4	75.4	297.88	3.83	85.4	341.88	3.90	95.4	385.88	3.95	105.4	429.88	3.99
4.6	76.1	313.03	3.99	86.1	359.03	4.05	96.1	405.03	4.11	106.1	451.03	4.16
4.8	76.8	328.32	4.14	86.8	376.32	4.21	96.8	424.32	4.27	106.8	472.32	4.32
5.0	77.5	343.75	4.29	87.5	393.75	4.37	97.5	443.75	4.43	107.5	493.75	4.48
5.2	78.2	359.32	4.44	88.2	411.32	4.52	98.2	463.32	4.59	108.2	515.32	4.64
5.4	78.9	375.03	4.59	88.9	429.03	4.68	98.9	483.03	4.75	108.9	537.03	4.80
5.6	79.6	390.88	4.73	89.6	446.88	4.83	99.6	502.88	4.90	109.6	558.88	4.96
5.8	80.3	406.87	4.88	90.3	464.87	4.98	100.3	522.87	5.06	110.3	580.87	5.12
6.0	81.0	423.00	5.02	91.0	483.00	5.13	101.0	543.00	5.21	111.0	603.00	5.28
6.2	81.7	439.27	5.17	91.7	501.27	5.28	101.7	563.27	5.36	111.7	625.27	5.44
6.4	82.4	455.68	5.31	92.4	519.68	5.42	102.4	583.68	5.52	112.4	647.68	5.59
6.6	83.1	472.23	5.45	93.1	538.23	5.57	103.1	604.23	5.67	113.1	670.23	5.75
6.8	83.8	488.92	5.59	93.8	556.92	5.72	103.8	624.92	5.82	113.8	692.92	5.90
7.0	84.5	505.75	5.73	94.5	575.75	5.86	104.5	645.75	5.97	114.5	715.75	6.05
7.2	85.2	522.72	5.87	95.2	594.72	6.01	105.2	666.72	6.12	115.2	738.72	6.21
7.4	85.9	539.83	6.01	95.9	613.83	6.15	105.9	687.83	6.26	115.9	761.83	6.36
7.6	86.6	557.08	6.15	96.6	633.08	6.29	106.6	709.08	6.41	116.6	785.08	6.51
7.8	87.3	574.47	6.28	97.3	652.47	6.43	107.3	730.47	6.55	117.3	808.47	6.66
8.0	88.0	592.00	6.42	98.0	672.00	6.57	108.0	752.00	6.70	118.0	832.00	6.81
8.2	88.7	609.67	6.55	98.7	691.67	6.71	108.7	773.67	6.84	118.7	855.67	6.95
8.4	89.4	627.48	6.69	99.4	711.48	6.85	109.4	795.48	6.99	119.4	879.48	7.10
8.6	90.1	645.43	6.82	100.1	731.43	6.99	110.1	817.43	7.13	120.1	903.43	7.25
8.8	90.8	663.52	6.95	100.8	751.52	7.13	110.8	839.52	7.27	120.8	927.52	7.39
9.0	91.5	681.75	7.08	101.5	771.75	7.26	111.5	861.75	7.41	121.5	951.75	7.54
9.2	92.2	700.12	7.21	102.2	792.12	7.40	112.2	884.12	7.55	122.2	976.12	7.68
9.4	92.9	718.63	7.34	102.9	812.63	7.53	112.9	906.63	7.69	122.9	1,000.63	7.82
9.6	93.6	737.28	7.47	103.6	833.28	7.67	113.6	929.28	7.83	123.6	1,025.28	7.97
9.8	94.3	756.07	7.60	104.3	854.07	7.80	114.3	952.07	7.97	124.3	1,050.07	8.11

Table 32.—*Area in square feet, A, top width in feet, T, and hydraulic radius in feet, r, of trapezoidal channels,* **side slopes 1¾ to 1**—Continued

Depth	Bottom width 60 feet			Bottom width 70 feet			Bottom width 80 feet			Bottom width 90 feet		
	T	A	r	T	A	r	T	A	r	T	A	r
10. 0	95. 00	775. 00	7. 73	105. 00	875. 00	7. 93	115. 00	975. 00	8. 10	125. 00	1,075.00	8. 25
10. 5	96. 75	822. 94	8. 04	106. 75	927. 94	8. 26	116. 75	1,032.94	8. 44	126. 75	1,137.94	8. 60
11. 0	98. 50	871. 8	8. 35	108. 50	981. 8	8. 59	118. 50	1,091. 8	8. 78	128. 50	1, 201. 8	8. 95
11. 5	100. 25	921. 4	8. 66	110. 25	1, 036. 4	8. 91	120. 25	1, 151. 4	9. 11	130. 25	1, 266. 4	9. 29
12. 0	102. 00	972. 0	8. 97	112. 00	1, 092. 0	9. 23	122. 00	1, 212. 0	9. 44	132. 00	1, 332. 0	9. 63
12. 5	103. 75	1, 023. 4	9. 27	113. 75	1, 148. 4	9. 54	123. 75	1, 273. 4	9. 77	133. 75	1, 398. 4	9. 96
13. 0	105. 50	1, 075. 8	9. 57	115. 50	1, 205. 8	9. 85	125. 50	1, 335. 8	10. 09	135. 50	1, 465. 8	10. 29
13. 5	107. 25	1, 128. 9	9. 87	117. 25	1, 263. 9	10. 16	127. 25	1, 398. 9	10. 41	137. 25	1, 533. 9	10. 62
14. 0	109. 00	1, 183. 0	10. 16	119. 00	1, 323. 0	10. 46	129. 00	1, 463. 0	10. 72	139. 00	1, 603. 0	10. 95
14. 5	110. 75	1, 237. 9	10. 45	120. 75	1, 382. 9	10. 77	130. 75	1, 527. 9	11. 04	140. 75	1, 672. 9	11. 27
15. 0	112. 50	1, 293. 8	10. 74	122. 50	1, 443. 8	11. 07	132. 50	1, 593. 8	11. 35	142. 50	1, 743. 8	11. 59
15. 5	114. 25	1, 350. 4	11. 03	124. 25	1, 505. 4	11. 36	134. 25	1, 660. 4	11. 65	144. 25	1, 815. 4	11. 91
16. 0	116. 00	1, 408. 0	11. 31	126. 00	1, 568. 0	11. 66	136. 00	1, 728. 0	11. 96	146. 00	1, 888. 0	12. 22
16. 5	117. 75	1, 466. 4	11. 59	127. 75	1, 631. 4	11. 95	137. 75	1, 796. 4	12. 26	147. 75	1, 961. 4	12. 53
17. 0	119. 50	1, 525. 8	11. 87	129. 50	1, 695. 8	12. 24	139. 50	1, 865. 8	12. 56	149. 50	2, 035. 8	12. 84
17. 5	121. 25	1, 585. 9	12. 15	131. 25	1, 760. 9	12. 53	141. 25	1, 935. 9	12. 86	151. 25	2, 110. 9	13. 15
18. 0	123. 00	1, 647. 0	12. 42	133. 00	1, 827. 0	12. 82	143. 00	2, 007. 0	13. 16	153. 00	2, 187. 0	13. 45
18. 5	124. 75	1, 708. 9	12. 70	134. 75	1, 893. 9	13. 10	144. 75	2, 078. 9	13. 45	154. 75	2, 263. 9	13. 76
19. 0	126. 50	1, 771. 8	12. 97	136. 50	1, 961. 8	13. 38	146. 50	2, 151. 8	13. 74	156. 50	2, 341. 8	14. 06
19. 5	128. 25	1, 835. 4	13. 24	138. 25	2, 030. 4	13. 66	148. 25	2, 225. 4	14. 03	158. 25	2, 420. 4	14. 36
20. 0	130. 00	1, 900. 0	13. 51	140. 00	2, 100. 0	13. 94	150. 00	2, 300. 0	14. 32	160. 00	2, 500. 0	14. 65

Table 33.—*Area in square feet, A, top width in feet, T, and hydraulic radius in feet, r, of trapezoidal channels,*
side slopes 2 to 1

Depth	Bottom width 2 feet			Bottom width 3 feet			Bottom width 4 feet			Bottom width 5 feet		
	T	A	r	T	A	r	T	A	r	T	A	r
0.4	3.6	1.12	.30	4.6	1.52	.32	5.6	1.92	.33	6.6	2.32	.34
0.6	4.4	1.92	.41	5.4	2.52	.44	6.4	3.12	.47	7.4	3.72	.48
0.8	5.2	2.88	.52	6.2	3.68	.56	7.2	4.48	.59	8.2	5.28	.62
1.0	6.0	4.00	.62	7.0	5.00	.67	8.0	6.00	.71	9.0	7.00	.74
1.2	6.8	5.28	.72	7.8	6.48	.77	8.8	7.68	.82	9.8	8.88	.86
1.4	7.6	6.72	.81	8.6	8.12	.88	9.6	9.52	.93	10.6	10.92	.97
1.6	8.4	8.32	.91	9.4	9.92	.98	10.4	11.52	1.03	11.4	13.12	1.08
1.8	9.2	10.08	1.00	10.2	11.88	1.08	11.2	13.68	1.14	12.2	15.48	1.19
2.0	10.0	12.00	1.10	11.0	14.00	1.17	12.0	16.00	1.24	13.0	18.00	1.29
2.2	10.8	14.08	1.19	11.8	16.28	1.27	12.8	18.48	1.34	13.8	20.68	1.39
2.4	11.6	16.32	1.28	12.6	18.72	1.36	13.6	21.12	1.43	14.6	23.52	1.49
2.6	12.4	18.72	1.37	13.4	21.32	1.46	14.4	23.92	1.53	15.4	26.52	1.59
2.8	13.2	21.28	1.47	14.2	24.08	1.55	15.2	26.88	1.63	16.2	29.68	1.69
3.0	14.0	24.00	1.56	15.0	27.00	1.64	16.0	30.00	1.72	17.0	33.00	1.79
3.2	14.8	26.88	1.65	15.8	30.08	1.74	16.8	33.28	1.82	17.8	36.48	1.89
3.4	15.6	29.92	1.74	16.6	33.32	1.83	17.6	36.72	1.91	18.6	40.12	1.99
3.6	16.4	33.12	1.83	17.4	36.72	1.92	18.4	40.32	2.01	19.4	43.92	2.08
3.8	17.2	36.48	1.92	18.2	40.28	2.01	19.2	44.08	2.10	20.2	47.88	2.18
4.0	18.0	40.00	2.01	19.0	44.00	2.11	20.0	48.00	2.19	21.0	52.00	2.27
4.2	18.8	43.68	2.10	19.8	47.88	2.20	20.8	52.08	2.29	21.8	56.28	2.37
4.4	19.6	47.52	2.19	20.6	51.92	2.29	21.6	56.32	2.38	22.6	60.72	2.46
4.6	20.4	51.52	2.28	21.4	56.12	2.38	22.4	60.72	2.47	23.4	65.32	2.55
4.8	21.2	55.68	2.37	22.2	60.48	2.47	23.2	65.28	2.56	24.2	70.08	2.65
5.0	22.0	60.00	2.46	23.0	65.00	2.56	24.0	70.00	2.66	25.0	75.00	2.74
5.2	22.8	64.48	2.55	23.8	69.68	2.65	24.8	74.88	2.75	25.8	80.08	2.83
5.4	23.6	69.12	2.64	24.6	74.52	2.74	25.6	79.92	2.84	26.6	85.32	2.93
5.6	24.4	73.92	2.73	25.4	79.52	2.84	26.4	85.12	2.93	27.4	90.72	3.02
5.8	25.2	78.88	2.82	26.2	84.68	2.93	27.2	90.48	3.02	28.2	96.28	3.11
6.0	26.0	84.00	2.91	27.0	90.00	3.02	28.0	96.00	3.11	29.0	102.00	3.20
6.2	26.8	89.28	3.00	27.8	95.48	3.11	28.8	101.68	3.20	29.8	107.88	3.30
6.4	27.6	94.72	3.09	28.6	101.12	3.20	29.6	107.52	3.30	30.6	113.92	3.39
6.6	28.4	100.32	3.18	29.4	106.92	3.29	30.4	113.52	3.39	31.4	120.12	3.48
6.8	29.2	106.08	3.27	30.2	112.88	3.38	31.2	119.68	3.48	32.2	126.48	3.57
7.0	30.0	112.00	3.36	31.0	119.00	3.47	32.0	126.00	3.57	33.0	133.00	3.66
7.2	30.8	118.08	3.45	31.8	125.28	3.56	32.8	132.48	3.66	33.8	139.68	3.75
7.4	31.6	124.32	3.54	32.6	131.72	3.65	33.6	139.12	3.75	34.6	146.52	3.85
7.6	32.4	130.72	3.63	33.4	138.32	3.74	34.4	145.92	3.84	35.4	153.52	3.94
7.8	33.2	137.28	3.72	34.2	145.08	3.83	35.2	152.88	3.93	36.2	160.68	4.03
8.0	34.0	144.00	3.81	35.0	152.00	3.92	36.0	160.00	4.02	37.0	168.00	4.12
8.2	----	------	----	35.8	159.08	4.01	36.8	167.28	4.11	37.8	175.48	4.21
8.4	----	------	----	36.6	166.32	4.10	37.6	174.72	4.20	38.6	183.12	4.30
8.6	----	------	----	37.4	173.72	4.19	38.4	182.32	4.29	39.4	190.92	4.39
8.8	----	------	----	38.2	181.28	4.28	39.2	190.08	4.38	40.2	198.88	4.48
9.0	----	------	----	39.0	189.00	4.37	40.0	198.00	4.47	41.0	207.00	4.57
9.2	----	------	----	39.8	196.88	4.46	40.8	206.08	4.56	41.8	215.28	4.67
9.4	----	------	----	40.6	204.92	4.55	41.6	214.32	4.66	42.6	223.72	4.76
9.6	----	------	----	41.4	213.12	4.64	42.4	222.72	4.75	43.4	232.32	4.85
9.8	----	------	----	42.2	221.48	4.73	43.2	231.28	4.84	44.2	241.08	4.94

Table 33.—*Area in square feet, A, top width in feet, T, and hydraulic radius in feet, r, of trapezoidal channels,*
side slopes 2 to 1—Continued

Depth	Bottom width 2 feet			Bottom width 3 feet			Bottom width 4 feet			Bottom width 5 feet		
	T	A	r	T	A	r	T	A	r	T	A	r
10.0	----	------	----	43.0	230.00	4.82	44.0	240.00	4.93	45.0	250.00	5.03
10.5	----	------	----	45.0	252.00	5.04	46.0	262.50	5.15	47.0	273.00	5.25
11.0	----	------	----	47.0	275.0	5.27	48.0	286.0	5.38	49.0	297.0	5.48
11.5	----	------	----	49.0	299.0	5.49	50.0	310.5	5.60	51.0	322.0	5.71
12.0	----	------	----	51.0	324.0	5.72	52.0	336.0	5.83	53.0	348.0	5.93
12.5	----	------	----	53.0	350.0	5.94	54.0	362.5	6.05	55.0	375.0	6.16
13.0				----	------	----	56.0	390.0	6.28	57.0	403.0	6.38
13.5				----	------	----	58.0	418.5	6.50	59.0	432.0	6.61
14.0				----	------	----	60.0	448.0	6.73	61.0	462.0	6.83
14.5				----	------	----	62.0	478.5	6.95	63.0	493.0	7.06
15.0				----	------	----	64.0	510.0	7.17	65.0	525.0	7.28
15.5				----	------	----	66.0	542.5	7.40	67.0	558.0	7.51
16.0				----	------	----	68.0	576.0	7.62	69.0	592.0	7.73
16.5				----	------	----	----	------	----	71.0	627.0	7.96
17.0				----	------	----	----	------	----	73.0	663.0	8.18
17.5				----	------	----	----	------	----	75.0	700.0	8.41
18.0				----	------	----	----	------	----	77.0	738.0	8.63
18.5				----	------	----	----	------	----	79.0	777.0	8.86
19.0				----	------	----	----	------	----	81.0	817.0	9.08
19.5				----	------	----	----	------	----	83.0	858.0	9.31
20.0				----	------	----	----	------	----	85.0	900.0	9.53

Depth	Bottom width 6 feet			Bottom width 7 feet			Bottom width 8 feet			Bottom width 9 feet		
	T	A	r	T	A	r	T	A	r	T	A	r
0.4	7.6	2.72	0.35	8.6	3.12	0.35	9.6	3.52	0.36	10.6	3.92	0.36
0.6	8.4	4.32	.50	9.4	4.92	.51	10.4	5.52	.52	11.4	6.12	.52
0.8	9.2	6.08	.63	10.2	6.88	.65	11.2	7.68	.66	12.2	8.48	.67
1.0	10.0	8.00	.76	11.0	9.00	.78	12.0	10.00	.80	13.0	11.00	.82
1.2	10.8	10.08	.89	11.8	11.28	.91	12.8	12.48	.93	13.8	13.68	.95
1.4	11.6	12.32	1.00	12.6	13.72	1.03	13.6	15.12	1.06	14.6	16.52	1.08
1.6	12.4	14.72	1.12	13.4	16.32	1.15	14.4	17.92	1.18	15.4	19.52	1.21
1.8	13.2	17.28	1.23	14.2	19.08	1.27	15.2	20.88	1.30	16.2	22.68	1.33
2.0	14.0	20.00	1.34	15.0	22.00	1.38	16.0	24.00	1.42	17.0	26.00	1.45
2.2	14.8	22.88	1.44	15.8	25.08	1.49	16.8	27.28	1.53	17.8	29.48	1.56
2.4	15.6	25.92	1.55	16.6	28.32	1.60	17.6	30.72	1.64	18.6	33.12	1.68
2.6	16.4	29.12	1.65	17.4	31.72	1.70	18.4	34.32	1.75	19.4	36.92	1.79
2.8	17.2	32.48	1.75	18.2	35.28	1.81	19.2	38.08	1.86	20.2	40.88	1.90
3.0	18.0	36.00	1.85	19.0	39.00	1.91	20.0	42.00	1.96	21.0	45.00	2.01
3.2	18.8	39.68	1.95	19.8	42.88	2.01	20.8	46.08	2.07	21.8	49.28	2.11
3.4	19.6	43.52	2.05	20.6	46.92	2.11	21.6	50.32	2.17	22.6	53.72	2.22
3.6	20.4	47.52	2.15	21.4	51.12	2.21	22.4	54.72	2.27	23.4	58.32	2.32
3.8	21.2	51.68	2.25	22.2	55.48	2.31	23.2	59.28	2.37	24.2	63.08	2.43
4.0	22.0	56.00	2.34	23.0	60.00	2.41	24.0	64.00	2.47	25.0	68.00	2.53
4.2	22.8	60.48	2.44	23.8	64.68	2.51	24.8	68.88	2.57	25.8	73.08	2.63
4.4	23.6	65.12	2.54	24.6	69.52	2.61	25.6	73.92	2.67	26.6	78.32	2.73
4.6	24.4	69.92	2.63	25.4	74.52	2.70	26.4	79.12	2.77	27.4	83.72	2.83
4.8	25.2	74.88	2.73	26.2	79.68	2.80	27.2	84.48	2.87	28.2	89.28	2.93

Table 33.—*Area in square feet, A, top width in feet, T, and hydraulic radius in feet, r, of trapezoidal channels,* **side slopes 2 to 1**—Continued

Depth	Bottom width 6 feet			Bottom width 7 feet			Bottom width 8 feet			Bottom width 9 feet		
	T	A	r	T	A	r	T	A	r	T	A	r
5.0	26.0	80.00	2.82	27.0	85.00	2.90	28.0	90.00	2.96	29.0	95.00	3.03
5.2	26.8	85.28	2.92	27.8	90.48	2.99	28.8	95.68	3.06	29.8	100.88	3.13
5.4	27.6	90.72	3.01	28.6	96.12	3.09	29.6	101.52	3.16	30.6	106.92	3.23
5.6	28.4	96.32	3.10	29.4	101.92	3.18	30.4	107.52	3.25	31.4	113.12	3.32
5.8	29.2	102.08	3.20	30.2	107.88	3.28	31.2	113.68	3.35	32.2	119.48	3.42
6.0	30.0	108.00	3.29	31.0	114.00	3.37	32.0	120.00	3.45	33.0	126.00	3.52
6.2	30.8	114.08	3.38	31.8	120.28	3.46	32.8	126.48	3.54	33.8	132.68	3.61
6.4	31.6	120.32	3.48	32.6	126.72	3.56	33.6	133.12	3.64	34.6	139.52	3.71
6.6	32.4	126.72	3.57	33.4	133.32	3.65	34.4	139.92	3.73	35.4	146.52	3.80
6.8	33.2	133.28	3.66	34.2	140.08	3.74	35.2	146.88	3.82	36.2	153.68	3.90
7.0	34.0	140.00	3.75	35.0	147.00	3.84	36.0	154.00	3.92	37.0	161.00	3.99
7.2	34.8	146.88	3.85	35.8	154.08	3.93	36.8	161.28	4.01	37.8	168.48	4.09
7.4	35.6	153.92	3.94	36.6	161.32	4.02	37.6	168.72	4.11	38.6	176.12	4.18
7.6	36.4	161.12	4.03	37.4	168.72	4.12	38.4	176.32	4.20	39.4	183.92	4.28
7.8	37.2	168.48	4.12	38.2	176.28	4.21	39.2	184.08	4.29	40.2	191.88	4.37
8.0	38.0	176.00	4.21	39.0	184.00	4.30	40.0	192.00	4.39	41.0	200.00	4.47
8.2	38.8	183.68	4.30	39.8	191.88	4.39	40.8	200.08	4.48	41.8	208.28	4.56
8.4	39.6	191.52	4.40	40.6	199.92	4.49	41.6	208.32	4.57	42.6	216.72	4.65
8.6	40.4	199.52	4.49	41.4	208.12	4.58	42.4	216.72	4.66	43.4	225.32	4.75
8.8	41.2	207.68	4.58	42.2	216.48	4.67	43.2	225.28	4.76	44.2	234.08	4.84
9.0	42.0	216.00	4.67	43.0	225.00	4.76	44.0	234.00	4.85	45.0	243.00	4.93
9.2	42.8	224.48	4.76	43.8	233.68	4.85	44.8	242.88	4.94	45.8	252.08	5.03
9.4	43.6	233.12	4.85	44.6	242.52	4.95	45.6	251.92	5.03	46.6	261.32	5.12
9.6	44.4	241.92	4.94	45.4	251.52	5.04	46.4	261.12	5.13	47.4	270.72	5.21
9.8	45.2	250.88	5.04	46.2	260.68	5.13	47.2	270.48	5.22	48.2	280.28	5.31
10.0	46.0	260.00	5.13	47.0	270.00	5.22	48.0	280.00	5.31	49.0	290.00	5.40
10.5	48.0	283.50	5.35	49.0	294.00	5.45	50.0	304.50	5.54	51.0	315.00	5.63
11.0	50.0	308.0	5.58	51.0	319.0	5.68	52.0	330.0	5.77	53.0	341.0	5.86
11.5	52.0	333.5	5.81	53.0	345.0	5.90	54.0	356.5	6.00	55.0	368.0	6.09
12.0	54.0	360.0	6.03	55.0	372.0	6.13	56.0	384.0	6.23	57.0	396.0	6.32
12.5	56.0	387.5	6.26	57.0	400.0	6.36	58.0	412.5	6.46	59.0	425.0	6.55
13.0	58.0	416.0	6.49	59.0	429.0	6.59	60.0	442.0	6.68	61.0	455.0	6.78
13.5	60.0	445.5	6.71	61.0	459.0	6.81	62.0	472.5	6.91	63.0	486.0	7.01
14.0	62.0	476.0	6.94	63.0	490.0	7.04	64.0	504.0	7.14	65.0	518.0	7.23
14.5	64.0	507.5	7.16	65.0	522.0	7.27	66.0	536.5	7.36	67.0	551.0	7.46
15.0	66.0	540.0	7.39	67.0	555.0	7.49	68.0	570.0	7.59	69.0	585.0	7.69
15.5	68.0	573.5	7.61	69.0	589.0	7.72	70.0	604.5	7.82	71.0	620.0	7.92
16.0	70.0	608.0	7.84	71.0	624.0	7.94	72.0	640.0	8.04	73.0	656.0	8.14
16.5	72.0	643.5	8.06	73.0	660.0	8.17	74.0	676.5	8.27	75.0	693.0	8.37
17.0	74.0	680.0	8.29	75.0	697.0	8.39	76.0	714.0	8.50	77.0	731.0	8.60
17.5	76.0	717.5	8.52	77.0	735.0	8.62	78.0	752.5	8.72	79.0	770.0	8.82
18.0	78.0	756.0	8.74	79.0	774.0	8.85	80.0	792.0	8.95	81.0	810.0	9.05
18.5	80.0	795.5	8.96	81.0	814.0	9.07	82.0	832.5	9.18	83.0	851.0	9.28
19.0	82.0	836.0	9.19	83.0	855.0	9.30	84.0	874.0	9.40	85.0	893.0	9.50
19.5	84.0	877.5	9.41	85.0	897.0	9.52	86.0	916.5	9.63	87.0	936.0	9.73
20.0	86.0	920.0	9.64	87.0	940.0	9.75	88.0	960.0	9.85	89.0	980.0	9.96

Table 33.—*Area in square feet, A, top width in feet, T, and hydraulic radius in feet, r, of trapezoidal channels,*
side slopes 2 to 1—Continued

Depth	Bottom width 10 feet			Bottom width 12 feet			Bottom width 14 feet			Bottom width 16 feet		
	T	A	r	T	A	r	T	A	r	T	A	r
0.4	11.6	4.32	0.37	13.6	5.12	0.37	15.6	5.92	0.37	17.6	6.72	0.38
0.6	12.4	6.72	.53	14.4	7.92	.54	16.4	9.12	.55	18.4	10.32	.55
0.8	13.2	9.28	.68	15.2	10.88	.70	17.2	12.48	.71	19.2	14.08	.72
1.0	14.0	12.00	.83	16.0	14.00	.85	18.0	16.00	.87	20.0	18.00	.88
1.2	14.8	14.88	.97	16.8	17.28	1.00	18.8	19.68	1.02	20.8	22.08	1.03
1.4	15.6	17.92	1.10	17.6	20.72	1.13	19.6	23.52	1.16	21.6	26.32	1.18
1.6	16.4	21.12	1.23	18.4	24.32	1.27	20.4	27.52	1.30	22.4	30.72	1.33
1.8	17.2	24.48	1.36	19.2	28.08	1.40	21.2	31.68	1.44	23.2	35.28	1.47
2.0	18.0	28.00	1.48	20.0	32.00	1.53	22.0	36.00	1.57	24.0	40.00	1.60
2.2	18.8	31.68	1.60	20.8	36.08	1.65	22.8	40.48	1.70	24.8	44.88	1.74
2.4	19.6	35.52	1.71	21.6	40.32	1.77	23.6	45.12	1.82	25.6	49.92	1.87
2.6	20.4	39.52	1.83	22.4	44.72	1.89	24.4	49.92	1.95	26.4	55.12	2.00
2.8	21.2	43.68	1.94	23.2	49.28	2.01	25.2	54.88	2.07	27.2	60.48	2.12
3.0	22.0	48.00	2.05	24.0	54.00	2.12	26.0	60.00	2.19	28.0	66.00	2.24
3.2	22.8	52.48	2.16	24.8	58.88	2.24	26.8	65.28	2.31	28.8	71.68	2.36
3.4	23.6	57.12	2.27	25.6	63.92	2.35	27.6	70.72	2.42	29.6	77.52	2.48
3.6	24.4	61.92	2.37	26.4	69.12	2.46	28.4	76.32	2.54	30.4	83.52	2.60
3.8	25.2	66.88	2.48	27.2	74.48	2.57	29.2	82.08	2.65	31.2	89.68	2.72
4.0	26.0	72.00	2.58	28.0	80.00	2.68	30.0	88.00	2.76	32.0	96.00	2.83
4.2	26.8	77.28	2.68	28.8	85.68	2.78	30.8	94.08	2.87	32.8	102.48	2.95
4.4	27.6	82.72	2.79	29.6	91.52	2.89	31.6	100.32	2.98	33.6	109.12	3.06
4.6	28.4	88.32	2.89	30.4	97.52	2.99	32.4	106.72	3.09	34.4	115.92	3.17
4.8	29.2	94.08	2.99	31.2	103.68	3.10	33.2	113.28	3.19	35.2	122.88	3.28
5.0	30.0	100.00	3.09	32.0	110.00	3.20	34.0	120.00	3.30	36.0	130.00	3.39
5.2	30.8	106.08	3.19	32.8	116.48	3.30	34.8	126.88	3.41	36.8	137.28	3.50
5.4	31.6	112.32	3.29	33.6	123.12	3.41	35.6	133.92	3.51	37.6	144.72	3.60
5.6	32.4	118.72	3.39	34.4	129.92	3.51	36.4	141.12	3.61	38.4	152.32	3.71
5.8	33.2	125.28	3.49	35.2	136.88	3.61	37.2	148.48	3.72	39.2	160.08	3.82
6.0	34.0	132.00	3.58	36.0	144.00	3.71	38.0	156.00	3.82	40.0	168.00	3.92
6.2	34.8	138.88	3.68	36.8	151.28	3.81	38.8	163.68	3.92	40.8	176.08	4.03
6.4	35.6	145.92	3.78	37.6	158.72	3.91	39.6	171.52	4.02	41.6	184.32	4.13
6.6	36.4	153.12	3.87	38.4	166.32	4.01	40.4	179.52	4.13	42.4	192.72	4.23
6.8	37.2	160.48	3.97	39.2	174.08	4.10	41.2	187.68	4.23	43.2	201.28	4.34
7.0	38.0	168.00	4.07	40.0	182.00	4.20	42.0	196.00	4.33	44.0	210.00	4.44
7.2	38.8	175.68	4.16	40.8	190.08	4.30	42.8	204.48	4.43	44.8	218.88	4.54
7.4	39.6	183.52	4.26	41.6	198.32	4.40	43.6	213.12	4.53	45.6	227.92	4.64
7.6	40.4	191.52	4.35	42.4	206.72	4.50	44.4	221.92	4.62	46.4	237.12	4.74
7.8	41.2	199.68	4.45	43.2	215.28	4.59	45.2	230.88	4.72	47.2	246.48	4.84
8.0	42.0	208.00	4.54	44.0	224.00	4.69	46.0	240.00	4.82	48.0	256.00	4.94
8.2	42.8	216.48	4.64	44.8	232.88	4.78	46.8	249.28	4.92	48.8	265.68	5.04
8.4	43.6	225.12	4.73	45.6	241.92	4.88	47.6	258.72	5.02	49.6	275.52	5.14
8.6	44.4	233.92	4.83	46.4	251.12	4.98	48.4	268.32	5.11	50.4	285.52	5.24
8.8	45.2	242.88	4.92	47.2	260.48	5.07	49.2	278.08	5.21	51.2	295.68	5.34
9.0	46.0	252.00	5.02	48.0	270.00	5.17	50.0	288.00	5.31	52.0	306.00	5.44
9.2	46.8	261.28	5.11	48.8	279.68	5.26	50.8	298.08	5.41	52.8	316.48	5.54
9.4	47.6	270.72	5.20	49.6	289.52	5.36	51.6	308.32	5.50	53.6	327.12	5.64
9.6	48.4	280.32	5.30	50.4	299.52	5.45	52.4	318.72	5.60	54.4	337.92	5.73
9.8	49.2	290.08	5.39	51.2	309.68	5.55	53.2	329.28	5.69	55.2	348.88	5.83

Table 33.—*Area in square feet, A, top width in feet, T, and hydraulic radius in feet, r, of trapezoidal channels,* **side slopes 2 to 1**—Continued

Depth	Bottom width 10 feet			Bottom width 12 feet			Bottom width 14 feet			Bottom width 16 feet		
	T	A	r	T	A	r	T	A	r	T	A	r
10.0	50.0	300.00	5.48	52.0	320.00	5.64	54.0	340.00	5.79	56.0	360.00	5.93
10.5	52.0	325.50	5.71	54.0	346.50	5.88	56.0	367.50	6.03	58.0	388.50	6.17
11.0	54.0	352.0	5.95	56.0	374.0	6.11	58.0	396.0	6.27	60.0	418.0	6.41
11.5	56.0	379.5	6.18	58.0	402.5	6.35	60.0	425.5	6.50	62.0	448.5	6.65
12.0	58.0	408.0	6.41	60.0	432.0	6.58	62.0	456.0	6.74	64.0	480.0	6.89
12.5	60.0	437.5	6.64	62.0	462.5	6.81	64.0	487.5	6.97	66.0	512.5	7.13
13.0	62.0	468.0	6.87	64.0	494.0	7.04	66.0	520.0	7.21	68.0	546.0	7.36
13.5	64.0	499.5	7.10	66.0	526.5	7.27	68.0	553.5	7.44	70.0	580.5	7.60
14.0	66.0	532.0	7.33	68.0	560.0	7.51	70.0	588.0	7.68	72.0	616.0	7.84
14.5	68.0	565.5	7.56	70.0	594.5	7.74	72.0	623.5	7.91	74.0	652.5	8.07
15.0	70.0	600.0	7.78	72.0	630.0	7.97	74.0	660.0	8.14	76.0	690.0	8.31
15.5	72.0	635.5	8.01	74.0	666.5	8.20	76.0	697.5	8.37	78.0	728.5	8.54
16.0	74.0	672.0	8.24	76.0	704.0	8.43	78.0	736.0	8.60	80.0	768.0	8.77
16.5	76.0	709.5	8.47	78.0	742.5	8.65	80.0	775.5	8.83	82.0	808.5	9.00
17.0	78.0	748.0	8.70	80.0	782.0	8.88	82.0	816.0	9.06	84.0	850.0	9.24
17.5	80.0	787.5	8.92	82.0	822.5	9.11	84.0	857.5	9.29	86.0	892.5	9.47
18.0	82.0	828.0	9.15	84.0	864.0	9.34	86.0	900.0	9.52	88.0	936.0	9.70
18.5	84.0	869.5	9.38	86.0	906.5	9.57	88.0	943.5	9.75	90.0	980.5	9.93
19.0	86.0	912.0	9.60	88.0	950.0	9.80	90.0	988.0	9.98	92.0	1,026.0	10.16
19.5	88.0	955.5	9.83	90.0	994.5	10.02	92.0	1,033.5	10.21	94.0	1,072.5	10.39
20.0	90.0	1,000.0	10.06	92.0	1,040.0	10.25	94.0	1,080.0	10.44	96.0	1,120.0	10.62

Depth	Bottom width 18 feet			Bottom width 20 feet			Bottom width 22 feet			Bottom width 24 feet		
	T	A	r	T	A	r	T	A	r	T	A	r
0.4	19.6	7.52	.38	21.6	8.32	0.38	23.6	9.12	0.38	25.6	9.92	0.38
0.6	20.4	11.52	.56	22.4	12.72	.56	24.4	13.92	.56	26.4	15.12	.57
0.8	21.2	15.68	.73	23.2	17.28	.73	25.2	18.88	.74	27.2	20.48	.74
1.0	22.0	20.00	.89	24.0	22.00	.90	26.0	24.00	.91	28.0	26.00	.91
1.2	22.8	24.48	1.05	24.8	26.88	1.06	26.8	29.28	1.07	28.8	31.68	1.08
1.4	23.6	29.12	1.20	25.6	31.92	1.22	27.6	34.72	1.23	29.6	37.52	1.24
1.6	24.4	33.92	1.35	26.4	37.12	1.37	28.4	40.32	1.38	30.4	43.52	1.40
1.8	25.2	38.88	1.49	27.2	42.48	1.51	29.2	46.08	1.53	31.2	49.68	1.55
2.0	26.0	44.00	1.63	28.0	48.00	1.66	30.0	52.00	1.68	32.0	56.00	1.70
2.2	26.8	49.28	1.77	28.8	53.68	1.80	30.8	58.08	1.82	32.8	62.48	1.85
2.4	27.6	54.72	1.90	29.6	59.52	1.94	31.6	64.32	1.96	33.6	69.12	1.99
2.6	28.4	60.32	2.04	30.4	65.52	2.07	32.4	70.72	2.10	34.4	75.92	2.13
2.8	29.2	66.08	2.16	31.2	71.68	2.20	33.2	77.28	2.24	35.2	82.88	2.27
3.0	30.0	72.00	2.29	32.0	78.00	2.33	34.0	84.00	2.37	36.0	90.00	2.41
3.2	30.8	78.08	2.42	32.8	84.48	2.46	34.8	90.88	2.50	36.8	97.28	2.54
3.4	31.6	84.32	2.54	33.6	91.12	2.59	35.6	97.92	2.63	37.6	104.72	2.67
3.6	32.4	90.72	2.66	34.4	97.92	2.71	36.4	105.12	2.76	38.4	112.32	2.80
3.8	33.2	97.28	2.78	35.2	104.88	2.84	37.2	112.48	2.88	39.2	120.08	2.93
4.0	34.0	104.00	2.90	36.0	112.00	2.96	38.0	120.00	3.01	40.0	128.00	3.06
4.2	34.8	110.88	3.01	36.8	119.28	3.08	38.8	127.68	3.13	40.8	136.08	3.18
4.4	35.6	117.92	3.13	37.6	126.72	3.19	39.6	135.52	3.25	41.6	144.32	3.30
4.6	36.4	125.12	3.24	38.4	134.32	3.31	40.4	143.52	3.37	42.4	152.72	3.43
4.8	37.2	132.48	3.36	39.2	142.08	3.43	41.2	151.68	3.49	43.2	161.28	3.55

Table 33.—*Area in square feet, A, top width in feet, T, and hydraulic radius in feet, r, of trapezoidal channels,*
side slopes 2 to 1—Continued

Depth	Bottom width 18 feet			Bottom width 20 feet			Bottom width 22 feet			Bottom width 24 feet		
	T	A	r	T	A	r	T	A	r	T	A	r
5.0	38.0	140.00	3.47	40.0	150.00	3.54	42.0	160.00	3.61	44.0	170.00	3.67
5.2	38.8	147.68	3.58	40.8	158.08	3.65	42.8	168.48	3.72	44.8	178.88	3.79
5.4	39.6	155.52	3.69	41.6	166.32	3.77	43.6	177.12	3.84	45.6	187.92	3.90
5.6	40.4	163.52	3.80	42.4	174.72	3.88	44.4	185.92	3.95	46.4	197.12	4.02
5.8	41.2	171.68	3.91	43.2	183.28	3.99	45.2	194.88	4.07	47.2	206.48	4.13
6.0	42.0	180.00	4.01	44.0	192.00	4.10	46.0	204.00	4.18	48.0	216.00	4.25
6.2	42.8	188.48	4.12	44.8	200.88	4.21	46.8	213.28	4.29	48.8	225.68	4.36
6.4	43.6	197.12	4.23	45.6	209.92	4.32	47.6	222.72	4.40	49.6	235.52	4.48
6.6	44.4	205.92	4.33	46.4	219.12	4.43	48.4	232.32	4.51	50.4	245.52	4.59
6.8	45.2	214.88	4.44	47.2	228.48	4.53	49.2	242.08	4.62	51.2	255.68	4.70
7.0	46.0	224.00	4.54	48.0	238.00	4.64	50.0	252.00	4.73	52.0	266.00	4.81
7.2	46.8	233.28	4.65	48.8	247.68	4.74	50.8	262.08	4.84	52.8	276.48	4.92
7.4	47.6	242.72	4.75	49.6	257.52	4.85	51.6	272.32	4.94	53.6	287.12	5.03
7.6	48.4	252.32	4.85	50.4	267.52	4.96	52.4	282.72	5.05	54.4	297.92	5.14
7.8	49.2	262.08	4.96	51.2	277.68	5.05	53.2	293.28	5.16	55.2	308.88	5.25
8.0	50.0	272.00	5.06	52.0	288.00	5.16	54.0	304.00	5.26	56.0	320.00	5.35
8.2	50.8	282.08	5.16	52.8	298.48	5.27	54.8	314.88	5.37	56.8	331.28	5.46
8.4	51.6	292.32	5.26	53.6	309.12	5.37	55.6	325.92	5.47	57.6	342.72	5.57
8.6	52.4	302.72	5.36	54.4	319.92	5.47	56.4	337.12	5.58	58.4	354.32	5.67
8.8	53.2	313.28	5.46	55.2	330.88	5.57	57.2	348.48	5.68	59.2	366.08	5.78
9.0	54.0	324.00	5.56	56.0	342.00	5.68	58.0	360.00	5.78	60.0	378.00	5.88
9.2	54.8	334.88	5.66	56.8	353.28	5.78	58.8	371.68	5.89	60.8	390.08	5.99
9.4	55.6	345.92	5.76	57.6	364.72	5.88	59.6	383.52	5.99	61.6	402.32	6.09
9.6	56.4	357.12	5.86	58.4	376.32	5.98	60.4	395.52	6.09	62.4	414.72	6.20
9.8	57.2	368.48	5.96	59.2	388.08	6.08	61.2	407.68	6.19	63.2	427.28	6.30
10.0	58.0	380.00	6.06	60.0	400.00	6.18	62.0	420.00	6.29	64.0	440.00	6.40
10.5	60.0	409.50	6.30	62.0	430.50	6.43	64.0	451.50	6.55	66.0	472.50	6.66
11.0	62.0	440.0	6.55	64.0	462.0	6.68	66.0	484.0	6.80	68.0	506.0	6.91
11.5	64.0	471.5	6.79	66.0	494.5	6.92	68.0	517.5	7.05	70.0	540.5	7.17
12.0	66.0	504.0	7.03	68.0	528.0	7.17	70.0	552.0	7.30	72.0	576.0	7.42
12.5	68.0	537.5	7.27	70.0	562.5	7.41	72.0	587.5	7.54	74.0	612.5	7.67
13.0	70.0	572.0	7.51	72.0	598.0	7.65	74.0	624.0	7.79	76.0	650.0	7.91
13.5	72.0	607.5	7.75	74.0	634.5	7.89	76.0	661.5	8.03	78.0	688.5	8.16
14.0	74.0	644.0	7.99	76.0	672.0	8.13	78.0	700.0	8.27	80.0	728.0	8.41
14.5	76.0	681.5	8.23	78.0	710.5	8.37	80.0	739.5	8.52	82.0	768.5	8.65
15.0	78.0	720.0	8.46	80.0	750.0	8.61	82.0	780.0	8.76	84.0	810.0	8.89
15.5	80.0	759.5	8.70	82.0	790.5	8.85	84.0	821.5	9.00	86.0	852.5	9.14
16.0	82.0	800.0	8.93	84.0	832.0	9.09	86.0	864.0	9.24	88.0	896.0	9.38
16.5	84.0	841.5	9.17	86.0	874.5	9.32	88.0	907.5	9.47	90.0	940.5	9.62
17.0	86.0	884.0	9.40	88.0	918.0	9.56	90.0	952.0	9.71	92.0	986.0	9.86
17.5	88.0	927.5	9.64	90.0	962.5	9.80	92.0	997.5	9.95	94.0	1,032.5	10.10
18.0	90.0	972.0	9.87	92.0	1,008.0	10.03	94.0	1,044.0	10.19	96.0	1,080.0	10.34
18.5	92.0	1,017.5	10.10	94.0	1,054.5	10.26	96.0	1,091.5	10.42	98.0	1,128.5	10.57
19.0	94.0	1,064.0	10.33	96.0	1,102.0	10.50	98.0	1,140.0	10.66	100.0	1,178.0	10.81
19.5	96.0	1,111.5	10.56	98.0	1,150.5	10.73	100.0	1,189.5	10.89	102.0	1,228.5	11.05
20.0	98.0	1,160.0	10.80	100.0	1,200.0	10.96	102.0	1,240.0	11.13	104.0	1,280.0	11.28

Table 33.—*Area in square feet, A, top width in feet, T, and hydraulic radius in feet, r, of trapezoidal channels,* **side slopes 2 to 1**—Continued

Depth	Bottom width 26 feet			Bottom width 28 feet			Bottom width 30 feet			Bottom width 32 feet		
	T	A	r	T	A	r	T	A	r	T	A	r
0.4	27.6	10.72	0.39	29.6	11.52	0.39	31.6	12.32	0.39	33.6	13.12	0.39
0.6	28.4	16.32	.57	30.4	17.52	.57	32.4	18.72	.57	34.4	19.92	.57
0.8	29.2	22.08	.75	31.2	23.68	.75	33.2	25.28	.75	35.2	26.88	.76
1.0	30.0	28.00	.92	32.0	30.00	.92	34.0	32.00	.93	36.0	34.00	.93
1.2	30.8	34.08	1.09	32.8	36.48	1.09	34.8	38.88	1.10	36.8	41.28	1.10
1.4	31.6	40.32	1.25	33.6	43.12	1.26	35.6	45.92	1.27	37.6	48.72	1.27
1.6	32.4	46.72	1.41	34.4	49.92	1.42	36.4	53.12	1.43	38.4	56.32	1.44
1.8	33.2	53.28	1.56	35.2	56.88	1.58	37.2	60.48	1.59	39.2	64.08	1.60
2.0	34.0	60.00	1.72	36.0	64.00	1.73	38.0	68.00	1.75	40.0	72.00	1.76
2.2	34.8	66.88	1.87	36.8	71.28	1.88	38.8	75.68	1.90	40.8	80.08	1.91
2.4	35.6	73.92	2.01	37.6	78.72	2.03	39.6	83.52	2.05	41.6	88.32	2.07
2.6	36.4	81.12	2.16	38.4	86.32	2.18	40.4	91.52	2.20	42.4	96.72	2.22
2.8	37.2	88.48	2.30	39.2	94.08	2.32	41.2	99.68	2.34	43.2	105.28	2.36
3.0	38.0	96.00	2.44	40.0	102.00	2.46	42.0	108.00	2.49	44.0	114.00	2.51
3.2	38.8	103.68	2.57	40.8	110.08	2.60	42.8	116.48	2.63	44.8	122.88	2.65
3.4	39.6	111.52	2.71	41.6	118.32	2.74	43.6	125.12	2.77	45.6	131.92	2.79
3.6	40.4	119.52	2.84	42.4	126.72	2.87	44.4	133.92	2.91	46.4	141.12	2.93
3.8	41.2	127.68	2.97	43.2	135.28	3.01	45.2	142.88	3.04	47.2	150.48	3.07
4.0	42.0	136.00	3.10	44.0	144.00	3.14	46.0	152.00	3.17	48.0	160.00	3.21
4.2	42.8	144.48	3.23	44.8	152.88	3.27	46.8	161.28	3.31	48.8	169.68	3.34
4.4	43.6	153.12	3.35	45.6	161.92	3.40	47.6	170.72	3.44	49.6	179.52	3.47
4.6	44.4	161.92	3.48	46.4	171.12	3.52	48.4	180.32	3.57	50.4	189.52	3.60
4.8	45.2	170.88	3.60	47.2	180.48	3.65	49.2	190.08	3.69	51.2	199.68	3.73
5.0	46.0	180.00	3.72	48.0	190.00	3.77	50.0	200.00	3.82	52.0	210.00	3.86
5.2	46.8	189.28	3.84	48.8	199.68	3.90	50.8	210.08	3.94	52.8	220.48	3.99
5.4	47.6	198.72	3.96	49.6	209.52	4.02	51.6	220.32	4.07	53.6	231.12	4.12
5.6	48.4	208.32	4.08	50.4	219.52	4.14	52.4	230.72	4.19	54.4	241.92	4.24
5.8	49.2	218.08	4.20	51.2	229.68	4.26	53.2	241.28	4.31	55.2	252.88	4.36
6.0	50.0	228.00	4.32	52.0	240.00	4.38	54.0	252.00	4.43	56.0	264.00	4.49
6.2	50.8	238.08	4.43	52.8	250.48	4.49	54.8	262.88	4.55	56.8	275.28	4.61
6.4	51.6	248.32	4.55	53.6	261.12	4.61	55.6	273.92	4.67	57.6	286.72	4.73
6.6	52.4	258.72	4.66	54.4	271.92	4.73	56.4	285.12	4.79	58.4	298.32	4.85
6.8	53.2	269.28	4.77	55.2	282.88	4.84	57.2	296.48	4.91	59.2	310.08	4.97
7.0	54.0	280.00	4.89	56.0	294.00	4.96	58.0	308.00	5.02	60.0	322.00	5.09
7.2	54.8	290.88	5.00	56.8	305.28	5.07	58.8	319.68	5.14	60.8	334.08	5.20
7.4	55.6	301.92	5.11	57.6	316.72	5.18	59.6	331.52	5.25	61.6	346.32	5.32
7.6	56.4	313.12	5.22	58.4	328.32	5.30	60.4	343.52	5.37	62.4	358.72	5.44
7.8	57.2	324.48	5.33	59.2	340.08	5.41	61.2	355.68	5.48	63.2	371.28	5.55
8.0	58.0	336.00	5.44	60.0	352.00	5.52	62.0	368.00	5.59	64.0	384.00	5.67
8.2	58.8	347.68	5.55	60.8	364.08	5.63	62.8	380.48	5.71	64.8	396.88	5.78
8.4	59.6	359.52	5.66	61.6	376.32	5.74	63.6	393.12	5.82	65.6	409.92	5.89
8.6	60.4	371.52	5.76	62.4	388.72	5.85	64.4	405.92	5.93	66.4	423.12	6.01
8.8	61.2	383.68	5.87	63.2	401.28	5.96	65.2	418.88	6.04	67.2	436.48	6.12
9.0	62.0	396.00	5.98	64.0	414.00	6.07	66.0	432.00	6.15	68.0	450.00	6.23
9.2	62.8	408.48	6.08	64.8	426.88	6.17	66.8	445.28	6.26	68.8	463.68	6.34
9.4	63.6	421.12	6.19	65.6	439.92	6.28	67.6	458.72	6.37	69.6	477.52	6.45
9.6	64.4	433.92	6.29	66.4	453.12	6.39	68.4	472.32	6.48	70.4	491.52	6.56
9.8	65.2	446.88	6.40	67.2	466.48	6.49	69.2	486.08	6.58	71.2	505.68	6.67

Table 33.—*Area in square feet, A, top width in feet, T, and hydraulic radius in feet, r, of trapezoidal channels,* **side slopes 2 to 1**—Continued

Depth	Bottom width 26 feet			Bottom width 28 feet			Bottom width 30 feet			Bottom width 32 feet		
	T	A	r	T	A	r	T	A	r	T	A	r
10.0	66.0	460.00	6.50	68.0	480.00	6.60	70.0	500.00	6.69	72.0	520.00	6.78
10.5	68.0	493.50	6.76	70.0	514.50	6.86	72.0	535.50	6.96	74.0	556.50	7.05
11.0	70.0	528.0	7.02	72.0	550.0	7.12	74.0	572.0	7.22	76.0	594.0	7.32
11.5	72.0	563.5	7.28	74.0	586.5	7.38	76.0	609.5	7.48	78.0	632.5	7.58
12.0	74.0	600.0	7.53	76.0	624.0	7.64	78.0	648.0	7.75	80.0	672.0	7.84
12.5	76.0	637.5	7.78	78.0	662.5	7.90	80.0	687.5	8.00	82.0	712.5	8.11
13.0	78.0	676.0	8.03	80.0	702.0	8.15	82.0	728.0	8.26	84.0	754.0	8.36
13.5	80.0	715.5	8.28	82.0	742.5	8.40	84.0	769.5	8.51	86.0	796.5	8.62
14.0	82.0	756.0	8.53	84.0	784.0	8.65	86.0	812.0	8.77	88.0	840.0	8.88
14.5	84.0	797.5	8.78	86.0	826.5	8.90	88.0	855.5	9.02	90.0	884.5	9.13
15.0	86.0	840.0	9.02	88.0	870.0	9.15	90.0	900.0	9.27	92.0	930.0	9.39
15.5	88.0	883.5	9.27	90.0	914.5	9.40	92.0	945.5	9.52	94.0	976.5	9.64
16.0	90.0	928.0	9.51	92.0	960.0	9.64	94.0	992.0	9.77	96.0	1,024.0	9.89
16.5	92.0	973.5	9.76	94.0	1,006.5	9.89	96.0	1,039.5	10.02	98.0	1,072.5	10.14
17.0	94.0	1,020.0	10.00	96.0	1,054.0	10.13	98.0	1,088.0	10.26	100.0	1,122.0	10.39
17.5	96.0	1,067.5	10.24	98.0	1,102.5	10.38	100.0	1,137.5	10.51	102.0	1,172.5	10.63
18.0	98.0	1,116.0	10.48	100.0	1,152.0	10.62	102.0	1,188.0	10.75	104.0	1,224.0	10.88
18.5	100.0	1,165.5	10.72	102.0	1,202.5	10.86	104.0	1,239.5	10.99	106.0	1,276.5	11.13
19.0	102.0	1,216.0	10.96	104.0	1,254.0	11.10	106.0	1,292.0	11.24	108.0	1,330.0	11.37
19.5	104.0	1,267.5	11.20	106.0	1,306.5	11.34	108.0	1,345.5	11.48	110.0	1,384.5	11.61
20.0	106.0	1,320.0	11.43	108.0	1,360.0	11.58	110.0	1,400.0	11.72	112.0	1,440.0	11.86

Depth	Bottom width 35 feet			Bottom width 40 feet			Bottom width 45 feet			Bottom width 50 feet		
	T	A	r	T	A	r	T	A	r	T	A	r
0.4	36.6	14.32	0.39	41.6	16.32	0.39	46.6	18.32	0.39	51.6	20.32	0.39
0.6	37.4	21.72	.58	42.4	24.72	.58	47.4	27.72	.58	52.4	30.72	.58
0.8	38.2	29.28	.76	43.2	33.28	.76	48.2	37.28	.77	53.2	41.28	.77
1.0	39.0	37.00	.94	44.0	42.00	.94	49.0	47.00	.95	54.0	52.00	.95
1.2	39.8	44.88	1.11	44.8	50.88	1.12	49.8	56.88	1.13	54.8	62.88	1.14
1.4	40.6	52.92	1.28	45.6	59.92	1.30	50.6	66.92	1.31	55.6	73.92	1.31
1.6	41.4	61.12	1.45	46.4	69.12	1.47	51.4	77.12	1.48	56.4	85.12	1.49
1.8	42.2	69.48	1.61	47.2	78.48	1.63	52.2	87.48	1.65	57.2	96.48	1.66
2.0	43.0	78.00	1.77	48.0	88.00	1.80	53.0	98.00	1.82	58.0	108.00	1.83
2.2	43.8	86.68	1.93	48.8	97.68	1.96	53.8	108.68	1.98	58.8	119.68	2.00
2.4	44.6	95.52	2.09	49.6	107.52	2.12	54.6	119.52	2.14	59.6	131.52	2.17
2.6	45.4	104.52	2.24	50.4	117.52	2.28	55.4	130.52	2.30	60.4	143.52	2.33
2.8	46.2	113.68	2.39	51.2	127.68	2.43	56.2	141.68	2.46	61.2	155.68	2.49
3.0	47.0	123.00	2.54	52.0	138.00	2.58	57.0	153.00	2.62	62.0	168.00	2.65
3.2	47.8	132.48	2.69	52.8	148.48	2.73	57.8	164.48	2.77	62.8	180.48	2.81
3.4	48.6	142.12	2.83	53.6	159.12	2.88	58.6	176.12	2.93	63.6	193.12	2.96
3.6	49.4	151.92	2.97	54.4	169.92	3.03	59.4	187.92	3.08	64.4	205.92	3.12
3.8	50.2	161.88	3.11	55.2	180.88	3.17	60.2	199.88	3.22	65.2	218.88	3.27
4.0	51.0	172.00	3.25	56.0	192.00	3.32	61.0	212.00	3.37	66.0	232.00	3.42
4.2	51.8	182.28	3.39	56.8	203.28	3.46	61.8	224.28	3.52	66.8	245.28	3.57
4.4	52.6	192.72	3.52	57.6	214.72	3.60	62.6	236.72	3.66	67.6	258.72	3.71
4.6	53.4	203.32	3.66	58.4	226.32	3.74	63.4	249.32	3.80	68.4	272.32	3.86
4.8	54.2	214.08	3.79	59.2	238.08	3.87	64.2	262.08	3.94	69.2	286.08	4.00

Table 33.—*Area in square feet, A, top width in feet, T, and hydraulic radius in feet, r, of trapezoidal channels,* **side slopes 2 to 1**—Continued

Depth	Bottom width 35 feet			Bottom width 40 feet			Bottom width 45 feet			Bottom width 50 feet		
	T	A	r	T	A	r	T	A	r	T	A	r
5.0	55.0	225.00	3.92	60.0	250.00	4.01	65.0	275.00	4.08	70.0	300.00	4.15
5.2	55.8	236.08	4.05	60.8	262.08	4.14	65.8	288.08	4.22	70.8	314.08	4.29
5.4	56.6	247.32	4.18	61.6	274.32	4.28	66.6	301.32	4.36	71.6	328.32	4.43
5.6	57.4	258.72	4.31	62.4	286.72	4.41	67.4	314.72	4.49	72.4	342.72	4.57
5.8	58.2	270.28	4.44	63.2	299.28	4.54	68.2	328.28	4.63	73.2	357.28	4.70
6.0	59.0	282.00	4.56	64.0	312.00	4.67	69.0	342.00	4.76	74.0	372.00	4.84
6.2	59.8	293.88	4.69	64.8	324.88	4.80	69.8	355.88	4.89	74.8	386.88	4.98
6.4	60.6	305.92	4.81	65.6	337.92	4.92	70.6	369.92	5.02	75.6	401.92	5.11
6.6	61.4	318.12	4.93	66.4	351.12	5.05	71.4	384.12	5.15	76.4	417.12	5.25
6.8	62.2	330.48	5.05	67.2	364.48	5.18	72.2	398.48	5.28	77.2	432.48	5.38
7.0	63.0	343.00	5.17	68.0	378.00	5.30	73.0	413.00	5.41	78.0	448.00	5.51
7.2	63.8	355.68	5.29	68.8	391.68	5.42	73.8	427.68	5.54	78.8	463.68	5.64
7.4	64.6	368.52	5.41	69.6	405.52	5.55	74.6	442.52	5.67	79.6	479.52	5.77
7.6	65.4	381.52	5.53	70.4	419.52	5.67	75.4	457.52	5.79	80.4	495.52	5.90
7.8	66.2	394.68	5.65	71.2	433.68	5.79	76.2	472.68	5.92	81.2	511.68	6.03
8.0	67.0	408.00	5.76	72.0	448.00	5.91	77.0	488.00	6.04	82.0	528.00	6.16
8.2	67.8	421.48	5.88	72.8	462.48	6.03	77.8	503.48	6.16	82.8	544.48	6.28
8.4	68.6	435.12	6.00	73.6	477.12	6.15	78.6	519.12	6.29	83.6	561.12	6.41
8.6	69.4	448.92	6.11	74.4	491.92	6.27	79.4	534.92	6.41	84.4	577.92	6.53
8.8	70.2	462.88	6.23	75.2	506.88	6.39	80.2	550.88	6.53	85.2	594.88	6.66
9.0	71.0	477.00	6.34	76.0	522.00	6.50	81.0	567.00	6.65	86.0	612.00	6.78
9.2	71.8	491.28	6.45	76.8	537.28	6.62	81.8	583.28	6.77	86.8	629.28	6.90
9.4	72.6	505.72	6.56	77.6	552.72	6.74	82.6	599.72	6.89	87.6	646.72	7.03
9.6	73.4	520.32	6.68	78.4	568.32	6.85	83.4	616.32	7.01	88.4	664.32	7.15
9.8	74.2	535.08	6.79	79.2	584.08	6.97	84.2	633.08	7.13	89.2	682.08	7.27
10.0	75.0	550.00	6.90	80.0	600.00	7.08	85.0	650.00	7.24	90.0	700.00	7.39
10.5	77.0	588.00	7.17	82.0	640.50	7.37	87.0	693.00	7.54	92.0	745.50	7.69
11.0	79.0	627.0	7.45	84.0	682.0	7.65	89.0	737.0	7.82	94.0	792.0	7.98
11.5	81.0	667.0	7.72	86.0	724.5	7.92	91.0	782.0	8.11	96.0	839.5	8.28
12.0	83.0	708.0	7.99	88.0	768.0	8.20	93.0	828.0	8.39	98.0	888.0	8.57
12.5	85.0	750.0	8.25	90.0	812.5	8.47	95.0	875.0	8.67	100.0	937.5	8.85
13.0	87.0	793.0	8.51	92.0	858.0	8.74	97.0	923.0	8.95	102.0	988.0	9.14
13.5	89.0	837.0	8.78	94.0	904.5	9.01	99.0	972.0	9.22	104.0	1,039.5	9.42
14.0	91.0	882.0	9.04	96.0	952.0	9.28	101.0	1,022.0	9.50	106.0	1,092.0	9.70
14.5	93.0	928.0	9.29	98.0	1,000.5	9.54	103.0	1,073.0	9.77	108.0	1,145.5	9.97
15.0	95.0	975.0	9.55	100.0	1,050.0	9.81	105.0	1,125.0	10.04	110.0	1,200.0	10.25
15.5	97.0	1,023.0	9.81	102.0	1,100.5	10.07	107.0	1,178.0	10.30	112.0	1,255.5	10.52
16.0	99.0	1,072.0	10.06	104.0	1,152.0	10.33	109.0	1,232.0	10.57	114.0	1,312.0	10.79
16.5	101.0	1,122.0	10.31	106.0	1,204.5	10.59	111.0	1,287.0	10.83	116.0	1,369.5	11.06
17.0	103.0	1,173.0	10.57	108.0	1,258.0	10.84	113.0	1,343.0	11.10	118.0	1,428.0	11.33
17.5	105.0	1,225.0	10.82	110.0	1,312.5	11.10	115.0	1,400.0	11.36	120.0	1,487.5	11.60
18.0	107.0	1,278.0	11.07	112.0	1,368.0	11.35	117.0	1,458.0	11.62	122.0	1,548.0	11.86
18.5	109.0	1,332.0	11.31	114.0	1,424.5	11.61	119.0	1,517.0	11.88	124.0	1,609.5	12.13
19.0	111.0	1,387.0	11.56	116.0	1,482.0	11.86	121.0	1,577.0	12.13	126.0	1,672.0	12.39
19.5	113.0	1,443.0	11.81	118.0	1,540.5	12.11	123.0	1,638.0	12.39	128.0	1,735.5	12.65
20.0	115.0	1,500.0	12.05	120.0	1,600.0	12.36	125.0	1,700.0	12.64	130.0	1,800.0	12.91

Table 33.—*Area in square feet, A, top width in feet, T, and hydraulic radius in feet, r, of trapezoidal channels,*
side slopes 2 to 1—Continued

Depth	Bottom width 60 feet			Bottom width 70 feet			Bottom width 80 feet			Bottom width 90 feet		
	T	A	r	T	A	r	T	A	r	T	A	r
0.4	61.6	24.32	0.39	71.6	28.32	0.39	81.6	32.32	0.40	91.6	36.32	0.40
0.6	62.4	36.72	.59	72.4	42.72	.59	82.4	48.72	.59	92.4	54.72	.59
0.8	63.2	49.28	.78	73.2	57.28	.78	83.2	65.28	.78	93.2	73.28	.78
1.0	64.0	62.00	.96	74.0	72.00	.97	84.0	82.00	.97	94.0	92.00	.97
1.2	64.8	74.88	1.15	74.8	86.88	1.15	84.8	98.88	1.16	94.8	110.88	1.16
1.4	65.6	87.92	1.33	75.6	101.92	1.34	85.6	115.92	1.34	95.6	129.92	1.35
1.6	66.4	101.12	1.51	76.4	117.12	1.52	86.4	133.12	1.53	96.4	149.12	1.53
1.8	67.2	114.48	1.68	77.2	132.48	1.70	87.2	150.48	1.71	97.2	168.48	1.72
2.0	68.0	128.00	1.86	78.0	148.00	1.87	88.0	168.00	1.89	98.0	188.00	1.90
2.2	68.8	141.68	2.03	78.8	163.68	2.05	88.8	185.68	2.07	98.8	207.68	2.08
2.4	69.6	155.52	2.20	79.6	179.52	2.22	89.6	203.52	2.24	99.6	227.52	2.26
2.6	70.4	169.52	2.37	80.4	195.52	2.40	90.4	221.52	2.42	100.4	247.52	2.44
2.8	71.2	183.68	2.53	81.2	211.68	2.57	91.2	239.68	2.59	101.2	267.68	2.61
3.0	72.0	198.00	2.70	82.0	228.00	2.73	92.0	258.00	2.76	102.0	288.00	2.78
3.2	72.8	212.48	2.86	82.8	244.48	2.90	92.8	276.48	2.93	102.8	308.48	2.96
3.4	73.6	227.12	3.02	83.6	261.12	3.06	93.6	295.12	3.10	103.6	329.12	3.13
3.6	74.4	241.92	3.18	84.4	277.92	3.23	94.4	313.92	3.27	104.4	349.92	3.30
3.8	75.2	256.88	3.34	85.2	294.88	3.39	95.2	332.88	3.43	105.2	370.88	3.47
4.0	76.0	272.00	3.49	86.0	312.00	3.55	96.0	352.00	3.60	106.0	392.00	3.63
4.2	76.8	287.28	3.65	86.8	329.28	3.71	96.8	371.28	3.76	106.8	413.28	3.80
4.4	77.6	302.72	3.80	87.6	346.72	3.87	97.6	390.72	3.92	107.6	434.72	3.96
4.6	78.4	318.32	3.95	88.4	364.32	4.02	98.4	410.32	4.08	108.4	456.32	4.13
4.8	79.2	334.08	4.10	89.2	382.08	4.18	99.2	430.08	4.24	109.2	478.08	4.29
5.0	80.0	350.00	4.25	90.0	400.00	4.33	100.0	450.00	4.40	110.0	500.00	4.45
5.2	80.8	366.08	4.40	90.8	418.08	4.48	100.8	470.08	4.55	110.8	522.08	4.61
5.4	81.6	382.32	4.54	91.6	436.32	4.63	101.6	490.32	4.71	111.6	544.32	4.77
5.6	82.4	398.72	4.69	92.4	454.72	4.78	102.4	510.72	4.86	112.4	566.72	4.93
5.8	83.2	415.28	4.83	93.2	473.28	4.93	103.2	531.28	5.01	113.2	589.28	5.08
6.0	84.0	432.00	4.98	94.0	492.00	5.08	104.0	552.00	5.17	114.0	612.00	5.24
6.2	84.8	448.88	5.12	94.8	510.88	5.23	104.8	572.88	5.32	114.8	634.88	5.39
6.4	85.6	465.92	5.26	95.6	529.92	5.37	105.6	593.92	5.47	115.6	657.92	5.55
6.6	86.4	483.12	5.40	96.4	549.12	5.52	106.4	615.12	5.62	116.4	681.12	5.70
6.8	87.2	500.48	5.54	97.2	568.48	5.66	107.2	636.48	5.76	117.2	704.48	5.85
7.0	88.0	518.00	5.67	98.0	588.00	5.80	108.0	658.00	5.91	118.0	728.00	6.00
7.2	88.8	535.68	5.81	98.8	607.68	5.95	108.8	679.68	6.06	118.8	751.68	6.15
7.4	89.6	553.52	5.95	99.6	627.52	6.09	109.6	701.52	6.20	119.6	775.52	6.30
7.6	90.4	571.52	6.08	100.4	647.52	6.23	110.4	723.52	6.35	120.4	799.52	6.45
7.8	91.2	589.68	6.21	101.2	667.68	6.37	111.2	745.68	6.49	121.2	823.68	6.60
8.0	92.0	608.00	6.35	102.0	688.00	6.50	112.0	768.00	6.63	122.0	848.00	6.74
8.2	92.8	626.48	6.48	102.8	708.48	6.64	112.8	790.48	6.78	122.8	872.48	6.89
8.4	93.6	645.12	6.61	103.6	729.12	6.78	113.6	813.12	6.92	123.6	897.12	7.03
8.6	94.4	663.92	6.74	104.4	749.92	6.91	114.4	835.92	7.06	124.4	921.92	7.18
8.8	95.2	682.88	6.87	105.2	770.88	7.05	115.2	858.88	7.20	125.2	946.88	7.32
9.0	96.0	702.00	7.00	106.0	792.00	7.18	116.0	882.00	7.33	126.0	972.00	7.46
9.2	96.8	721.28	7.13	106.8	813.28	7.32	116.8	905.28	7.47	126.8	997.28	7.60
9.4	97.6	740.72	7.26	107.6	834.72	7.45	117.6	928.72	7.61	127.6	1,022.72	7.75
9.6	98.4	760.32	7.39	108.4	856.32	7.58	118.4	952.32	7.75	128.4	1,048.32	7.89
9.8	99.2	780.08	7.51	109.2	878.08	7.71	119.2	976.08	7.88	129.2	1,074.08	8.03

Table 33.—*Area in square feet, A, top width in feet, T, and hydraulic radius in feet, r, of trapezoidal channels,* **side slopes 2 to 1**—Continued

Depth	Bottom width 60 feet			Bottom width 70 feet			Bottom width 80 feet			Bottom width 90 feet		
	T	A	r	T	A	r	T	A	r	T	A	r
10.0	100.0	800.00	7.64	110.0	900.00	7.85	120.0	1,000.00	8.02	130.0	1,100.00	8.17
10.5	102.0	850.50	7.95	112.0	955.50	8.17	122.0	1,060.50	8.35	132.0	1,165.50	8.51
11.0	104.0	902.0	8.26	114.0	1,012.0	8.49	124.0	1,122.0	8.68	134.0	1,232.0	8.85
11.5	106.0	954.5	8.57	116.0	1,069.5	8.81	126.0	1,184.5	9.01	136.0	1,299.5	9.19
12.0	108.0	1,008.0	8.87	118.0	1,128.0	9.12	128.0	1,248.0	9.34	138.0	1,368.0	9.52
12.5	110.0	1,062.5	9.17	120.0	1,187.5	9.43	130.0	1,312.5	9.66	140.0	1,437.5	9.85
13.0	112.0	1,118.0	9.46	122.0	1,248.0	9.74	132.0	1,378.0	9.98	142.0	1,508.0	10.18
13.5	114.0	1,174.5	9.76	124.0	1,309.5	10.04	134.0	1,444.5	10.29	144.0	1,579.5	10.50
14.0	116.0	1,232.0	10.05	126.0	1,372.0	10.35	136.0	1,512.0	10.60	146.0	1,652.0	10.82
14.5	118.0	1,290.5	10.34	128.0	1,435.5	10.65	138.0	1,580.5	10.91	148.0	1,725.5	11.14
15.0	120.0	1,350.0	10.62	130.0	1,500.0	10.94	140.0	1,650.0	11.22	150.0	1,800.0	11.46
15.5	122.0	1,410.5	10.91	132.0	1,565.5	11.24	142.0	1,720.5	11.52	152.0	1,875.5	11.77
16.0	124.0	1,472.0	11.19	134.0	1,632.0	11.53	144.0	1,792.0	11.82	154.0	1,952.0	12.08
16.5	126.0	1,534.5	11.47	136.0	1,699.5	11.82	146.0	1,864.5	12.12	156.0	2,029.5	12.39
17.0	128.0	1,598.0	11.75	138.0	1,768.0	12.11	148.0	1,938.0	12.42	158.0	2,108.0	12.70
17.5	130.0	1,662.5	12.02	140.0	1,837.5	12.39	150.0	2,012.5	12.72	160.0	2,187.5	13.00
18.0	132.0	1,728.0	12.30	142.0	1,908.0	12.68	152.0	2,088.0	13.01	162.0	2,268.0	13.30
18.5	134.0	1,794.5	12.57	144.0	1,979.5	12.96	154.0	2,164.5	13.30	164.0	2,349.5	13.60
19.0	136.0	1,862.0	12.84	146.0	2,052.0	13.24	156.0	2,242.0	13.59	166.0	2,432.0	13.90
19.5	138.0	1,930.5	13.11	148.0	2,125.5	13.52	158.0	2,320.5	13.88	168.0	2,515.5	14.20
20.0	140.0	2,000.0	13.38	150.0	2,200.0	13.80	160.0	2,400.0	14.16	170.0	2,600.0	14.49

Table 34.—*Area in square feet, A, top width in feet, T, and hydraulic radius in feet, r, of trapezoidal channels,*
side slopes 2½ to 1

Depth	Bottom width 2 feet			Bottom width 3 feet			Bottom width 4 feet			Bottom width 5 feet		
	T	A	r	T	A	r	T	A	r	T	A	r
0.4	4.0	1.20	0.29	5.0	1.60	0.31	6.0	2.00	0.32	7.0	2.40	0.34
0.6	5.0	2.10	.40	6.0	2.70	.43	7.0	3.30	.46	8.0	3.90	.47
0.8	6.0	3.20	.51	7.0	4.00	.55	8.0	4.80	.58	9.0	5.60	.60
1.0	7.0	4.50	.61	8.0	5.50	.66	9.0	6.50	.69	10.0	7.50	.72
1.2	8.0	6.00	.71	9.0	7.20	.76	10.0	8.40	.80	11.0	9.60	.84
1.4	9.0	7.70	.81	10.0	9.10	.86	11.0	10.50	.91	12.0	11.90	.95
1.6	10.0	9.60	.90	11.0	11.20	.96	12.0	12.80	1.01	13.0	14.40	1.06
1.8	11.0	11.70	1.00	12.0	13.50	1.06	13.0	15.30	1.12	14.0	17.10	1.16
2.0	12.0	14.00	1.10	13.0	16.00	1.16	14.0	18.00	1.22	15.0	20.00	1.27
2.2	13.0	16.50	1.19	14.0	18.70	1.26	15.0	20.90	1.32	16.0	23.10	1.37
2.4	14.0	19.20	1.29	15.0	21.60	1.36	16.0	24.00	1.42	17.0	26.40	1.47
2.6	15.0	22.10	1.38	16.0	24.70	1.45	17.0	27.30	1.52	18.0	29.90	1.57
2.8	16.0	25.20	1.48	17.0	28.00	1.55	18.0	30.80	1.61	19.0	33.60	1.67
3.0	17.0	28.50	1.57	18.0	31.50	1.64	19.0	34.50	1.71	20.0	37.50	1.77
3.2	18.0	32.00	1.66	19.0	35.20	1.74	20.0	38.40	1.81	21.0	41.60	1.87
3.4	19.0	35.70	1.76	20.0	39.10	1.83	21.0	42.50	1.91	22.0	45.90	1.97
3.6	20.0	39.60	1.85	21.0	43.20	1.93	22.0	46.80	2.00	23.0	50.40	2.07
3.8	21.0	43.70	1.95	22.0	47.50	2.02	23.0	51.30	2.10	24.0	55.10	2.16
4.0	22.0	48.00	2.04	23.0	52.00	2.12	24.0	56.00	2.19	25.0	60.00	2.26
4.2	23.0	52.50	2.13	24.0	56.70	2.21	25.0	60.90	2.29	26.0	65.10	2.36
4.4	24.0	57.20	2.23	25.0	61.60	2.31	26.0	66.00	2.38	27.0	70.40	2.45
4.6	25.0	62.10	2.32	26.0	66.70	2.40	27.0	71.30	2.48	28.0	75.90	2.55
4.8	26.0	67.20	2.41	27.0	72.00	2.50	28.0	76.80	2.57	29.0	81.60	2.65
5.0	27.0	72.50	2.51	28.0	77.50	2.59	29.0	82.50	2.67	30.0	87.50	2.74
5.2	28.0	78.00	2.60	29.0	83.20	2.68	30.0	88.40	2.76	31.0	93.60	2.84
5.4	29.0	83.70	2.69	30.0	89.10	2.78	31.0	94.50	2.86	32.0	99.90	2.93
5.6	30.0	89.60	2.79	31.0	95.20	2.87	32.0	100.80	2.95	33.0	106.40	3.03
5.8	31.0	95.70	2.88	32.0	101.50	2.96	33.0	107.30	3.05	34.0	113.10	3.12
6.0	32.0	102.00	2.97	33.0	108.00	3.06	34.0	114.00	3.14	35.0	120.00	3.22
6.2	33.0	108.50	3.07	34.0	114.70	3.15	35.0	120.90	3.23	36.0	127.10	3.31
6.4	34.0	115.20	3.16	35.0	121.60	3.25	36.0	128.00	3.33	37.0	134.40	3.41
6.6	35.0	122.10	3.25	36.0	128.70	3.34	37.0	135.30	3.42	38.0	141.90	3.50
6.8	36.0	129.20	3.35	37.0	136.00	3.43	38.0	142.80	3.52	39.0	149.60	3.59
7.0	37.0	136.50	3.44	38.0	143.50	3.53	39.0	150.50	3.61	40.0	157.50	3.69
7.2	38.0	144.00	3.53	39.0	151.20	3.62	40.0	158.40	3.70	41.0	165.60	3.78
7.4	39.0	151.70	3.62	40.0	159.10	3.71	41.0	166.50	3.80	42.0	173.90	3.88
7.6	40.0	159.60	3.72	41.0	167.20	3.81	42.0	174.80	3.89	43.0	182.40	3.97
7.8	41.0	167.70	3.81	42.0	175.50	3.90	43.0	183.30	3.98	44.0	191.10	4.07
8.0	42.0	176.00	3.90	43.0	184.00	3.99	44.0	192.00	4.08	45.0	200.00	4.16
8.2	-----	-------	----	44.0	192.70	4.09	45.0	200.90	4.17	46.0	209.10	4.25
8.4	-----	-------	----	45.0	201.60	4.18	46.0	210.00	4.27	47.0	218.40	4.35
8.6	-----	-------	----	46.0	210.70	4.27	47.0	219.30	4.36	48.0	227.90	4.44
8.8	-----	-------	----	47.0	220.00	4.37	48.0	228.80	4.45	49.0	237.60	4.54
9.0	-----	-------	----	48.0	229.50	4.46	49.0	238.50	4.55	50.0	247.50	4.63
9.2	-----	-------	----	49.0	239.20	4.55	50.0	248.40	4.64	51.0	257.60	4.72
9.4	-----	-------	----	50.0	249.10	4.65	51.0	258.50	4.73	52.0	267.90	4.82
9.6	-----	-------	----	51.0	259.20	4.74	52.0	268.80	4.83	53.0	278.40	4.91
9.8	-----	-------	----	52.0	269.50	4.83	53.0	279.30	4.92	54.0	289.10	5.00

Table 34.—*Area in square feet, A, top width in feet, T, and hydraulic radius in feet, r, of trapezoidal channels,*
side slopes 2½ to 1—Continued

Depth	Bottom width 2 feet			Bottom width 3 feet			Bottom width 4 feet			Bottom width 5 feet		
	T	A	r	T	A	r	T	A	r	T	A	r
10.0	-----	--------	----	53.0	280.00	4.93	54.0	290.00	5.01	55.0	300.00	5.10
10.5	-----	--------	----	55.5	307.12	5.16	56.5	317.62	5.25	57.5	328.12	5.33
11.0	-----	--------	----	58.0	335.50	5.39	59.0	346.50	5.48	60.0	357.50	5.57
11.5	-----	--------	----	60.5	365.12	5.62	61.5	376.62	5.71	62.5	388.12	5.80
12.0	-----	--------	----	63.0	396.00	5.86	64.0	408.00	5.95	65.0	420.00	6.03
12.5	-----	--------	----	65.5	428.12	6.09	66.5	440.62	6.18	67.5	453.12	6.27
13.0	-----	--------	----	-----	--------	----	69.0	474.50	6.41	70.0	487.50	6.50
13.5	-----	--------	----	-----	--------	----	71.5	509.62	6.64	72.5	523.12	6.73
14.0	-----	--------	----	-----	--------	----	74.0	546.00	6.88	75.0	560.00	6.97
14.5	-----	--------	----	-----	--------	----	76.5	583.62	7.11	77.5	598.12	7.20
15.0	-----	--------	----	-----	--------	----	79.0	622.50	7.34	80.0	637.50	7.43
15.5	-----	--------	----	-----	--------	----	81.5	662.62	7.58	82.5	678.12	7.67
16.0	-----	--------	----	-----	--------	----	84.0	704.00	7.81	85.0	720.00	7.90
16.5	-----	--------	----	-----	--------	----	-----	--------	----	87.5	763.12	8.13
17.0	-----	--------	----	-----	--------	----	-----	--------	----	90.0	807.50	8.36
17.5	-----	--------	----	-----	--------	----	-----	--------	----	92.5	853.12	8.60
18.0	-----	--------	----	-----	--------	----	-----	--------	----	95.0	900.00	8.83
18.5	-----	--------	----	-----	--------	----	-----	--------	----	97.5	948.12	9.06
19.0	-----	--------	----	-----	--------	----	-----	--------	----	100.0	997.50	9.29
19.5	-----	--------	----	-----	--------	----	-----	--------	----	102.5	1,048.12	9.53
20.0	-----	--------	----	-----	--------	----	-----	--------	----	105.0	1,100.00	9.76

Depth	Bottom width 6 feet			Bottom width 7 feet			Bottom width 8 feet			Bottom width 9 feet		
	T	A	r	T	A	r	T	A	r	T	A	r
0.4	8.0	2.80	0.34	9.0	3.20	0.35	10.0	3.60	0.35	11.0	4.00	0.36
0.6	9.0	4.50	.49	10.0	5.10	.50	11.0	5.70	.51	12.0	6.30	.52
0.8	10.0	6.40	.62	11.0	7.20	.64	12.0	8.00	.65	13.0	8.80	.66
1.0	11.0	8.50	.75	12.0	9.50	.77	13.0	10.50	.78	14.0	11.50	.80
1.2	12.0	10.80	.87	13.0	12.00	.89	14.0	13.20	.91	15.0	14.40	.93
1.4	13.0	13.30	.98	14.0	14.70	1.01	15.0	16.10	1.04	16.0	17.50	1.06
1.6	14.0	16.00	1.09	15.0	17.60	1.13	16.0	19.20	1.16	17.0	20.80	1.18
1.8	15.0	18.90	1.20	16.0	20.70	1.24	17.0	22.50	1.27	18.0	24.30	1.30
2.0	16.0	22.00	1.31	17.0	24.00	1.35	18.0	26.00	1.39	19.0	28.00	1.42
2.2	17.0	25.30	1.42	18.0	27.50	1.46	19.0	29.70	1.50	20.0	31.90	1.53
2.4	18.0	28.80	1.52	19.0	31.20	1.57	20.0	33.60	1.61	21.0	36.00	1.64
2.6	19.0	32.50	1.62	20.0	35.10	1.67	21.0	37.70	1.71	22.0	40.30	1.75
2.8	20.0	36.40	1.73	21.0	39.20	1.78	22.0	42.00	1.82	23.0	44.80	1.86
3.0	21.0	40.50	1.83	22.0	43.50	1.88	23.0	46.50	1.93	24.0	49.50	1.97
3.2	22.0	44.80	1.93	23.0	48.00	1.98	24.0	51.20	2.03	25.0	54.40	2.07
3.4	23.0	49.30	2.03	24.0	52.70	2.08	25.0	56.10	2.13	26.0	59.50	2.18
3.6	24.0	54.00	2.13	25.0	57.60	2.18	26.0	61.20	2.23	27.0	64.80	2.28
3.8	25.0	58.90	2.23	26.0	62.70	2.28	27.0	66.50	2.34	28.0	70.30	2.39
4.0	26.0	64.00	2.32	27.0	68.00	2.38	28.0	72.00	2.44	29.0	76.00	2.49
4.2	27.0	69.30	2.42	28.0	73.50	2.48	29.0	77.70	2.54	30.0	81.90	2.59
4.4	28.0	74.80	2.52	29.0	79.20	2.58	30.0	83.60	2.64	31.0	88.00	2.69
4.6	29.0	80.50	2.62	30.0	85.10	2.68	31.0	89.70	2.74	32.0	94.30	2.79
4.8	30.0	86.40	2.71	31.0	91.20	2.78	32.0	96.00	2.84	33.0	100.80	2.89

Table 34.—*Area in square feet, A, top width in feet, T, and hydraulic radius in feet, r, of trapezoidal channels,* **side slopes 2½ to 1**—Continued

Depth	Bottom width 6 feet			Bottom width 7 feet			Bottom width 8 feet			Bottom width 9 feet		
	T	A	r	T	A	r	T	A	r	T	A	r
5.0	31.0	92.50	2.81	32.0	97.50	2.87	33.0	102.50	2.93	34.0	107.50	2.99
5.2	32.0	98.80	2.91	33.0	104.00	2.97	34.0	109.20	3.03	35.0	114.40	3.09
5.4	33.0	105.30	3.00	34.0	110.70	3.07	35.0	116.10	3.13	36.0	121.50	3.19
5.6	34.0	112.00	3.10	35.0	117.60	3.16	36.0	123.20	3.23	37.0	128.80	3.29
5.8	35.0	118.90	3.19	36.0	124.70	3.26	37.0	130.50	3.33	38.0	136.30	3.39
6.0	36.0	126.00	3.29	37.0	132.00	3.36	38.0	138.00	3.42	39.0	144.00	3.49
6.2	37.0	133.30	3.38	38.0	139.50	3.45	39.0	145.70	3.52	40.0	151.90	3.58
6.4	38.0	140.80	3.48	39.0	147.20	3.55	40.0	153.60	3.62	41.0	160.00	3.68
6.6	39.0	148.50	3.57	40.0	155.10	3.65	41.0	161.70	3.71	42.0	168.30	3.78
6.8	40.0	156.40	3.67	41.0	163.20	3.74	42.0	170.00	3.81	43.0	176.80	3.88
7.0	41.0	164.50	3.76	42.0	171.50	3.84	43.0	178.50	3.91	44.0	185.50	3.97
7.2	42.0	172.80	3.86	43.0	180.00	3.93	44.0	187.20	4.00	45.0	194.40	4.07
7.4	43.0	181.30	3.95	44.0	188.70	4.03	45.0	196.10	4.10	46.0	203.50	4.17
7.6	44.0	190.00	4.05	45.0	197.60	4.12	46.0	205.20	4.19	47.0	212.80	4.26
7.8	45.0	198.90	4.14	46.0	206.70	4.22	47.0	214.50	4.29	48.0	222.30	4.36
8.0	46.0	208.00	4.24	47.0	216.00	4.31	48.0	224.00	4.39	49.0	232.00	4.45
8.2	47.0	217.30	4.33	48.0	225.50	4.41	49.0	233.70	4.48	50.0	241.90	4.55
8.4	48.0	226.80	4.43	49.0	235.20	4.50	50.0	243.60	4.58	51.0	252.00	4.65
8.6	49.0	236.50	4.52	50.0	245.10	4.60	51.0	253.70	4.67	52.0	262.30	4.74
8.8	50.0	246.40	4.62	51.0	255.20	4.69	52.0	264.00	4.77	53.0	272.80	4.84
9.0	51.0	256.50	4.71	52.0	265.50	4.79	53.0	274.50	4.86	54.0	283.50	4.93
9.2	52.0	266.80	4.80	53.0	276.00	4.88	54.0	285.20	4.96	55.0	294.40	5.03
9.4	53.0	277.30	4.90	54.0	286.70	4.98	55.0	296.10	5.05	56.0	305.50	5.12
9.6	54.0	288.00	4.99	55.0	297.60	5.07	56.0	307.20	5.15	57.0	316.80	5.22
9.8	55.0	298.90	5.09	56.0	308.70	5.16	57.0	318.50	5.24	58.0	328.30	5.31
10.0	56.0	310.00	5.18	57.0	320.00	5.26	58.0	330.00	5.34	59.0	340.00	5.41
10.5	58.5	338.62	5.41	59.5	349.12	5.49	60.5	359.62	5.57	61.5	370.12	5.65
11.0	61.0	368.5	5.65	62.0	379.5	5.73	63.0	390.5	5.81	64.0	401.5	5.88
11.5	63.5	399.6	5.88	64.5	411.1	5.96	65.5	422.6	6.04	66.5	434.1	6.12
12.0	66.0	432.0	6.12	67.0	444.0	6.20	68.0	456.0	6.28	69.0	468.0	6.36
12.5	68.5	465.6	6.35	69.5	478.1	6.43	70.5	490.6	6.51	71.5	503.1	6.59
13.0	71.0	500.5	6.58	72.0	513.5	6.67	73.0	526.5	6.75	74.0	539.5	6.83
13.5	73.5	536.6	6.82	74.5	550.1	6.90	75.5	563.6	6.98	76.5	577.1	7.06
14.0	76.0	574.0	7.05	77.0	588.0	7.14	78.0	602.0	7.22	79.0	616.0	7.30
14.5	78.5	612.6	7.29	79.5	627.1	7.37	80.5	641.6	7.45	81.5	656.1	7.53
15.0	81.0	652.5	7.52	82.0	667.5	7.60	83.0	682.5	7.69	84.0	697.5	7.77
15.5	83.5	693.6	7.75	84.5	709.1	7.84	85.5	724.6	7.92	86.5	740.1	8.00
16.0	86.0	736.0	7.99	87.0	752.0	8.07	88.0	768.0	8.16	89.0	784.0	8.24
16.5	88.5	779.6	8.22	89.5	796.1	8.31	90.5	812.6	8.39	91.5	829.1	8.47
17.0	91.0	824.5	8.45	92.0	841.5	8.54	93.0	858.5	8.62	94.0	875.5	8.71
17.5	93.5	870.6	8.69	94.5	888.1	8.77	95.5	905.6	8.86	96.5	923.1	8.94
18.0	96.0	918.0	8.92	97.0	936.0	9.01	98.0	954.0	9.09	99.0	972.0	9.18
18.5	98.5	966.6	9.15	99.5	985.1	9.24	100.5	1,003.6	9.33	101.5	1,022.1	9.41
19.0	101.0	1,016.5	9.38	102.0	1,035.5	9.47	103.0	1,054.5	9.56	104.0	1,073.5	9.64
19.5	103.5	1,067.6	9.62	104.5	1,087.1	9.71	105.5	1,106.6	9.79	106.5	1,126.1	9.88
20.0	106.0	1,120.0	9.85	107.0	1,140.0	9.94	108.0	1,160.0	10.03	109.0	1,180.0	10.11

Table 94.—*Area in square feet, A, top width in feet, T, and hydraulic radius in feet, r, of trapezoidal channels,*
side slopes 2½ to 1—Continued

Depth	Bottom width 10 feet			Bottom width 12 feet			Bottom width 14 feet			Bottom width 16 feet		
	T	A	r	T	A	r	T	A	r	T	A	r
0.4	12.0	4.40	0.36	14.0	5.20	0.37	16.0	6.00	0.37	18.0	6.80	0.37
0.6	13.0	6.90	.52	15.0	8.10	.53	17.0	9.30	.54	19.0	10.50	.55
0.8	14.0	9.60	.67	16.0	11.20	.69	18.0	12.80	.70	20.0	14.40	.71
1.0	15.0	12.50	.81	17.0	14.50	.83	19.0	16.50	.85	21.0	18.50	.87
1.2	16.0	15.60	.95	18.0	18.00	.97	20.0	20.40	1.00	22.9	22.80	1.02
1.4	17.0	18.90	1.08	19.0	21.70	1.11	21.0	24.50	1.14	23.0	27.30	1.16
1.6	18.0	22.40	1.20	20.0	25.60	1.24	22.0	28.80	1.27	24.0	32.00	1.30
1.8	19.0	26.10	1.33	21.0	29.70	1.37	23.0	33.30	1.41	25.0	36.90	1.44
2.0	20.0	30.00	1.44	22.0	34.00	1.49	24.0	38.00	1.53	26.0	42.00	1.57
2.2	21.0	34.10	1.56	23.0	38.50	1.61	25.0	42.90	1.66	27.0	47.30	1.70
2.4	22.0	38.40	1.68	24.0	43.20	1.73	26.0	48.00	1.78	28.0	52.80	1.83
2.6	23.0	42.90	1.79	25.0	48.10	1.85	27.0	53.30	1.90	29.0	58.50	1.95
2.8	24.0	47.60	1.90	26.0	53.20	1.96	28.0	58.80	2.02	30.0	64.40	2.07
3.0	25.0	52.50	2.01	27.0	58.50	2.08	29.0	64.50	2.14	31.0	70.50	2.19
3.2	26.0	57.60	2.12	28.0	64.00	2.19	30.0	70.40	2.25	32.0	76.80	2.31
3.4	27.0	62.90	2.22	29.0	69.70	2.30	31.0	76.50	2.37	33.0	83.30	2.43
3.6	28.0	68.40	2.33	30.0	75.60	2.41	32.0	82.80	2.48	34.0	90.00	2.54
3.8	29.0	74.10	2.43	31.0	81.70	2.52	33.0	89.30	2.59	35.0	96.90	2.66
4.0	30.0	80.00	2.54	32.0	88.00	2.62	34.0	96.00	2.70	36.0	104.00	2.77
4.2	31.0	86.10	2.64	33.0	94.50	2.73	35.0	102.90	2.81	37.0	111.30	2.88
4.4	32.0	92.40	2.74	34.0	101.20	2.84	36.0	110.00	2.92	38.0	118.80	2.99
4.6	33.0	98.90	2.84	35.0	108.10	2.94	37.0	117.30	3.03	39.0	126.50	3.10
4.8	34.0	105.60	2.95	36.0	115.20	3.04	38.0	124.80	3.13	40.0	134.40	3.21
5.0	35.0	112.50	3.05	37.0	122.50	3.15	39.0	132.50	3.24	41.0	142.50	3.32
5.2	36.0	119.60	3.15	38.0	130.00	3.25	40.0	140.40	3.34	42.0	150.80	3.43
5.4	37.0	126.90	3.25	39.0	137.70	3.35	41.0	148.50	3.45	43.0	159.30	3.53
5.6	38.0	134.40	3.35	40.0	145.60	3.45	42.0	156.80	3.55	44.0	168.00	3.64
5.8	39.0	142.10	3.45	41.0	153.70	3.56	43.0	165.30	3.65	45.0	176.90	3.75
6.0	40.0	150.00	3.55	42.0	162.00	3.66	44.0	174.00	3.76	46.0	186.00	3.85
6.2	41.0	158.10	3.64	43.0	170.50	3.76	45.0	182.90	3.86	47.0	195.30	3.95
6.4	42.0	166.40	3.74	44.0	179.20	3.86	46.0	192.00	3.96	48.0	204.80	4.06
6.6	43.0	174.90	3.84	45.0	188.10	3.96	47.0	201.30	4.06	49.0	214.50	4.16
6.8	44.0	183.60	3.94	46.0	197.20	4.06	48.0	210.80	4.16	50.0	224.40	4.26
7.0	45.0	192.50	4.04	47.0	206.50	4.16	49.0	220.50	4.27	51.0	234.50	4.37
7.2	46.0	201.60	4.13	48.0	216.00	4.25	50.0	230.40	4.37	52.0	244.80	4.47
7.4	47.0	210.90	4.23	49.0	225.70	4.35	51.0	240.50	4.47	53.0	255.30	4.57
7.6	48.0	220.40	4.33	50.0	235.60	4.45	52.0	250.80	4.57	54.0	266.00	4.67
7.8	49.0	230.10	4.42	51.0	245.70	4.55	53.0	261.30	4.67	55.0	276.90	4.77
8.0	50.0	240.00	4.52	52.0	256.00	4.65	54.0	272.00	4.77	56.0	288.00	4.87
8.2	51.0	250.10	4.62	53.0	266.50	4.75	55.0	282.90	4.86	57.0	299.30	4.98
8.4	52.0	260.40	4.71	54.0	277.20	4.84	56.0	294.00	4.96	58.0	310.80	5.08
8.6	53.0	270.90	4.81	55.0	288.10	4.94	57.0	305.30	5.06	59.0	322.50	5.18
8.8	54.0	281.60	4.91	56.0	299.20	5.04	58.0	316.80	5.16	60.0	334.40	5.28
9.0	55.0	292.50	5.00	57.0	310.50	5.14	59.0	328.50	5.26	61.0	346.50	5.37
9.2	56.0	303.60	5.10	58.0	322.00	5.23	60.0	340.40	5.36	62.0	358.80	5.47
9.4	57.0	314.90	5.19	59.0	333.70	5.33	61.0	352.50	5.45	63.0	371.30	5.57
9.6	58.0	326.40	5.29	60.0	345.60	5.43	62.0	364.80	5.55	64.0	384.00	5.67
9.8	59.0	338.10	5.39	61.0	357.70	5.52	63.0	377.30	5.65	65.0	396.90	5.77

Table 34.—*Area in square feet, A, top width in feet, T, and hydraulic radius in feet, r, of trapezoidal channels,*
side slopes 2½ to 1—Continued

Depth	Bottom width 10 feet			Bottom width 12 feet			Bottom width 14 feet			Bottom width 16 feet		
	T	A	r	T	A	r	T	A	r	T	A	r
10. 0	60. 0	350. 00	5. 48	62. 0	370. 00	5. 62	64. 0	390. 00	5. 75	66. 0	410. 00	5. 87
10. 5	62. 5	380. 62	5. 72	64. 5	401. 62	5. 86	66. 5	422. 62	5. 99	68. 5	443. 62	6. 12
11. 0	65. 0	412. 5	5. 96	67. 0	434. 5	6. 10	69. 0	456. 5	6. 23	71. 0	478. 5	6. 36
11. 5	67. 5	445. 6	6. 20	69. 5	468. 6	6. 34	71. 5	491. 6	6. 47	73. 5	514. 6	6. 60
12. 0	70. 0	480. 0	6. 43	72. 0	504. 0	6. 58	74. 0	528. 0	6. 72	76. 0	552. 0	6. 85
12. 5	72. 5	515. 6	6. 67	74. 5	540. 6	6. 82	76. 5	565. 6	6. 96	78. 5	590. 6	7. 09
13. 0	75. 0	552. 5	6. 91	77. 0	578. 5	7. 05	79. 0	604. 5	7. 20	81. 0	630. 5	7. 33
13. 5	77. 5	590. 6	7. 14	79. 5	617. 6	7. 29	81. 5	644. 6	7. 44	83. 5	671. 6	7. 57
14. 0	80. 0	630. 0	7. 38	82. 0	658. 0	7. 53	84. 0	686. 0	7. 67	86. 0	714. 0	7. 81
14. 5	82. 5	670. 6	7. 61	84. 5	699. 6	7. 77	86. 5	728. 6	7. 91	88. 5	757. 6	8. 05
15. 0	85. 0	712. 5	7. 85	87. 0	742. 5	8. 00	89. 0	772. 5	8. 15	91. 0	802. 5	8. 29
15. 5	87. 5	755. 6	8. 08	89. 5	786. 6	8. 24	91. 5	817. 6	8. 39	93. 5	848. 6	8. 53
16. 0	90. 0	800. 0	8. 32	92. 0	832. 0	8. 48	94. 0	864. 0	8. 63	96. 0	896. 0	8. 77
16. 5	92. 5	845. 6	8. 55	94. 5	878. 6	8. 71	96. 5	911. 6	8. 86	98. 5	944. 6	9. 01
17. 0	95. 0	892. 5	8. 79	97. 0	926. 5	8. 95	99. 0	960. 5	9. 10	101. 0	994. 5	9. 25
17. 5	97. 5	940. 6	9. 02	99. 5	975. 6	9. 18	101. 5	1, 010. 6	9. 34	103. 5	1, 045. 6	9. 48
18. 0	100. 0	990. 0	9. 26	102. 0	1, 026. 0	9. 42	104. 0	1, 062. 0	9. 57	106. 0	1, 098. 0	9. 72
18. 5	102. 5	1, 040. 6	9. 49	104. 5	1, 077. 6	9. 65	106. 5	1, 114. 6	9. 81	108. 5	1, 151. 6	9. 96
19. 0	105. 0	1, 092. 5	9. 73	107. 0	1, 130. 5	9. 89	109. 0	1, 168. 5	10. 05	111. 0	1, 206. 5	10. 20
19. 5	107. 5	1, 145. 6	9. 96	109. 5	1, 184. 6	10. 12	111. 5	1, 223. 6	10. 28	113. 5	1, 262. 6	10. 43
20. 0	110. 0	1, 200. 0	10. 20	112. 0	1, 240. 0	10. 36	114. 0	1, 280. 0	10. 52	116. 0	1, 320. 0	10. 67

Depth	Bottom width 18 feet			Bottom width 20 feet			Bottom width 22 feet			Bottom width 24 feet		
	T	A	r	T	A	r	T	A	r	T	A	r
0. 4	20. 0	7. 60	0. 38	22. 0	8. 40	0. 38	24. 0	9. 20	0. 38	26. 0	10. 00	0. 38
0. 6	21. 0	11. 70	. 55	23. 0	12. 90	. 56	25. 0	14. 10	. 56	27. 0	15. 30	. 56
0. 8	22. 0	16. 00	. 72	24. 0	17. 60	. 72	26. 0	19. 20	. 73	28. 0	20. 80	. 73
1. 0	23. 0	20. 50	. 88	25. 0	22. 50	. 89	27. 0	24. 50	. 89	29. 0	26. 50	. 90
1. 2	24. 0	25. 20	1. 03	26. 0	27. 60	1. 04	28. 0	30. 00	1. 05	30. 0	32. 40	1. 06
1. 4	25. 0	30. 10	1. 18	27. 0	32. 90	1. 19	29. 0	35. 70	1. 21	31. 0	38. 50	1. 22
1. 6	26. 0	35. 20	1. 32	28. 0	38. 40	1. 34	30. 0	41. 60	1. 36	32. 0	44. 80	1. 37
1. 8	27. 0	40. 50	1. 46	29. 0	44. 10	1. 49	31. 0	47. 70	1. 51	33. 0	51. 30	1. 52
2. 0	28. 0	46. 00	1. 60	30. 0	50. 00	1. 62	32. 0	54. 00	1. 65	34. 0	58. 00	1. 67
2. 2	29. 0	51. 70	1. 73	31. 0	56. 10	1. 76	33. 0	60. 50	1. 79	35. 0	64. 90	1. 81
2. 4	30. 0	57. 60	1. 86	32. 0	62. 40	1. 90	34. 0	67. 20	1. 92	36. 0	72. 00	1. 95
2. 6	31. 0	63. 70	1. 99	33. 0	68. 90	2. 03	35. 0	74. 10	2. 06	37. 0	79. 30	2. 09
2. 8	32. 0	70. 00	2. 12	34. 0	75. 60	2. 16	36. 0	81. 20	2. 19	38. 0	86. 80	2. 22
3. 0	33. 0	76. 50	2. 24	35. 0	82. 50	2. 28	37. 0	88. 50	2. 32	39. 0	94. 50	2. 35
3. 2	34. 0	83. 20	2. 36	36. 0	89. 60	2. 41	38. 0	96. 00	2. 45	40. 0	102. 40	2. 48
3. 4	35. 0	90. 10	2. 48	37. 0	96. 90	2. 53	39. 0	103. 70	2. 57	41. 0	110. 50	2. 61
3. 6	36. 0	97. 20	2. 60	38. 0	104. 40	2. 65	40. 0	111. 60	2. 70	42. 0	118. 80	2. 74
3. 8	37. 0	104. 50	2. 72	39. 0	112. 10	2. 77	41. 0	119. 70	2. 82	43. 0	127. 30	2. 86
4. 0	38. 0	112. 00	2. 83	40. 0	120. 00	2. 89	42. 0	128. 00	2. 94	44. 0	136. 00	2. 99
4. 2	39. 0	119. 70	2. 95	41. 0	128. 10	3. 01	43. 0	136. 50	3. 06	45. 0	144. 90	3. 11
4. 4	40. 0	127. 60	3. 06	42. 0	136. 40	3. 12	44. 0	145. 20	3. 18	46. 0	154. 00	3. 23
4. 6	41. 0	135. 70	3. 17	43. 0	144. 90	3. 24	45. 0	154. 10	3. 29	47. 0	163. 30	3. 35
4. 8	42. 0	144. 00	3. 28	44. 0	153. 60	3. 35	46. 0	163. 20	3. 41	48. 0	172. 80	3. 47

Table 94.—*Area in square feet, A, top width in feet, T, and hydraulic radius in feet, r, of trapezoidal channels,* **side slopes 2½ to 1**—Continued

Depth	Bottom width 18 feet			Bottom width 20 feet			Bottom width 22 feet			Bottom width 24 feet		
	T	A	r	T	A	r	T	A	r	T	A	r
5.0	43.0	152.50	3.39	45.0	162.50	3.46	47.0	172.50	3.53	49.0	182.50	3.58
5.2	44.0	161.20	3.50	46.0	171.60	3.57	48.0	182.00	3.64	50.0	192.40	3.70
5.4	45.0	170.10	3.61	47.0	180.90	3.69	49.0	191.70	3.75	51.0	202.50	3.82
5.6	46.0	179.20	3.72	48.0	190.40	3.80	50.0	201.60	3.87	52.0	212.80	3.93
5.8	47.0	188.50	3.83	49.0	200.10	3.91	51.0	211.70	3.98	53.0	223.30	4.04
6.0	48.0	198.00	3.94	50.0	210.00	4.01	52.0	222.00	4.09	54.0	234.00	4.16
6.2	49.0	207.70	4.04	51.0	220.10	4.12	53.0	232.50	4.20	55.0	244.90	4.27
6.4	50.0	217.60	4.15	52.0	230.40	4.23	54.0	243.20	4.31	56.0	256.00	4.38
6.6	51.0	227.70	4.25	53.0	240.90	4.34	55.0	254.10	4.42	57.0	267.30	4.49
6.8	52.0	238.00	4.36	54.0	251.60	4.44	56.0	265.20	4.52	58.0	278.80	4.60
7.0	53.0	248.50	4.46	55.0	262.50	4.55	57.0	276.50	4.63	59.0	290.50	4.71
7.2	54.0	259.20	4.57	56.0	273.60	4.66	58.0	288.00	4.74	60.0	302.40	4.82
7.4	55.0	270.10	4.67	57.0	284.90	4.76	59.0	299.70	4.85	61.0	314.50	4.93
7.6	56.0	281.20	4.77	58.0	296.40	4.86	60.0	311.60	4.95	62.0	326.80	5.03
7.8	57.0	292.50	4.87	59.0	308.10	4.97	61.0	323.70	5.06	63.0	339.30	5.14
8.0	58.0	304.00	4.98	60.0	320.00	5.07	62.0	336.00	5.16	64.0	352.00	5.25
8.2	59.0	315.70	5.08	61.0	332.10	5.18	63.0	348.50	5.27	65.0	364.90	5.35
8.4	60.0	327.60	5.18	62.0	344.40	5.28	64.0	361.20	5.37	66.0	378.00	5.46
8.6	61.0	339.70	5.28	63.0	356.90	5.38	65.0	374.10	5.48	67.0	391.30	5.57
8.8	62.0	352.00	5.38	64.0	369.60	5.48	66.0	387.20	5.58	68.0	404.80	5.67
9.0	63.0	364.50	5.48	65.0	382.50	5.59	67.0	400.50	5.68	69.0	418.50	5.78
9.2	64.0	377.20	5.58	66.0	395.60	5.69	68.0	414.00	5.79	70.0	432.40	5.88
9.4	65.0	390.10	5.68	67.0	408.90	5.79	69.0	427.70	5.89	71.0	446.50	5.98
9.6	66.0	403.20	5.78	68.0	422.40	5.89	70.0	441.60	5.99	72.0	460.80	6.09
9.8	67.0	416.50	5.88	69.0	436.10	5.99	71.0	455.70	6.09	73.0	475.30	6.19
10.0	68.0	430.00	5.98	70.0	450.00	6.09	72.0	470.00	6.20	74.0	490.00	6.29
10.5	70.5	464.62	6.23	72.5	485.62	6.34	74.5	506.62	6.45	76.5	527.62	6.55
11.0	73.0	500.5	6.48	75.0	522.5	6.59	77.0	544.5	6.70	79.0	566.5	6.81
11.5	75.5	537.6	6.73	77.5	560.6	6.84	79.5	583.6	6.95	81.5	606.6	7.06
12.0	78.0	576.0	6.97	80.0	600.0	7.09	82.0	624.0	7.20	84.0	648.0	7.31
12.5	80.5	615.6	7.22	82.5	640.6	7.34	84.5	665.6	7.45	86.5	690.6	7.56
13.0	83.0	656.5	7.46	85.0	682.5	7.58	87.0	708.5	7.70	89.0	734.5	7.81
13.5	85.5	698.6	7.70	87.5	725.6	7.83	89.5	752.6	7.95	91.5	779.6	8.06
14.0	88.0	742.0	7.94	90.0	770.0	8.07	92.0	798.0	8.19	94.0	826.0	8.31
14.5	90.5	786.6	8.19	92.5	815.6	8.32	94.5	844.6	8.44	96.5	873.6	8.56
15.0	93.0	832.5	8.43	95.0	862.5	8.56	97.0	892.5	8.68	99.0	922.5	8.80
15.5	95.5	879.6	8.67	97.5	910.6	8.80	99.5	941.6	8.93	101.5	972.6	9.05
16.0	98.0	928.0	8.91	100.0	960.0	9.04	102.0	992.0	9.17	104.0	1,024.0	9.30
16.5	100.5	977.6	9.15	102.5	1,010.6	9.28	104.5	1,043.6	9.41	106.5	1,076.6	9.54
17.0	103.0	1,028.5	9.39	105.0	1,062.5	9.53	107.0	1,096.5	9.66	109.0	1,130.5	9.78
17.5	105.5	1,080.6	9.62	107.5	1,115.6	9.77	109.5	1,150.6	9.90	111.5	1,185.6	10.03
18.0	108.0	1,134.0	9.87	110.0	1,170.0	10.01	112.0	1,206.0	10.14	114.0	1,242.0	10.27
18.5	110.5	1,188.6	10.11	112.5	1,225.6	10.25	114.5	1,262.6	10.38	116.5	1,299.6	10.51
19.0	113.0	1,244.5	10.34	115.0	1,282.5	10.48	117.0	1,320.5	10.62	119.0	1,358.5	10.75
19.5	115.5	1,301.6	10.58	117.5	1,340.6	10.72	119.5	1,379.6	10.86	121.5	1,418.6	11.00
20.0	118.0	1,360.0	10.82	120.0	1,400.0	10.96	122.0	1,440.0	11.10	124.0	1,480.0	11.24

Table 34.—*Area in square feet, A, top width in feet, T, and hydraulic radius in feet, r, of trapezoidal channels,*
side slopes 2½ to 1—Continued

Depth	Bottom width 26 feet			Bottom width 28 feet			Bottom width 30 feet			Bottom width 32 feet		
	T	A	r	T	A	r	T	A	r	T	A	r
0.4	28.0	10.80	.38	30.0	11.60	.38	32.0	12.40	.39	34.0	13.20	.39
0.6	29.0	16.50	.56	31.0	17.70	.57	33.0	18.90	.57	35.0	20.10	.57
0.8	30.0	22.40	.74	32.0	24.00	.74	34.0	25.60	.75	36.0	27.20	.75
1.0	31.0	28.50	.91	33.0	30.50	.91	35.0	32.50	.92	37.0	34.50	.92
1.2	32.0	34.80	1.07	34.0	37.20	1.08	36.0	39.60	1.09	38.0	42.00	1.09
1.4	33.0	41.30	1.23	35.0	44.10	1.24	37.0	46.90	1.25	39.0	49.70	1.26
1.6	34.0	48.00	1.39	36.0	51.20	1.40	38.0	54.40	1.41	40.0	57.60	1.42
1.8	35.0	54.90	1.54	37.0	58.50	1.55	39.0	62.10	1.56	41.0	65.70	1.58
2.0	36.0	62.00	1.69	38.0	66.00	1.70	40.0	70.00	1.72	42.0	74.00	1.73
2.2	37.0	69.30	1.83	39.0	73.70	1.85	41.0	78.10	1.87	43.0	82.50	1.88
2.4	38.0	76.80	1.97	40.0	81.60	1.99	42.0	86.40	2.01	44.0	91.20	2.03
2.6	39.0	84.50	2.11	41.0	89.70	2.14	43.0	94.90	2.16	45.0	100.10	2.18
2.8	40.0	92.40	2.25	42.0	98.00	2.27	44.0	103.60	2.30	46.0	109.20	2.32
3.0	41.0	100.50	2.38	43.0	106.50	2.41	45.0	112.50	2.44	47.0	118.50	2.46
3.2	42.0	108.80	2.52	44.0	115.20	2.55	46.0	121.60	2.57	48.0	128.00	2.60
3.4	43.0	117.30	2.65	45.0	124.10	2.68	47.0	130.90	2.71	49.0	137.70	2.74
3.6	44.0	126.00	2.78	46.0	133.20	2.81	48.0	140.40	2.84	50.0	147.60	2.87
3.8	45.0	134.90	2.90	47.0	142.50	2.94	49.0	150.10	2.97	51.0	157.70	3.01
4.0	46.0	144.00	3.03	48.0	152.00	3.07	50.0	160.00	3.10	52.0	168.00	3.14
4.2	47.0	153.30	3.15	49.0	161.70	3.19	51.0	170.10	3.23	53.0	178.50	3.27
4.4	48.0	162.80	3.28	50.0	171.60	3.32	52.0	180.40	3.36	54.0	189.20	3.40
4.6	49.0	172.50	3.40	51.0	181.70	3.44	53.0	190.90	3.49	55.0	200.10	3.52
4.8	50.0	182.40	3.52	52.0	192.00	3.57	54.0	201.60	3.61	56.0	211.20	3.65
5.0	51.0	192.50	3.64	53.0	202.50	3.69	55.0	212.50	3.73	57.0	222.50	3.78
5.2	52.0	202.80	3.76	54.0	213.20	3.81	56.0	223.60	3.85	58.0	234.00	3.90
5.4	53.0	213.30	3.87	55.0	224.10	3.93	57.0	234.90	3.98	59.0	245.70	4.02
5.6	54.0	224.00	3.99	56.0	235.20	4.04	58.0	246.40	4.10	60.0	257.60	4.14
5.8	55.0	234.90	4.10	57.0	246.50	4.16	59.0	258.10	4.21	61.0	269.70	4.27
6.0	56.0	246.00	4.22	58.0	258.00	4.28	60.0	270.00	4.33	62.0	282.00	4.38
6.2	57.0	257.30	4.33	59.0	269.70	4.39	61.0	282.10	4.45	63.0	294.50	4.50
6.4	58.0	268.80	4.45	60.0	281.60	4.51	62.0	294.40	4.57	64.0	307.20	4.62
6.6	59.0	280.50	4.56	61.0	293.70	4.62	63.0	306.90	4.68	65.0	320.10	4.74
6.8	60.0	292.40	4.67	62.0	306.00	4.74	64.0	319.60	4.80	66.0	333.20	4.86
7.0	61.0	304.50	4.78	63.0	318.50	4.85	65.0	332.50	4.91	67.0	346.50	4.97
7.2	62.0	316.80	4.89	64.0	331.20	4.96	66.0	345.60	5.03	68.0	360.00	5.09
7.4	63.0	329.30	5.00	65.0	344.10	5.07	67.0	358.90	5.14	69.0	373.70	5.20
7.6	64.0	342.00	5.11	66.0	357.20	5.18	68.0	372.40	5.25	70.0	387.60	5.31
7.8	65.0	354.90	5.22	67.0	370.50	5.29	69.0	386.10	5.36	71.0	401.70	5.43
8.0	66.0	368.00	5.33	68.0	384.00	5.40	70.0	400.00	5.47	72.0	416.00	5.54
8.2	67.0	381.30	5.43	69.0	397.70	5.51	71.0	414.10	5.58	73.0	430.50	5.65
8.4	68.0	394.80	5.54	70.0	411.60	5.62	72.0	428.40	5.69	74.0	445.20	5.76
8.6	69.0	408.50	5.76	71.0	425.70	5.73	73.0	442.90	5.80	75.0	460.10	5.88
8.8	70.0	422.40	5.76	72.0	440.00	5.84	74.0	457.60	5.91	76.0	475.20	5.99
9.0	71.0	436.50	5.86	73.0	454.50	5.94	75.0	472.50	6.02	77.0	490.50	6.10
9.2	72.0	450.80	5.97	74.0	469.20	6.05	76.0	487.60	6.13	78.0	506.00	6.21
9.4	73.0	465.30	6.07	75.0	484.10	6.16	77.0	502.90	6.24	79.0	521.70	6.31
9.6	74.0	480.00	6.18	76.0	499.20	6.26	78.0	518.40	6.35	80.0	537.60	6.42
9.8	75.0	494.90	6.28	77.0	514.50	6.37	79.0	534.10	6.45	81.0	553.70	6.53

Table 34.—*Area in square feet, A, top width in feet, T, and hydraulic radius in feet, r, of trapezoidal channels,*
side slopes 2½ to 1—Continued

Depth	Bottom width 26 feet			Bottom width 28 feet			Bottom width 30 feet			Bottom width 32 feet		
	T	A	r	T	A	r	T	A	r	T	A	r
10.0	76.0	510.00	6.39	78.0	530.00	6.48	80.0	550.00	6.56	82.0	570.00	6.64
10.5	78.5	548.62	6.65	80.5	569.62	6.74	82.5	590.62	6.82	84.5	611.62	6.91
11.0	81.0	588.5	6.90	83.0	610.5	7.00	85.0	632.5	7.09	87.0	654.5	7.17
11.5	83.5	629.6	7.16	85.5	652.6	7.26	87.5	675.6	7.35	89.5	698.6	7.44
12.0	86.0	672.0	7.42	88.0	696.0	7.51	90.0	720.0	7.61	92.0	744.0	7.70
12.5	88.5	715.6	7.67	90.5	740.6	7.77	92.5	765.6	7.87	94.5	790.6	7.96
13.0	91.0	760.5	7.92	93.0	786.5	8.02	95.0	812.5	8.12	97.0	838.5	8.22
13.5	93.5	806.6	8.17	95.5	833.6	8.28	97.5	860.6	8.38	99.5	887.6	8.48
14.0	96.0	854.0	8.42	98.0	882.0	8.53	100.0	910.0	8.63	102.0	938.0	8.73
14.5	98.5	902.6	8.67	100.5	931.6	8.78	102.5	960.6	8.89	104.5	989.6	8.99
15.0	101.0	952.5	8.92	103.0	982.5	9.03	105.0	1,012.5	9.14	107.0	1,042.5	9.24
15.5	103.5	1,003.6	9.17	105.5	1,034.6	9.28	107.5	1,065.6	9.39	109.5	1,096.6	9.50
16.0	106.0	1,056.0	9.41	108.0	1,088.0	9.53	110.0	1,120.0	9.64	112.0	1,152.0	9.75
16.5	108.5	1,109.6	9.66	110.5	1,142.6	9.78	112.5	1,175.6	9.89	114.5	1,208.6	10.00
17.0	111.0	1,164.5	9.91	113.0	1,198.5	10.03	115.0	1,232.5	10.14	117.0	1,266.5	10.25
17.5	113.5	1,220.6	10.15	115.5	1,255.6	10.27	117.5	1,290.6	10.39	119.5	1,325.6	10.50
18.0	116.0	1,278.0	10.40	118.0	1,314.0	10.52	120.0	1,350.0	10.64	122.0	1,386.0	10.75
18.5	118.5	1,336.6	10.64	120.5	1,373.6	10.76	122.5	1,410.6	10.88	124.5	1,447.6	11.00
19.0	121.0	1,396.5	10.88	123.0	1,434.5	11.01	125.0	1,472.5	11.13	127.0	1,510.5	11.25
19.5	123.5	1,457.6	11.13	125.5	1,496.6	11.25	127.5	1,535.6	11.37	129.5	1,574.6	11.49
20.0	126.0	1,520.0	11.37	128.0	1,560.0	11.50	130.0	1,600.0	11.62	132.0	1,640.0	11.74

Depth	Bottom width 35 feet			Bottom width 40 feet			Bottom width 45 feet			Bottom width 50 feet		
	T	A	r	T	A	r	T	A	r	T	A	r
0.4	37.0	14.40	.39	42.0	16.40	.39	47.0	18.40	.39	52.0	20.40	.39
0.6	38.0	21.90	.57	43.0	24.90	.58	48.0	27.90	.58	53.0	30.90	.58
0.8	39.0	29.60	.75	44.0	33.60	.76	49.0	37.60	.76	54.0	41.60	.77
1.0	40.0	37.50	.93	45.0	42.50	.94	50.0	47.50	.94	55.0	52.50	.95
1.2	41.0	45.60	1.10	46.0	51.60	1.11	51.0	57.60	1.12	56.0	63.60	1.13
1.4	42.0	53.90	1.27	47.0	60.90	1.28	52.0	67.90	1.29	57.0	74.90	1.30
1.6	43.0	62.40	1.43	48.0	70.40	1.45	53.0	78.40	1.46	58.0	86.40	1.47
1.8	44.0	71.10	1.59	49.0	80.10	1.61	54.0	89.10	1.63	59.0	98.10	1.64
2.0	45.0	80.00	1.75	50.0	90.00	1.77	55.0	100.00	1.79	60.0	110.00	1.81
2.2	46.0	89.10	1.90	51.0	100.10	1.93	56.0	111.10	1.95	61.0	122.10	1.97
2.4	47.0	98.40	2.05	52.0	110.40	2.09	57.0	122.40	2.11	62.0	134.40	2.14
2.6	48.0	107.90	2.20	53.0	120.90	2.24	58.0	133.90	2.27	63.0	146.90	2.30
2.8	49.0	117.60	2.35	54.0	131.60	2.39	59.0	145.60	2.42	64.0	159.60	2.45
3.0	50.0	127.50	2.49	55.0	142.50	2.54	60.0	157.50	2.58	65.0	172.50	2.61
3.2	51.0	137.60	2.63	56.0	153.60	2.68	61.0	169.60	2.73	66.0	185.60	2.76
3.4	52.0	147.90	2.77	57.0	164.90	2.83	62.0	181.90	2.87	67.0	198.90	2.91
3.6	53.0	158.40	2.91	58.0	176.40	2.97	63.0	194.40	3.02	68.0	212.40	3.06
3.8	54.0	169.10	3.05	59.0	188.10	3.11	64.0	207.10	3.16	69.0	226.10	3.21
4.0	55.0	180.00	3.18	60.0	200.00	3.25	65.0	220.00	3.31	70.0	240.00	3.35
4.2	56.0	191.10	3.32	61.0	212.10	3.39	66.0	233.10	3.45	71.0	254.10	3.50
4.4	57.0	202.40	3.45	62.0	224.40	3.52	67.0	246.40	3.59	72.0	268.40	3.64
4.6	58.0	213.90	3.58	63.0	236.90	3.66	68.0	259.90	3.73	73.0	282.90	3.78
4.8	59.0	225.60	3.71	64.0	249.60	3.79	69.0	273.60	3.86	74.0	297.60	3.92

Table 34.—*Area in square feet, A, top width in feet, T, and hydraulic radius in feet, r, of trapezoidal channels,* **side slopes 2½ to 1**—Continued

Depth	Bottom width 35 feet			Bottom width 40 feet			Bottom width 45 feet			Bottom width 50 feet		
	T	A	r	T	A	r	T	A	r	T	A	r
5.0	60.0	237.50	3.84	65.0	262.50	3.92	70.0	287.50	4.00	75.0	312.50	4.06
5.2	61.0	249.60	3.96	66.0	275.60	4.05	71.0	301.60	4.13	76.0	327.60	4.20
5.4	62.0	261.90	4.09	67.0	288.90	4.18	72.0	315.90	4.26	77.0	342.90	4.34
5.6	63.0	274.40	4.21	68.0	302.40	4.31	73.0	330.40	4.40	78.0	358.40	4.47
5.8	64.0	287.10	4.33	69.0	316.10	4.44	74.0	345.10	4.53	79.0	374.10	4.61
6.0	65.0	300.00	4.46	70.0	330.00	4.56	75.0	360.00	4.66	80.0	390.00	4.74
6.2	66.0	313.10	4.58	71.0	344.10	4.69	76.0	375.10	4.79	81.0	406.10	4.87
6.4	67.0	326.40	4.70	72.0	358.40	4.81	77.0	390.40	4.91	82.0	422.40	5.00
6.6	68.0	339.90	4.82	73.0	372.90	4.94	78.0	405.90	5.04	83.0	438.90	5.13
6.8	69.0	353.60	4.94	74.0	387.60	5.06	79.0	421.60	5.17	84.0	455.60	5.26
7.0	70.0	367.50	5.06	75.0	402.50	5.18	80.0	437.50	5.29	85.0	472.50	5.39
7.2	71.0	381.60	5.17	76.0	417.60	5.30	81.0	453.60	5.41	86.0	489.60	5.52
7.4	72.0	395.90	5.29	77.0	432.90	5.42	82.0	469.90	5.54	87.0	506.90	5.64
7.6	73.0	410.40	5.41	78.0	448.40	5.54	83.0	486.40	5.66	88.0	524.40	5.77
7.8	74.0	425.10	5.52	79.0	464.10	5.66	84.0	503.10	5.78	89.0	542.10	5.89
8.0	75.0	440.00	5.64	80.0	480.00	5.78	85.0	520.00	5.90	90.0	560.00	6.02
8.2	76.0	455.10	5.75	81.0	496.10	5.89	86.0	537.10	6.02	91.0	578.10	6.14
8.4	77.0	470.40	5.86	82.0	512.40	6.01	87.0	554.40	6.14	92.0	596.40	6.26
8.6	78.0	485.90	5.98	83.0	528.90	6.13	88.0	571.90	6.26	93.0	614.90	6.38
8.8	79.0	501.60	6.09	84.0	545.60	6.24	89.0	589.60	6.38	94.0	633.60	6.51
9.0	80.0	517.50	6.20	85.0	562.50	6.36	90.0	607.50	6.50	95.0	652.50	6.63
9.2	81.0	533.60	6.31	86.0	579.60	6.47	91.0	625.60	6.62	96.0	671.60	6.75
9.4	82.0	549.90	6.42	87.0	596.90	6.59	92.0	643.90	6.73	97.0	690.90	6.87
9.6	83.0	566.40	6.53	88.0	614.40	6.70	93.0	662.40	6.85	98.0	710.40	6.99
9.8	84.0	583.10	6.64	89.0	632.10	6.81	94.0	681.10	6.97	99.0	730.10	7.10
10.0	85.0	600.00	6.75	90.0	650.00	6.93	95.0	700.00	7.08	100.0	750.00	7.22
10.5	87.5	643.12	7.03	92.5	695.62	7.21	97.5	748.12	7.37	102.5	800.62	7.51
11.0	90.0	687.5	7.30	95.0	742.5	7.48	100.0	797.5	7.65	105.0	852.5	7.80
11.5	92.5	733.1	7.56	97.5	790.6	7.76	102.5	848.1	7.93	107.5	905.6	8.09
12.0	95.0	780.0	7.83	100.0	840.0	8.03	105.0	900.0	8.21	110.0	960.0	8.38
12.5	97.5	828.1	8.09	102.5	890.6	8.30	107.5	953.1	8.49	112.5	1,015.6	8.66
13.0	100.0	877.5	8.36	105.0	942.5	8.57	110.0	1,007.5	8.76	115.0	1,072.5	8.94
13.5	102.5	928.1	8.62	107.5	995.6	8.83	112.5	1,063.1	9.03	117.5	1,130.6	9.21
14.0	105.0	980.0	8.88	110.0	1,050.0	9.10	115.0	1,120.0	9.30	120.0	1,190.0	9.49
14.5	107.5	1,033.1	9.14	112.5	1,105.6	9.36	117.5	1,178.1	9.57	122.5	1,250.6	9.76
15.0	110.0	1,087.5	9.39	115.0	1,162.5	9.63	120.0	1,237.5	9.84	125.0	1,312.5	10.04
15.5	112.5	1,143.1	9.65	117.5	1,220.6	9.89	122.5	1,298.1	10.10	127.5	1,375.6	10.31
16.0	115.0	1,200.0	9.90	120.0	1,280.0	10.15	125.0	1,360.0	10.37	130.0	1,440.0	10.58
16.5	117.5	1,258.1	10.16	122.5	1,340.6	10.40	127.5	1,423.1	10.63	132.5	1,505.6	10.84
17.0	120.0	1,317.5	10.41	125.0	1,402.5	10.66	130.0	1,487.5	10.89	135.0	1,572.5	11.11
17.5	122.5	1,378.1	10.66	127.5	1,465.6	10.92	132.5	1,553.1	11.15	137.5	1,640.6	11.37
18.0	125.0	1,440.0	10.91	130.0	1,530.0	11.17	135.0	1,620.0	11.41	140.0	1,710.0	11.64
18.5	127.5	1,503.1	11.17	132.5	1,595.6	11.43	137.5	1,688.1	11.67	142.5	1,780.6	11.90
19.0	130.0	1,567.5	11.42	135.0	1,662.5	11.68	140.0	1,757.5	11.93	145.0	1,852.5	12.16
19.5	132.5	1,633.1	11.66	137.5	1,730.6	11.93	142.5	1,828.1	12.19	147.5	1,925.6	12.42
20.0	135.0	1,700.0	11.91	140.0	1,800.0	12.19	145.0	1,900.0	12.44	150.0	2,000.0	12.68

Table 34.—*Area in square feet, A, top width in feet, T, and hydraulic radius in feet, r, of trapezoidal channels,* **side slopes 2½ to 1**—Continued

Depth	Bottom width 60 feet			Bottom width 70 feet			Bottom width 80 feet			Bottom width 90 feet		
	T	A	r	T	A	r	T	A	r	T	A	r
0.4	62.0	24.40	0.39	72.0	28.40	0.39	82.0	32.40	0.39	92.0	36.40	0.39
0.6	63.0	36.90	.58	73.0	42.90	.59	83.0	48.90	.59	93.0	54.90	.59
0.8	64.0	49.60	.77	74.0	57.60	.78	84.0	65.60	.78	94.0	73.60	.78
1.0	65.0	62.50	.96	75.0	72.50	.96	85.0	82.50	.97	95.0	92.50	.97
1.2	66.0	75.60	1.14	76.0	87.60	1.15	86.0	99.60	1.15	96.0	111.60	1.16
1.4	67.0	88.90	1.32	77.0	102.90	1.33	87.0	116.90	1.34	97.0	130.90	1.34
1.6	68.0	102.40	1.49	78.0	118.40	1.51	88.0	134.40	1.52	98.0	150.40	1.53
1.8	69.0	116.10	1.67	79.0	134.10	1.68	89.0	152.10	1.70	99.0	170.10	1.71
2.0	70.0	130.00	1.84	80.0	150.00	1.86	90.0	170.00	1.87	100.0	190.00	1.89
2.2	71.0	144.10	2.01	81.0	166.10	2.03	91.0	188.10	2.05	101.0	210.10	2.06
2.4	72.0	158.40	2.17	82.0	182.40	2.20	92.0	206.40	2.22	102.0	230.40	2.24
2.6	73.0	172.90	2.34	83.0	198.90	2.37	93.0	224.90	2.39	103.0	250.90	2.41
2.8	74.0	187.60	2.50	84.0	215.60	2.53	94.0	243.60	2.56	104.0	271.60	2.58
3.0	75.0	202.50	2.66	85.0	232.50	2.70	95.0	262.50	2.73	105.0	292.50	2.76
3.2	76.0	217.60	2.82	86.0	249.60	2.86	96.0	281.60	2.90	106.0	313.60	2.92
3.4	77.0	232.90	2.97	87.0	266.90	3.02	97.0	300.90	3.06	107.0	334.90	3.09
3.6	78.0	248.40	3.13	88.0	284.40	3.18	98.0	320.40	3.22	108.0	356.40	3.26
3.8	79.0	264.10	3.28	89.0	302.10	3.34	99.0	340.10	3.39	109.0	378.10	3.42
4.0	80.0	280.00	3.43	90.0	320.00	3.50	100.0	360.00	3.55	110.0	400.00	3.59
4.2	81.0	296.10	3.58	91.0	338.10	3.65	101.0	380.10	3.70	111.0	422.10	3.75
4.4	82.0	312.40	3.73	92.0	356.40	3.80	102.0	400.40	3.86	112.0	444.40	3.91
4.6	83.0	328.90	3.88	93.0	374.90	3.96	103.0	420.90	4.02	113.0	466.90	4.07
4.8	84.0	345.60	4.03	94.0	393.60	4.11	104.0	441.60	4.17	114.0	489.60	4.23
5.0	85.0	362.50	4.17	95.0	412.50	4.26	105.0	462.50	4.33	115.0	512.50	4.38
5.2	86.0	379.60	4.31	96.0	431.60	4.40	106.0	483.60	4.48	116.0	535.60	4.54
5.4	87.0	396.90	4.46	97.0	450.90	4.55	107.0	504.90	4.63	117.0	558.90	4.69
5.6	88.0	414.40	4.60	98.0	470.40	4.70	108.0	526.40	4.78	118.0	582.40	4.85
5.8	89.0	432.10	4.74	99.0	490.10	4.84	109.0	548.10	4.93	119.0	606.10	5.00
6.0	90.0	450.00	4.87	100.0	510.00	4.98	110.0	570.00	5.08	120.0	630.00	5.15
6.2	91.0	468.10	5.01	101.0	530.10	5.13	111.0	592.10	5.22	121.0	654.10	5.30
6.4	92.0	486.40	5.15	102.0	550.40	5.27	112.0	614.40	5.37	122.0	678.40	5.45
6.6	93.0	504.90	5.28	103.0	570.90	5.41	113.0	636.90	5.51	123.0	702.90	5.60
6.8	94.0	523.60	5.42	104.0	591.60	5.55	114.0	659.60	5.66	124.0	727.60	5.75
7.0	95.0	542.50	5.55	105.0	612.50	5.69	115.0	682.50	5.80	125.0	752.50	5.89
7.2	96.0	561.60	5.69	106.0	633.60	5.82	116.0	705.60	5.94	126.0	777.60	6.04
7.4	97.0	580.90	5.82	107.0	654.90	5.96	117.0	728.90	6.08	127.0	802.90	6.18
7.6	98.0	600.40	5.95	108.0	676.40	6.10	118.0	752.40	6.22	128.0	828.40	6.33
7.8	99.0	620.10	6.08	109.0	698.10	6.23	119.0	776.10	6.36	129.0	854.10	6.47
8.0	100.0	640.00	6.21	110.0	720.00	6.37	120.0	800.00	6.50	130.0	880.00	6.61
8.2	101.0	660.10	6.34	111.0	742.10	6.50	121.0	824.10	6.64	131.0	906.10	6.75
8.4	102.0	680.40	6.47	112.0	764.40	6.63	122.0	848.40	6.77	132.0	932.40	6.89
8.6	103.0	700.90	6.59	113.0	786.90	6.77	123.0	872.90	6.91	133.0	958.90	7.03
8.8	104.0	721.60	6.72	114.0	809.60	6.90	124.0	897.60	7.05	134.0	985.60	7.17
9.0	105.0	742.50	6.85	115.0	832.50	7.03	125.0	922.50	7.18	135.0	1,012.50	7.31
9.2	106.0	763.60	6.97	116.0	855.60	7.16	126.0	947.60	7.31	136.0	1,039.60	7.45
9.4	107.0	784.90	7.10	117.0	878.90	7.29	127.0	972.90	7.45	137.0	1,066.90	7.59
9.6	108.0	806.40	7.22	118.0	902.40	7.42	128.0	998.40	7.58	138.0	1,094.40	7.72
9.8	109.0	828.10	7.34	119.0	926.10	7.54	129.0	1,024.10	7.71	139.0	1,122.10	7.86

Table 34.—*Area in square feet, A, top width in feet, T, and hydraulic radius in feet, r, of trapezoidal channels,* **side slopes 2½ to 1**—Continued

Depth	Bottom width 60 feet			Bottom width 70 feet			Bottom width 80 feet			Bottom width 90 feet		
	T	A	r	T	A	r	T	A	r	T	A	r
10.0	110.0	850.00	7.47	120.0	950.00	7.67	130.0	1,050.00	7.84	140.0	1,150.00	7.99
10.5	112.5	905.62	7.77	122.5	1,010.62	7.99	132.5	1,115.62	8.17	142.5	1,220.62	8.33
11.0	115.0	962.5	8.07	125.0	1,072.5	8.30	135.0	1,182.5	8.49	145.0	1,292.5	8.66
11.5	117.5	1,020.6	8.37	127.5	1,135.6	8.61	137.5	1,250.6	8.81	147.5	1,365.6	8.99
12.0	120.0	1,080.0	8.67	130.0	1,200.0	8.91	140.0	1,320.0	9.13	150.0	1,440.0	9.31
12.5	122.5	1,140.6	8.96	132.5	1,265.6	9.22	142.5	1,390.6	9.44	152.5	1,515.6	9.63
13.0	125.0	1,202.5	9.25	135.0	1,332.5	9.52	145.0	1,462.5	9.75	155.0	1,592.5	9.95
13.5	127.5	1,265.6	9.54	137.5	1,400.6	9.82	147.5	1,535.6	10.06	157.5	1,670.6	10.27
14.0	130.0	1,330.0	9.82	140.0	1,470.0	10.11	150.0	1,610.0	10.36	160.0	1,750.0	10.58
14.5	132.5	1,395.6	10.11	142.5	1,540.6	10.40	152.5	1,685.6	10.66	162.5	1,830.6	10.89
15.0	135.0	1,462.5	10.39	145.0	1,612.5	10.69	155.0	1,762.5	10.96	165.0	1,912.5	11.20
15.5	137.5	1,530.6	10.67	147.5	1,685.6	10.98	157.5	1,840.6	11.26	167.5	1,995.6	11.50
16.0	140.0	1,600.0	10.95	150.0	1,760.0	11.27	160.0	1,920.0	11.55	170.0	2,080.0	11.81
16.5	142.5	1,670.6	11.22	152.5	1,835.6	11.56	162.5	2,000.6	11.85	172.5	2,165.6	12.11
17.0	145.0	1,742.5	11.50	155.0	1,912.5	11.84	165.0	2,082.5	12.14	175.0	2,252.5	12.41
17.5	147.5	1,815.6	11.77	157.5	1,990.6	12.12	167.5	2,165.6	12.43	177.5	2,340.6	12.70
18.0	150.0	1,890.0	12.04	160.0	2,070.0	12.40	170.0	2,250.0	12.72	180.0	2,430.0	13.00
18.5	152.5	1,965.6	12.31	162.5	2,150.6	12.68	172.5	2,335.6	13.00	182.5	2,520.6	13.29
19.0	155.0	2,042.5	12.58	165.0	2,232.5	12.96	175.0	2,422.5	13.29	185.0	2,612.5	13.58
19.5	157.5	2,120.6	12.85	167.5	2,315.6	13.23	177.5	2,510.6	13.57	187.5	2,705.6	13.87
20.0	160.0	2,200.0	13.12	170.0	2,400.0	13.51	180.0	2,600.0	13.85	190.0	2,800.0	14.16

Table 35.—*Area in square feet, A, top width in feet, T, and hydraulic radius in feet, r, of trapezoidal channels,*
side slopes 3 to 1

Depth	Bottom width 2 feet			Bottom width 3 feet			Bottom width 4 feet			Bottom width 5 feet		
	T	A	r	T	A	r	T	A	r	T	A	r
0.4	4.4	1.28	0.28	5.4	1.68	0.30	6.4	2.08	0.32	7.4	2.48	0.33
0.6	5.6	2.28	.39	6.6	2.88	.42	7.6	3.48	.45	8.6	4.08	.46
0.8	6.8	3.52	.50	7.8	4.32	.54	8.8	5.12	.57	9.8	5.92	.59
1.0	8.0	5.00	.60	9.0	6.00	.64	10.0	7.00	.68	11.0	8.00	.71
1.2	9.2	6.72	.70	10.2	7.92	.75	11.2	9.12	.79	12.2	10.32	.82
1.4	10.4	8.68	.80	11.4	10.08	.85	12.4	11.48	.89	13.4	12.88	.93
1.6	11.6	10.88	.90	12.6	12.48	.95	13.6	14.08	1.00	14.6	15.68	1.04
1.8	12.8	13.32	1.00	13.8	15.12	1.05	14.8	16.92	1.10	15.8	18.72	1.14
2.0	14.0	16.00	1.09	15.0	18.00	1.15	16.0	20.00	1.20	17.0	22.00	1.25
2.2	15.2	18.92	1.19	16.2	21.12	1.25	17.2	23.32	1.30	18.2	25.52	1.35
2.4	16.4	22.08	1.29	17.4	24.48	1.35	18.4	26.88	1.40	19.4	29.28	1.45
2.6	17.6	25.48	1.38	18.6	28.08	1.44	19.6	30.68	1.50	20.6	33.28	1.55
2.8	18.8	29.12	1.48	19.8	31.92	1.54	20.8	34.72	1.60	21.8	37.52	1.65
3.0	20.0	33.00	1.57	21.0	36.00	1.64	22.0	39.00	1.70	23.0	42.00	1.75
3.2	21.2	37.12	1.67	22.2	40.32	1.74	23.2	43.52	1.80	24.2	46.72	1.85
3.4	22.4	41.48	1.76	23.4	44.88	1.83	24.4	48.28	1.89	25.4	51.68	1.95
3.6	23.6	46.08	1.86	24.6	49.68	1.93	25.6	53.28	1.99	26.6	56.88	2.05
3.8	24.8	50.92	1.96	25.8	54.72	2.02	26.8	58.52	2.09	27.8	62.32	2.15
4.0	26.0	56.00	2.05	27.0	60.00	2.12	28.0	64.00	2.18	29.0	68.00	2.24
4.2	27.2	61.32	2.15	28.2	65.52	2.22	29.2	69.72	2.28	30.2	73.92	2.34
4.4	28.4	66.88	2.24	29.4	71.28	2.31	30.4	75.68	2.38	31.4	80.08	2.44
4.6	29.6	72.68	2.34	30.6	77.28	2.41	31.6	81.88	2.47	32.6	86.48	2.54
4.8	30.8	78.72	2.43	31.8	83.52	2.50	32.8	88.32	2.57	33.8	93.12	2.63
5.0	32.0	85.00	2.53	33.0	90.00	2.60	34.0	95.00	2.67	35.0	100.00	2.73
5.2	33.2	91.52	2.62	34.2	96.72	2.70	35.2	101.92	2.76	36.2	107.12	2.83
5.4	34.4	98.28	2.72	35.4	103.68	2.79	36.4	109.08	2.86	37.4	114.48	2.92
5.6	35.6	105.28	2.81	36.6	110.88	2.89	37.6	116.48	2.96	38.6	122.08	3.02
5.8	36.8	112.52	2.91	37.8	118.32	2.98	38.8	124.12	3.05	39.8	129.92	3.12
6.0	38.0	120.00	3.00	39.0	126.00	3.08	40.0	132.00	3.15	41.0	138.00	3.21
6.2	39.2	127.72	3.10	40.2	133.92	3.17	41.2	140.12	3.24	42.2	146.32	3.31
6.4	40.4	135.68	3.19	41.4	142.08	3.27	42.4	148.48	3.34	43.4	154.88	3.41
6.6	41.6	143.88	3.29	42.6	150.48	3.36	43.6	157.08	3.43	44.6	163.68	3.50
6.8	42.8	152.32	3.38	43.8	159.12	3.46	44.8	165.92	3.53	45.8	172.72	3.60
7.0	44.0	161.00	3.48	45.0	168.00	3.55	46.0	175.00	3.63	47.0	182.00	3.69
7.2	45.2	169.92	3.57	46.2	177.12	3.65	47.2	184.32	3.72	48.2	191.52	3.79
7.4	46.4	179.08	3.67	47.4	186.48	3.74	48.4	193.88	3.82	49.4	201.28	3.89
7.6	47.6	188.48	3.76	48.6	196.08	3.84	49.6	203.68	3.91	50.6	211.28	3.98
7.8	48.8	198.12	3.86	49.8	205.92	3.93	50.8	213.72	4.01	51.8	221.52	4.08
8.0	50.0	208.00	3.95	51.0	216.00	4.03	52.0	224.00	4.10	53.0	232.00	4.17
8.2	----	----	----	52.2	226.32	4.13	53.2	234.52	4.20	54.2	242.72	4.27
8.4	----	----	----	53.4	236.88	4.22	54.4	245.28	4.29	55.4	253.68	4.36
8.6	----	----	----	54.6	247.68	4.32	55.6	256.28	4.39	56.6	264.88	4.46
8.8	----	----	----	55.8	258.72	4.41	56.8	267.52	4.48	57.8	276.32	4.56
9.0	----	----	----	57.0	270.00	4.51	58.0	279.00	4.58	59.0	288.00	4.65
9.2	----	----	----	58.2	281.52	4.60	59.2	290.72	4.68	60.2	299.92	4.75
9.4	----	----	----	59.4	293.28	4.70	60.4	302.68	4.77	61.4	312.08	4.84
9.6	----	----	----	60.6	305.28	4.79	61.6	314.88	4.87	62.6	324.48	4.94
9.8	----	----	----	61.8	317.52	4.89	62.8	327.32	4.96	63.8	337.12	5.03

Table 35.—*Area in square feet, A, top width in feet, T, and hydraulic radius in feet, r, of trapezoidal channels,* **side slopes 3 to 1**—Continued

Depth	Bottom width 2 feet			Bottom width 3 feet			Bottom width 4 feet			Bottom width 5 feet		
	T	A	r	T	A	r	T	A	r	T	A	r
10.0	----	--------	----	63.0	330.00	4.98	64.0	340.00	5.06	65.0	350.00	5.13
10.5	----	--------	----	66.0	362.25	5.22	67.0	372.75	5.29	68.0	383.25	5.37
11.0	----	--------	----	69.0	396.0	5.46	70.0	407.0	5.53	71.0	418.0	5.61
11.5	----	--------	----	72.0	431.3	5.69	73.0	442.8	5.77	74.0	454.3	5.84
12.0	----	--------	----	75.0	468.0	5.93	76.0	480.0	6.01	77.0	492.0	6.08
12.5	----	--------	----	78.0	506.3	6.17	79.0	518.8	6.25	80.0	531.3	6.32
13.0							82.0	559.0	6.48	83.0	572.0	6.56
13.5							85.0	600.8	6.72	86.0	614.3	6.80
14.0							88.0	644.0	6.96	89.0	658.0	7.03
14.5							91.0	688.8	7.20	92.0	703.3	7.27
15.0							94.0	735.0	7.43	95.0	750.0	7.51
15.5							97.0	782.8	7.67	98.0	798.3	7.75
16.0							100.0	832.0	7.91	101.0	848.0	7.99
16.5										104.0	899.3	8.22
17.0										107.0	952.0	8.46
17.5										110.0	1,006.3	8.70
18.0										113.0	1,062.0	8.94
18.5										116.0	1,119.3	9.17
19.0										119.0	1,178.0	9.41
19.5										122.0	1,238.3	9.65
20.0										125.0	1,300.0	9.89

Depth	Bottom width 6 feet			Bottom width 7 feet			Bottom width 8 feet			Bottom width 9 feet		
	T	A	r	T	A	r	T.	A	r	T	A	r
0.4	8.4	2.88	.34	9.4	3.28	.34	10.4	3.68	.35	11.4	4.08	.35
0.6	9.6	4.68	.48	10.6	5.28	.49	11.6	5.88	.50	12.6	6.48	.51
0.8	10.8	6.72	.61	11.8	7.52	.62	12.8	8.32	.64	13.8	9.12	.65
1.0	12.0	9.00	.73	13.0	10.00	.75	14.0	11.00	.77	15.0	12.00	.78
1.2	13.2	11.52	.85	14.2	12.72	.87	15.2	13.92	.89	16.2	15.12	.91
1.4	14.4	14.28	.96	15.4	15.68	.99	16.4	17.08	1.01	17.4	18.48	1.04
1.6	15.6	17.28	1.07	16.6	18.88	1.10	17.6	20.48	1.13	18.6	22.08	1.15
1.8	16.8	20.52	1.18	17.8	22.32	1.21	18.8	24.12	1.24	19.8	25.92	1.27
2.0	18.0	24.00	1.29	19.0	26.00	1.32	20.0	28.00	1.36	21.0	30.00	1.39
2.2	19.2	27.72	1.39	20.2	29.92	1.43	21.2	32.12	1.47	22.2	34.32	1.50
2.4	20.4	31.68	1.50	21.4	34.08	1.54	22.4	36.48	1.57	23.4	38.88	1.61
2.6	21.6	35.88	1.60	22.6	38.48	1.64	23.6	41.08	1.68	24.6	43.68	1.72
2.8	22.8	40.32	1.70	23.8	43.12	1.75	24.8	45.92	1.79	25.8	48.72	1.82
3.0	24.0	45.00	1.80	25.0	48.00	1.85	26.0	51.00	1.89	27.0	54.00	1.93
3.2	25.2	49.92	1.90	26.2	53.12	1.95	27.2	56.32	1.99	28.2	59.52	2.04
3.4	26.4	55.08	2.00	27.4	58.48	2.05	28.4	61.88	2.10	29.4	65.28	2.14
3.6	27.6	60.48	2.10	28.6	64.08	2.15	29.6	67.68	2.20	30.6	71.28	2.24
3.8	28.8	66.12	2.20	29.8	69.92	2.25	30.8	73.72	2.30	31.8	77.52	2.35
4.0	30.0	72.00	2.30	31.0	76.00	2.35	32.0	80.00	2.40	33.0	84.00	2.45
4.2	31.2	78.12	2.40	32.2	82.32	2.45	33.2	86.52	2.50	34.2	90.72	2.55
4.4	32.4	84.48	2.50	33.4	88.88	2.55	34.4	93.28	2.60	35.4	97.68	2.65
4.6	33.6	91.08	2.60	34.6	95.68	2.65	35.6	100.28	2.70	36.6	104.88	2.75
4.8	34.8	97.92	2.69	35.8	102.72	2.75	36.8	107.52	2.80	37.8	112.32	2.85

Table 35.—*Area in square feet, A, top width in feet, T, and hydraulic radius in feet, r, of trapezoidal channels,*
side slopes 3 to 1—Continued

Depth	Bottom width 6 feet			Bottom width 7 feet			Bottom width 8 feet			Bottom width 9 feet		
	T	A	r	T	A	r	T	A	r	T	A	r
5.0	36.0	105.00	2.79	37.0	110.00	2.85	38.0	115.00	2.90	39.0	120.00	2.95
5.2	37.2	112.32	2.89	38.2	117.52	2.95	39.2	122.72	3.00	40.2	127.92	3.05
5.4	38.4	119.88	2.99	39.4	125.28	3.04	40.4	130.68	3.10	41.4	136.08	3.15
5.6	39.6	127.68	3.08	40.6	133.28	3.14	41.6	138.88	3.20	42.6	144.48	3.25
5.8	40.8	135.72	3.18	41.8	141.52	3.24	42.8	147.32	3.30	43.8	153.12	3.35
6.0	42.0	144.00	3.28	43.0	150.00	3.34	44.0	156.00	3.40	45.0	162.00	3.45
6.2	43.2	152.52	3.37	44.2	158.72	3.43	45.2	164.92	3.49	46.2	171.12	3.55
6.4	44.4	161.28	3.47	45.4	167.68	3.53	46.4	174.08	3.59	47.4	180.48	3.65
6.6	45.6	170.28	3.57	46.6	176.88	3.63	47.6	183.48	3.69	48.6	190.08	3.75
6.8	46.8	179.52	3.66	47.8	186.32	3.73	48.8	193.12	3.79	49.8	199.92	3.84
7.0	48.0	189.00	3.76	49.0	196.00	3.82	50.0	203.00	3.88	51.0	210.00	3.94
7.2	49.2	198.72	3.86	50.2	205.92	3.92	51.2	213.12	3.98	52.2	220.32	4.04
7.4	50.4	208.68	3.95	51.4	216.08	4.02	52.4	223.48	4.08	53.4	230.88	4.14
7.6	51.6	218.88	4.05	52.6	226.48	4.11	53.6	234.08	4.18	54.6	241.68	4.24
7.8	52.8	229.32	4.14	53.8	237.12	4.21	54.8	244.92	4.27	55.8	252.72	4.33
8.0	54.0	240.00	4.24	55.0	248.00	4.31	56.0	256.00	4.37	57.0	264.00	4.43
8.2	55.2	250.92	4.34	56.2	259.12	4.40	57.2	267.32	4.47	58.2	275.52	4.53
8.4	56.4	262.08	4.43	57.4	270.48	4.50	58.4	278.88	4.56	59.4	287.28	4.62
8.6	57.6	273.48	4.53	58.6	282.08	4.59	59.6	290.68	4.66	60.6	299.28	4.72
8.8	58.8	285.12	4.62	59.8	293.92	4.69	60.8	302.72	4.76	61.8	311.52	4.82
9.0	60.0	297.00	4.72	61.0	306.00	4.79	62.0	315.00	4.85	63.0	324.00	4.91
9.2	61.2	309.12	4.82	62.2	318.32	4.88	63.2	327.52	4.95	64.2	336.72	5.01
9.4	62.4	321.48	4.91	63.4	330.88	4.98	64.4	340.28	5.04	65.4	349.68	5.11
9.6	63.6	334.08	5.01	64.6	343.68	5.08	65.6	353.28	5.14	66.6	362.88	5.21
9.8	64.8	346.92	5.10	65.8	356.72	5.17	66.8	366.52	5.24	67.8	376.32	5.30
10.0	66.0	360.00	5.20	67.0	370.00	5.27	68.0	380.00	5.33	69.0	390.00	5.40
10.5	69.0	393.75	5.44	70.0	404.25	5.51	71.0	414.75	5.57	72.0	425.25	5.64
11.0	72.0	429.0	5.68	73.0	440.0	5.75	74.0	451.0	5.81	75.0	462.0	5.88
11.5	75.0	465.8	5.92	76.0	477.3	5.99	77.0	488.8	6.05	78.0	500.3	6.12
12.0	78.0	504.0	6.15	79.0	516.0	6.22	80.0	528.0	6.29	81.0	540.0	6.36
12.5	81.0	543.8	6.39	82.0	556.3	6.46	83.0	568.8	6.53	84.0	581.3	6.60
13.0	84.0	585.0	6.63	85.0	598.0	6.70	86.0	611.0	6.77	87.0	624.0	6.84
13.5	87.0	627.8	6.87	88.0	641.3	6.94	89.0	654.8	7.01	90.0	668.3	7.08
14.0	90.0	672.0	7.11	91.0	686.0	7.18	92.0	700.0	7.25	93.0	714.0	7.32
14.5	93.0	717.8	7.35	94.0	732.3	7.42	95.0	746.8	7.49	96.0	761.3	7.56
15.0	96.0	765.0	7.58	97.0	780.0	7.66	98.0	795.0	7.73	99.0	810.0	7.80
15.5	99.0	813.8	7.82	100.0	829.3	7.90	101.0	844.8	7.97	102.0	860.3	8.04
16.0	102.0	864.0	8.06	103.0	880.0	8.13	104.0	896.0	8.21	105.0	912.0	8.28
16.5	105.0	915.8	8.30	106.0	932.3	8.37	107.0	948.8	8.44	108.0	965.3	8.52
17.0	108.0	969.0	8.54	109.0	986.0	8.61	110.0	1,003.0	8.68	111.0	1,020.0	8.75
17.5	111.0	1,023.8	8.77	112.0	1,041.3	8.85	113.0	1,058.8	8.92	114.0	1,076.3	8.99
18.0	114.0	1,080.0	9.01	115.0	1,098.0	9.09	116.0	1,116.0	9.16	117.0	1,134.0	9.23
18.5	117.0	1,137.8	9.25	118.0	1,156.3	9.32	119.0	1,174.8	9.40	120.0	1,193.3	9.47
19.0	120.0	1,197.0	9.49	121.0	1,216.0	9.56	122.0	1,235.0	9.64	123.0	1,254.0	9.71
19.5	123.0	1,257.8	9.73	124.0	1,277.3	9.80	125.0	1,296.8	9.87	126.0	1,316.3	9.95
20.0	126.0	1,320.0	9.96	127.0	1,340.0	10.04	128.0	1,360.0	10.11	129.0	1,380.0	10.19

Table 35.—*Area in square feet, A, top width in feet, T, and hydraulic radius in feet, r, of trapezoidal channels,* **side slopes 3 to 1**—Continued

Depth	Bottom width 10 feet			Bottom width 12 feet			Bottom width 14 feet			Bottom width 16 feet		
	T	A	r	T	A	r	T	A	r	T	A	r
0.4	12.4	4.48	0.36	14.4	5.28	0.36	16.4	6.08	0.37	18.4	6.88	0.37
0.6	13.6	7.08	.51	15.6	8.28	.52	17.6	9.48	.53	19.6	10.68	.54
0.8	14.8	9.92	.66	16.8	11.52	.68	18.8	13.12	.69	20.8	14.72	.70
1.0	16.0	13.00	.80	18.0	15.00	.82	20.0	17.00	.84	22.0	19.00	.85
1.2	17.2	16.32	.93	19.2	18.72	.96	21.2	21.12	.98	23.2	23.52	1.00
1.4	18.4	19.88	1.05	20.4	22.68	1.09	22.4	25.48	1.11	24.4	28.28	1.14
1.6	19.6	23.68	1.18	21.6	26.88	1.22	23.6	30.08	1.25	25.6	33.28	1.27
1.8	20.8	27.72	1.30	22.8	31.32	1.34	24.8	34.92	1.38	26.8	38.52	1.41
2.0	22.0	32.00	1.41	24.0	36.00	1.46	26.0	40.00	1.50	28.0	44.00	1.54
2.2	23.2	36.52	1.53	25.2	40.92	1.58	27.2	45.32	1.62	29.2	49.72	1.66
2.4	24.4	41.28	1.64	26.4	46.08	1.70	28.4	50.88	1.74	30.4	55.68	1.79
2.6	25.6	46.28	1.75	27.6	51.48	1.81	29.6	56.68	1.86	31.6	61.88	1.91
2.8	26.8	51.52	1.86	28.8	57.12	1.92	30.8	62.72	1.98	32.8	68.32	2.03
3.0	28.0	57.00	1.97	30.0	63.00	2.03	32.0	69.00	2.09	34.0	75.00	2.14
3.2	29.2	62.72	2.07	31.2	69.12	2.14	33.2	75.52	2.21	35.2	81.92	2.26
3.4	30.4	68.68	2.18	32.4	75.48	2.25	34.4	82.28	2.32	36.4	89.08	2.38
3.6	31.6	74.88	2.29	33.6	82.08	2.36	35.6	89.28	2.43	37.6	96.48	2.49
3.8	32.8	81.32	2.39	34.8	88.92	2.47	36.8	96.52	2.54	38.8	104.12	2.60
4.0	34.0	88.00	2.49	36.0	96.00	2.57	38.0	104.00	2.65	40.0	112.00	2.71
4.2	35.2	94.92	2.60	37.2	103.32	2.68	39.2	111.72	2.75	41.2	120.12	2.82
4.4	36.4	102.08	2.70	38.4	110.88	2.78	40.4	119.68	2.86	42.4	128.48	2.93
4.6	37.6	109.48	2.80	39.6	118.68	2.89	41.6	127.88	2.97	43.6	137.08	3.04
4.8	38.8	117.12	2.90	40.8	126.72	2.99	42.8	136.32	3.07	44.8	145.92	3.15
5.0	40.0	125.00	3.00	42.0	135.00	3.09	44.0	145.00	3.18	46.0	155.00	3.25
5.2	41.2	133.12	3.10	43.2	143.52	3.20	45.2	153.92	3.28	47.2	164.32	3.36
5.4	42.4	141.48	3.20	44.4	152.28	3.30	46.4	163.08	3.39	48.4	173.88	3.47
5.6	43.6	150.08	3.30	45.6	161.28	3.40	47.6	172.48	3.49	49.6	183.68	3.57
5.8	44.8	158.92	3.40	46.8	170.52	3.50	48.8	182.12	3.59	50.8	193.72	3.68
6.0	46.0	168.00	3.50	48.0	180.00	3.60	50.0	192.00	3.70	52.0	204.00	3.78
6.2	47.2	177.32	3.60	49.2	189.72	3.70	51.2	202.12	3.80	53.2	214.52	3.89
6.4	48.4	186.88	3.70	50.4	199.68	3.81	52.4	212.48	3.90	54.4	225.28	3.99
6.6	49.6	196.68	3.80	51.6	209.88	3.91	53.6	223.08	4.00	55.6	236.28	4.09
6.8	50.8	206.72	3.90	52.8	220.32	4.01	54.8	233.92	4.10	56.8	247.52	4.19
7.0	52.0	217.00	4.00	54.0	231.00	4.11	56.0	245.00	4.20	58.0	259.00	4.30
7.2	53.2	227.52	4.10	55.2	241.92	4.20	57.2	256.32	4.31	59.2	270.72	4.40
7.4	54.4	238.28	4.19	56.4	253.08	4.30	58.4	267.88	4.41	60.4	282.68	4.50
7.6	55.6	249.28	4.29	57.6	264.48	4.40	59.6	279.68	4.51	61.6	294.88	4.60
7.8	56.8	260.52	4.39	58.8	276.12	4.50	60.8	291.72	4.61	62.8	307.32	4.70
8.0	58.0	272.00	4.49	60.0	288.00	4.60	62.0	304.00	4.71	64.0	320.00	4.81
8.2	59.2	283.72	4.59	61.2	300.12	4.70	63.2	316.52	4.81	65.2	332.92	4.91
8.4	60.4	295.68	4.68	62.4	312.48	4.80	64.4	329.28	4.91	66.4	346.08	5.01
8.6	61.6	307.88	4.78	63.6	325.08	4.90	65.6	342.28	5.00	67.6	359.48	5.11
8.8	62.8	320.32	4.88	64.8	337.92	4.99	66.8	355.52	5.10	68.8	373.12	5.21
9.0	64.0	333.00	4.98	66.0	351.00	5.09	68.0	369.00	5.20	70.0	387.00	5.31
9.2	65.2	345.92	5.07	67.2	364.32	5.19	69.2	382.72	5.30	71.2	401.12	5.41
9.4	66.4	359.08	5.17	68.4	377.88	5.29	70.4	396.68	5.40	72.4	415.48	5.51
9.6	67.6	372.48	5.27	69.6	391.68	5.39	71.6	410.88	5.50	73.6	430.08	5.61
9.8	68.8	386.12	5.36	70.8	405.72	5.48	72.8	425.32	5.60	74.8	444.92	5.71

Table 35.—*Area in square feet, A, top width in feet, T, and hydraulic radius in feet, r, of trapezoidal channels,*
side slopes 3 to 1—Continued

Depth	Bottom width 10 feet			Bottom width 12 feet			Bottom width 14 feet			Bottom width 16 feet		
	T	A	r	T	A	r	T	A	r	T	A	r
10.0	70.0	400.00	5.46	72.0	420.00	5.58	74.0	440.00	5.70	76.0	460.00	5.80
10.5	73.0	435.75	5.70	75.0	456.75	5.83	77.0	477.75	5.94	79.0	498.75	6.05
11.0	76.0	473.0	5.94	78.0	495.0	6.07	80.0	517.0	6.19	82.0	539.0	6.30
11.5	79.0	511.8	6.19	81.0	534.8	6.31	83.0	557.8	6.43	85.0	580.8	6.54
12.0	82.0	552.0	6.43	84.0	576.0	6.55	86.0	600.0	6.67	88.0	624.0	6.79
12.5	85.0	593.8	6.67	87.0	618.8	6.80	89.0	643.8	6.92	91.0	668.8	7.04
13.0	88.0	637.0	6.91	90.0	663.0	7.04	92.0	689.0	7.16	94.0	715.0	7.28
13.5	91.0	681.8	7.15	93.0	708.8	7.28	95.0	735.8	7.40	97.0	762.8	7.52
14.0	94.0	728.0	7.39	96.0	756.0	7.52	98.0	784.0	7.65	100.0	812.0	7.77
14.5	97.0	775.8	7.63	99.0	804.8	7.76	101.0	833.8	7.89	103.0	862.8	8.01
15.0	100.0	825.0	7.87	102.0	855.0	8.00	104.0	885.0	8.13	106.0	915.0	8.25
15.5	103.0	875.8	8.11	105.0	906.8	8.24	107.0	937.8	8.37	109.0	968.8	8.50
16.0	106.0	928.0	8.35	108.0	960.0	8.48	110.0	992.0	8.61	112.0	1,024.0	8.74
16.5	109.0	981.8	8.59	111.0	1,014.8	8.72	113.0	1,047.8	8.85	115.0	1,080.8	8.98
17.0	112.0	1,037.0	8.82	114.0	1,071.0	8.96	116.0	1,105.0	9.09	118.0	1,139.0	9.22
17.5	115.0	1,093.8	9.06	117.0	1,128.8	9.20	119.0	1,163.8	9.33	121.0	1,198.8	9.46
18.0	118.0	1,152.0	9.30	120.0	1,188.0	9.44	122.0	1,224.0	9.57	124.0	1,260.0	9.70
18.5	121.0	1,211.8	9.54	123.0	1,248.8	9.68	125.0	1,285.8	9.81	127.0	1,322.8	9.95
19.0	124.0	1,273.0	9.78	126.0	1,311.0	9.92	128.0	1,349.0	10.05	130.0	1,387.0	10.19
19.5	127.0	1,335.8	10.02	129.0	1,374.8	10.16	131.0	1,413.8	10.29	133.0	1,452.8	10.43
20.0	130.0	1,400.0	10.26	132.0	1,440.0	10.40	134.0	1,480.0	10.53	136.0	1,520.0	10.67

Depth	Bottom width 18 feet			Bottom width 20 feet			Bottom width 22 feet			Bottom width 24 feet		
	T	A	r	T	A	r	T	A	r	T	A	r
0.4	20.4	7.68	0.37	22.4	8.48	0.38	24.4	9.28	0.38	26.4	10.08	.38
0.6	21.6	11.88	.55	23.6	13.08	.55	25.6	14.28	.55	27.6	15.48	.56
0.8	22.8	16.32	.71	24.8	17.92	.72	26.8	19.52	.72	28.8	21.12	.73
1.0	24.0	21.00	.86	26.0	23.00	.87	28.0	25.00	.88	30.0	27.00	.89
1.2	25.2	25.92	1.01	27.2	28.32	1.03	29.2	30.72	1.04	31.2	33.12	1.05
1.4	26.4	31.08	1.16	28.4	33.88	1.17	30.4	36.68	1.19	32.4	39.48	1.20
1.6	27.6	36.48	1.30	29.6	39.68	1.32	31.6	42.88	1.34	33.6	46.08	1.35
1.8	28.8	42.12	1.43	30.8	45.72	1.46	32.8	49.32	1.48	34.8	52.92	1.50
2.0	30.0	48.00	1.57	32.0	52.00	1.59	34.0	56.00	1.62	36.0	60.00	1.64
2.2	31.2	54.12	1.70	33.2	58.52	1.73	35.2	62.92	1.75	37.2	67.32	1.78
2.4	32.4	60.48	1.82	34.4	65.28	1.86	36.4	70.08	1.88	38.4	74.88	1.91
2.6	33.6	67.08	1.95	35.6	72.28	1.98	37.6	77.48	2.02	39.6	82.68	2.04
2.8	34.8	73.92	2.07	36.8	79.52	2.11	38.8	85.12	2.14	40.8	90.72	2.18
3.0	36.0	81.00	2.19	38.0	87.00	2.23	40.0	93.00	2.27	42.0	99.00	2.30
3.2	37.2	88.32	2.31	39.2	94.72	2.35	41.2	101.12	2.39	43.2	107.52	2.43
3.4	38.4	95.88	2.43	40.4	102.68	2.47	42.4	109.48	2.52	44.4	116.28	2.56
3.6	39.6	103.68	2.54	41.6	110.88	2.59	43.6	118.08	2.64	45.6	125.28	2.68
3.8	40.8	111.72	2.66	42.8	119.32	2.71	44.8	126.92	2.76	46.8	134.52	2.80
4.0	42.0	120.00	2.77	44.0	128.00	2.83	46.0	136.00	2.88	48.0	144.00	2.92
4.2	43.2	128.52	2.88	45.2	136.92	2.94	47.2	145.32	2.99	49.2	153.72	3.04
4.4	44.4	137.28	3.00	46.4	146.08	3.05	48.4	154.88	3.11	50.4	163.68	3.16
4.6	45.6	146.28	3.11	47.6	155.48	3.17	49.6	164.68	3.22	51.6	173.88	3.28
4.8	46.8	155.52	3.22	48.8	165.12	3.28	50.8	174.72	3.34	52.8	184.32	3.39

Table 35.—*Area in square feet, A, top width in feet, T, and hydraulic radius in feet, r, of trapezoidal channels,* **side slopes 3 to 1**—Continued

Depth	Bottom width 18 feet			Bottom width 20 feet			Bottom width 22 feet			Bottom width 24 feet		
	T	A	r	T	A	r	T	A	r	T	A	r
5.0	48.0	165.00	3.33	50.0	175.00	3.39	52.0	185.00	3.45	54.0	195.00	3.51
5.2	49.2	174.72	3.43	51.2	185.12	3.50	53.2	195.52	3.56	55.2	205.92	3.62
5.4	50.4	184.68	3.54	52.4	195.48	3.61	54.4	206.28	3.67	56.4	217.08	3.73
5.6	51.6	194.88	3.65	53.6	206.08	3.72	55.6	217.28	3.78	57.6	228.48	3.85
5.8	52.8	205.32	3.75	54.8	216.92	3.83	56.8	228.52	3.89	58.8	240.12	3.96
6.0	54.0	216.00	3.86	56.0	228.00	3.93	58.0	240.00	4.00	60.0	252.00	4.07
6.2	55.2	226.92	3.97	57.2	239.32	4.04	59.2	251.72	4.11	61.2	264.12	4.18
6.4	56.4	238.08	4.07	58.4	250.88	4.15	60.4	263.68	4.22	62.4	276.48	4.29
6.6	57.6	249.48	4.18	59.6	262.68	4.25	61.6	275.88	4.33	63.6	289.08	4.40
6.8	58.8	261.12	4.28	60.8	274.72	4.36	62.8	288.32	4.44	64.8	301.92	4.51
7.0	60.0	273.00	4.38	62.0	287.00	4.47	64.0	301.00	4.54	66.0	315.00	4.61
7.2	61.2	285.12	4.49	63.2	299.52	4.57	65.2	313.92	4.65	67.2	328.32	4.72
7.4	62.4	297.48	4.59	64.4	312.28	4.67	66.4	327.08	4.75	68.4	341.88	4.83
7.6	63.6	310.08	4.69	65.6	325.28	4.78	67.6	340.48	4.86	69.6	355.68	4.94
7.8	64.8	322.92	4.80	66.8	338.52	4.88	68.8	354.12	4.96	70.8	369.72	5.04
8.0	66.0	336.00	4.90	68.0	352.00	4.99	70.0	368.00	5.07	72.0	384.00	5.15
8.2	67.2	349.32	5.00	69.2	365.72	5.09	71.2	382.12	5.17	73.2	398.52	5.25
8.4	68.4	362.88	5.10	70.4	379.68	5.19	72.4	396.48	5.28	74.4	413.28	5.36
8.6	69.6	376.68	5.20	71.6	393.88	5.29	73.6	411.08	5.38	75.6	428.28	5.46
8.8	70.8	390.72	5.30	72.8	408.32	5.40	74.8	425.92	5.48	76.8	443.52	5.57
9.0	72.0	405.00	5.41	74.0	423.00	5.50	76.0	441.00	5.59	78.0	459.00	5.67
9.2	73.2	419.52	5.51	75.2	437.92	5.60	77.2	456.32	5.69	79.2	474.72	5.78
9.4	74.4	434.28	5.61	76.4	453.08	5.70	78.4	471.88	5.79	80.4	490.68	5.88
9.6	75.6	449.28	5.71	77.6	468.48	5.80	79.6	487.68	5.90	81.6	506.88	5.98
9.8	76.8	464.52	5.81	78.8	484.12	5.91	80.8	503.72	6.00	82.8	523.32	6.09
10.0	78.0	480.00	5.91	80.0	500.00	6.01	82.0	520.00	6.10	84.0	540.00	6.19
10.5	81.0	519.75	6.16	83.0	540.75	6.26	85.0	561.75	6.35	87.0	582.75	6.45
11.0	84.0	561.0	6.41	86.0	583.0	6.51	88.0	605.0	6.61	90.0	627.0	6.70
11.5	87.0	603.8	6.65	89.0	626.8	6.76	91.0	649.8	6.86	93.0	672.8	6.95
12.0	90.0	648.0	6.90	92.0	672.0	7.01	94.0	696.0	7.11	96.0	720.0	7.21
12.5	93.0	693.8	7.15	95.0	718.8	7.26	97.0	743.8	7.36	99.0	768.8	7.46
13.0	96.0	741.0	7.39	98.0	767.0	7.50	100.0	793.0	7.61	102.0	819.0	7.71
13.5	99.0	789.8	7.64	101.0	816.8	7.75	103.0	843.8	7.86	105.0	870.8	7.96
14.0	102.0	840.0	7.88	104.0	868.0	8.00	106.0	896.0	8.11	108.0	924.0	8.21
14.5	105.0	891.8	8.13	107.0	920.8	8.24	109.0	949.8	8.35	111.0	978.8	8.46
15.0	108.0	945.0	8.37	110.0	975.0	8.49	112.0	1,005.0	8.60	114.0	1,035.0	8.71
15.5	111.0	999.8	8.62	113.0	1,030.8	8.73	115.0	1,061.8	8.85	117.0	1,092.8	8.95
16.0	114.0	1,056.0	8.86	116.0	1,088.0	8.98	118.0	1,120.0	9.09	120.0	1,152.0	9.20
16.5	117.0	1,113.8	9.10	119.0	1,146.8	9.22	121.0	1,179.8	9.34	123.0	1,212.8	9.45
17.0	120.0	1,173.0	9.35	122.0	1,207.0	9.47	124.0	1,241.0	9.58	126.0	1,275.0	9.69
17.5	123.0	1,233.8	9.59	125.0	1,268.8	9.71	127.0	1,303.8	9.83	129.0	1,338.8	9.94
18.0	126.0	1,296.0	9.83	128.0	1,332.0	9.95	130.0	1,368.0	10.07	132.0	1,404.0	10.19
18.5	129.0	1,359.8	10.07	131.0	1,396.8	10.19	133.0	1,433.8	10.31	135.0	1,470.8	10.43
19.0	132.0	1,425.0	10.31	134.0	1,463.0	10.44	136.0	1,501.0	10.56	138.0	1,539.0	10.68
19.5	135.0	1,491.8	10.56	137.0	1,530.8	10.68	139.0	1,569.8	10.80	141.0	1,608.8	10.92
20.0	138.0	1,560.0	10.80	140.0	1,600.0	10.92	142.0	1,640.0	11.04	144.0	1,680.0	11.16

Table 35.—*Area in square feet, A, top width in feet, T, and hydraulic radius in feet, r, of trapezoidal channels,* side slopes 3 to 1—Continued

Depth	Bottom width 26 feet			Bottom width 28 feet			Bottom width 30 feet			Bottom width 32 feet		
	T	A	r	T	A	r	T	A	r	T	A	r
0.4	28.4	10.88	0.38	30.4	11.68	0.38	32.4	12.48	0.38	34.4	13.28	0.38
0.6	29.6	16.68	.56	31.6	17.88	.56	33.6	19.08	.56	35.6	20.28	.57
0.8	30.8	22.72	.73	32.8	24.32	.74	34.8	25.92	.74	36.8	27.52	.74
1.0	32.0	29.00	.90	34.0	31.00	.90	36.0	33.00	.91	38.0	35.00	.91
1.2	33.2	35.52	1.06	35.2	37.92	1.07	37.2	40.32	1.07	39.2	42.72	1.08
1.4	34.4	42.28	1.21	36.4	45.08	1.22	38.4	47.88	1.23	40.4	50.68	1.24
1.6	35.6	49.28	1.36	37.6	52.48	1.38	39.6	55.68	1.39	41.6	58.88	1.40
1.8	36.8	56.52	1.51	38.8	60.12	1.53	40.8	63.72	1.54	42.8	67.32	1.55
2.0	38.0	64.00	1.66	40.0	68.00	1.67	42.0	72.00	1.69	44.0	76.00	1.70
2.2	39.2	71.72	1.80	41.2	76.12	1.82	43.2	80.52	1.83	45.2	84.92	1.85
2.4	40.4	79.68	1.93	42.4	84.48	1.96	44.4	89.28	1.98	46.4	94.08	1.99
2.6	41.6	87.88	2.07	43.6	93.08	2.09	45.6	98.28	2.12	47.6	103.48	2.14
2.8	42.8	96.32	2.20	44.8	101.92	2.23	46.8	107.52	2.25	48.8	113.12	2.28
3.0	44.0	105.00	2.33	46.0	111.00	2.36	48.0	117.00	2.39	50.0	123.00	2.41
3.2	45.2	113.92	2.46	47.2	120.32	2.49	49.2	126.72	2.52	51.2	133.12	2.55
3.4	46.4	123.08	2.59	48.4	129.88	2.62	50.4	136.68	2.65	52.4	143.48	2.68
3.6	47.6	132.48	2.72	49.6	139.68	2.75	51.6	146.88	2.78	53.6	154.08	2.81
3.8	48.8	142.12	2.84	50.8	149.72	2.88	52.8	157.32	2.91	54.8	164.92	2.94
4.0	50.0	152.00	2.96	52.0	160.00	3.00	54.0	168.00	3.04	56.0	176.00	3.07
4.2	51.2	162.12	3.08	53.2	170.52	3.13	55.2	178.92	3.16	57.2	187.32	3.20
4.4	52.4	172.48	3.20	54.4	181.28	3.25	56.4	190.08	3.29	58.4	198.88	3.32
4.6	53.6	183.08	3.32	55.6	192.28	3.37	57.6	201.48	3.41	59.6	210.68	3.45
4.8	54.8	193.92	3.44	56.8	203.52	3.49	58.8	213.12	3.53	60.8	222.72	3.57
5.0	56.0	205.00	3.56	58.0	215.00	3.61	60.0	225.00	3.65	62.0	235.00	3.69
5.2	57.2	216.32	3.67	59.2	226.72	3.72	61.2	237.12	3.77	63.2	247.52	3.81
5.4	58.4	227.88	3.79	60.4	238.68	3.84	62.4	249.48	3.89	64.4	260.28	3.93
5.6	59.6	239.68	3.90	61.6	250.88	3.96	63.6	262.08	4.01	65.6	273.28	4.05
5.8	60.8	251.72	4.02	62.8	263.32	4.07	64.8	274.92	4.12	66.8	286.52	4.17
6.0	62.0	264.00	4.13	64.0	276.00	4.19	66.0	288.00	4.24	68.0	300.00	4.29
6.2	63.2	276.52	4.24	65.2	288.92	4.30	67.2	301.32	4.35	69.2	313.72	4.41
6.4	64.4	289.28	4.35	66.4	302.08	4.41	68.4	314.88	4.47	70.4	327.68	4.52
6.6	65.6	302.28	4.46	67.6	315.48	4.52	69.6	328.68	4.58	71.6	341.88	4.64
6.8	66.8	315.52	4.57	68.8	329.12	4.64	70.8	342.72	4.69	72.8	356.32	4.75
7.0	68.0	329.00	4.68	70.0	343.00	4.75	72.0	357.00	4.81	74.0	371.00	4.86
7.2	69.2	342.72	4.79	71.2	357.12	4.86	73.2	371.52	4.92	75.2	385.92	4.98
7.4	70.4	356.68	4.90	72.4	371.48	4.97	74.4	386.28	5.03	76.4	401.08	5.09
7.6	71.6	370.88	5.01	73.6	386.08	5.08	75.6	401.28	5.14	77.6	416.48	5.20
7.8	72.8	385.32	5.11	74.8	400.92	5.18	76.8	416.52	5.25	78.8	432.12	5.31
8.0	74.0	400.00	5.22	76.0	416.00	5.29	78.0	432.00	5.36	80.0	448.00	5.42
8.2	75.2	414.92	5.33	77.2	431.32	5.40	79.2	447.72	5.47	81.2	464.12	5.53
8.4	76.4	430.08	5.44	78.4	446.88	5.51	80.4	463.68	5.58	82.4	480.48	5.64
8.6	77.6	445.48	5.54	79.6	462.68	5.62	81.6	479.88	5.69	83.6	497.08	5.75
8.8	78.8	461.12	5.65	80.8	478.72	5.72	82.8	496.32	5.79	84.8	513.92	5.86
9.0	80.0	477.00	5.75	82.0	495.00	5.83	84.0	513.00	5.90	86.0	531.00	5.97
9.2	81.2	493.12	5.86	83.2	511.52	5.94	85.2	529.92	6.01	87.2	548.32	6.08
9.4	82.4	509.48	5.96	84.4	528.28	6.04	86.4	547.08	6.12	88.4	565.88	6.19
9.6	83.6	526.08	6.07	85.6	545.28	6.15	87.6	564.48	6.22	89.6	583.68	6.30
9.8	84.8	542.92	6.17	86.8	562.52	6.25	88.8	582.12	6.33	90.8	601.72	6.40

Table 35.—*Area in square feet, A, top width in feet, T, and hydraulic radius in feet, r, of trapezoidal channels,*
side slopes 3 to 1—Continued

Depth	Bottom width 26 feet			Bottom width 28 feet			Bottom width 30 feet			Bottom width 32 feet		
	T	A	r	T	A	r	T	A	r	T	A	r
10.0	86.0	560.00	6.27	88.0	580.00	6.36	90.0	600.00	6.43	92.0	620.00	6.51
10.5	89.0	603.75	6.53	91.0	624.75	6.62	93.0	645.75	6.70	95.0	666.75	6.78
11.0	92.0	649.0	6.79	94.0	671.0	6.88	96.0	693.0	6.96	98.0	715.0	7.04
11.5	95.0	695.8	7.05	97.0	718.8	7.14	99.0	741.8	7.22	101.0	764.8	7.30
12.0	98.0	744.0	7.30	100.0	768.0	7.39	102.0	792.0	7.48	104.0	816.0	7.56
12.5	101.0	793.8	7.56	103.0	818.8	7.65	105.0	843.8	7.74	107.0	868.8	7.82
13.0	104.0	845.0	7.81	106.0	871.0	7.90	108.0	897.0	7.99	110.0	923.0	8.08
13.5	107.0	897.8	8.06	109.0	924.8	8.16	111.0	951.8	8.25	113.0	978.8	8.34
14.0	110.0	952.0	8.31	112.0	980.0	8.41	114.0	1,008.0	8.50	116.0	1,036.0	8.59
14.5	113.0	1,007.8	8.56	115.0	1,036.8	8.66	117.0	1,065.8	8.76	119.0	1,094.8	8.85
15.0	116.0	1,065.0	8.81	118.0	1,095.0	8.91	120.0	1,125.0	9.01	122.0	1,155.0	9.10
15.5	119.0	1,123.8	9.06	121.0	1,154.8	9.16	123.0	1,185.8	9.26	125.0	1,216.8	9.36
16.0	122.0	1,184.0	9.31	124.0	1,216.0	9.41	126.0	1,248.0	9.51	128.0	1,280.0	9.61
16.5	125.0	1,245.8	9.56	127.0	1,278.8	9.66	129.0	1,311.8	9.76	131.0	1,344.8	9.86
17.0	128.0	1,309.0	9.80	130.0	1,343.0	9.91	132.0	1,377.0	10.01	134.0	1,411.0	10.11
17.5	131.0	1,373.8	10.05	133.0	1,408.8	10.16	135.0	1,443.8	10.26	137.0	1,478.8	10.36
18.0	134.0	1,440.0	10.30	136.0	1,476.0	10.41	138.0	1,512.0	10.51	140.0	1,548.0	10.61
18.5	137.0	1,507.8	10.54	139.0	1,544.8	10.65	141.0	1,581.8	10.76	143.0	1,618.8	10.86
19.0	140.0	1,577.0	10.79	142.0	1,615.0	10.90	144.0	1,653.0	11.01	146.0	1,691.0	11.11
19.5	143.0	1,647.8	11.03	145.0	1,686.8	11.15	147.0	1,725.8	11.26	149.0	1,764.8	11.36
20.0	146.0	1,720.0	11.28	148.0	1,760.0	11.39	150.0	1,800.0	11.50	152.0	1,840.0	11.61

Depth	Bottom width 35 feet			Bottom width 40 feet			Bottom width 45 feet			Bottom width 50 feet		
	T	A	r	T	A	r	T	A	r	T	A	r
0.4	37.4	14.48	0.39	42.4	16.48	0.39	47.4	18.48	0.39	52.4	20.48	0.39
0.6	38.6	22.08	.57	43.6	25.08	.57	48.6	28.08	.58	53.6	31.08	.58
0.8	39.8	29.92	.75	44.8	33.92	.75	49.8	37.92	.76	54.8	41.92	.76
1.0	41.0	38.00	.92	46.0	43.00	.93	51.0	48.00	.94	56.0	53.00	.94
1.2	42.2	46.32	1.09	47.2	52.32	1.10	52.2	58.32	1.11	57.2	64.32	1.12
1.4	43.4	54.88	1.25	48.4	61.88	1.27	53.4	68.88	1.28	58.4	75.88	1.29
1.6	44.6	63.68	1.41	49.6	71.68	1.43	54.6	79.68	1.45	59.6	87.68	1.46
1.8	45.8	72.72	1.57	50.8	81.72	1.59	55.8	90.72	1.61	60.8	99.72	1.62
2.0	47.0	82.00	1.72	52.0	92.00	1.75	57.0	102.00	1.77	62.0	112.00	1.79
2.2	48.2	91.52	1.87	53.2	102.52	1.90	58.2	113.52	1.93	63.2	124.52	1.95
2.4	49.4	101.28	2.02	54.4	113.28	2.05	59.4	125.28	2.08	64.4	137.28	2.11
2.6	50.6	111.28	2.16	55.6	124.28	2.20	60.6	137.28	2.23	65.6	150.28	2.26
2.8	51.8	121.52	2.31	56.8	135.52	2.35	61.8	149.52	2.38	66.8	163.52	2.42
3.0	53.0	132.00	2.45	58.0	147.00	2.49	63.0	162.00	2.53	68.0	177.00	2.57
3.2	54.2	142.72	2.58	59.2	158.72	2.63	64.2	174.72	2.68	69.2	190.72	2.72
3.4	55.4	153.68	2.72	60.4	170.68	2.78	65.4	187.68	2.82	70.4	204.68	2.86
3.6	56.6	164.88	2.85	61.6	182.88	2.91	66.6	200.88	2.96	71.6	218.88	3.01
3.8	57.8	176.32	2.99	62.8	195.32	3.05	67.8	214.32	3.10	72.8	233.32	3.15
4.0	59.0	188.00	3.12	64.0	208.00	3.19	69.0	228.00	3.24	74.0	248.00	3.29
4.2	60.2	199.92	3.25	65.2	220.92	3.32	70.2	241.92	3.38	75.2	262.92	3.43
4.4	61.4	212.08	3.38	66.4	234.08	3.45	71.4	256.08	3.52	76.4	278.08	3.57
4.6	62.6	224.48	3.50	67.6	247.48	3.58	72.6	270.48	3.65	77.6	293.48	3.71
4.8	63.8	237.12	3.63	68.8	261.12	3.71	73.8	285.12	3.78	78.8	309.12	3.85

Table 35.—*Area in square feet, A, top width in feet, T, and hydraulic radius in feet, r, of trapezoidal channels,* **side slopes 3 to 1**—Continued

Depth	Bottom width 35 feet			Bottom width 40 feet			Bottom width 45 feet			Bottom width 50 feet		
	T	A	r	T	A	r	T	A	r	T	A	r
5.0	65.0	250.00	3.75	70.0	275.00	3.84	75.0	300.00	3.92	80.0	325.00	3.98
5.2	66.2	263.12	3.88	71.2	289.12	3.97	76.2	315.12	4.05	81.2	341.12	4.12
5.4	67.4	276.48	4.00	72.4	303.48	4.09	77.4	330.48	4.18	82.4	357.48	4.25
5.6	68.6	290.08	4.12	73.6	318.08	4.22	78.6	346.08	4.30	83.6	374.08	4.38
5.8	69.8	303.92	4.24	74.8	332.92	4.34	79.8	361.92	4.43	84.8	390.92	4.51
6.0	71.0	318.00	4.36	76.0	348.00	4.46	81.0	378.00	4.56	86.0	408.00	4.64
6.2	72.2	332.32	4.48	77.2	363.32	4.59	82.2	394.32	4.68	87.2	425.32	4.77
6.4	73.4	346.88	4.60	78.4	378.88	4.71	83.4	410.88	4.81	88.4	442.88	4.89
6.6	74.6	361.68	4.71	79.6	394.68	4.83	84.6	427.68	4.93	89.6	460.68	5.02
6.8	75.8	376.72	4.83	80.8	410.72	4.95	85.8	444.72	5.05	90.8	478.72	5.15
7.0	77.0	392.00	4.95	82.0	427.00	5.07	87.0	462.00	5.18	92.0	497.00	5.27
7.2	78.2	407.52	5.06	83.2	443.52	5.19	88.2	479.52	5.30	93.2	515.52	5.40
7.4	79.4	423.28	5.17	84.4	460.28	5.30	89.4	497.28	5.42	94.4	534.28	5.52
7.6	80.6	439.28	5.29	85.6	477.28	5.42	90.6	515.28	5.54	95.6	553.28	5.64
7.8	81.8	455.52	5.40	86.8	494.52	5.54	91.8	533.52	5.66	96.8	572.52	5.76
8.0	83.0	472.00	5.51	88.0	512.00	5.65	93.0	552.00	5.77	98.0	592.00	5.88
8.2	84.2	488.72	5.63	89.2	529.72	5.77	94.2	570.72	5.89	99.2	611.72	6.01
8.4	85.4	505.68	5.74	90.4	547.68	5.88	95.4	589.68	6.01	100.4	631.68	6.13
8.6	86.6	522.88	5.85	91.6	565.88	6.00	96.6	608.88	6.13	101.6	651.88	6.24
8.8	87.8	540.32	5.96	92.8	584.32	6.11	97.8	628.32	6.24	102.8	672.32	6.36
9.0	89.0	558.00	6.07	94.0	603.00	6.22	99.0	648.00	6.36	104.0	693.00	6.48
9.2	90.2	575.92	6.18	95.2	621.92	6.33	100.2	667.92	6.47	105.2	713.92	6.60
9.4	91.4	594.08	6.29	96.4	641.08	6.45	101.4	688.08	6.59	106.4	735.08	6.72
9.6	92.6	612.48	6.40	97.6	660.48	6.56	102.6	708.48	6.70	107.6	756.48	6.83
9.8	93.8	631.12	6.51	98.8	680.12	6.67	103.8	729.12	6.82	108.8	778.12	6.95
10.0	95.0	650.00	6.62	100.0	700.00	6.78	105.0	750.00	6.93	110.0	800.00	7.06
10.5	98.0	698.25	6.89	103.0	750.75	7.06	108.0	803.25	7.21	113.0	855.75	7.35
11.0	101.0	748.0	7.15	106.0	803.0	7.33	111.0	858.0	7.49	116.0	913.0	7.64
11.5	104.0	799.3	7.42	109.0	856.8	7.60	114.0	914.3	7.77	119.0	971.8	7.92
12.0	107.0	852.0	7.68	112.0	912.0	7.87	117.0	972.0	8.04	122.0	1,032.0	8.20
12.5	110.0	906.3	7.95	115.0	968.8	8.14	120.0	1,031.3	8.31	125.0	1,093.8	8.47
13.0	113.0	962.0	8.21	118.0	1,027.0	8.40	123.0	1,092.0	8.58	128.0	1,157.0	8.75
13.5	116.0	1,019.3	8.47	121.0	1,086.8	8.67	126.0	1,154.3	8.85	131.0	1,221.8	9.02
14.0	119.0	1,078.0	8.73	124.0	1,148.0	8.93	129.0	1,218.0	9.12	134.0	1,288.0	9.30
14.5	122.0	1,138.3	8.98	127.0	1,210.8	9.19	132.0	1,283.3	9.39	137.0	1,355.8	9.57
15.0	125.0	1,200.0	9.24	130.0	1,275.0	9.45	135.0	1,350.0	9.65	140.0	1,425.0	9.84
15.5	128.0	1,263.3	9.50	133.0	1,340.8	9.71	138.0	1,418.3	9.92	143.0	1,495.8	10.10
16.0	131.0	1,328.0	9.75	136.0	1,408.0	9.97	141.0	1,488.0	10.18	146.0	1,568.0	10.37
16.5	134.0	1,394.3	10.01	139.0	1,476.8	10.23	144.0	1,559.3	10.44	149.0	1,641.8	10.64
17.0	137.0	1,462.0	10.26	142.0	1,547.0	10.49	147.0	1,632.0	10.70	152.0	1,717.0	10.90
17.5	140.0	1,531.3	10.51	145.0	1,618.8	10.74	150.0	1,706.3	10.96	155.0	1,793.8	11.16
18.0	143.0	1,602.0	10.76	148.0	1,692.0	11.00	153.0	1,782.0	11.22	158.0	1,872.0	11.43
18.5	146.0	1,674.3	11.01	151.0	1,766.8	11.25	156.0	1,859.3	11.48	161.0	1,951.8	11.69
19.0	149.0	1,748.0	11.27	154.0	1,843.0	11.51	159.0	1,938.0	11.73	164.0	2,033.0	11.95
19.5	152.0	1,823.3	11.52	157.0	1,920.8	11.76	162.0	2,018.3	11.99	167.0	2,115.8	12.21
20.0	155.0	1,900.0	11.77	160.0	2,000.0	12.01	165.0	2,100.0	12.25	170.0	2,200.0	12.47

Table 35.—*Area in square feet, A, top width in feet, T, and hydraulic radius in feet, r, of trapezoidal channels,*
side slopes 3 to 1—Continued

Depth	Bottom width 60 feet			Bottom width 70 feet			Bottom width 80 feet			Bottom width 90 feet		
	T	A	r	T	A	r	T	A	r	T	A	r
0.4	62.4	24.48	.39	72.4	28.48	.39	82.4	32.48	.39	92.4	36.48	.39
0.6	63.6	37.08	.58	73.6	43.08	.58	83.6	49.08	.59	93.6	55.08	.59
0.8	64.8	49.92	.77	74.8	57.92	.77	84.8	65.92	.77	94.8	73.92	.78
1.0	66.0	63.00	.95	76.0	73.00	.96	86.0	83.00	.96	96.0	93.00	.97
1.2	67.2	76.32	1.13	77.2	88.32	1.14	87.2	100.32	1.15	97.2	112.32	1.15
1.4	68.4	89.88	1.31	78.4	103.88	1.32	88.4	117.88	1.33	98.4	131.88	1.33
1.6	69.6	103.68	1.48	79.6	119.68	1.49	89.6	135.68	1.51	99.6	151.68	1.51
1.8	70.8	117.72	1.65	80.8	135.72	1.67	90.8	153.72	1.68	100.8	171.72	1.69
2.0	72.0	132.00	1.82	82.0	152.00	1.84	92.0	172.00	1.86	102.0	192.00	1.87
2.2	73.2	146.52	1.98	83.2	168.52	2.01	93.2	190.52	2.03	103.2	212.52	2.05
2.4	74.4	161.28	2.15	84.4	185.28	2.18	94.4	209.28	2.20	104.4	233.28	2.22
2.6	75.6	176.28	2.31	85.6	202.28	2.34	95.6	228.28	2.37	105.6	254.28	2.39
2.8	76.8	191.52	2.46	86.8	219.52	2.50	96.8	247.52	2.53	106.8	275.52	2.56
3.0	78.0	207.00	2.62	88.0	237.00	2.66	98.0	267.00	2.70	108.0	297.00	2.73
3.2	79.2	222.72	2.78	89.2	254.72	2.82	99.2	286.72	2.86	109.2	318.72	2.89
3.4	80.4	238.68	2.93	90.4	272.68	2.98	100.4	306.68	3.02	110.4	340.68	3.06
3.6	81.6	254.88	3.08	91.6	290.88	3.14	101.6	326.88	3.18	111.6	362.88	3.22
3.8	82.8	271.32	3.23	92.8	309.32	3.29	102.8	347.32	3.34	112.8	385.32	3.38
4.0	84.0	288.00	3.38	94.0	328.00	3.44	104.0	368.00	3.49	114.0	408.00	3.54
4.2	85.2	304.92	3.52	95.2	346.92	3.59	105.2	388.92	3.65	115.2	430.92	3.70
4.4	86.4	322.08	3.67	96.4	366.08	3.74	106.4	410.08	3.80	116.4	454.08	3.85
4.6	87.6	339.48	3.81	97.6	385.48	3.89	107.6	431.48	3.96	117.6	477.48	4.01
4.8	88.8	357.12	3.95	98.8	405.12	4.04	108.8	453.12	4.11	118.8	501.12	4.16
5.0	90.0	375.00	4.09	100.0	425.00	4.18	110.0	475.00	4.26	120.0	525.00	4.32
5.2	91.2	393.12	4.23	101.2	445.12	4.33	111.2	497.12	4.40	121.2	549.12	4.47
5.4	92.4	411.48	4.37	102.4	465.48	4.47	112.4	519.48	4.55	122.4	573.48	4.62
5.6	93.6	430.08	4.51	103.6	486.08	4.61	113.6	542.08	4.70	123.6	598.08	4.77
5.8	94.8	448.92	4.64	104.8	506.92	4.75	114.8	564.92	4.84	124.8	622.92	4.92
6.0	96.0	468.00	4.78	106.0	528.00	4.89	116.0	588.00	4.99	126.0	648.00	5.06
6.2	97.2	487.32	4.91	107.2	549.32	5.03	117.2	611.32	5.13	127.2	673.32	5.21
6.4	98.4	506.88	5.04	108.4	570.88	5.17	118.4	634.88	5.27	128.4	698.88	5.36
6.6	99.6	526.68	5.18	109.6	592.68	5.30	119.6	658.68	5.41	129.6	724.68	5.50
6.8	100.8	546.72	5.31	110.8	614.72	5.44	120.8	682.72	5.55	130.8	750.72	5.64
7.0	102.0	567.00	5.44	112.0	637.00	5.57	122.0	707.00	5.69	132.0	777.00	5.79
7.2	103.2	587.52	5.57	113.2	659.52	5.71	123.2	731.52	5.83	133.2	803.52	5.93
7.4	104.4	608.28	5.70	114.4	682.28	5.84	124.4	756.28	5.96	134.4	830.28	6.07
7.6	105.6	629.28	5.82	115.6	705.28	5.97	125.6	781.28	6.10	135.6	857.28	6.21
7.8	106.8	650.52	5.95	116.8	728.52	6.11	126.8	806.52	6.24	136.8	884.52	6.35
8.0	108.0	672.00	6.08	118.0	752.00	6.24	128.0	832.00	6.37	138.0	912.00	6.49
8.2	109.2	693.72	6.20	119.2	775.72	6.37	129.2	857.72	6.50	139.2	939.72	6.62
8.4	110.4	715.68	6.33	120.4	799.68	6.49	130.4	883.68	6.64	140.4	967.68	6.76
8.6	111.6	737.88	6.45	121.6	823.88	6.62	131.6	909.88	6.77	141.6	995.88	6.90
8.8	112.8	760.32	6.57	122.8	848.32	6.75	132.8	936.32	6.90	142.8	1,024.32	7.03
9.0	114.0	783.00	6.70	124.0	873.00	6.88	134.0	963.00	7.03	144.0	1,053.00	7.17
9.2	115.2	805.92	6.82	125.2	897.92	7.00	135.2	989.92	7.16	145.2	1,081.92	7.30
9.4	116.4	829.08	6.94	126.4	923.08	7.13	136.4	1,017.08	7.29	146.4	1,111.08	7.43
9.6	117.6	852.48	7.06	127.6	948.48	7.26	137.6	1,044.48	7.42	147.6	1,140.48	7.57
9.8	118.8	876.12	7.18	128.8	974.12	7.38	138.8	1,072.12	7.55	148.8	1,170.12	7.70

Table 35.—*Area in square feet, A, top width in feet, T, and hydraulic radius in feet, r, of trapezoidal channels,* **side slopes 3 to 1**—Continued

Depth	Bottom width 60 feet			Bottom width 70 feet			Bottom width 80 feet			Bottom width 90 feet		
	T	A	r	T	A	r	T	A	r	T	A	r
10.0	120.0	900.00	7.30	130.0	1,000.00	7.50	140.0	1,100.00	7.68	150.0	1,200.00	7.83
10.5	123.0	960.75	7.60	133.0	1,065.75	7.81	143.0	1,170.75	8.00	153.0	1,275.75	8.16
11.0	126.0	1,023.0	7.90	136.0	1,133.0	8.12	146.0	1,243.0	8.31	156.0	1,353.0	8.48
11.5	129.0	1,086.8	8.19	139.0	1,201.8	8.42	149.0	1,316.8	8.62	159.0	1,431.8	8.80
12.0	132.0	1,152.0	8.48	142.0	1,272.0	8.72	152.0	1,392.0	8.93	162.0	1,512.0	9.11
12.5	135.0	1,218.8	8.76	145.0	1,343.8	9.02	155.0	1,468.8	9.23	165.0	1,593.8	9.43
13.0	138.0	1,287.0	9.05	148.0	1,417.0	9.31	158.0	1,547.0	9.54	168.0	1,677.0	9.74
13.5	141.0	1,356.8	9.33	151.0	1,491.8	9.60	161.0	1,626.8	9.84	171.0	1,761.8	10.05
14.0	144.0	1,428.0	9.61	154.0	1,568.0	9.89	164.0	1,708.0	10.13	174.0	1,848.0	10.35
14.5	147.0	1,500.8	9.89	157.0	1,645.8	10.18	167.0	1,790.8	10.43	177.0	1,935.8	10.65
15.0	150.0	1,575.0	10.17	160.0	1,725.0	10.46	170.0	1,875.0	10.72	180.0	2,025.0	10.95
15.5	153.0	1,650.8	10.45	163.0	1,805.8	10.75	173.0	1,960.8	11.01	183.0	2,115.8	11.25
16.0	156.0	1,728.0	10.72	166.0	1,888.0	11.03	176.0	2,048.0	11.30	186.0	2,208.0	11.55
16.5	159.0	1,806.8	10.99	169.0	1,971.8	11.31	179.0	2,136.8	11.59	189.0	2,301.8	11.84
17.0	162.0	1,887.0	11.26	172.0	2,057.0	11.59	182.0	2,227.0	11.88	192.0	2,397.0	12.14
17.5	165.0	1,968.8	11.53	175.0	2,143.8	11.86	185.0	2,318.8	12.16	195.0	2,493.8	12.43
18.0	168.0	2,052.0	11.80	178.0	2,232.0	12.14	188.0	2,412.0	12.44	198.0	2,592.0	12.72
18.5	171.0	2,136.8	12.07	181.0	2,321.8	12.42	191.0	2,506.8	12.72	201.0	2,691.8	13.00
19.0	174.0	2,223.0	12.34	184.0	2,413.0	12.69	194.0	2,603.0	13.00	204.0	2,793.0	13.29
19.5	177.0	2,310.8	12.60	187.0	2,505.8	12.96	197.0	2,700.8	13.28	207.0	2,895.8	13.57
20.0	180.0	2,400.0	12.87	190.0	2,600.0	13.23	200.0	2,800.0	13.56	210.0	3,000.0	13.86

Table 36.—*Discharge in second-feet of Cipolletti weirs, computed from $Q = 3.367\ LH^{3/2}$*

Depth on crest (feet)	Length of weir in feet									
	100	150	200	300	400	500	600	700	800	900
0.01	0.3	1	1	1	1	2	2	2	3	3
.02	1.0	1	2	3	4	5	6	7	8	9
.03	1.8	3	4	5	7	9	11	12	14	16
.04	2.7	4	5	8	11	13	16	19	22	24
.05	3.8	6	8	11	15	19	23	26	30	34
.06	5.0	7	10	15	20	25	30	35	40	45
.07	6.2	9	12	19	25	31	37	44	50	56
.08	7.6	11	15	23	30	38	46	53	61	69
.09	9.1	14	18	27	36	45	55	64	73	82
.10	10.7	16	21	32	43	53	64	75	85	96
.11	12.3	18	25	37	49	61	74	86	98	111
.12	14.0	21	28	42	56	70	84	98	112	126
.13	15.8	24	32	47	63	79	95	110	126	142
.14	17.6	26	35	53	71	88	106	123	141	159
.15	19.6	29	39	59	78	98	117	137	156	176
.16	21.6	32	43	65	86	108	129	151	172	194
.17	23.6	35	47	71	94	118	142	165	189	212
.18	25.7	39	51	77	103	129	154	180	206	231
.19	27.9	42	56	84	112	139	167	195	223	251
.20	30.1	45	60	90	120	151	181	211	241	271
.21	32.4	49	65	97	130	162	194	227	259	292
.22	34.7	52	69	104	139	174	208	243	278	313
.23	37.1	56	74	111	149	186	223	260	297	334
.24	39.6	59	79	119	158	198	238	277	317	356
.25	42.1	63	84	126	168	210	253	295	337	379
.26	44.6	67	89	134	179	223	268	312	357	402
.27	47.2	71	94	142	189	236	283	331	378	425
.28	49.9	75	100	150	200	249	299	349	399	449
.29	52.6	79	105	158	210	263	315	368	421	473
.30	55.3	83	111	166	221	277	332	387	443	498
.31	58.1	87	116	174	232	291	349	407	465	523
.32	60.9	91	122	183	244	305	366	427	488	548
.33	63.8	96	128	191	255	319	383	447	511	574
.34	66.7	100	133	200	267	334	400	467	534	601
.35	69.7	105	139	209	279	349	418	488	558	627
.36	72.7	109	145	218	291	364	436	509	582	654
.37	75.8	114	152	227	303	379	455	530	606	682
.38	78.9	118	158	237	315	394	473	552	631	710
.39	82.0	123	164	246	328	410	492	574	656	738
.40	85.2	128	170	256	341	426	511	596	681	767
.41	88.4	133	177	265	354	442	530	619	707	795
.42	91.6	137	183	275	367	458	550	641	733	825
.43	94.9	142	190	285	380	475	570	665	759	854
.44	98.3	147	197	295	393	491	590	688	786	884
.45	101.6	152	203	305	407	508	610	711	813	915
.46	105.0	158	210	315	420	525	630	735	840	945
.47	108.5	163	217	325	434	542	651	759	868	976
.48	112.0	168	224	336	448	560	672	784	896	1,008
.49	115.5	173	231	346	462	577	693	808	924	1,039
.50	119.0	179	238	357	476	595	714	833	952	1,071

Table 36.—*Discharge in second-feet of Cipolletti weirs, computed from $Q=3.367\ LH^{3/2}$*—Continued

Depth on crest (feet)	Length of weir in feet									
	100	150	200	300	400	500	600	700	800	900
0.51	122.6	184	245	368	490	613	736	858	981	1,104
.52	126.2	189	252	379	505	631	757	884	1,010	1,136
.53	129.9	195	260	390	520	650	779	909	1,039	1,169
.54	133.6	200	267	401	534	668	802	935	1,069	1,202
.55	137.3	206	275	412	549	687	824	961	1,099	1,236
.56	141.1	212	282	423	564	705	847	988	1,129	1,270
.57	144.9	217	290	435	580	724	869	1,014	1,159	1,304
.58	148.7	223	297	446	595	744	892	1,041	1,190	1,338
.59	152.6	229	305	458	610	763	915	1,068	1,221	1,373
.60	156.5	235	313	469	626	782	939	1,095	1,252	1,408
.61	160.4	241	321	481	642	802	962	1,123	1,283	1,444
.62	164.4	247	329	493	657	822	986	1,151	1,315	1,479
.63	168.3	253	337	505	673	842	1,010	1,178	1,347	1,515
.64	172.4	259	345	517	689	862	1,034	1,207	1,379	1,551
.65	176.4	265	353	529	706	882	1,059	1,235	1,411	1,588
.66	180.5	271	361	542	722	903	1,083	1,264	1,444	1,625
.67	184.6	277	369	554	739	923	1,108	1,292	1,477	1,662
.68	188.8	283	378	566	755	944	1,133	1,321	1,510	1,699
.69	193.0	289	386	579	772	965	1,158	1,351	1,544	1,737
.70	197.2	296	394	592	789	986	1,183	1,380	1,577	1,775
.71	201.4	302	403	604	806	1,007	1,208	1,410	1,611	1,813
.72	205.7	309	411	617	823	1,028	1,234	1,440	1,645	1,851
.73	210.0	315	420	630	840	1,050	1,260	1,470	1,680	1,890
.74	214.3	321	429	643	857	1,072	1,286	1,500	1,715	1,929
.75	218.7	328	437	656	875	1,093	1,312	1,531	1,749	1,968
.76	223.1	334	446	669	892	1,115	1,338	1,561	1,784	2,008
.77	227.5	341	455	682	910	1,137	1,365	1,592	1,820	2,047
.78	231.9	348	464	696	928	1,160	1,392	1,623	1,855	2,087
.79	236.4	356	473	709	946	1,182	1,418	1,655	1,891	2,128
.80	240.9	361	482	723	964	1,205	1,445	1,686	1,927	2,168
.81	245.4	368	491	736	982	1,227	1,473	1,718	1,963	2,209
.82	250.0	375	500	750	1,000	1,250	1,500	1,750	2,000	2,250
.83	254.6	382	509	764	1,018	1,273	1,527	1,782	2,037	2,291
.84	259.2	389	518	778	1,037	1,296	1,555	1,814	2,074	2,333
.85	263.8	396	528	792	1,055	1,319	1,583	1,847	2,111	2,374
.86	268.5	403	537	806	1,074	1,343	1,611	1,880	2,148	2,417
.87	273.2	410	546	820	1,093	1,366	1,639	1,912	2,186	2,459
.88	277.9	417	556	834	1,112	1,390	1,668	1,945	2,223	2,501
.89	282.7	424	565	848	1,131	1,413	1,696	1,979	2,261	2,544
.90	287.5	431	575	862	1,150	1,437	1,725	2,012	2,300	2,587
.91	292.3	438	585	877	1,169	1,461	1,754	2,046	2,338	2,630
.92	297.1	446	594	891	1,188	1,485	1,783	2,080	2,377	2,674
.93	301.9	453	604	906	1,208	1,510	1,812	2,114	2,416	2,717
.94	306.8	460	614	920	1,227	1,534	1,841	2,148	2,455	2,761
.95	311.7	468	623	935	1,247	1,559	1,870	2,182	2,494	2,806
.96	316.7	475	633	950	1,267	1,583	1,900	2,217	2,533	2,850
.97	321.6	482	643	965	1,287	1,608	1,930	2,251	2,573	2,895
.98	326.6	490	653	980	1,306	1,633	1,960	2,286	2,613	2,940
.99	331.6	497	663	995	1,327	1,658	1,990	2,321	2,653	2,985
1.00	336.7	505	673	1,010	1,347	1,683	2,020	2,357	2,693	3,030

Table 36.—*Discharge in second-feet of Cipolletti weirs, computed from $Q = 3.367\ LH^{3/2}$*—Continued

Depth on crest (feet)	Length of weir in feet								
	100	200	300	400	500	600	700	800	900
1.01	341.7	683	1,025	1,367	1,709	2,050	2,392	2,734	3,076
1.02	346.8	694	1,040	1,387	1,734	2,081	2,428	2,775	3,121
1.03	351.9	704	1,056	1,408	1,760	2,112	2,464	2,815	3,167
1.04	357.1	714	1,071	1,428	1,785	2,142	2,499	2,857	3,214
1.05	362.2	724	1,087	1,449	1,811	2,173	2,536	2,898	3,260
1.06	367.4	735	1,102	1,470	1,837	2,205	2,572	2,939	3,307
1.07	372.6	745	1,118	1,491	1,863	2,236	2,608	2,981	3,354
1.08	377.9	756	1,134	1,511	1,889	2,267	2,645	3,023	3,401
1.09	383.1	766	1,149	1,532	1,916	2,299	2,682	3,065	3,448
1.10	388.4	777	1,165	1,554	1,942	2,330	2,719	3,107	3,496
1.11	393.7	787	1,181	1,575	1,969	2,362	2,756	3,150	3,543
1.12	399.0	798	1,197	1,596	1,995	2,394	2,793	3,192	3,591
1.13	404.4	809	1,213	1,618	2,022	2,426	2,831	3,235	3,640
1.14	409.8	820	1,229	1,639	2,049	2,459	2,869	3,278	3,688
1.15	415.2	830	1,246	1,661	2,076	2,491	2,906	3,322	3,737
1.16	420.6	841	1,262	1,682	2,103	2,524	2,944	3,365	3,786
1.17	426.1	852	1,278	1,704	2,130	2,556	2,982	3,409	3,835
1.18	431.5	863	1,295	1,726	2,158	2,589	3,021	3,452	3,884
1.19	437.0	874	1,311	1,748	2,185	2,622	3,059	3,496	3,933
1.20	442.6	885	1,328	1,770	2,213	2,655	3,098	3,540	3,983
1.21	448.1	896	1,344	1,792	2,240	2,689	3,137	3,585	4,033
1.22	453.7	907	1,361	1,815	2,268	2,722	3,176	3,629	4,083
1.23	459.3	919	1,378	1,837	2,296	2,756	3,215	3,674	4,133
1.24	464.9	930	1,395	1,859	2,324	2,789	3,254	3,719	4,184
1.25	470.5	941	1,412	1,882	2,353	2,823	3,294	3,764	4,235
1.26	476.2	952	1,428	1,905	2,381	2,857	3,333	3,809	4,285
1.27	481.8	964	1,446	1,927	2,409	2,891	3,373	3,855	4,337
1.28	487.5	975	1,463	1,950	2,438	2,925	3,413	3,900	4,388
1.29	493.3	987	1,480	1,973	2,466	2,960	3,453	3,946	4,439
1.30	499.0	998	1,497	1,996	2,495	2,994	3,493	3,992	4,491
1.31	504.8	1,010	1,514	2,019	2,524	3,029	3,534	4,038	4,543
1.32	510.6	1,021	1,532	2,042	2,553	3,063	3,574	4,085	4,595
1.33	516.4	1,033	1,549	2,066	2,582	3,098	3,615	4,131	4,648
1.34	522.2	1,044	1,567	2,089	2,611	3,133	3,656	4,178	4,700
1.35	528.1	1,056	1,584	2,112	2,640	3,168	3,697	4,225	4,753
1.36	534.0	1,068	1,602	2,136	2,670	3,204	3,738	4,272	4,806
1.37	539.9	1,080	1,620	2,159	2,699	3,239	3,779	4,319	4,859
1.38	545.8	1,092	1,637	2,183	2,729	3,275	3,820	4,366	4,912
1.39	551.7	1,103	1,655	2,207	2,759	3,310	3,862	4,414	4,965
1.40	557.7	1,115	1,673	2,231	2,788	3,346	3,904	4,462	5,019
1.41	563.7	1,127	1,691	2,255	2,818	3,382	3,946	4,509	5,073
1.42	569.7	1,139	1,709	2,279	2,848	3,418	3,988	4,557	5,127
1.43	575.7	1,151	1,727	2,303	2,879	3,454	4,030	4,606	5,181
1.44	581.8	1,164	1,745	2,327	2,909	3,491	4,072	4,654	5,236
1.45	587.8	1,176	1,763	2,351	2,939	3,527	4,115	4,703	5,290
1.46	593.9	1,188	1,782	2,376	2,970	3,564	4,157	4,751	5,345
1.47	600.0	1,200	1,800	2,400	3,000	3,600	4,200	4,800	5,400
1.48	606.2	1,212	1,819	2,425	3,031	3,637	4,243	4,849	5,456
1.49	612.3	1,225	1,837	2,449	3,062	3,674	4,286	4,899	5,511
1.50	618.5	1,237	1,856	2,474	3,092	3,711	4,330	4,948	5,566

Table 36.—*Discharge in second-feet of Cipolletti weirs, computed from* $Q = 3.367\ LH^{3/2}$—Continued

Depth on crest (feet)	Length of weir in feet								
	100	200	300	400	500	600	700	800	900
1.51	624.7	1,249	1,874	2,499	3,123	3,748	4,373	4,998	5,622
1.52	630.9	1,262	1,893	2,524	3,155	3,785	4,416	5,047	5,678
1.53	637.1	1,274	1,911	2,549	3,186	3,823	4,460	5,097	5,734
1.54	643.4	1,287	1,930	2,574	3,217	3,860	4,504	5,147	5,791
1.55	649.7	1,299	1,949	2,599	3,248	3,898	4,548	5,197	5,847
1.56	656.0	1,312	1,968	2,624	3,280	3,936	4,592	5,248	5,904
1.57	662.3	1,325	1,987	2,649	3,311	3,974	4,636	5,298	5,961
1.58	668.6	1,337	2,006	2,675	3,343	4,012	4,680	5,349	6,018
1.59	675.0	1,350	2,025	2,700	3,375	4,050	4,725	5,400	6,075
1.60	681.4	1,363	2,044	2,725	3,407	4,088	4,770	5,451	6,132
1.61	687.8	1,376	2,063	2,751	3,439	4,127	4,814	5,502	6,190
1.62	694.2	1,388	2,083	2,777	3,471	4,165	4,859	5,553	6,248
1.63	700.6	1,401	2,102	2,802	3,503	4,204	4,904	5,605	6,306
1.64	707.1	1,414	2,121	2,828	3,535	4,242	4,950	5,657	6,364
1.65	713.6	1,427	2,141	2,854	3,568	4,281	4,995	5,708	6,422
1.66	720.0	1,440	2,160	2,880	3,600	4,320	5,040	5,760	6,480
1.67	726.6	1,453	2,180	2,906	3,633	4,359	5,086	5,813	6,539
1.68	733.1	1,466	2,199	2,932	3,666	4,399	5,132	5,865	6,598
1.69	739.7	1,479	2,219	2,959	3,698	4,438	5,178	5,917	6,657
1.70	746.2	1,492	2,239	2,985	3,731	4,477	5,224	5,970	6,716
1.71	752.8	1,506	2,258	3,011	3,764	4,517	5,270	6,023	6,775
1.72	759.4	1,519	2,278	3,038	3,797	4,557	5,316	6,076	6,835
1.73	766.1	1,532	2,298	3,064	3,830	4,596	5,362	6,129	6,895
1.74	772.7	1,545	2,318	3,091	3,864	4,636	5,409	6,182	6,954
1.75	779.4	1,559	2,338	3,118	3,897	4,676	5,456	6,235	7,015
1.76	786.1	1,572	2,358	3,144	3,930	4,716	5,503	6,289	7,075
1.77	792.8	1,586	2,378	3,171	3,964	4,757	5,550	6,342	7,135
1.78	799.5	1,599	2,399	3,198	3,998	4,797	5,597	6,396	7,196
1.79	806.3	1,613	2,419	3,225	4,031	4,838	5,644	6,450	7,256
1.80	813.0	1,626	2,439	3,252	4,065	4,878	5,691	6,504	7,317
1.81	819.8	1,640	2,459	3,279	4,099	4,919	5,739	6,559	7,378
1.82	826.6	1,653	2,480	3,306	4,133	4,960	5,786	6,613	7,440
1.83	833.4	1,667	2,500	3,334	4,167	5,001	5,834	6,668	7,501
1.84	840.3	1,681	2,521	3,361	4,201	5,042	5,882	6,722	7,563
1.85	847.1	1,694	2,541	3,389	4,236	5,083	5,930	6,777	7,624
1.86	854.0	1,708	2,562	3,416	4,270	5,124	5,978	6,832	7,686
1.87	860.9	1,722	2,583	3,444	4,305	5,166	6,026	6,887	7,748
1.88	867.8	1,736	2,603	3,471	4,339	5,207	6,075	6,943	7,810
1.89	874.8	1,750	2,624	3,499	4,374	5,249	6,123	6,998	7,873
1.90	881.7	1,763	2,645	3,527	4,409	5,290	6,172	7,054	7,935
1.91	888.7	1,777	2,666	3,555	4,443	5,332	6,221	7,110	7,998
1.92	895.7	1,791	2,687	3,583	4,478	5,374	6,270	7,165	8,061
1.93	902.7	1,805	2,708	3,611	4,513	5,416	6,319	7,221	8,124
1.94	909.7	1,819	2,729	3,639	4,549	5,458	6,368	7,278	8,187
1.95	916.8	1,834	2,750	3,667	4,584	5,500	6,417	7,334	8,251
1.96	923.8	1,848	2,771	3,695	4,619	5,543	6,467	7,390	8,314
1.97	930.9	1,862	2,793	3,724	4,654	5,585	6,516	7,447	8,378
1.98	938.0	1,876	2,814	3,752	4,690	5,628	6,566	7,504	8,442
1.99	945.1	1,890	2,835	3,780	4,726	5,671	6,616	7,561	8,506
2.00	952.2	1,904	2,857	3,809	4,761	5,713	6,666	7,618	8,570

Table 36.—*Discharge in second-feet of Cipolletti weirs, computed from $Q=3.367\ LH^{3/2}$*—Continued

Depth on crest (feet)	Length of weir in feet								
	100	200	300	400	500	600	700	800	900
2.1	1,024.5	2,049	3,074	4,098	5,123	6,147	7,172	8,196	9,221
2.2	1,098.6	2,197	3,296	4,394	5,493	6,592	7,690	8,789	9,887
2.3	1,174.3	2,349	3,523	4,697	5,872	7,046	8,220	9,395	10,569
2.4	1,251.7	2,504	3,755	5,007	6,259	7,510	8,762	10,014	11,266
2.5	1,330.8	2,662	3,992	5,323	6,654	7,985	9,316	10,646	11,977
2.6	1,411.4	2,823	4,234	5,646	7,057	8,469	9,880	11,291	12,703
2.7	1,493.6	2,987	4,481	5,975	7,468	8,962	10,455	11,949	13,443
2.8	1,577.4	3,155	4,732	6,310	7,887	9,464	11,042	12,619	14,196
2.9	1,662.6	3,325	4,988	6,651	8,313	9,976	11,638	13,301	14,964
3.0	1,749.4	3,499	5,248	6,997	8,747	10,496	12,246	13,995	15,744
3.1	1,837.6	3,675	5,513	7,350	9,188	11,025	12,863	14,700	16,538
3.2	1,927.2	3,854	5,782	7,709	9,636	11,563	13,490	15,418	17,345
3.3	2,018.2	4,036	6,055	8,073	10,091	12,109	14,128	16,146	18,164
3.4	2,110.7	4,221	6,332	8,443	10,553	12,664	14,775	16,885	18,996
3.5	2,204.5	4,409	6,613	8,818	11,022	13,227	15,431	17,636	19,840
3.6	2,299.6	4,599	6,899	9,198	11,498	13,798	16,097	18,397	20,696
3.7	2,396.1	4,792	7,188	9,584	11,980	14,377	16,773	19,169	21,565
3.8	2,493.9	4,988	7,482	9,975	12,469	14,963	17,457	19,951	22,445
3.9	2,593.0	5,186	7,779	10,372	12,965	15,558	18,151	20,744	23,337
4.0	2,693.3	5,387	8,080	10,773	13,467	16,160	18,853	21,547	24,240
4.1	2,795.0	5,590	8,385	11,180	13,975	16,770	19,565	22,360	25,155
4.2	2,897.8	5,796	8,694	11,591	14,489	17,387	20,285	23,183	26,081
4.3	3,001.9	6,004	9,006	12,008	15,010	18,012	21,014	24,016	27,018
4.4	3,107.3	6,215	9,322	12,429	15,536	18,644	21,751	24,858	27,965
4.5	3,213.8	6,428	9,641	12,855	16,069	19,283	22,497	25,710	28,924
4.6	3,321.5	6,643	9,965	13,286	16,608	19,929	23,251	26,572	29,894
4.7	3,430.4	6,861	10,291	13,722	17,152	20,583	24,013	27,443	30,874
4.8	3,540.5	7,081	10,621	14,162	17,702	21,243	24,783	28,324	31,864
4.9	3,651.7	7,303	10,955	14,607	18,258	21,910	25,562	29,214	32,865
5.0	3,764.0	7,528	11,292	15,056	18,820	22,584	26,348	30,112	33,876
5.5	4,342.5	8,685	13,028	17,370	21,713	26,055	30,398	34,740	39,083
6.0	4,948.0	9,896	14,844	19,792	24,740	29,688	34,636	39,584	44,532
6.5	5,579.2	11,158	16,738	22,317	27,896	33,475	39,054	44,633	50,213
7.0	6,235.2	12,470	18,705	24,941	31,176	37,411	43,646	49,881	56,116
7.5	6,915.0	13,830	20,745	27,660	34,575	41,490	48,405	55,320	62,235
8.0	7,617.9	15,236	22,854	30,472	38,090	45,707	53,325	60,943	68,561
8.5	8,343.1	16,686	25,029	33,372	41,716	50,059	58,402	66,745	75,088
9.0	9,090.0	18,180	27,270	36,360	45,450	54,540	63,630	72,720	81,810
9.5	9,857.9	19,716	29,574	39,432	49,290	59,148	69,006	78,863	88,721
10	10,646.3	21,293	31,939	42,585	53,232	63,878	74,524	85,171	95,817
11	12,282.6	24,565	36,848	49,130	61,413	73,695	85,978	98,260	110,543
12	13,995.0	27,990	41,985	55,980	69,975	83,970	97,965	111,960	125,955
13	15,780.3	31,561	47,341	63,121	78,902	94,682	110,462	126,242	142,023
14	17,635.7	35,271	52,907	70,543	88,178	105,814	123,450	141,085	158,721
15	19,558.6	39,117	58,676	78,234	97,793	117,351	136,910	156,469	176,027
16	21,546.7	43,093	64,640	86,187	107,733	129,280	150,827	172,373	193,920
17	23,597.9	47,196	70,794	94,392	117,990	141,587	165,185	188,783	212,381
18	25,710.4	51,421	77,131	102,842	128,552	154,262	179,973	205,683	231,394
19	27,882.4	55,765	83,647	111,530	139,412	167,295	195,177	223,059	250,942
20	30,112.4	60,225	90,337	120,450	150,562	180,674	210,787	240,899	271,011

Table 36 is not accurate for heads of water on the weir crest greater than one-third its length.   Also, owing to inability to measure the depth accurately and the effect of wind, etc., the tabulated discharge is likely to be in error for depths less than 0.2 foot.   Where velocity of approach exists, before taking out the discharge the measured head on the weir crest in table 36 should be increased by 1.5 times h, the velocity of approach head computed from $v^2 \div 2g$ where v is the velocity of approach in feet per second and g is the acceleration due to gravity.

The discharges for suppressed, thin-edged rectangular weirs can be obtained from table 36 by multiplying the appropriate tabular discharges therein by the factor 0.989.   The discharges for thin-edged rectangular weirs with end contractions can be obtained from table 36 by multiplying the appropriate tabular discharges therein by $0.989(L-0.2H) \div L$ where L is the crest length and H is the water depth on the crest, each in feet.

The discharges for thin-edged suppressed submerged weirs can be obtained from table 36 with fair accuracy by multiplying the appropriate value of the depth of water on the weir crest therein by the proper value of the coefficient n selected from the tabulation below before taking out the discharge.   In this tabulation D equals the head of water on the weir crest on the upstream side thereof and d equals the head of water thereon on the downstream side thereof.

Table 37.—*Values of n for use in Herschel's formula for submerged weirs, $Q = 3.367 \, L(nH)^{3/2}$*

$\dfrac{d}{D}$	0.0	0.01	0.02	0.03	0.04	0.05	0.06	0.07	0.08	0.09
0.0	1.000	1.004	1.006	1.006	1.007	1.007	1.007	1.006	1.006	1.005
.1	1.005	1.003	1.002	1.000	.998	.996	.994	.992	.989	.987
.2	.985	.982	.980	.977	.975	.972	.970	.967	.964	.961
.3	.959	.956	.953	.950	.947	.944	.941	.938	.935	.932
.4	.929	.926	.922	.919	.915	.912	.908	.904	.900	.896
.5	.892	.888	.884	.880	.875	.871	.866	.861	.856	.851
.6	.846	.841	.836	.830	.824	.818	.813	.806	.800	.794
.7	.787	.780	.773	.766	.758	.750	.742	.732	.723	.714
.8	.703	.692	.681	.669	.656	.644	.631	.618	.604	.590
.9	.574	.557	.539	.520	.498	.471	.441	.402	.352	.275

Table 38.—*Discharge in second-feet per foot of length over sharp-crested vertical weirs without end contractions*[1]

[Computed from the formula $Q=\left(0.405+\dfrac{.00984}{h}\right)\left(1+0.55\dfrac{h^2}{(p+h)^2}\right)Lh\sqrt{2gh}$
(h=observed head, in feet; p=height of weir, in feet; L=length of crest, in feet;
Q=discharge in second-feet.)]

p / h	2	4	6	8	10	20	30
0.1	0.13	0.13	0.13	0.13	0.13	0.13	0.13
0.2	.33	.33	.33	.33	.33	.33	.33
0.3	.58	.58	.58	.58	.58	.58	.58
0.4	.88	.88	.87	.87	.87	.87	.87
0.5	1.23	1.21	1.21	1.21	1.21	1.20	1.20
0.6	1.62	1.59	1.58	1.58	1.57	1.57	1.57
0.7	2.04	1.99	1.98	1.98	1.97	1.97	1.97
0.8	2.50	2.43	2.41	2.41	2.40	2.40	2.40
0.9	3.00	2.90	2.88	2.86	2.86	2.85	2.85
1.0	3.53	3.40	3.36	3.35	3.34	3.33	3.33
1.1	4.10	3.93	3.88	3.86	3.85	3.84	3.83
1.2	4.69	4.48	4.42	4.40	4.38	4.36	4.36
1.3	5.32	5.07	4.99	4.96	4.94	4.92	4.91
1.4	5.99	5.68	5.58	5.55	5.52	5.49	5.48
1.5	6.69	6.30	6.20	6.16	6.13	6.08	6.07
1.6	7.40	6.97	6.84	6.78	6.75	6.69	6.68
1.7	8.15	7.66	7.50	7.43	7.39	7.33	7.31
1.8	8.93	8.37	8.18	8.09	8.05	7.98	7.96
1.9	9.74	9.11	8.89	8.79	8.74	8.65	8.63
2.0	10.58	9.87	9.62	9.51	9.44	9.34	9.32
2.1	11.44	10.65	10.37	10.24	10.17	10.05	10.02
2.2	12.33	11.46	11.14	10.99	10.91	10.78	10.75
2.3	13.25	12.29	11.93	11.77	11.67	11.52	11.48
2.4	14.20	13.15	12.75	12.56	12.45	12.28	12.24
2.5	15.18	14.03	13.59	13.37	13.25	13.06	13.02
2.6	16.17	14.92	14.44	14.20	14.07	13.85	13.80
2.7	17.19	15.84	15.31	15.05	14.90	14.65	14.60
2.8	18.23	16.79	16.21	15.92	15.76	15.48	15.42
2.9	19.29	17.75	17.12	16.81	16.63	16.32	16.25
3.0	20.38	18.74	18.06	17.71	17.52	17.18	17.10
3.1	21.50	19.74	19.01	18.64	18.42	18.05	17.96
3.2	22.64	20.77	19.98	19.58	19.34	18.93	18.83
3.3	23.80	21.82	20.98	20.54	20.28	19.83	19.72
3.4	24.98	22.89	21.99	21.52	21.24	20.75	20.63
3.5	26.20	23.98	23.01	22.51	22.22	21.69	21.55
3.6	27.42	25.09	24.06	23.52	23.20	22.62	22.48
3.7	28.67	26.23	25.13	24.55	24.21	23.58	23.43
3.8	29.94	27.38	26.22	25.60	25.23	24.56	24.39
3.9	31.23	28.55	27.32	26.66	26.27	25.54	25.37
4.0	32.54	29.74	28.45	27.74	27.32	26.55	26.35
4.1	33.87	30.96	29.59	28.84	28.39	27.56	27.34
4.2	35.22	32.18	30.75	29.96	29.48	28.59	28.35
4.3	36.59	33.43	31.92	31.09	30.58	29.63	29.38
4.4	37.99	34.70	33.12	32.24	31.70	30.68	30.42

[1] This table should not be used where the weir is submerged, nor unless the over-falling sheet is aerated on the downstream face of the weir. If a vacuum forms under the falling sheet, the discharge may be 5 percent greater than given in this table. This table is not accurate for values of h greater than one-third L.

Table 38.—*Discharge in second-feet per foot of length over sharp-crested vertical weirs without end contractions*[1]—Continued

h \ p	2	4	6	8	10	20	30
4.5	39.40	35.98	34.33	33.40	32.83	31.74	31.47
4.6	40.83	37.29	35.56	34.58	33.98	32.82	32.53
4.7	42.28	38.61	36.80	35.78	35.14	33.92	33.61
4.8	43.75	39.96	38.07	37.00	36.32	35.04	34.70
4.9	45.23	41.32	39.35	38.23	37.52	36.17	35.80
5.0	46.73	42.69	40.65	39.48	38.74	37.21	36.91
5.1	48.25	44.09	41.96	40.73	39.97	38.45	38.03
5.2	49.79	45.50	43.29	42.01	41.20	39.61	39.17
5.3	51.36	46.93	44.64	43.30	42.45	40.78	40.31
5.4	52.94	48.38	46.00	44.60	43.71	41.96	41.47
5.5	54.54	49.85	47.38	45.93	45.00	43.16	42.64
5.6	56.15	51.34	48.79	47.27	46.31	44.38	43.83
5.7	57.78	52.83	50.19	48.62	47.62	45.60	45.02
5.8	59.42	54.34	51.62	49.99	48.94	46.83	46.22
5.9	61.09	55.88	53.07	51.38	50.29	48.08	47.44
6.0	62.77	57.43	54.53	52.78	51.64	49.34	48.67
6.1	64.46	59.00	56.00	54.20	53.02	50.61	49.91
6.2	66.18	60.58	57.50	55.63	54.40	51.90	51.16
6.3	67.91	62.18	59.01	57.07	55.80	53.20	52.42
6.4	69.65	63.79	60.53	58.53	57.22	54.50	53.70
6.5	71.42	65.42	62.07	60.01	58.65	55.82	54.98
6.6	73.19	67.07	63.63	61.50	60.09	57.16	56.27
6.7	74.99	68.74	65.20	63.00	61.55	58.50	57.58
6.8	76.80	70.42	66.78	64.53	63.02	59.96	58.90
6.9	78.62	72.11	68.38	66.06	64.50	61.23	60.22
7.0	80.46	73.82	70.00	67.60	66.00	62.61	61.56
7.1	82.32	75.55	71.63	69.17	67.52	64.00	62.91
7.2	84.18	77.29	73.28	70.74	69.04	65.40	64.27
7.3	86.07	79.04	74.94	72.34	70.58	66.81	65.64
7.4	87.97	80.81	76.61	73.94	72.14	68.24	67.02
7.5	89.89	82.60	78.30	75.56	73.70	69.68	68.41
7.6	91.82	84.40	80.01	77.19	75.28	71.13	69.81
7.7	93.76	86.22	81.73	78.84	76.88	72.59	71.23
7.8	95.72	88.05	83.46	80.50	78.48	74.06	72.65
7.9	97.70	89.90	85.21	82.18	80.11	75.55	74.09
8.0	99.68	91.75	86.97	83.87	81.74	77.04	75.53
8.1	101.69	93.63	88.75	85.57	83.39	78.55	76.98
8.2	103.70	95.51	90.54	87.29	85.25	80.06	78.44
8.3	105.73	97.42	92.34	89.02	86.72	81.59	79.92
8.4	107.78	99.34	94.16	90.76	88.41	83.13	81.40
8.5	109.84	101.27	96.00	92.52	90.11	84.69	82.90
8.6	111.91	103.21	97.84	94.29	91.82	86.25	84.41
8.7	113.99	105.17	99.70	96.07	93.55	87.82	85.92
8.8	116.09	107.14	101.57	97.87	95.28	89.40	87.44
8.9	118.20	109.13	103.46	99.68	97.04	91.00	88.98
9.0	120.33	111.13	105.36	101.50	98.80	92.61	90.52
9.1	122.47	113.15	107.28	103.34	100.58	94.23	92.08
9.2	124.62	115.18	109.21	105.19	102.37	95.86	93.65
9.3	126.79	117.22	111.15	107.06	104.17	97.49	95.22
9.4	128.97	119.27	113.10	108.93	105.99	99.14	96.80
9.5	131.16	121.34	115.07	110.82	107.82	100.80	98.40
9.6	133.36	123.42	117.05	112.72	109.65	102.48	100.00
9.7	135.58	125.51	119.04	114.64	111.50	104.16	101.62
9.8	137.82	127.63	121.05	116.57	113.37	105.85	103.25
9.9	140.06	129.74	123.07	118.51	115.25	107.56	104.88
10.0	142.31	131.87	125.10	120.46	117.14	109.27	106.52

Table 39.—*Multipliers for broad-crested weirs of rectangular cross section (Type a, fig. 2)*

[*p*=height of weir; *c*=width of crest; *h*=observed head; all in feet.]

p	4.6	4.6	11.25	11.25	11.25	11.25	11.25	11.25	11.25	11.25
c	2.6	6.6	.48	.93	1.65	3.17	5.88	8.98	12.24	16.30
h										
0.5	....	....	.821	.792	.806	.792	.799	.801	.786	.790
1.0	.765	.708	.997	.899	.808	.795	.791	.794	.815	.790
1.5	.789	.709	1.00	.982	.878	.796	.796	.793	.814	.792
2.0	.814	.710	1.00	1.00	.906	.815	.797	.792	.797	.793
2.5	.835	.711	1.00	1.00	.985	.844	.797	.790	.796	.793
3.0	.857	.711	1.00	1.00	1.00	.870	.797	.788	.794	.791
3.5	.878	.712	1.00	1.00	1.00	.90	.812	.787	.794	.791
4.0	.899	.714	1.00	1.00	1.00	.93	.834	.786	.792	.789
5.0	.940	.716	1.00	1.00	1.00	.97	(a)	.78	.79	.78
6.0	.986	.718	1.00	1.00	1.00	.98	(a)	.78	.78	.78
7.0	....	....	1.00	1.00	1.00	(a)	(a)	.77	.78	.77
8.0	....	....	1.00	1.00	1.00	(a)	(a)	.77	.77	.77
9.0	....	....	1.00	1.00	1.00	(a)	(a)	.77	.77	.77
10.0	....	....	1.00	1.00	1.00	(a)	(a)	.77	.77	.77

a Value doubtful.

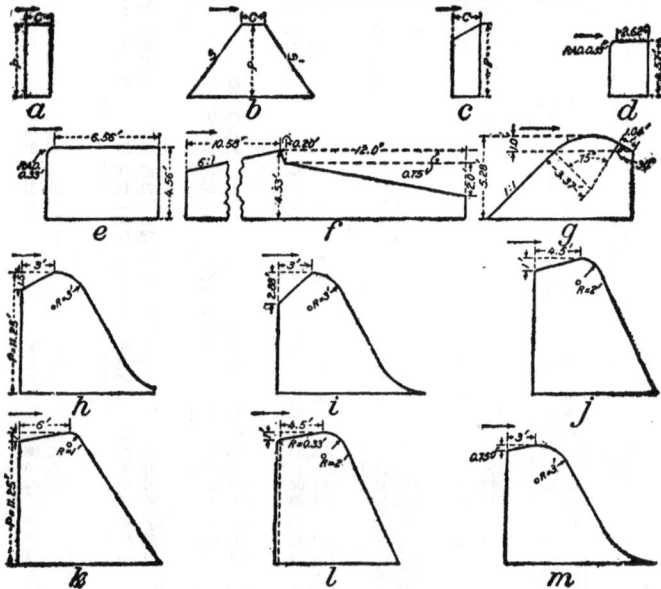

FIG. 2.—Types of weirs.

Table 40.—*Multipliers for weirs of trapezoidal cross section*

[p=height of weir, in feet; c=width of crest, in feet; s=upstream slope; s'=downstream slope; h=observed head, in feet.]

	Type b, fig. 2							Type c, fig. 2	
p c s s'	4.9 .33 2:1 0	4.9 .66 2:1 0	4.9 .66 3:1 0	4.9 .66 4:1 0	4.9 .66 5:1 0	4.9 .33 2:1 5:1	4.9 .66 2:1 2:1	4.65 7.00 4.67:1	11.25 6.00 6:1 ......
h									
1.0	1.137	1.048	1.066	1.039	1.009	1.095	1.071	1.042	1.060
1.5	1.131	1.068	1.066	1.039	1.009	1.071	1.066	1.033	1.069
2.0	1.120	1.080	1.061	1.033	1.005	1.044	1.053	1.024	1.054
2.5	1.106	1.085	1.052	1.026	.997	1.024	1.047	1.012	1.012
3.0	1.094	1.088	1.047	1.020	.991	1.009	1.047	.995	.985
3.5	1.085	1.087	1.043	1.017	.988	1.003	1.050	.983	.979
4.0	1.072	1.084	1.038	1.012	.984	1.014	1.052	.977	.976
4.5	1.064	1.081	1.035	1.009	.980	1.023	1.055	.974	.973
5.0	......	......	......	......	......	......	......	.97	.97
6.0	......	......	......	......	......	......	......	.97	.96
7.0	......	......	......	......	......	......	......	.97	.96
8.0	......	......	......	......	......	......	......	.96	.95
9.0	......	......	......	......	......	......	......	.96	.95
10.0	......	......	......	......	......	......	......	.96	.95

Table 41.—*Multipliers for compound weirs*

[p=height of weir, in feet; h=observed head, in feet.]

p	4.57	4.56	4.53	5.28	11.25	11.25	11.25	11.25	11.25	11.25
Type, fig. 2.	d	e	f	g	h	i	j	k	l	m
h										
0.5	......	......	......	......	.941	.924	.933	.962	.971	.947
1.0	.842	.836	.929	.976	1.039	1.033	.988	1.045	1.033	1.000
1.5	.866	.834	.950	.979	1.087	1.093	1.018	1.066	1.042	1.036
2.0	.888	.831	.953	.988	1.109	1.133	1.033	1.063	1.035	1.063
2.5	.906	.826	.947	1.000	1.118	1.153	1.045	1.020	1.033	1.085
3.0	.927	.822	.942	1.016	1.120	1.163	1.054	.997	1.045	1.096
3.5	.945	.817	.936	1.032	1.127	1.169	1.060	.994	1.054	1.108
4.0	.965	.812	.931	1.044	1.123	1.165	1.060	.991	1.057	1.110
5.0	1.00	.80	.92	1.05	1.11	1.16	1.05	.98	1.05	1.10
6.0	......	......	......	......	1.11	1.15	1.04	.98	1.04	1.10
7.0	......	......	......	......	1.10	1.14	1.04	.97	1.04	1.09
8.0	......	......	......	......	1.10	1.14	1.04	.97	1.03	1.09
9.0	......	......	......	......	1.09	1.14	1.03	.97	1.03	1.08
10.0	......	......	......	......	1.09	1.13	1.03	.97	1.03	1.08

Table 42.—*Discharge of sharp-edged rectangular submerged orifices in second-feet, computed from the formula*

$$Q = 0.61 \sqrt{2gH} \; A$$

Head H, feet	Cross-sectional area A of orifice, square feet							
	0.25	**0.5**	**0.75**	**1.0**	**1.25**	**1.5**	**1.75**	**2.0**
0.01	0.122	0.245	0.367	0.489	0.611	0.734	0.856	0.978
.02	0.173	0.346	0.518	0.691	0.864	1.037	1.210	1.382
.03	0.212	0.424	0.635	0.847	1.059	1.271	1.483	1.694
.04	0.245	0.489	0.734	0.978	1.223	1.468	1.712	1.957
.05	0.273	0.547	0.820	1.093	1.367	1.640	1.913	2.186
.06	0.300	0.599	0.899	1.198	1.497	1.797	2.097	2.396
.07	0.324	0.647	0.971	1.294	1.617	1.941	2.265	2.588
.08	0.346	0.691	1.037	1.383	1.729	2.074	2.420	2.766
.09	0.367	0.734	1.101	1.468	1.835	2.201	2.638	2.935
.10	0.387	0.773	1.160	1.557	1.933	2.320	2.707	3.094
.11	0.406	0.811	1.217	1.622	2.027	2.433	2.839	3.244
.12	0.424	0.847	1.271	1.694	2.118	2.542	2.965	3.389
.13	0.441	0.882	1.323	1.764	2.205	2.645	3.086	3.527
.14	0.458	0.915	1.373	1.830	2.287	2.745	3.203	3.660
.15	0.474	0.947	1.421	1.895	2.369	2.842	3.316	3.790
.16	0.489	0.978	1.467	1.956	2.445	2.934	3.423	3.912
.17	0.504	1.008	1.512	2.016	2.520	3.024	3.528	4.032
.18	0.519	1.037	1.556	2.075	2.593	3.112	3.631	4.150
.19	0.533	1.066	1.599	2.132	2.665	3.198	3.731	4.264
.20	0.547	1.094	1.641	2.188	2.735	3.282	3.829	4.376
.21	0.561	1.120	1.681	2.241	2.801	3.361	3.921	4.482
.22	0.574	1.148	1.722	2.296	2.870	3.464	4.018	4.592
.23	0.587	1.172	1.759	2.345	2.931	3.517	4.103	4.690
.24	0.600	1.198	1.797	2.396	2.995	3.599	4.193	4.792
.25	0.612	1.223	1.834	2.446	3.057	3.668	4.280	4.891
.26	0.624	1.247	1.871	2.494	3.117	3.741	4.365	4.988
.27	0.636	1.270	1.906	2.541	3.176	3.811	4.446	5.082
.28	0.646	1.294	1.942	2.589	3.236	3.883	4.530	5.178
.29	0.659	1.319	1.978	2.638	3.297	3.956	4.616	5.276
.30	0.670	1.339	2.009	2.678	3.347	4.017	4.687	5.356
.31	0.681	1.363	2.045	2.726	3.407	4.089	4.771	5.452
.32	0.692	1.382	2.073	2.764	3.455	4.146	4.837	5.528
.33	0.703	1.405	2.107	2.810	3.513	4.215	4.917	5.620
.34	0.713	1.426	2.139	2.852	3.565	4.278	4.991	5.704
.35	0.724	1.446	2.169	2.892	3.615	4.338	5.061	5.784
.36	0.734	1.467	2.201	2.934	3.667	4.401	5.135	5.868
.37	0.745	1.488	2.232	2.976	3.720	4.464	5.208	5.952
.38	0.754	1.508	2.262	3.016	3.770	4.524	5.278	6.032
.39	0.764	1.527	2.291	3.054	3.818	4.582	5.345	6.109
0.40	0.774	1.547	2.321	3.094	3.867	4.641	5.415	6.188

Table 42.—*Discharge of sharp-edged rectangular submerged orifices in second-feet, computed from the formula $Q=0.61 \sqrt{2gH} A$—* Continued

Head H, feet	Cross-sectional area A of orifice, square feet							
	0.25	0.5	0.75	1.0	1.25	1.5	1.75	2.0
0.41	0.783	1.567	2.350	3.133	3.917	4.700	5.483	6.266
.42	0.792	1.585	2.377	3.170	3.962	4.754	5.547	6.339
.43	0.802	1.604	2.406	3.208	4.010	4.812	5.614	6.416
.44	0.811	1.622	2.433	3.244	4.055	4.866	5.677	6.488
.45	0.820	1.640	2.461	3.281	4.101	4.921	5.741	6.562
.46	0.829	1.659	2.489	3.318	4.147	4.977	5.807	6.636
.47	0.839	1.678	2.517	3.356	4.195	5.035	5.874	6.713
.48	0.847	1.695	2.542	3.389	4.237	5.084	5.931	6.778
.49	0.856	1.712	2.568	3.424	4.280	5.136	5.992	6.848
.50	0.865	1.729	2.594	3.458	4.323	5.188	6.052	6.917
.51	0.873	1.746	2.620	3.493	4.366	5.239	6.112	6.986
.52	0.882	1.763	2.645	3.527	4.409	5.290	6.172	7.054
.53	0.890	1.780	2.670	3.560	4.451	5.341	6.231	7.121
.54	0.898	1.797	2.695	3.593	4.491	5.390	6.288	7.186
.55	0.907	1.813	2.719	3.626	4.533	5.439	6.345	7.252
.56	0.915	1.830	2.745	3.660	4.575	5.490	6.405	7.320
.57	0.923	1.846	2.769	3.692	4.615	5.538	6.461	7.384
.58	0.931	1.862	2.794	3.725	4.656	5.587	6.518	7.450
.59	0.939	1.879	2.818	3.757	4.697	5.636	6.575	7.514
.60	0.947	1.895	2.842	3.790	4.737	5.684	6.632	7.579
.61	0.955	1.910	2.865	3.820	4.775	5.730	6.685	7.640
.62	0.963	1.925	2.887	3.850	4.812	5.775	6.737	7.700
.63	0.971	1.941	2.911	3.882	4.853	5.823	6.793	7.764
.64	0.978	1.956	2.934	3.912	4.890	5.868	6.846	7.824
.65	0.986	1.972	2.958	3.944	4.930	5.916	6.902	7.888
.66	0.993	1.987	2.980	3.974	4.967	5.960	6.954	7.947
.67	1.001	2.002	3.003	4.004	5.005	6.006	7.007	8.008
.68	1.008	2.016	3.024	4.032	5.040	6.048	7.056	8.064
.69	1.016	2.032	3.048	4.064	5.080	6.096	7.112	8.128
.70	1.023	2.046	3.069	4.092	5.115	6.138	7.161	8.184
.71	1.031	2.062	3.093	4.124	5.155	6.186	7.217	8.248
.72	1.038	2.076	3.114	4.152	5.190	6.228	7.266	8.304
.73	1.045	2.090	3.135	4.180	5.225	6.270	7.315	8.360
.74	1.052	2.104	3.158	4.210	5.260	6.311	7.369	8.421
.75	1.059	2.118	3.178	4.237	5.296	6.355	7.413	8.475
.76	1.066	2.132	3.198	4.264	5.330	6.396	7.462	8.528
.77	1.072	2.145	3.217	4.290	5.362	6.434	7.507	8.579
.78	1.080	2.160	3.240	4.320	5.400	6.480	7.560	8.640
.79	1.087	2.174	3.261	4.348	5.435	6.522	7.609	8.696
0.80	1.094	2.188	3.282	4.376	5.470	6.564	7.658	8.752

Table 43.—*Coefficients C to be applied to a discharge given by table 42 to give the discharge of the same orifice suppressed, computed from the formula* $C = 1 + 0.15\ r$.

d = height of orifice, in feet.
l = length of orifice, in feet.
r = ratio of suppressed perimeter to total perimeter.

Size of orifice			Bottom suppressed		Bottom and sides suppressed	
d, feet	l, feet	A, square feet	r	C	r	C
0.25	1.0	0.25	0.40	1.06	0.60	1.09
	2.0	.50	.44	1.07	.56	1.08
	3.0	.75	.46	1.07	.54	1.08
0.5	1.0	.50	.33	1.05	.67	1.10
	1.5	.75	.37	1.06	.63	1.09
	2.0	1.00	.40	1.06	.60	1.09
	2.5	1.25	.42	1.06	.58	1.09
	3.0	1.50	.43	1.06	.57	1.09
0.75	1.33	1.00	.32	1.05	.68	1.10
	1.67	1.25	.34	1.05	.66	1.10
	2.00	1.50	.36	1.05	.64	1.10
	2.33	1.75	.38	1.06	.62	1.09
	2.67	2.00	0.39	1.06	0.61	1.09

EXAMPLE: To find the discharge of a standard submerged rectangular orifice 0.5 by 2.5 feet with bottom and side suppressions under a head of 0.18 feet.

For an area of 1.25 square feet ($= 0.5 \times 2.5$) and a head of 0.18 feet, table 42 gives a discharge of 2.593 second-feet. For a height, d, of 0.5 feet and a length, l, of 2.5 feet, with bottom and sides suppressed, table 43 gives a coefficient of 1.09. Then $2.593 \times 1.09 = 2.826$ second-feet, the discharge desired.

Table 44.—*Factors for solution of Scobey's formula for flow in concrete pipe.* $Q = 0.00546 \ C_s \ d^{2.625} \ H^{0.5}$

Q = Discharge in second-feet
C_s = Scobey's coefficient

d = Diameter in inches
H = Friction head per 1,000 feet

d	0.00546 C_s $d^{2.625}$				
	$C_s = 0.310$	$C_s = 0.345$	$C_s = 0.370$	$C_s = 0.380$	$C_s = 0.400$
6	0.1867	0.2078	0.2229	0.2289	0.2409
8	0.3973	0.4422	0.4742	0.4870	0.5127
10	0.7138	0.7944	0.8519	0.8749	0.9210
12	1.1519	1.2819	1.3748	1.4120	1.4863
15	2.0692	2.3028	2.4696	2.5364	2.6699
18	3.3392	3.7162	3.9855	4.0932	4.3087
21	5.0047	5.5698	5.9734	6.1348	6.4577
24	7.1057	7.9080	8.4810	8.7103	9.1687
27	9.6802	10.7731	11.5538	11.8661	12.4906
30	12.764	14.205	15.235	15.646	16.470
33	16.393	18.244	19.566	20.094	21.152
36	20.599	22.925	24.586	25.250	26.579
39	25.416	28.285	30.335	31.155	32.794
42	30.873	34.359	36.849	37.844	39.837
45	37.003	41.181	44.165	45.359	47.746
48	43.834	48.783	52.318	53.732	56.560
51	51.396	57.198	61.343	63.002	66.317
54	59.716	66.458	71.273	73.200	77.052
57	68.822	76.592	82.142	84.362	88.802
60	78.741	87.631	93.981	96.521	101.601
63	89.500	99.605	106.823	109.710	115.484
66	101.125	112.542	120.697	123.959	130.483
69	113.641	126.471	135.635	139.301	146.633
72	127.073	141.420	151.667	155.767	163.965
78	156.784	174.486	187.130	192.187	202.302
84	190.453	211.956	227.315	233.459	245.746
90	228.266	254.038	272.446	279.810	294.537
96	270.406	300.936	322.743	331.465	348.911
102	317.052	352.848	378.416	388.644	409.099
108	368.376	409.967	439.675	451.558	475.324
114	424.551	472.484	506.722	520.417	547.808
120	485.741	540.583	579.756	595.425	626.763

$C_s = 0.310$ for concrete pipe lines 21 inches and less in diameter with mortar joints.
$C_s = 0.345$ for concrete pipe lines 21 inches and less in diameter with rubber gasket joints.
$C_s = 0.370$ for concrete pipe lines 24 inches and larger in diameter with rubber gasket or smooth troweled mortar joints.
$C_s = 0.380$ for concrete pipe 36 inches or larger in diameter with very best dense smooth surface obtained by placing concrete against metal forms where pipe is not subject to interior surface deterioration and is carrying clear water.
$C_s = 0.400$ for special considerations involving best possible flow conditions and alinement, and for establishment of minimum design water surface.

Table 45.—*Theoretical velocity of water in feet per second for heads of 0 to 2.6 feet*

$$V = \sqrt{2gh}. \quad g = 32.16$$

Head, in feet	0.000	0.001	0.002	0.003	0.004	0.005	0.006	0.007	0.008	0.009
0.00	------	0.254	0.358	0.439	0.507	0.567	0.621	0.671	0.717	0.761
.01	0.802	0.841	0.878	0.914	0.949	0.982	1.014	1.046	1.076	1.105
.02	1.134	1.162	1.190	1.216	1.242	1.268	1.293	1.318	1.342	1.366
.03	1.388	1.412	1.435	1.457	1.479	1.500	1.522	1.543	1.563	1.584
.04	1.604	1.624	1.644	1.663	1.682	1.701	1.720	1.739	1.757	1.775
.05	1.793	1.811	1.829	1.846	1.864	1.881	1.898	1.915	1.931	1.948
.06	1.964	1.981	1.997	2.013	2.028	2.045	2.060	2.076	2.091	2.107
.07	2.122	2.137	2.152	2.167	2.182	2.196	2.211	2.225	2.240	2.254
.08	2.268	2.283	2.297	2.310	2.324	2.338	2.352	2.366	2.379	2.393
.09	2.406	2.419	2.433	2.446	2.459	2.472	2.485	2.498	2.511	2.523
.10	2.536	2.549	2.561	2.574	2.586	2.599	2.611	2.623	2.636	2.648
.11	2.660	2.672	2.684	2.696	2.708	2.720	2.732	2.743	2.755	2.767
.12	2.778	2.790	2.801	2.813	2.824	2.835	2.847	2.858	2.869	2.880
.13	2.892	2.903	2.914	2.925	2.936	2.947	2.958	2.968	2.979	2.990
.14	3.001	3.011	3.022	3.033	3.043	3.054	3.064	3.075	3.085	3.096
.15	3.106	3.116	3.127	3.137	3.147	3.157	3.168	3.178	3.188	3.198
.16	3.208	3.218	3.228	3.238	3.248	3.258	3.267	3.277	3.287	3.297
.17	3.307	3.316	3.326	3.336	3.345	3.355	3.365	3.374	3.384	3.393
.18	3.402	3.412	3.421	3.431	3.440	3.450	3.459	3.468	3.477	3.486
.19	3.496	3.505	3.514	3.523	3.532	3.541	3.551	3.560	3.569	3.578

Head, in feet	0.00	0.01	0.02	0.03	0.04	0.05	0.06	0.07	0.08	0.09
0.2	3.586	3.675	3.762	3.846	3.929	4.010	4.089	4.167	4.244	4.319
0.3	4.393	4.465	4.536	4.607	4.676	4.745	4.812	4.878	4.944	5.008
0.4	5.072	5.135	5.197	5.259	5.320	5.380	5.439	5.498	5.556	5.614
0.5	5.671	5.727	5.783	5.838	5.893	5.947	6.001	6.054	6.107	6.160
0.6	6.212	6.263	6.315	6.365	6.416	6.465	6.515	6.564	6.613	6.662
0.7	6.710	6.757	6.805	6.852	6.899	6.946	6.992	7.038	7.083	7.128
0.8	7.173	7.218	7.262	7.306	7.350	7.394	7.438	7.481	7.523	7.566
0.9	7.608	7.650	7.692	7.734	7.776	7.817	7.858	7.898	7.939	7.979
1.0	8.020	8.060	8.099	8.139	8.179	8.218	8.257	8.296	8.335	8.373
1.1	8.412	8.450	8.487	8.525	8.563	8.600	8.638	8.675	8.712	8.749
1.2	8.785	8.822	8.858	8.894	8.930	8.967	9.002	9.038	9.073	9.108
1.3	9.144	9.179	9.214	9.249	9.284	9.318	9.353	9.387	9.421	9.455
1.4	9.489	9.523	9.557	9.590	9.624	9.657	9.690	9.724	9.757	9.790
1.5	9.822	9.855	9.888	9.920	9.953	9.985	10.017	10.049	10.081	10.113
1.6	10.145	10.176	10.208	10.239	10.271	10.302	10.333	10.364	10.395	10.425
1.7	10.457	10.487	10.518	10.549	10.579	10.611	10.640	10.670	10.700	10.730
1.8	10.760	10.790	10.820	10.849	10.879	10.908	10.938	10.967	10.996	11.026
1.9	11.055	11.084	11.113	11.142	11.171	11.199	11.228	11.257	11.285	11.314
2.0	11.342	11.370	11.399	11.427	11.455	11.483	11.511	11.539	11.567	11.594
2.1	11.622	11.650	11.677	11.705	11.732	11.760	11.787	11.814	11.841	11.868
2.2	11.896	11.923	11.949	11.976	12.003	12.030	12.057	12.083	12.110	12.137
2.3	12.163	12.189	12.216	12.242	12.268	12.294	12.321	12.347	12.373	12.399
2.4	12.424	12.450	12.476	12.502	12.528	12.553	12.579	12.604	12.630	12.655
2.5	12.681	12.706	12.731	12.757	12.782	12.807	12.832	12.857	12.882	12.907

Table 46.—*Theoretical velocity of water in feet per second for heads of 0 to 50 feet*

$$V = \sqrt{2gh}. \quad g = 32.16$$

Head, in feet	0.0	0.1	0.2	0.3	0.4	0.5	0.6	0.7	0.8	0.9
0	0.0	2.5	3.6	4.4	5.1	5.7	6.2	6.7	7.2	7.6
1	8.0	8.4	8.8	9.1	9.5	9.8	10.1	10.5	10.8	11.1
2	11.3	11.6	11.9	12.2	12.4	12.7	12.9	13.2	13.4	13.7
3	13.9	14.1	14.3	14.6	14.8	15.0	15.2	15.4	15.6	15.8
4	16.0	16.2	16.4	16.6	16.8	17.0	17.2	17.4	17.6	17.8
5	17.9	18.1	18.3	18.5	18.6	18.8	19.0	19.2	19.3	19.5
6	19.6	19.8	20.0	20.1	20.3	20.5	20.6	20.8	20.9	21.1
7	21.2	21.4	21.5	21.7	21.8	22.0	22.1	22.3	22.4	22.5
8	22.7	22.8	23.0	23.1	23.3	23.4	23.5	23.7	23.8	23.9
9	24.1	24.2	24.3	24.5	24.6	24.7	24.8	25.0	25.1	25.2
10	25.4	25.5	25.6	25.7	25.9	26.0	26.1	26.2	26.4	26.5
11	26.6	26.7	26.8	27.0	27.1	27.2	27.3	27.4	27.5	27.7
12	27.8	27.9	28.0	28.1	28.2	28.4	28.5	28.6	28.7	28.8
13	28.9	29.0	29.1	29.2	29.4	29.5	29.6	29.7	29.8	29.9
14	30.0	30.1	30.2	30.3	30.4	30.5	30.6	30.7	30.9	31.0
15	31.1	31.2	31.3	31.4	31.5	31.6	31.7	31.8	31.9	32.0
16	32.1	32.2	32.3	32.4	32.5	32.6	32.7	32.8	32.9	33.0
17	33.1	33.2	33.3	33.4	33.5	33.5	33.6	33.7	33.8	33.9
18	34.0	34.1	34.2	34.3	34.4	34.5	34.6	34.7	34.8	34.9
19	35.0	35.0	35.2	35.2	35.3	35.4	35.4	35.5	35.6	35.8
20	35.9	36.0	36.0	36.1	36.2	36.3	36.4	36.5	36.6	36.7
21	36.8	36.8	36.9	37.0	37.1	37.2	37.3	37.4	37.4	37.5
22	37.6	37.7	37.8	37.9	38.0	38.0	38.1	38.2	38.3	38.4
23	38.5	38.5	38.6	38.7	38.8	38.9	39.0	39.0	39.1	39.2
24	39.3	39.4	39.5	39.5	39.6	39.7	39.8	39.9	39.9	40.0
25	40.1	40.2	40.3	40.3	40.4	40.5	40.6	40.7	40.7	40.8
26	40.9	41.0	41.1	41.1	41.2	41.3	41.4	41.4	41.5	41.6
27	41.7	41.8	41.8	41.9	42.0	42.1	42.1	42.2	42.3	42.4
28	42.4	42.5	42.6	42.7	42.7	42.8	42.9	43.0	43.1	43.2
29	43.2	43.3	43.3	43.4	43.5	43.6	43.6	43.7	43.8	43.9
30	43.9	44.0	44.1	44.2	44.2	44.3	44.4	44.4	44.5	44.6
31	44.7	44.7	44.8	44.9	44.9	45.0	45.1	45.2	45.2	45.3
32	45.4	45.4	45.5	45.5	45.6	45.7	45.8	45.9	45.9	46.0
33	46.1	46.1	46.2	46.3	46.3	46.4	46.5	46.6	46.6	46.7
34	46.8	46.8	46.9	47.0	47.0	47.1	47.2	47.2	47.3	47.4
35	47.4	47.5	47.6	47.6	47.7	47.8	47.9	47.9	48.0	48.1
36	48.1	48.2	48.3	48.3	48.4	48.5	48.5	48.6	48.6	48.7
37	48.8	48.8	48.9	49.0	49.1	49.1	49.2	49.2	49.3	49.4
38	49.4	49.5	49.6	49.6	49.7	49.8	49.8	49.9	50.0	50.0
39	50.1	50.1	50.2	50.3	50.3	50.4	50.5	50.5	50.6	50.7
40	50.7	50.8	50.8	50.9	51.0	51.0	51.1	51.2	51.2	51.3
41	51.4	51.4	51.5	51.5	51.6	51.7	51.7	51.8	51.9	51.9
42	52.0	52.0	52.1	52.2	52.2	52.3	52.3	52.4	52.5	52.5
43	52.6	52.7	52.7	52.8	52.8	52.9	53.0	53.0	53.1	53.1
44	53.2	53.3	53.3	53.4	53.4	53.5	53.6	53.6	53.7	53.7
45	53.8	53.9	53.9	54.0	54.0	54.1	54.2	54.2	54.3	54.3
46	54.4	54.5	54.5	54.6	54.6	54.7	54.7	54.8	54.9	54.9
47	55.0	55.0	55.1	55.2	55.2	55.3	55.3	55.4	55.5	55.5
48	55.6	55.6	55.7	55.7	55.8	55.9	55.9	56.0	56.0	56.1
49	56.1	56.2	56.3	56.3	56.4	56.4	56.5	56.5	56.6	56.7

Table 47.—*Theoretical heads in feet for velocities of 0 to 38 feet per second*

$$h = \frac{V^2}{2g}$$

V	0.00	0.01	0.02	0.03	0.04	0.05	0.06	0.07	0.08	0.09
0.0	0.000	0.000	0.000	0.000	0.000	0.000	0.000	0.000	0.000	0.000
0.1	.000	.000	.000	.000	.000	.000	.000	.000	.001	.001
0.2	.001	.001	.001	.001	.001	.001	.001	.001	.001	.001
0.3	.001	.002	.002	.002	.002	.002	.002	.002	.002	.002
0.4	.003	.003	.003	.003	.003	.003	.003	.003	.004	.004
0.5	.004	.004	.004	.004	.005	.005	.005	.005	.005	.005
0.6	.006	.006	.006	.006	.006	.007	.007	.007	.007	.007
0.7	.008	.008	.008	.008	.009	.009	.009	.009	.010	.010
0.8	.010	.010	.011	.011	.011	.011	.012	.012	.012	.012
0.9	.013	.013	.013	.013	.014	.014	.014	.015	.015	.015
1.0	.016	.016	.016	.017	.017	.017	.018	.018	.018	.019
1.1	.019	.019	.020	.020	.020	.021	.021	.021	.022	.022
1.2	.022	.023	.023	.024	.024	.024	.025	.025	.026	.026
1.3	.026	.027	.027	.028	.028	.028	.029	.029	.030	.030
1.4	.031	.031	.031	.032	.032	.033	.033	.034	.034	.035
1.5	.035	.035	.036	.036	.037	.037	.038	.038	.039	.039
1.6	.040	.040	.041	.041	.042	.042	.043	.043	.044	.044
1.7	.045	.046	.046	.047	.047	.048	.048	.049	.049	.050
1.8	.050	.051	.052	.052	.053	.053	.054	.054	.055	.056
1.9	.056	.057	.057	.058	.059	.059	.060	.060	.061	.062
2.0	.062	.063	.064	.064	.065	.065	.066	.067	.067	.068
2.1	.069	.069	.070	.071	.071	.072	.073	.073	.074	.075
2.2	.075	.076	.077	.077	.078	.079	.079	.080	.081	.082
2.3	.082	.083	.084	.084	.085	.086	.087	.087	.088	.089
2.4	.090	.090	.091	.092	.093	.093	.094	.095	.096	.096
2.5	.097	.098	.099	.100	.100	.101	.102	.103	.104	.104
2.6	.105	.106	.107	.108	.108	.109	.110	.111	.112	.113
2.7	.113	.114	.115	.116	.117	.118	.118	.119	.120	.121
2.8	.122	.123	.124	.125	.125	.126	.127	.128	.129	.130
2.9	.131	.132	.133	.134	.134	.136	.136	.137	.138	.139
3.0	.140	.141	.142	.143	.144	.145	.146	.147	.148	.148
3.1	.149	.150	.151	.152	.153	.154	.155	.156	.157	.158
3.2	.159	.160	.161	.162	.163	.164	.165	.166	.167	.168
3.3	.169	.170	.171	.172	.173	.175	.176	.177	.178	.179
3.4	.180	.181	.182	.183	.184	.185	.186	.187	.188	.189
3.5	.190	.192	.193	.194	.195	.196	.197	.198	.199	.200
3.6	.202	.203	.204	.205	.206	.207	.208	.209	.211	.212
3.7	.213	.214	.215	.216	.218	.219	.220	.221	.222	.223
3.8	.225	.226	.227	.228	.229	.230	.232	.233	.234	.235
3.9	.237	.238	.239	.240	.241	.243	.244	.245	.246	.248
4.0	.249	.250	.251	.253	.254	.255	.256	.258	.259	.260
4.1	.261	.263	.264	.265	.267	.268	.269	.270	.272	.273
4.2	.274	.276	.277	.278	.280	.281	.282	.284	.285	.286
4.3	.288	.289	.290	.292	.293	.294	.296	.297	.298	.300
4.4	.301	.302	.304	.305	.307	.308	.309	.311	.312	.313
4.5	.315	.316	.318	.319	.320	.322	.323	.325	.326	.328
4.6	.329	.330	.332	.333	.335	.336	.338	.339	.341	.342
4.7	.343	.345	.346	.348	.349	.351	.352	.354	.355	.357
4.8	.358	.360	.361	.363	.364	.366	.367	.369	.370	.372
4.9	.373	.375	.376	.378	.379	.381	.382	.384	.385	.387

Table 47.—*Theoretical heads in feet for velocities of 0 to 38 feet per second*—Continued

$$h = \frac{V^2}{2g}$$

V	0.00	0.01	0.02	0.03	0.04	0.05	0.06	0.07	0.08	0.09
5.0	0.389	0.391	0.392	0.394	0.395	0.397	0.398	0.400	0.401	0.403
5.1	.404	.406	.407	.409	.410	.412	.414	.415	.417	.418
5.2	.420	.422	.423	.425	.427	.429	.430	.432	.434	.435
5.3	.437	.439	.440	.442	.443	.445	.447	.448	.450	.451
5.4	.453	.455	.456	.458	.460	.462	.463	.465	.467	.468
5.5	.470	.472	.474	.475	.477	.479	.481	.483	.484	.486
5.6	.488	.490	.491	.493	.495	.497	.498	.500	.502	.503
5.7	.505	.507	.509	.510	.512	.514	.516	.518	.519	.521
5.8	.523	.525	.527	.528	.530	.532	.534	.536	.537	.539
5.9	.541	.543	.545	.547	.549	.551	.552	.554	.556	.558
6.0	.560	.562	.564	.565	.567	.569	.571	.573	.574	.576
6.1	.578	.580	.582	.584	.586	.588	.590	.592	.594	.596
6.2	.598	.600	.602	.604	.606	.608	.610	.612	.614	.616
6.3	.618	.620	.622	.624	.626	.628	.629	.631	.633	.635
6.4	.637	.639	.641	.643	.645	.647	.649	.651	.653	.655
6.5	.657	.659	.661	.663	.665	.667	.669	.671	.673	.675
6.6	.677	.679	.681	.683	.685	.688	.690	.692	.694	.696
6.7	.698	.700	.702	.704	.706	.709	.711	.713	.715	.717
6.8	.719	.721	.723	.725	.727	.730	.732	.734	.736	.738
6.9	.740	.742	.744	.747	.749	.751	.753	.755	.757	.760
7.0	.762	.764	.766	.769	.771	.773	.775	.777	.779	.782
7.1	.784	.786	.788	.791	.793	.795	.797	.799	.801	.804
7.2	.806	.808	.810	.812	.814	.817	.819	.821	.823	.825
7.3	.827	.829	.832	.834	.837	.839	.841	.844	.846	.849
7.4	.851	.853	.856	.858	.861	.863	.865	.868	.870	.873
7.5	.875	.877	.880	.882	.884	.887	.889	.891	.893	.896
7.6	.898	.900	.903	.905	.908	.910	.912	.915	.917	.920
7.7	.922	.924	.927	.929	.932	.934	.936	.939	.941	.943
7.8	.946	.948	.951	.953	.956	.958	.960	.963	.965	.967
7.9	.970	.973	.975	.978	.980	.983	.985	.988	.990	.993
8.0	.995	.998	1.000	1.003	1.005	1.008	1.010	1.013	1.015	1.018
8.1	1.020	1.023	.025	.028	.030	.033	.035	.038	.040	.043
8.2	.045	.048	.050	.053	.055	.058	.061	.063	.066	.068
8.3	.071	.074	.076	.079	.081	.084	.087	.089	.092	.094
8.4	.097	.100	.102	.105	.107	.110	.113	.115	.118	.120
8.5	.123	.126	.128	.131	.134	.137	.139	.142	.145	.147
8.6	.150	.153	.155	.158	.161	.164	.166	.169	.172	.174
8.7	.177	.180	.182	.185	.188	.191	.193	.196	.199	.201
8.8	.204	.207	.209	.212	.215	.218	.220	.223	.226	.228
8.9	.231	.234	.237	.239	.242	.245	.248	.251	.253	.256
9.0	.259	.262	.265	.267	.270	.273	.276	.279	.281	.284
9.1	.287	.290	.293	.296	.299	.302	.304	.307	.310	.313
9.2	.316	.319	.322	.325	.328	.331	.333	.336	.339	.342
9.3	.345	.348	.351	.354	.357	.360	.362	.365	.368	.371
9.4	.374	.377	.380	.383	.386	.389	.391	.394	.397	.400
9.5	.403	.406	.409	.412	.415	.418	.421	.424	.427	.430
9.6	.433	.436	.439	.442	.445	.448	.451	.454	.457	.460
9.7	.463	.466	.469	.472	.475	.478	.481	.484	.487	.490
9.8	.493	.496	.499	.502	.505	.509	.512	.515	.518	.521
9.9	.524	.527	.530	.533	.536	.540	.543	.546	.549	.552

Table 47.—*Theoretical heads in feet for velocities of 0 to 38 feet per second*—Continued

$$h = \frac{V^2}{2g}$$

V	0.00	0.01	0.02	0.03	0.04	0.05	0.06	0.07	0.08	0.09
10.0	1.555	1.558	1.561	1.564	1.567	1.571	1.574	1.577	1.580	1.583
10.1	.586	.589	.592	.596	.599	.602	.605	.608	.612	.615
10.2	.618	.621	.624	.627	.630	.634	.637	.640	.643	.646
10.3	.649	.652	.656	.659	.662	.666	.669	.672	.675	.679
10.4	.682	.685	.688	.692	.695	.698	.701	.704	.708	.711
10.5	.714	.717	.720	.723	.726	.730	.733	.736	.739	.742
10.6	.745	.749	.752	.756	.759	.763	.766	.770	.773	.777
10.7	.780	.783	.787	.790	.793	.797	.800	.803	.806	.810
10.8	.813	.816	.820	.823	.827	.830	.833	.837	.840	.844
10.9	.847	.850	.854	.857	.861	.864	.867	.871	.874	.878
11.0	.881	.884	.888	.891	.895	.898	.901	.905	.908	.912
11.1	.915	.919	.922	.926	.929	.933	.936	.940	.943	.947
11.2	.950	.954	.957	.961	.964	.968	.971	.975	.978	.982
11.3	1.985	1.989	1.992	1.996	1.999	2.003	2.007	2.010	2.014	2.017
11.4	2.021	2.025	2.028	2.032	2.035	2.039	2.042	2.046	2.049	2.053
11.5	.056	.060	.063	.067	.070	.074	.078	.081	.085	.088
11.6	.092	.096	.099	.103	.106	.110	.114	.117	.121	.124
11.7	.128	.132	.135	.139	.143	.147	.150	.154	.158	.161
11.8	.165	.169	.172	.176	.180	.184	.187	.191	.195	.198
11.9	.202	.206	.209	.213	.217	.221	.224	.228	.232	.235
12.0	.239	.243	.246	.250	.254	.258	.261	.265	.269	.272
12.1	.276	.280	.284	.287	.291	.295	.299	.303	.306	.310
12.2	.314	.318	.322	.325	.329	.333	.337	.341	.344	.348
12.3	.352	.356	.360	.364	.368	.372	.375	.379	.383	.387
12.4	.391	.395	.399	.402	.406	.410	.414	.418	.421	.425
12.5	2.429	2.433	2.437	2.441	2.445	2.449	2.452	2.456	2.460	2.464
12.6	.468	.472	.476	.480	.484	.488	.492	.496	.500	.504
12.7	.508	.512	.516	.520	.524	.528	.531	.535	.539	.543
12.8	.547	.551	.555	.559	.563	.567	.571	.575	.579	.583
12.9	.587	.591	.595	.599	.603	.607	.611	.615	.619	.623
13.0	.627	.631	.635	.639	.643	.648	.652	.656	.660	.664
13.1	.668	.672	.676	.680	.684	.689	.693	.697	.701	.705
13.2	.709	.713	.717	.721	.725	.730	.734	.738	.742	.746
13.3	.750	.754	.758	.763	.767	.771	.775	.779	.784	.788
13.4	.792	.796	.800	.804	.808	.813	.817	.821	.825	.829
13.5	.833	.837	.842	.846	.850	.855	.859	.863	.867	.872
13.6	.876	.880	.884	.888	.892	.897	.901	.905	.909	.913
13.7	.917	.921	.926	.930	.935	.939	.943	.948	.952	.957
13.8	2.961	2.965	2.970	2.974	2.978	2.983	2.987	2.991	2.995	3.000
13.9	3.004	3.008	3.013	3.017	3.021	3.026	3.030	3.034	3.038	3.043
14.0	.047	.051	.056	.060	.065	.069	.073	.078	.082	.087
14.1	.091	.095	.100	.104	.109	.113	.117	.122	.126	.131
14.2	.135	.139	.144	.148	.153	.157	.161	.166	.170	.175
14.3	.179	.184	.188	.193	.197	.202	.206	.211	.215	.220
14.4	.224	.229	.233	.238	.242	.247	.251	.256	.260	.265
14.5	.269	.274	.278	.283	.287	.292	.296	.301	.305	.310
14.6	.314	.319	.323	.328	.332	.337	.341	.346	.350	.355
14.7	.359	.364	.368	.373	.377	.382	.387	.391	.396	.400
14.8	.405	.410	.414	.419	.424	.429	.433	.438	.443	.447
14.9	.452	.457	.461	.466	.470	.475	.480	.484	.489	.493

Table 47.—*Theoretical heads in feet for velocities of 0 to 38 feet per second*—Continued

$$h = \frac{V^2}{2g}$$

V	0.00	0.01	0.02	0.03	0.04	0.05	0.06	0.07	0.08	0.09
15.0	3.498	3.503	3.507	3.512	3.517	3.522	3.526	3.531	3.536	3.540
15.1	.545	.550	.554	.559	.564	.569	.573	.578	.583	.587
15.2	.592	.597	.601	.606	.611	.616	.620	.625	.630	.634
15.3	.639	.644	.649	.653	.658	.663	.668	.673	.677	.682
15.4	.687	.692	.696	.701	.706	.711	.715	.720	.725	.729
15.5	.734	.739	.744	.749	.754	.759	.764	.769	.774	.779
15.6	.784	.789	.794	.798	.803	.808	.813	.818	.822	.827
15.7	.832	.837	.841	.846	.851	.856	.860	.865	.870	.874
15.8	.879	.884	.889	.895	.900	.905	.910	.915	.921	.926
15.9	.931	.936	.941	.946	.951	.956	.960	.964	.969	.974
16.0	3.980	3.985	3.990	3.995	4.000	4.005	4.010	4.015	4.020	4.025
16.1	4.030	4.035	4.040	4.045	4.050	4.055	4.060	4.065	4.070	4.075
16.2	.080	.085	.090	.095	.100	.106	.111	.116	.121	.126
16.3	.131	.136	.141	.146	.151	.157	.162	.167	.172	.177
16.4	.182	.187	.192	.197	.202	.208	.213	.218	.223	.228
16.5	.233	.238	.243	.248	.253	.259	.264	.269	.274	.279
16.6	.284	.289	.294	.300	.305	.310	.315	.320	.326	.331
16.7	.336	.341	.346	.352	.357	.362	.367	.372	.378	.383
16.8	.388	.393	.398	.404	.409	.414	.419	.424	.430	.435
16.9	.440	.445	.451	.456	.461	.467	.472	.477	.482	.488
17.0	.493	.498	.504	.509	.514	.520	.525	.530	.535	.541
17.1	.546	.551	.557	.562	.567	.573	.578	.583	.588	.594
17.2	.599	.604	.610	.615	.621	.626	.631	.637	.642	.648
17.3	.653	.658	.664	.669	.675	.680	.685	.691	.696	.702
17.4	.707	.712	.718	.723	.729	.734	.739	.745	.750	.756
17.5	.761	.767	.772	.778	.783	.789	.794	.800	.805	.811
17.6	.816	.822	.827	.833	.838	.844	.849	.855	.860	.866
17.7	.871	.877	.882	.888	.893	.899	.904	.910	.915	.921
17.8	.926	.932	.937	.943	.948	.954	.959	.965	.970	.976
17.9	4.981	4.987	4.992	4.998	5.003	5.009	5.015	5.020	5.026	5.031
18.0	5.037	5.043	5.048	5.054	5.059	5.065	5.071	5.076	5.082	5.087
18.1	.093	.099	.104	.110	.115	.121	.127	.132	.138	.143
18.2	.149	.155	.161	.166	.172	.178	.184	.190	.195	.201
18.3	.207	.213	.218	.224	.229	.235	.241	.246	.252	.257
18.4	.263	.269	.275	.280	.286	.292	.298	.304	.309	.315
18.5	.321	.327	.333	.338	.344	.350	.356	.362	.367	.373
18.6	.379	.385	.391	.396	.402	.408	.414	.420	.425	.431
18.7	.437	.443	.449	.454	.460	.466	.472	.478	.483	.489
18.8	.495	.501	.507	.513	.519	.525	.530	.536	.542	.548
18.9	.554	.560	.566	.572	.578	.584	.589	.595	.601	.607
19.0	5.613	5.619	5.625	5.631	5.639	5.643	5.648	5.654	5.660	5.666
19.1	.672	.678	.684	.690	.696	.702	.707	.713	.719	.725
19.2	.731	.737	.743	.749	.755	.761	.767	.773	.779	.785
19.3	.791	.797	.803	.809	.815	.821	.827	.833	.839	.845
19.4	.851	.857	.863	.869	.875	.882	.888	.894	.900	.906
19.5	.912	.918	.924	.930	.936	.943	.949	.955	.961	.967
19.6	5.973	5.979	5.985	5.991	5.997	6.004	6.010	6.016	6.022	6.028
19.7	6.034	6.040	6.046	6.052	6.058	6.065	6.071	6.077	6.083	6.089
19.8	.095	.101	.107	.114	.120	.126	.132	.138	.145	.151
19.9	.157	.163	.169	.176	.182	.188	.198	.200	.207	.213

Table 47.—*Theoretical heads in feet for velocities of 0 to 38 feet per second*—Continued

$$h = \frac{V^2}{2g}$$

V	0.00	0.01	0.02	0.03	0.04	0.05	0.06	0.07	0.08	0.09
20.0	6.219	6.225	6.231	6.238	6.244	6.250	6.256	6.262	6.269	6.275
20.1	.281	.287	.294	.300	.306	.313	.319	.325	.331	.338
20.2	.344	.350	.357	.363	.369	.376	.382	.388	.394	.401
20.3	.407	.413	.420	.426	.432	.439	.445	.451	.457	.464
20.4	.470	.476	.483	.489	.496	.502	.508	.515	.521	.528
20.5	.534	.540	.547	.553	.560	.566	.572	.579	.585	.592
20.6	.598	.604	.611	.617	.624	.630	.636	.643	.649	.656
20.7	.662	.668	.675	.681	.688	.694	.700	.707	.713	.720
20.8	.726	.733	.739	.746	.752	.759	.765	.772	.778	.785
20.9	.791	.798	.804	.811	.817	.824	.830	.837	.843	.850
21.0	.856	.863	.869	.876	.882	.889	.896	.902	.909	.915
21.1	.922	.929	.935	.942	.948	.955	.962	.968	.975	.981
21.2	6.988	6.995	7.001	7.008	7.014	7.021	7.028	7.034	7.041	7.047
21.3	7.054	7.061	7.067	7.074	7.080	7.087	7.094	7.100	7.107	7.113
21.4	.120	.127	.133	.140	.147	.154	.160	.167	.174	.180
21.5	.187	.194	.200	.207	.214	.221	.227	.234	.241	.247
21.6	.254	.261	.267	.274	.281	.288	.294	.301	.308	.314
21.7	.321	.328	.335	.341	.348	.355	.362	.369	.375	.382
21.8	.389	.396	.403	.409	.416	.423	.430	.437	.443	.450
21.9	.457	.464	.471	.477	.484	.491	.498	.505	.511	.518
22.0	.525	.532	.539	.545	.552	.559	.566	.573	.579	.586
22.1	.593	.600	.607	.614	.621	.628	.634	.641	.648	.655
22.2	.662	.669	.676	.683	.690	.697	.703	.710	.717	.724
22.3	.731	.738	.745	.752	.759	.766	.773	.780	.787	.794
22.4	.801	.808	.815	.822	.829	.836	.843	.850	.857	.864
22.5	.871	.878	.885	.892	.899	.906	.913	.920	.927	.934
22.6	7.941	7.948	7.955	7.962	7.969	7.976	7.983	7.990	7.997	8.004
22.7	8.011	8.018	8.025	8.032	8.039	8.047	8.054	8.061	8.068	8.075
22.8	.082	.089	.096	.103	.110	.118	.125	.132	.139	.146
22.9	.153	.160	.167	.174	.181	.189	.196	.203	.210	.217
23.0	.224	.231	.238	.246	.253	.260	.267	.274	.282	.289
23.1	.296	.303	.310	.318	.325	.332	.339	.346	.354	.361
23.2	.368	.375	.382	.390	.397	.404	.411	.418	.426	.433
23.3	.440	.447	.455	.462	.469	.477	.484	.491	.498	.506
23.4	.513	.520	.528	.535	.542	.550	.557	.564	.571	.579
23.5	.586	.593	.601	.608	.615	.623	.630	.637	.644	.622
23.6	.659	.666	.674	.681	.689	.696	.703	.711	.718	.726
23.7	.733	.740	.748	.755	.763	.770	.777	.785	.792	.800
23.8	.807	.814	.822	.829	.837	.844	.851	.859	.866	.874
23.9	.881	.888	.896	.903	.911	.918	.925	.933	.940	.948
24.0	8.955	8.963	8.970	8.978	8.985	8.993	9.000	9.008	9.015	9.023
24.1	9.030	9.038	9.045	9.053	9.060	9.068	9.075	9.083	9.090	9.098
24.2	.105	.113	.120	.128	.135	.143	.151	.158	.166	.173
24.3	.181	.189	.196	.204	.211	.219	.226	.234	.241	.249
24.4	.256	.264	.271	.279	.286	.294	.302	.309	.317	.324
24.5	.332	.340	.347	.355	.363	.371	.378	.386	.394	.401
24.6	.409	.417	.424	.432	.439	.447	.455	.462	.470	.477
24.7	.485	.493	.500	.508	.516	.524	.531	.539	.547	.554
24.8	.562	.570	.577	.585	.593	.601	.608	.616	.624	.631
24.9	.639	.647	.655	.662	.670	.678	.686	.694	.701	.709

Table 47.—*Theoretical heads in feet for velocities of 0 to 38 feet per second*—Continued

$$h = \frac{V^2}{2g}$$

V	0.00	0.01	0.02	0.03	0.04	0.05	0.06	0.07	0.08	0.09
25.0	9.717	9.725	9.733	9.740	9.748	9.756	9.764	9.772	9.779	9.787
25.1	.795	.803	.811	.818	.826	.834	.842	.850	.857	.865
25.2	.873	.881	.889	.897	.905	.913	.920	.928	.936	.944
25.3	9.952	9.960	9.968	9.975	9.983	9.991	9.999	10.007	10.014	10.022
25.4	10.030	10.038	10.046	10.054	10.062	10.070	10.077	10.085	10.093	10.101
25.5	10.109	10.117	10.125	10.133	10.141	10.149	10.157	10.165	10.173	10.181
25.6	10.189	10.197	10.205	10.213	10.221	10.229	10.237	10.245	10.253	10.261
25.7	10.269	10.277	10.285	10.293	10.301	10.309	10.317	10.325	10.333	10.341
25.8	10.349	10.357	10.365	10.373	10.381	10.389	10.397	10.405	10.413	10.421
25.9	10.429	10.437	10.445	10.453	10.461	10.470	10.478	10.486	10.494	10.502
26.0	10.510	10.518	10.526	10.534	10.542	10.551	10.559	10.567	10.575	10.583
26.1	10.591	10.599	10.607	10.615	10.623	10.632	10.640	10.648	10.656	10.664
26.2	10.672	10.680	10.689	10.697	10.706	10.714	10.722	10.731	10.739	10.748
26.3	10.756	10.764	10.772	10.780	10.788	10.796	10.804	10.812	10.820	10.828
26.4	10.836	10.844	10.852	10.861	10.869	10.877	10.885	10.893	10.902	10.910
26.5	10.918	10.926	10.935	10.943	10.951	10.960	10.968	10.976	10.984	10.993
26.6	11.001	11.009	11.017	11.025	11.033	11.042	11.050	11.058	11.066	11.074
26.7	11.082	11.090	11.099	11.107	11.116	11.124	11.132	11.141	11.149	11.158
26.8	11.166	11.174	11.183	11.191	11.200	11.208	11.216	11.225	11.233	11.242
26.9	11.250	11.258	11.267	11.275	11.284	11.292	11.300	11.309	11.317	11.326
27.0	11.334	11.342	11.351	11.359	11.368	11.376	11.384	11.393	11.401	11.410
27.1	11.418	11.426	11.435	11.443	11.452	11.460	11.468	11.477	11.485	11.494
27.2	11.502	11.511	11.519	11.528	11.536	11.545	11.553	11.562	11.570	11.579
27.3	11.587	11.596	11.604	11.613	11.621	11.630	11.638	11.646	11.655	11.664
27.4	11.672	11.681	11.689	11.698	11.706	11.715	11.724	11.732	11.741	11.749
27.5	11.758	11.766	11.775	11.783	11.792	11.800	11.808	11.817	11.825	11.834
27.6	11.842	11.851	11.859	11.868	11.877	11.886	11.894	11.903	11.912	11.920
27.7	11.929	11.938	11.946	11.955	11.964	11.973	11.981	11.990	11.999	12.007
27.8	12.016	12.025	12.033	12.042	12.050	12.059	12.068	12.076	12.085	12.093
27.9	12.102	12.111	12.119	12.128	12.137	12.146	12.154	12.163	12.172	12.180
28.0	12.189	12.198	12.206	12.215	12.224	12.233	12.241	12.250	12.259	12.267
28.1	12.276	12.285	12.294	12.302	12.311	12.320	12.329	12.338	12.346	12.355
28.2	12.364	12.373	12.382	12.390	12.399	12.408	12.417	12.426	12.434	12.443
28.3	12.452	12.461	12.470	12.478	12.487	12.496	12.505	12.514	12.522	12.531
28.4	12.540	12.549	12.558	12.566	12.575	12.584	12.593	12.602	12.610	12.619
28.5	12.628	12.637	12.646	12.655	12.664	12.673	12.681	12.690	12.699	12.708
28.6	12.717	12.726	12.735	12.744	12.753	12.762	12.770	12.779	12.788	12.797
28.7	12.806	12.815	12.824	12.833	12.842	12.851	12.860	12.869	12.878	12.887
28.8	12.896	12.905	12.914	12.923	12.932	12.941	12.949	12.958	12.967	12.976
28.9	12.985	12.994	13.003	13.012	13.021	13.030	13.039	13.048	13.057	13.066
29.0	13.075	13.084	13.093	13.102	13.111	13.120	13.129	13.138	13.147	13.156
29.1	13.165	13.174	13.183	13.192	13.201	13.211	13.220	13.229	13.238	13.247
29.2	13.256	13.265	13.274	13.283	13.292	13.302	13.311	13.320	13.329	13.338
29.3	13.347	13.356	13.365	13.374	13.383	13.393	13.402	13.411	13.420	13.429
29.4	13.438	13.447	13.456	13.466	13.475	13.484	13.493	13.502	13.512	13.521
29.5	13.530	13.539	13.548	13.558	13.567	13.576	13.585	13.594	13.604	13.613
29.6	13.622	13.631	13.640	13.650	13.659	13.668	13.677	13.686	13.696	13.705
29.7	13.714	13.723	13.733	13.742	13.751	13.761	13.770	13.779	13.788	13.798
29.8	13.807	13.816	13.825	13.835	13.844	13.855	13.862	13.871	13.881	13.890
29.9	13.899	13.908	13.918	13.927	13.937	13.946	13.955	13.965	13.974	13.984

Table 47.—*Theoretical heads in feet for velocities of 0 to 38 feet per second*—Continued

$$h = \frac{V^2}{2g}$$

V	0.00	0.01	0.02	0.03	0.04	0.05	0.06	0.07	0.08	0.09
30.0	13.993	14.002	14.012	14.021	14.030	14.040	14.049	14.028	14.067	14.077
30.1	14.086	14.095	14.105	14.114	14.124	14.133	14.142	14.152	14.161	14.171
30.2	14.180	14.189	14.199	14.208	14.218	14.228	14.236	14.246	14.255	14.265
30.3	14.274	14.283	14.296	14.302	14.312	14.321	14.330	14.340	14.349	14.359
30.4	14.366	14.376	14.385	14.395	14.404	14.414	14.423	14.433	14.442	14.452
30.5	14.461	14.471	14.480	14.490	14.499	14.509	14.518	14.528	14.537	14.547
30.6	14.556	14.566	14.575	14.585	14.594	14.604	14.613	14.623	14.632	14.642
30.7	14.651	14.661	14.670	14.680	14.690	14.699	14.709	14.718	14.728	14.737
30.8	14.747	14.757	14.766	14.776	14.785	14.795	14.804	14.814	14.824	14.833
30.9	14.843	14.852	14.862	14.872	14.881	14.891	14.900	14.910	14.920	14.930
31.0	14.939	14.949	14.958	14.968	14.978	14.987	14.997	15.007	15.016	15.026
31.1	15.036	15.045	15.055	15.065	15.074	15.084	15.094	15.103	15.113	15.123
31.2	15.133	15.142	15.152	15.162	15.171	15.181	15.191	15.201	15.210	15.220
31.3	15.230	15.239	15.249	15.259	15.269	15.278	15.288	15.298	15.308	15.317
31.4	15.327	15.337	15.347	15.356	15.366	15.376	15.386	15.396	15.405	15.415
31.5	15.425	15.435	15.445	15.454	15.464	15.474	15.484	15.494	15.503	15.513
31.6	15.523	15.533	15.543	15.552	15.562	15.572	15.582	15.592	15.602	15.612
31.7	15.622	15.631	15.641	15.651	15.661	15.671	15.681	15.691	15.700	15.710
31.8	15.720	15.730	15.740	15.750	15.760	15.770	15.779	15.789	15.799	15.809
31.9	15.819	15.829	15.839	15.849	15.859	15.869	15.879	15.889	15.899	15.909
32.0	15.919	15.929	15.939	15.948	15.958	15.968	15.978	15.988	15.998	16.008
32.1	16.018	16.028	16.038	16.048	16.058	16.068	16.078	16.088	16.098	16.108
32.2	16.118	16.128	16.138	16.148	16.158	16.168	16.178	16.188	16.198	16.208
32.3	16.218	16.228	16.239	16.248	16.259	16.269	16.279	16.289	16.299	16.309
32.4	16.319	16.329	16.339	16.349	16.359	16.369	16.380	16.390	16.400	16.410
32.5	16.420	16.430	16.440	16.450	16.460	16.470	16.480	16.491	16.501	16.511
32.6	16.521	16.531	16.541	16.551	16.562	16.572	16.582	16.592	16.602	16.612
32.7	16.623	16.633	16.643	16.653	16.663	16.674	16.684	16.694	16.704	16.714
32.8	16.725	16.735	16.745	16.755	16.765	16.775	16.786	16.796	16.806	16.816
32.9	16.827	16.837	16.847	16.857	16.868	16.878	16.888	16.898	16.909	16.919
33.0	16.929	16.939	16.950	16.960	16.970	16.980	16.991	17.001	17.011	17.021
33.1	17.032	17.042	17.052	17.063	17.073	17.083	17.094	17.104	17.114	17.125
33.2	17.135	17.145	17.155	17.166	17.176	17.187	17.197	17.207	17.217	17.228
33.3	17.238	17.249	17.259	17.269	17.280	17.290	17.300	17.311	17.321	17.332
33.4	17.342	17.352	17.363	17.373	17.384	17.394	17.404	17.415	17.425	17.436
33.5	17.446	17.456	17.467	17.477	17.488	17.498	17.508	17.519	17.529	17.540
33.6	17.550	17.561	17.571	17.582	17.592	17.603	17.613	17.623	17.634	17.644
33.7	17.655	17.665	17.675	17.686	17.697	17.707	17.718	17.728	17.739	17.749
33.8	17.760	17.770	17.781	17.791	17.802	17.812	17.823	17.833	17.844	17.855
33.9	17.865	17.876	17.886	17.896	17.907	17.918	17.928	17.939	17.950	17.960
34.0	17.971	17.981	17.992	18.002	18.013	18.024	18.034	18.045	18.055	18.066
34.1	18.077	18.087	18.098	18.108	18.119	18.130	18.140	18.151	18.161	18.172
34.2	18.183	18.193	18.204	18.215	18.225	18.236	18.246	18.257	18.268	18.278
34.3	18.289	18.300	18.311	18.321	18.332	18.343	18.353	18.364	18.375	18.385
34.4	18.396	18.407	18.417	18.428	18.439	18.449	18.460	18.471	18.482	18.492
34.5	18.503	18.514	18.525	18.535	18.546	18.557	18.567	18.578	18.589	18.600
34.6	18.611	18.621	18.632	18.643	18.653	18.664	18.675	18.686	18.697	18.707
34.7	18.718	18.729	18.740	18.751	18.761	18.772	18.783	18.794	18.805	18.815
34.8	18.826	18.837	18.848	18.859	18.870	18.880	18.891	18.902	18.913	18.924
34.9	18.935	18.945	18.956	18.967	18.978	18.989	19.000	19.011	19.022	19.032

Table 47.—*Theoretical heads in feet for velocities of 0 to 38 feet per second*—Continued

$$h = \frac{V^2}{2g}$$

V	0.00	0.01	0.02	0.03	0.04	0.05	0.06	0.07	0.08	0.09
35.0	19.043	19.054	19.065	19.076	19.087	19.098	19.109	19.120	19.130	19.141
35.1	19.152	19.163	19.174	19.185	19.196	19.207	19.218	19.229	19.240	19.251
35.2	19.262	19.272	19.283	19.294	19.305	19.316	19.327	19.338	19.349	19.360
35.3	19.371	19.382	19.393	19.404	19.415	19.426	19.437	19.448	19.459	19.470
35.4	19.481	19.492	19.503	19.514	19.525	19.536	19.547	19.558	19.569	19.580
35.5	19.591	19.602	19.613	19.624	19.635	19.647	19.658	19.669	19.680	19.691
35.6	19.702	19.713	19.724	19.735	19.746	19.757	19.768	19.779	19.790	19.802
35.7	19.813	19.824	19.835	19.846	19.857	19.868	19.879	19.890	19.901	19.913
35.8	19.924	19.935	19.946	19.957	19.968	19.980	19.991	20.002	20.013	20.024
35.9	20.035	20.046	20.058	20.069	20.080	20.091	20.102	20.114	20.125	20.136
36.0	20.147	20.158	20.170	20.181	20.192	20.203	20.214	20.225	20.237	20.248
36.1	20.259	20.270	20.282	20.293	20.304	20.315	20.327	20.338	20.349	20.360
36.2	20.372	20.383	20.394	20.405	20.417	20.428	20.439	20.450	20.462	20.473
36.3	20.484	20.496	20.507	20.518	20.529	20.541	20.552	20.563	20.575	20.586
36.4	20.597	20.609	20.620	20.631	20.643	20.654	20.665	20.677	20.688	20.699
36.5	20.711	20.722	20.733	20.745	20.756	20.767	20.779	20.790	20.802	20.813
36.6	20.824	20.836	20.847	20.858	20.870	20.881	20.893	20.904	20.915	20.927
36.7	20.938	20.950	20.961	20.972	20.984	20.995	21.007	21.018	21.030	21.041
36.8	21.053	21.064	21.075	21.087	21.098	21.110	21.121	21.133	21.144	21.156
36.9	21.167	21.178	21.190	21.201	21.213	21.224	21.236	21.247	21.259	21.270
37.0	21.282	21.293	21.305	21.316	21.328	21.339	21.351	21.363	21.374	21.386
37.1	21.397	21.409	21.420	21.432	21.443	21.455	21.466	21.478	21.490	21.501
37.2	21.513	21.524	21.536	21.547	21.559	21.571	21.582	21.594	21.605	21.617
37.3	21.628	21.640	21.652	21.663	21.675	21.686	21.698	21.710	21.721	21.733
37.4	21.745	21.756	21.768	21.779	21.791	21.803	21.814	21.826	21.838	21.842
37.5	21.861	21.873	21.884	21.889	21.908	21.919	21.931	21.936	21.954	21.966

Table 48.—*Amount of material in cubic yards per 100 linear feet of level cut,*

side slopes ½ to 1

Depth of center cut in feet	.0	.1	.2	.3	.4	.5	.6	.7	.8	.9
0	0.0	0.0	0.1	0.2	0.3	0.5	0.7	0.9	1.2	1.5
1	1.9	2.2	2.7	3.1	3.6	4.2	4.7	5.4	6.0	6.7
2	7.4	8.2	9.0	9.8	10.7	11.6	12.5	13.5	14.5	15.6
3	17	18	19	20	21	23	24	25	27	28
4	30	31	33	34	36	37	39	41	43	44
5	46	48	50	52	54	56	58	60	62	64
6	67	69	71	73	76	78	81	83	86	88
7	91	93	96	99	101	104	107	110	113	116
8	119	121	125	128	131	134	137	140	143	147
9	150	153	157	160	164	167	171	174	178	181
10	185	189	193	196	200	204	208	212	216	220
11	224	228	232	236	241	245	249	253	258	262
12	267	271	276	280	285	289	294	299	303	308
13	313	318	323	328	333	337	343	348	353	358
14	363	368	373	379	384	389	395	400	406	411
15	417	422	428	433	439	445	451	456	462	468
16	474	480	486	492	498	504	510	516	523	529
17	535	541	548	554	561	567	574	580	587	593
18	600	607	613	620	627	634	641	648	655	661
19	669	676	683	690	697	704	711	719	726	733
20	741	748	756	763	771	778	786	793	801	809
21	817	824	832	840	848	856	864	872	880	888
22	896	904	913	921	929	937	946	954	963	971
23	980	988	997	1,005	1,014	1,023	1,031	1,040	1,049	1,058
24	1,067	1,076	1,085	1,093	1,103	1,112	1,121	1,130	1,139	1,148
25	1,157	1,167	1,176	1,185	1,195	1,204	1,214	1,223	1,233	1,242
26	1,252	1,261	1,271	1,281	1,291	1,300	1,310	1,320	1,330	1,340
27	1,350	1,360	1,370	1,380	1,390	1,400	1,411	1,421	1,431	1,441
28	1,452	1,462	1,473	1,483	1,494	1,504	1,515	1,525	1,536	1,547
29	1,557	1,568	1,579	1,590	1,601	1,612	1,623	1,633	1,645	1,656
30	1,667	1,678	1,689	1,700	1,711	1,723	1,734	1,745	1,757	1,768
31	1,780	1,791	1,803	1,814	1,826	1,837	1,849	1,861	1,873	1,884
32	1,896	1,908	1,920	1,932	1,944	1,956	1,968	1,980	1,992	2,004
33	2,017	2,029	2,041	2,053	2,066	2,078	2,091	2,103	2,116	2,128
34	2,141	2,153	2,166	2,179	2,191	2,204	2,217	2,230	2,243	2,256

Table 48.—*Amount of material in cubic yards per 100 linear feet of level cut,*
side slopes ½ to 1—Continued

Depth of center cut in feet	.0	.1	.2	.3	.4	.5	.6	.7	.8	.9
35	2,269	2,281	2,295	2,308	2,321	2,334	2,347	2,360	2,373	2,387
36	2,400	2,413	2,427	2,440	2,454	2,467	2,481	2,494	2,508	2,521
37	2,535	2,549	2,563	2,576	2,590	2,604	2,618	2,632	2,646	2,660
38	2,674	2,688	2,702	2,716	2,731	2,745	2,759	2,773	2,788	2,802
39	2,817	2,831	2,846	2,860	2,875	2,889	2,904	2,919	2,933	2,948
40	2,963	2,978	2,993	3,008	3,023	3,037	3,053	3,068	3,083	3,098
41	3,113	3,128	3,143	3,159	3,174	3,189	3,205	3,220	3,236	3,251
42	3,267	3,282	3,298	3,313	3,329	3,345	3,361	3,376	3,392	3,408
43	3,424	3,440	3,456	3,472	3,488	3,504	3,520	3,536	3,553	3,569
44	3,585	3,601	3,618	3,634	3,651	3,667	3,684	3,700	3,717	3,733
45	3,750	3,767	3,783	3,800	3,817	3,834	3,851	3,868	3,885	3,901
46	3,919	3,936	3,953	3,970	3,987	4,004	4,021	4,039	4,056	4,073
47	4,091	4,108	4,126	4,143	4,161	4,178	4,196	4,213	4,231	4,249
48	4,267	4,284	4,302	4,320	4,338	4,356	4,374	4,392	4,410	4,428
49	4,446	4,464	4,483	4,501	4,519	4,537	4,556	4,574	4,593	4,611
50	4,630	4,648	4,667	4,685	4,704	4,723	4,741	4,760	4,779	4,798
51	4,817	4,836	4,855	4,873	4,893	4,912	4,931	4,950	4,969	4,988
52	5,007	5,027	5,046	5,065	5,085	5,104	5,124	5,143	5,163	5,182
53	5,202	5,221	5,241	5,261	5,281	5,300	5,320	5,340	5,360	5,380
54	5,400	5,420	5,440	5,460	5,480	5,500	5,521	5,541	5,561	5,581
55	5,602	5,622	5,643	5,663	5,684	5,704	5,725	5,745	5,766	5,787
56	5,807	5,828	5,849	5,870	5,891	5,912	5,933	5,953	5,975	5,996
57	6,017	6,038	6,059	6,080	6,101	6,123	6,144	6,165	6,187	6,208
58	6,230	6,251	6,273	6,294	6,316	6,337	6,359	6,381	6,403	6,424
59	6,446	6,468	6,490	6,512	6,534	6,556	6,578	6,600	6,622	6,644
60	6,667	-------	-------	-------	-------	-------	-------	-------	-------	-------

Table 49.—*Amount of material in cubic yards per 100 linear feet of level cut,*

side slopes 1 to 1.

Depth of center cut in feet	.0	.1	.2	.3	.4	.5	.6	.7	.8	.9
0	0.0	0.0	0.1	0.3	0.6	0.9	1.3	1.8	2.4	3.0
1	3.7	4.5	5.3	6.3	7.3	8.3	9.5	10.7	12.0	13.4
2	15	16	18	20	21	23	25	27	29	31
3	33	36	38	40	43	45	48	51	54	56
4	59	62	65	68	72	75	78	82	85	89
5	93	96	100	104	108	112	116	120	125	129
6	133	138	142	147	152	156	161	166	171	176
7	181	187	192	197	203	208	214	220	225	231
8	237	243	249	255	261	268	274	280	287	293
9	300	307	313	320	327	334	341	349	356	363
10	370	378	385	393	401	408	416	424	432	440
11	448	456	465	473	481	490	498	507	516	524
12	533	542	551	560	569	579	588	597	607	616
13	626	636	645	655	665	675	685	695	705	716
14	726	736	747	757	768	779	789	800	811	822
15	833	844	856	867	878	890	901	913	925	936
16	948	960	972	984	996	1,008	1,021	1,033	1,045	1,058
17	1,070	1,083	1,096	1,108	1,121	1,134	1,147	1,160	1,173	1,187
18	1,200	1,213	1,227	1,240	1,254	1,268	1,281	1,295	1,309	1,323
19	1,337	1,351	1,365	1,380	1,394	1,408	1,423	1,437	1,452	1,467
20	1,481	1,496	1,511	1,526	1,541	1,556	1,572	1,587	1,602	1,618
21	1,633	1,649	1,665	1,680	1,696	1,712	1,728	1,744	1,760	1,776
22	1,793	1,809	1,825	1,842	1,858	1,875	1,892	1,908	1,925	1,942
23	1,959	1,976	1,993	2,011	2,028	2,045	2,063	2,080	2,098	2,116
24	2,133	2,151	2,169	2,187	2,205	2,223	2,241	2,260	2,278	2,296
25	2,315	2,333	2,352	2,371	2,389	2,408	2,427	2,446	2,465	2,484
26	2,504	2,523	2,542	2,562	2,581	2,601	2,621	2,640	2,660	2,680
27	2,700	2,720	2,740	2,760	2,781	2,801	2,821	2,842	2,862	2,883
28	2,904	2,924	2,945	2,966	2,987	3,008	3,029	3,051	3,072	3,093
29	3,115	3,136	3,158	3,180	3,201	3,223	3,245	3,267	3,289	3,311
30	3,333	3,356	3,378	3,400	3,423	3,445	3,468	3,491	3,513	3,536
31	3,559	3,582	3,605	3,628	3,652	3,675	3,698	3,722	3,745	3,769
32	3,793	3,816	3,840	3,864	3,888	3,912	3,936	3,960	3,985	4,009
33	4,033	4,058	4,082	4,107	4,132	4,156	4,181	4,206	4,231	4,256
34	4,281	4,307	4,332	4,357	4,383	4,408	4,434	4,460	4,485	4,511
35	4,537	4,563	4,589	4,615	4,641	4,668	4,694	4,720	4,747	4,773
36	4,800	4,827	4,853	4,880	4,907	4,934	4,961	4,988	5,016	5,043
37	5,070	5,098	5,125	5,153	5,181	5,208	5,236	5,264	5,292	5,320
38	5,348	5,376	5,405	5,433	5,461	5,490	5,518	5,547	5,576	5,604
39	5,633	5,662	5,691	5,720	5,749	5,779	5,808	5,837	5,867	5,896
40	5,926	5,956	5,985	6,015	6,045	6,075	6,105	6,135	6,165	6,196
41	6,226	6,256	6,287	6,317	6,348	6,379	6,409	6,440	6,471	6,502
42	6,533	6,564	6,596	6,627	6,658	6,690	6,721	6,763	6,785	6,816
43	6,848	6,880	6,912	6,944	6,976	7,008	7,041	7,073	7,105	7,138
44	7,170	7,203	7,236	7,268	7,301	7,334	7,367	7,400	7,433	7,467

Table 49.—*Amount of material in cubic yards per 100 linear feet of level cut,*

side slopes 1 to 1—Continued.

Depth of center cut in feet	.0	.1	.2	.3	.4	.5	.6	.7	.8	.9
45	7,500	7,533	7,567	7,600	7,634	7,668	7,701	7,735	7,769	7,803
46	7,837	7,871	7,905	7,940	7,974	8,008	8,043	8,077	8,112	8,147
47	8,181	8,216	8,251	8,286	8,321	8,356	8,392	8,427	8,462	8,498
48	8,533	8,569	8,605	8,640	8,676	8,712	8,748	8,784	8,820	8,856
49	8,893	8,929	8,965	9,002	9,038	9,075	9,112	9,148	9,185	9,222
50	9,259	9,296	9,333	9,371	9,408	9,445	9,483	9,520	9,558	9,596
51	9,633	9,671	9,709	9,747	9,785	9,823	9,861	9,900	9,938	9,976
52	10,015	10,053	10,092	10,131	10,169	10,208	10,247	10,286	10,325	10,364
53	10,404	10,443	10,482	10,522	10,561	10,601	10,641	10,680	10,720	10,760
54	10,800	10,840	10,880	10,920	10,961	11,001	11,041	11,082	11,122	11,163
55	11,204	11,244	11,285	11,326	11,367	11,408	11,449	11,491	11,532	11,573
56	11,615	11,656	11,698	11,740	11,781	11,823	11,865	11,907	11,949	11,991
57	12,033	12,076	12,118	12,160	12,203	12,245	12,288	12,331	12,373	12,416
58	12,459	12,502	12,545	12,588	12,632	12,675	12,718	12,762	12,805	12,849
59	12,893	12,936	12,980	13,024	13,068	13,112	13,156	13,200	13,245	13,289
60	13,333	......	......	......	......	......	......	......	......	......

Table 50.—*Amount of material in cubic yards per 100 linear feet of level cut,*

side slopes 1½ to 1.

Depth of center cut in feet	.0	.1	.2	.3	.4	.5	.6	.7	.8	.9
0	0.0	0.0	0.2	0.5	0.9	1.4	2.0	2.7	3.6	4.5
1	5.6	6.7	8.0	9.4	10.9	12.5	14.2	16.1	18.0	20.1
2	22	24	27	29	32	35	38	41	44	47
3	50	53	57	60	64	68	72	76	80	84
4	89	93	98	103	108	112	118	123	128	133
5	139	144	150	156	162	168	174	180	187	193
6	200	207	214	222	228	235	242	249	257	264
7	272	280	288	296	304	312	321	329	338	347
8	356	364	374	383	392	401	411	420	430	440
9	450	460	470	480	491	501	512	522	533	544
10	556	567	577	589	601	612	624	636	648	660
11	672	684	697	709	722	735	748	760	774	787
12	800	813	827	840	854	868	882	896	910	924
13	939	953	968	983	998	1,012	1,028	1,043	1,058	1,073
14	1,089	1,104	1,120	1,136	1,152	1,168	1,184	1,200	1,217	1,233

Table 50.—*Amount of material in cubic yards per 100 linear feet of level cut,*

side slopes 1½ to 1—Continued.

Depth of center cut in feet	.0	.1	.2	.3	.4	.5	.6	.7	.8	.9
15	1,250	1,267	1,284	1,300	1,318	1,335	1,352	1,369	1,387	1,404
16	1,422	1,440	1,458	1,476	1,494	1,512	1,531	1,549	1,568	1,587
17	1,606	1,624	1,644	1,663	1,682	1,701	1,721	1,740	1,760	1,780
18	1,800	1,820	1,840	1,860	1,881	1,901	1,922	1,943	1,964	1,984
19	2,006	2,027	2,048	2,069	2,091	2,112	2,134	2,156	2,178	2,200
20	2,222	2,244	2,267	2,289	2,311	2,335	2,358	2,380	2,404	2,427
21	2,450	2,473	2,497	2,520	2,544	2,568	2,592	2,616	2,640	2,664
22	2,689	2,713	2,738	2,763	2,788	2,812	2,838	2,863	2,888	2,913
23	2,939	2,964	2,990	3,016	3,042	3,068	3,094	3,120	3,147	3,173
24	3,200	3,227	3,254	3,280	3,308	3,335	3,362	3,389	3,417	3,444
25	3,472	3,500	3,528	3,556	3,584	3,612	3,641	3,669	3,698	3,727
26	3,756	3,784	3,814	3,843	3,872	3,901	3,931	3,960	3,990	4,020
27	4,050	4,080	4,110	4,140	4,171	4,201	4,232	4,263	4,294	4,324
28	4,356	4,387	4,418	4,449	4,481	4,512	4,544	4,576	4,608	4,640
29	4,672	4,704	4,737	4,769	4,802	4,835	4,868	4,900	4,934	4,967
30	5,000	5,033	5,067	5,100	5,134	5,168	5,202	5,236	5,270	5,304
31	5,339	5,373	5,408	5,443	5,478	5,512	5,548	5,583	5,618	5,653
32	5,689	5,724	5,760	5,796	5,832	5,868	5,904	5,940	5,977	6,013
33	6,050	6,087	6,124	6,160	6,198	6,235	6,272	6,309	6,347	6,384
34	6,422	6,460	6,498	6,536	6,574	6,612	6,651	6,689	6,728	6,767
35	6,806	6,844	6,884	6,923	6,962	7,001	7,041	7,080	7,120	7,160
36	7,200	7,240	7,280	7,320	7,361	7,401	7,442	7,483	7,524	7,564
37	7,606	7,647	7,688	7,729	7,771	7,812	7,854	7,896	7,938	7,980
38	8,022	8,064	8,107	8,149	8,192	8,235	8,278	8,320	8,364	8,407
39	8,450	8,493	8,537	8,580	8,624	8,668	8,712	8,756	8,800	8,844
40	8,889	8,933	8,978	9,023	9,068	9,112	9,158	9,203	9,248	9,293
41	9,339	9,384	9,430	9,476	9,522	9,568	9,614	9,660	9,707	9,753
42	9,800	9,847	9,894	9,940	9,988	10,035	10,082	10,129	10,177	10,224
43	10,272	10,320	10,368	10,416	10,464	10,512	10,561	10,609	10,658	10,707
44	10,756	10,804	10,854	10,903	10,952	11,001	11,051	11,100	11,150	11,200
45	11,250	11,300	11,350	11,400	11,451	11,501	11,552	11,603	11,654	11,704
46	11,756	11,807	11,858	11,909	11,961	12,012	12,064	12,116	12,168	12,220
47	12,272	12,324	12,377	12,429	12,482	12,535	12,588	12,640	12,694	12,747
48	12,800	12,853	12,907	12,960	13,014	13,068	13,122	13,176	13,230	13,284
49	13,339	13,393	13,448	13,503	13,558	13,612	13,668	13,723	13,778	13,833
50	13,889	13,944	14,000	14,056	14,112	14,168	14,224	14,280	14,337	14,392
51	14,450	14,507	14,564	14,620	14,678	14,735	14,792	14,849	14,907	14,964
52	15,022	15,080	15,138	15,196	15,254	15,312	15,371	15,430	15,489	15,548
53	15,606	15,664	15,724	15,783	15,842	15,901	15,961	16,020	16,080	16,140
54	16,200	16,260	16,320	16,380	16,441	16,501	16,562	16,623	16,684	16,744
55	16,806	16,867	16,928	16,989	17,051	17,112	17,174	17,236	17,298	17,360
56	17,422	17,484	17,547	17,609	17,672	17,735	17,798	17,860	17,924	17,987
57	18,050	18,113	18,177	18,240	18,304	18,368	18,432	18,496	18,560	18,624
58	18,689	18,753	18,818	18,883	18,948	19,012	19,078	19,143	19,208	19,273
59	19,339	19,404	19,470	19,536	19,602	19,668	19,734	19,800	19,867	19,933
60	20,000									

Table 51.—*Amount of material in cubic yards per 100 linear feet of level cut,*

side slopes 2 to 1.

Depth of center cut in feet	.0	.1	.2	.3	.4	.5	.6	.7	.8	.9
0	0.0	0.1	0.3	0.7	1.2	1.9	2.7	3.6	4.7	6.0
1	7.4	9.0	10.7	12.5	14.5	16.7	19.0	21.4	24.0	26.7
2	30	33	36	39	43	46	50	54	58	62
3	67	71	76	81	86	91	96	101	107	113
4	119	125	131	137	143	150	157	164	171	178
5	185	193	200	208	216	224	232	241	249	258
6	267	276	285	294	303	313	323	333	343	353
7	363	373	384	395	406	417	428	439	451	462
8	474	486	498	510	523	535	548	561	574	587
9	600	613	627	641	655	669	683	697	711	726
10	741	756	771	786	801	817	832	848	864	880
11	896	913	929	946	963	980	997	1,014	1,031	1,049
12	1,067	1,084	1,103	1,121	1,139	1,157	1,176	1,195	1,214	1,233
13	1,252	1,271	1,291	1,310	1,330	1,350	1,370	1,390	1,411	1,431
14	1,452	1,473	1,494	1,515	1,536	1,557	1,579	1,601	1,623	1,645
15	1,667	1,689	1,711	1,734	1,757	1,780	1,803	1,826	1,849	1,873
16	1,896	1,920	1,944	1,968	1,992	2,017	2,041	2,066	2,091	2,116
17	2,141	2,166	2,191	2,217	2,243	2,269	2,295	2,321	2,347	2,373
18	2,400	2,427	2,454	2,481	2,508	2,535	2,563	2,590	2,618	2,646
19	2,674	2,702	2,731	2,759	2,788	2,817	2,846	2,875	2,904	2,938
20	2,963	2,993	3,023	3,053	3,083	3,113	3,143	3,174	3,205	3,236
21	3,267	3,298	3,329	3,361	3,392	3,424	3,456	3,488	3,520	3,553
22	3,585	3,618	3,651	3,684	3,717	3,750	3,783	3,817	3,851	3,885
23	3,919	3,953	3,987	4,021	4,056	4,091	4,126	4,161	4,196	4,231
24	4,267	4,302	4,338	4,374	4,410	4,446	4,483	4,519	4,556	4,593
25	4,630	4,667	4,704	4,741	4,779	4,817	4,855	4,893	4,931	4,969
26	5,007	5,046	5,085	5,124	5,163	5,202	5,241	5,281	5,320	5,360
27	5,400	5,440	5,480	5,521	5,561	5,602	5,643	5,684	5,725	5,766
28	5,807	5,849	5,891	5,933	5,975	6,017	6,059	6,101	6,144	6,187
29	6,230	6,273	6,316	6,359	6,403	6,446	6,490	6,534	6,578	6,622
30	6,667	6,711	6,756	6,801	6,846	6,891	6,936	6,981	7,027	7,073
31	7,119	7,165	7,211	7,257	7,303	7,350	7,397	7,444	7,491	7,538
32	7,585	7,633	7,680	7,728	7,776	7,824	7,872	7,921	7,969	8,018
33	8,067	8,116	8,165	8,214	8,263	8,313	8,363	8,413	8,463	8,513
34	8,563	8,613	8,664	8,715	8,766	8,817	8,868	8,919	8,971	9,022
35	9,074	9,126	9,178	9,230	9,283	9,335	9,388	9,441	9,494	9,547
36	9,600	9,653	9,707	9,761	9,815	9,869	9,923	9,977	10,031	10,086
37	10,141	10,196	10,251	10,306	10,361	10,417	10,472	10,528	10,584	10,640
38	10,696	10,753	10,809	10,866	10,923	10,980	11,037	11,094	11,151	11,209
39	11,267	11,325	11,383	11,441	11,499	11,557	11,616	11,675	11,734	11,793
40	11,852	11,911	11,971	12,030	12,090	12,150	12,210	12,270	12,331	12,391
41	12,452	12,513	12,574	12,635	12,696	12,757	12,819	12,881	12,943	13,005
42	13,067	13,129	13,191	13,254	13,317	13,380	13,443	13,506	13,569	13,633
43	13,696	13,760	13,824	13,888	13,952	14,017	14,081	14,146	14,211	14,276
44	14,341	14,406	14,471	14,537	14,603	14,669	14,735	14,801	14,867	14,933

Table 51.—*Amount of material in cubic yards per 100 linear feet of level cut,*

side slopes 2 to 1 — Continued.

Depth of center cut in feet	.0	.1	.2	.3	.4	.5	.6	.7	.8	.9
45	15,000	15,067	15,134	15,201	15,268	15,335	15,403	15,470	15,538	15,606
46	15,674	15,742	15,811	15,879	15,948	16,017	16,086	16,155	16,224	16,293
47	16,363	16,433	16,503	16,573	16,643	16,713	16,783	16,854	16,925	16,996
48	17,067	17,138	17,209	17,281	17,352	17,424	17,496	17,568	17,640	17,713
49	17,785	17,858	17,931	18,004	18,077	18,150	18,223	18,297	18,371	18,445
50	18,519	18,593	18,667	18,741	18,816	18,891	18,966	19,041	19,116	19,191
51	19,267	19,342	19,418	19,494	19,570	19,646	19,723	19,799	19,876	19,953
52	20,030	20,107	20,184	20,261	20,339	20,417	20,495	20,573	20,651	20,729
53	20,807	20,886	20,965	21,044	21,123	21,202	21,281	21,361	21,440	21,520
54	21,600	21,680	21,760	21,841	21,921	22,002	22,083	22,164	22,245	22,326
55	22,407	22,489	22,571	22,653	22,735	22,817	22,899	22,981	23,064	23,147
56	23,230	23,313	23,396	23,479	23,563	23,646	23,730	23,814	23,898	23,982
57	24,067	24,151	24,236	24,321	24,406	24,491	24,576	24,661	24,747	24,833
58	24,919	25,005	25,091	25,177	25,263	25,350	25,447	25,524	25,611	25,698
59	25,785	25,873	25,960	26,048	26,136	26,224	26,312	26,401	26,489	26,578
60	26,667	......	......	......	......	......	......	......	......	......

Table 52.—*Amount of material in cubic yards per 100 linear feet of level cut,*

side slopes 3 to 1.

Depth of center cut in feet	.0	.1	.2	.3	.4	.5	.6	.7	.8	.9
0	0.0	0.1	0.4	1.0	1.8	2.8	4.0	5.4	7.1	9.0
1	11.1	13.4	16.0	18.8	21.8	25.0	28.4	32.2	36.1	40.1
2	44	49	54	59	64	69	75	81	87	93
3	100	106	114	121	128	136	144	152	160	168
4	178	187	196	205	215	225	235	245	256	267
5	278	289	300	312	324	336	348	361	373	387
6	400	413	427	441	455	469	484	499	514	529
7	544	560	576	592	608	625	642	659	676	693
8	711	729	747	765	784	803	822	841	860	880
9	900	920	940	961	982	1,003	1,024	1,045	1,067	1,089
10	1,111	1,133	1,156	1,179	1,202	1,225	1,248	1,272	1,296	1,320
11	1,344	1,369	1,394	1,419	1,444	1,469	1,495	1,521	1,547	1,573
12	1,600	1,627	1,654	1,681	1,708	1,736	1,764	1,792	1,820	1,849
13	1,878	1,907	1,936	1,965	1,995	2,025	2,055	2,085	2,116	2,147
14	2,178	2,209	2,240	2,272	2,304	2,336	2,368	2,401	2,434	2,467

Table 52.—*Amount of material in cubic yards per 100 linear feet of level cut,*

side slopes 3 to 1—Continued.

Depth of center cut in feet	.0	.1	.2	.3	.4	.5	.6	.7	.8	.9
15	2,500	2,533	2,567	2,601	2,635	2,669	2,704	2,739	2,774	2,809
16	2,844	2,880	2,916	2,952	2,988	3,025	3,062	3,099	3,136	3,173
17	3,211	3,249	3,287	3,325	3,364	3,403	3,442	3,481	3,520	3,560
18	3,600	3,640	3,680	3,721	3,762	3,803	3,844	3,885	3,927	3,969
19	4,011	4,053	4,096	4,139	4,182	4,225	4,268	4,312	4,356	4,400
20	4,444	4,489	4,534	4,579	4,624	4,669	4,715	4,761	4,807	4,853
21	4,900	4,947	4,994	5,041	5,088	5,137	5,184	5,232	5,280	5,329
22	5,378	5,427	5,476	5,525	5,575	5,625	5,675	5,725	5,776	5,827
23	5,878	5,929	5,980	6,032	6,084	6,136	6,188	6,240	6,294	6,346
24	6,400	6,453	6,507	6,561	6,615	6,669	6,724	6,779	6,834	6,889
25	6,944	7,000	7,056	7,112	7,168	7,225	7,282	7,339	7.396	7,453
26	7,511	7,569	7,627	7,685	7,744	7,803	7,862	7,921	7 980	8,040
27	8,100	8,160	8,220	8,281	8,342	8,403	8,464	8,525	8 587	8,649
28	8,711	8,773	8,836	8,899	8,962	9,025	9,088	9,152	9,216	9,280
29	9,344	9,409	9,474	9,539	9,604	9,669	9,735	9,801	9,867	9,993
30	10,000	10,067	10,134	10,201	10,268	10,336	10,404	10,472	10,540	10,609
31	10,678	10,747	10,816	10,885	10,955	11,025	11,095	11,165	11,236	11,307
32	11,378	11,449	11,520	11,592	11,664	11,736	11,808	11,881	11,954	12,027
33	12,100	12,173	12,247	12,321	12,395	12,469	12,544	12,619	12,694	12,769
34	12,844	12,920	12,996	13,072	13,148	13,225	13,302	13,379	13,456	13,533
35	13,611	13,689	13,767	13,845	13,924	14,003	14,082	14,161	14,240	14,320
36	14,400	14,480	14,560	14,641	14,722	14,803	14,884	14,965	15,047	15,129
37	15,211	15,293	15,376	15,459	15,542	15,625	15,708	15,792	15,876	15,960
38	16,044	16,129	16,214	16,299	16,384	16,469	16,555	16,641	16,727	16,813
39	16,900	16,987	17,074	17,161	17,248	17,336	17,424	17,512	17,600	17,689
40	17,778	17,867	17,956	18,045	18,135	18,225	18,315	18,405	18,496	18,587
41	18,678	18,769	18,860	18,952	19,044	19,136	19,228	19,321	19.414	19,507
42	19,600	19,693	19,787	19,881	19,975	20,069	20,164	20,259	20,354	20,449
43	20,544	20,640	20,736	20,832	20,928	21,025	21,122	21,219	21,316	21,413
44	21,511	21,609	21,707	21,805	21,904	22,003	22,102	22,201	22,300	22,400
45	22,500	22,600	22,700	22,801	22,902	23,003	23,104	23,205	23,307	23,409
46	23,511	23,613	23,716	23,819	23,922	24,025	24,128	24,232	24,336	24,440
47	24,544	24,649	24,754	24,859	24,964	25,069	25,175	25,281	25,387	25,493
48	25,600	25,707	25,814	25,921	26,029	26,136	26,244	26,352	26,460	26,569
49	26,678	26,787	26,896	27,005	27,115	27,225	27,335	27,445	27,556	27,667
50	27,778	27,889	28,000	28,112	28,224	28,336	28,448	28,561	28,674	28,787
51	28,900	29,013	29,127	29,241	29,355	29,469	29,584	29,699	29,814	29,929
52	30,044	30,160	30,276	30,392	30,508	30,625	30,742	30,859	30,976	31,093
53	31,211	31,329	31,447	31,565	31,684	31,803	31,922	32,041	32,160	32,280
54	32,400	32,520	32,640	32,761	32,882	33,003	33,124	33,245	33,367	33,489
55	33,611	33,733	33,856	33,979	34,102	34,225	34,348	34,472	34,596	34,720
56	34,844	34,969	35,094	35,219	35,344	35,459	35,595	35,721	35,847	35,973
57	36,100	36,227	36,354	36,481	36,608	36,736	36,864	36,992	37,120	37,249
58	37,378	37,507	37,636	37,765	37,895	38,025	38,155	38,285	38,416	38,547
59	38,678	38,809	38,940	39,072	39,204	39,336	39,468	39,601	39,734	39,867
60	40,000	......	......	......	......	......	......	......	......	......

Table 53.—*Amount of material in cubic yards per 100 linear feet of cut on sloping ground,*
side slopes ½ to 1

Depth of center cut in feet	Surface slope of ground in percent										
	10	15	20	25	30	35	40	45	50	55	60
1.0	2	2	2	2	2	2	2	2	2	2	2
1.5	4	4	4	4	4	4	4	4	4	5	5
2.0	7	7	7	8	8	8	8	8	8	8	8
2.5	12	12	12	12	12	12	12	12	12	13	13
3.0	17	17	17	17	17	17	17	18	18	18	18
3.5	23	23	23	23	23	23	24	24	24	25	25
4.0	30	30	30	30	30	31	31	31	32	32	33
4.5	38	38	38	38	38	39	39	39	40	41	41
5.0	46	47	47	47	47	48	48	49	49	50	51
5.5	56	56	57	57	57	58	58	59	60	61	62
6.0	67	67	67	68	68	69	69	70	71	72	73
6.5	78	79	79	79	80	81	82	82	83	85	86
7.0	91	91	92	92	93	94	95	96	97	98	100
7.5	104	105	105	106	107	107	109	110	111	113	114
8.0	119	119	120	120	121	122	123	125	126	128	130
8.5	134	135	135	136	137	138	139	141	143	145	147
9.0	150	151	152	152	153	155	156	158	160	162	165
9.5	168	168	169	170	171	172	174	176	178	181	184
10.0	186	186	187	188	189	191	193	195	198	200	204
10.5	205	205	206	207	209	211	213	215	218	221	224
11.0	225	225	226	228	229	231	233	236	239	242	246
11.5	246	246	247	249	251	253	255	258	261	265	269
12.0	267	268	269	271	273	275	278	281	284	288	293
12.5	290	291	292	294	296	298	301	305	309	313	318
13.0	314	315	316	318	320	323	326	330	334	339	344
13.5	338	339	341	343	345	348	352	355	360	365	371
14.0	364	365	367	369	371	374	378	382	387	393	399
14.5	390	392	393	396	398	402	406	410	415	421	428
15.0	418	419	421	423	426	430	434	439	444	451	458
15.5	446	447	449	452	455	459	463	469	475	481	489
16.0	475	477	479	482	485	489	494	499	506	513	521
16.5	505	507	509	512	516	520	525	531	538	545	554
17.0	537	538	541	544	548	552	557	564	571	579	588
17.5	569	570	573	576	580	585	591	597	605	614	623
18.0	602	603	606	610	614	619	625	632	640	649	659
18.5	635	637	640	644	648	654	660	668	676	686	696
19.0	670	672	675	679	684	690	696	704	713	723	735
19.5	706	708	711	715	720	726	734	742	751	762	774
20.0	743	745	748	752	758	764	772	780	790	801	814
20.5	780	783	786	791	796	803	811	820	830	842	855
21.0	819	821	825	830	835	842	851	860	871	883	897
21.5	858	861	865	870	876	883	892	902	913	926	941
22.0	899	901	905	911	917	925	934	944	956	970	985
22.5	940	943	947	952	959	967	977	987	1,000	1,014	1,030
23.0	982	985	990	995	1,002	1,011	1,020	1,032	1,045	1,060	1,077

Table 53.—*Amount of material in cubic yards per 100 linear feet of cut on sloping ground,*
side slopes ½ to 1—Continued

Depth of center cut in feet	\multicolumn{11}{Surface slope of ground in percent}
	10	15	20	25	30	35	40	45	50	55	60
23.5	1,025	1,028	1,033	1,039	1,046	1,055	1,065	1,077	1 091	1,106	1,124
24.0	1,069	1,073	1,077	1,084	1,091	1,100	1,111	1,124	1,138	1,154	1,172
24.5	1,114	1,118	1,123	1,129	1,137	1,147	1,158	1,171	1,186	1,203	1,222
25.0	1,160	1,164	1,169	1,176	1,184	1,194	1,206	1,219	1,235	1,252	1,272
25.5	1,207	1,211	1,216	1,223	1,232	1,242	1,254	1,268	1,284	1,303	1,323
26.0	1,255	1,259	1,264	1,272	1,281	1,291	1,304	1,319	1,335	1,354	1,376
26.5	1,304	1,308	1,314	1,321	1,330	1,342	1,355	1,370	1,387	1,407	1,429
27.0	1,353	1,358	1,364	1,371	1,381	1,393	1,406	1,422	1,440	1,460	1,484
27.5	1,404	1,408	1,415	1,423	1,433	1,445	1,459	1,475	1,494	1,515	1,539
28.0	1,455	1,460	1,467	1,475	1,485	1,498	1,512	1,529	1,549	1,571	1,595
28.5	1,508	1,513	1,519	1,528	1,539	1,552	1,567	1,584	1,604	1,627	1,653
29.0	1,561	1,566	1,573	1,582	1,593	1,607	1,622	1,640	1,661	1,685	1,711
29.5	1,616	1,621	1,628	1,637	1,649	1,662	1,679	1,698	1,719	1,743	1,771
30.0	1,671	1,676	1,684	1,693	1,705	1,719	1,736	1,756	1,778	1,803	1,832
30.5	1,727	1,732	1,740	1,750	1,762	1,777	1,794	1,815	1,838	1,864	1,893
31.0	1,784	1,790	1,798	1,808	1,821	1,836	1,854	1,875	1,898	1,925	1,956
31.5	1,842	1,848	1,856	1,867	1,880	1,896	1,914	1,935	1,960	1,988	2,019
32.0	1,901	1,907	1,915	1,926	1,940	1,956	1,975	1,997	2,023	2,051	2,084
32.5	1,961	1,967	1,976	1,987	2,001	2,018	2,038	2,060	2,086	2,116	2,149
33.0	2,022	2,028	2,037	2,049	2,063	2,080	2,101	2,124	2,151	2,182	2,216
33.5	2,083	2,090	2,099	2,111	2,126	2,144	2,165	2,189	2,217	2,248	2,284
34.0	2,146	2,153	2,162	2,175	2,190	2,208	2,230	2,255	2,283	2,316	2,352
34.5	2,210	2,217	2,226	2,239	2,255	2,274	2,296	2,322	2,351	2,384	2,422
35.0	2,274	2,281	2,291	2,305	2,321	2,340	2,363	2,389	2,420	2,454	2,493
35.5	2,340	2,347	2,357	2,371	2,388	2,408	2,431	2,458	2,489	2,525	2,565
36.0	2,406	2,414	2,424	2,438	2,455	2,476	2,500	2,528	2,560	2,596	2,637
36.5	2,473	2,481	2,492	2,506	2,524	2,545	2,570	2,599	2,632	2,669	2,711
37.0	2,542	2,550	2,561	2,575	2,594	2,615	2,641	2,670	2,704	2,743	2,786
37.5	2,611	2,619	2,630	2,646	2,664	2,686	2,713	2,743	2,778	2,817	2,862
38.0	2,681	2,689	2,701	2,717	2,736	2,759	2,785	2,817	2,852	2,893	2,939
38.5	2,752	2,760	2,773	2,788	2,808	2,832	2,859	2,891	2,928	2,969	3,016
39.0	2,824	2,833	2,845	2,861	2,882	2,906	2,934	2,967	3,004	3,047	3,095
39.5	2,897	2,906	2,919	2,935	2,956	2,981	3,010	3,043	3,082	3,126	3,175
40.0	2,970	2,980	2,993	3,010	3,031	3,057	3,086	3,121	3,160	3,205	3,256
40.5	3,045	3,055	3,068	3,086	3,107	3,133	3,164	3,199	3,240	3,286	3,338
41.0	3,121	3,131	3,144	3,162	3,185	3,211	3,243	3,279	3,320	3,368	3,421
41.5	3,197	3,207	3,222	3,240	3,263	3,290	3,322	3,359	3,402	3,450	3,505
42.0	3,275	3,285	3,300	3,319	3,342	3,370	3,403	3,441	3,484	3,534	3,590
42.5	3,353	3,364	3,379	3,398	3,422	3,451	3,484	3,523	3,568	3,619	3,676
43.0	3,433	3,443	3,459	3,478	3,503	3,532	3,567	3,607	3,652	3,704	3,763
43.5	3,513	3,524	3,540	3,560	3,585	3,615	3,650	3,691	3,738	3,791	3,851
44.0	3,594	3,605	3,621	3,642	3,668	3,698	3,735	3,776	3,824	3,878	3,940

Table 54.—*Amount of material in cubic yards per 100 linear feet of cut on sloping ground,*

side slopes 1 to 1.

Depth of center cut in feet	Surface slope of ground in per cent										
	10	15	20	25	30	35	40	45	50	55	60
1.0	4	4	4	4	4	4	5	5	5	6	6
1.5	8	8	9	9	9	9	10	10	11	12	13
2.0	15	15	16	16	16	17	18	19	20	21	23
2.5	23	24	24	25	25	27	27	29	31	33	36
3.0	33	33	34	35	36	38	39	42	44	47	52
3.5	46	46	47	48	49	51	54	57	60	65	70
4.0	59	60	61	63	65	67	70	74	79	85	92
4.5	76	77	78	80	83	85	89	94	100	107	117
5.0	94	95	97	99	102	106	111	117	124	133	145
5.5	113	114	117	120	123	128	133	141	149	161	175
6.0	134	136	139	142	146	152	158	167	177	191	208
6.5	157	160	163	166	172	178	186	196	208	224	244
7.0	183	185	189	193	199	206	215	227	242	260	283
7.5	210	212	217	222	229	237	248	261	278	299	325
8.0	239	242	247	253	261	270	282	297	316	340	370
8.5	270	274	279	286	295	305	319	336	357	384	418
9.0	303	307	312	320	330	342	357	376	400	430	468
9.5	338	342	348	356	367	381	398	419	446	479	522
10.0	374	378	385	395	406	422	441	464	494	531	578
10.5	412	417	425	436	448	465	486	512	545	585	637
11.0	453	458	467	478	492	510	533	562	598	642	700
11.5	495	501	510	523	538	558	583	615	653	702	765
12.0	539	545	555	569	586	607	634	669	711	764	833
12.5	585	592	603	618	637	659	689	726	772	830	904
13.0	632	640	652	668	689	713	745	785	835	897	978
13.5	681	691	703	720	743	769	803	847	900	967	1,054
14.0	733	743	756	774	799	827	864	911	968	1,040	1,134
14.5	787	797	811	831	857	887	927	977	1,039	1,116	1,216
15.0	841	852	868	888	916	949	994	1,045	1,111	1,194	1,301
15.5	898	910	927	949	978	1,014	1,059	1,116	1,187	1,276	1,390
16.0	957	970	987	1,011	1,042	1,080	1,128	1,189	1,264	1,359	1,480
16.5	1,018	1,031	1,050	1,075	1,108	1,148	1,199	1,265	1,344	1,445	1,573
17.0	1,080	1,095	1,115	1,141	1,176	1,219	1,273	1,343	1,427	1,534	1,669
17.5	1,145	1,160	1,182	1,209	1,246	1,292	1,349	1,423	1,512	1,626	1,770
18.0	1,212	1,227	1,250	1,280	1,319	1,368	1,428	1,506	1,600	1,720	1,874
18.5	1,281	1,297	1,321	1,353	1,394	1,445	1,509	1,591	1,691	1,817	1,980
19.0	1,351	1,368	1,393	1,426	1,470	1,523	1,591	1,678	1,783	1,916	2,088
19.5	1,422	1,440	1,467	1,502	1,548	1,604	1,676	1,767	1,878	2,018	2,199
20.0	1,496	1,515	1,542	1,580	1,628	1,687	1,763	1,859	1,975	2,123	2,313
20.5	1,572	1,592	1,620	1,660	1,710	1,773	1,852	1,953	2,075	2,230	2,430
21.0	1,649	1,670	1,701	1,742	1,795	1,861	1,943	2,049	2,178	2,340	2,550
21.5	1,729	1,751	1,783	1,826	1,882	1,951	2,037	2,148	2,283	2,453	2,673
22.0	1,811	1,834	1,868	1,913	1,971	2,043	2,134	2,250	2,391	2,569	2,800
22.5	1,894	1,918	1,953	2,001	2,061	2,136	2,231	2,353	2,501	2,687	2,928
23.0	1,979	2,004	2,041	2,090	2,153	2,232	2,331	2,458	2,613	2,808	3,059

Table 54.—*Amount of material in cubic yards per 100 linear feet of cut on sloping ground,*

side slopes 1 to 1 — Continued.

Depth of center cut in feet	Surface slope of ground in per cent										
	10	15	20	25	30	35	40	45	50	55	60
23.5	2,065	2,091	2,130	2,181	2,247	2,330	2,434	2,566	2,728	2,931	3,194
24.0	2,154	2,181	2,221	2,275	2,344	2,430	2,539	2,677	2,845	3,057	3,331
24.5	2,245	2,274	2,315	2,371	2,443	2,533	2,646	2,790	2,965	3,186	3,472
25.0	2,338	2,368	2,411	2,469	2,545	2,637	2,755	2,905	3,088	3,318	3,615
25.5	2,432	2,463	2,508	2,568	2,647	2,743	2,866	3,022	3,212	3,451	3,761
26.0	2,529	2,561	2,608	2,670	2,752	2,852	2,980	3,142	3,340	3,588	3,910
26.5	2,627	2,661	2,709	2,774	2,859	2,963	3,095	3,264	3,469	3,727	4,062
27.0	2,727	2,762	2,813	2,880	2,968	3,076	3,212	3,388	3,601	3,869	4,217
27.5	2,829	2,865	2,918	2,988	3,079	3,191	3,332	3,515	3,736	4,014	4,374
28.0	2,932	2,970	3,024	3,097	3,191	3,308	3,454	3,643	3,872	4,161	4,534
28.5	3,038	3,077	3,133	3,208	3,306	3,427	3,579	3,775	4,012	4,311	4,698
29.0	3,146	3,187	3,245	3,322	3,423	3,548	3,706	3,909	4,154	4,464	4,864
29.5	3,255	3,297	3,357	3,438	3,542	3,671	3,835	4,045	4,298	4,619	5,033
30.0	3,367	3,409	3,471	3,555	3,663	3,797	3,967	4,183	4,445	4,777	5,205
30.5	3,480	3,524	3,588	3,675	3,786	3,924	4,100	4,323	4,595	4,937	5,380
31.0	3,595	3,641	3,707	3,796	3,911	4,054	4,236	4,466	4,747	5,100	5,558
31.5	3,712	3,759	3,828	3,920	4,039	4,187	4,374	4,612	4,901	5,266	5,739
32.0	3,831	3,880	3,951	4,046	4,169	4,322	4,514	4,760	5,058	5,435	5,923
32.5	3,952	4,002	4,075	4,173	4,300	4,457	4,656	4,909	5,217	5,606	6,109
33.0	4,074	4,126	4,201	4,302	4,433	4,595	4,800	5,061	5,379	5,780	6,298
33.5	4,198	4,252	4,329	4,433	4,568	4,735	4,946	5,215	5,543	5,956	6,491
34.0	4,324	4,379	4,459	4,566	4,705	4,877	5,095	5,372	5,710	6,135	6,686
34.5	4,452	4,509	4,592	4,702	4,845	5,022	5,246	5,531	5,879	6,317	6,884
35.0	4,583	4,641	4,726	4,839	4,987	5,169	5,399	5,693	6,051	6,502	7,085
35.5	4,714	4,774	4,861	4,978	5,130	5,317	5,555	5,856	6,225	6,689	7,288
36.0	4,848	4,910	5,000	5,120	5,276	5,469	5,712	6,023	6,402	6,879	7,496
36.5	4,984	5,048	5,140	5,263	5,423	5,621	5,872	6,191	6,581	7,071	7,705
37.0	5,122	5,187	5,282	5,408	5,573	5,776	6,034	6,362	6,762	7,266	7,918
37.5	5,261	5,328	5,426	5,555	5,725	5,933	6,198	6,535	6,946	7,464	8,132
38.0	5,402	5,471	5,571	5,705	5,879	6,093	6,365	6,711	7,133	7,665	8,353
38.5	5,545	5,615	5,718	5,855	6,033	6,254	6,532	6,888	7,321	7,867	8,572
39.0	5,690	5,763	5,868	6,008	6,191	6,418	6,703	7,069	7,513	8,073	8,797
39.5	5,837	5,912	6,020	6,164	6,351	6,584	6,877	7,252	7,707	8,282	9,024
40.0	5,986	6,062	6,173	6,321	6,513	6,752	7,052	7,436	7,903	8,493	9,254
40.5	6,137	6,215	6,328	6,480	6,677	6,921	7,230	7,623	8,102	8,706	9,487
41.0	6,289	6,369	6,485	6,641	6,843	7,093	7,410	7,813	8,304	8,922	9,722
41.5	6,442	6,524	6,644	6,803	7,011	7,266	7,591	8,004	8,507	9,140	9,961
42.0	6,599	6,683	6,806	6,969	7,181	7,443	7,775	8,198	8,713	9,362	10,203
42.5	6,758	6,844	6,969	7,136	7,353	7,622	7,962	8,395	8,922	9,587	10,447
43.0	6,917	7,006	7,134	7,305	7,527	7,802	8,150	8,593	9,133	9,814	10,694
43.5	7,079	7,170	7,300	7,476	7,703	7,984	8,341	8,794	9,347	10,043	10,944
44.0	7,243	7,335	7,469	7,648	7,880	8,169	8,533	8,997	9,563	10,175	11,197

Table 55.—*Amount of material in cubic yards per 100 linear feet of cut on sloping ground,*

side slopes 1½ to 1.

Depth of center cut in feet	Surface slope of ground in per cent										
	10	15	20	25	30	35	40	45	50	55	60
0.5	1	1	1	1	1	1	1	2	2	2	6
1.0	6	6	7	7	7	8	9	11	13	18	29
1.5	12	13	13	14	15	17	19	22	28	39	65
2.0	23	23	24	26	28	31	34	41	51	70	117
2.5	36	37	38	41	44	48	55	64	80	109	183
3.0	51	53	55	58	63	69	78	92	114	157	263
3.5	70	72	75	79	85	94	106	125	155	213	357
4.0	91	94	98	104	112	123	139	163	203	278	467
4.5	113	118	124	132	141	155	176	206	257	352	590
5.0	142	146	153	162	174	192	217	255	318	435	730
5.5	172	177	185	195	211	232	262	309	384	526	882
6.0	205	211	220	233	251	276	312	368	457	624	1,051
6.5	240	248	258	273	295	324	367	431	537	735	1,233
7.0	278	287	299	317	341	375	425	500	622	852	1,430
7.5	319	329	343	363	391	430	488	574	714	978	1,641
8.0	364	375	391	414	446	491	556	654	813	1,113	1,870
8.5	411	423	441	467	503	555	627	738	918	1,257	2,107
9.0	460	474	495	524	564	622	703	827	1,029	1,409	2,364
9.5	513	528	552	583	628	691	783	922	1,146	1,569	2,633
10.0	569	585	611	647	697	765	868	1,021	1,271	1,740	2,919
10.5	627	645	673	712	768	844	956	1,125	1,401	1,918	3,217
11.0	687	708	739	781	843	927	1,049	1,235	1,537	2,104	3,531
11.5	752	774	808	855	922	1,013	1,149	1,350	1,680	2,301	3,860
12.0	819	843	879	931	1,003	1,103	1,250	1,470	1,829	2,504	4,203
12.5	888	914	954	1,010	1,089	1,197	1,356	1,595	1,985	2,717	4,560
13.0	961	989	1,032	1,093	1,178	1,295	1,467	1,725	2,147	2,939	4,933
13.5	1,036	1,066	1,112	1,178	1,269	1,396	1,581	1,860	2,316	3,170	5,318
14.0	1,114	1,147	1,196	1,267	1,365	1,502	1,701	2,001	2,489	3,410	5,721
14.5	1,195	1,230	1,284	1,359	1,465	1,612	1,825	2,146	2,669	3,657	6,136
15.0	1,279	1,316	1,374	1,454	1,568	1,724	1,952	2,297	2,857	3,914	6,567
15.5	1,366	1,406	1,467	1,553	1,674	1,841	2,085	2,453	3,051	4,179	7,012
16.0	1,455	1,498	1,563	1,654	1,784	1,961	2,221	2,613	3,250	4,453	7,472
16.5	1,547	1,593	1,662	1,759	1,897	2,085	2,362	2,779	3,456	4,735	7,945
17.0	1,643	1,691	1,765	1,868	2,014	2,214	2,507	2,951	3,670	5,027	8,435
17.5	1,741	1,792	1,870	1,979	2,134	2,346	2,656	3,126	3,889	5,326	8,937
18.0	1,841	1,896	1,979	2,094	2,258	2,482	2,809	3,308	4,114	5,636	9,456
18.5	1,945	2,002	2,090	2,212	2,385	2,622	2,967	3,494	4,346	5,953	9,988
19.0	2,051	2,111	2,205	2,334	2,516	2,766	3,130	3,686	4,585	6,279	10,535
19.5	2,160	2,225	2,322	2,458	2,650	2,913	3,299	3,881	4,828	6,614	11,097
20.0	2,272	2,341	2,442	2,586	2,787	3,064	3,472	4,083	5,079	6,957	11,673
20.5	2,387	2,460	2,566	2,717	2,929	3,220	3,648	4,289	5,337	7,310	12,265
21.0	2,506	2,581	2,692	2,851	3,073	3,379	3,828	4,502	5,600	7,670	12,871
21.5	2,627	2,705	2,822	2,988	3,221	3,541	4,013	4,719	5,870	8,040	13,491
22.0	2,751	2,832	2,955	3,129	3,373	3,708	4,201	4,941	6,147	8,417	14,127
22.5	2,877	2,962	3,090	3,272	3,527	3,878	4,394	5,168	6,429	8,804	14,775

Table 55.—*Amount of material in cubic yards per 100 linear feet of cut on sloping ground,*

side slopes 1½ to 1. — Continued.

Depth of center cut in feet	Surface slope of ground in per cent										
	10	15	20	25	30	35	40	45	50	55	60
23.0	3,007	3,096	3,229	3,420	3,686	4,053	4,592	5,400	6,718	9,201	15,440
23.5	3,139	3,232	3,372	3,570	3,848	4,231	4,794	5,638	7,013	9,606	16,118
24.0	3,274	3,371	3,517	3,724	4,014	4,413	5,000	5,881	7,314	10,019	16,812
24.5	3,412	3,513	3,665	3,881	4,183	4,599	5,211	6,129	7,622	10,441	17,519
25.0	3,552	3,657	3,816	4,040	4,355	4,788	5,425	6,382	7,936	10,871	18,242
25.5	3,695	3,804	3,970	4,203	4,531	4,981	5,644	6,639	8,256	11,310	18,978
26.0	3,842	3,954	4,128	4,370	4,711	5,178	5,868	6,902	8,584	11,758	19,731
26.5	3,991	4,109	4,288	4,539	4,892	5,380	6,095	7,169	8,917	12,215	20,497
27.0	4,144	4,266	4,451	4,712	5,080	5,585	6,328	7,443	9,257	12,680	21,277
27.5	4,298	4,425	4,617	4,888	5,270	5,793	6,564	7,721	9,603	13,153	22,072
28.0	4,456	4,588	4,786	5,068	5,464	6,006	6,805	8,005	9,956	13,637	22,881
28.5	4,616	4,753	4,958	5,250	5,661	6,223	7,050	8,292	10,314	14,128	23,706
29.0	4,779	4,921	5,134	5,436	5,860	6,443	7,300	8,586	10,680	14,627	24,546
29.5	4,946	5,093	5,313	5,626	6,064	6,667	7,555	8,885	11,052	15,136	25,399
30.0	5,115	5,267	5,495	5,818	6,272	6,895	7,813	9,189	11,429	15,654	26,268
30.5	5,287	5,444	5,680	6,014	6,482	7,127	8,076	9,497	11,813	16,181	27,150
31.0	5,462	5,624	5,868	6,213	6,697	7,363	8,342	9,811	12,203	16,715	28,047
31.5	5,639	5,806	6,058	6,414	6,914	7,602	8,613	10,130	12,600	17,259	28,958
32.0	5,820	5,992	6,252	6,619	7,136	7,845	8,889	10,455	13,004	17,811	29,885
32.5	6,003	6,180	6,449	6,828	7,360	8,092	9,169	10,784	13,413	18,372	30,826
33.0	6,189	6,372	6,649	7,040	7,589	8,343	9,453	11,119	13,829	18,941	31,782
33.5	6,378	6,567	6,852	7,255	7,821	8,598	9,742	11,458	14,251	19,520	32,753
34.0	6,570	6,764	7,057	7,472	8,055	8,856	10,034	11,802	14,680	20,105	33,738
34.5	6,764	6,964	7,266	7,693	8,294	9,118	10,331	12,151	15,115	20,701	34,738
35.0	6,962	7,168	7,479	7,919	8,537	9,385	10,634	12,506	15,557	21,307	35,754
35.5	7,162	7,374	7,694	8,147	8,783	9,655	10,940	12,865	16,004	21,921	36,782
36.0	7,366	7,584	7,913	8,378	9,032	9,929	11,250	13,230	16,458	22,542	37,826
36.5	7,572	7,796	8,134	8,612	9,284	10,206	11,565	13,601	16,919	23,172	38,884
37.0	7,780	8,011	8,359	8,850	9,540	10,482	11,883	13,977	17,386	23,812	39,958
37.5	7,991	8,229	8,585	9,090	9,799	10,773	12,206	14,356	17,857	24,461	41,045
38.0	8,206	8,450	8,816	9,334	10,062	11,062	12,535	14,742	18,337	25,116	42,148
38.5	8,424	8,674	9,050	9,582	10,329	11,356	12,867	15,133	18,823	25,781	43,266
39.0	8,644	8,900	9,286	9,832	10,599	11,652	13,203	15,528	19,315	26,455	44,398
39.5	8,867	9,130	9,526	10,086	10,873	11,952	13,544	15,929	19,814	27,137	45,545
40.0	9,093	9,363	9,769	10,343	11,150	12,258	13,889	16,335	20,319	27,829	46,699
40.5	9,322	9,598	10,014	10,603	11,430	12,567	14,236	16,745	20,829	28,529	47,873
41.0	9,554	9,836	10,263	10,867	11,714	12,879	14,590	17,163	21,346	29,238	49,062
41.5	9,788	10,078	10,515	11,133	12,002	13,195	14,950	17,584	21,870	29,955	50,265
42.0	10,025	10,322	10,770	11,403	12,293	13,515	15,313	18,010	22,401	30,682	51,483
42.5	10,266	10,569	11,028	11,677	12,587	13,838	15,679	18,441	22,937	31,417	52,716
43.0	10,509	10,819	11,289	11,953	12,885	14,166	16,049	18,877	23,480	32,160	53,963
43.5	10,754	11,072	11,553	12,233	13,186	14,497	16,425	19,319	24,029	32,912	55,225
44.0	11,003	11,329	11,821	12,516	13,492	14,833	16,805	19,766	24,586	33,674	56,506

Table 56.—*Amount of material in cubic yards per 100 linear feet of cut on sloping ground,*

side slopes 2 to 1.

Depth of center cut in feet	Surface slope of ground in per cent							
	10	**15**	**20**	**25**	**30**	**35**	**40**	**45**
0.5	2	2	2	3	3	4	5	10
1.0	7	8	8	9	11	14	20	38
1.5	18	19	20	23	26	33	47	87
2.0	31	33	36	40	47	58	83	156
2.5	48	51	55	61	72	90	128	244
3.0	70	74	80	89	104	131	186	352
3.5	95	100	109	121	142	178	252	479
4.0	124	131	142	158	186	233	330	623
4.5	157	165	179	200	235	294	417	788
5.0	193	203	221	247	289	363	514	972
5.5	233	246	267	299	350	439	622	1,176
6.0	278	293	318	356	417	523	741	1,400
6.5	326	344	373	417	489	614	869	1,643
7.0	378	399	432	484	568	712	1,008	1,906
7.5	434	458	496	556	652	817	1,158	2,189
8.0	493	521	564	632	741	929	1,317	2,491
8.5	557	588	637	713	837	1,049	1,486	2,819
9.0	625	659	715	800	938	1,176	1,667	3,160
9.5	697	735	797	892	1,046	1,312	1,857	3,521
10.0	772	814	883	988	1,159	1,453	2,058	3,903
10.5	851	897	973	1,089	1,278	1,601	2,269	4,304
11.0	933	984	1,067	1,095	1,401	1,754	2,489	4,722
11.5	1,020	1,076	1,167	1,307	1,532	1,920	2,721	5,162
12.0	1,111	1,172	1,270	1,423	1,668	2,091	2,963	5,621
12.5	1,205	1,271	1,377	1,543	1,810	2,268	3,215	6,099
13.0	1,304	1,375	1,490	1,669	1,959	2,453	3,478	6,597
13.5	1,406	1,483	1,607	1,800	2,112	2,644	3,750	7,113
14.0	1,513	1,595	1,729	1,936	2,271	2,846	4,033	7,649
14.5	1,622	1,711	1,854	2,076	2,436	3,053	4,325	8,203
15.0	1,736	1,832	1,985	2,223	2,608	3,268	4,630	8,779
15.5	1,854	1,956	2,119	2,374	2,784	3,489	4,944	9,378
16.0	1,975	2,084	2,257	2,529	2,966	3,718	5,268	9,981
16.5	2,101	2,217	2,401	2,690	3,155	3,954	5,603	10,625
17.0	2,230	2,353	2,549	2,856	3,349	4,197	5,946	11,282
17.5	2,364	2,493	2,701	3,027	3,549	4,448	6,302	11,954
18.0	2,500	2,637	2,857	3,202	3,754	4,706	6,667	12,645
18.5	2,641	2,785	3,018	3,382	3,965	4,971	7,043	13,358
19.0	2,785	2,938	3,183	3,568	4,183	5,243	7,429	14,091
19.5	2,934	3,095	3,353	3,759	4,406	5,621	7,825	14,842
20.0	3,087	3,255	3,527	3,953	4,634	5,809	8,231	15,613
20.5	3,243	3,420	3,706	4,151	4,869	6,103	8,648	16,403
21.0	3,403	3,589	3,889	4,356	5,109	6,405	9,075	17,213
21.5	3,567	3,762	4,076	4,565	5,355	6,713	9,512	18,042
22.0	3,734	3,939	4,268	4,780	5,608	7,029	9,959	18,891
22.5	3,906	4,120	4,464	5,000	5,866	7,352	10,417	19,760

Table 56.—*Amount of material in cubic yards per 100 linear feet of cut on sloping ground,*

side slopes 2 to 1—Continued.

Depth of center cut in feet	Surface slope of ground in per cent							
	10	15	20	25	30	35	40	45
23.0	4,082	4,306	4,665	5,225	6,130	7,683	10,886	20,648
23.5	4,262	4,495	4,879	5,454	6,399	8,021	11,364	21,555
24.0	4,445	4,688	5,080	5,689	6,675	8,365	11,853	22,482
24.5	4,631	4,885	5,293	5,928	6,955	8,715	12,352	23,428
25.0	4,823	5,087	5,512	6,174	7,242	9,075	12,861	24,395
25.5	5,018	5,292	5,734	6,424	7,533	9,442	13,380	25,381
26.0	5,216	5,500	5,960	6,678	7,830	9,817	13,909	26,385
26.5	5,419	5,714	6,192	6,938	8,135	10,199	14,450	27,410
27.0	5,625	5,932	6,428	7,202	8,445	10,587	15,000	28,454
27.5	5,835	6,154	6,669	7,471	8,762	10,983	15,561	29,518
28.0	6,049	6,380	6,813	7,746	9,083	11,386	16,132	30,600
28.5	6,268	6,611	7,163	8,027	9,411	11,798	16,714	31,704
29.0	6,490	6,845	7,417	8,311	9,744	12,215	17,305	32,826
29.5	6,715	7,083	7,674	8,598	10,082	12,638	17,906	33,967
30.0	6,945	7,328	7,937	8,891	10,428	13,071	18,519	35,129
30.5	7,178	7,572	8,204	9,188	10,779	13,510	19,141	36,309
31.0	7,415	7,821	8,475	9,491	11,135	13,954	19,773	37,509
31.5	7,657	8,075	8,750	9,801	11,497	14,410	20,417	38,729
32.0	7,902	8,333	9,030	10,115	11,865	14,871	21,071	39,968
32.5	8,150	8,596	9,314	10,434	12,238	15,339	21,735	41,227
33.0	8,403	8,863	9,603	10,758	12,617	15,815	22,409	42,506
33.5	8,660	9,133	9,896	11,086	13,002	16,298	23,093	43,803
34.0	8,920	9,408	10,194	11,419	13,393	16,788	23,787	45,120
34.5	9,184	9,687	10,496	11,757	13,791	17,286	24,492	46,457
35.0	9,452	9,970	10,802	12,100	14,194	17,791	25,207	47,813
35.5	9,724	10,257	11,113	12,447	14,602	18,302	25,932	49,199
36.0	10,000	10,548	11,429	12,800	15,016	18,820	26,668	50,585
36.5	10,280	10,843	11,749	13,158	15,436	19,346	27,414	52,000
37.0	10,563	11,142	12,073	13,522	15,861	19,880	28,170	53,434
37.5	10,850	11,445	12,401	13,891	16,293	20,422	28,937	54,888
38.0	11,142	11,752	12,733	14,264	16,730	20,971	29,713	56,361
38.5	11,437	12,063	13,071	14,642	17,174	21,527	30,500	57,855
39.0	11,737	12,378	13,413	15,025	17,623	22,190	31,297	59,368
39.5	12,039	12,697	13,759	15,413	18,078	22,660	32,104	60,906
40.0	12,346	13,021	14,110	15,805	18,539	23,237	32,923	62,451
40.5	12,656	13,349	14,465	16,202	19,006	23,821	33,752	64,021
41.0	12,971	13,681	14,824	16,605	19,479	24,414	34,590	65,611
41.5	13,290	14,017	15,187	17,013	19,957	25,012	35,438	67,221
42.0	13,612	14,357	15,556	17,425	20,441	25,619	36,298	68,851
42.5	13,938	14,701	15,929	17,842	20,930	26,231	37,168	70,501
43.0	14,267	15,049	16,306	18,264	21,424	26,852	38,047	72,170
43.5	14,601	15,401	16,687	18,691	21,925	27,481	38,937	73,858
44.0	14,939	15,757	17,073	19,124	22,432	28,116	39,837	75,565

Table 57.—*Amount of material in cubic yards per 100 linear feet of cut on sloping ground,*

side slopes 3 to 1

Depth of center cut in feet	Surface slope of ground in percent				
	10	15	20	25	30
1.0	12	14	17	25	58
1.5	27	31	39	57	132
2.0	49	56	69	102	234
2.5	76	87	109	159	365
3.0	110	125	156	229	526
3.5	150	171	213	311	716
4.0	195	223	278	406	936
4.5	247	282	352	514	1,184
5.0	305	348	434	635	1,462
5.5	369	421	525	768	1,769
6.0	440	502	625	914	2,105
6.5	516	589	734	1,073	2,471
7.0	598	683	851	1,244	2,865
7.5	687	784	977	1,429	3,289
8.0	781	892	1,111	1,625	3,743
8.5	882	1,007	1,254	1,835	4,225
9.0	989	1.129	1,406	2,057	4,737
9.5	1,102	1,257	1,567	2.292	5,278
10.0	1,221	1,393	1,736	2,540	5,848
10.5	1,346	1.536	1,914	2,800	6,447
11.0	1,477	1,686	2,101	3,073	7,076
11.5	1,615	1,843	2,296	3,359	7,734
12.0	1,758	2,006	2,500	3,657	8,421
12.5	1,908	2,177	2,713	3,968	9,137
13.0	2,063	2,355	2,934	4,292	9,883
13.5	2,225	2,539	3,164	4,629	10,658
14.0	2,393	2,731	3,403	4,978	11,462
14.5	2,567	2,929	3,650	5,340	12,295
15.0	2,747	3,135	3,906	5,714	13,158
15.5	2,933	3,347	4,171	6,102	14,050
16.0	3,126	3,567	4,444	6,502	14,971
16.5	3,324	3,793	4,727	6,914	15,921
17.0	3,529	4,026	5,017	7,340	16,901
17.5	3,739	4,267	5,317	7,778	17,909
18.0	3,956	4,514	5,625	8,229	18,947
18.5	4,179	4,768	5,942	8,692	20,015
19.0	4,408	5,030	6,267	9,168	21,111
19.5	4,643	5,298	6,602	9,657	22,237
20.0	4,884	5,573	6,944	10,159	23,392
20.5	5,131	5,855	7,296	10,673	24,576
21.0	5,385	6,144	7,656	11,200	25,789
21.5	5,644	6,440	8,025	11,740	27,032
22.0	5,910	6,743	8,403	12,292	28,304
22.5	6,181	7,053	8,789	12,857	29,605
23.0	6,459	7,370	9,184	13,435	30,936

Table 57.—*Amount of material in cubic yards per 100 linear feet of cut on sloping ground,*

side slopes 3 to 1—Continued

Depth of center cut in feet	Surface slope of ground in percent				
	10	15	20	25	30
23.5	6,743	7,694	9,588	14,025	32,295
24.0	7,033	8,025	10,000	14,629	33,684
24.5	7,329	8,363	10,421	15,244	35,102
25.0	7,631	8,708	10,851	15,873	36,550
25.5	7,940	9,060	11,289	16,514	38,026
26.0	8,254	9,418	11,736	17,168	39,532
26.5	8,574	9,784	12,192	17,835	41,067
27.0	8,901	10,157	12,656	18,514	42,632
27.5	9,234	10,536	13,129	19,206	44,225
28.0	9,573	10,923	13,611	19,911	45,848
28.5	9,918	11,317	14,102	20,629	47,500
29.0	10,269	11,717	14,601	21,359	49,181
29.5	10,626	12,125	15,109	22,102	50,892
30.0	10,989	12,539	15,625	22,857	52,632
30.5	11,358	12,961	16,150	23,625	54,401
31.0	11,734	13,389	16,684	24,406	56,199
31.5	12,115	13,824	17,227	25,200	58,026
32.0	12,503	14,267	17,778	26,006	59,883
32.5	12,897	14,716	18,338	26,825	61,769
33.0	13,297	15,172	18,906	27,657	63,684
33.5	13,703	15,636	19,484	28,502	65,629
34.0	14,115	16,106	20,069	29,359	67,602
34.5	14,533	16,583	20,664	30,229	69,605
35.0	14,957	17,067	21,267	31,111	71,637
35.5	15,388	17,558	21,879	32,006	73,699
36.0	15,824	18,056	22,500	32,914	75,789
36.5	16,267	18,561	23,129	33,835	77,909
37.0	16,716	19,073	23,767	34,768	80,058
37.5	17,170	19,592	24,414	35,714	82,237
38.0	17,631	20.118	25,069	36,673	84,444
38.5	18,098	20,651	25,734	37,644	86,681
39.0	18,571	21,191	26,406	38,629	88,947
39.5	19,051	21,738	27,088	39,625	91,243
40.0	19,536	22,292	27,778	40,635	93,567
40.5	20,027	22,853	28,477	41,657	95,921
41.0	20,525	23,420	29,184	42,692	98,304
41.5	21,029	23,995	29,900	43,740	100,716
42.0	21,538	24,577	30,625	44,800	103,158
42.5	22,054	25,165	31,359	45,873	105,629
43.0	22,576	25,761	32,101	46,959	108,129
43.5	23,104	26,364	32,852	48,057	110,658
44.0	23,639	26,973	33,611	49,168	113,216

Table 58.—*Five-halves powers of numbers*

Number	.00	.01	.02	.03	.04	.05	.06	.07	.08	.09
0.0	0.0000	0.0000	0.0001	0.0002	0.0003	0.0006	0.0009	0.0013	0.0018	0.0024
0.1	.0032	.0040	.0050	.0061	.0073	.0087	.0102	.0119	.0137	.0157
0.2	.0179	.0202	.0227	.0254	.0282	.0313	.0345	.0379	.0415	.0453
0.3	.0493	.0535	.0579	.0626	.0674	.0725	.0778	.0833	.0890	.0950
0.4	.1012	.1076	.1143	.1212	.1284	.1358	.1435	.1514	.1596	.1681
0.5	.1768	.1857	.1950	.2045	.2143	.2243	.2347	.2453	.2562	.2674
0.6	.2789	.2906	.3027	.3150	.3277	.3406	.3539	.3674	.3813	.3955
0.7	.4100	.4248	.4399	.4553	.4711	.4871	.5035	.5203	.5373	.5547
0.8	.5724	.5905	.6089	.6276	.6467	.6661	.6859	.7060	.7265	.7473
0.9	.7684	.7900	.8118	.8341	.8567	.8796	.9030	.9267	.9507	.9752
1.0	1.000	1.025	1.051	1.077	1.103	1.130	1.157	1.184	1.212	1.240
1.1	1.269	1.298	1.328	1.357	1.388	1.418	1.449	1.481	1.513	1.545
1.2	1.577	1.611	1.644	1.678	1.712	1.747	1.782	1.818	1.854	1.890
1.3	1.927	1.964	2.002	2.040	2.079	2.118	2.157	2.197	2.237	2.278
1.4	2.319	2.361	2.403	2.445	2.488	2.532	2.576	2.620	2.665	2.710
1.5	2.756	2.802	2.848	2.896	2.943	2.991	3.040	3.089	3.138	3.188
1.6	3.238	3.289	3.340	3.392	3.444	3.497	3.550	3.604	3.658	3.713
1.7	3.768	3.824	3.880	3.937	3.994	4.051	4.109	4.168	4.227	4.287
1.8	4.347	4.408	4.469	4.530	4.592	4.655	4.718	4.782	4.846	4.911
1.9	4.976	5.042	5.108	5.175	5.242	5.310	5.378	5.447	5.516	5.586
2.0	5.657	5.728	5.799	5.871	5.944	6.017	6.091	6.165	6.240	6.315
2.1	6.391	6.467	6.544	6.621	6.699	6.778	6.857	6.937	7.017	7.098
2.2	7.179	7.261	7.343	7.426	7.510	7.594	7.678	7.764	7.849	7.936
2.3	8.023	8.110	8.198	8.287	8.376	8.466	8.556	8.647	8.739	8.831
2.4	8.923	9.017	9.110	9.205	9.300	9.395	9.492	9.588	9.686	9.784
2.5	9.882	9.981	10.08	10.18	10.28	10.38	10.49	10.59	10.69	10.80
2.6	10.90	11.01	11.11	11.22	11.32	11.43	11.54	11.65	11.76	11.87
2.7	11.98	12.09	12.20	12.31	12.43	12.54	12.66	12.77	12.89	13.00
2.8	13.12	13.24	13.35	13.47	13.59	13.71	13.83	13.95	14.08	14.20
2.9	14.32	14.45	14.57	14.69	14.82	14.95	15.07	15.20	15.33	15.46
3.0	15.59	15.72	15.85	15.98	16.11	16.25	16.38	16.51	16.65	16.78
3.1	16.92	17.06	17.19	17.33	17.47	17.61	17.75	17.89	18.03	18.18
3.2	18.32	18.46	18.61	18.75	18.90	19.04	19.19	19.34	19.48	19.63
3.3	19.78	19.93	20.08	20.24	20.39	20.54	20.69	20.85	21.00	21.16
3.4	21.32	21.47	21.63	21.79	21.95	22.11	22.27	22.43	22.59	22.75
3.5	22.92	23.08	23.25	23.41	23.58	23.74	23.91	24.08	24.25	24.42
3.6	24.59	24.76	24.93	25.11	25.28	25.45	25.63	25.80	25.98	26.16
3.7	26.33	26.51	26.69	26.87	27.05	27.23	27.41	27.60	27.78	27.96
3.8	28.15	28.33	28.52	28.71	28.90	29.08	29.27	29.46	29.65	29.85
3.9	30.04	30.23	30.42	30.62	30.81	31.01	31.21	31.40	31.60	31.80
4.0	32.00	32.20	32.40	32.60	32.81	33.01	33.21	33.42	33.62	33.83
4.1	34.04	34.25	34.45	34.66	34.87	35.08	35.30	35.51	35.72	35.94
4.2	36.15	36.37	36.58	36.80	37.02	37.24	37.46	37.68	37.90	38.12
4.3	38.34	38.56	38.79	39.01	39.24	39.47	39.69	39.92	40.15	40.38
4.4	40.61	40.84	41.07	41.31	41.54	41.77	42.01	42.24	42.48	42.72
4.5	42.96	43.20	43.44	43.68	43.92	44.16	44.40	44.65	44.89	45.14
4.6	45.38	45.63	45.88	46.13	46.38	46.63	46.88	47.13	47.38	47.64
4.7	47.89	48.15	48.40	48.66	48.92	49.17	49.43	49.69	49.95	50.22
4.8	50.48	50.74	51.01	51.27	51.54	51.80	52.07	52.34	52.61	52.88
4.9	53.15	53.42	53.69	53.97	54.24	54.51	54.79	55.07	55.34	55.62

Table 58.—*Five-halves powers of numbers*—Continued

Number	.00	.01	.02	.03	.04	.05	.06	.07	.08	.09
5.0	55.90	56.18	56.46	56.74	57.03	57.31	57.59	57.88	58.16	58.45
5.1	58.74	59.03	59.32	59.61	59.90	60.19	60.48	60.78	61.07	61.36
5.2	61.66	61.96	62.26	62.55	62.85	63.15	63.45	63.76	64.06	64.36
5.3	64.67	64.97	65.28	65.59	65.90	66.20	66.51	66.82	67.14	67.45
5.4	67.76	68.08	68.39	68.71	69.02	69.34	69.66	69.98	70.30	70.62
5.5	70.94	71.27	71.59	71.91	72.24	72.57	72.89	73.22	73.55	73.88
5.6	74.21	74.54	74.88	75.21	75.54	75.88	76.22	76.55	76.89	77.23
5.7	77.57	77.91	78.25	78.59	78.94	79.28	79.63	79.97	80.32	80.67
5.8	81.02	81.37	81.72	82.07	82.42	82.77	83.13	83.48	83.84	84.20
5.9	84.55	84.91	85.27	85.63	85.99	86.36	86.72	87.08	87.45	87.81
6.0	88.18	88.55	88.92	89.29	89.66	90.03	90.40	90.78	91.15	91.53
6.1	91.90	92.28	92.66	93.04	93.42	93.80	94.18	94.56	94.94	95.33
6.2	95.71	96.10	96.49	96.88	97.27	97.66	98.05	98.44	98.83	99.23
6.3	99.62	100.0	100.4	100.8	101.2	101.6	102.0	102.4	102.8	103.2
6.4	103.6	104.0	104.4	104.8	105.2	105.7	106.1	106.5	106.9	107.3
6.5	107.7	108.1	108.5	109.0	109.4	109.8	110.2	110.6	111.1	111.5
6.6	111.9	112.3	112.8	113.2	113.6	114.0	114.5	114.9	115.3	115.8
6.7	116.2	116.6	117.1	117.5	117.9	118.4	118.8	119.3	119.7	120.1
6.8	120.6	121.0	121.5	121.9	122.4	122.8	123.3	123.7	124.2	124.6
6.9	125.1	125.5	126.0	126.4	126.9	127.3	127.8	128.3	128.7	129.2
7.0	129.6	130.1	130.6	131.0	131.5	132.0	132.4	132.9	133.4	133.8
7.1	134.3	134.8	135.3	135.7	136.2	136.7	137.2	137.7	138.1	138.6
7.2	139.1	139.6	140.1	140.6	141.0	141.5	142.0	142.5	143.0	143.5
7.3	144.0	144.5	145.0	145.5	146.0	146.5	147.0	147.5	148.0	148.5
7.4	149.0	149.5	150.0	150.5	151.0	151.5	152.0	152.5	153.0	153.5
7.5	154.0	154.6	155.1	155.6	156.1	156.6	157.1	157.7	158.2	158.7
7.6	159.2	159.8	160.3	160.8	161.3	161.9	162.4	162.9	163.5	164.0
7.7	164.5	165.1	165.6	166.1	166.7	167.2	167.7	168.3	168.8	169.4
7.8	169.9	170.5	171.0	171.6	172.1	172.7	173.2	173.8	174.3	174.9
7.9	175.4	176.0	176.5	177.1	177.6	178.2	178.8	179.3	179.9	180.5
8.0	181.0	181.6	182.2	182.7	183.3	183.9	184.4	185.0	185.6	186.2
8.1	186.7	187.3	187.9	188.5	189.0	189.6	190.2	190.8	191.4	192.0
8.2	192.5	193.1	193.7	194.3	194.9	195.5	196.1	196.7	197.3	197.9
8.3	198.5	199.1	199.7	200.3	200.9	201.5	202.1	202.7	203.3	203.9
8.4	204.5	205.1	205.7	206.3	206.9	207.6	208.2	208.8	209.4	210.0
8.5	210.6	211.3	211.9	212.5	213.1	213.8	214.4	215.0	215.6	216.3
8.6	216.9	217.5	218.2	218.8	219.4	220.1	220.7	221.3	222.0	222.6
8.7	223.3	223.9	224.5	225.2	225.8	226.5	227.1	227.8	228.4	229.1
8.8	229.7	230.4	231.0	231.7	232.3	233.0	233.7	234.3	235.0	235.6
8.9	236.3	237.0	237.6	238.3	239.0	239.6	240.3	241.0	241.7	242.3
9.0	243.0	243.7	244.4	245.0	245.7	246.4	247.1	247.8	248.4	249.1
9.1	249.8	250.5	251.2	251.9	252.6	253.3	253.9	254.6	255.3	256.0
9.2	256.7	257.4	258.1	258.8	259.5	260.2	260.9	261.6	262.3	263.1
9.3	263.8	264.5	265.2	265.9	266.6	267.3	268.0	268.8	269.5	270.2
9.4	270.9	271.6	272.3	273.1	273.8	274.5	275.3	276.0	276.7	277.4
9.5	278.2	278.9	279.6	280.4	281.1	281.8	282.6	283.3	284.1	284.8
9.6	285.5	286.3	287.0	287.8	288.5	289.3	290.0	290.8	291.5	292.3
9.7	293.0	293.8	294.6	295.3	296.1	296.8	297.6	298.4	299.1	299.9
9.8	300.7	301.4	302.2	303.0	303.7	304.5	305.3	306.1	306.8	307.6
9.9	308.4	309.2	309.9	310.7	311.5	312.3	313.1	313.9	314.6	315.4

Table 58.—*Five-halves powers of numbers*—Continued

Number	.00	.01	.02	.03	.04	.05	.06	.07	.08	.09
10. 0	316. 2	317. 0	317. 8	318. 6	319. 4	320. 2	321. 0	321. 8	322. 6	323. 4
10. 1	324. 2	325. 0	325. 8	326. 6	327. 4	328. 2	329. 0	329. 8	330. 7	331. 5
10. 2	332. 3	333. 1	333. 9	334. 7	335. 5	336. 4	337. 2	338. 0	338. 8	339. 7
10. 3	340. 5	341. 3	342. 1	343. 0	343. 8	344. 6	345. 5	346. 3	347. 1	348. 0
10. 4	348. 8	349. 6	350. 5	351. 3	352. 2	353. 0	353. 9	354. 7	355. 6	356. 4
10. 5	357. 3	358. 1	359. 0	359. 8	360. 7	361. 5	362. 4	363. 2	364. 1	365. 0
10. 6	365. 8	366. 7	367. 5	368. 4	369. 3	370. 1	371. 0	371. 9	372. 8	373. 6
10. 7	374. 5	375. 4	376. 3	377. 1	378. 0	378. 9	379. 8	380. 7	381. 5	382. 4
10. 8	383. 3	384. 2	385. 1	386. 0	386. 9	387. 8	388. 7	389. 6	390. 5	391. 4
10. 9	392. 3	393. 2	394. 1	395. 0	395. 9	396. 8	397. 7	398. 6	399. 5	400. 4
11. 0	401. 3	402. 2	403. 1	404. 1	405. 0	405. 9	406. 8	407. 7	408. 6	409. 6
11. 1	410. 5	411. 4	412. 3	413. 3	414. 2	415. 1	416. 1	417. 0	417. 9	418. 9
11. 2	419. 8	420. 7	421. 7	422. 6	423. 6	424. 5	425. 4	426. 4	427. 3	428. 3
11. 3	429. 2	430. 2	431. 1	432. 1	433. 0	434. 0	435. 0	435. 9	436. 9	437. 8
11. 4	438. 8	439. 8	440. 7	441. 7	442. 7	443. 6	444. 6	445. 6	446. 5	447. 5
11. 5	448. 5	449. 5	450. 4	451. 4	452. 4	453. 4	454. 4	455. 3	456. 3	457. 3
11. 6	458. 3	459. 3	460. 3	461. 3	462. 3	463. 2	464. 2	465. 2	466. 2	467. 2
11. 7	468. 2	469. 2	470. 2	471. 2	472. 2	473. 3	474. 3	475. 3	476. 3	477. 3
11. 8	478. 3	479. 3	480. 3	481. 4	482. 4	483. 4	484. 4	485. 4	486. 5	487. 5
11. 9	488. 5	489. 5	490. 6	491. 6	492. 6	493. 7	494. 7	495. 7	496. 8	497. 8
12. 0	498. 8	499. 9	500. 9	502. 0	503. 0	504. 0	505. 1	506. 1	507. 2	508. 2
12. 1	509. 3	510. 3	511. 4	512. 5	513. 5	514. 6	515. 6	516. 7	517. 7	518. 8
12. 2	519. 9	520. 9	522. 0	523. 1	524. 1	525. 2	526. 3	527. 4	528. 4	529. 5
12. 3	530. 6	531. 7	532. 8	533. 8	534. 9	536. 0	537. 1	538. 2	539. 3	540. 4
12. 4	541. 4	542. 5	543. 6	544. 7	545. 8	546. 9	548. 0	549. 1	550. 2	551. 3
12. 5	552. 4	553. 5	554. 6	555. 7	556. 9	558. 0	559. 1	560. 2	561. 3	562. 4
12. 6	563. 5	564. 7	565. 8	566. 9	568. 0	569. 1	570. 3	571. 4	572. 5	573. 7
12. 7	574. 8	575. 9	577. 1	578. 2	579. 3	580. 5	581. 6	582. 7	583. 9	585. 0
12. 8	586. 2	587. 3	588. 5	589. 6	590. 8	591. 9	593. 1	594. 2	595. 4	596. 5
12. 9	597. 7	598. 8	600. 0	601. 2	602. 3	603. 5	604. 7	605. 8	607. 0	608. 2
13. 0	609. 3	610. 5	611. 7	612. 9	614. 0	615. 2	616. 4	617. 6	618. 8	619. 9
13. 1	621. 1	622. 3	623. 5	624. 7	625. 9	627. 1	628. 3	629. 5	630. 7	631. 8
13. 2	633. 0	634. 2	635. 4	636. 6	637. 9	639. 1	640. 3	641. 5	642. 7	643. 9
13. 3	645. 1	646. 3	647. 5	648. 7	650. 0	651. 2	652. 4	653. 6	654. 8	656. 1
13. 4	657. 3	658. 5	659. 8	661. 0	662. 2	663. 4	664. 7	665. 9	667. 2	668. 4
13. 5	669. 6	670. 9	672. 1	673. 4	674. 6	675. 8	677. 1	678. 3	679. 6	680. 8
13. 6	682. 1	683. 4	684. 6	685. 9	687. 1	688. 4	689. 6	690. 9	692. 2	693. 4
13. 7	694. 7	696. 0	697. 2	698. 5	699. 8	701. 1	702. 3	703. 6	704. 9	706. 2
13. 8	707. 5	708. 7	710. 0	711. 3	712. 6	713. 9	715. 2	716. 5	717. 8	719. 0
13. 9	720. 3	721. 6	722. 9	724. 2	725. 5	726. 8	728. 1	729. 4	730. 7	732. 1
14. 0	733. 4	734. 7	736. 0	737. 3	738. 6	739. 9	741. 2	742. 6	743. 9	745. 2
14. 1	746. 5	747. 9	749. 2	750. 5	751. 8	753. 2	754. 5	755. 8	757. 2	758. 5
14. 2	759. 8	761. 2	762. 5	763. 9	765. 2	766. 5	767. 9	769. 2	770. 6	771. 9
14. 3	773. 3	774. 6	776. 0	777. 3	778. 7	780. 1	781. 4	782. 8	784. 1	785. 5
14. 4	786. 9	788. 2	789. 6	791. 0	792. 4	793. 7	795. 1	796. 5	797. 9	799. 2
14. 5	800. 6	802. 0	803. 4	804. 8	806. 1	807. 5	808. 9	810. 3	811. 7	813. 1
14. 6	814. 5	815. 9	817. 3	818. 7	820. 1	821. 5	822. 9	824. 3	825. 7	827. 1
14. 7	828. 5	829. 9	831. 3	832. 7	834. 1	835. 6	837. 0	838. 4	839. 8	841. 2
14. 8	842. 7	844. 1	845. 5	846. 9	848. 4	849. 8	851. 2	852. 7	854. 1	855. 5
14. 9	857. 0	858. 4	859. 8	861. 3	862. 7	864. 2	865. 6	867. 1	868. 5	870. 0

Table 58.—*Five-halves powers of numbers*—Continued

Number	.00	.01	.02	.03	.04	.05	.06	.07	.08	.09
15.0	871.4	872.9	874.3	875.8	877.2	878.7	880.2	881.6	883.1	884.6
15.1	886.0	887.5	889.0	890.4	891.9	893.4	894.8	896.3	897.8	899.3
15.2	900.8	902.2	903.7	905.2	906.7	908.2	909.7	911.2	912.7	914.2
15.3	915.6	917.1	918.6	920.1	921.6	923.1	924.7	926.2	927.7	929.2
15.4	930.7	932.2	933.7	935.2	936.7	938.3	939.8	941.3	942.8	944.3
15.5	945.9	947.4	948.9	950.4	952.0	953.5	955.0	956.6	958.1	959.7
15.6	961.2	962.7	964.3	965.8	967.4	968.9	970.5	972.0	973.6	975.1
15.7	976.7	978.2	979.8	981.3	982.9	984.5	986.0	987.6	989.2	990.7
15.8	992.3	993.9	995.4	997.0	998.6	1,000	1,002	1,003	1,005	1,006
15.9	1,008	1,010	1,011	1,013	1,014	1,016	1,018	1,019	1,021	1,022
16.0	1,024	1,026	1,027	1,029	1,030	1,032	1,034	1,035	1,037	1,038
16.1	1,040	1,042	1,043	1,045	1,047	1,048	1,050	1,051	1,053	1,055
16.2	1,056	1,058	1,060	1,061	1,063	1,064	1,066	1,068	1,069	1,071
16.3	1,073	1,074	1,076	1,078	1,079	1,081	1,083	1,084	1,086	1,088
16.4	1,089	1,091	1,093	1,094	1,096	1,098	1,099	1,101	1,103	1,104
16.5	1,106	1,108	1,109	1,111	1,113	1,114	1,116	1,118	1,119	1,121
16.6	1,123	1,124	1,126	1,128	1,129	1,131	1,133	1,135	1,136	1,138
16.7	1,140	1,141	1,143	1,145	1,147	1,148	1,150	1,152	1,153	1,155
16.8	1,157	1,159	1,160	1,162	1,164	1,165	1,167	1,169	1,171	1,172
16.9	1,174	1,176	1,178	1,179	1,181	1,183	1,185	1,186	1,188	1,190
17.0	1,192	1,193	1,195	1,197	1,199	1,200	1,202	1,204	1,206	1,207
17.1	1,209	1,211	1,213	1,214	1,216	1,218	1,220	1,222	1,223	1,225
17.2	1,227	1,229	1,231	1,232	1,234	1,236	1,238	1,239	1,241	1,243
17.3	1,245	1,247	1,248	1,250	1,252	1,254	1,256	1,257	1,259	1,261
17.4	1,263	1,265	1,267	1,268	1,270	1,272	1,274	1,276	1,277	1,279
17.5	1,281	1,283	1,285	1,287	1,288	1,290	1,292	1,294	1,296	1,298
17.6	1,300	1,301	1,303	1,305	1,307	1,309	1,311	1,312	1,314	1,316
17.7	1,318	1,320	1,322	1,324	1,326	1,327	1,329	1,331	1,333	1,335
17.8	1,337	1,339	1,341	1,342	1,344	1,346	1,348	1,350	1,352	1,354
17.9	1,356	1,357	1,359	1,361	1,363	1,365	1,367	1,369	1,371	1,373
18.0	1,375	1,377	1,378	1,380	1,382	1,384	1,386	1,388	1,390	1,392
18.1	1,394	1,396	1,398	1,400	1,402	1,403	1,405	1,407	1,409	1,411
18.2	1,413	1,415	1,417	1,419	1,421	1,423	1,425	1,427	1,429	1,431
18.3	1,433	1,435	1,437	1,438	1,440	1,442	1,444	1,446	1,448	1,450
18.4	1,452	1,454	1,456	1,458	1,460	1,462	1,464	1,466	1,468	1.470
18.5	1,472	1,474	1,476	1,478	1,480	1,482	1,484	1,486	1,488	1,490
18.6	1,492	1,494	1,496	1,498	1,500	1,502	1,504	1,506	1,508	1,510
18.7	1,512	1,514	1,516	1,518	1,520	1,522	1,524	1,526	1,528	1,530
18.8	1,532	1,535	1,537	1,539	1,541	1,543	1,545	1,547	1,549	1,551
18.9	1,553	1,555	1,557	1,559	1,561	1,563	1,565	1,567	1,569	1,571
19.0	1,574	1,576	1,578	1,580	1,582	1,584	1,586	1,588	1,590	1,592
19.1	1,594	1,596	1,599	1,601	1,603	1,605	1,607	1,609	1,611	1,613
19.2	1,615	1,617	1,620	1,622	1,624	1,626	1,628	1,630	1,632	1,634
19.3	1,636	1,639	1,641	1,643	1,645	1,647	1,649	1,651	1,653	1,656
19.4	1,658	1,660	1,662	1,664	1,666	1,668	1,671	1,673	1,675	1,677
19.5	1,679	1,681	1,683	1,686	1,688	1,690	1,692	1,694	1,696	1,699
19.6	1,701	1,703	1,705	1,707	1,709	1,712	1,714	1,716	1,718	1,720
19.7	1,723	1,725	1,727	1,729	1,731	1,733	1,736	1,738	1,740	1,742
19.8	1,744	1,747	1,749	1,751	1,753	1,756	1,758	1,760	1,762	1,764
19.9	1,767	1,769	1,771	1,773	1,775	1,778	1,780	1,782	1,784	1,787

Table 58.—*Five-halves powers of numbers*—Continued

Number	.00	.01	.02	.03	.04	.05	.06	.07	.08	.09
20.0	1,789	1,791	1,793	1,796	1,798	1,800	1,802	1,805	1,807	1,809
20.1	1,811	1,814	1,816	1,818	1,820	1,823	1,825	1,827	1,829	1,832
20.2	1,834	1,836	1,838	1,841	1,843	1,845	1,848	1,850	1,852	1,854
20.3	1,857	1,859	1,861	1,864	1,866	1,868	1,870	1,873	1,875	1,877
20.4	1,880	1,882	1,884	1,887	1,889	1,891	1,893	1,896	1,898	1,900
20.5	1,903	1,905	1,907	1,910	1,912	1,914	1,917	1,919	1,921	1,924
20.6	1,926	1,928	1,931	1,933	1,935	1,938	1,940	1,942	1,945	1,947
20.7	1,950	1,952	1,954	1,957	1,959	1,961	1,964	1,966	1,968	1,971
20.8	1,973	1,976	1,978	1,980	1,983	1,985	1,987	1,990	1,992	1,995
20.9	1,997	1,999	2,002	2,004	2,007	2,009	2,011	2,014	2,016	2,019
21.0	2,021	2,023	2,026	2,028	2,031	2,033	2,035	2,038	2,040	2,043
21.1	2,045	2,047	2,050	2,052	2,055	2,057	2,060	2,062	2,065	2,067
21.2	2,069	2,072	2,074	2,077	2,079	2,082	2,084	2,087	2,089	2,091
21.3	2,094	2,096	2,099	2,101	2,104	2,106	2,109	2,111	2,114	2,116
21.4	2,119	2,121	2,123	2,126	2,128	2,131	2,133	2,136	2,138	2,141
21.5	2,143	2,146	2,148	2,151	2,153	2,156	2,158	2,161	2,163	2,166
21.6	2,168	2,171	2,173	2,176	2,178	2,181	2,183	2,186	2,189	2,191
21.7	2,194	2,196	2,199	2,201	2,204	2,206	2,209	2,211	2,214	2,216
21.8	2,219	2,221	2,224	2,227	2,229	2,232	2,234	2,237	2,239	2,242
21.9	2,244	2,247	2,250	2,252	2,255	2,257	2,260	2,262	2,265	2,268
22.0	2,270	2,273	2,275	2,278	2,280	2,283	2,286	2,288	2,291	2,293
22.1	2,296	2,299	2,301	2,304	2,306	2,309	2,312	2,314	2,317	2,319
22.2	2,322	2,325	2,327	2,330	2,333	2,335	2,338	2,340	2,343	2,346
22.3	2,348	2,351	2,354	2,356	2,359	2,362	2,364	2,367	2,369	2,372
22.4	2,375	2,377	2,380	2,383	2,385	2,388	2,391	2,393	2,396	2,399
22.5	2,401	2,404	2,407	2,409	2,412	2,415	2,417	2,420	2,423	2,425
22.6	2,428	2,431	2,434	2,436	2,439	2,442	2,444	2,447	2,450	2,452
22.7	2,455	2,458	2,460	2,463	2,466	2,469	2,471	2,474	2,477	2,479
22.8	2,482	2,485	2,488	2,490	2,493	2,496	2,499	2,501	2,504	2,507
22.9	2,510	2,512	2,515	2,518	2,520	2,523	2,526	2,529	2,531	2,534
23.0	2,537	2,540	2,543	2,545	2,548	2,551	2,554	2,556	2,559	2,562
23.1	2,565	2,567	2,570	2,573	2,576	2,579	2,581	2,584	2,587	2,590
23.2	2,593	2,595	2,598	2,601	2,604	2,606	2,609	2,612	2,615	2,618
23.3	2,621	2,623	2,626	2,629	2,632	2,635	2,637	2,640	2,643	2,646
23.4	2,649	2,652	2,654	2,657	2,660	2,663	2,666	2,669	2,671	2,674
23.5	2,677	2,680	2,683	2,686	2,689	2,691	2,694	2,697	2,700	2,703
23.6	2,706	2,709	2,711	2,714	2,717	2,720	2,723	2,726	2,729	2,732
23.7	2,734	2,737	2,740	2,743	2,746	2,749	2,752	2,755	2,758	2,760
23.8	2,763	2,766	2,769	2,772	2,775	2,778	2,781	2,784	2,787	2,790
23.9	2,793	2,795	2,798	2,801	2,804	2,807	2,810	2,813	2,816	2,819
24.0	2,822	2,825	2,828	2,831	2,834	2,837	2,839	2,842	2,845	2,848
24.1	2,851	2,854	2,857	2,860	2,863	2,866	2,869	2,872	2,875	2,878
24.2	2,881	2,884	2,887	2,890	2,893	2,896	2,899	2,902	2,905	2,908
24.3	2,911	2,914	2,917	2,920	2,923	2,926	2,929	2,932	2,935	2,938
24.4	2,941	2,944	2,947	2,950	2,953	2,956	2,959	2,962	2,965	2,968
24.5	2,971	2,974	2,977	2,980	2,983	2,986	2,989	2,992	2,995	2,998
24.6	3,001	3,005	3,008	3,011	3,014	3,017	3,020	3,023	3,026	3,029
24.7	3,032	3,035	3,038	3,041	3,044	3,047	3,051	3,054	3,057	3,060
24.8	3,063	3,066	3,069	3,072	3,075	3,078	3,081	3,085	3,088	3,091
24.9	3,094	3,097	3,100	3,103	3,106	3,109	3,113	3,116	3,119	3,122

Table 58.—*Five-halves powers of numbers*—Continued

Number	.00	.01	.02	.03	.04	.05	.06	.07	.08	.09
25.0	3,125	3,128	3,131	3,134	3,138	3,141	3,144	3,147	3,150	3,153
25.1	3,156	3,159	3,163	3,166	3,169	3,172	3,175	3,178	3,182	3,185
25.2	3,188	3,191	3,194	3,197	3,201	3,204	3,207	3,210	3,213	3,216
25.3	3,220	3,223	3,226	3,229	3,232	3,236	3,239	3,242	3,245	3,248
25.4	3,252	3,255	3,258	3,261	3,264	3,268	3,271	3,274	3,277	3,280
25.5	3,284	3,287	3,290	3,293	3,296	3,300	3,303	3,306	3,309	3,313
25.6	3,316	3,319	3,322	3,326	3,329	3,332	3,335	3,339	3,342	3,345
25.7	3,348	3,352	3,355	3,358	3,361	3,365	3,368	3,371	3,374	3,378
25.8	3,381	3,384	3,388	3,391	3,394	3,397	3,401	3,404	3,407	3,411
25.9	3,414	3,417	3,420	3,424	3,427	3,430	3,434	3,437	3,440	3,444
26.0	3,447	3,450	3,454	3,457	3,460	3,464	3,467	3,470	3,474	3,477
26.1	3,480	3,484	3,487	3,490	3,494	3,497	3,500	3,504	3,507	3,510
26.2	3,514	3,517	3,520	3,524	3,527	3,530	3,534	3,537	3,540	3,544
26.3	3,547	3,551	3,554	3,557	3,561	3,564	3,567	3,571	3,574	3,578
26.4	3,581	3,584	3,588	3,591	3,595	3,598	3,601	3,605	3,608	3,612
26.5	3,615	3,618	3,622	3,625	3,629	3,632	3,636	3,639	3,642	3,646
26.6	3,649	3,653	3,656	3,660	3,663	3,666	3,670	3,673	3,677	3,680
26.7	3,684	3,687	3,691	3,694	3,697	3,701	3,704	3,708	3,711	3,715
26.8	3,718	3,722	3,725	3,729	3,732	3,736	3,739	3,743	3,746	3,750
26.9	3,753	3,757	3,760	3,763	3,767	3,770	3,774	3,777	3,781	3,784
27.0	3,788	3,792	3,795	3,799	3,802	3,806	3,809	3,813	3,816	3,820
27.1	3,823	3,827	3,830	3,834	3,837	3,841	3,844	3,848	3,851	3,855
27.2	3,859	3,862	3,866	3,869	3,873	3,876	3,880	3,883	3,887	3,891
27.3	3,894	3,898	3,901	3,905	3,908	3,912	3,916	3,919	3,923	3,926
27.4	3,930	3,933	3,937	3,941	3,944	3,948	3,951	3,955	3,959	3,962
27.5	3,966	3,969	3,973	3,977	3,980	3,984	3,987	3,991	3,995	3,998
27.6	4,002	4,006	4,009	4,013	4,016	4,020	4,024	4,027	4,031	4,035
27.7	4,038	4,042	4,046	4,049	4,053	4,057	4,060	4,064	4,068	4,071
27.8	4,075	4,079	4,082	4,086	4,090	4,093	4,097	4,101	4,104	4,108
27.9	4,112	4,115	4,119	4,123	4,126	4,130	4,134	4,137	4,141	4,145
28.0	4,149	4,152	4,156	4,160	4,163	4,167	4,171	4,175	4,178	4,182
28.1	4,186	4,189	4,193	4,197	4,201	4,204	4,208	4,212	4,216	4,219
28.2	4,223	4,227	4,231	4,234	4,238	4,242	4,246	4,249	4,253	4,257
28.3	4,261	4,264	4,268	4,272	4,276	4,279	4,283	4,287	4,291	4,295
28.4	4,298	4,302	4,306	4,310	4,313	4,317	4,321	4,325	4,329	4,332
28.5	4,336	4,340	4,344	4,348	4,351	4,355	4,359	4,363	4,367	4,371
28.6	4,374	4,378	4,382	4,386	4,390	4,394	4,397	4,401	4,405	4,409
28.7	4,413	4,417	4,420	4,424	4,428	4,432	4,436	4,440	4,444	4,447
28.8	4,451	4,455	4,459	4,463	4,467	4,471	4,474	4,478	4,482	4,486
28.9	4,490	4,494	4,498	4,502	4,506	4,509	4,513	4,517	4,521	4,525
29.0	4,529	4,533	4,537	4,541	4,545	4,548	4,552	4,556	4,560	4,564
29.1	4,568	4,572	4,576	4,580	4,584	4,588	4,592	4,596	4,600	4,603
29.2	4,607	4,611	4,615	4,619	4,623	4,627	4,631	4,635	4,639	4,643
29.3	4,647	4,651	4,655	4,659	4,663	4,667	4,671	4,675	4,679	4,683
29.4	4,687	4,691	4,695	4,699	4,703	4,707	4,711	4,715	4,719	4,723
29.5	4,727	4,731	4,735	4,739	4,743	4,747	4,751	4,755	4,759	4,763
29.6	4,767	4,771	4,775	4,779	4,783	4,787	4,791	4,795	4,799	4,803
29.7	4,807	4,811	4,815	4,819	4,823	4,827	4,832	4,836	4,840	4,844
29.8	4,848	4,852	4,856	4,860	4,864	4,868	4,872	4,876	4,880	4,884
29.9	4,889	4,893	4,897	4,901	4,905	4,909	4,913	4,917	4,921	4,925

Table 58.—*Five-halves powers of numbers*—Continued

Number	.0	.1	.2	.3	.4	.5	.6	.7	.8	.9
30	4,930	4,971	5,012	5,054	5,095	5,137	5,180	5,222	5,265	5,308
31	5,351	5,394	5,437	5,481	5,525	5,569	5,613	5,658	5,703	5,747
32	5,793	5,838	5,884	5,929	5,975	6,022	6,068	6,115	6,161	6,209
33	6,256	6,303	6,351	6,399	6,447	6,495	6,544	6,593	6,642	6,691
34	6,741	6,790	6,840	6,890	6,941	6,991	7,042	7,093	7,144	7,196
35	7,247	7,299	7,351	7,403	7,456	7,509	7,562	7,615	7,668	7,722
36	7,776	7,830	7,884	7,939	7,994	8,049	8,104	8,160	8,215	8,271
37	8,327	8,384	8,440	8,497	8,554	8,611	8,669	8,727	8,785	8,843
38	8,901	8,960	9,019	9,078	9,138	9,197	9,257	9,317	9,377	9,438
39	9,499	9,560	9,621	9,682	9,744	9,806	9,868	9,931	9,993	10,056
40	10,119	10,183	10,246	10,310	10,374	10,438	10,503	10,568	10,633	10,698
41	10,764	10,829	10,895	10,962	11,028	11,095	11,162	11,229	11,296	11,364
42	11,432	11,500	11,569	11,637	11,706	11,775	11,845	11,914	11,984	12,054
43	12,125	12,195	12,266	12,337	12,409	12,480	12,552	12,624	12,697	12,769
44	12,842	12,915	12,988	13,062	13,136	13,210	13,284	13,359	13,434	13,509
45	13,584	13,660	13,736	13,812	13,888	13,965	14,041	14,119	14,196	14,274
46	14,351	14,430	14,508	14,587	14,665	14,745	14,824	14,904	14,984	15,064
47	15,144	15,225	15,306	15,387	15,468	15,550	15,632	15,714	15,797	15,880
48	15,963	16,046	16,129	16,213	16,297	16,382	16,466	16,551	16,636	16,721
49	16,807	16,893	16,979	17,065	17,152	17,239	17,326	17,414	17,501	17,589
50	17,678	17,766	17,855	17,944	18,033	18,123	18,213	18,303	18,393	18,484
51	18,575	18,666	18,757	18,849	18,941	19,033	19,126	19,219	19,312	19,405
52	19,499	19,593	19,687	19,781	19,876	19,971	20,066	20,162	20,257	20,354
53	20,450	20,546	20,643	20,740	20,838	20,936	21,034	21,132	21,230	21,329
54	21,428	21,527	21,627	21,727	21,827	21,928	22,028	22,129	22,231	22,332
55	22,434	22,536	22,639	22,741	22,844	22,947	23,051	23,155	23,259	23,363
56	23,468	23,573	23,678	23,783	23,889	23,995	24,101	24,208	24,315	24,422
57	24,529	24,637	24,745	24,853	24,962	25,071	25,180	25,289	25,399	25,509
58	25,619	25,730	25,841	25,952	26,063	26,175	26,287	26,399	26,512	26,625
59	26,738	26,852	26,965	27,079	27,194	27,308	27,423	27,538	27,654	27,769
60	27,885	28,002	28,118	28,235	28,353	28,470	28,588	28,706	28,824	28,943
61	29,062	29,181	29,301	29,421	29,541	29,661	29,782	29,903	30,024	30,146
62	30,268	30,390	30,512	30,635	30,758	30,882	31,005	31,129	31,254	31,378
63	31,503	31,628	31,754	31,879	32,005	32,132	32,258	32,385	32,513	32,640
64	32,768	32,896	33,025	33,153	33,282	33,412	33,541	33,671	33,802	33,932
65	34,063	34,194	34,326	34,457	34,590	34,722	34,855	34,988	35,121	35,254
66	35,388	35,523	35,657	35,792	35,927	36,062	36,198	36,334	36,470	36,607
67	36,744	36,881	37,019	37,157	37,295	37,433	37,572	37,711	37,851	37,990
68	38,130	38,271	38,411	38,552	38,694	38,835	38,977	39,119	39,262	39,405
69	39,548	39,691	39,835	39,979	40,123	40,268	40,413	40,559	40,704	40,850
70	40,996	41,143	41,290	41,437	41,585	41,732	41,880	42,029	42,178	42,327
71	42,476	42,626	42,776	42,926	43,077	43,228	43,379	43,531	43,683	43,835
72	43,988	44,141	44,294	44,447	44,601	44,755	44,910	45,065	45,220	45,375
73	45,531	45,687	45,843	46,000	46,157	46,315	46,472	46,630	46,789	46,947
74	47,106	47,266	47,425	47,585	47,745	47,906	48,067	48,228	48,390	48,552
75	48,714	48,876	49,039	49,203	49,366	49,530	49,694	49,859	50,023	50,189
76	50,354	50,520	50,686	50,852	51,019	51,186	51,354	51,521	51,690	51,858
77	52,027	52,196	52,365	52,535	52,705	52,875	53,046	53,217	53,389	53,560
78	53,732	53,905	54,078	54,251	54,424	54,598	54,772	54,946	55,121	55,296
79	55,471	55,647	55,823	55,999	56,176	56,353	56,530	56,708	56,886	57,065

Table 58.—*Five-halves powers of numbers*—Continued

Number	.0	.1	.2	.3	.4	.5	.6	.7	.8	.9
80	57,243	57,422	57,602	57,782	57,962	58,142	58,323	58,504	58,685	58,867
81	59,049	59,231	59,414	59,597	59,781	59,964	60,149	60,333	60,518	60,703
82	60,888	61,074	61,260	61,447	61,634	61,821	62,008	62,196	62,384	62,573
83	62,762	62,951	63,141	63,330	63,521	63,711	63,902	64,093	64,285	64,477
84	64,669	64,862	65,055	65,248	65,442	65,636	65,830	66,025	66,220	66,415
85	66,611	66,807	67,004	67,201	67,398	67,595	67,793	67,991	68,190	68,388
86	68,588	68,787	68,987	69,187	69,388	69,589	69,790	69,992	70,194	70,396
87	70,599	70,802	71,005	71,209	71,413	71,618	71,822	72,028	72,233	72,439
88	72,645	72,852	73,059	73,266	73,473	73,681	73,890	74,098	74,307	74,517
89	74,727	74,937	75,147	75,358	75,569	75,781	75,992	76,205	76,417	76,630
90	76,843	77,057	77,271	77,485	77,700	77,915	78,130	78,346	78,562	78,779
91	78,996	79,213	79,430	79,648	79,867	80,085	80,304	80,524	80,743	80,963
92	81,184	81,405	81,626	81,847	82,069	82,291	82,514	82,737	82,960	83,184
93	83,408	83,632	83,857	84,082	84,308	84,534	84,760	84,986	85,213	85,441
94	85,668	85,896	86,125	86,353	86,582	86,812	87,042	87,272	87,503	87,734
95	87,965	88,196	88,429	88,661	88,894	89,127	89,360	89,594	89,828	90,063
96	90,298	90,533	90,769	91,005	91,242	91,478	91,716	91,953	92,191	92,429
97	92,668	92,907	93,146	93,386	93,626	93,867	94,108	94,349	94,590	94,832
98	95,075	95,317	95,561	95,804	96,048	96,292	96,537	96,782	97,027	97,273
99	97,519	97,765	98,012	98,259	98,507	98,755	99,003	99,252	99,501	99,750

Table 59.—Three-halves powers of numbers

Number	.00	.01	.02	.03	.04	.05	.06	.07	.08	.09
0.0	0.0000	0.0010	0.0028	0.0052	0.0080	0.0112	0.0147	0.0185	0.0226	0.0270
0.1	.0316	.0365	.0416	.0469	.0524	.0581	.0640	.0701	.0764	.0828
0.2	.0894	.0962	.1032	.1103	.1176	.1250	.1326	.1403	.1482	.1562
0.3	.1643	.1726	.1810	.1896	.1983	.2071	.2160	.2251	.2342	.2436
0.4	.2530	.2625	.2722	.2820	.2919	.3019	.3120	.3222	.3326	.3430
0.5	.3536	.3642	.3750	.3858	.3968	.4079	.4191	.4303	.4417	.4532
0.6	.4648	.4764	.4882	.5000	.5120	.5240	.5362	.5484	.5607	.5732
0.7	.5857	.5983	.6109	.6237	.6366	.6495	.6626	.6757	.6889	.7022
0.8	.7155	.7290	.7425	.7562	.7699	.7837	.7975	.8115	.8255	.8396
0.9	.8538	.8681	.8824	.8969	.9114	.9259	.9406	.9553	.9702	.9850
1.0	1.000	1.015	1.030	1.045	1.061	1.076	1.091	1.107	1.122	1.138
1.1	1.154	1.169	1.185	1.201	1.217	1.233	1.249	1.266	1.282	1.298
1.2	1.315	1.331	1.348	1.364	1.381	1.398	1.414	1.431	1.448	1.465
1.3	1.482	1.499	1.517	1.534	1.551	1.569	1.586	1.604	1.621	1.639
1.4	1.657	1.674	1.692	1.710	1.728	1.746	1.764	1.782	1.800	1.819
1.5	1.837	1.856	1.874	1.893	1.911	1.930	1.948	1.967	1.986	2.005
1.6	2.024	2.043	2.062	2.081	2.100	2.119	2.139	2.158	2.178	2.197
1.7	2.217	2.236	2.256	2.275	2.295	2.315	2.335	2.355	2.375	2.395
1.8	2.415	2.435	2.455	2.476	2.496	2.516	2.537	2.557	2.578	2.598
1.9	2.619	2.640	2.660	2.681	2.702	2.723	2.744	2.765	2.786	2.807
2.0	2.828	2.850	2.871	2.892	2.914	2.935	2.957	2.978	3.000	3.021
2.1	3.043	3.065	3.087	3.109	3.131	3.153	3.175	3.197	3.219	3.241
2.2	3.263	3.285	3.308	3.330	3.353	3.375	3.398	3.420	3.443	3.465
2.3	3.488	3.511	3.534	3.557	3.580	3.602	3.626	3.649	3.672	3.695
2.4	3.718	3.741	3.765	3.788	3.811	3.835	3.858	3.882	3.906	3.929
2.5	3.953	3.977	4.000	4.024	4.048	4.072	4.096	4.120	4.144	4.168
2.6	4.192	4.217	4.241	4.265	4.289	4.314	4.338	4.363	4.387	4.412
2.7	4.437	4.461	4.486	4.511	4.536	4.560	4.585	4.610	4.635	4.660
2.8	4.685	4.710	4.736	4.761	4.786	4.811	4.837	4.862	4.888	4.913
2.9	4.939	4.964	4.990	5.015	5.041	5.067	5.093	5.118	5.144	5.170
3.0	5.196	5.222	5.248	5.274	5.300	5.327	5.353	5.379	5.405	5.432
3.1	5.458	5.485	5.511	5.538	5.564	5.591	5.617	5.644	5.671	5.698
3.2	5.724	5.751	5.778	5.805	5.832	5.859	5.886	5.913	5.940	5.968
3.3	5.995	6.022	6.049	6.077	6.104	6.132	6.159	6.186	6.214	6.242
3.4	6.269	6.297	6.325	6.352	6.380	6.408	6.436	6.464	6.492	6.520
3.5	6.548	6.576	6.604	6.632	6.660	6.689	6.717	6.745	6.774	6.802
3.6	6.831	6.859	6.888	6.916	6.945	6.973	7.002	7.031	7.059	7.088
3.7	7.117	7.146	7.175	7.204	7.233	7.262	7.291	7.320	7.349	7.378
3.8	7.408	7.437	7.466	7.495	7.525	7.554	7.584	7.613	7.643	7.672
3.9	7.702	7.732	7.761	7.791	7.821	7.850	7.880	7.910	7.940	7.970
4.0	8.000	8.030	8.060	8.090	8.120	8.150	8.181	8.211	8.241	8.272
4.1	8.302	8.332	8.363	8.393	8.424	8.454	8.485	8.515	8.546	8.577
4.2	8.607	8.638	8.669	8.700	8.731	8.762	8.793	8.824	8.855	8.886
4.3	8.917	8.948	8.979	9.010	9.041	9.073	9.104	9.135	9.167	9.198
4.4	9.230	9.261	9.293	9.324	9.356	9.387	9.419	9.451	9.482	9.514
4.5	9.546	9.578	9.610	9.642	9.674	9.705	9.737	9.770	9.802	9.834
4.6	9.866	9.898	9.930	9.963	9.995	10.03	10.06	10.09	10.12	10.16
4.7	10.19	10.22	10.25	10.29	10.32	10.35	10.39	10.42	10.45	10.48
4.8	10.52	10.55	10.58	10.62	10.65	10.68	10.71	10.75	10.78	10.81
4.9	10.85	10.88	10.91	10.95	10.98	11.01	11.05	11.08	11.11	11.15

Table 59.—*Three-halves powers of numbers*—Continued

Number	.00	.01	.02	.03	.04	.05	.06	.07	.08	.09
5.0	11.18	11.21	11.25	11.28	11.31	11.35	11.38	11.42	11.45	11.48
5.1	11.52	11.55	11.59	11.62	11.65	11.69	11.72	11.76	11.79	11.82
5.2	11.86	11.89	11.93	11.96	11.99	12.03	12.06	12.10	12.13	12.17
5.3	12.20	12.24	12.27	12.31	12.34	12.37	12.41	12.44	12.48	12.51
5.4	12.55	12.58	12.62	12.65	12.69	12.72	12.76	12.79	12.83	12.86
5.5	12.90	12.93	12.97	13.00	13.04	13.07	13.11	13.15	13.18	13.22
5.6	13.25	13.29	13.32	13.36	13.39	13.43	13.47	13.50	13.54	13.57
5.7	13.61	13.64	13.68	13.72	13.75	13.79	13.82	13.86	13.90	13.93
5.8	13.97	14.00	14.04	14.08	14.11	14.15	14.19	14.22	14.26	14.29
5.9	14.33	14.37	14.40	14.44	14.48	14.51	14.55	14.59	14.62	14.66
6.0	14.70	14.73	14.77	14.81	14.84	14.88	14.92	14.95	14.99	15.03
6.1	15.07	15.10	15.14	15.18	15.21	15.25	15.29	15.33	15.36	15.40
6.2	15.44	15.48	15.51	15.55	15.59	15.62	15.66	15.70	15.74	15.78
6.3	15.81	15.85	15.89	15.93	15.96	16.00	16.04	16.08	16.12	16.15
6.4	16.19	16.23	16.27	16.30	16.34	16.38	16.42	16.46	16.50	16.53
6.5	16.57	16.61	16.65	16.69	16.73	16.76	16.80	16.84	16.88	16.92
6.6	16.96	16.99	17.03	17.07	17.11	17.15	17.19	17.23	17.26	17.30
6.7	17.34	17.38	17.42	17.46	17.50	17.54	17.58	17.62	17.65	17.69
6.8	17.73	17.77	17.81	17.85	17.89	17.93	17.97	18.01	18.05	18.09
6.9	18.12	18.16	18.20	18.24	18.28	18.32	18.36	18.40	18.44	18.48
7.0	18.52	18.56	18.60	18.64	18.68	18.72	18.76	18.80	18.84	18.88
7.1	18.92	18.96	19.00	19.04	19.08	19.12	19.16	19.20	19.24	19.28
7.2	19.32	19.36	19.40	19.44	19.48	19.52	19.56	19.60	19.64	19.68
7.3	19.72	19.76	19.80	19.85	19.89	19.93	19.97	20.01	20.05	20.09
7.4	20.13	20.17	20.21	20.25	20.29	20.33	20.38	20.42	20.46	20.50
7.5	20.54	20.58	20.62	20.66	20.70	20.75	20.79	20.83	20.87	20.91
7.6	20.95	20.99	21.03	21.08	21.12	21.16	21.20	21.24	21.28	21.33
7.7	21.37	21.41	21.45	21.49	21.53	21.58	21.62	21.66	21.70	21.74
7.8	21.78	21.83	21.87	21.91	21.95	21.99	22.04	22.08	22.12	22.16
7.9	22.20	22.25	22.29	22.33	22.37	22.42	22.46	22.50	22.54	22.59
8.0	22.63	22.67	22.71	22.75	22.80	22.84	22.88	22.93	22.97	23.01
8.1	23.05	23.10	23.14	23.18	23.22	23.27	23.31	23.35	23.40	23.44
8.2	23.48	23.52	23.57	23.61	23.65	23.70	23.74	23.78	23.83	23.87
8.3	23.91	23.96	24.00	24.04	24.09	24.13	24.17	24.22	24.26	24.30
8.4	24.35	24.39	24.43	24.48	24.52	24.56	24.61	24.65	24.69	24.74
8.5	24.78	24.83	24.87	24.91	24.96	25.00	25.04	25.09	25.13	25.18
8.6	25.22	25.26	25.31	25.35	25.40	25.44	25.48	25.53	25.57	25.62
8.7	25.66	25.71	25.75	25.79	25.84	25.88	25.93	25.97	26.02	26.06
8.8	26.11	26.15	26.19	26.24	26.28	26.33	26.37	26.42	26.46	26.51
8.9	26.55	26.60	26.64	26.69	26.73	26.78	26.82	26.87	26.91	26.96
9.0	27.00	27.05	27.09	27.14	27.18	27.23	27.27	27.32	27.36	27.41
9.1	27.45	27.50	27.54	27.59	27.63	27.68	27.72	27.77	27.81	27.86
9.2	27.90	27.95	27.95	28.04	28.09	28.13	28.18	28.22	28.27	28.32
9.3	28.36	28.41	28.45	28.50	28.54	28.59	28.64	28.68	28.73	28.77
9.4	28.82	28.87	28.91	28.96	29.00	29.05	29.10	29.14	29.19	29.23
9.5	29.28	29.33	29.37	29.42	29.47	29.51	29.56	29.61	29.65	29.70
9.6	29.74	29.79	29.84	29.88	29.93	29.98	30.02	30.07	30.12	30.16
9.7	30.21	30.26	30.30	30.35	30.40	30.44	30.49	30.54	30.58	30.63
9.8	30.68	30.73	30.77	30.82	30.87	30.91	30.96	31.01	31.06	31.10
9.9	31.15	31.20	31.24	31.29	31.34	31.39	31.43	31.48	31.53	31.58

Table 59.—*Three-halves powers of numbers*—Continued

Number	.00	.01	.02	.03	.04	.05	.06	.07	.08	.09
10.0	31.62	31.67	31.72	31.77	31.81	31.86	31.91	31.96	32.00	32.05
10.1	32.10	32.15	32.19	32.24	32.29	32.34	32.38	32.43	32.48	32.53
10.2	32.58	32.62	32.67	32.72	32.77	32.82	32.86	32.91	32.96	33.01
10.3	33.06	33.10	33.15	33.20	33.25	33.30	33.35	33.39	33.44	33.49
10.4	33.54	33.59	33.64	33.68	33.73	33.78	33.83	33.88	33.93	33.98
10.5	34.02	34.07	34.12	34.17	34.22	34.27	34.32	34.36	34.41	34.46
10.6	34.51	34.56	34.61	34.66	34.71	34.76	34.80	34.85	34.90	34.95
10.7	35.00	35.05	35.10	35.15	35.20	35.25	35.30	35.34	35.39	35.44
10.8	35.49	35.54	35.59	35.64	35.69	35.74	35.79	35.84	35.89	35.94
10.9	35.99	36.04	36.09	36.14	36.18	36.23	36.28	36.33	36.38	36.43
11.0	36.48	36.53	36.58	36.63	36.68	36.73	36.78	36.83	36.88	36.93
11.1	36.98	37.03	37.08	37.13	37.18	37.23	37.28	37.33	37.38	37.43
11.2	37.48	37.53	37.58	37.63	37.68	37.73	37.78	37.83	37.88	37.94
11.3	37.99	38.04	38.09	38.14	38.19	38.24	38.29	38.34	38.39	38.44
11.4	38.49	38.54	38.59	38.64	38.69	38.74	38.80	38.85	38.90	38.95
11.5	39.00	39.05	39.10	39.15	39.20	39.25	39.30	39.36	39.41	39.46
11.6	39.51	39.56	39.61	39.66	39.71	39.76	39.82	39.87	39.92	39.97
11.7	40.02	40.07	40.12	40.17	40.23	40.28	40.33	40.38	40.43	40.48
11.8	40.53	40.59	40.64	40.69	40.74	40.79	40.84	40.90	40.95	41.00
11.9	41.05	41.10	41.15	41.21	41.26	41.31	41.36	41.41	41.47	41.52
12.0	41.57	41.62	41.67	41.73	41.78	41.83	41.88	41.93	41.99	42.04
12.1	42.09	42.14	42.19	42.25	42.30	42.35	42.40	42.46	42.51	42.56
12.2	42.61	42.67	42.72	42.77	42.82	42.88	42.93	42.98	43.03	43.09
12.3	43.14	43.19	43.24	43.30	43.35	43.40	43.45	43.51	43.56	43.61
12.4	43.66	43.72	43.77	43.82	43.88	43.93	43.98	44.04	44.09	44.14
12.5	44.19	44.25	44.30	44.35	44.41	44.46	44.51	44.57	44.62	44.67
12.6	44.73	44.78	44.83	44.89	44.94	44.99	45.05	45.10	45.15	45.21
12.7	45.26	45.31	45.37	45.42	45.47	45.53	45.58	45.63	45.69	45.74
12.8	45.79	45.85	45.90	45.96	46.01	46.06	46.12	46.17	46.22	46.28
12.9	46.33	46.39	46.44	46.49	46.55	46.60	46.66	46.71	46.76	46.82
13.0	46.87	46.93	46.98	47.03	47.09	47.14	47.20	47.25	47.31	47.36
13.1	47.41	47.47	47.52	47.58	47.63	47.69	47.74	47.79	47.85	47.90
13.2	47.96	48.01	48.07	48.12	48.18	48.23	48.29	48.34	48.39	48.45
13.3	48.50	48.56	48.61	48.67	48.72	48.78	48.83	48.89	48.94	49.00
13.4	49.05	49.11	49.16	49.22	49.27	49.33	49.38	49.44	49.49	49.55
13.5	49.60	49.66	49.71	49.77	49.82	49.88	49.93	49.99	50.04	50.10
13.6	50.15	50.21	50.26	50.32	50.38	50.43	50.49	50.54	50.60	50.65
13.7	50.71	50.76	50.82	50.88	50.93	50.99	51.04	51.10	51.15	51.21
13.8	51.26	51.32	51.38	51.43	51.49	51.54	51.60	51.66	51.71	51.77
13.9	51.82	51.88	51.93	51.99	52.05	52.10	52.16	52.21	52.27	52.33
14.0	52.38	52.44	52.50	52.55	52.61	52.66	52.72	52.78	52.83	52.89
14.1	52.95	53.00	53.06	53.11	53.17	53.23	53.28	53.34	53.40	53.45
14.2	53.51	53.57	53.62	53.68	53.74	53.79	53.85	53.91	53.96	54.02
14.3	54.08	54.13	54.19	54.25	54.30	54.36	54.42	54.47	54.53	54.59
14.4	54.64	54.70	54.76	54.82	54.87	54.93	54.99	55.04	55.10	55.16
14.5	55.21	55.27	55.33	55.39	55.44	55.50	55.56	55.61	55.67	55.73
14.6	55.79	55.84	55.90	55.96	56.02	56.07	56.13	56.19	56.25	56.30
14.7	56.36	56.42	56.48	56.53	56.59	56.65	56.71	56.76	56.82	56.88
14.8	56.94	56.99	57.05	57.11	57.17	57.23	57.28	57.34	57.40	57.46
14.9	57.51	57.57	57.63	57.69	57.75	57.80	57.86	57.92	57.98	58.04

Table 59.—*Three-halves powers of numbers*—Continued

Number	.00	.01	.02	.03	.04	.05	.06	.07	.08	.09
15.0	58.09	58.15	58.21	58.27	58.33	58.39	58.44	58.50	58.56	58.62
15.1	58.68	58.73	58.79	58.85	58.91	58.97	59.03	59.09	59.14	59.20
15.2	59.26	59.32	59.38	59.44	59.49	59.55	59.61	59.67	59.73	59.79
15.3	59.85	59.90	59.96	60.02	60.08	60.14	60.20	60.26	60.32	60.38
15.4	60.43	60.49	60.55	60.61	60.67	60.73	60.79	60.85	60.91	60.96
15.5	61.02	61.08	61.14	61.20	61.26	61.32	61.38	61.44	61.50	61.56
15.6	61.62	61.67	61.73	61.79	61.85	61.91	61.97	62.03	62.09	62.15
15.7	62.21	62.27	62.33	62.39	62.45	62.51	62.57	62.62	62.68	62.74
15.8	62.80	62.86	62.92	62.98	63.04	63.10	63.16	63.22	63.28	63.34
15.9	63.40	63.46	63.52	63.58	63.64	63.70	63.76	63.82	63.88	63.94
16.0	64.00	64.06	64.12	64.18	64.24	64.30	64.36	64.42	64.48	64.54
16.1	64.60	64.66	64.72	64.78	64.84	64.90	64.96	65.02	65.08	65.14
16.2	65.20	65.26	65.32	65.38	65.45	65.51	65.57	65.63	65.69	65.75
16.3	65.81	65.87	65.93	65.99	66.05	66.11	66.17	66.23	66.29	66.35
16.4	66.41	66.48	66.54	66.60	66.66	66.72	66.78	66.84	66.90	66.96
16.5	67.02	67.08	67.15	67.21	67.27	67.33	67.39	67.45	67.51	67.57
16.6	67.63	67.69	67.76	67.82	67.88	67.94	68.00	68.06	68.12	68.18
16.7	68.25	68.31	68.37	68.43	68.49	68.55	68.61	68.68	68.74	68.80
16.8	68.86	68.92	68.98	69.04	69.11	69.17	69.23	69.29	69.35	69.41
16.9	69.48	69.54	69.60	69.66	69.72	69.78	69.85	69.91	69.97	70.03
17.0	70.09	70.15	70.22	70.28	70.34	70.40	70.46	70.53	70.59	70.65
17.1	70.71	70.77	70.84	70.90	70.96	71.02	71.08	71.15	71.21	71.27
17.2	71.33	71.40	71.46	71.52	71.58	71.64	71.71	71.77	71.83	71.89
17.3	71.96	72.02	72.08	72.14	72.21	72.27	72.33	72.39	72.46	72.52
17.4	72.58	72.64	72.71	72.77	72.83	72.89	72.96	73.02	73.08	73.15
17.5	73.21	73.27	73.33	73.40	73.46	73.52	73.58	73.65	73.71	73.77
17.6	73.84	73.90	73.96	74.03	74.09	74.15	74.21	74.28	74.34	74.40
17.7	74.47	74.53	74.59	74.66	74.72	74.78	74.85	74.91	75.04	75.04
17.8	75.10	75.16	75.22	75.29	75.35	75.41	75.48	75.54	75.61	75.67
17.9	75.73	75.80	75.86	75.92	75.99	76.05	76.11	76.18	76.24	76.30
18.0	76.37	76.43	76.49	76.56	76.62	76.69	76.75	76.81	76.88	76.94
18.1	77.00	77.07	77.13	77.20	77.26	77.32	77.39	77.45	77.52	77.58
18.2	77.64	77.71	77.77	77.84	77.90	77.96	78.03	78.09	78.16	78.22
18.3	78.28	78.35	78.41	78.48	78.54	78.61	78.67	78.73	78.80	78.86
18.4	78.93	78.99	79.06	79.12	79.18	79.25	79.31	79.38	79.44	79.51
18.5	79.57	79.64	79.70	79.77	79.83	79.89	79.96	80.02	80.09	80.15
18.6	80.22	80.28	80.35	80.41	80.48	80.54	80.61	80.67	80.74	80.80
18.7	80.87	80.93	81.00	81.06	81.12	81.19	81.25	81.32	81.38	81.45
18.8	81.51	81.58	81.64	81.71	81.78	81.84	81.91	81.97	82.04	82.10
18.9	82.17	82.23	82.30	82.36	82.43	82.49	82.56	82.62	82.69	82.75
19.0	82.82	82.88	82.95	83.02	83.08	83.15	83.21	83.28	83.34	83.41
19.1	83.47	83.54	83.60	83.67	83.74	83.80	83.87	83.93	84.00	84.06
19.2	84.13	84.20	84.26	84.33	84.39	84.46	84.52	84.59	84.66	84.72
19.3	84.79	84.85	84.92	84.99	85.05	85.12	85.18	85.25	85.32	85.38
19.4	85.45	85.51	85.58	85.65	85.71	85.78	85.84	85.91	85.98	86.04
19.5	86.11	86.18	86.24	86.31	86.37	86.44	86.51	86.57	86.64	86.71
19.6	86.77	86.84	86.91	86.97	87.04	87.11	87.17	87.24	87.30	87.37
19.7	87.44	87.50	87.57	87.64	87.70	87.77	87.84	87.90	87.97	88.04
19.8	88.10	88.17	88.24	88.30	88.37	88.44	88.51	88.57	88.64	88.71
19.9	88.77	88.84	88.91	88.97	89.04	89.11	89.17	89.24	89.31	89.38

Table 59.—*Three-halves powers of numbers*—Continued

Number	.00	.01	.02	.03	.04	.05	.06	.07	.08	.09
20.0	89.44	89.51	89.58	89.64	89.71	89.78	89.85	89.91	89.98	90.05
20.1	90.11	90.18	90.25	90.32	90.38	90.45	90.52	90.59	90.65	90.72
20.2	90.79	90.86	90.92	90.99	91.06	91.12	91.19	91.26	91.33	91.40
20.3	91.46	91.53	91.60	91.67	91.73	91.80	91.87	91.94	92.00	92.07
20.4	92.14	92.21	92.27	92.34	92.41	92.48	92.55	92.61	92.68	92.75
20.5	92.82	92.89	92.95	93.02	93.09	93.16	93.23	93.29	93.36	93.43
20.6	93.50	93.57	93.63	93.70	93.77	93.84	93.91	93.97	94.04	94.11
20.7	94.18	94.25	94.32	94.38	94.45	94.52	94.59	94.66	94.73	94.79
20.8	94.86	94.93	95.00	95.07	95.14	95.20	95.27	95.34	95.41	95.48
20.9	95.55	95.62	95.68	95.75	95.82	95.89	95.96	96.03	96.10	96.17
21.0	96.23	96.30	96.37	96.44	96.51	96.58	96.65	96.72	96.78	96.85
21.1	96.92	96.99	97.06	97.13	97.20	97.27	97.34	97.41	97.47	97.54
21.2	97.61	97.68	97.75	97.82	97.89	97.96	98.03	98.10	98.17	98.23
21.3	98.30	98.37	98.44	98.51	98.58	98.65	98.72	98.79	98.86	98.93
21.4	99.00	99.07	99.14	99.20	99.27	99.34	99.41	99.48	99.55	99.62
21.5	99.69	99.76	99.83	99.90	99.97	100.0	100.1	100.2	100.2	100.3
21.6	100.4	100.5	100.5	100.6	100.7	100.7	100.8	100.9	100.9	101.0
21.7	101.1	101.2	101.2	101.3	101.4	101.4	101.5	101.6	101.6	101.7
21.8	101.8	101.9	101.9	102.0	102.1	102.1	102.2	102.3	102.3	102.4
21.9	102.5	102.6	102.6	102.7	102.8	102.8	102.9	103.0	103.0	103.1
22.0	103.2	103.3	103.3	103.4	103.5	103.5	103.6	103.7	103.8	103.8
22.1	103.9	104.0	104.0	104.1	104.2	104.2	104.3	104.4	104.5	104.5
22.2	104.6	104.7	104.7	104.8	104.9	105.0	105.0	105.1	105.2	105.2
22.3	105.3	105.4	105.4	105.5	105.6	105.7	105.7	105.8	105.9	105.9
22.4	106.0	106.1	106.2	106.2	106.3	106.4	106.4	106.5	106.6	106.7
22.5	106.7	106.8	106.9	106.9	107.0	107.1	107.2	107.2	107.3	107.4
22.6	107.4	107.5	107.6	107.7	107.7	107.8	107.9	107.9	108.0	108.1
22.7	108.2	108.2	108.3	108.4	108.4	108.5	108.6	108.7	108.7	108.8
22.8	108.9	108.9	109.0	109.1	109.2	109.2	109.3	109.4	109.4	109.5
22.9	109.6	109.7	109.7	109.8	109.9	109.9	110.0	110.1	110.2	110.2
23.0	110.3	110.4	110.4	110.5	110.6	110.7	110.7	110.8	110.9	111.0
23.1	111.0	111.1	111.2	111.2	111.3	111.4	111.5	111.5	111.6	111.7
23.2	111.7	111.8	111.9	112.0	112.0	112.1	112.2	112.3	112.3	112.4
23.3	112.5	112.5	112.6	112.7	112.8	112.8	112.9	113.0	113.0	113.1
23.4	113.2	113.3	113.3	113.4	113.5	113.6	113.6	113.7	113.8	113.8
23.5	113.9	114.0	114.1	114.1	114.2	114.3	114.4	114.4	114.5	114.6
23.6	114.6	114.7	114.8	114.9	114.9	115.0	115.1	115.2	115.2	115.3
23.7	115.4	115.5	115.5	115.6	115.7	115.7	115.8	115.9	116.0	116.0
23.8	116.1	116.2	116.3	116.3	116.4	116.5	116.5	116.6	116.7	116.8
23.9	116.8	116.9	117.0	117.0	117.1	117.2	117.3	117.4	117.4	117.5
24.0	117.6	117.6	117.7	117.8	117.9	117.9	118.0	118.1	118.2	118.2
24.1	118.3	118.4	118.5	118.5	118.6	118.7	118.8	118.8	118.9	119.0
24.2	119.0	119.1	119.2	119.3	119.3	119.4	119.5	119.6	119.6	119.7
24.3	119.8	119.9	119.9	120.0	120.1	120.2	120.2	120.3	120.4	120.5
24.4	120.5	120.6	120.7	120.7	120.8	120.9	121.0	121.0	121.1	121.2
24.5	121.3	121.3	121.4	121.5	121.6	121.6	121.7	121.8	121.9	121.9
24.6	122.0	122.1	122.2	122.2	122.3	122.4	122.5	122.5	122.6	122.7
24.7	122.8	122.8	122.9	123.0	123.1	123.1	123.2	123.3	123.4	123.4
24.8	123.5	123.6	123.7	123.7	123.8	123.9	124.0	124.0	124.1	124.2
24.9	124.3	124.3	124.4	124.5	124.6	124.6	124.7	124.8	124.9	124.9

Table 59.—*Three-halves powers of numbers*—Continued

Num-ber	.00	.01	.02	.03	.04	.05	.06	.07	.08	.09
25.0	125.0	125.1	125.2	125.2	125.3	125.4	125.5	125.5	125.6	125.7
25.1	125.8	125.8	125.9	126.0	126.1	126.1	126.2	126.3	126.4	126.4
25.2	126.5	126.6	126.7	126.7	126.8	126.9	127.0	127.0	127.1	127.2
25.3	127.3	127.3	127.4	127.5	127.6	127.6	127.7	127.8	127.9	127.9
25.4	128.0	128.1	128.2	128.2	128.3	128.4	128.5	128.5	128.6	128.7
25.5	128.8	128.8	128.9	129.0	129.1	129.1	129.2	129.3	129.4	129.5
25.6	129.5	129.6	129.7	129.8	129.8	129.9	130.0	130.1	130.1	130.2
25.7	130.3	130.4	130.4	130.5	130.6	130.7	130.7	130.8	130.9	131.0
25.8	131.0	131.1	131.2	131.3	131.4	131.4	131.5	131.6	131.7	131.7
25.9	131.8	131.9	132.0	132.0	132.1	132.2	132.3	132.3	132.4	132.5
26.0	132.6	132.7	132.7	132.8	132.9	133.0	133.0	133.1	133.2	133.3
26.1	133.3	133.4	133.5	133.6	133.6	133.7	133.8	133.9	134.0	134.0
26.2	134.1	134.2	134.3	134.3	134.4	134.5	134.6	134.6	134.7	134.8
26.3	134.9	135.0	135.0	135.1	135.2	135.3	135.3	135.4	135.5	135.6
26.4	135.6	135.7	135.8	135.9	136.0	136.0	136.1	136.2	136.3	136.3
26.5	136.4	136.5	136.6	136.6	136.7	136.8	136.9	137.0	137.0	137.1
26.6	137.2	137.3	137.3	137.4	137.5	137.6	137.7	137.7	137.8	137.9
26.7	138.0	138.0	138.1	138.2	138.3	138.4	138.4	138.5	138.6	138.7
26.8	138.7	138.8	138.9	139.0	139.1	139.1	139.2	139.3	139.4	139.4
26.9	139.5	139.6	139.7	139.8	139.8	139.9	140.0	140.1	140.1	140.2
27.0	140.3	140.4	140.5	140.5	140.6	140.7	140.8	140.8	140.9	141.0
27.1	141.1	141.2	141.2	141.3	141.4	141.5	141.5	141.6	141.7	141.8
27.2	141.9	141.9	142.0	142.1	142.2	142.2	142.3	142.4	142.5	142.6
27.3	142.6	142.7	142.8	142.9	143.0	143.0	143.1	143.2	143.3	143.3
27.4	143.4	143.5	143.6	143.7	143.7	143.8	143.9	144.0	144.1	144.1
27.5	144.2	144.3	144.4	144.4	144.5	144.6	144.7	144.8	144.8	144.9
27.6	145.0	145.1	145.2	145.2	145.3	145.4	145.5	145.5	145.6	145.7
27.7	145.8	145.9	145.9	146.0	146.1	146.2	146.3	146.3	146.4	146.5
27.8	146.6	146.7	146.7	146.8	146.9	147.0	147.1	147.1	147.2	147.3
27.9	147.4	147.4	147.5	147.6	147.7	147.8	147.8	147.9	148.0	148.1
28.0	148.2	148.2	148.3	148.4	148.5	148.6	148.6	148.7	148.8	148.9
28.1	149.0	149.0	149.1	149.2	149.3	149.4	149.4	149.5	149.6	149.7
28.2	149.8	149.8	149.9	150.0	150.1	150.2	150.2	150.3	150.4	150.5
28.3	150.5	150.6	150.7	150.8	150.9	150.9	151.0	151.1	151.2	151.3
28.4	151.3	151.4	151.5	151.6	151.7	151.7	151.8	151.9	152.0	152.1
28.5	152.1	152.2	152.3	152.4	152.5	152.5	152.6	152.7	152.8	152.9
28.6	152.9	153.0	153.1	153.1	153.2	153.3	153.4	153.4	153.5	153.7
28.7	153.8	153.8	153.9	154.0	154.1	154.2	154.2	154.3	154.4	154.5
28.8	154.6	154.6	154.7	154.8	154.9	155.0	155.0	155.1	155.2	155.3
28.9	155.4	155.4	155.5	155.6	155.7	155.7	155.8	155.9	156.0	156.1
29.0	156.2	156.3	156.3	156.4	156.5	156.6	156.7	156.7	156.8	156.9
29.1	157.0	157.1	157.1	157.2	157.3	157.4	157.5	157.5	157.6	157.7
29.2	157.8	157.9	158.0	158.0	158.1	158.2	158.3	158.4	158.4	158.5
29.3	158.6	158.7	158.8	158.8	158.9	159.0	159.1	159.2	159.2	159.3
29.4	159.4	159.5	159.6	159.7	159.7	159.8	159.9	160.0	160.1	160.1
29.5	160.2	160.3	160.4	160.5	160.6	160.6	160.7	160.8	160.9	161.0
29.6	161.0	161.1	161.2	161.3	161.4	161.4	161.5	161.6	161.7	161.8
29.7	161.9	161.9	162.0	162.1	162.2	162.3	162.3	162.4	162.5	162.6
29.8	162.7	162.8	162.8	162.9	163.0	163.1	163.2	163.2	163.3	163.4
29.9	163.5	163.6	163.7	163.7	163.8	163.9	164.0	164.1	164.2	164.2

Table 59.—*Three-halves powers of numbers*—Continued

Number	.0	.1	.2	.3	.4	.5	.6	.7	.8	.9
30	164.3	165.1	166.0	166.8	167.6	168.4	169.3	170.1	170.9	171.8
31	172.6	173.4	174.3	175.1	176.0	176.8	177.6	178.5	179.3	180.2
32	181.0	181.9	182.7	183.6	184.4	185.3	186.1	187.0	187.8	188.7
33	189.6	190.4	191.3	192.2	193.0	193.9	194.8	195.6	196.5	197.4
34	198.3	199.1	200.0	200.9	201.8	202.6	203.5	204.4	205.3	206.2
35	207.1	208.0	208.8	209.7	210.6	211.5	212.4	213.3	214.2	215.1
36	216.0	216.9	217.8	218.7	219.6	220.5	221.4	222.3	223.2	224.2
37	225.1	226.0	226.9	227.8	228.7	229.6	230.6	231.5	232.4	233.3
38	234.2	235.2	236.1	237.0	238.0	238.9	239.8	240.8	241.7	242.6
39	243.6	244.5	245.4	246.4	247.3	248.3	249.2	250.1	251.1	252.0
40	253.0	253.9	254.9	255.8	256.8	257.7	258.7	259.7	260.6	261.6
41	262.5	263.5	264.5	265.4	266.4	267.3	268.3	269.3	270.2	271.2
42	272.2	273.2	274.1	275.1	276.1	277.1	278.0	279.0	280.0	281.0
43	282.0	283.0	283.9	284.9	285.9	286.9	287.9	288.9	289.9	290.9
44	291.9	292.9	293.9	294.9	295.9	296.9	297.9	298.9	299.9	300.9
45	301.9	302.9	303.9	304.9	305.9	306.9	307.9	308.9	310.0	311.0
46	312.0	313.0	314.0	315.0	316.1	317.1	318.1	319.1	320.2	321.2
47	322.2	323.2	324.3	325.3	326.3	327.4	328.4	329.4	330.5	331.5
48	332.6	333.6	334.6	335.7	336.7	337.8	338.8	339.9	340.9	342.0
49	343.0	344.1	345.1	346.2	347.2	348.3	349.3	350.4	351.4	352.5
50	353.6	354.6	355.7	356.7	357.8	358.9	359.9	361.0	362.1	363.1
51	364.2	365.3	366.4	367.4	368.5	369.6	370.7	371.7	372.8	373.9
52	375.0	376.1	377.1	378.2	379.3	380.4	381.5	382.6	383.7	384.8
53	385.8	386.9	388.0	389.1	390.2	391.3	392.4	393.5	394.6	395.7
54	396.8	397.9	399.0	400.1	401.2	402.3	403.4	404.6	405.7	406.8
55	407.9	409.0	410.1	411.2	412.3	413.5	414.6	415.7	416.8	417.9
56	419.1	420.2	421.3	422.4	423.6	424.7	425.8	426.9	428.1	429.2
57	430.3	431.5	432.6	433.7	434.9	436.0	437.2	438.3	439.4	440.6
58	441.7	442.9	444.0	445.1	446.3	447.4	448.6	449.7	450.9	452.0
59	453.2	454.3	455.5	456.6	457.8	459.0	460.1	461.3	462.4	463.6
60	464.8	465.9	467.1	468.2	469.4	470.6	471.7	472.9	474.1	475.3
61	476.4	477.6	478.8	479.9	481.1	482.3	483.5	484.6	485.8	487.0
62	488.2	489.4	490.6	491.7	492.9	494.1	495.3	496.5	497.7	498.9
63	500.0	501.2	502.4	503.6	504.8	506.0	507.2	508.4	509.6	510.8
64	512.0	513.2	514.4	515.6	516.8	518.0	519.2	520.4	521.6	522.8
65	524.0	525.3	526.5	527.7	528.9	530.1	531.3	532.5	533.8	535.0
66	536.2	537.4	538.6	539.8	541.1	542.3	543.5	544.7	546.0	547.2
67	548.4	549.6	550.9	552.1	553.3	554.6	555.8	557.0	558.3	559.5
68	560.7	562.0	563.2	564.5	565.7	566.9	568.2	569.4	570.7	571.9
69	573.2	574.4	575.7	576.9	578.1	579.4	580.6	581.9	583.2	584.4
70	585.7	586.9	588.2	589.4	590.7	591.9	593.2	594.5	595.7	597.0
71	598.3	599.5	600.8	602.1	603.3	604.6	605.9	607.1	608.4	609.7
72	610.9	612.2	613.5	614.8	616.0	617.3	618.6	619.9	621.2	622.4
73	623.7	625.0	626.3	627.6	628.8	630.1	631.4	632.7	634.0	635.3
74	636.6	637.9	639.2	640.4	641.7	643.0	644.3	645.6	646.9	648.2
75	649.5	650.8	652.1	653.4	654.7	656.0	657.3	658.6	659.9	661.2
76	662.6	663.9	665.2	666.5	667.8	669.1	670.4	671.7	673.0	674.4
77	675.7	677.0	678.3	679.6	680.9	682.3	683.6	684.9	686.2	687.6
78	688.9	690.2	691.5	692.9	694.2	695.5	696.8	698.2	699.5	700.8
79	702.2	703.5	704.8	706.2	707.5	708.8	710.2	711.5	712.9	714.2

Table 59.—*Three-halves powers of numbers*—Continued

Number	.0	.1	.2	.3	.4	.5	.6	.7	.8	.9
80	715.5	716.9	718.2	719.6	720.9	722.3	723.6	725.0	726.3	727.7
81	729.0	730.4	731.7	733.1	734.4	735.8	737.1	738.5	739.8	741.2
82	742.5	743.9	745.3	746.6	748.0	749.3	750.7	752.1	753.4	754.8
83	756.2	757.5	758.9	760.3	761.6	763.0	764.4	765.8	767.1	768.5
84	769.9	771.2	772.6	774.0	775.4	776.8	778.1	779.5	780.9	782.3
85	783.7	785.0	786.4	787.8	789.2	790.6	792.0	793.4	794.8	796.1
86	797.5	798.9	800.3	801.7	803.1	804.5	805.9	807.3	808.7	810.1
87	811.5	812.9	814.3	815.7	817.1	818.5	819.9	821.3	822.7	824.1
88	825.5	826.9	828.3	829.7	831.1	832.6	834.0	835.4	836.8	838.2
89	839.6	841.0	842.5	843.9	845.3	846.7	848.1	849.5	851.0	852.4
90	853.8	855.2	856.7	858.1	859.5	860.9	862.3	863.8	865.2	866.7
91	868.1	869.5	870.9	872.4	873.8	875.2	876.7	878.1	879.6	881.0
92	882.4	883.9	885.3	886.8	888.2	889.6	891.1	892.5	894.0	895.4
93	896.9	898.3	899.8	901.2	902.7	904.1	905.6	907.0	908.5	909.9
94	911.4	912.8	914.3	915.7	917.2	918.6	920.1	921.6	923.0	924.5
95	925.9	927.4	928.9	930.3	931.8	933.3	934.7	936.2	937.7	939.1
96	940.6	942.1	943.5	945.0	946.5	948.0	949.4	950.9	952.4	953.9
97	955.3	956.8	958.3	959.8	961.3	962.7	964.2	965.7	967.2	968.7
98	970.2	971.6	973.1	974.6	976.1	977.6	979.1	980.6	982.1	983.5
99	985.0	986.5	988.0	989.5	991.0	992.5	994.0	995.5	997.0	998.5

Table 60.—*Eight-thirds powers of numbers*

Number	.00	.01	.02	.03	.04	.05	.06	.07	.08	.09
0.0	0.0000	0.0000	0.0000	0.0001	0.0002	0.0003	0.0006	0.0008	0.0012	0.0016
0.1	.0022	.0028	.0035	.0043	.0053	.0064	.0075	.0089	.0103	.0119
0.2	.0137	.0156	.0176	.0199	.0222	.0248	.0275	.0305	.0336	.0368
0.3	.0403	.0440	.0479	.0520	.0563	.0608	.0656	.0706	.0758	.0812
0.4	.0869	.0928	.0989	.1053	.1120	.1189	.1261	.1335	.1412	.1492
0.5	.1575	.1660	.1749	.1840	.1934	.2031	.2131	.2234	.2340	.2449
0.6	.2561	.2676	.2795	.2917	.3042	.3170	.3302	.3437	.3576	.3718
0.7	.3863	.4012	.4164	.4320	.4480	.4643	.4810	.4981	.5155	.5333
0.8	.5515	.5701	.5891	.6084	.6282	.6483	.6689	.6898	.7111	.7329
0.9	.7551	.7776	.8006	.8241	.8479	.8722	.8969	.9220	.9476	.9736
1.0	1.0000	1.027	1.054	1.082	1.110	1.139	1.168	1.198	1.228	1.258
1.1	1.289	1.321	1.353	1.385	1.418	1.452	1.486	1.520	1.555	1.590
1.2	1.626	1.662	1.699	1.737	1.775	1.813	1.852	1.892	1.931	1.972
1.3	2.013	2.055	2.097	2.139	2.182	2.226	2.270	2.315	2.361	2.406
1.4	2.453	2.500	2.547	2.596	2.644	2.693	2.743	2.794	2.845	2.896
1.5	2.948	3.001	3.054	3.108	3.163	3.218	3.273	3.330	3.387	3.444
1.6	3.502	3.561	3.620	3.680	3.740	3.802	3.863	3.926	3.989	4.052
1.7	4.117	4.181	4.247	4.313	4.380	4.447	4.515	4.584	4.654	4.724
1.8	4.794	4.866	4.938	5.010	5.084	5.158	5.232	5.308	5.384	5.460
1.9	5.538	5.616	5.695	5.774	5.854	5.935	6.017	6.099	6.182	6.265
2.0	6.350	6.435	6.520	6.607	6.694	6.782	6.870	6.960	7.050	7.140
2.1	7.232	7.324	7.417	7.511	7.605	7.700	7.796	7.893	7.990	8.088
2.2	8.187	8.287	8.387	8.488	8.590	8.693	8.796	8.900	9.005	9.111
2.3	9.217	9.325	9.433	9.541	9.651	9.761	9.873	9.985	10.10	10.21
2.4	10.33	10.44	10.56	10.67	10.79	10.91	11.03	11.15	11.27	11.39
2.5	11.51	11.64	11.76	11.88	12.01	12.14	12.26	12.39	12.52	12.65
2.6	12.78	12.91	13.05	13.18	13.31	13.45	13.58	13.72	13.86	14.00
2.7	14.14	14.28	14.42	14.56	14.70	14.84	14.99	15.13	15.28	15.43
2.8	15.57	15.72	15.87	16.02	16.18	16.33	16.48	16.63	16.79	16.95
2.9	17.10	17.26	17.42	17.58	17.74	17.90	18.06	18.23	18.39	18.55
3.0	18.72	18.89	19.06	19.22	19.39	19.56	19.74	19.91	20.08	20.26
3.1	20.43	20.61	20.78	20.96	21.14	21.32	21.50	21.68	21.87	22.05
3.2	22.24	22.42	22.61	22.80	22.99	23.18	23.37	23.56	23.75	23.94
3.3	24.14	24.33	24.53	24.73	24.93	25.13	25.33	25.53	25.73	25.93
3.4	26.14	26.34	26.55	26.76	26.97	27.18	27.39	27.60	27.81	28.02
3.5	28.24	28.45	28.67	28.89	29.11	29.33	29.55	29.77	29.99	30.22
3.6	30.44	30.67	30.90	31.12	31.35	31.58	31.81	32.05	32.28	32.51
3.7	32.75	32.99	33.22	33.46	33.70	33.94	34.18	34.43	34.67	34.92
3.8	35.16	35.41	35.66	35.91	36.16	36.41	36.66	36.92	37.17	37.43
3.9	37.69	37.94	38.20	38.46	38.72	38.99	39.25	39.52	39.78	40.05
4.0	40.32	40.59	40.86	41.13	41.40	41.68	41.95	42.23	42.50	42.78
4.1	43.06	43.34	43.62	43.91	44.19	44.48	44.76	45.05	45.34	45.63
4.2	45.92	46.21	46.51	46.80	47.10	47.39	47.69	47.99	48.29	48.59
4.3	48.89	49.20	49.50	49.81	50.12	50.42	50.73	51.04	51.36	51.67
4.4	51.98	52.30	52.62	52.94	53.25	53.57	53.90	54.22	54.54	54.87
4.5	55.20	55.52	55.85	56.18	56.51	56.85	57.18	57.51	57.85	58.19
4.6	58.53	58.87	59.21	59.55	59.89	60.24	60.58	60.93	61.28	61.63
4.7	61.98	62.33	62.69	63.04	63.40	63.76	64.11	64.47	64.83	65.20
4.8	65.56	65.93	66.29	66.66	67.03	67.40	67.77	68.14	68.52	68.89
4.9	69.27	69.64	70.02	70.40	70.78	71.17	71.55	71.94	72.32	72.71

Table 60.—*Eight-thirds powers of numbers*—Continued

Number	.00	.01	.02	.03	.04	.05	.06	.07	.08	.09
5.0	73.10	73.49	73.88	74.28	74.67	75.07	75.46	75.86	76.26	76.66
5.1	77.06	77.47	77.87	78.28	78.69	79.10	79.51	79.92	80.33	80.74
5.2	81.16	81.58	82.00	82.41	82.84	83.26	83.68	84.11	84.53	84.96
5.3	85.39	85.82	86.25	86.68	87.12	87.55	87.99	88.43	88.87	89.31
5.4	89.75	90.20	90.64	91.09	91.54	91.99	92.44	92.89	93.34	93.80
5.5	94.25	94.71	95.17	95.63	96.09	96.56	97.02	97.49	97.95	98.42
5.6	98.89	99.37	99.84	100.3	100.8	101.3	101.7	102.2	102.7	103.2
5.7	103.7	104.2	104.6	105.1	105.6	106.1	106.6	107.1	107.6	108.1
5.8	108.6	109.1	109.6	110.1	110.6	111.1	111.6	112.1	112.6	113.1
5.9	113.7	114.2	114.7	115.2	115.7	116.2	116.8	117.3	117.8	118.3
6.0	118.9	119.4	119.9	120.5	121.0	121.5	122.1	122.6	123.1	123.7
6.1	124.2	124.8	125.3	125.9	126.4	127.0	127.5	128.1	128.6	129.2
6.2	129.7	130.3	130.9	131.4	132.0	132.5	133.1	133.7	134.2	134.8
6.3	135.4	136.0	136.5	137.1	137.7	138.3	138.9	139.4	140.0	140.6
6.4	141.2	141.8	142.4	143.0	143.6	144.2	144.8	145.3	145.9	146.6
6.5	147.2	147.8	148.4	149.0	149.6	150.2	150.8	151.4	152.0	152.6
6.6	153.3	153.9	154.5	155.1	155.8	156.4	157.0	157.6	158.3	158.9
6.7	159.5	160.2	160.8	161.5	162.1	162.7	163.4	164.0	164.7	165.3
6.8	166.0	166.6	167.3	167.9	168.6	169.2	169.9	170.6	171.2	171.9
6.9	172.6	173.2	173.9	174.6	175.2	175.9	176.6	177.3	177.9	178.6
7.0	179.3	180.0	180.7	181.4	182.1	182.7	183.4	184.1	184.8	185.5
7.1	186.2	186.9	187.6	188.3	189.0	189.7	190.4	191.2	191.9	192.6
7.2	193.3	194.0	194.7	195.4	196.2	196.9	197.6	198.3	199.1	199.8
7.3	200.5	201.3	202.0	202.7	203.5	204.2	205.0	205.7	206.5	207.2
7.4	207.9	208.7	209.4	210.2	211.0	211.7	212.5	213.2	214.0	214.8
7.5	215.5	216.3	217.1	217.8	218.6	219.4	220.2	220.9	221.7	222.5
7.6	223.3	224.1	224.8	225.6	226.4	227.2	228.0	228.8	229.6	230.4
7.7	231.2	232.0	232.8	233.6	234.4	235.2	236.0	236.8	237.7	238.5
7.8	239.3	240.1	240.9	241.7	242.6	243.4	244.2	245.1	245.9	246.7
7.9	247.6	248.4	249.2	250.1	250.9	251.8	252.6	253.4	254.3	255.1
8.0	256.0	256.9	257.7	258.6	259.4	260.3	261.2	262.0	262.9	263.8
8.1	264.6	265.5	266.4	267.2	268.1	269.0	269.9	270.8	271.6	272.5
8.2	273.4	274.3	275.2	276.1	277.0	277.9	278.8	279.7	280.6	281.5
8.3	282.4	283.3	284.2	285.1	286.1	287.0	287.9	288.8	289.7	290.6
8.4	291.6	292.5	293.4	294.4	295.3	296.2	297.2	298.1	299.0	300.0
8.5	300.9	301.9	302.8	303.8	304.7	305.7	306.6	307.6	308.5	309.5
8.6	310.5	311.4	312.4	313.4	314.3	315.3	316.3	317.2	318.2	319.2
8.7	320.2	321.2	322.1	323.1	324.1	325.1	326.1	327.1	328.1	329.1
8.8	330.1	331.1	332.1	333.1	334.1	335.1	336.1	337.1	338.1	339.2
8.9	340.2	341.2	342.2	343.2	344.3	345.3	346.3	347.4	348.4	349.4
9.0	350.5	351.5	352.5	353.6	354.6	355.7	356.7	357.8	358.8	359.9
9.1	360.9	362.0	363.1	364.1	365.2	366.3	367.3	368.4	369.5	370.5
9.2	371.6	372.7	373.8	374.9	375.9	377.0	378.1	379.2	380.3	381.4
9.3	382.5	383.6	384.7	385.8	386.9	388.0	389.1	390.2	391.3	392.4
9.4	393.6	394.7	395.8	396.9	398.0	399.2	400.3	401.4	402.6	403.7
9.5	404.8	406.0	407.1	408.2	409.4	410.5	411.7	412.8	414.0	415.1
9.6	416.3	417.4	418.6	419.8	420.9	422.1	423.3	424.4	425.6	426.8
9.7	427.9	429.1	430.3	431.5	432.7	433.9	435.0	436.2	437.4	438.6
9.8	439.8	441.0	442.2	443.4	444.6	445.8	447.0	448.2	449.5	450.7
9.9	451.9	453.1	454.3	455.5	456.8	458.0	459.2	460.5	461.7	462.9

Table 60.—*Eight-thirds powers of numbers*—Continued

Num-ber	.00	.01	.02	.03	.04	.05	.06	.07	.08	.09
10. 0	464. 2	465. 4	466. 6	467. 9	469. 1	470. 4	471. 6	472. 9	474. 1	475. 4
10. 1	476. 6	477. 9	479. 2	480. 4	481. 7	483. 0	484. 2	485. 5	486. 8	488. 1
10. 2	489. 3	490. 6	491. 9	493. 2	494. 5	495. 8	497. 0	498. 3	499. 6	500. 9
10. 3	502. 2	503. 5	504. 8	506. 1	507. 4	508. 8	510. 1	511. 4	512. 7	514. 0
10. 4	515. 3	516. 7	518. 0	519. 3	520. 6	522. 0	523. 3	524. 6	526. 0	527. 3
10. 5	528. 7	530. 0	531. 3	532. 7	534. 0	535. 4	536. 7	538. 1	539. 5	540. 8
10. 6	542. 2	543. 6	544. 9	546. 3	547. 7	549. 0	550. 4	551. 8	553. 2	554. 5
10. 7	555. 9	557. 3	558. 7	560. 1	561. 5	562. 9	564. 3	565. 7	567. 1	568. 5
10. 8	569. 9	571. 3	572. 7	574. 1	575. 5	577. 0	578. 4	579. 8	581. 2	582. 6
10. 9	584. 1	585. 5	586. 9	588. 4	589. 8	591. 2	592. 7	594. 1	595. 6	597. 0
11. 0	598. 5	599. 9	601. 4	602. 8	604. 3	605 8	607. 2	608. 7	610. 2	611. 6
11. 1	613. 1	614. 6	616. 0	617. 5	619. 0	620. 5	622. 0	623. 5	624. 9	626. 4
11. 2	627. 9	629. 4	630. 9	632. 4	633. 9	635. 4	636. 9	638. 5	640. 0	641. 5
11. 3	643. 0	644. 5	646. 0	647. 6	649. 1	650. 6	652. 1	653. 7	655. 2	656. 7
11. 4	658. 3	659. 8	661. 4	662. 9	664. 5	666. 0	667. 6	669. 1	670. 7	672. 2
11. 5	673. 8	675. 4	676. 9	678. 5	680. 1	681. 6	683. 2	684. 8	686. 4	687. 9
11. 6	689. 5	691. 1	692. 7	694. 3	695. 9	697. 5	699. 1	700. 7	702. 3	703. 9
11. 7	705. 5	707. 1	708. 7	710. 3	711. 9	713. 6	715. 2	716. 8	718. 4	720. 1
11. 8	721. 7	723. 3	725. 0	726. 6	728. 2	729. 9	731. 5	733. 2	734. 8	736. 5
11. 9	738. 1	739. 8	741. 4	743. 1	744. 8	746. 4	748. 1	749. 8	751. 4	753. 1
12. 0	754. 8	756. 5	758. 1	759. 8	761. 5	763. 2	764. 9	766. 6	768. 3	770. 0
12. 1	771. 7	773. 4	775. 1	776. 8	778. 5	780. 2	781. 9	783. 6	785. 3	787. 1
12. 2	788. 8	790. 5	792. 2	794. 0	795. 7	797. 4	799. 2	800. 9	802. 7	804. 4
12. 3	806. 1	807. 9	809. 6	811. 4	813. 2	814. 9	816. 7	818. 4	820. 2	822. 0
12. 4	823. 7	825. 5	827. 3	829. 1	830. 8	832. 6	834. 4	836. 2	838. 0	839. 8
12. 5	841. 6	843. 4	845. 2	847. 0	848. 8	850. 6	852. 4	854. 2	856. 0	857. 8
12. 6	859. 6	861. 5	863. 3	865. 1	866. 9	868. 8	870. 6	872. 4	874. 3	876. 1
12. 7	878. 0	879. 8	881. 7	883. 5	885. 4	887. 2	889. 1	890. 9	892. 8	894. 7
12. 8	896. 5	898. 4	900. 3	902. 1	904. 0	905. 9	907. 8	909. 7	911. 5	913. 4
12. 9	915. 3	917. 2	919. 1	921. 0	922. 9	924. 8	926. 7	928. 6	930. 5	932. 4
13. 0	934. 4	936. 3	938. 2	940. 1	942. 1	944. 0	945. 9	947. 8	949. 8	951. 7
13. 1	953. 7	955. 6	957. 5	959. 5	961. 4	963. 4	965. 3	967. 3	969. 3	971. 2
13. 2	973. 2	975. 2	977. 1	979. 1	981. 1	983. 1	985. 0	987. 0	989. 0	991. 0
13. 3	993. 0	995. 0	997. 0	999. 0	1, 001	1, 003	1, 005	1, 007	1, 009	1, 011
13. 4	1, 013	1, 015	1, 017	1, 019	1, 021	1, 023	1, 025	1, 027	1, 029	1, 031
13. 5	1, 033	1, 035	1, 037	1, 039	1, 041	1, 044	1, 046	1, 048	1, 050	1, 052
13. 6	1, 054	1, 056	1, 058	1, 060	1, 062	1, 064	1, 066	1, 068	1, 070	1, 073
13. 7	1, 075	1, 077	1, 079	1, 081	1, 083	1, 085	1, 087	1, 089	1, 091	1, 094
13. 8	1, 096	1, 098	1, 100	1, 102	1, 104	1, 106	1, 108	1, 111	1, 113	1, 115
13. 9	1, 117	1, 119	1, 121	1, 123	1, 126	1, 128	1, 130	1, 132	1, 134	1, 136
14. 0	1, 139	1, 141	1, 143	1, 145	1, 147	1, 149	1, 152	1, 154	1, 156	1, 158
14. 1	1, 160	1, 163	1, 165	1, 167	1, 169	1, 171	1, 174	1, 176	1, 178	1, 180
14. 2	1, 182	1, 185	1, 187	1, 189	1, 191	1, 194	1, 196	1, 198	1, 200	1, 203
14. 3	1, 205	1, 207	1, 209	1, 211	1, 214	1, 216	1, 218	1, 221	1, 223	1, 225
14. 4	1, 227	1, 230	1, 232	1, 234	1, 236	1, 239	1, 241	1, 243	1, 246	1, 248
14. 5	1, 250	1, 253	1, 255	1, 257	1, 259	1, 262	1, 264	1, 266	1, 269	1, 271
14. 6	1, 273	1, 276	1, 278	1, 280	1, 283	1, 285	1, 287	1, 290	1, 292	1, 294
14. 7	1, 297	1, 299	1, 301	1, 304	1, 306	1, 309	1, 311	1, 313	1, 316	1, 318
14. 8	1, 320	1, 323	1, 325	1, 328	1, 330	1, 332	1, 335	1, 337	1, 339	1, 342
14. 9	1, 344	1, 347	1, 349	1, 352	1, 354	1, 356	1, 359	1, 361	1, 364	1, 366

Table 60.—*Eight-thirds powers of numbers*—Continued

Number	.00	.01	.02	.03	.04	.05	.06	.07	.08	.09
15.0	1,368	1,371	1,373	1,376	1,378	1,381	1,383	1,386	1,388	1,391
15.1	1,393	1,395	1,398	1,400	1,403	1,405	1,408	1,410	1,413	1,415
15.2	1,418	1,420	1,423	1,425	1,428	1,430	1,433	1,435	1,438	1,440
15.3	1,443	1,445	1,448	1,450	1,453	1,455	1,458	1,460	1,463	1,465
15.4	1,468	1,471	1,473	1,476	1,478	1,481	1,483	1,486	1,488	1,491
15.5	1,494	1,496	1,499	1,501	1,504	1,506	1,509	1,512	1,514	1,517
15.6	1,519	1,522	1,525	1,527	1,530	1,532	1,535	1,538	1,540	1,543
15.7	1,545	1,548	1,551	1,553	1,556	1,559	1,561	1,564	1,567	1,569
15.8	1,572	1,575	1,577	1,580	1,583	1,585	1,588	1,591	1,593	1,596
15.9	1,599	1,601	1,604	1,607	1,609	1,612	1,615	1,617	1,620	1,623
16.0	1,625	1,628	1,631	1,634	1,636	1,639	1,642	1,645	1,647	1,650
16.1	1,653	1,655	1,658	1,661	1,664	1,666	1,669	1,672	1,675	1,677
16.2	1,680	1,683	1,686	1,689	1,691	1,694	1,697	1,700	1,702	1,705
16.3	1,708	1,711	1,714	1,716	1,719	1,722	1,725	1,728	1,730	1,733
16.4	1,736	1,739	1,742	1,745	1,747	1,750	1,753	1,756	1,759	1,762
16.5	1,765	1,767	1,770	1,773	1,776	1,779	1,782	1,785	1,787	1,790
16.6	1,793	1,796	1,799	1,802	1,805	1,808	1,811	1,813	1,816	1,819
16.7	1,822	1,825	1,828	1,831	1,834	1,837	1,840	1,843	1,846	1,848
16.8	1,851	1,854	1,857	1,860	1,863	1,866	1,869	1,872	1,875	1,878
16.9	1,881	1,884	1,887	1,890	1,893	1,896	1,899	1,902	1,905	1,908
17.0	1,911	1,914	1,917	1,920	1,923	1,926	1,929	1,932	1,935	1,938
17.1	1,941	1,944	1,947	1,950	1,953	1,956	1,959	1,962	1,965	1,968
17.2	1,971	1,974	1,977	1,981	1,984	1,987	1,990	1,993	1,996	1,999
17.3	2,002	2,005	2,008	2,011	2,014	2,017	2,021	2,024	2,027	2,030
17.4	2,033	2,036	2.039	2,042	2,045	2,049	2,052	2,055	2,058	2,061
17.5	2,064	2,067	2,071	2,074	2,077	2,080	2,083	2,086	2,090	2,093
17.6	2,096	2,099	2,102	2,105	2,109	2,112	2,115	2,118	2,121	2,125
17.7	2,128	2,131	2,134	2,137	2,141	2,144	2,147	2,150	2,154	2,157
17.8	2,160	2,163	2,166	2,170	2,173	2,176	2,180	2,183	2,186	2,189
17.9	2,193	2,196	2,199	2,202	2,206	2,209	2,212	2,215	2,219	2,222
18.0	2,225	2,229	2,232	2,235	2,239	2,242	2,245	2,248	2,252	2,255
18.1	2,258	2,262	2,265	2,268	2,272	2,275	2,278	2,282	2,285	2,289
18.2	2,292	2,295	2,299	2,302	2,305	2,309	2,312	2,315	2,319	2,322
18.3	2,326	2,329	2,332	2,336	2,339	2,343	2,346	2,349	2,353	2,356
18.4	2,360	2,363	2,366	2,370	2,373	2,377	2,380	2,384	2,387	2,391
18.5	2,394	2,397	2,401	2,404	2,408	2,411	2,415	2,418	2,422	2,425
18.6	2,429	2,432	2,436	2,439	2,443	2,446	2,450	2,453	2,457	2,460
18.7	2,464	2,467	2,471	2,474	2,478	2,481	2,485	2,488	2,492	2,495
18.8	2,499	2,502	2,506	2,510	2,513	2,517	2,520	2,524	2,527	2,531
18.9	2,535	2,538	2,542	2,545	2,549	2,553	2,556	2,560	2,563	2,567
19.0	2,570	2,574	2,578	2,581	2,585	2,589	2,592	2,596	2,599	2,603
19.1	2,607	2,610	2,614	2,618	2,621	2,625	2,629	2,632	2,636	2,640
19.2	2,643	2,647	2,651	2,654	2,658	2,662	2,665	2,669	2,673	2,676
19.3	2,680	2,684	2,688	2,691	2,695	2,699	2,702	2,706	2,710	2,714
19.4	2,717	2,721	2,725	2,729	2,732	2,736	2,740	2,744	2,747	2,751
19.5	2,755	2,759	2,762	2,766	2,770	2,774	2,777	2,781	2,785	2,789
19.6	2,793	2,796	2,800	2,804	2,808	2,812	2,816	2,819	2,823	2,827
19.7	2,831	2,835	2,838	2,842	2,846	2,850	2,854	2,858	2,862	2,865
19.8	2,869	2,873	2,877	2,881	2,885	2,889	2,893	2,896	2,900	2,904
19.9	2,908	2,912	2,916	2,920	2,924	2,928	2,932	2,935	2,939	2,943

Table 60.—*Eight-thirds powers of numbers*—Continued

Number	.00	.01	.02	.03	.04	.05	.06	.07	.08	.09
20.0	2,947	2,951	2,955	2,959	2,963	2,967	2,971	2,975	2,979	2.983
20.1	2,987	2,991	2,995	2,999	3,003	3,007	3,011	3,015	3,018	3,022
20.2	3,026	3,030	3,034	3,038	3,043	3,047	3,051	3,055	3,059	3,063
20.3	3,067	3,071	3,075	3,079	3,083	3,087	3,091	3,095	3,099	3.103
20.4	3,107	3,111	3,115	3,119	3,123	3,127	3,131	3,136	3,140	3,144
20.5	3,148	3,152	3,156	3,160	3,164	3,168	3,173	3,177	3,181	3,185
20.6	3,189	3,193	3,197	3,201	3,206	3,210	3,214	3,218	3,222	3,226
20.7	3,230	3,235	3,239	3,243	3,247	3,251	3,255	3,260	3,264	3,268
20.8	3,272	3,276	3,281	3,285	3,289	3,293	3,297	3,302	3,306	3,310
20.9	3,314	3,319	3,323	3,327	3,331	3,336	3,340	3,344	3,348	3,352
21.0	3,357	3,361	3,365	3,370	3,374	3,378	3,382	3,387	3,391	3,395
21.1	3,400	3,404	3,408	3,412	3,417	3,421	3,425	3,430	3,434	3,438
21.2	3,443	3,447	3,451	3,456	3,460	3.464	3,469	3,473	3,477	3,482
21.3	3,486	3,491	3,495	3,499	3,504	3,508	3,512	3,517	3,521	3,526
21.4	3,530	3,534	3,539	3,543	3,548	3,552	3,556	3,561	3,565	3,570
21.5	3,574	3,579	3,583	3,587	3,592	3,596	3,601	3,605	3,610	3,614
21.6	3,619	3,623	3,628	3,632	3,637	3,641	3,646	3,650	3,654	3,659
21.7	3,663	3,668	3,673	3,677	3,682	3,686	3,691	3,695	3,700	3,704
21.8	3,709	3,713	3,718	3,722	3,727	3,731	3,736	3,741	3,745	3,750
21.9	3,754	3,759	3,763	3,768	3,773	3,777	3,782	3,786	3,791	3,796
22.0	3,800	3,805	3,809	3,814	3,819	3,823	3,828	3,832	3,837	3,842
22.1	3,846	3,851	3,856	3,860	3,865	3,870	3,874	3,879	3,884	3,888
22.2	3,893	3,898	3,902	3,907	3,912	3,916	3,921	3,926	3,930	3,935
22.3	3,940	3,945	3,949	3,954	3,959	3,963	3,968	3,973	3,978	3,982
22.4	3,987	3,992	3,997	4,001	4,006	4,011	4,016	4,020	4,025	4,030
22.5	4,035	4,040	4,044	4,049	4,054	4,059	4,064	4,068	4,073	4,078
22.6	4,083	4,088	4,092	4,097	4,102	4,107	4,112	4,117	4,121	4,126
22.7	4,131	4,136	4,141	4,146	4,151	4,155	4,160	4,165	4,170	4,175
22.8	4,180	4,185	4,190	4,195	4,199	4,204	4,209	4,214	4,219	4,224
22.9	4,229	4,234	4,239	4,244	4,249	4,254	4,259	4,263	4,268	4,273
23.0	4,278	4,283	4,288	4,293	4,298	4,303	4,308	4,313	4,318	4,323
23.1	4,328	4,333	4,338	4.343	4,348	4,353	4,358	4,363	4,368	4,373
23.2	4,378	4,383	4,388	4,393	4,398	4,404	4,409	4,414	4,419	4,424
23.3	4,429	4,434	4,439	4,444	4,449	4,454	4,459	4,464	4,469	4,475
23.4	4,480	4,485	4,490	4,495	4,500	4,505	4,510	4,515	4,521	4,526
23.5	4,531	4,536	4,541	4,546	4,552	4,557	4,562	4,567	4,572	4,577
23.6	4,582	4,588	4,593	4,598	4,603	4,608	4,614	4,619	4,624	4,629
23.7	4,634	4,640	4,645	4,650	4,655	4,661	4,666	4,671	4,676	4,682
23.8	4,687	4,692	4,697	4,703	4,708	4,713	4,718	4,724	4,729	4,734
23.9	4,739	4,745	4,750	4,755	4,761	4,766	4,771	4,777	4,782	4,787
24.0	4,793	4,798	4,803	4,809	4,814	4,819	4,825	4,830	4,835	4,841
24.1	4,846	4,851	4,857	4,862	4,867	4,873	4,878	4,884	4,889	4,894
24.2	4,900	4,905	4,911	4,916	4,921	4,927	4,932	4,938	4,943	4,948
24.3	4,954	4,959	4,965	4,970	4,976	4,981	4,987	4,992	4,998	5,003
24.4	5,008	5,014	5,019	5,025	5,030	5,036	5.041	5,047	5,052	5,058
24.5	5,063	5,069	5,074	5,080	5,086	5,091	5,097	5,102	5,108	5,113
24.6	5,119	5,124	5,130	5.135	5,141	5,147	5,152	5,158	5,163	5,169
24.7	5,174	5,180	5,186	5,191	5,197	5,202	5,208	5,214	5,219	5,225
24.8	5,230	5,236	5,242	5,247	5,253	5,259	5,264	5,270	5,276	5,281
24.9	5,287	5,293	5,298	5,304	5,310	5,315	5,321	5,327	5,332	5,338

Table 60.—*Eight-thirds powers of numbers*—Continued

Number	.00	.01	.02	.03	.04	.05	.06	.07	.08	.09
25.0	5,344	5,349	5,355	5,361	5,367	5,372	5,378	5,384	5,389	5.395
25.1	5,401	5.407	5,412	5,418	5,424	5,430	5,435	5,441	5,447	5,453
25.2	5,458	5,464	5,470	5,476	5,482	5,487	5,493	5,499	5,505	5,511
25.3	5,516	5,522	5,528	5,534	5,540	5,546	5,551	5,557	5,563	5.569
25.4	5,575	5,581	5,586	5,592	5,598	5,604	5,610	5,616	5,622	5,628
25.5	5,633	5,639	5.645	5,651	5,657	5,663	5,669	5,675	5,681	5,687
25.6	5,693	5,698	5,704	5,710	5,716	5,722	5,728	5,734	5,740	5,746
25.7	5,752	5,758	5,764	5,770	5,776	5,782	5,788	5,794	5,800	5,806
25.8	5,812	5,818	5,824	5,830	5,836	5,842	5,848	5,854	5,860	5,866
25.9	5,872	5,878	5,884	5,890	5,896	5,903	5,909	5,915	5,921	5,927
26.0	5,933	5,939	5,945	5,951	5,957	5,963	5,969	5,976	5,982	5,988
26.1	5,994	6,000	6,006	6,012	6,018	6,025	6,031	6,037	6,043	6,049
26.2	6,055	6,061	6,068	6,074	6,080	6,086	6,092	6.099	6,105	6,111
26.3	6,117	6,123	6,130	6,136	6,142	6,148	6,154	6,161	6,167	6,173
26.4	6,179	6,186	6,192	6,198	6,204	6,211	6,217	6,223	6,229	6,236
26.5	6,242	6,248	6,255	6,261	6,267	6.274	6,280	6,286	6,292	6,299
26.6	6,305	6,311	6,318	6,324	6.330	6.337	6,343	6,349	6,356	6,362
26.7	6,368	6,375	6,381	6,388	6,394	6,400	6,407	6,413	6,419	6,426
26.8	6,432	6,439	6,445	6,451	6,458	6,464	6,471	6,477	6,484	6,490
26.9	6,496	6,503	6,509	6,516	6,522	6,529	6.535	6,542	6,548	6,555
27.0	6,561	6,568	6,574	6,581	6,587	6,594	6,600	6,607	6,613	6,620
27.1	6,626	6,633	6,639	6,646	6.652	6,659	6,665	6,672	6,678	6,685
27.2	6,691	6,698	6,705	6,711	6,718	6,724	6,731	6,737	6,744	6,751
27.3	6,757	6,764	6,770	6,777	6,784	6,790	6,797	6,804	6,810	6,817
27.4	6,823	6,830	6,837	6,843	6,850	6,857	6,863	6,870	6,877	6,883
27.5	6,890	6.897	6,903	6,910	6,917	6,924	6,930	6,937	6,944	6,950
27.6	6,957	6.964	6,971	6,977	6,984	6,991	6,997	7,004	7,011	7,018
27.7	7,024	7,031	7,038	7,045	7,052	7,058	7,065	7,072	7,079	7,085
27.8	7,092	7,099	7,106	7,113	7,120	7,126	7,133	7,140	7,147	7,154
27.9	7,161	7,167	7,174	7,181	7,188	7,195	7,202	7,209	7,215	7,222
28.0	7,229	7,236	7,243	7,250	7,257	7,264	7,271	7,277	7,284	7,291
28.1	7,298	7,305	7,312	7,319	7,326	7,333	7,340	7,347	7,354	7,361
28.2	7,368	7,375	7,382	7,389	7,396	7,403	7,410	7,417	7.424	7,431
28.3	7,438	7,445	7,452	7,459	7,466	7,473	7,480	7,487	7,494	7,501
28.4	7,508	7,515	7,522	7,529	7,536	7,543	7,550	7,557	7,564	7,572
28.5	7,579	7,586	7,593	7,600	7,607	7,614	7,621	7,628	7,635	7,643
28.6	7,650	7,657	7,664	7,671	7,678	7,685	7,693	7,700	7,707	7,714
28.7	7,721	7,728	7,736	7,743	7,750	7,757	7,764	7,772	7,779	7,786
28.8	7,793	7,800	7,808	7,815	7,822	7,829	7,837	7,844	7,851	7,858
28.9	7,866	7,873	7,880	7,887	7,895	7,902	7,909	7,916	7,924	7,931
29.0	7,938	7,946	7,953	7,960	7,968	7,975	7,982	7,990	7,997	8,004
29.1	8,012	8,019	8,026	8,034	8,041	8,048	8,056	8,063	8,070	8,078
29.2	8,085	8,093	8,100	8,107	8,115	8,122	8,130	8,137	8,144	8,152
29.3	8,159	8,167	8,174	8,182	8,189	8,196	8,204	8,211	8,219	8,226
29.4	8,234	8,241	8,249	8,256	8,264	8,271	8,279	8,286	8,294	8,301
29.5	8,309	8,316	8,324	8,331	8,339	8,346	8,354	8,361	8,369	8,376
29.6	8,384	8,391	8,399	8,407	8,414	8,422	8,429	8,437	8,444	8,452
29.7	8,460	8,467	8,475	8,482	8,490	8,498	8,505	8,513	8,521	8,528
29.8	8,536	8,543	8,551	8,559	8,566	8,574	8,582	8,589	8,597	8,605
29.9	8,612	8,620	8,628	8,635	8,643	8,651	8,659	8,666	8,674	8,682

Table 61.—*Five-thirds powers of numbers*

Number	.00	.01	.02	.03	.04	.05	.06	.07	.08	.09
0.0	.0000	.0005	.0015	.0029	.0047	.0068	.0092	.0119	.0149	.0181
0.1	.0215	.0253	.0292	.0334	.0377	.0423	.0472	.0522	.0574	.0628
0.2	.0684	.0742	.0802	.0863	.0927	.0992	.1059	.1128	.1198	.1271
0.3	.1344	.1420	.1497	.1576	.1656	.1738	.1822	.1907	.1994	.2082
0.4	.2172	.2263	.2355	.2450	.2545	.2643	.2741	.2841	.2943	.3046
0.5	.3150	.3255	.3363	.3471	.3581	.3692	.3805	.3919	.4034	.4150
0.6	.4268	.4387	.4508	.4630	.4753	.4877	.5003	.5130	.5258	.5388
0.7	.5519	.5651	.5784	.5918	.6054	.6191	.6329	.6469	.6609	.6751
0.8	.6894	.7038	.7184	.7330	.7478	.7627	.7777	.7929	.8081	.8235
0.9	.8390	.8545	.8703	.8861	.9020	.9181	.9342	.9505	.9669	.9834
1.0	1.000	1.017	1.034	1.050	1.068	1.085	1.102	1.119	1.137	1.154
1.1	1.172	1.190	1.208	1.226	1.244	1.262	1.281	1.299	1.318	1.336
1.2	1.355	1.374	1.393	1.412	1.431	1.450	1.470	1.489	1.509	1.529
1.3	1.548	1.568	1.588	1.608	1.629	1.649	1.669	1.690	1.711	1.731
1.4	1.752	1.773	1.794	1.815	1.836	1.858	1.879	1.900	1.922	1.944
1.5	1.966	1.987	2.009	2.032	2.054	2.076	2.098	2.121	2.143	2.166
1.6	2.189	2.212	2.235	2.258	2.281	2.304	2.327	2.351	2.374	2.398
1.7	2.421	2.445	2.469	2.493	2.517	2.541	2.566	2.590	2.614	2.639
1.8	2.664	2.688	2.713	2.738	2.763	2.788	2.813	2.838	2.864	2.889
1.9	2.915	2.940	2.966	2.992	3.018	3.044	3.070	3.096	3.122	3.148
2.0	3.175	3.201	3.228	3.255	3.281	3.308	3.335	3.362	3.389	3.416
2.1	3.444	3.471	3.499	3.526	3.554	3.581	3.609	3.637	3.665	3.693
2.2	3.721	3.750	3.778	3.806	3.835	3.863	3.892	3.921	3.950	3.979
2.3	4.008	4.037	4.066	4.095	4.124	4.154	4.183	4.213	4.243	4.272
2.4	4.302	4.332	4.362	4.392	4.422	4.453	4.483	4.513	4.544	4.574
2.5	4.605	4.636	4.667	4.698	4.728	4.760	4.791	4.822	4.853	4.885
2.6	4.916	4.948	4.979	5.011	5.043	5.075	5.107	5.139	5.171	5.203
2.7	5.235	5.268	5.300	5.333	5.365	5.398	5.431	5.463	5.496	5.529
2.8	5.562	5.596	5.629	5.662	5.695	5.729	5.762	5.796	5.830	5.864
2.9	5.897	5.931	5.965	5.999	6.034	6.068	6.102	6.137	6.171	6.206
3.0	6.240	6.275	6.310	6.345	6.380	6.415	6.450	6.485	6.520	6.555
3.1	6.591	6.626	6.662	6.697	6.733	6.769	6.805	6.841	6.877	6.913
3.2	6.949	6.985	7.021	7.058	7.094	7.131	7.167	7.204	7.241	7.278
3.3	7.315	7.352	7.389	7.426	7.463	7.500	7.538	7.575	7.613	7.650
3.4	7.688	7.725	7.763	7.801	7.839	7.877	7.915	7.953	7.992	8.030
3.5	8.068	8.107	8.145	8.184	8.223	8.261	8.300	8.339	8.378	8.417
3.6	8.456	8.495	8.535	8.574	8.613	8.653	8.692	8.732	8.772	8.811
3.7	8.851	8.891	8.931	8.971	9.011	9.051	9.092	9.132	9.172	9.213
3.8	9.253	9.294	9.335	9.376	9.416	9.457	9.498	9.539	9.580	9.622
3.9	9.663	9.704	9.746	9.787	9.829	9.870	9.912	9.954	9.996	10.04
4.0	10.08	10.12	10.16	10.21	10.25	10.29	10.33	10.38	10.42	10.46
4.1	10.50	10.55	10.59	10.63	10.67	10.72	10.76	10.80	10.85	10.89
4.2	10.93	10.98	11.02	11.06	11.11	11.15	11.19	11.24	11.28	11.33
4.3	11.37	11.41	11.46	11.50	11.55	11.59	11.64	11.68	11.73	11.77
4.4	11.81	11.86	11.90	11.95	11.99	12.04	12.08	12.13	12.17	12.22
4.5	12.27	12.31	12.36	12.40	12.45	12.49	12.54	12.59	12.63	12.68
4.6	12.72	12.77	12.82	12.86	12.91	12.95	13.00	13.05	13.09	13.14
4.7	13.19	13.23	13.28	13.33	13.38	13.42	13.47	13.52	13.56	13.61
4.8	13.66	13.71	13.75	13.80	13.85	13.90	13.94	13.99	14.04	14.09
4.9	14.14	14.18	14.23	14.28	14.33	14.38	14.43	14.47	14.52	14.57

Table 61.—*Five-thirds powers of numbers*—Continued

Number	.00	.01	.02	.03	.04	.05	.06	.07	.08	.09
5.0	14.62	14.67	14.72	14.77	14.82	14.86	14.91	14.96	15.01	15.06
5.1	15.11	15.16	15.21	15.26	15.31	15.36	15.41	15.46	15.51	15.56
5.2	15.61	15.66	15.71	15.76	15.81	15.86	15.91	15.96	16.01	16.06
5.3	16.11	16.16	16.21	16.26	16.31	16.37	16.42	16.47	16.52	16.57
5.4	16.62	16.67	16.72	16.78	16.83	16.88	16.93	16.98	17.03	17.09
5.5	17.14	17.19	17.24	17.29	17.35	17.40	17.45	17.50	17.55	17.61
5.6	17.66	17.71	17.76	17.82	17.87	17.92	17.98	18.03	18.08	18.14
5.7	18.19	18.24	18.29	18.35	18.40	18.45	18.51	18.56	18.62	18.67
5.8	18.72	18.78	18.83	18.88	18.94	18.99	19.05	19.10	19.16	19.21
5.9	19.26	19.32	19.37	19.43	19.48	19.54	19.59	19.65	19.70	19.76
6.0	19.81	19.87	19.92	19.98	20.03	20.09	20.14	20.20	20.25	20.31
6.1	20.36	20.42	20.48	20.53	20.59	20.64	20.70	20.76	20.81	20.87
6.2	20.92	20.98	21.04	21.09	21.15	21.21	21.26	21.32	21.38	21.43
6.3	21.49	21.55	21.60	21.66	21.72	21.77	21.83	21.89	21.95	22.00
6.4	22.06	22.12	22.18	22.23	22.29	22.35	22.41	22.47	22.52	22.58
6.5	22.64	22.70	22.76	22.81	22.87	22.93	22.99	23.05	23.11	23.16
6.6	23.22	23.28	23.34	23.40	23.46	23.52	23.58	23.63	23.69	23.75
6.7	23.81	23.87	23.93	23.99	24.05	24.11	24.17	24.23	24.29	24.35
6.8	24.41	24.47	24.53	24.59	24.65	24.71	24.77	24.83	24.89	24.95
6.9	25.01	25.07	25.13	25.19	25.25	25.31	25.37	25.43	25.49	25.55
7.0	25.62	25.68	25.74	25.80	25.86	25.92	25.98	26.04	26.10	26.17
7.1	26.23	26.29	26.35	26.41	26.47	26.54	26.60	26.66	26.72	26.78
7.2	26.85	26.91	26.97	27.03	27.10	27.16	27.22	27.28	27.35	27.41
7.3	27.47	27.53	27.60	27.66	27.72	27.79	27.85	27.91	27.97	28.04
7.4	28.10	28.16	28.23	28.29	28.35	28.42	28.48	28.55	28.61	28.67
7.5	28.74	28.80	28.86	28.93	28.99	29.06	29.12	29.19	29.25	29.31
7.6	29.38	29.44	29.51	29.57	29.64	29.70	29.77	29.83	29.90	29.96
7.7	30.03	30.09	30.16	30.22	30.29	30.35	30.42	30.48	30.55	30.61
7.8	30.68	30.74	30.81	30.87	30.94	31.01	31.07	31.14	31.20	31.27
7.9	31.34	31.40	31.47	31.53	31.60	31.67	31.73	31.80	31.87	31.93
8.0	32.00	32.07	32.13	32.20	32.27	32.33	32.40	32.47	32.54	32.60
8.1	32.67	32.74	32.80	32.87	32.94	33.01	33.07	33.14	33.21	33.28
8.2	33.34	33.41	33.48	33.55	33.62	33.68	33.75	33.82	33.89	33.96
8.3	34.02	34.09	34.16	34.23	34.30	34.37	34.44	34.50	34.57	34.64
8.4	34.71	34.78	34.85	34.92	34.99	35.06	35.13	35.19	35.26	35.33
8.5	35.40	35.47	35.54	35.61	35.68	35.75	35.82	35.89	35.96	36.03
8.6	36.10	36.17	36.24	36.31	36.38	36.45	36.52	36.59	36.66	36.73
8.7	36.80	36.87	36.94	37.01	37.08	37.15	37.23	37.30	37.37	37.44
8.8	37.51	37.58	37.65	37.72	37.79	37.87	37.94	38.01	38.08	38.15
8.9	38.22	38.29	38.37	38.44	38.51	38.58	38.65	38.72	38.80	38.87
9.0	38.94	39.01	39.09	39.16	39.23	39.30	39.37	39.45	39.52	39.59
9.1	39.66	39.74	39.81	39.88	39.96	40.03	40.10	40.17	40.25	40.32
9.2	40.39	40.47	40.54	40.61	40.69	40.76	40.83	40.91	40.98	41.05
9.3	41.13	41.20	41.28	41.35	41.42	41.50	41.57	41.65	41.72	41.79
9.4	41.87	41.94	42.02	42.09	42.17	42.24	42.31	42.39	42.46	42.54
9.5	42.61	42.69	42.76	42.84	42.91	42.99	43.06	43.14	43.21	43.29
9.6	43.36	43.44	43.51	43.59	43.66	43.74	43.82	43.89	43.97	44.04
9.7	44.12	44.19	44.27	44.35	44.42	44.50	44.57	44.65	44.73	44.80
9.8	44.88	44.96	45.03	45.11	45.18	45.26	45.34	45.41	45.49	45.57
9.9	45.64	45.72	45.80	45.88	45.95	46.03	46.11	46.18	46.26	46.34

Table 61.—*Five-thirds powers of numbers*—Continued

Number	.00	.01	.02	.03	.04	.05	.06	.07	.08	.09
10.0	46.42	46.49	46.57	46.65	46.73	46.80	46.88	46.96	47.04	47.11
10.1	47.19	47.27	47.35	47.43	47.50	47.58	47.66	47.74	47.82	47.90
10.2	47.97	48.05	48.13	48.21	48.29	48.37	48.44	48.52	48.60	48.68
10.3	48.76	48.84	48.92	49.00	49.08	49.15	49.23	49.31	49.39	49.47
10.4	49.55	49.63	49.71	49.79	49.87	49.95	50.03	50.11	50.19	50.27
10.5	50.35	50.43	50.51	50.59	50.67	50.75	50.83	50.91	50.99	51.07
10.6	51.15	51.23	51.31	51.39	51.47	51.55	51.63	51.71	51.79	51.88
10.7	51.96	52.04	52.12	52.20	52.28	52.36	52.44	52.52	52.61	52.69
10.8	52.77	52.85	52.93	53.01	53.09	53.18	53.26	53.34	53.42	53.50
10.9	53.59	53.67	53.75	53.83	53.91	54.00	54.08	54.16	54.24	54.32
11.0	54.41	54.49	54.57	54.65	54.74	54.82	54.90	54.99	55.07	55.15
11.1	55.23	55.32	55.40	55.48	55.57	55.65	55.73	55.82	55.90	55.98
11.2	56.07	56.15	56.23	56.32	56.40	56.48	56.57	56.65	56.73	56.82
11.3	56.90	56.99	57.07	57.15	57.24	57.32	57.41	57.49	57.58	57.66
11.4	57.74	57.83	57.91	58.00	58.08	58.17	58.25	58.34	58.42	58.51
11.5	58.59	58.68	58.76	58.85	58.93	59.02	59.10	59.19	59.27	59.36
11.6	59.44	59.53	59.61	59.70	59.78	59.87	59.96	60.04	60.13	60.21
11.7	60.30	60.38	60.47	60.56	60.64	60.73	60.82	60.90	60.99	61.07
11.8	61.16	61.25	61.33	61.42	61.51	61.59	61.68	61.77	61.85	61.94
11.9	62.03	62.11	62.20	62.29	62.37	62.46	62.55	62.64	62.72	62.81
12.0	62.90	62.99	63.07	63.16	63.25	63.34	63.42	63.51	63.60	63.69
12.1	63.77	63.86	63.95	64.04	64.13	64.21	64.30	64.39	64.48	64.57
12.2	64.65	64.74	64.83	64.92	65.01	65.10	65.19	65.27	65.36	65.45
12.3	65.54	65.63	65.72	65.81	65.90	65.98	66.07	66.16	66.25	66.34
12.4	66.43	66.52	66.61	66.70	66.79	66.88	66.97	67.06	67.15	67.24
12.5	67.33	67.42	67.51	67.60	67.69	67.78	67.87	67.96	68.05	68.14
12.6	68.23	68.32	68.41	68.50	68.59	68.68	68.77	68.86	68.95	69.04
12.7	69.13	69.22	69.31	69.40	69.49	69.59	69.68	69.77	69.86	69.95
12.8	70.04	70.13	70.22	70.31	70.41	70.50	70.59	70.68	70.77	70.86
12.9	70.95	71.05	71.14	71.23	71.32	71.41	71.51	71.60	71.69	71.78
13.0	71.87	71.97	72.06	72.15	72.24	72.34	72.43	72.52	72.61	72.71
13.1	72.80	72.89	72.98	73.08	73.17	73.26	73.35	73.45	73.54	73.63
13.2	73.73	73.82	73.91	74.01	74.10	74.19	74.29	74.38	74.47	74.57
13.3	74.66	74.75	74.85	74.94	75.03	75.13	75.22	75.32	75.41	75.50
13.4	75.60	75.69	75.79	75.88	75.97	76.07	76.16	76.26	76.35	76.45
13.5	76.54	76.63	76.73	76.82	76.92	77.01	77.11	77.20	77.30	77.39
13.6	77.49	77.58	77.68	77.77	77.87	77.96	78.06	78.15	78.25	78.34
13.7	78.44	78.53	78.63	78.73	78.82	78.92	79.01	79.11	79.20	79.30
13.8	79.40	79.49	79.59	79.68	79.78	79.88	79.97	80.07	80.16	80.26
13.9	80.36	80.45	80.55	80.65	80.74	80.84	80.94	81.03	81.13	81.23
14.0	81.32	81.42	81.52	81.61	81.71	81.81	81.90	82.00	82.10	82.20
14.1	82.29	82.39	82.49	82.59	82.68	82.78	82.88	82.98	83.07	83.17
14.2	83.27	83.37	83.46	83.56	83.66	83.76	83.86	83.95	84.05	84.15
14.3	84.25	84.35	84.44	84.54	84.64	84.74	84.84	84.94	85.04	85.13
14.4	85.23	85.33	85.43	85.53	85.63	85.73	85.82	85.92	86.02	86.12
14.5	86.22	86.32	86.42	86.52	86.62	86.72	86.82	86.92	87.02	87.11
14.6	87.21	87.31	87.41	87.51	87.61	87.71	87.81	87.91	88.01	88.11
14.7	88.21	88.31	88.41	88.51	88.61	88.71	88.81	88.91	89.01	89.11
14.8	89.21	89.32	89.42	89.52	89.62	89.72	89.82	89.92	90.02	90.12
14.9	90.22	90 32	90.42	90.52	90.63	90.73	90.83	90.93	91.03	91.13

Table 61.—*Five-thirds powers of numbers*—Continued

Number	.00	.01	.02	.03	.04	.05	.06	.07	.08	.09
15.0	91.23	91.33	91.44	91.54	91.64	91.74	91.84	91.94	92.05	92.15
15.1	92.25	92.35	92.45	92.55	92.66	92.76	92.86	92.96	93.06	93.17
15.2	93.27	93.37	93.47	93.58	93.68	93.78	93.88	93.99	94.09	94.19
15.3	94.29	94.40	94.50	94.60	94.71	94.81	94.91	95.01	95.12	95.22
15.4	95.32	95.43	95.53	95.63	95.74	95.84	95.94	96.05	96.15	96.25
15.5	96.36	96.46	96.56	96.67	96.77	96.88	96.98	97.08	97.19	97.29
15.6	97.40	97.50	97.60	97.71	97.81	97.92	98.02	98.13	98.23	98.33
15.7	98.44	98.54	98.65	98.75	98.86	98.96	99.07	99.17	99.28	99.38
15.8	99.49	99.59	99.70	99.80	99.91	100.0	100.1	100.2	100.3	100.4
15.9	100.5	100.6	100.7	100.9	101.0	101.1	101.2	101.3	101.4	101.5
16.0	101.6	101.7	101.8	101.9	102.0	102.1	102.2	102.3	102.4	102.5
16.1	102.7	102.8	102.9	103.0	103.1	103.2	103.3	103.4	103.5	103.6
16.2	103.7	103.8	103.9	104.0	104.1	104.3	104.4	104.5	104.6	104.7
16.3	104.8	104.9	105.0	105.1	105.2	105.3	105.4	105.5	105.6	105.8
16.4	105.9	106.0	106.1	106.2	106.3	106.4	106.5	106.6	106.7	106.8
16.5	106.9	107.0	107.2	107.3	107.4	107.5	107.6	107.7	107.8	107.9
16.6	108.0	108.1	108.2	108.3	108.5	108.6	108.7	108.8	108.9	109.0
16.7	109.1	109.2	109.3	109.4	109.5	109.7	109.8	109.9	110.0	110.1
16.8	110.2	110.3	110.4	110.5	110.6	110.7	110.9	111.0	111.1	111.2
16.9	111.3	111.4	111.5	111.6	111.7	111.8	112.0	112.1	112.2	112.3
17.0	112.4	112.5	112.6	112.7	112.8	112.9	113.1	113.2	113.3	113.4
17.1	113.5	113.6	113.7	113.8	113.9	114.1	114.2	114.3	114.4	114.5
17.2	114.6	114.7	114.8	114.9	115.1	115.2	115.3	115.4	115.5	115.6
17.3	115.7	115.8	115.9	116.1	116.2	116.3	116.4	116.5	116.6	116.7
17.4	116.8	116.9	117.1	117.2	117.3	117.4	117.5	117.6	117.7	117.8
17.5	118.0	118.1	118.2	118.3	118.4	118.5	118.6	118.7	118.9	119.0
17.6	119.1	119.2	119.3	119.4	119.5	119.6	119.8	119.9	120.0	120.1
17.7	120.2	120.3	120.4	120.6	120.7	120.8	120.9	121.0	121.1	121.2
17.8	121.3	121.5	121.6	121.7	121.8	121.9	122.0	122.1	122.3	122.4
17.9	122.5	122.6	122.7	122.8	122.9	123.1	123.2	123.3	123.4	123.5
18.0	123.6	123.7	123.9	124.0	124.1	124.2	124.3	124.4	124.5	124.7
18.1	124.8	124.9	125.0	125.1	125.2	125.4	125.5	125.6	125.7	125.8
18.2	125.9	126.0	126.2	126.3	126.4	126.5	126.6	126.7	126.9	127.0
18.3	127.1	127.2	127.3	127.4	127.5	127.7	127.8	127.9	128.0	128.1
18.4	128.2	128.4	128.5	128.6	128.7	128.8	128.9	129.1	129.2	129.3
18.5	129.4	129.5	129.6	129.8	129.9	130.0	130.1	130.2	130.3	130.5
18.6	130.6	130.7	130.8	130.9	131.0	131.2	131.3	131.4	131.5	131.6
18.7	131.7	131.9	132.0	132.1	132.2	132.3	132.5	132.6	132.7	132.8
18.8	132.9	133.0	133.2	133.3	133.4	133.5	133.6	133.7	133.9	134.0
18.9	134.1	134.2	134.3	134.5	134.6	134.7	134.8	134.9	135.0	135.2
19.0	135.3	135.4	135.5	135.6	135.8	135.9	136.0	136.1	136.2	136.4
19.1	136.5	136.6	136.7	136.8	137.0	137.1	137.2	137.3	137.4	137.5
19.2	137.7	137.8	137.9	138.0	138.1	138.3	138.4	138.5	138.6	138.7
19.3	138.9	139.0	139.1	139.2	139.3	139.5	139.6	139.7	139.9	140.0
19.4	140.1	140.2	140.3	140.4	140.5	140.7	140.8	140.9	141.0	141.2
19.5	141.3	141.4	141.5	141.6	141.8	141.9	142.0	142.1	142.2	142.4
19.6	142.5	142.6	142.7	142.8	143.0	143.1	143.2	143.3	143.5	143.6
19.7	143.7	143.8	143.9	144.1	144.2	144.3	144.4	144.5	144.7	144.8
19.8	144.9	145.0	145.2	145.3	145.4	145.5	145.6	145.8	145.9	146.0
19.9	146.1	146.3	146.4	146.5	146.6	146.7	146.9	147.0	147.1	147.2

Table 61.—*Five-thirds powers of numbers*—Continued

Number	.00	.01	.02	.03	.04	.05	.06	.07	.08	.09
20.0	147.4	147.5	147.6	147.7	147.9	148.0	148.1	148.2	148.3	148.5
20.1	148.6	148.7	148.8	149.0	149.1	149.2	149.3	149.5	149.6	149.7
20.2	149.8	149.9	150.1	150.2	150.3	150.4	150.6	150.7	150.8	150.9
20.3	151.1	151.2	151.3	151.4	151.6	151.7	151.8	151.9	152.1	152.2
20.4	152.3	152.4	152.6	152.7	152.8	152.9	153.1	153.2	153.3	153.4
20.5	153.6	153.7	153.8	153.9	154.1	154.2	154.3	154.4	154.6	154.7
20.6	154.8	154.9	155.1	155.2	155.3	155.4	155.6	155.7	155.8	155.9
20.7	156.1	156.2	156.3	156.4	156.6	156.7	156.8	156.9	157.1	157.2
20.8	157.3	157.4	157.6	157.7	157.8	157.9	158.1	158.2	158.3	158.5
20.9	158.6	158.7	158.8	159.0	159.1	159.2	159.3	159.5	159.6	159.7
21.0	159.8	160.0	160.1	160.2	160.4	160.5	160.6	160.7	160.9	161.0
21.1	161.1	161.2	161.4	161.5	161.6	161.8	161.9	162.0	162.1	162.3
21.2	162.4	162.5	162.6	162.8	162.9	163.0	163.2	163.3	163.4	163.5
21.3	163.7	163.8	163.9	164.1	164.2	164.3	164.4	164.6	164.7	164.8
21.4	165.0	165.1	165.2	165.3	165.5	165.6	165.7	165.9	166.0	166.1
21.5	166.2	166.4	166.5	166.6	166.8	166.9	167.0	167.1	167.3	167.4
21.6	167.5	167.7	167.8	167.9	168.0	168.2	168.3	168.4	168.6	168.7
21.7	168.8	169.0	169.1	169.2	169.3	169.5	169.6	169.7	169.9	170.0
21.8	170.1	170.3	170.4	170.5	170.6	170.8	170.9	171.0	171.2	171.3
21.9	171.4	171.6	171.7	171.8	171.9	172.1	172.2	172.3	172.5	172.6
22.0	172.7	172.9	173.0	173.1	173.3	173.4	173.5	173.6	173.8	173.9
22.1	174.0	174.2	174.3	174.4	174.6	174.7	174.8	175.0	175.1	175.2
22.2	175.4	175.5	175.6	175.8	175.9	176.0	176.1	176.3	176.4	176.5
22.3	176.7	176.8	176.9	177.1	177.2	177.3	177.5	177.6	177.7	177.9
22.4	178.0	178.1	178.3	178.4	178.5	178.7	178.8	178.9	179.1	179.2
22.5	179.3	179.5	179.6	179.7	179.9	180.0	180.1	180.3	180.4	180.5
22.6	180.7	180.8	180.9	181.1	181.2	181.3	181.5	181.6	181.7	181.9
22.7	182.0	182.1	182.3	182.4	182.5	182.7	182.8	182.9	183.1	183.2
22.8	183.3	183.5	183.6	183.7	183.9	184.0	184.1	184.3	184.4	184.5
22.9	184.7	184.8	184.9	185.1	185.2	185.3	185.5	185.6	185.7	185.9
23.0	186.0	186.1	186.3	186.4	186.6	186.7	186.8	187.0	187.1	187.2
23.1	187.4	187.5	187.6	187.8	187.9	188.0	188.2	188.3	188.4	188.6
23.2	188.7	188.9	189.0	189.1	189.3	189.4	189.5	189.7	189.8	189.9
23.3	190.1	190.2	190.3	190.5	190.6	190.8	190.9	191.0	191.2	191.3
23.4	191.4	191.6	191.7	191.8	192.0	192.1	192.3	192.4	192.5	192.7
23.5	192.8	192.9	193.1	193.2	193.3	193.5	193.6	193.8	193.9	194.0
23.6	194.2	194.3	194.4	194.6	194.7	194.9	195.0	195.1	195.3	195.4
23.7	195.5	195.7	195.8	196.0	196.1	196.2	196.4	196.5	196.6	196.8
23.8	196.9	197.1	197.2	197.3	197.5	197.6	197.8	197.9	198.0	198.2
23.9	198.3	198.4	198.6	198.7	198.9	199.0	199.1	199.3	199.4	199.5
24.0	199.7	199.8	200.0	200.1	200.2	200.4	200.5	200.7	200.8	200.9
24.1	201.1	201.2	201.4	201.5	201.6	201.8	201.9	202.1	202.2	202.3
24.2	202.5	202.6	202.7	202.9	203.0	203.2	203.3	203.4	203.6	203.7
24.3	203.9	204.0	204.1	204.3	204.4	204.6	204.7	204.8	205.0	205.1
24.4	205.3	205.4	205.5	205.7	205.8	206.0	206.1	206.2	206.4	206.5
24.5	206.7	206.8	207.0	207.1	207.2	207.4	207.5	207.7	207.8	207.9
24.6	208.1	208.2	208.4	208.5	208.6	208.8	208.9	209.1	209.2	209.3
24.7	209.5	209.6	209.8	209.9	210.1	210.2	210.3	210.5	210.6	210.8
24.8	210.9	211.0	211.2	211.3	211.5	211.6	211.8	211.9	212.0	212.2
24.9	212.3	212.5	212.6	212.8	212.9	213.0	213.2	213.3	213.5	213.6

Table 61.—*Five-thirds powers of numbers*—Continued

Number	.00	.01	.02	.03	.04	.05	.06	.07	.08	.09
25.0	213.7	213.9	214.0	214.2	214.3	214.5	214.6	214.7	214.9	215.0
25.1	215.2	215.3	215.5	215.6	215.7	215.9	216.0	216.2	216.3	216.5
25.2	216.6	216.7	216.9	217.0	217.2	217.3	217.5	217.6	217.8	217.9
25.3	218.0	218.2	218.3	218.5	218.6	218.8	218.9	219.0	219.2	219.3
25.4	219.5	219.6	219.8	219.9	220.1	220.2	220.3	220.5	220.6	220.8
25.5	220.9	221.1	221.2	221.4	221.5	221.6	221.8	221.9	222.1	222.2
25.6	222.4	222.5	222.7	222.8	222.9	223.1	223.2	223.4	223.5	223.7
25.7	223.8	224.0	224.1	224.3	224.4	224.5	224.7	224.8	225.0	225.1
25.8	225.3	225.4	225.6	225.7	225.9	226.0	226.1	226.3	226.4	226.6
25.9	226.7	226.9	227.0	227.2	227.3	227.5	227.6	227.7	227.9	228.0
26.0	228.2	228.3	228.5	228.6	228.8	228.9	229.1	229.2	229.4	229.5
26.1	229.7	229.8	229.9	230.1	230.2	230.4	230.5	230.7	230.8	231.0
26.2	231.1	231.3	231.4	231.6	231.7	231.9	232.0	232.1	232.3	232.4
26.3	232.6	232.7	232.9	233.0	233.2	233.3	233.5	233.6	233.8	233.9
26.4	234.1	234.2	234.4	234.5	234.7	234.8	235.0	235.1	235.3	235.4
26.5	235.5	235.7	235.8	236.0	236.1	236.3	236.4	236.6	236.7	236.9
26.6	237.0	237.2	237.3	237.5	237.6	237.8	237.9	238.1	238.2	238.4
26.7	238.5	238.7	238.8	239.0	239.1	239.3	239.4	239.6	239.7	239.9
26.8	240.0	240.2	240.3	240.5	240.6	240.8	240.9	241.1	241.2	241.4
26.9	241.5	241.7	241.8	242.0	242.1	242.3	242.4	242.6	242.7	242.9
27.0	243.0	243.2	243.3	243.5	243.6	243.8	243.9	244.1	244.2	244.4
27.1	244.5	244.7	244.8	245.0	245.1	245.3	245.4	245.6	245.7	245.9
27.2	246.0	246.2	246.3	246.5	246.6	246.8	246.9	247.1	247.2	247.4
27.3	247.5	247.7	247.8	248.0	248.1	248.3	248.4	248.6	248.7	248.9
27.4	249.0	249.2	249.3	249.5	249.6	249.8	249.9	250.1	250.2	250.4
27.5	250.5	250.7	250.8	251.0	251.2	251.3	251.5	251.6	251.8	251.9
27.6	252.1	252.2	252.4	252.5	252.7	252.8	253.0	253.1	253.3	253.4
27.7	253.6	253.7	253.9	254.0	254.2	254.4	254.5	254.7	254.8	255.0
27.8	255.1	255.3	255.5	255.6	255.7	255.9	256.0	256.2	256.3	256.5
27.9	256.6	256.8	257.0	257.1	257.3	257.4	257.6	257.7	257.9	258.0
28.0	258.2	258.3	258.5	258.6	258.8	259.0	259.1	259.3	259.4	259.6
28.1	259.7	259.9	260.0	260.2	260.3	260.5	260.6	260.8	261.0	261.1
28.2	261.3	261.4	261.6	261.7	261.9	262.0	262.2	262.3	262.5	262.7
28.3	262.8	263.0	263.1	263.3	263.4	263.6	263.7	263.9	264.1	264.2
28.4	264.4	264.5	264.7	264.8	265.0	265.1	265.3	265.4	265.6	265.8
28.5	265.9	266.1	266.2	266.4	266.5	266.7	266.8	267.0	267.2	267.3
28.6	267.5	267.6	267.8	267.9	268.1	268.3	268.4	268.6	268.7	268.9
28.7	269.0	269.2	269.3	269.5	269.7	269.8	270.0	270.1	270.3	270.4
28.8	270.6	270.8	270.9	271.1	271.2	271.4	271.5	271.7	271.8	272.0
28.9	272.2	272.3	272.5	272.6	272.8	272.9	273.1	273.3	273.4	273.6
29.0	273.7	273.9	274.0	274.2	274.4	274.5	274.7	274.8	275.0	275.2
29.1	275.3	275.5	275.6	275.8	275.9	276.1	276.3	276.4	276.6	276.7
29.2	276.9	277.0	277.2	277.4	277.5	277.7	277.8	278.0	278.2	278.3
29.3	278.5	278.6	278.8	278.9	279.1	279.3	279.4	279.6	279.7	279.9
29.4	280.1	280.2	280.4	280.5	280.7	280.9	281.0	281.2	281.3	281.5
29.5	281.6	281.8	282.0	282.1	282.3	282.4	282.6	282.8	282.9	283.1
29.6	283.2	283.4	283.6	283.7	283.9	284.0	284.2	284.4	284.5	284.7
29.7	284.8	285.0	285.2	285.3	285.5	285.6	285.8	286.0	286.1	286.3
29.8	286.4	286.6	286.8	286.9	287.1	287.2	287.4	287.6	287.7	287.9
29.9	288.0	288.2	288.4	288.5	288.7	288.8	289.0	289.2	289.3	289.5

Table 61.—*Five-thirds powers of numbers*—Continued

Number	.0	.1	.2	.3	.4	.5	.6	.7	.8	.9
30	289.6	291.3	292.9	294.5	296.1	297.7	299.4	301.0	302.6	304.3
31	305.9	307.6	309.2	310.9	312.5	314.2	315.8	317.5	319.2	320.9
32	322.5	324.2	325.9	327.6	329.3	331.0	332.7	334.4	336.1	337.8
33	339.5	341.2	342.9	344.7	346.4	348.1	349.9	351.6	353.3	355.1
34	356.8	358.6	360.3	362.1	363.9	365.6	367.4	369.2	370.9	372.7
35	374.5	376.3	378.1	379.9	381.7	383.5	385.3	387.1	388.9	390.7
36	392.5	394.3	396.1	398.0	399.8	401.6	403.5	405.3	407.1	409.0
37	410.8	412.7	414.5	416.4	418.3	420.1	422.0	423.9	425.7	427.6
38	429.5	431.4	433.3	435.2	437.1	439.0	440.9	442.8	444.7	446.6
39	448.5	450.4	452.4	454.3	456.2	458.1	460.1	462.0	464.0	465.9
40	467.8	469.8	471.7	473.7	475.7	477.6	479.6	481.6	483.5	485.5
41	487.5	489.5	491.5	493.5	495.5	497.4	499.4	501.4	503.5	505.5
42	507.5	509.5	511.5	513.5	515.6	517.6	519.6	521.7	523.7	525.7
43	527.8	529.8	531.9	533.9	536.0	538.0	540.1	542.2	544.2	546.3
44	548.4	550.5	552.5	554.6	556.7	558.8	560.9	563.0	565.1	567.2
45	569.3	571.4	573.5	575.7	577.8	579.9	582.0	584.2	586.3	588.4
46	590.6	592.7	594.8	597.0	599.1	601.3	603.5	605.6	607.8	609.9
47	612.1	614.3	616.5	618.6	620.8	623.0	625.2	627.4	629.6	631.8
48	634.0	636.2	638.4	640.6	642.8	645.0	647.2	649.5	651.7	653.9
49	656.1	658.4	660.6	662.8	665.1	667.3	669.6	671.8	674.1	676.3
50	678.6	680.9	683.1	685.4	687.7	690.0	692.2	694.5	696.8	699.1
51	701.4	703.7	706.0	708.3	710.6	712.9	715.2	717.5	719.8	722.1
52	724.4	726.8	729.1	731.4	733.8	736.1	738.4	740.8	743.1	745.5
53	747.8	750.2	752.5	754.9	757.2	759.6	762.0	764.3	766.7	769.1
54	771.5	773.9	776.2	778.6	781.0	783.4	785.8	788.2	790.6	793.0
55	795.4	797.8	800.3	802.7	805.1	807.5	809.9	812.4	814.8	817.2
56	819.7	822.1	824.6	827.0	829.5	831.9	834.4	836.8	839.3	841.8
57	844.2	846.7	849.2	851.6	854.1	856.6	859.1	861.6	864.1	866.6
58	869.1	871.6	874.1	876.6	879.1	881.6	884.1	886.6	889.1	891.6
59	894.2	896.7	899.2	901.8	904.3	906.8	909.4	911.9	914.5	917.0
60	919.6	922.1	924.7	927.2	929.8	932.4	934.9	937.5	940.1	942.7
61	945.3	947.8	950.4	953.0	955.6	958.2	960.8	963.4	966.0	968.6
62	971.2	973.8	976.5	979.1	981.7	984.3	986.9	989.6	992.2	994.8
63	997.5	1,000	1,003	1,005	1,008	1,011	1,013	1,016	1,019	1,021
64	1,024	1,027	1,029	1,032	1,035	1,037	1,040	1,043	1,045	1,048
65	1,051	1,054	1,056	1,059	1,062	1,064	1,067	1,070	1,072	1,075
66	1,078	1,081	1,083	1,086	1,089	1,092	1,094	1,097	1,100	1,102
67	1,105	1,108	1,111	1,114	1,116	1,119	1,122	1,125	1,127	1,130
68	1,133	1,136	1,138	1,141	1,144	1,147	1,150	1,152	1,155	1,158
69	1,161	1,164	1,166	1,169	1,172	1,175	1,178	1,180	1,183	1,186
70	1,189	1,192	1,195	1,197	1,200	1,203	1,206	1,209	1,212	1,215
71	1,217	1,220	1,223	1,226	1,229	1,232	1,235	1,237	1,240	1,243
72	1,246	1,249	1,252	1,255	1,258	1,261	1,263	1,266	1,269	1,272
73	1,275	1,278	1,281	1,284	1,287	1,290	1,293	1,296	1,298	1,301
74	1,304	1,307	1,310	1,313	1,316	1,319	1,322	1,325	1,328	1,331
75	1,334	1,337	1,340	1,343	1,346	1,349	1,352	1,355	1,358	1,361
76	1,364	1,367	1,370	1,373	1,376	1,379	1,382	1,385	1,388	1,391
77	1,394	1,397	1,400	1,403	1,406	1,409	1,412	1,415	1,418	1,421
78	1,424	1,427	1,430	1,433	1,436	1,439	1,442	1,445	1,448	1,451
79	1,454	1,458	1,461	1,464	1,467	1,470	1,473	1,476	1,479	1,482

Table 61.—*Five-thirds powers of numbers*—Continued

Number	.0	.1	.2	.3	.4	.5	.6	.7	.8	.9
80	1,485	1,488	1,492	1,495	1,498	1,501	1,504	1,507	1,510	1,513
81	1,516	1,520	1,523	1,526	1,529	1,532	1,535	1,538	1,541	1,545
82	1,548	1,551	1,554	1,557	1,560	1,563	1,567	1,570	1,573	1,576
83	1,579	1,582	1,586	1,589	1,592	1,595	1,598	1,602	1,605	1,608
84	1,611	1,614	1,618	1,621	1,624	1,627	1,630	1,634	1,637	1,640
85	1,643	1,646	1,650	1,653	1,656	1,659	1,663	1,666	1,669	1,672
86	1,676	1,679	1,682	1,685	1,689	1,692	1,695	1,698	1,702	1,705
87	1,708	1,711	1,715	1,718	1,721	1,725	1,728	1,731	1,734	1,738
88	1,741	1,744	1,748	1,751	1,754	1,758	1,761	1,764	1,767	1,771
89	1,774	1,777	1,781	1,784	1,787	1,791	1,794	1,797	1,801	1,804
90	1,807	1,811	1,814	1,818	1,821	1,824	1,828	1,831	1,834	1,838
91	1,841	1,844	1,848	1,851	1,855	1,858	1,861	1,865	1,868	1,872
92	1,875	1,878	1,882	1,885	1,889	1,892	1,895	1,899	1,902	1,906
93	1,909	1,912	1,916	1,919	1,923	1,926	1,930	1,933	1,936	1,940
94	1,943	1,947	1,950	1,954	1,957	1,961	1,964	1,968	1,971	1,974
95	1,978	1,981	1,985	1,988	1,992	1,995	1,999	2,002	2,006	2,009
96	2,013	2,016	2,020	2,023	2,027	2,030	2,034	2,037	2,041	2,044
97	2,048	2,051	2,055	2,058	2,062	2,065	2,069	2,072	2,076	2,080
98	2,083	2,087	2,090	2,094	2,097	2,101	2,104	2,108	2,112	2,115
99	2,119	2,122	2,126	2,129	2,133	2,137	2,140	2,144	2,147	2,151

Table 62.—*Four-thirds powers of numbers*

Number	.00	.01	.02	.03	.04	.05	.06	.07	.08	.09
0.0	0.0000	0.0022	0.0054	0.0093	0.0137	0.0184	0.0235	0.0288	0.0345	0.0403
0.1	.0464	.0527	.0592	.0659	.0727	.0797	.0869	.0942	.1016	.1092
0.2	.1170	.1248	.1328	.1409	.1491	.1575	.1659	.1745	.1832	.1920
0.3	.2008	.2098	.2189	.2280	.2373	.2467	.2561	.2656	.2752	.2849
0.4	.2947	.3046	.3145	.3246	.3347	.3448	.3551	.3654	.3758	.3863
0.5	.3969	.4075	.4182	.4289	.4397	.4506	.4616	.4726	.4837	.4948
0.6	.5061	.5173	.5287	.5401	.5515	.5631	.5746	.5863	.5980	.6097
0.7	.6215	.6334	.6453	.6573	.6693	.6814	.6936	.7058	.7180	.7303
0.8	.7427	.7551	.7675	.7800	.7926	.8052	.8178	.8305	.8433	.8561
0.9	.8689	.8818	.8948	.9078	.9208	.9339	.9470	.9602	.9734	.9867
1.0	1.000	1.013	1.027	1.040	1.054	1.067	1.081	1.094	1.108	1.122
1.1	1.136	1.149	1.163	1.177	1.191	1.205	1.219	1.233	1.247	1.261
1.2	1.275	1.289	1.304	1.318	1.332	1.347	1.361	1.375	1.390	1.404
1.3	1.419	1.433	1.448	1.463	1.477	1.492	1.507	1.522	1.586	1.551
1.4	1.566	1.581	1.596	1.611	1.626	1.641	1.656	1.671	1.687	1.702
1.5	1.717	1.732	1.748	1.763	1.778	1.794	1.809	1.825	1.840	1.856
1.6	1.871	1.887	1.903	1.918	1.934	1.950	1.966	1.981	1.997	2.013
1.7	2.029	2.045	2.061	2.077	2.093	2.109	2.125	2.141	2.157	2.173
1.8	2.190	2.206	2.222	2.238	2.255	2.271	2.287	2.304	2.320	2.337
1.9	2.353	2.370	2.386	2.403	2.420	2.436	2.453	2.470	2.486	2.503
2.0	2.520	2.537	2.553	2.570	2.587	2.604	2.621	2.638	2.655	2.672
2.1	2.689	2.706	2.723	2.741	2.758	2.775	2.792	2.809	2.827	2.844
2.2	2.861	2.879	2.896	2.913	2.931	2.948	2.966	2.983	3.001	3.018
2.3	3.036	3.054	3.071	3.089	3.107	3.124	3.142	3.160	3.178	3.195
2.4	3.213	3.231	3.249	3.267	3.285	3.303	3.321	3.339	3.357	3.375
2.5	3.393	3.411	3.429	3.447	3.466	3.484	3.502	3.520	3.539	3.557
2.6	3.575	3.594	3.612	3.630	3.649	3.667	3.686	3.704	3.723	3.741
2.7	3.760	3.778	3.797	3.815	3.834	3.853	3.871	3.890	3.909	3.928
2.8	3.946	3.965	3.984	4.003	4.022	4.041	4.060	4.079	4.098	4.117
2.9	4.136	4.155	4.174	4.193	4.212	4.231	4.250	4.269	4.288	4.308
3.0	4.327	4.346	4.365	4.385	4.404	4.423	4.443	4.462	4.481	4.501
3.1	4.520	4.540	4.559	4.579	4.598	4.618	4.637	4.657	4.676	4.696
3.2	4.716	4.735	4.755	4.775	4.794	4.814	4.834	4.854	4.873	4.893
3.3	4.913	4.933	4.953	4.973	4.993	5.013	5.033	5.053	5.073	5.093
3.4	5.113	5.133	5.153	5.153	5.173	5.193	5.213	5.233	5.253	5.274
3.5	5.314	5.334	5.355	5.375	5.395	5.415	5.436	5.456	5.477	5.497
3.6	5.517	5.538	5.558	5.579	5.599	5.620	5.640	5.661	5.682	5.702
3.7	5.723	5.743	5.764	5.785	5.805	5.826	5.847	5.868	5.888	5.909
3.8	5.930	5.951	5.972	5.992	6.013	6.034	6.055	6.076	6.097	6.118
3.9	6.139	6.160	6.181	6.202	6.223	6.244	6.265	6.286	6.307	6.328
4.0	6.350	6.371	6.392	6.413	6.434	6.456	6.477	6.498	6.519	6.541
4.1	6.562	6.583	6.605	6.626	6.648	6.669	6.690	6.712	6.733	6.755
4.2	6.776	6.798	6.819	6.841	6.863	6.884	6.906	6.927	6.949	6.971
4.3	6.992	7.014	7.036	7.057	7.079	7.101	7.123	7.145	7.166	7.188
4.4	7.210	7.232	7.254	7.276	7.298	7.319	7.341	7.363	7.385	7.407
4.5	7.429	7.451	7.473	7.495	7.518	7.540	7.562	7.584	7.606	7.628
4.6	7.650	7.672	7.695	7.717	7.739	7.761	7.784	7.806	7.828	7.850
4.7	7.873	7.895	7.918	7.940	7.962	7.985	8.007	8.030	8.052	8.074
4.8	8.097	8.119	8.142	8.164	8.187	8.210	8.232	8.255	8.277	8.300
4.9	8.323	8.345	8.368	8.391	8.413	8.436	8.459	8.482	8.504	8.527

Table 62.—*Four-thirds powers of numbers*—Continued

Number	.00	.01	.02	.03	.04	.05	.06	.07	.08	.09
5.0	8.550	8.573	8.596	8.618	8.641	8.664	8.687	8.710	8.733	8.756
5.1	8.779	8.802	8.825	8.848	8.871	8.894	8.917	8.940	8.963	8.986
5.2	9.009	9.032	9.055	9.078	9.101	9.125	9.148	9.171	9.194	9.217
5.3	9.241	9.264	9.287	9.310	9.334	9.357	9.380	9.404	9.427	9.450
5.4	9.474	9.497	9.521	9.544	9.568	9.591	9.614	9.638	9.661	9.685
5.5	9.708	9.732	9.756	9.779	9.803	9.826	9.850	9.874	9.897	9.921
5.6	9.945	9.968	9.992	10.02	10.04	10.06	10.09	10.11	10.13	10.16
5.7	10.18	10.21	10.23	10.25	10.28	10.30	10.33	10.35	10.37	10.40
5.8	10.42	10.44	10.47	10.49	10.52	10.54	10.56	10.59	10.61	10.64
5.9	10.66	10.69	10.71	10.73	10.76	10.78	10.81	10.83	10.85	10.88
6.0	10.90	10.93	10.95	10.98	11.00	11.02	11.05	11.07	11.10	11.12
6.1	11.15	11.17	11.19	11.22	11.24	11.27	11.29	11.32	11.34	11.37
6.2	11.39	11.41	11.44	11.46	11.49	11.51	11.54	11.56	11.59	11.61
6.3	11.64	11.66	11.68	11.71	11.73	11.76	11.78	11.81	11.83	11.86
6.4	11.88	11.91	11.93	11.96	11.98	12.01	12.03	12.06	12.08	12.11
6.5	12.13	12.16	12.18	12.21	12.23	12.26	12.28	12.31	12.33	12.36
6.6	12.38	12.41	12.43	12.46	12.48	12.51	12.53	12.56	12.58	12.61
6.7	12.63	12.66	12.68	12.71	12.73	12.76	12.78	12.81	12.83	12.86
6.8	12.88	12.91	12.93	12.96	12.98	13.01	13.03	13.06	13.09	13.11
6.9	13.14	13.16	13.19	13.21	13.24	13.26	13.29	13.31	13.34	13.37
7.0	13.39	13.42	13.44	13.47	13.49	13.52	13.54	13.57	13.59	13.62
7.1	13.65	13.67	13.70	13.72	13.75	13.77	13.80	13.83	13.85	13.88
7.2	13.90	13.93	13.95	13.98	14.01	14.03	14.06	14.08	14.11	14.14
7.3	14.16	14.19	14.21	14.24	14.26	14.29	14.32	14.34	14.37	14.39
7.4	14.42	14.45	14.47	14.50	14.52	14.55	14.58	14.60	14.63	14.65
7.5	14.68	14.71	14.73	14.76	14.79	14.81	14.84	14.86	14.89	14.92
7.6	14.94	14.97	14.99	15.02	15.05	15.07	15.10	15.13	15.15	15.18
7.7	15.21	15.23	15.26	15.28	15.31	15.34	15.36	15.39	15.42	15.44
7.8	15.47	15.50	15.52	15.55	15.57	15.60	15.63	15.65	15.68	15.71
7.9	15.73	15.76	15.79	15.81	15.84	15.87	15.89	15.92	15.95	15.97
8.0	16.00	16.03	16.05	16.08	16.11	16.13	16.16	16.19	16.21	16.24
8.1	16.27	16.29	16.32	16.35	16.37	16.40	16.43	16.45	16.48	16.51
8.2	16.54	16.56	16.59	16.62	16.64	16.67	16.70	16.72	16.75	16.78
8.3	16.80	16.83	16.86	16.89	16.91	16.94	16.97	16.99	17.02	17.05
8.4	17.08	17.10	17.13	17.16	17.18	17.21	17.24	17.27	17.29	17.32
8.5	17.35	17.37	17.40	17.43	17.46	17.48	17.51	17.54	17.57	17.59
8.6	17.62	17.65	17.67	17.70	17.73	17.76	17.78	17.81	17.84	17.87
8.7	17.89	17.92	17.95	17.98	18.00	18.03	18.06	18.09	18.11	18.14
8.8	18.17	18.20	18.22	18.25	18.28	18.31	18.33	18.36	18.39	18.42
8.9	18.44	18.47	18.50	18.53	18.55	18.58	18.61	18.64	18.67	18.69
9.0	18.72	18.75	18.78	18.80	18.83	18.86	18.89	18.92	18.94	18.97
9.1	19.00	19.03	19.05	19.08	19.11	19.14	19.17	19.19	19.22	19.25
9.2	19.28	19.31	19.33	19.36	19.39	19.42	19.45	19.47	19.50	19.53
9.3	19.56	19.59	19.61	19.64	19.67	19.70	19.73	19.76	19.78	19.81
9.4	19.84	19.87	19.89	19.92	19.95	19.98	20.01	20.04	20.06	20.09
9.5	20.12	20.15	20.18	20.20	20.23	20.26	20.29	20.32	20.35	20.37
9.6	20.40	20.43	20.46	20.49	20.52	20.54	20.57	20.60	20.63	20.66
9.7	20.69	20.72	20.74	20.77	20.80	20.83	20.86	20.89	20.91	20.94
9.8	20.97	21.00	21.03	21.06	21.09	21.11	21.14	21.17	21.20	21.23
9.9	21.26	21.29	21.31	21.34	21.37	21.40	21.43	21.46	21.49	21.52

Table 62.—*Four-thirds powers of numbers*—Continued

Number	.00	.01	.02	.03	.04	.05	.06	.07	.08	.09
10.0	21.54	21.57	21.60	21.63	21.66	21.69	21.72	21.75	21.77	21.80
10.1	21.83	21.86	21.89	21.92	21.95	21.98	22.01	22.03	22.06	22.09
10.2	22.12	22.15	22.18	22.21	22.24	22.27	22.29	22.32	22.35	22.38
10.3	22.41	22.44	22.47	22.50	22.53	22.56	22.58	22.61	22.64	22.67
10.4	22.70	22.73	22.76	22.79	22.82	22.85	22.88	22.90	22.93	22.96
10.5	22.99	23.02	23.05	23.08	23.11	23.14	23.17	23.20	23.23	23.26
10.6	23.28	23.31	23.34	23.37	23.40	23.43	23.46	23.49	23.52	23.55
10.7	23.58	23.61	23.64	23.67	23.70	23.73	23.75	23.78	23.81	23.84
10.8	23.87	23.90	23.93	23.96	23.99	24.02	24.05	24.08	24.11	24.14
10.9	24.17	24.20	24.23	24.26	24.29	24.32	24.35	24.37	24.40	24.43
11.0	24.46	24.49	24.52	24.55	24.58	24.61	24.64	24.67	24.70	24.73
11.1	24.76	24.79	24.83	24.86	24.88	24.91	24.94	24.97	25.00	25.03
11.2	25.06	25.09	25.12	25.15	25.18	25.21	25.24	25.27	25.30	25.33
11.3	25.36	25.39	25.42	25.45	25.48	25.51	25.54	25.57	25.60	25.63
11.4	25.66	25.69	25.72	25.75	25.78	25.81	25.84	25.87	25.90	25.93
11.5	25.96	25.99	26.02	26.05	26.08	26.11	26.14	26.17	26.20	26.23
11.6	26.26	26.29	26.32	26.35	26.38	26.41	26.44	26.47	26.50	26.53
11.7	26.56	26.59	26.62	26.65	26.68	26.71	26.74	26.77	26.80	26.83
11.8	26.86	26.89	26.93	26.96	26.99	27.02	27.05	27.08	27.11	27.14
11.9	27.17	27.20	27.23	27.26	27.29	27.32	27.35	27.38	27.41	27.44
12.0	27.47	27.50	27.53	27.56	27.60	27.63	27.66	27.69	27.72	27.75
12.1	27.78	27.81	27.84	27.87	27.90	27.93	27.96	27.99	28.02	28.05
12.2	28.09	28.12	28.15	28.18	28.21	28.24	28.27	28.30	28.33	28.36
12.3	28.39	28.42	28.45	28.49	28.52	28.55	28.58	28.61	28.64	28.67
12.4	28.70	28.73	28.76	28.79	28.82	28.86	28.89	28.92	28.95	28.98
12.5	29.01	29.04	29.07	29.10	29.13	29.16	29.20	29.23	29.26	29.29
12.6	29.32	29.35	29.38	29.41	29.44	29.48	29.51	29.54	29.57	29.60
12.7	29.63	29.66	29.69	29.72	29.75	29.79	29.82	29.85	29.88	29.91
12.8	29.94	29.97	30.00	30.04	30.07	30.10	30.13	30.16	30.19	30.22
12.9	30.25	30.29	30.32	30.35	30.38	30.41	30.44	30.47	30.50	30.54
13.0	30.57	30.60	30.63	30.66	30.69	30.72	30.76	30.79	30.82	30.85
13.1	30.88	30.91	30.94	30.98	31.01	31.04	31.07	31.10	31.13	31.16
13.2	31.20	31.23	31.26	31.29	31.32	31.35	31.39	31.42	31.45	31.48
13.3	31.51	31.54	31.57	31.61	31.64	31.67	31.70	31.73	31.76	31.80
13.4	31.83	31.86	31.89	31.92	31.95	31.99	32.02	32.05	32.08	32.11
13.5	32.14	32.18	32.21	32.24	32.27	32.30	32.34	32.37	32.40	32.43
13.6	32.46	32.49	32.53	32.56	32.59	32.62	32.65	32.69	32.72	32.75
13.7	32.78	32.81	32.85	32.88	32.91	32.94	32.97	33.00	33.04	33.07
13.8	33.10	33.13	33.16	33.20	33.23	33.26	33.29	33.32	33.36	33.39
13.9	33.42	33.45	33.49	33.52	33.55	33.58	33.61	33.65	33.68	33.71
14.0	33.74	33.77	33.81	33.84	33.87	33.90	33.93	33.97	34.00	34.03
14.1	34.06	34.10	34.13	34.16	34.19	34.22	34.26	34.29	34.32	34.35
14.2	34.39	34.42	34.45	34.48	34.52	34.55	34.58	34.61	34.64	34.68
14.3	34.71	34.74	34.77	34.81	34.84	34.87	34.90	34.94	34.97	35.00
14.4	35.03	35.07	35.10	35.13	35.16	35.20	35.23	35.26	35.29	35.33
14.5	35.36	35.39	35.42	35.46	35.49	35.52	35.55	35.59	35.62	35.65
14.6	35.68	35.72	35.75	35.78	35.81	35.85	35.88	35.91	35.94	35.98
14.7	36.01	36.04	36.08	36.11	36.14	36.17	36.21	36.24	36.27	36.30
14.8	36.34	36.37	36.40	36.44	36.47	36.50	36.53	36.57	36.60	36.63
14.9	36.66	36.70	36.73	36.76	36.80	36.83	36.86	36.89	36.93	36.96

Table 62.—*Four-thirds powers of numbers*—Continued

Number	.00	.01	.02	.03	.04	.05	.06	.07	.08	.09
15.0	36.99	37.03	37.06	37.09	37.12	37.16	37.19	37.22	37.26	37.29
15.1	37.32	37.36	37.39	37.42	37.45	37.49	37.52	37.55	37.59	37.62
15.2	37.65	37.69	37.72	37.75	37.78	37.82	37.85	37.88	37.92	37.95
15.3	37.98	38.02	38.05	38.08	38.12	38.15	38.18	38.21	38.25	38.28
15.4	38.31	38.35	38.38	38.41	38.45	38.48	38.51	38.55	38.58	38.61
15.5	38.65	38.68	38.71	38.75	38.78	38.81	38.85	38.88	38.91	38.95
15.6	38.98	39.01	39.05	39.08	39.11	39.15	39.18	39.21	39.25	39.28
15.7	39.31	39.35	39.38	39.41	39.45	39.48	39.51	39.55	39.58	39.61
15.8	39.65	39.68	39.71	39.75	39.78	39.81	39.85	39.88	39.91	39.95
15.9	39.98	40.02	40.05	40.08	40.12	40.15	40.18	40.22	40.25	40.28
16.0	40.32	40.35	40.38	40.42	40.45	40.49	40.52	40.55	40.59	40.62
16.1	40.65	40.69	40.72	40.75	40.79	40.82	40.86	40.89	40.92	40.96
16.2	40.99	41.02	41.06	41.09	41.13	41.16	41.19	41.23	41.26	41.29
16.3	41.33	41.36	41.40	41.43	41.46	41.50	41.53	41.57	41.60	41.63
16.4	41.67	41.70	41.73	41.77	41.80	41.84	41.87	41.90	41.94	41.97
16.5	42.01	42.04	42.07	42.11	42.14	42.18	42.21	42.24	42.28	42.31
16.6	42.35	42.38	42.41	42.45	42.48	42.52	42.55	42.58	42.62	42.65
16.7	42.69	42.72	42.75	42.79	42.82	42.86	42.89	42.93	42.96	42.99
16.8	43.03	43.06	43.10	43.13	43.16	43.20	43.23	43.27	43.30	43.34
16.9	43.37	43.40	43.44	43.47	43.51	43.54	43.57	43.61	43.64	43.68
17.0	43.71	43.75	43.78	43.81	43.85	43.88	43.92	43.95	43.99	44.02
17.1	44.05	44.09	44.12	44.16	44.19	44.23	44.26	44.30	44.33	44.36
17.2	44.40	44.43	44.47	44.50	44.54	44.57	44.61	44.64	44.67	44.71
17.3	44.74	44.78	44.81	44.85	44.88	44.92	44.95	44.98	45.02	45.05
17.4	45.09	45.12	45.16	45.19	45.23	45.26	45.30	45.33	45.37	45.40
17.5	45.43	45.47	45.50	45.54	45.57	45.61	45.64	45.68	45.71	45.75
17.6	45.78	45.82	45.85	45.88	45.92	45.95	45.99	46.02	46.06	46.09
17.7	46.13	46.16	46.20	46.23	46.27	46.30	46.34	46.37	46.41	46.44
17.8	46.48	46.51	46.55	46.58	46.62	46.65	46.68	46.72	46.75	46.79
17.9	46.82	46.86	46.89	46.93	46.96	47.00	47.03	47.07	47.10	47.14
18.0	47.17	47.21	47.24	47.28	47.31	47.35	47.38	47.42	47.45	47.49
18.1	47.52	47.56	47.59	47.63	47.66	47.70	47.73	47.77	47.80	47.84
18.2	47.87	47.91	47.94	47.98	48.01	48.05	48.08	48.12	48.15	48.19
18.3	48.22	48.26	48.29	48.33	48.37	48.40	48.44	48.47	48.51	48.54
18.4	48.58	48.61	48.65	48.68	48.72	48.75	48.79	48.82	48.86	48.89
18.5	48.93	48.96	49.00	49.03	49.07	49.10	49.14	49.18	49.21	49.25
18.6	49.28	49.32	49.35	49.39	49.42	49.46	49.49	49.53	49.56	49.60
18.7	49.64	49.67	49.71	49.74	49.78	49.81	49.85	49.88	49.92	49.95
18.8	49.99	50.02	50.06	50.10	50.13	50.17	50.20	50.24	50.27	50.31
18.9	50.34	50.38	50.42	50.45	50.49	50.52	50.56	50.59	50.63	50.66
19.0	50.70	50.74	50.77	50.81	50.84	50.88	50.91	50.95	50.98	51.02
19.1	51.06	51.09	51.13	51.16	51.20	51.23	51.27	51.31	51.34	51.38
19.2	51.41	51.45	51.48	51.52	51.56	51.59	51.63	51.66	51.70	51.73
19.3	51.77	51.81	51.84	51.88	51.91	51.95	51.98	52.02	52.06	52.09
19.4	52.13	52.16	52.20	52.24	52.27	52.31	52.34	52.38	52.41	52.45
19.5	52.49	52.52	52.56	52.59	52.63	52.67	52.70	52.74	52.77	52.81
19.6	52.85	52.88	52.92	52.95	52.99	53.03	53.06	53.10	53.13	53.17
19.7	53.21	53.24	53.28	53.31	53.35	53.39	53.42	53.46	53.49	53.53
19.8	53.57	53.60	53.64	53.67	53.71	53.75	53.78	53.82	53.85	53.89
19.9	53.93	53.96	54.00	54.04	54.07	54.11	54.14	54.18	54.22	54.25

Table 62.—*Four-thirds powers of numbers*—Continued

Number	.00	.01	.02	.03	.04	.05	.06	.07	.08	.09
20.0	54.29	54.32	54.36	54.40	54.43	54.47	54.51	54.54	54.58	54.61
20.1	54.65	54.69	54.72	54.76	54.80	54.83	54.87	54.90	54.94	54.98
20.2	55.01	55.05	55.09	55.12	55.16	55.20	55.23	55.27	55.30	55.34
20.3	55.38	55.41	55.45	55.49	55.52	55.56	55.60	55.63	55.67	55.70
20.4	55.74	55.78	55.81	55.85	55.89	55.92	55.96	56.00	56.03	56.07
20.5	56.11	56.14	56.18	56.21	56.25	56.29	56.32	56.36	56.40	56.43
20.6	56.47	56.51	56.54	56.58	56.62	56.65	56.69	56.76	56.76	56.80
20.7	56.84	56.87	56.91	56.95	56.98	57.02	57.06	57.09	57.13	57.17
20.8	57.20	57.24	57.28	57.31	57.35	57.39	57.42	57.46	57.50	57.53
20.9	57.57	57.61	57.64	57.68	57.72	57.75	57.79	57.83	57.86	57.90
21.0	57.94	57.97	58.01	58.05	58.08	58.12	58.16	58.20	58.23	58.27
21.1	58.31	58.34	58.38	58.42	58.45	58.49	58.53	58.56	58.60	58.64
21.2	58.67	58.71	58.75	58.79	58.82	58.86	58.90	58.93	58.97	59.01
21.3	59.04	59.08	59.12	59.15	59.19	59.23	59.27	59.30	59.34	59.38
21.4	59.41	59.45	59.49	59.52	59.56	59.60	59.64	59.67	59.71	59.75
21.5	59.78	59.82	59.86	59.90	59.93	59.97	60.01	60.04	60.08	60.12
21.6	60.15	60.19	60.23	60.27	60.30	60.34	60.38	60.42	60.45	60.49
21.7	60.53	60.56	60.60	60.64	60.68	60.71	60.75	60.79	60.82	60.86
21.8	60.90	60.94	60.97	61.01	61.05	61.09	61.12	61.16	61.20	61.23
21.9	61.27	61.31	61.35	61.38	61.42	61.46	61.50	61.53	61.57	61.61
22.0	61.64	61.68	61.72	61.76	61.79	61.83	61.87	61.91	61.94	61.98
22.1	62.02	62.06	62.09	62.13	62.17	62.21	62.24	62.28	62.32	62.36
22.2	62.39	62.43	62.47	62.51	62.54	62.58	62.62	62.66	62.69	62.73
22.3	62.77	62.81	62.84	62.88	62.92	62.96	62.99	63.03	63.07	63.11
22.4	63.14	63.18	63.22	63.26	63.29	63.33	63.37	63.41	63.44	63.48
22.5	63.52	63.56	63.60	63.63	63.67	63.71	63.75	63.78	63.82	63.86
22.6	63.90	63.93	63.97	64.01	64.05	64.09	64.12	64.16	64.20	64.24
22.7	64.27	64.31	64.35	64.39	64.42	64.46	64.50	64.54	64.58	64.61
22.8	64.65	64.69	64.73	64.77	64.80	64.84	64.88	64.92	64.95	64.99
22.9	65.03	65.07	65.11	65.14	65.18	65.22	65.26	65.30	65.33	65.37
23.0	65.41	65.45	65.48	65.52	65.56	65.60	65.64	65.67	65.71	65.75
23.1	65.79	65.83	65.86	65.90	65.94	65.98	66.02	66.05	66.09	66.13
23.2	66.17	66.21	66.24	66.28	66.32	66.36	66.40	66.43	66.47	66.51
23.3	66.55	66.59	66.63	66.66	66.70	66.74	66.78	66.82	66.85	66.89
23.4	66.93	66.97	67.01	67.04	67.08	67.12	67.16	67.20	67.24	67.27
23.5	67.31	67.35	67.39	67.43	67.46	67.50	67.54	67.58	67.62	67.66
23.6	67.69	67.73	67.77	67.81	67.85	67.89	67.92	67.96	68.00	68.04
23.7	68.08	68.11	68.15	68.19	68.23	68.27	68.31	68.34	68.38	68.42
23.8	68.46	68.50	68.54	68.57	68.61	68.65	68.69	68.73	68.77	68.81
23.9	68.84	68.88	68.92	68.96	69.00	69.04	69.07	69.11	69.15	69.19
24.0	69.23	69.27	69.30	69.34	69.38	69.42	69.46	69.50	69.54	69.57
24.1	69.61	69.65	69.69	69.73	69.77	69.81	69.84	69.88	69.92	69.96
24.2	70.00	70.04	70.08	70.11	70.15	70.19	70.23	70.27	70.31	70.35
24.3	70.38	70.42	70.46	70.50	70.54	70.58	70.62	70.65	70.69	70.73
24.4	70.77	70.81	70.85	70.89	70.93	70.96	71.00	71.04	71.08	71.12
24.5	71.16	71.20	71.24	71.27	71.31	71.35	71.39	71.43	71.47	71.51
24.6	71.55	71.58	71.62	71.66	71.70	71.74	71.78	71.82	71.86	71.89
24.7	71.93	71.97	72.01	72.05	72.09	72.13	72.17	72.21	72.24	72.28
24.8	72.32	72.36	72.40	72.44	72.48	72.52	72.56	72.59	72.63	72.67
24.9	72.71	72.75	72.79	72.83	72.87	72.91	72.94	72.98	73.02	73.06

Table 62.—*Four-thirds powers of numbers*—Continued

Number	.00	.01	.02	.03	.04	.05	.06	.07	.08	.09
25.0	73.10	73.14	73.18	73.22	73.26	73.30	73.33	73.37	73.41	73.45
25.1	73.49	73.53	73.57	73.61	73.65	73.69	73.72	73.76	73.80	73.84
25.2	73.88	73.92	73.96	74.00	74.04	74.08	74.12	74.15	74.19	74.23
25.3	74.27	74.31	74.35	74.39	74.43	74.47	74.51	74.55	74.59	74.62
25.4	74.66	74.70	74.74	74.78	74.82	74.86	74.90	74.94	74.98	75.02
25.5	75.06	75.10	75.13	75.17	75.21	75.25	75.29	75.33	75.37	75.41
25.6	75.45	75.49	75.53	75.57	75.61	75.65	75.68	75.72	75.76	75.80
25.7	75.84	75.88	75.92	75.96	76.00	76.04	76.08	76.12	76.16	76.20
25.8	76.24	76.28	76.31	76.35	76.39	76.43	76.47	76.51	76.55	76.59
25.9	76.63	76.67	76.71	76.75	76.79	76.83	76.87	76.91	76.95	76.99
26.0	77.02	77.06	77.10	77.14	77.18	77.22	77.26	77.30	77.34	77.38
26.1	77.42	77.46	77.50	77.54	77.58	77.62	77.66	77.70	77.74	77.78
26.2	77.82	77.86	77.90	77.93	77.97	78.01	78.05	78.09	78.13	78.17
26.3	78.21	78.25	78.29	78.33	78.37	78.41	78.45	78.49	78.53	78.57
26.4	78.61	78.65	78.69	78.73	78.77	78.81	78.85	78.89	78.93	78.97
26.5	79.01	79.05	79.09	79.13	79.17	79.21	79.24	79.28	79.32	79.36
26.6	79.40	79.44	79.48	79.52	79.56	79.60	79.64	79.68	79.72	79.76
26.7	79.80	79.84	79.88	79.92	79.96	80.00	80.04	80.08	80.12	80.16
26.8	80.20	80.24	80.28	80.32	80.36	80.40	80.44	80.48	80.52	80.56
26.9	80.60	80.64	80.68	80.72	80.76	80.80	80.84	80.88	80.92	80.96
27.0	81.00	81.04	81.08	81.12	81.16	81.20	81.24	81.28	81.32	81.36
27.1	81.40	81.44	81.48	81.52	81.56	81.60	81.64	81.68	81.72	81.76
27.2	81.80	81.84	81.88	81.92	81.96	82.00	82.04	82.08	82.12	82.16
27.3	82.20	82.24	82.28	82.32	82.36	82.40	82.44	82.48	82.52	82.56
27.4	82.60	82.64	82.68	82.72	82.76	82.80	82.85	82.89	82.93	82.97
27.5	83.01	83.05	83.09	83.13	83.17	83.21	83.25	83.29	83.33	83.37
27.6	83.41	83.45	83.49	83.53	83.57	83.61	83.65	83.69	83.73	83.77
27.7	83.81	83.85	83.89	83.93	83.97	84.01	84.05	84.09	84.13	84.18
27.8	84.22	84.26	84.30	84.34	84.38	84.42	84.46	84.50	84.54	84.58
27.9	84.62	84.66	84.70	84.74	84.78	84.82	84.86	84.90	84.94	84.98
28.0	85.02	85.06	85.11	85.15	85.19	85.23	85.27	85.31	85.35	85.39
28.1	85.43	85.47	85.51	85.55	85.59	85.63	85.67	85.71	85.75	85.79
28.2	85.84	85.88	85.92	85.96	86.00	86.04	86.08	86.12	86.16	86.20
28.3	86.24	86.28	86.32	86.36	86.40	86.44	86.49	86.53	86.57	86.61
28.4	86.65	86.69	86.73	86.77	86.81	86.85	86.89	86.93	86.97	87.01
28.5	87.05	87.10	87.14	87.18	87.22	87.26	87.30	87.34	87.38	87.42
28.6	87.46	87.50	87.54	87.58	87.63	87.67	87.71	87.75	87.79	87.83
28.7	87.87	87.91	87.95	87.99	88.03	88.07	88.12	88.16	88.20	88.24
28.8	88.28	88.32	88.36	88.40	88.44	88.48	88.52	88.57	88.61	88.65
28.9	88.69	88.73	88.77	88.81	88.85	88.89	88.93	88.97	89.02	89.06
29.0	89.10	89.14	89.18	89.22	89.26	89.30	89.34	89.38	89.43	89.47
29.1	89.51	89.55	89.59	89.63	89.67	89.71	89.75	89.79	89.84	89.88
29.2	89.92	89.96	90.00	90.04	90.08	90.12	90.16	90.20	90.25	90.29
29.3	90.33	90.37	90.41	90.45	90.49	90.53	90.57	90.62	90.66	90.70
29.4	90.74	90.78	90.82	90.86	90.90	90.95	90.99	91.03	91.07	91.11
29.5	91.15	91.19	91.23	91.27	91.32	91.36	91.40	91.44	91.48	91.52
29.6	91.56	91.60	91.65	91.69	91.73	91.77	91.81	91.85	91.89	91.93
29.7	91.98	92.02	92.06	92.10	92.14	92.18	92.22	92.27	92.31	92.35
29.8	92.39	92.43	92.47	92.51	92.55	92.60	92.64	92.68	92.72	92.76
29.9	92.80	92.84	92.89	92.93	92.97	93.01	93.05	93.09	93.13	93.18

Table 62.—*Four-thirds powers of numbers*—Continued

Number	.0	.1	.2	.3	.4	.5	.6	.7	.8	.9
30	93.22	93.63	94.05	94.46	94.88	95.29	95.71	96.13	96.55	96.96
31	97.38	97.80	98.22	98.64	99.06	99.48	99.90	100.3	100.7	101.2
32	101.6	102.0	102.4	102.9	103.3	103.7	104.1	104.6	105.0	105.4
33	105.8	106.3	106.7	107.1	107.6	108.0	108.4	108.9	109.3	109.7
34	110.1	110.6	111.0	111.4	111.9	112.3	112.7	113.2	113.6	114.1
35	114.5	114.9	115.4	115.8	116.2	116.7	117.1	117.6	118.0	118.4
36	118.9	119.3	119.8	120.2	120.6	121.1	121.5	122.0	122.4	122.8
37	123.3	123.7	124.2	124.6	125.1	125.5	126.0	126.4	126.9	127.3
38	127.8	128.2	128.7	129.1	129.6	130.0	130.5	130.9	131.4	131.8
39	132.3	132.7	133.2	133.6	134.1	134.5	135.0	135.4	135.9	136.3
40	136.8	137.3	137.7	138.2	138.6	139.1	139.5	140.0	140.5	140.9
41	141.4	141.8	142.3	142.8	143.2	143.7	144.1	144.6	145.1	145.5
42	146.0	146.5	146.9	147.4	147.8	148.3	148.8	149.2	149.7	150.2
43	150.6	151.1	151.6	152.0	152.5	153.0	153.5	153.9	154.4	154.9
44	155.3	155.8	156.3	156.7	157.2	157.7	158.2	158.6	159.1	159.6
45	160.1	160.5	161.0	161.5	162.0	162.4	162.9	163.4	163.9	164.3
46	164.8	165.3	165.8	166.3	166.7	167.2	167.7	168.2	168.6	169.1
47	169.6	170.1	170.6	171.1	171.5	172.0	172.5	173.0	173.5	174.0
48	174.4	174.9	175.4	175.9	176.4	176.9	177.4	177.8	178.3	178.8
49	179.3	179.8	180.3	180.8	181.3	181.7	182.2	182.7	183.2	183.7
50	184.2	184.7	185.2	185.7	186.2	186.7	187.2	187.6	188.1	188.6
51	189.1	189.6	190.1	190.6	191.1	191.6	192.1	192.6	193.1	193.6
52	194.1	194.6	195.1	195.6	196.1	196.6	197.1	197.6	198.1	198.6
53	199.1	199.6	200.1	200.6	201.1	201.6	202.1	202.6	203.1	203.6
54	204.1	204.6	205.1	205.6	206.1	206.6	207.1	207.6	208.1	208.7
55	209.2	209.7	210.2	210.7	211.2	211.7	212.2	212.7	213.2	213.7
56	214.2	214.8	215.3	215.8	216.3	216.8	217.3	217.8	218.3	218.9
57	219.4	219.9	220.4	220.9	221.4	221.9	222.4	223.0	223.5	224.0
58	224.5	225.0	225.5	226.1	226.6	227.1	227.6	228.1	228.6	229.2
59	229.7	230.2	230.7	231.2	231.8	232.3	232.8	233.3	233.8	234.4
60	234.9	235.4	235.9	236.5	237.0	237.5	238.0	238.6	239.1	239.6
61	240.1	240.7	241.2	241.7	242.2	242.8	243.3	243.8	244.3	244.9
62	245.4	245.9	246.4	247.0	247.5	248.0	248.6	249.1	249.6	250.2
63	250.7	251.2	251.7	252.3	252.8	253.3	253.9	254.4	254.9	255.5
64	256.0	256.5	257.1	257.6	258.1	258.7	259.2	259.7	260.3	260.8
65	261.3	261.9	262.4	263.0	263.5	264.0	264.6	265.1	265.6	266.2
66	266.7	267.3	267.8	268.3	268.9	269.4	270.0	270.5	271.0	271.6
67	272.1	272.7	273.2	273.7	274.3	274.8	275.4	275.9	276.5	277.0
68	277.6	278.1	278.6	279.2	279.7	280.3	280.8	281.4	281.9	282.5
69	283.0	283.6	284.1	284.6	285.2	285.7	286.3	286.8	287.4	287.9
70	288.5	289.0	289.6	290.1	290.7	291.2	291.8	292.3	292.9	293.4
71	294.0	294.6	295.1	295.7	296.2	296.8	297.3	297.9	298.4	299.0
72	299.5	300.1	300.6	301.2	301.8	302.3	302.9	303.4	304.0	304.5
73	305.1	305.6	306.2	306.8	307.3	307.9	308.4	309.0	309.6	310.1
74	310.7	311.2	311.8	312.4	312.9	313.5	314.0	314.6	315.2	315.7
75	316.3	316.8	317.4	318.0	318.5	319.1	319.7	320.2	320.8	321.4
76	321.9	322.5	323.1	323.6	324.2	324.7	325.3	325.9	326.4	327.0
77	327.6	328.2	328.7	329.3	329.9	330.4	331.0	331.6	332.1	332.7
78	333.3	333.8	334.4	335.0	335.5	336.1	336.7	337.3	337.8	338.4
79	339.0	339.5	340.1	340.7	341.3	341.8	342.4	343.0	343.6	344.1

Table 62.—*Four-thirds powers of numbers*—Continued

Number	.0	.1	.2	.3	.4	.5	.6	.7	.8	.9
80	344.7	345.3	345.9	346.4	347.0	347.6	348.2	348.7	349.3	349.9
81	350.5	351.0	351.6	352.2	352.8	353.4	353.9	354.5	355.1	355.7
82	356.2	356.8	357.4	358.0	358.6	359.1	359.7	360.3	360.9	361.5
83	362.1	362.6	363.2	363.8	364.4	365.0	365.5	366.1	366.7	367.3
84	367.9	368.5	369.0	369.6	370.2	370.8	371.4	372.0	372.6	373.1
85	373.7	374.3	374.9	375.5	376.1	376.7	377.3	377.8	378.4	379.0
86	379.6	380.2	380.8	381.4	382.0	382.5	383.1	383.7	384.3	384.9
87	385.5	386.1	386.7	387.3	387.9	388.5	389.1	389.6	390.2	390.8
88	391.4	392.0	392.6	393.2	393.8	394.4	395.0	395.6	396.2	396.8
89	397.4	398.0	398.6	399.1	399.7	400.3	400.9	401.5	402.1	402.7
90	403.3	403.9	404.5	405.1	405.7	406.3	406.9	407.5	408.1	408.7
91	409.3	409.9	410.5	411.1	411.7	412.3	412.9	413.5	414.1	414.7
92	415.3	415.9	416.5	417.1	417.7	418.3	418.9	419.5	420.1	420.7
93	421.4	422.0	422.6	423.2	423.8	424.4	425.0	425.6	426.2	426.8
94	427.4	428.0	428.6	429.2	429.8	430.4	431.0	431.7	432.3	432.9
95	433.5	434.1	434.7	435.3	435.9	436.5	437.1	437.7	438.3	439.0
96	439.6	440.2	440.8	441.4	442.0	442.6	443.2	443.8	444.5	445.1
97	445.7	446.3	446.9	447.5	448.1	448.8	449.4	450.0	450.6	451.2
98	451.8	452.4	453.1	453.7	454.3	454.9	455.5	456.1	456.7	457.4
99	458.0	458.6	459.2	459.8	460.4	461.1	461.7	462.3	462.9	463.5

Table 63.—*Two-thirds powers of numbers*

Number	.00	.01	.02	.03	.04	.05	.06	.07	.08	.09
0.0	0.000	0.046	0.074	0.097	0.117	0.136	0.153	0.170	0.186	0.201
0.1	.215	.229	.243	.256	.269	.282	.295	.307	.319	.331
0.2	.342	.353	.364	.375	.386	.397	.407	.418	.428	.438
0.3	.448	.458	.468	.477	.487	.497	.506	.515	.525	.534
0.4	.543	.552	.561	.570	.578	.587	.596	.604	.613	.622
0.5	.630	.638	.647	.655	.663	.671	.679	.687	.695	.703
0.6	.711	.719	.727	.735	.743	.750	.758	.765	.773	.781
0.7	.788	.796	.803	.811	.818	.825	.832	.840	.847	.855
0.8	.862	.869	.876	.883	.890	.897	.904	.911	.918	.925
0.9	.932	.939	.946	.953	.960	.966	.973	.980	.987	.993
1.0	1.000	1.007	1.013	1.020	1.027	1.033	1.040	1.046	1.053	1.059
1.1	1.065	1.072	1.078	1.085	1.091	1.097	1.104	1.110	1.117	1.123
1.2	1.129	1.136	1.142	1.148	1.154	1.160	1.167	1.173	1.179	1.185
1.3	1.191	1.197	1.203	1.209	1.215	1.221	1.227	1.233	1.239	1.245
1.4	1.251	1.257	1.263	1.269	1.275	1.281	1.287	1.293	1.299	1.305
1.5	1.310	1.316	1.322	1.328	1.334	1.339	1.345	1.351	1.357	1.362
1.6	1.368	1.374	1.379	1.385	1.391	1.396	1.402	1.408	1.413	1.419
1.7	1.424	1.430	1.436	1.441	1.447	1.452	1.458	1.463	1.469	1.474
1.8	1.480	1.485	1.491	1.496	1.502	1.507	1.513	1.518	1.523	1.529
1.9	1.534	1.539	1.545	1.550	1.556	1.561	1.566	1.571	1.577	1.582
2.0	1.587	1.593	1.598	1.603	1.608	1.613	1.619	1.624	1.629	1.634
2.1	1.639	1.645	1.650	1.655	1.660	1.665	1.671	1.676	1.681	1.686
2.2	1.691	1.697	1.702	1.707	1.712	1.717	1.722	1.727	1.732	1.737
2.3	1.742	1.747	1.752	1.757	1.762	1.767	1.772	1.777	1.782	1.787
2.4	1.792	1.797	1.802	1.807	1.812	1.817	1.822	1.827	1.832	1.837
2.5	1.842	1.847	1.852	1.857	1.862	1.867	1.871	1.876	1.881	1.886
2.6	1.891	1.896	1.900	1.905	1.910	1.915	1.920	1.925	1.929	1.934
2.7	1.939	1.944	1.949	1.953	1.958	1.963	1.968	1.972	1.977	1.982
2.8	1.987	1.992	1.996	2.001	2.006	2.010	2.015	2.020	2.024	2.029
2.9	2.034	2.038	2.043	2.048	2.052	2.057	2.062	2.066	2.071	2.075
3.0	2.080	2.085	2.089	2.094	2.099	2.103	2.108	2.112	2.117	2.122
3.1	2.126	2.131	2.135	2.140	2.144	2.149	2.153	2.158	2.163	2.167
3.2	2.172	2.176	2.180	2.185	2.190	2.194	2.199	2.203	2.208	2.212
3.3	2.217	2.221	2.226	2.230	2.234	2.239	2.243	2.248	2.252	2.257
3.4	2.261	2.265	2.270	2.274	2.279	2.283	2.288	2.292	2.296	2.301
3.5	2.305	2.310	2.314	2.318	2.323	2.327	2.331	2.336	2.340	2.345
3.6	2.349	2.353	2.358	2.362	2.366	2.371	2.375	2.379	2.384	2.388
3.7	2.392	2.397	2.401	2.405	2.409	2.414	2.418	2.422	2.427	2.431
3.8	2.435	2.439	2.444	2.448	2.452	2.457	2.461	2.465	2.469	2.474
3.9	2.478	2.482	2.486	2.490	2.495	2.499	2.503	2.507	2.511	2.516
4.0	2.520	2.524	2.528	2.532	2.537	2.541	2.545	2.549	2.553	2.558
4.1	2.562	2.566	2.570	2.574	2.579	2.583	2.587	2.591	2.595	2.599
4.2	2.603	2.607	2.611	2.616	2.620	2.624	2.628	2.632	2.636	2.640
4.3	2.644	2.648	2.653	2.657	2.661	2.665	2.669	2.673	2.677	2.681
4.4	2.685	2.689	2.693	2.698	2.702	2.706	2.710	2.714	2.718	2.722
4.5	2.726	2.730	2.734	2.738	2.742	2.746	2.750	2.754	2.758	2.762
4.6	2.766	2.770	2.774	2.778	2.782	2.786	2.790	2.794	2.798	2.802
4.7	2.806	2.810	2.814	2.818	2.822	2.826	2.830	2.834	2.838	2.842
4.8	2.846	2.850	2.854	2.858	2.862	2.865	2.869	2.873	2.877	2.881
4.9	2.885	2.889	2.893	2.897	2.901	2.904	2.908	2.912	2.916	2.920

Table 63.—*Two-thirds powers of numbers*—Continued

Number	.00	.01	.02	.03	.04	.05	.06	.07	.08	.09
5.0	2.924	2.928	2.932	2.936	2.940	2.944	2.947	2.951	2.955	2.959
5.1	2.963	2.967	2.971	2.975	2.979	2.982	2.986	2.990	2.994	2.998
5.2	3.001	3.005	3.009	3.013	3.017	3.021	3.024	3.028	3.032	3.036
5.3	3.040	3.044	3.047	3.051	3.055	3.059	3.063	3.067	3.070	3.074
5.4	3.078	3.082	3.086	3.089	3.093	3.097	3.101	3.105	3.108	3.112
5.5	3.116	3.120	3.123	3.127	3.131	3.135	3.138	3.142	3.146	3.150
5.6	3.154	3.157	3.161	3.165	3.169	3.172	3.176	3.180	3.184	3.187
5.7	3.191	3.195	3.198	3.202	3.206	3.210	3.213	3.217	3.221	3.224
5.8	3.228	3.232	3.236	3.239	3.243	3.247	3.250	3.254	3.258	3.261
5.9	3.265	3.269	3.273	3.276	3.280	3.284	3.287	3.291	3.295	3.298
6.0	3.302	3.306	3.309	3.313	3.317	3.320	3.324	3.328	3.331	3.335
6.1	3.339	3.342	3.346	3.350	3.353	3.357	3.360	3.364	3.368	3.371
6.2	3.375	3.379	3.382	3.386	3.389	3.393	3.397	3.400	3.404	3.408
6.3	3.411	3.415	3.418	3.422	3.426	3.429	3.433	3.436	3.440	3.444
6.4	3.447	3.451	3.454	3.458	3.461	3.465	3.469	3.472	3.476	3.479
6.5	3.483	3.486	3.490	3.494	3.497	3.501	3.504	3.508	3.511	3.515
6.6	3.519	3.522	3.526	3.529	3.533	3.536	3.540	3.543	3.547	3.550
6.7	3.554	3.558	3.561	3.565	3.568	3.572	3.575	3.579	3.582	3.586
6.8	3.589	3.593	3.596	3.600	3.603	3.607	3.610	3.614	3.617	3.621
6.9	3.624	3.628	3.631	3.635	3.638	3.642	3.645	3.649	3.652	3.656
7.0	3.659	3.663	3.666	3.670	3.673	3.677	3.680	3.684	3.687	3.691
7.1	3.694	3.698	3.701	3.705	3.708	3.712	3.715	3.718	3.722	3.725
7.2	3.729	3.732	3.736	3.739	3.742	3.746	3.749	3.753	3.756	3.760
7.3	3.763	3.767	3.770	3.773	3.777	3.780	3.784	3.787	3.791	3.794
7.4	3.797	3.801	3.804	3.808	3.811	3.814	3.818	3.821	3.825	3.828
7.5	3.832	3.835	3.838	3.842	3.845	3.849	3.852	3.855	3.859	3.862
7.6	3.866	3.869	3.872	3.876	3.879	3.883	3.886	3.889	3.893	3.896
7.7	3.899	3.903	3.906	3.910	3.913	3.916	3.920	3.923	3.926	3.930
7.8	3.933	3.937	3.940	3.943	3.947	3.950	3.953	3.957	3.960	3.963
7.9	3.967	3.970	3.973	3.977	3.980	3.983	3.987	3.990	3.993	3.997
8.0	4.000	4.003	4.007	4.010	4.013	4.017	4.020	4.023	4.027	4.030
8.1	4.033	4.037	4.040	4.043	4.047	4.050	4.053	4.057	4.060	4.063
8.2	4.066	4.070	4.073	4.076	4.080	4.083	4.086	4.090	4.093	4.096
8.3	4.099	4.103	4.106	4.109	4.113	4.116	4.119	4.122	4.126	4.129
8.4	4.132	4.136	4.139	4.142	4.145	4.149	4.152	4.155	4.159	4.162
8.5	4.165	4.168	4.172	4.175	4.178	4.181	4.185	4.188	4.191	4.194
8.6	4.198	4.201	4.204	4.207	4.211	4.214	4.217	4.220	4.224	4.227
8.7	4.230	4.233	4.237	4.240	4.243	4.246	4.249	4.253	4.256	4.259
8.8	4.262	4.266	4.269	4.272	4.275	4.279	4.282	4.285	4.288	4.291
8.9	4.295	4.298	4.301	4.304	4.307	4.311	4.314	4.317	4.320	4.324
9.0	4.327	4.330	4.333	4.336	4.340	4.343	4.346	4.349	4.352	4.356
9.1	4.359	4.362	4.365	4.368	4.372	4.375	4.378	4.381	4.384	4.387
9.2	4.391	4.394	4.397	4.400	4.403	4.407	4.410	4.413	4.416	4.419
9.3	4.422	4.426	4.429	4.432	4.435	4.438	4.441	4.445	4.448	4.451
9.4	4.454	4.457	4.460	4.464	4.467	4.470	4.473	4.476	4.479	4.482
9.5	4.486	4.489	4.492	4.495	4.498	4.501	4.504	4.508	4.511	4.514
9.6	4.517	4.520	4.523	4.526	4.530	4.533	4.536	4.539	4.542	4.545
9.7	4.548	4.551	4.555	4.558	4.561	4.564	4.567	4.570	4.573	4.576
9.8	4.580	4.583	4.586	4.589	4.592	4.595	4.598	4.601	4.604	4.608
9.9	4.611	4.614	4.617	4.620	4.623	4.626	4.629	4.632	4.635	4.639

Table 63.—*Two-thirds powers of numbers*—Continued

Number	.00	.01	.02	.03	.04	.05	.06	.07	.08	.09
10.0	4.642	4.645	4.648	4.651	4.654	4.657	4.660	4.663	4.666	4.669
10.1	4.672	4.676	4.679	4.682	4.685	4.688	4.691	4.694	4.697	4.700
10.2	4.703	4.706	4.709	4.712	4.716	4.719	4.722	4.725	4.728	4.731
10.3	4.734	4.737	4.740	4.743	4.746	4.749	4.752	4.755	4.758	4.762
10.4	4.765	4.768	4.771	4.774	4.777	4.780	4.783	4.786	4.789	4.792
10.5	4.795	4.798	4.801	4.804	4.807	4.810	4.813	4.816	4.819	4.822
10.6	4.825	4.828	4.832	4.835	4.838	4.841	4.844	4.847	4.850	4.853
10.7	4.856	4.859	4.862	4.865	4.868	4.871	4.874	4.877	4.880	4.883
10.8	4.886	4.889	4.892	4.895	4.898	4.901	4.904	4.907	4.910	4.913
10.9	4.916	4.919	4.922	4.925	4.928	4.931	4.934	4.937	4.940	4.943
11.0	4.946	4.949	4.952	4.955	4.958	4.961	4.964	4.967	4.970	4.973
11.1	4.976	4.979	4.982	4.985	4.988	4.991	4.994	4.997	5.000	5.003
11.2	5.006	5.009	5.012	5.015	5.018	5.021	5.024	5.027	5.030	5.033
11.3	5.036	5.039	5.042	5.044	5.047	5.050	5.053	5.056	5.059	5.062
11.4	5.065	5.068	5.071	5.074	5.077	5.080	5.083	5.086	5.089	5.092
11.5	5.095	5.098	5.101	5.104	5.107	5.110	5.113	5.115	5.118	5.121
11.6	5.124	5.127	5.130	5.133	5.136	5.139	5.142	5.145	5.148	5.151
11.7	5.154	5.157	5.160	5.163	5.166	5.168	5.171	5.174	5.177	5.180
11.8	5.183	5.186	5.189	5.192	5.195	5.198	5.201	5.204	5.207	5.209
11.9	5.212	5.215	5.218	5.221	5.224	5.227	5.230	5.233	5.236	5.238
12.0	5.241	5.244	5.247	5.250	5.253	5.256	5.259	5.262	5.265	5.268
12.1	5.271	5.273	5.276	5.279	5.282	5.285	5.288	5.291	5.294	5.297
12.2	5.300	5.302	5.305	5.308	5.311	5.314	5.317	5.320	5.323	5.326
12.3	5.329	5.331	5.334	5.337	5.340	5.343	5.346	5.349	5.352	5.354
12.4	5.357	5.360	5.363	5.366	5.369	5.372	5.375	5.377	5.380	5.383
12.5	5.386	5.389	5.392	5.395	5.398	5.400	5.403	5.406	5.409	5.412
12.6	5.415	5.418	5.421	5.423	5.426	5.429	5.432	5.435	5.438	5.440
12.7	5.443	5.446	5.449	5.452	5.455	5.458	5.461	5.463	5.466	5.469
12.8	5.472	5.475	5.478	5.480	5.483	5.486	5.489	5.492	5.495	5.498
12.9	5.500	5.503	5.506	5.509	5.512	5.515	5.517	5.520	5.523	5.526
13.0	5.529	5.532	5.534	5.537	5.540	5.543	5.546	5.549	5.551	5.554
13.1	5.557	5.560	5.563	5.566	5.568	5.571	5.574	5.577	5.580	5.582
13.2	5.585	5.588	5.591	5.594	5.597	5.600	5.602	5.605	5.608	5.611
13.3	5.614	5.616	5.619	5.622	5.625	5.628	5.630	5.633	5.636	5.639
13.4	5.642	5.644	5.647	5.650	5.653	5.656	5.658	5.661	5.664	5.667
13.5	5.670	5.672	5.675	5.678	5.681	5.684	5.686	5.689	5.692	5.695
13.6	5.698	5.700	5.703	5.706	5.709	5.712	5.714	5.717	5.720	5.723
13.7	5.725	5.728	5.731	5.734	5.737	5.739	5.742	5.745	5.748	5.751
13.8	5.753	5.756	5.759	5.762	5.765	5.767	5.770	5.773	5.776	5.778
13.9	5.781	5.784	5.787	5.789	5.792	5.795	5.798	5.801	5.803	5.806
14.0	5.809	5.812	5.814	5.817	5.820	5.823	5.825	5.828	5.831	5.834
14.1	5.836	5.839	5.842	5.845	5.847	5.850	5.853	5.856	5.859	5.861
14.2	5.864	5.867	5.870	5.872	5.875	5.878	5.830	5.883	5.886	5.889
14.3	5.892	5.894	5.897	5.900	5.902	5.905	5.908	5.911	5.913	5.916
14.4	5.919	5.922	5.924	5.927	5.930	5.933	5.935	5.938	5.941	5.944
14.5	5.946	5.949	5.952	5.955	5.957	5.960	5.963	5.965	5.968	5.971
14.6	5.974	5.976	5.979	5.982	5.985	5.987	5.990	5.993	5.995	5.998
14.7	6.001	6.004	6.006	6.009	6.012	6.014	6.017	6.020	6.023	6.025
14.8	6.028	6.031	6.034	6.036	6.039	6.042	6.044	6.047	6.050	6.052
14.9	6.055	6.058	6.061	6.063	6.066	6.069	6.071	6.074	6.077	6.080

Table 63.—*Two-thirds powers of numbers*—Continued

Number	.00	.01	.02	.03	.04	.05	.06	.07	.08	.09
15.0	6.082	6.085	6.088	6.090	6.093	6.096	6.098	6.101	6.104	6.106
15.1	6.109	6.112	6.115	6.117	6.120	6.123	6.125	6.128	6.131	6.133
15.2	6.136	6.139	6.141	6.144	6.147	6.150	6.152	6.155	6.158	6.160
15.3	6.163	6.166	6.168	6.171	6.174	6.176	6.179	6.182	6.184	6.187
15.4	6.190	6.193	6.195	6.198	6.200	6.203	6.206	6.209	6.211	6.214
15.5	6.217	6.219	6.222	6.225	6.227	6.230	6.233	6.235	6.238	6.241
15.6	6.243	6.246	6.249	6.251	6.254	6.257	6.259	6.262	6.265	6.267
15.7	6.270	6.273	6.275	6.278	6.281	6.283	6.286	6.289	6.291	6.294
15.8	6.297	6.299	6.302	6.305	6.307	6.310	6.312	6.315	6.318	6.320
15.9	6.323	6.326	6.328	6.331	6.334	6.336	6.339	6.342	6.344	6.347
16.0	6.350	6.352	6.355	6.358	6.360	6.363	6.365	6.368	6.371	6.373
16.1	6.376	6.379	6.381	6.384	6.387	6.389	6.392	6.394	6.397	6.400
16.2	6.402	6.405	6.408	6.410	6.413	6.416	6.418	6.421	6.423	6.426
16.3	6.429	6.431	6.434	6.437	6.439	6.442	6.444	6.447	6.450	6.452
16.4	6.455	6.458	6.460	6.463	6.465	6.468	6.471	6.473	6.476	6.479
16.5	6.481	6.484	6.486	6.489	6.492	6.494	6.497	6.500	6.502	6.505
16.6	6.507	6.510	6.513	6.515	6.518	6.520	6.523	6.526	6.528	6.531
16.7	6.533	6.536	6.539	6.541	6.544	6.546	6.549	6.552	6.554	6.557
16.8	6.560	6.562	6.565	6.567	6.570	6.573	6.575	6.578	6.580	6.583
16.9	6.586	6.588	6.591	6.593	6.596	6.599	6.601	6.604	6.606	6.609
17.0	6.611	6.614	6.617	6.619	6.622	6.624	6.627	6.630	6.632	6.635
17.1	6.637	6.640	6.643	6.645	6.648	6.650	6.653	6.655	6.658	6.661
17.2	6.663	6.666	6.668	6.671	6.674	6.676	6.679	6.681	6.684	6.686
17.3	6.689	6.692	6.694	6.697	6.699	6.702	6.705	6.707	6.710	6.712
17.4	6.715	6.717	6.720	6.723	6.725	6.728	6.730	6.733	6.735	6.738
17.5	6.740	6.743	6.746	6.748	6.751	6.753	6.756	6.758	6.761	6.764
17.6	6.766	6.769	6.771	6.774	6.776	6.779	6.782	6.784	6.787	6.789
17.7	6.792	6.794	6.797	6.799	6.802	6.805	6.807	6.810	6.812	6.815
17.8	6.817	6.820	6.822	6.825	6.828	6.830	6.833	6.835	6.838	6.840
17.9	6.843	6.845	6.848	6.850	6.853	6.856	6.858	6.861	6.863	6.866
18.0	6.868	6.871	6.873	6.876	6.878	6.881	6.884	6.886	6.889	6.891
18.1	6.894	6.896	6.899	6.901	6.904	6.906	6.909	6.911	6.914	6.917
18.2	6.919	6.922	6.924	6.927	6.929	6.932	6.934	6.937	6.939	6.942
18.3	6.944	6.947	6.949	6.952	6.954	6.957	6.960	6.962	6.965	6.967
18.4	6.970	6.972	6.975	6.977	6.980	6.982	6.985	6.987	6.990	6.992
18.5	6.995	6.997	7.000	7.002	7.005	7.007	7.010	7.013	7.015	7.018
18.6	7.020	7.023	7.025	7.028	7.030	7.032	7.035	7.038	7.040	7.043
18.7	7.045	7.048	7.050	7.053	7.055	7.058	7.060	7.063	7.065	7.068
18.8	7.070	7.073	7.075	7.078	7.080	7.083	7.085	7.088	7.090	7.093
18.9	7.095	7.098	7.100	7.103	7.105	7.108	7.110	7.113	7.115	7.118
19.0	7.120	7.123	7.125	7.128	7.130	7.133	7.135	7.138	7.140	7.143
19.1	7.145	7.148	7.150	7.153	7.155	7.158	7.160	7.163	7.165	7.168
19.2	7.170	7.173	7.175	7.178	7.180	7.183	7.185	7.188	7.190	7.193
19.3	7.195	7.198	7.200	7.203	7.205	7.208	7.210	7.212	7.215	7.217
19.4	7.220	7.222	7.225	7.227	7.230	7.232	7.235	7.237	7.240	7.242
19.5	7.245	7.247	7.250	7.252	7.255	7.257	7.260	7.262	7.265	7.267
19.6	7.270	7.272	7.274	7.277	7.279	7.282	7.284	7.287	7.289	7.292
19.7	7.294	7.297	7.299	7.302	7.304	7.307	7.309	7.311	7.314	7.316
19.8	7.319	7.321	7.324	7.326	7.329	7.331	7.334	7.336	7.339	7.341
19.9	7.343	7.346	7.348	7.351	7.353	7.356	7.358	7.361	7.363	7.366

Table 63.—*Two-thirds powers of numbers*—Continued

Number	.00	.01	.02	.03	.04	.05	.06	.07	.08	.09
20.0	7.368	7.371	7.373	7.375	7.378	7.380	7.383	7.385	7.388	7.390
20.1	7.393	7.395	7.397	7.400	7.402	7.405	7.407	7.410	7.412	7.415
20.2	7.417	7.420	7.422	7.424	7.427	7.429	7.432	7.434	7.437	7.439
20.3	7.442	7.444	7.446	7.449	7.451	7.454	7.456	7.459	7.461	7.464
20.4	7.466	7.468	7.471	7.473	7.476	7.478	7.481	7.483	7.485	7.488
20.5	7.490	7.493	7.495	7.498	7.500	7.503	7.505	7.507	7.510	7.512
20.6	7.515	7.517	7.520	7.522	7.524	7.527	7.529	7.532	7.534	7.537
20.7	7.539	7.541	7.544	7.546	7.549	7.551	7.554	7.556	7.558	7.561
20.8	7.563	7.566	7.568	7.571	7.573	7.575	7.578	7.580	7.583	7.585
20.9	7.587	7.590	7.592	7.595	7.597	7.600	7.602	7.604	7.607	7.609
21.0	7.612	7.614	7.617	7.619	7.621	7.624	7.626	7.629	7.631	7.633
21.1	7.636	7.638	7.641	7.643	7.645	7.648	7.650	7.653	7.655	7.658
21.2	7.660	7.662	7.665	7.667	7.670	7.672	7.674	7.677	7.679	7.682
21.3	7.684	7.686	7.689	7.691	7.694	7.696	7.698	7.701	7.703	7.706
21.4	7.708	7.710	7.713	7.715	7.718	7.720	7.722	7.725	7.727	7.730
21.5	7.732	7.734	7.737	7.739	7.742	7.744	7.746	7.749	7.751	7.754
21.6	7.756	7.758	7.761	7.763	7.766	7.768	7.770	7.773	7.775	7.778
21.7	7.780	7.782	7.785	7.787	7.789	7.792	7.794	7.797	7.799	7.801
21.8	7.804	7.806	7.809	7.811	7.813	7.816	7.818	7.820	7.823	7.825
21.9	7.828	7.830	7.832	7.835	7.837	7.840	7.842	7.844	7.847	7.849
22.0	7.851	7.854	7.856	7.859	7.861	7.863	7.866	7.868	7.870	7.873
22.1	7.875	7.878	7.880	7.882	7.885	7.887	7.889	7.892	7.894	7.897
22.2	7.899	7.901	7.904	7.906	7.908	7.911	7.913	7.916	7.918	7.920
22.3	7.923	7.925	7.927	7.930	7.932	7.934	7.937	7.939	7.942	7.944
22.4	7.946	7.949	7.951	7.953	7.956	7.958	7.960	7.963	7.965	7.968
22.5	7.970	7.972	7.975	7.977	7.979	7.982	7.984	7.986	7.989	7.991
22.6	7.994	7.996	7.998	8.001	8.003	8.005	8.008	8.010	8.012	8.015
22.7	8.017	8.019	8.022	8.024	8.027	8.029	8.031	8.034	8.036	8.038
22.8	8.041	8.043	8.045	8.048	8.050	8.052	8.055	8.057	8.059	8.062
22.9	8.064	8.066	8.069	8.071	8.074	8.076	8.078	8.081	8.083	8.085
23.0	8.088	8.090	8.092	8.095	8.097	8.099	8.102	8.104	8.106	8.109
23.1	8.111	8.113	8.116	8.118	8.120	8.123	8.125	8.127	8.130	8.132
23.2	8.134	8.137	8.139	8.141	8.144	8.146	8.148	8.151	8.153	8.155
23.3	8.158	8.160	8.162	8.165	8.167	8.169	8.172	8.174	8.176	8.179
23.4	8.181	8.183	8.186	8.188	8.190	8.193	8.195	8.197	8.200	8.202
23.5	8.204	8.207	8.209	8.211	8.214	8.216	8.218	8.221	8.223	8.225
23.6	8.228	8.230	8.232	8.235	8.237	8.239	8.242	8.244	8.246	8.249
23.7	8.251	8.253	8.255	8.258	8.260	8.262	8.265	8.267	8.269	8.272
23.8	8.274	8.276	8.279	8.281	8.283	8.286	8.288	8.290	8.293	8.295
23.9	8.297	8.300	8.302	8.304	8.306	8.309	8.311	8.313	8.316	8.318
24.0	8.320	8.323	8.325	8.327	8.330	8.332	8.334	8.337	8.339	8.341
24.1	8.343	8.346	8.348	8.350	8.353	8.355	8.357	8.360	8.362	8.364
24.2	8.366	8.369	8.371	8.374	8.376	8.378	8.380	8.383	8.385	8.387
24.3	8.390	8.392	8.394	8.396	8.399	8.401	8.403	8.406	8.408	8.410
24.4	8.413	8.415	8.417	8.419	8.422	8.424	8.426	8.429	8.431	8.433
24.5	8.436	8.438	8.440	8.442	8.445	8.447	8.449	8.452	8.454	8.456
24.6	8.458	8.461	8.463	8.465	8.468	8.470	8.472	8.474	8.477	8.479
24.7	8.481	8.484	8.486	8.488	8.490	8.493	8.495	8.497	8.500	8.502
24.8	8.504	8.507	8.509	8.511	8.513	8.516	8.518	8.520	8.522	8.525
24.9	8.527	8.529	8.532	8.534	8.536	8.538	8.541	8.543	8.545	8.548

Table 63.—*Two-thirds powers of numbers*—Continued

Number	.00	.01	.02	.03	.04	.05	.06	.07	.08	.09
25.0	8.550	8.552	8.554	8.557	8.559	8.561	8.564	8.566	8.568	8.570
25.1	8.573	8.575	8.577	8.580	8.582	8.584	8.586	8.589	8.591	8.593
25.2	8.595	8.598	8.600	8.602	8.605	8.607	8.609	8.611	8.614	8.616
25.3	8.618	8.620	8.623	8.625	8.627	8.629	8.632	8.634	8.636	8.639
25.4	8.641	8.643	8.645	8.648	8.650	8.652	8.654	8.657	8.659	8.661
25.5	8.664	8.666	8.668	8.670	8.673	8.675	8.677	8.679	8.682	8.684
25.6	8.686	8.688	8.691	8.693	8.695	8.697	8.700	8.702	8.704	8.706
25.7	8.709	8.711	8.713	8.716	8.718	8.720	8.722	8.725	8.727	8.729
25.8	8.731	8.734	8.736	8.738	8.740	8.743	8.745	8.747	8.749	8.752
25.9	8.754	8.756	8.758	8.761	8.763	8.765	8.767	8.770	8.772	8.774
26.0	8.776	8.779	8.781	8.783	8.785	8.788	8.790	8.792	8.794	8.797
26.1	8.799	8.801	8.803	8.806	8.808	8.810	8.812	8.815	8.817	8.819
26.2	8.821	8.824	8.826	8.828	8.830	8.833	8.835	8.837	8.839	8.842
26.3	8.844	8.846	8.848	8.850	8.853	8.855	8.857	8.859	8.862	8.864
26.4	8.866	8.868	8.871	8.873	8.875	8.877	8.880	8.882	8.884	8.886
26.5	8.889	8.891	8.893	8.895	8.897	8.900	8.902	8.904	8.906	8.909
26.6	8.911	8.913	8.915	8.918	8.920	8.922	8.924	8.927	8.929	8.931
26.7	8.933	8.935	8.938	8.940	8.942	8.944	8.947	8.949	8.951	8.953
26.8	8.956	8.958	8.960	8.962	8.964	8.967	8.969	8.971	8.973	8.976
26.9	8.978	8.980	8.982	8.984	8.987	8.989	8.991	8.993	8.996	8.998
27.0	9.000	9.002	9.004	9.007	9.009	9.011	9.013	9.016	9.018	9.020
27.1	9.022	9.024	9.027	9.029	9.031	9.033	9.036	9.038	9.040	9.042
27.2	9.044	9.047	9.049	9.051	9.053	9.055	9.058	9.060	9.062	9.064
27.3	9.067	9.069	9.071	9.073	9.075	9.078	9.080	9.082	9.084	9.086
27.4	9.089	9.091	9.093	9.095	9.098	9.100	9.102	9.104	9.106	9.109
27.5	9.111	9.113	9.115	9.117	9.120	9.122	9.124	9.126	9.128	9.131
27.6	9.133	9.135	9.137	9.139	9.142	9.144	9.146	9.148	9.150	9.153
27.7	9.155	9.157	9.159	9.162	9.164	9.166	9.168	9.170	9.173	9.175
27.8	9.177	9.179	9.181	9.184	9.186	9.188	9.190	9.192	9.195	9.197
27.9	9.199	9.201	9.203	9.206	9.208	9.210	9.212	9.214	9.216	9.219
28.0	9.221	9.223	9.225	9.227	9.230	9.232	9.234	9.236	9.238	9.241
28.1	9.243	9.245	9.247	9.249	9.252	9.254	9.256	9.258	9.260	9.263
28.2	9.265	9.267	9.269	9.271	9.274	9.276	9.278	9.280	9.282	9.284
28.3	9.287	9.289	9.291	9.293	9.295	9.298	9.300	9.302	9.304	9.306
28.4	9.308	9.311	9.313	9.315	9.317	9.319	9.322	9.324	9.326	9.328
28.5	9.330	9.333	9.335	9.337	9.339	9.341	9.343	9.346	9.348	9.350
28.6	9.352	9.354	9.356	9.359	9.361	9.363	9.365	9.367	9.370	9.372
28.7	9.374	9.376	9.378	9.380	9.383	9.385	9.387	9.389	9.391	9.393
28.8	9.396	9.398	9.400	9.402	9.404	9.407	9.409	9.411	9.413	9.415
28.9	9.417	9.420	9.422	9.424	9.426	9.428	9.430	9.433	9.435	9.437
29.0	9.439	9.441	9.443	9.446	9.448	9.450	9.452	9.454	9.456	9.459
29.1	9.461	9.463	9.465	9.467	9.469	9.472	9.474	9.476	9.478	9.480
29.2	9.482	9.485	9.487	9.489	9.491	9.493	9.495	9.498	9.500	9.502
29.3	9.504	9.506	9.508	9.511	9.513	9.515	9.517	9.519	9.521	9.524
29.4	9.526	9.528	9.530	9.532	9.534	9.537	9.539	9.541	9.543	9.545
29.5	9.547	9.549	9.552	9.554	9.556	9.558	9.560	9.562	9.565	9.567
29.6	9.569	9.571	9.573	9.575	9.577	9.580	9.582	9.584	9.586	9.588
29.7	9.590	9.593	9.595	9.597	9.599	9.601	9.603	9.605	9.608	9.610
29.8	9.612	9.614	9.616	9.618	9.621	9.623	9.625	9.627	9.629	9.631
29.9	9.633	9.636	9.638	9.640	9.642	9.644	9.646	9.648	9.651	9.653

Table 63.—*Two-thirds powers of numbers*—Continued

Number	.0	.1	.2	.3	.4	.5	.6	.7	.8	.9
30	9. 655	9. 676	9. 698	9. 719	9. 741	9. 762	9. 783	9. 805	9. 826	9. 847
31	9. 868	9. 889	9. 911	9. 932	9. 953	9. 974	9. 995	10. 02	10. 04	10. 06
32	10. 08	10. 10	10. 12	10. 14	10. 16	10. 18	10. 20	10. 23	10. 25	10. 27
33	10. 29	10. 31	10. 33	10. 35	10. 37	10. 39	10. 41	10. 43	10. 45	10. 47
34	10. 50	10. 52	10. 54	10. 56	10. 58	10. 60	10. 62	10. 64	10. 66	10. 68
35	10. 70	10. 72	10. 74	10. 76	10. 78	10. 80	10. 82	10. 84	10. 86	10. 88
36	10. 90	10. 92	10. 94	10. 96	10. 98	11. 00	11. 02	11. 04	11. 06	11. 08
37	11. 10	11. 12	11. 14	11. 16	11. 18	11. 20	11. 22	11. 24	11. 26	11. 28
38	11. 30	11. 32	11. 34	11. 36	11. 38	11. 40	11. 42	11. 44	11. 46	11. 48
39	11. 50	11. 52	11. 54	11. 56	11. 58	11. 60	11. 62	11. 64	11. 66	11. 68
40	11. 70	11. 72	11. 74	11. 75	11. 77	11. 79	11. 81	11. 83	11. 85	11. 87
41	11. 89	11. 91	11. 93	11. 95	11. 97	11. 99	12. 01	12. 03	12. 04	12. 06
42	12. 08	12. 10	12. 12	12. 14	12. 16	12. 18	12. 20	12. 22	12. 24	12. 25
43	12. 27	12. 29	12. 31.	12. 33	12. 35	12. 37	12. 39	12. 41	12. 43	12. 44
44	12. 46	12. 48	12. 50	12. 52	12. 54	12. 56	12. 58	12. 60	12. 61	12. 63
45	12. 65	12. 67	12. 69	12. 71	12. 73	12. 75	12. 76	12. 78	12. 80	12. 82
46	12. 84	12. 86	12. 88	12. 89	12. 91	12. 93	12. 95	12. 97	12. 99	13. 01
47	13. 02	13. 04	13. 06	13. 08	13. 10	13. 12	13. 13	13. 15	13. 17	13. 19
48	13. 21	13. 23	13. 24	13. 26	13. 28	13. 30	13. 32	13. 34	13. 35	13. 37
49	13. 39	13. 41	13. 43	13. 45	13. 46	13. 48	13. 50	13. 52	13. 54	13. 55
50	13. 57	13. 59	13. 61	13. 63	13. 64	13. 66	13. 68	13. 70	13. 72	13. 73
51	13. 75	13. 77	13. 79	13. 81	13. 82	13. 84	13. 86	13. 88	13. 90	13. 91
52	13. 93	13. 95	13. 97	13. 99	14. 00	14. 02	14. 04	14. 06	14. 07	14. 09
53	14. 11	14. 13	14. 15	14. 16	14. 18	14. 20	14. 22	14. 23	14. 25	14. 27
54	14. 29	14. 30	14. 32	14. 34	14. 36	14. 37	14. 39	14. 41	14. 43	14. 44
55	14. 46	14. 48	14. 50	14. 51	14. 53	14. 55	14. 57	14. 58	14. 60	14. 62
56	14. 64	14. 65	14. 67	14. 69	14. 71	14. 72	14. 74	14. 76	14. 78	14. 79
57	14. 81	14. 83	14. 85	14. 86	14. 88	14. 90	14. 91	14. 93	14. 95	14. 97
58	14. 98	15. 00	15. 02	15. 04	15. 05	15. 07	15. 09	15. 10	15. 12	15. 14
59	15. 16	15. 17	15. 19	15. 21	15. 22	15. 24	15. 26	15. 28	15. 29	15. 31
60	15. 33	15. 34	15. 36	15. 38	15. 39	15. 41	15. 43	15. 45	15. 46	15. 48
61	15. 50	15. 51	15. 53	15. 55	15. 56	15. 58	15. 60	15. 61	15. 63	15. 65
62	15. 66	15. 68	15. 70	15. 72	15. 73	15. 75	15. 77	15. 78	15. 80	15. 82
63	15. 83	15. 85	15. 87	15. 88	15. 90	15. 92	15. 93	15. 95	15. 97	15. 98
64	16. 00	16. 02	16. 03	16. 05	16. 07	16. 08	16. 10	16. 12	16. 13	16. 15
65	16. 17	16. 18	16. 20	16. 22	16. 23	16. 25	16. 27	16. 28	16. 30	16. 32
66	16. 33	16. 35	16. 36	16. 38	16. 40	16. 41	16. 43	16. 45	16. 46	16. 48
67	16. 50	16. 51	16. 53	16. 55	16. 56	16. 58	16. 59	16. 61	16. 63	16. 64
68	16. 66	16. 68	16. 69	16. 71	16. 73	16. 74	16. 76	16. 77	16. 79	16. 81
69	16. 82	16. 84	16. 86	16. 87	16. 89	16. 90	16. 92	16. 94	16. 95	16. 97
70	16. 98	17. 00	17. 02	17. 03	17. 05	17. 07	17. 08	17. 10	17. 11	17. 13
71	17. 15	17. 16	17. 18	17. 19	17. 21	17. 23	17. 24	17. 26	17. 27	17. 29
72	17. 31	17. 32	17. 34	17. 36	17. 37	17. 39	17. 40	17. 42	17. 43	17. 45
73	17. 47	17. 48	17. 50	17. 51	17. 53	17. 55	17. 56	17. 58	17. 59	17. 61
74	17. 63	17. 64	17. 66	17. 67	17. 69	17. 71	17. 72	17. 74	17. 75	17. 77
75	17. 78	17. 80	17. 82	17. 83	17. 85	17. 86	17. 88	17. 89	17. 91	17. 93
76	17. 94	17. 96	17. 97	17. 99	18. 01	18. 02	18. 04	18. 05	18. 07	18. 08
77	18. 10	18. 11	18. 13	18. 15	18. 16	18. 18	18. 19	18. 21	18. 22	18. 24
78	18. 26	18. 27	18. 29	18. 30	18. 32	18. 33	18. 35	18. 36	18. 38	18. 40
79	18. 41	18. 43	18. 44	18. 46	18. 47	18. 49	18. 50	18. 52	18. 54	18. 55

Table 63.—*Two-thirds powers of numbers*—Continued

Number	.0	.1	.2	.3	.4	.5	.6	.7	.8	.9
80	18. 57	18. 58	18. 60	18. 61	18. 63	18. 64	18. 66	18. 67	18. 69	18. 71
81	18. 72	18. 74	18. 75	18. 77	18. 78	18. 80	18. 81	18. 83	18. 84	18. 86
82	18. 87	18. 89	18. 91	18. 92	18. 94	18. 95	18. 97	18. 98	19. 00	19. 01
83	19. 03	19. 04	19. 06	19. 07	19. 09	19. 10	19. 12	19. 13	19. 15	19. 16
84	19. 18	19. 20	19. 21	19. 23	19. 24	19. 26	19. 27	19. 29	19. 30	19. 32
85	19. 33	19. 35	19. 36	19. 38	19. 39	19. 41	19. 42	19. 44	19. 45	19. 47
86	19. 48	19. 50	19. 51	19. 53	19. 54	19. 56	19. 57	19. 59	19. 60	19. 62
87	19. 63	19. 65	19. 66	19. 68	19. 69	19. 71	19. 72	19. 74	19. 75	19. 77
88	19. 78	19. 80	19. 81	19. 83	19. 84	19. 86	19. 87	19. 89	19. 90	19. 92
89	19. 93	19. 95	19. 96	19. 98	19. 99	20. 01	20. 02	20. 04	20. 05	20. 07
90	20. 08	20. 10	20. 11	20. 13	20. 14	20. 16	20. 17	20. 19	20. 20	20. 22
91	20. 23	20. 25	20. 26	20. 28	20. 29	20. 31	20. 32	20. 34	20. 35	20. 36
92	20. 38	20. 39	20. 41	20. 42	20. 44	20. 45	20. 47	20. 48	20. 50	20. 51
93	20. 53	20. 54	20. 56	20. 57	20. 59	20. 60	20. 62	20. 63	20. 64	20. 66
94	20. 67	20. 69	20. 70	20. 72	20. 73	20. 75	20. 76	20. 78	20. 79	20. 81
95	20. 82	20. 83	20. 85	20. 86	20. 88	20. 89	20. 91	20. 92	20. 94	20. 95
96	20. 97	20. 98	21. 00	21. 01	21. 02	21. 04	21. 05	21. 07	21. 08	21. 10
97	21. 11	21. 13	21. 14	21. 15	21. 17	21. 18	21. 20	21. 21	21. 23	21. 24
98	21. 26	21. 27	21. 29	21. 30	21. 31	21. 33	21. 34	21. 36	21. 37	21. 39
99	21. 40	21. 41	21. 43	21. 44	21. 46	21. 47	21. 49	21. 50	21. 52	21. 53

Table 64.—*Three-fifths powers of numbers*

Number	.00	.01	.02	.03	.04	.05	.06	.07	.08	.09
0.0	0.0000	0.0631	0.0956	0.1220	0.1450	0.1657	0.1849	0.2028	0.2197	0.2358
0.1	.2512	.2660	.2802	.2940	.3074	.3204	.3330	.3454	.3574	.3692
0.2	.3807	.3920	.4031	.4140	.4247	.4353	.4456	.4558	.4659	.4758
0.3	.4856	.4952	.5048	.5142	.5235	.5326	.5417	.5507	.5596	.5684
0.4	.5771	.5857	.5942	.6027	.6110	.6193	.6276	.6357	.6438	.6518
0.5	.6598	.6676	.6755	.6832	.6909	.6986	.7062	.7137	.7212	.7286
0.6	.7360	.7434	.7506	.7579	.7651	.7722	.7793	.7864	.7934	.8004
0.7	.8073	.8142	.8211	.8279	.8347	.8415	.8482	.8549	.8615	.8681
0.8	.8747	.8812	.8877	.8942	.9007	.9071	.9135	.9198	.9262	.9325
0.9	.9387	.9450	.9512	.9574	.9636	.9697	.9758	.9819	.9880	.9940
1.0	1.000	1.006	1.012	1.018	1.024	1.030	1.036	1.041	1.047	1.053
1.1	1.059	1.065	1.070	1.076	1.082	1.087	1.093	1.099	1.104	1.110
1.2	1.116	1.121	1.127	1.132	1.138	1.143	1.149	1.154	1.160	1.165
1.3	1.170	1.176	1.181	1.187	1.192	1.197	1.203	1.208	1.213	1.218
1.4	1.224	1.229	1.234	1.239	1.245	1.250	1.255	1.260	1.265	1.270
1.5	1.275	1.281	1.286	1.291	1.296	1.301	1.306	1.311	1.316	1.321
1.6	1.326	1.331	1.336	1.341	1.346	1.350	1.355	1.360	1.365	1.370
1.7	1.375	1.380	1.385	1.389	1.394	1.399	1.404	1.409	1.413	1.418
1.8	1.423	1.428	1.432	1.437	1.442	1.446	1.451	1.456	1.460	1.465
1.9	1.470	1.474	1.479	1.484	1.488	1.493	1.497	1.502	1.507	1.511
2.0	1.516	1.520	1.525	1.529	1.534	1.538	1.543	1.547	1.552	1.556
2.1	1.561	1.565	1.570	1.574	1.579	1.583	1.587	1.592	1.596	1.601
2.2	1.605	1.609	1.614	1.618	1.622	1.627	1.631	1.635	1.640	1.644
2.3	1.648	1.653	1.657	1.661	1.665	1.670	1.674	1.678	1.682	1.687
2.4	1.691	1.695	1.699	1.704	1.708	1.712	1.716	1.720	1.725	1.729
2.5	1.733	1.737	1.741	1.745	1.749	1.754	1.758	1.762	1.766	1.770
2.6	1.774	1.778	1.782	1.786	1.790	1.795	1.799	1.803	1.807	1.811
2.7	1.815	1.819	1.823	1.827	1.831	1.835	1.839	1.843	1.847	1.851
2.8	1.855	1.859	1.863	1.867	1.871	1.875	1.879	1.882	1.886	1.890
2.9	1.894	1.898	1.902	1.906	1.910	1.914	1.918	1.922	1.925	1.929
3.0	1.933	1.937	1.941	1.945	1.949	1.952	1.956	1.960	1.964	1.968
3.1	1.972	1.975	1.979	1.983	1.987	1.991	1.994	1.998	2.002	2.006
3.2	2.010	2.013	2.017	2.021	2.025	2.028	2.032	2.036	2.040	2.043
3.3	2.047	2.051	2.054	2.058	2.062	2.066	2.069	2.073	2.077	2.080
3.4	2.084	2.088	2.091	2.095	2.099	2.102	2.106	2.110	2.113	2.117
3.5	2.121	2.124	2.128	2.131	2.135	2.139	2.142	2.146	2.149	2.153
3.6	2.157	2.160	2.164	2.167	2.171	2.175	2.178	2.182	2.185	2.189
3.7	2.192	2.196	2.200	2.203	2.207	2.210	2.214	2.217	2.221	2.224
3.8	2.228	2.231	2.235	2.238	2.242	2.245	2.249	2.252	2.256	2.259
3.9	2.263	2.266	2.270	2.273	2.277	2.280	2.284	2.287	2.290	2.294
4.0	2.297	2.301	2.304	2.308	2.311	2.315	2.318	2.321	2.325	2.328
4.1	2.332	2.335	2.339	2.342	2.345	2.349	2.352	2.355	2.359	2.362
4.2	2.366	2.369	2.372	2.376	2.379	2.383	2.386	2.389	2.393	2.396
4.3	2.399	2.403	2.406	2.409	2.413	2.416	2.419	2.423	2.426	2.429
4.4	2.433	2.436	2.439	2.443	2.446	2.449	2.452	2.456	2.459	2.462
4.5	2.466	2.469	2.472	2.475	2.479	2.482	2.485	2.489	2.492	2.495
4.6	2.498	2.502	2.505	2.508	2.511	2.515	2.518	2.521	2.524	2.528
4.7	2.531	2.534	2.537	2.540	2.544	2.547	2.550	2.553	2.557	2.560
4.8	2.563	2.566	2.569	2.573	2.576	2.579	2.582	2.585	2.589	2.592
4.9	2.595	2.598	2.601	2.604	2.608	2.611	2.614	2.617	2.620	2.623

Table 64.—*Three-fifths powers of numbers*—Continued

Number	.00	.01	.02	.03	.04	.05	.06	.07	.08	.09
5.0	2.627	2.630	2.633	2.636	2.639	2.642	2.645	2.649	2.652	2.655
5.1	2.658	2.661	2.664	2.667	2.670	2.674	2.677	2.680	2.683	2.686
5.2	2.689	2.692	2.695	2.698	2.701	2.705	2.708	2.711	2.714	2.717
5.3	2.720	2.723	2.726	2.729	2.732	2.735	2.738	2.741	2.745	2.748
5.4	2.751	2.754	2.757	2.760	2.763	2.766	2.769	2.772	2.775	2.778
5.5	2.781	2.784	2.787	2.790	2.793	2.796	2.799	2.802	2.805	2.808
5.6	2.811	2.814	2.817	2.820	2.823	2.826	2.829	2.832	2.835	2.838
5.7	2.841	2.844	2.847	2.850	2.853	2.856	2.859	2.862	2.865	2.868
5.8	2.871	2.874	2.877	2.880	2.883	2.886	2.889	2.892	2.895	2.898
5.9	2.901	2.904	2.907	2.910	2.913	2.915	2.918	2.921	2.924	2.927
6.0	2.930	2.933	2.936	2.939	2.942	2.945	2.948	2.951	2.954	2.956
6.1	2.959	2.962	2.965	2.968	2.971	2.974	2.977	2.980	2.983	2.985
6.2	2.988	2.991	2.994	2.997	3.000	3.003	3.006	3.009	3.011	3.014
6.3	3.017	3.020	3.023	3.026	3.029	3.032	3.034	3.037	3.040	3.043
6.4	3.046	3.049	3.052	3.054	3.057	3.060	3.063	3.066	3.069	3.071
6.5	3.074	3.077	3.080	3.083	3.086	3.088	3.091	3.094	3.097	3.100
6.6	3.103	3.105	3.108	3.111	3.114	3.117	3.119	3.122	3.125	3.128
6.7	3.131	3.134	3.136	3.139	3.142	3.145	3.148	3.150	3.153	3.156
6.8	3.159	3.161	3.164	3.167	3.170	3.173	3.175	3.178	3.181	3.184
6.9	3.186	3.189	3.192	3.195	3.198	3.200	3.203	3.206	3.209	3.211
7.0	3.214	3.217	3.220	3.222	3.225	3.228	3.231	3.233	3.236	3.239
7.1	3.242	3.244	3.247	3.250	3.253	3.255	3.258	3.261	3.263	3.266
7.2	3.269	3.272	3.274	3.277	3.280	3.282	3.285	3.288	3.291	3.293
7.3	3.296	3.299	3.301	3.304	3.307	3.310	3.312	3.315	3.318	3.320
7.4	3.323	3.326	3.328	3.331	3.334	3.337	3.339	3.342	3.345	3.347
7.5	3.350	3.353	3.355	3.358	3.361	3.363	3.366	3.369	3.371	3.374
7.6	3.377	3.379	3.382	3.385	3.387	3.390	3.393	3.395	3.398	3.401
7.7	3.403	3.406	3.409	3.411	3.414	3.416	3.419	3.422	3.424	3.427
7.8	3.430	3.432	3.435	3.438	3.440	3.443	3.446	3.448	3.451	3.453
7.9	3.456	3.459	3.461	3.464	3.467	3.469	3.472	3.474	3.477	3.480
8.0	3.482	3.485	3.487	3.490	3.493	3.495	3.498	3.500	3.503	3.506
8.1	3.508	3.511	3.513	3.516	3.519	3.521	3.524	3.526	3.529	3.532
8.2	3.534	3.537	3.539	3.542	3.545	3.547	3.550	3.552	3.555	3.557
8.3	3.560	3.563	3.565	3.568	3.570	3.573	3.575	3.578	3.581	3.583
8.4	3.586	3.588	3.591	3.593	3.596	3.598	3.601	3.604	3.606	3.609
8.5	3.611	3.614	3.616	3.619	3.621	3.624	3.626	3.629	3.632	3.634
8.6	3.637	3.639	3.642	3.644	3.647	3.649	3.652	3.654	3.657	3.659
8.7	3.662	3.664	3.667	3.670	3.672	3.675	3.677	3.680	3.682	3.685
8.8	3.687	3.690	3.692	3.695	3.697	3.700	3.702	3.705	3.707	3.710
8.9	3.712	3.715	3.717	3.720	3.722	3.725	3.727	3.730	3.732	3.735
9.0	3.737	3.740	3.742	3.745	3.747	3.750	3.752	3.755	3.757	3.760
9.1	3.762	3.765	3.767	3.769	3.772	3.774	3.777	3.779	3.782	3.784
9.2	3.787	3.789	3.792	3.794	3.797	3.799	3.802	3.804	3.807	3.809
9.3	3.811	3.814	3.816	3.819	3.821	3.824	3.826	3.829	3.831	3.834
9.4	3.836	3.838	3.841	3.843	3.846	3.848	3.851	3.853	3.856	3.858
9.5	3.860	3.863	3.865	3.868	3.870	3.873	3.875	3.877	3.880	3.882
9.6	3.885	3.887	3.890	3.892	3.894	3.897	3.899	3.902	3.904	3.907
9.7	3.909	3.911	3.914	3.916	3.919	3.921	3.923	3.926	3.928	3.931
9.8	3.933	3.936	3.938	3.940	3.943	3.945	3.948	3.950	3.952	3.955
9.9	3.957	3.960	3.962	3.964	3.967	3.969	3.972	3.974	3.976	3.979

Table 64.—*Three-fifths powers of numbers*—Continued

Number	.00	.01	.02	.03	.04	.05	.06	.07	.08	.09
10. 0	3. 981	3. 983	3. 986	3. 988	3. 991	3. 993	3. 995	3. 998	4. 000	4. 003
10. 1	4. 005	4. 007	4. 010	4. 012	4. 014	4. 017	4. 019	4. 022	4. 024	4. 026
10. 2	4. 029	4. 031	4. 033	4. 036	4. 038	4. 040	4. 043	4. 045	4. 048	4. 050
10. 3	4. 052	4. 055	4. 057	4. 059	4. 062	4. 064	4. 066	4. 069	4. 071	4. 074
10. 4	4. 076	4. 078	4. 081	4. 083	4. 085	4. 088	4. 090	4. 092	4. 095	4. 097
10. 5	4. 099	4. 102	4. 104	4. 106	4. 109	4. 111	4. 113	4. 116	4. 118	4. 120
10. 6	4. 123	4. 125	4. 127	4. 130	4. 132	4. 134	4. 137	4. 139	4. 141	4. 144
10. 7	4. 146	4. 148	4. 151	4. 153	4. 155	4. 158	4. 160	4. 162	4. 165	4. 167
10. 8	4. 169	4. 172	4. 174	4. 176	4. 178	4. 181	4. 183	4. 185	4. 188	4. 190
10. 9	4. 192	4. 195	4. 197	4. 199	4. 202	4. 204	4. 206	4. 208	4. 211	4. 213
11. 0	4. 215	4. 218	4. 220	4. 222	4. 225	4. 227	4. 229	4. 231	4. 234	4. 236
11. 1	4. 238	4. 241	4. 243	4. 245	4. 247	4. 250	4. 252	4. 254	4. 257	4. 259
11. 2	4. 261	4. 263	4. 266	4. 268	4. 270	4. 273	4. 275	4. 277	4. 279	4. 282
11. 3	4. 284	4. 286	4. 289	4. 291	4. 293	4. 295	4. 298	4. 300	4. 302	4. 304
11. 4	4. 307	4. 309	4. 311	4. 313	4. 316	4. 318	4. 320	4. 323	4. 325	4. 327
11. 5	4. 329	4. 332	4. 334	4. 336	4. 338	4. 341	4. 343	4. 345	4. 347	4. 350
11. 6	4. 352	4. 354	4. 356	4. 359	4. 361	4. 363	4. 365	4. 368	4. 370	4. 372
11. 7	4. 374	4. 377	4. 379	4. 381	4. 383	4. 386	4. 388	4. 390	4. 392	4. 394
11. 8	4. 397	4. 399	4. 401	4. 403	4. 406	4. 408	4. 410	4. 412	4. 415	4. 417
11. 9	4. 419	4. 421	4. 424	4. 426	4. 428	4. 430	4. 432	4. 435	4. 437	4. 439
12. 0	4. 441	4. 444	4. 446	4. 448	4. 450	4. 452	4. 455	4. 457	4. 459	4. 461
12. 1	4. 463	4. 466	4. 468	4. 470	4. 472	4. 475	4. 477	4. 479	4. 481	4. 483
12. 2	4. 486	4. 488	4. 490	4. 492	4. 494	4. 497	4. 499	4. 501	4. 503	4. 505
12. 3	4. 508	4. 510	4. 512	4. 514	4. 516	4. 519	4. 521	4. 523	4. 525	4. 527
12. 4	4. 530	4. 532	4. 534	4. 536	4. 538	4. 540	4. 543	4. 545	4. 547	4. 549
12. 5	4. 551	4. 554	4. 556	4. 558	4. 560	4. 562	4. 565	4. 567	4. 569	4. 571
12. 6	4. 573	4. 575	4. 578	4. 580	4. 582	4. 584	4. 586	4. 588	4. 591	4. 593
12. 7	4. 595	4. 597	4. 599	4. 601	4. 604	4. 606	4. 608	4. 610	4. 612	4. 614
12. 8	4. 617	4. 619	4. 621	4. 623	4. 625	4. 627	4. 630	4. 632	4. 634	4. 636
12. 9	4. 638	4. 640	4. 643	4. 645	4. 647	4. 649	4. 651	4. 653	4. 655	4. 658
13. 0	4. 660	4. 662	4. 664	4. 666	4. 668	4. 671	4. 673	4. 675	4. 677	4. 679
13. 1	4. 681	4. 683	4. 686	4. 688	4. 690	4. 692	4. 694	4. 696	4. 698	4. 701
13. 2	4. 703	4. 705	4. 707	4. 709	4. 711	4. 713	4. 715	4. 718	4. 720	4. 722
13. 3	4. 724	4. 726	4. 728	4. 730	4. 733	4. 735	4. 737	4. 739	4. 741	4. 743
13. 4	4. 745	4. 747	4. 750	4. 752	4. 754	4. 756	4. 758	4. 760	4. 762	4. 764
13. 5	4. 767	4. 769	4. 771	4. 773	4. 775	4. 777	4. 779	4. 781	4. 783	4. 786
13. 6	4. 788	4. 790	4. 792	4. 794	4. 796	4. 798	4. 800	4. 802	4. 805	4. 807
13. 7	4. 809	4. 811	4. 813	4. 815	4. 817	4. 819	4. 821	4. 823	4. 826	4. 828
13. 8	4. 830	4. 832	4. 834	4. 836	4. 838	4. 840	4. 842	4. 844	4. 847	4. 849
13. 9	4. 851	4. 853	4. 855	4. 857	4. 859	4. 861	4. 863	4. 865	4. 867	4. 870
14. 0	4. 872	4. 874	4. 876	4. 878	4. 880	4. 882	4. 884	4. 886	4. 888	4. 890
14. 1	4. 893	4. 895	4. 897	4. 899	4. 901	4. 903	4. 905	4. 907	4. 909	4. 911
14. 2	4. 913	4. 915	4. 917	4. 920	4. 922	4. 924	4. 926	4. 928	4. 930	4. 932
14. 3	4. 934	4. 936	4. 938	4. 940	4. 942	4. 944	4. 946	4. 949	4. 951	4. 953
14. 4	4. 955	4. 957	4. 959	4. 961	4. 963	4. 965	4. 967	4. 969	4. 971	4. 973

Table 64.—*Three-fifths powers of numbers*—Continued

Number	.00	.01	.02	.03	.04	.05	.06	.07	.08	.09
14.5	4.975	4.977	4.979	4.981	4.984	4.986	4.988	4.990	4.992	4.994
14.6	4.996	4.998	5.000	5.002	5.004	5.006	5.008	5.010	5.012	5.014
14.7	5.016	5.018	5.020	5.023	5.025	5.027	5.029	5.031	5.033	5.035
14.8	5.037	5.039	5.041	5.043	5.045	5.047	5.049	5.051	5.053	5.055
14.9	5.057	5.059	5.061	5.063	5.065	5.067	5.069	5.071	5.073	5.076
15.0	5.078	5.080	5.082	5.084	5.086	5.088	5.090	5.092	5.094	5.096
15.1	5.098	5.100	5.102	5.104	5.106	5.108	5.110	5.112	5.114	5.116
15.2	5.118	5.120	5.122	5.124	5.126	5.128	5.130	5.132	5.134	5.136
15.3	5.138	5.140	5.142	5.144	5.146	5.148	5.150	5.152	5.154	5.156
15.4	5.158	5.160	5.162	5.162	5.164	5.166	5.168	5.170	5.172	5.174	5.176
15.5	5.178	5.180	5.182	5.184	5.186	5.188	5.190	5.192	5.194	5.196
15.6	5.198	5.200	5.202	5.204	5.206	5.208	5.210	5.212	5.214	5.216
15.7	5.218	5.220	5.222	5.224	5.226	5.228	5.230	5.232	5.234	5.236
15.8	5.238	5.240	5.242	5.244	5.246	5.248	5.250	5.252	5.254	5.256
15.9	5.258	5.260	5.262	5.264	5.266	5.268	5.270	5.272	5.274	5.276
16.0	5.278	5.280	5.282	5.284	5.286	5.288	5.290	5.292	5.294	5.296
16.1	5.298	5.300	5.302	5.304	5.306	5.308	5.310	5.312	5.314	5.316
16.2	5.318	5.319	5.321	5.323	5.325	5.327	5.329	5.331	5.333	5.335
16.3	5.337	5.339	5.341	5.343	5.345	5.347	5.349	5.351	5.353	5.355
16.4	5.357	5.359	5.361	5.363	5.365	5.367	5.369	5.371	5.372	5.374
16.5	5.376	5.378	5.380	5.382	5.384	5.386	5.388	5.390	5.392	5.394
16.6	5.396	5.398	5.400	5.402	5.404	5.406	5.408	5.410	5.412	5.413
16.7	5.415	5.417	5.419	5.421	5.423	5.425	5.427	5.429	5.431	5.433
16.8	5.435	5.437	5.439	5.441	5.443	5.445	5.446	5.448	5.450	5.452
16.9	5.454	5.456	5.458	5.460	5.462	5.464	5.466	5.468	5.470	5.472
17.0	5.474	5.475	5.477	5.479	5.481	5.483	5.485	5.487	5.489	5.491
17.1	5.493	5.495	5.497	5.499	5.501	5.502	5.504	5.506	5.508	5.510
17.2	5.512	5.514	5.516	5.518	5.520	5.522	5.524	5.526	5.527	5.529
17.3	5.531	5.533	5.535	5.537	5.539	5.541	5.543	5.545	5.547	5.549
17.4	5.550	5.552	5.554	5.556	5.558	5.560	5.562	5.564	5.566	5.568
17.5	5.570	5.571	5.573	5.575	5.577	5.579	5.581	5.583	5.585	5.587
17.6	5.589	5.591	5.592	5.594	5.596	5.598	5.600	5.602	5.604	5.606
17.7	5.608	5.610	5.611	5.613	5.615	5.617	5.619	5.621	5.623	5.625
17.8	5.627	5.629	5.630	5.632	5.634	5.636	5.638	5.640	5.642	5.644
17.9	5.646	5.648	5.649	5.651	5.653	5.655	5.657	5.659	5.661	5.663
18.0	5.665	5.666	5.668	5.670	5.672	5.674	5.676	5.678	5.680	5.682
18.1	5.683	5.685	5.687	5.689	5.691	5.693	5.695	5.697	5.698	5.700
18.2	5.702	5.704	5.706	5.708	5.710	5.712	5.713	5.715	5.717	5.719
18.3	5.721	5.723	5.725	5.727	5.728	5.730	5.732	5.734	5.736	5.738
18.4	5.740	5.742	5.743	5.745	5.747	5.749	5.751	5.753	5.755	5.757
18.5	5.758	5.760	5.762	5.764	5.766	5.768	5.770	5.771	5.773	5.775
18.6	5.777	5.779	5.781	5.783	5.785	5.786	5.788	5.790	5.792	5.794
18.7	5.796	5.798	5.799	5.801	5.803	5.805	5.807	5.809	5.811	5.812
18.8	5.814	5.816	5.818	5.820	5.822	5.824	5.825	5.827	5.829	5.831
18.9	5.833	5.835	5.837	5.838	5.840	5.842	5.844	5.846	5.848	5.849

Table 64.—*Three-fifths powers of numbers*—Continued

Number	.00	.01	.02	.03	.04	.05	.06	.07	.08	.09
19.0	5.851	5.853	5.855	5.857	5.859	5.861	5.862	5.864	5.866	5.868
19.1	5.870	5.872	5.873	5.875	5.877	5.879	5.881	5.883	5.884	5.886
19.2	5.888	5.890	5.892	5.894	5.896	5.897	5.899	5.901	5.903	5.905
19.3	5.907	5.908	5.910	5.912	5.914	5.916	5.918	5.919	5.921	5.923
19.4	5.925	5.927	5.929	5.930	5.932	5.934	5.936	5.938	5.940	5.941
19.5	5.943	5.945	5.947	5.949	5.951	5.952	5.954	5.956	5.958	5.960
19.6	5.961	5.963	5.965	5.967	5.969	5.971	5.972	5.974	5.976	5.978
19.7	5.980	5.982	5.983	5.985	5.987	5.989	5.991	5.992	5.994	5.996
19.8	5.998	6.000	6.002	6.003	6.005	6.007	6.009	6.011	6.012	6.014
19.9	6.016	6.018	6.020	6.022	6.023	6.025	6.027	6.029	6.031	6.032
20.0	6.034	6.036	6.038	6.040	6.041	6.043	6.045	6.047	6.049	6.050
20.1	6.052	6.054	6.056	6.058	6.059	6.061	6.063	6.065	6.067	6.069
20.2	6.070	6.072	6.074	6.076	6.078	6.079	6.081	6.083	6.085	6.087
20.3	6.088	6.090	6.092	6.094	6.096	6.097	6.099	6.101	6.103	6.105
20.4	6.106	6.108	6.110	6.112	6.113	6.115	6.117	6.119	6.121	6.122
20.5	6.124	6.126	6.128	6.130	6.131	6.133	6.135	6.137	6.139	6.140
20.6	6.142	6.144	6.146	6.148	6.149	6.151	6.153	6.155	6.156	6.158
20.7	6.160	6.162	6.164	6.165	6.167	6.169	6.171	6.173	6.174	6.176
20.8	6.178	6.180	6.181	6.183	6.185	6.187	6.189	6.190	6.192	6.194
20.9	6.196	6.197	6.199	6.201	6.203	6.205	6.206	6.208	6.210	6.212
21.0	6.213	6.215	6.217	6.219	6.221	6.222	6.224	6.226	6.228	6.229
21.1	6.231	6.233	6.235	6.236	6.238	6.240	6.242	6.244	6.245	6.247
21.2	6.249	6.251	6.252	6.254	6.256	6.258	6.259	6.261	6.263	6.265
21.3	6.267	6.268	6.270	6.272	6.274	6.275	6.277	6.279	6.281	6.282
21.4	6.284	6.286	6.288	6.289	6.291	6.293	6.295	6.297	6.298	6.300
21.5	6.302	6.304	6.305	6.307	6.309	6.311	6.312	6.314	6.316	6.318
21.6	6.319	6.321	6.323	6.325	6.326	6.328	6.330	6.332	6.333	6.335
21.7	6.337	6.339	6.340	6.342	6.344	6.346	6.347	6.349	6.351	6.353
21.8	6.354	6.356	6.358	6.360	6.361	6.363	6.365	6.367	6.368	6.370
21.9	6.372	6.374	6.375	6.377	6.379	6.381	6.382	6.384	6.386	6.388
22.0	6.389	6.391	6.393	6.395	6.396	6.398	6.400	6.401	6.403	6.405
22.1	6.407	6.408	6.410	6.412	6.414	6.415	6.417	6.419	6.421	6.422
22.2	6.424	6.426	6.428	6.429	6.431	6.433	6.435	6.436	6.438	6.440
22.3	6.441	6.443	6.445	6.447	6.448	6.450	6.452	6.454	6.455	6.457
22.4	6.459	6.460	6.462	6.464	6.466	6.467	6.469	6.471	6.473	6.474
22.5	6.476	6.478	6.479	6.481	6.483	6.485	6.486	6.488	6.490	6.492
22.6	6.493	6.495	6.497	6.498	6.500	6.502	6.504	6.505	6.507	6.509
22.7	6.511	6.512	6.514	6.516	6.517	6.519	6.521	6.523	6.524	6.526
22.8	6.528	6.529	6.531	6.533	6.535	6.536	6.538	6.540	6.541	6.543
22.9	6.545	6.547	6.548	6.550	6.552	6.553	6.555	6.557	6.559	6.560
23.0	6.562	6.564	6.565	6.567	6.569	6.571	6.572	6.574	6.576	6.577
23.1	6.579	6.581	6.583	6.584	6.586	6.588	6.589	6.591	6.593	6.594
23.2	6.596	6.598	6.600	6.601	6.603	6.605	6.606	6.608	6.610	6.612
23.3	6.613	6.615	6.617	6.618	6.620	6.622	6.623	6.625	6.627	6.629
23.4	6.630	6.632	6.634	6.635	6.637	6.639	6.640	6.642	6.644	6.646

Table 64.—*Three-fifths powers of numbers*—Continued

Number	.00	.01	.02	.03	.04	.05	.06	.07	.08	.09
23.5	6.647	6.649	6.651	6.652	6.654	6.656	6.657	6.659	6.661	6.662
23.6	6.664	6.666	6.668	6.669	6.671	6.673	6.674	6.676	6.678	6.679
23.7	6.681	6.683	6.685	6.686	6.688	6.690	6.691	6.693	6.695	6.696
23.8	6.698	6.700	6.701	6.703	6.705	6.706	6.708	6.710	6.712	6.713
23.9	6.715	6.717	6.718	6.720	6.722	6.723	6.725	6.727	6.728	6.730
24.0	6.732	6.733	6.735	6.737	6.738	6.740	6.742	6.744	6.745	6.747
24.1	6.749	6.750	6.752	6.754	6.755	6.757	6.759	6.760	6.762	6.764
24.2	6.765	6.767	6.769	6.770	6.772	6.774	6.775	6.777	6.779	6.780
24.3	6.782	6.784	6.785	6.787	6.789	6.790	6.792	6.794	6.795	6.797
24.4	6.799	6.801	6.802	6.804	6.806	6.807	6.809	6.811	6.812	6.814
24.5	6.816	6.817	6.819	6.821	6.822	6.824	6.826	6.827	6.829	6.831
24.6	6.832	6.834	6.836	6.837	6.839	6.841	6.842	6.844	6.846	6.847
24.7	6.849	6.851	6.852	6.854	6.856	6.857	6.859	6.861	6.862	6.864
24.8	6.865	6.867	6.869	6.870	6.872	6.874	6.875	6.877	6.879	6.880
24.9	6.882	6.884	6.885	6.887	6.889	6.890	6.892	6.894	6.895	6.897
25.0	6.899	6.900	6.902	6.904	6.905	6.907	6.909	6.910	6.912	6.914
25.1	6.915	6.917	6.919	6.920	6.922	6.923	6.925	6.927	6.928	6.930
25.2	6.932	6.933	6.935	6.937	6.938	6.940	6.942	6.943	6.945	6.947
25.3	6.948	6.950	6.952	6.953	6.955	6.956	6.958	6.960	6.961	6.963
25.4	6.965	6.966	6.968	6.970	6.971	6.973	6.975	6.976	6.978	6.979
25.5	6.981	6.983	6.984	6.986	6.988	6.989	6.991	6.993	6.994	6.996
25.6	6.998	6.999	7.001	7.002	7.004	7.006	7.007	7.009	7.011	7.012
25.7	7.014	7.016	7.017	7.019	7.020	7.022	7.024	7.025	7.027	7.029
25.8	7.030	7.032	7.034	7.035	7.037	7.038	7.040	7.042	7.043	7.045
25.9	7.047	7.048	7.050	7.052	7.053	7.055	7.056	7.058	7.060	7.061
26.0	7.063	7.065	7.066	7.068	7.069	7.071	7.073	7.074	7.076	7.078
26.1	7.079	7.081	7.082	7.084	7.086	7.087	7.089	7.091	7.092	7.094
26.2	7.095	7.097	7.099	7.100	7.102	7.104	7.105	7.107	7.108	7.110
26.3	7.112	7.113	7.115	7.117	7.118	7.120	7.121	7.123	7.125	7.126
26.4	7.128	7.130	7.131	7.133	7.134	7.136	7.138	7.139	7.141	7.142
26.5	7.144	7.146	7.147	7.149	7.151	7.152	7.154	7.155	7.157	7.159
26.6	7.160	7.162	7.163	7.165	7.167	7.168	7.170	7.172	7.173	7.175
26.7	7.176	7.178	7.180	7.181	7.183	7.184	7.186	7.188	7.189	7.191
26.8	7.193	7.194	7.196	7.197	7.199	7.201	7.202	7.204	7.205	7.207
26.9	7.209	7.210	7.212	7.213	7.215	7.217	7.218	7.220	7.221	7.223
27.0	7.225	7.226	7.228	7.229	7.231	7.233	7.234	7.236	7.238	7.239
27.1	7.241	7.242	7.244	7.246	7.247	7.249	7.250	7.252	7.254	7.255
27.2	7.257	7.258	7.260	7.262	7.263	7.265	7.266	7.268	7.270	7.271
27.3	7.273	7.274	7.276	7.278	7.279	7.281	7.282	7.284	7.286	7.287
27.4	7.289	7.290	7.292	7.293	7.295	7.297	7.298	7.300	7.301	7.303
27.5	7.305	7.306	7.308	7.309	7.311	7.313	7.314	7.316	7.317	7.319
27.6	7.321	7.322	7.324	7.325	7.327	7.329	7.330	7.332	7.333	7.335
27.7	7.336	7.338	7.340	7.341	7.343	7.344	7.346	7.348	7.349	7.351
27.8	7.352	7.354	7.356	7.357	7.359	7.360	7.362	7.363	7.365	7.367
27.9	7.368	7.370	7.371	7.373	7.375	7.376	7.378	7.379	7.381	7.382

Table 64.—*Three-fifths powers of numbers*—Continued

Number	.00	.01	.02	.03	.04	.05	.06	.07	.08	.09
28.0	7.384	7.386	7.387	7.389	7.390	7.392	7.394	7.395	7.397	7.398
28.1	7.400	7.401	7.403	7.405	7.406	7.408	7.409	7.411	7.412	7.414
28.2	7.416	7.417	7.419	7.420	7.422	7.424	7.425	7.427	7.428	7.430
28.3	7.431	7.433	7.435	7.436	7.438	7.439	7.441	7.442	7.444	7.446
28.4	7.447	7.449	7.450	7.452	7.453	7.455	7.457	7.458	7.460	7.461
28.5	7.463	7.464	7.466	7.468	7.469	7.471	7.472	7.474	7.475	7.477
28.6	7.479	7.480	7.482	7.483	7.485	7.486	7.488	7.490	7.491	7.493
28.7	7.494	7.496	7.497	7.499	7.501	7.502	7.504	7.505	7.507	7.508
28.8	7.510	7.511	7.513	7.515	7.516	7.518	7.519	7.521	7.522	7.524
28.9	7.526	7.527	7.529	7.530	7.532	7.533	7.535	7.536	7.538	7.540
29.0	7.541	7.543	7.544	7.546	7.547	7.549	7.551	7.552	7.554	7.555
29.1	7.557	7.558	7.560	7.561	7.563	7.565	7.566	7.568	7.569	7.571
29.2	7.572	7.574	7.575	7.577	7.579	7.580	7.582	7.583	7.585	7.586
29.3	7.588	7.589	7.591	7.593	7.594	7.596	7.597	7.599	7.600	7.602
29.4	7.603	7.605	7.607	7.608	7.610	7.611	7.613	7.614	7.616	7.617
29.5	7.619	7.620	7.622	7.624	7.625	7.627	7.628	7.630	7.631	7.633
29.6	7.634	7.636	7.638	7.639	7.641	7.642	7.644	7.645	7.647	7.648
29.7	7.650	7.651	7.653	7.655	7.656	7.658	7.659	7.661	7.662	7.664
29.8	7.665	7.667	7.668	7.670	7.671	7.673	7.675	7.676	7.678	7.679
29.9	7.681	7.682	7.684	7.685	7.687	7.688	7.690	7.692	7.693	7.695

Number	.0	.1	.2	.3	.4	.5	.6	.7	.8	.9
30	7.696	7.712	7.727	7.742	7.758	7.773	7.788	7.803	7.819	7.834
31	7.849	7.864	7.879	7.895	7.910	7.925	7.940	7.955	7.970	7.985
32	8.000	8.015	8.030	8.045	8.060	8.075	8.090	8.105	8.119	8.134
33	8.149	8.164	8.179	8.193	8.208	8.223	8.238	8.252	8.267	8.282
34	8.296	8.311	8.326	8.340	8.355	8.369	8.384	8.398	8.413	8.427
35	8.442	8.456	8.471	8.485	8.500	8.514	8.528	8.543	8.557	8.571
36	8.586	8.600	8.614	8.629	8.643	8.657	8.671	8.686	8.700	8.714
37	8.728	8.742	8.756	8.771	8.785	8.799	8.813	8.827	8.841	8.855
38	8.869	8.883	8.897	8.911	8.925	8.939	8.953	8.967	8.980	8.994
39	9.008	9.022	9.036	9.050	9.064	9.077	9.091	9.105	9.119	9.132
40	9.146	9.160	9.174	9.187	9.201	9.215	9.228	9.242	9.255	9.269
41	9.283	9.296	9.310	9.323	9.337	9.350	9.364	9.377	9.391	9.404
42	9.418	9.431	9.445	9.458	9.472	9.485	9.498	9.512	9.525	9.538
43	9.552	9.565	9.578	9.592	9.605	9.618	9.631	9.645	9.658	9.671
44	9.684	9.698	9.711	9.724	9.737	9.750	9.763	9.777	9.790	9.803
45	9.816	9.829	9.842	9.855	9.868	9.881	9.894	9.907	9.920	9.933
46	9.946	9.959	9.972	9.985	9.998	10.01	10.02	10.04	10.05	10.06
47	10.08	10.09	10.10	10.11	10.13	10.14	10.15	10.17	10.18	10.19
48	10.20	10.22	10.23	10.24	10.25	10.27	10.28	10.29	10.31	10.32
49	10.33	10.34	10.36	10.37	10.38	10.39	10.41	10.42	10.43	10.44
50	10.46	10.47	10.48	10.49	10.51	10.52	10.53	10.54	10.56	10.57
51	10.58	10.59	10.61	10.62	10.63	10.64	10.66	10.67	10.68	10.69
52	10.71	10.72	10.73	10.74	10.75	10.77	10.78	10.79	10.80	10.82
53	10.83	10.84	10.85	10.87	10.88	10.89	10.90	10.91	10.93	10.94
54	10.95	10.96	10.97	10.99	11.00	11.01	11.02	11.04	11.05	11.06

Table 64.—*Three-fifths powers of numbers*—Continued

Number	.0	.1	.2	.3	.4	.5	.6	.7	.8	.9
55	11.07	11.08	11.10	11.11	11.12	11.13	11.14	11.16	11.17	11.18
56	11.19	11.20	11.22	11.23	11.24	11.25	11.26	11.28	11.29	11.30
57	11.31	11.32	11.34	11.35	11.36	11.37	11.38	11.39	11.41	11.42
58	11.43	11.44	11.45	11.47	11.48	11.49	11.50	11.51	11.52	11.54
59	11.55	11.56	11.57	11.58	11.60	11.61	11.62	11.63	11.64	11.65
60	11.67	11.68	11.69	11.70	11.71	11.72	11.74	11.75	11.76	11.77
61	11.78	11.79	11.80	11.82	11.83	11.84	11.85	11.86	11.87	11.89
62	11.90	11.91	11.92	11.93	11.94	11.95	11.97	11.98	11.99	12.00
63	12.01	12.02	12.03	12.05	12.06	12.07	12.08	12.09	12.10	12.11
64	12.13	12.14	12.15	12.16	12.17	12.18	12.19	12.21	12.22	12.23
65	12.24	12.25	12.26	12.27	12.28	12.30	12.31	12.32	12.33	12.34
66	12.35	12.36	12.37	12.39	12.40	12.41	12.42	12.43	12.44	12.45
67	12.46	12.47	12.49	12.50	12.51	12.52	12.53	12.54	12.55	12.56
68	12.57	12.59	12.60	12.61	12.62	12.63	12.64	12.65	12.66	12.67
69	12.69	12.70	12.71	12.72	12.73	12.74	12.75	12.76	12.77	12.78
70	12.80	12.81	12.82	12.83	12.84	12.85	12.86	12.87	12.88	12.89
71	12.90	12.92	12.93	12.94	12.95	12.96	12.97	12.98	12.99	13.00
72	13.01	13.02	13.04	13.05	13.06	13.07	13.08	13.09	13.10	13.11
73	13.12	13.13	13.14	13.15	13.16	13.18	13.19	13.20	13.21	13.22
74	13.23	13.24	13.25	13.26	13.27	13.28	13.29	13.30	13.32	13.33
75	13.34	13.35	13.36	13.37	13.38	13.39	13.40	13.41	13.42	13.43
76	13.44	13.45	13.46	13.47	13.49	13.50	13.51	13.52	13.53	13.54
77	13.55	13.56	13.57	13.58	13.59	13.60	13.61	13.62	13.63	13.64
78	13.65	13.66	13.67	13.69	13.70	13.71	13.72	13.73	13.74	13.75
79	13.76	13.77	13.78	13.79	13.80	13.81	13.82	13.83	13.84	13.85
80	13.86	13.87	13.88	13.89	13.90	13.91	13.93	13.94	13.95	13.96
81	13.97	13.98	13.99	14.00	14.01	14.02	14.03	14.04	14.05	14.06
82	14.07	14.08	14.09	14.10	14.11	14.12	14.13	14.14	14.15	14.16
83	14.17	14.18	14.19	14.20	14.21	14.22	14.23	14.24	14.25	14.26
84	14.27	14.28	14.30	14.31	14.32	14.33	14.34	14.35	14.36	14.37
85	14.38	14.39	14.40	14.41	14.42	14.43	14.44	14.45	14.46	14.47
86	14.48	14.49	14.50	14.51	14.52	14.53	14.54	14.55	14.56	14.57
87	14.58	14.59	14.60	14.61	14.62	14.63	14.64	14.65	14.66	14.67
88	14.68	14.69	14.70	14.71	14.72	14.73	14.74	14.75	14.76	14.77
89	14.78	14.79	14.80	14.81	14.82	14.83	14.84	14.85	14.86	14.87
90	14.88	14.89	14.90	14.91	14.92	14.93	14.94	14.95	14.96	14.97
91	14.98	14.99	15.00	15.01	15.02	15.03	15.04	15.05	15.06	15.07
92	15.08	15.09	15.10	15.10	15.11	15.12	15.13	15.14	15.15	15.16
93	15.17	15.18	15.19	15.20	15.21	15.22	15.23	15.24	15.25	15.26
94	15.27	15.28	15.29	15.30	15.31	15.32	15.33	15.34	15.35	15.36
95	15.37	15.38	15.39	15.40	15.41	15.42	15.43	15.44	15.45	15.46
96	15.47	15.48	15.48	15.49	15.50	15.51	15.52	15.53	15.54	15.55
97	15.56	15.57	15.58	15.59	15.60	15.61	15.62	15.63	15.64	15.65
98	15.66	15.67	15.68	15.69	15.70	15.71	15.72	15.72	15.73	15.74
99	15.75	15.76	15.77	15.78	15.79	15.80	15.81	15.82	15.83	15.84

Table 64.—*Three-fifths powers of numbers*—Continued

Number	0	1	2	3	4	5	6	7	8	9
100	15.85	15.94	16.04	16.13	16.23	16.32	16.41	16.51	16.60	16.69
110	16.78	16.87	16.96	17.05	17.15	17.24	17.33	17.41	17.50	17.59
120	17.68	17.77	17.86	17.94	18.03	18.12	18.21	18.29	18.38	18.47
130	18.55	18.64	18.72	18.81	18.89	18.98	19.06	19.14	19.23	19.31
140	19.39	19.48	19.56	19.64	19.73	19.81	19.89	19.97	20.05	20.13
150	20.21	20.29	20.38	20.46	20.54	20.62	20.70	20.77	20.85	20.93
160	21.01	21.09	21.17	21.25	21.33	21.40	21.48	21.56	21.64	21.71
170	21.79	21.87	21.94	22.02	22.10	22.17	22.25	22.32	22.40	22.48
180	22.55	22.63	22.70	22.78	22.85	22.92	23.00	23.07	23.15	23.22
190	23.29	23.37	23.44	23.51	23.59	23.66	23.73	23.81	23.88	23.95
200	24.02	24.09	24.17	24.24	24.31	24.38	24.45	24.52	24.59	24.67
210	24.74	24.81	24.88	24.95	25.02	25.09	25.16	25.23	25.30	25.37
220	25.44	25.51	25.57	25.64	25.71	25.78	25.85	25.92	25.99	26.06
230	26.12	26.19	26.26	26.33	26.40	26.46	26.53	26.60	26.67	26.73
240	26.80	26.87	26.93	27.00	27.07	27.13	27.20	27.27	27.33	27.40
250	27.46	27.53	27.60	27.66	27.73	27.79	27.86	27.92	27.99	28.05
260	28.12	28.18	28.25	28.31	28.38	28.44	28.51	28.57	28.63	28.70
270	28.76	28.83	28.89	28.95	29.02	29.08	29.14	29.21	29.27	29.33
280	29.40	29.46	29.52	29.59	29.65	29.71	29.77	29.84	29.90	29.96
290	30.02	30.08	30.15	30.21	30.27	30.33	30.39	30.45	30.52	30.58
300	30.64	30.70	30.76	30.82	30.88	30.94	31.01	31.07	31.13	31.19
310	31.25	31.31	31.37	31.43	31.49	31.55	31.61	31.67	31.73	31.79
320	31.85	31.91	31.97	32.03	32.09	32.15	32.21	32.26	32.32	32.38
330	32.44	32.50	32.56	32.62	32.68	32.74	32.79	32.85	32.91	32.97
340	33.03	33.09	33.14	33.20	33.26	33.32	33.38	33.43	33.49	33.55
350	33.61	33.67	33.72	33.78	33.84	33.90	33.95	34.01	34.07	34.12
360	34.18	34.24	34.29	34.35	34.41	34.46	34.52	34.58	34.63	34.69
370	34.75	34.80	34.86	34.92	34.97	35.03	35.08	35.14	35.20	35.25
380	35.31	35.36	35.42	35.47	35.53	35.59	35.64	35.70	35.75	35.81
390	35.86	35.92	35.97	36.03	36.08	36.14	36.19	36.25	36.30	36.36
400	36.41	36.47	36.52	36.57	36.63	36.68	36.74	36.79	36.85	36.90
410	36.95	37.01	37.06	37.12	37.17	37.22	37.28	37.33	37.39	37.44
420	37.49	37.55	37.60	37.65	37.71	37.76	37.81	37.87	37.92	37.97
430	38.03	38.08	38.13	38.18	38.24	38.29	38.34	38.40	38.45	38.50
440	38.55	38.61	38.66	38.71	38.76	38.82	38.87	38.92	38.97	39.03
450	39.08	39.13	39.18	39.23	39.29	39.34	39.39	39.44	39.49	39.54
460	39.60	39.65	39.70	39.75	39.80	39.85	39.91	39.96	40.01	40.06
470	40.11	40.16	40.21	40.26	40.32	40.37	40.42	40.47	40.52	40.57
480	40.62	40.67	40.72	40.77	40.82	40.87	40.92	40.97	41.03	41.08
490	41.13	41.18	41.23	41.28	41.33	41.38	41.43	41.48	41.53	41.58

Table 64.—*Three-fifths powers of numbers*—Continued

Number	0	1	2	3	4	5	6	7	8	9
500	41.63	41.68	41.73	41.78	41.83	41.88	41.93	41.98	42.03	42.08
510	42.13	42.17	42.22	42.27	42.32	42.37	42.42	42.47	42.52	42.57
520	42.62	42.67	42.72	42.77	42.82	42.86	42.91	42.96	43.01	43.06
530	43.11	43.16	43.21	43.26	43.30	43.35	43.40	43.45	43.50	43.55
540	43.59	43.64	43.69	43.74	43.79	43.84	43.88	43.93	43.98	44.03
550	44.08	44.13	44.17	44.22	44.27	44.32	44.37	44.41	44.46	44.51
560	44.56	44.60	44.65	44.70	44.75	44.79	44.84	44.89	44.94	44.98
570	45.03	45.08	45.13	45.17	45.22	45.27	45.32	45.36	45.41	45.46
580	45.50	45.55	45.60	45.65	45.69	45.74	45.79	45.83	45.88	45.93
590	45.97	46.02	46.07	46.11	46.16	46.21	46.25	46.30	46.35	46.39
600	46.44	46.49	46.53	46.58	46.63	46.67	46.72	46.76	46.81	46.86
610	46.90	46.95	46.99	47.04	47.09	47.13	47.18	47.22	47.27	47.32
620	47.36	47.41	47.45	47.50	47.55	47.59	47.64	47.68	47.73	47.77
630	47.82	47.86	47.91	47.96	48.00	48.05	48.09	48.14	48.18	48.23
640	48.27	48.32	48.36	48.41	48.45	48.50	48.54	48.59	48.63	48.68
650	48.72	48.77	48.81	48.86	48.90	48.95	48.99	49.04	49.08	49.13
660	49.17	49.22	49.26	49.31	49.35	49.40	49.44	49.49	49.53	49.57
670	49.62	49.66	49.71	49.75	49.80	49.84	49.88	49.93	49.97	50.02
680	50.06	50.11	50.15	50.19	50.24	50.28	50.33	50.37	50.41	50.46
690	50.50	50.55	50.59	50.63	50.68	50.72	50.77	50.81	50.85	50.90
700	50.94	50.98	51.03	51.07	51.11	51.16	51.20	51.25	51.29	51.33
710	51.38	51.42	51.46	51.51	51.55	51.59	51.64	51.68	51.72	51.77
720	51.81	51.85	51.89	51.94	51.98	52.02	52.07	52.11	52.15	52.20
730	52.24	52.28	52.32	52.37	52.41	52.45	52.50	52.54	52.58	52.62
740	52.67	52.71	52.75	52.80	52.84	52.88	52.92	52.97	53.01	53.05
750	53.09	53.14	53.18	53.22	53.26	53.31	53.35	53.39	53.43	53.47
760	53.52	53.56	53.60	53.64	53.69	53.73	53.77	53.81	53.85	53.90
770	53.94	53.98	54.02	54.06	54.11	54.15	54.19	54.23	54.27	54.32
780	54.36	54.40	54.44	54.48	54.52	54.57	54.61	54.65	54.69	54.73
790	54.77	54.82	54.86	54.90	54.94	54.98	55.02	55.06	55.11	55.15
800	55.19	55.23	55.27	55.31	55.35	55.40	55.44	55.48	55.52	55.56
810	55.60	55.64	55.68	55.73	55.77	55.81	55.85	55.89	55.93	55.97
820	56.01	56.05	56.09	56.14	56.18	56.22	56.26	56.30	56.34	56.38
830	56.42	56.46	56.50	56.54	56.58	56.63	56.67	56.71	56.75	56.79
840	56.83	56.87	56.91	56.95	56.99	57.03	57.07	57.11	57.15	57.19
850	57.23	57.27	57.31	57.35	57.40	57.44	57.48	57.52	57.56	57.60
860	57.64	57.68	57.72	57.76	57.80	57.84	57.88	57.92	57.96	58.00
870	58.04	58.08	58.12	58.16	58.20	58.24	58.28	58.32	58.36	58.40
880	58.44	58.48	58.52	58.56	58.60	58.64	58.68	58.72	58.76	58.80
890	58.83	58.87	58.91	58.95	58.99	59.03	59.07	59.11	59.15	59.19
900	59.23	59.27	59.31	59.35	59.39	59.43	59.47	59.51	59.55	59.59
910	59.62	59.66	59.70	59.74	59.78	59.82	59.86	59.90	59.94	59.98
920	60.02	60.06	60.10	60.13	60.17	60.21	60.25	60.29	60.33	60.37
930	60.41	60.45	60.49	60.52	60.56	60.60	60.64	60.68	60.72	60.76
940	60.80	60.84	60.87	60.91	60.95	60.99	61.03	61.07	61.11	61.14
950	61.18	61.22	61.26	61.30	61.34	61.38	61.42	61.45	61.49	61.53
960	61.57	61.61	61.65	61.68	61.72	61.76	61.80	61.84	61.88	61.91
970	61.95	61.99	62.03	62.07	62.11	62.14	62.18	62.22	62.26	62.30
980	62.34	62.37	62.41	62.45	62.49	62.53	62.56	62.60	62.64	62.68
990	62.72	62.75	62.79	62.83	62.87	62.91	62.94	62.98	63.02	63.06

Table 65.—*Squares, cubes, square roots, cube roots, reciprocals, and area and circumference of circles of radius N*

N	N^2	N^3	$N^{\frac{1}{2}}$	$N^{\frac{1}{3}}$	$\dfrac{1}{N}$	πN^2	$2\,\pi N$
1	1	1	1.0000	1.0000	1.000000	3.142	6.283
2	4	8	1.4142	1.2599	.500000	12.566	12.566
3	9	27	1.7321	1.4422	.333333	28.274	18.850
4	16	64	2.0000	1.5874	.250000	50.265	25.133
5	25	125	2.2361	1.7100	.200000	78.540	31.416
6	36	216	2.4495	1.8171	.166667	113.097	37.699
7	49	343	2.6458	1.9129	.142857	153.938	43.982
8	64	512	2.8284	2.0000	.125000	201.062	50.265
9	81	729	3.0000	2.0801	.111111	254.469	56.549
10	100	1,000	3.1623	2.1544	.100000	314.159	62.832
11	121	1,331	3.3166	2.2240	.090909	380.133	69.115
12	144	1,728	3.4641	2.2894	.083333	452.389	75.398
13	169	2,197	3.6056	2.3513	.076923	530.929	81.681
14	196	2,744	3.7417	2.4101	.071429	615.752	87.965
15	225	3,375	3.8730	2.4662	.066667	706.858	94.248
16	256	4,096	4.0000	2.5198	.062500	804.248	100.531
17	289	4,913	4.1231	2.5713	.058824	907.920	106.814
18	324	5,832	4.2426	2.6207	.055556	1,017.876	1.3.097
19	361	6,859	4.3589	2.6684	.052632	1,134.115	1.9.381
20	400	8,000	4.4721	2.7144	.050000	1,256.637	125.664
21	441	9,261	4.5826	2.7589	.047619	1,385.442	131.947
22	484	10,648	4.6904	2.8020	.045455	1,520.531	138.230
23	529	12,167	4.7958	2.8439	.043478	1,661.903	144.513
24	576	13,824	4.8990	2.8845	.041667	1,809.557	150.796
25	625	15,625	5.0000	2.9240	.040000	1,963.495	157,080
26	676	17,576	5.0990	2.9625	.038462	2,123.717	163.363
27	729	19,683	5.1962	3.0000	.037037	2,290.221	169.646
28	784	21,952	5.2915	3.0366	.035714	2,463.009	175.929
29	841	24,389	5.3852	3.0723	.034483	2,642.079	182.212
30	900	27,000	5.4772	3.1072	.033333	2,827.433	188.496
31	961	29,791	5.5678	3.1414	.032258	3,019.071	194.779
32	1,024	32,768	5.6569	3.1748	.031250	3,216.991	201.062
33	1,089	35,937	5.7446	3.2075	.030303	3,421.194	207.345
34	1,156	39,304	5.8310	3.2396	.029412	3,631.681	213.628
35	1,225	42,875	5.9161	3.2711	.028571	3,848.451	219.911
36	1,296	46,656	6.0000	3.3019	.027778	4,071.504	226.195
37	1,369	50,653	6.0828	3.3322	.027027	4,300.840	232.478
38	1,444	54,872	6.1644	3.3620	.026316	4,536.460	238.761
39	1,521	59,319	6.2450	3.3912	.025641	4,778.362	245.044
40	1,600	64,000	6.3246	3.4200	.025000	5,026.548	251.327
41	1,681	68,921	6.4031	3.4482	.024390	5,281.017	257.611
42	1,764	74,088	6.4807	3.4760	.023810	5,541.770	263.894
43	1,849	79,507	6.5574	3.5034	.023256	5,808.805	270.177
44	1,936	85,184	6.6332	3.5303	.022727	6,082.128	276.460
45	2,025	91,125	6.7082	3.5569	.022222	6,361.725	282.743
46	2,116	97,836	6.7823	3.5830	.021739	6,647.610	289.027
47	2,209	103,823	6.8557	3.6088	.021277	6,939.778	295.810
48	2,304	110,592	6.9282	3.6342	.020833	7,238.230	301.593
49	2,401	117,649	7.0000	3.6593	.020408	7,542.964	307.876
50	2,500	125,000	7.0711	3.6840	.020000	7,853.982	314.159

Table 65.—*Squares, cubes, square roots, cube roots, reciprocals, and area and circumference of circles of radius N*—Continued

N	N^2	N^3	$N^{\frac{1}{2}}$	$N^{\frac{1}{3}}$	$\frac{1}{N}$	πN^2	$2\pi N$
51	2,601	132,651	7.1414	3.7084	.019607	8,171.283	320.442
52	2,704	140,608	7.2111	3.7325	.019231	8,494.867	326.726
53	2,809	148,877	7.2801	3.7563	.018868	8,824.734	333.009
54	2,916	157,464	7.3485	3.7798	.018519	9,160.884	339.292
55	3,025	166,375	7.4162	3.8030	.018182	9,503.318	345.575
56	3,136	175,616	7.4833	3.8259	.017857	9,852.035	351.858
57	3,249	185,193	7.5498	3.8485	.017544	10,207.035	358.142
58	3,364	195,112	7.6158	3.8709	.017241	10,568.318	364.425
59	3,481	205,379	7.6811	3.8930	.016949	10,935.884	370.708
60	3,600	216,000	7.7460	3.9149	.016667	11,309.734	376.991
61	3,721	226,981	7.8102	3.9365	.016393	11,689.866	383.274
62	3,844	238,328	7.8740	3.9579	.016129	12,076.282	389.557
63	3,969	250,047	7.9373	3.9791	.015873	12,468.981	395.841
64	4,096	262,144	8.0000	4.0000	.015625	12,867.964	402.124
65	4,225	274,625	8.0623	4.0207	.015385	13,273.229	408.407
66	4,356	287,496	8.1240	4.0412	.015156	13,684.778	414.690
67	4,489	300,763	8.1854	4.0615	.014925	14,102.610	420.973
68	4,624	314,432	8.2462	4.0817	.014706	14,526.725	427.257
69	4,761	328,509	8.3066	4.1016	.014493	14,957.123	433.540
70	4,900	343,000	8.3666	4.1213	.014286	15,393.804	439.823
71	5,041	357,911	8.4261	4.1408	.014085	15,836.769	446.106
72	5,184	373,248	8.4853	4.1602	.013889	16,286.017	452.389
73	5,329	389,017	8.5440	4.1793	.013699	16,741.547	458.673
74	5,476	405,224	8.6023	4.1983	.013514	17,203.362	464.956
75	5,625	421,875	8.6603	4.2172	.013333	17,671.459	471.239
76	5,776	438,976	8.7178	4.2358	.013158	18,145.839	477.522
77	5,929	456,533	8.7750	4.2543	.012987	18,626.503	483.805
78	6,084	474,552	8.8318	4.2727	.012821	19,113.450	490.088
79	6,241	493,039	8.8882	4.2908	.012658	19,606.680	496.372
80	6,400	512,000	8.9443	4.3089	.012500	20,106.193	502.655
81	6,561	531,441	9.0000	4.3267	.012346	20,611.990	508.938
82	6,724	551,368	9.0554	4.3445	.012195	21,124.069	515.221
83	6,889	571,787	9.1104	4.3621	.012048	21,642.432	521.504
84	7,056	592,704	9.1652	4.3795	.011905	22,167.078	527.788
85	7,225	614,125	9.2195	4.3968	.011765	22,698.007	534.071
86	7,396	636,056	9.2736	4.4140	.011628	23,235.220	540.354
87	7,569	658,503	9.3274	4.4310	.011494	23,778.715	546.637
88	7,744	681,472	9.3808	4.4480	.011364	24,328.494	552.920
89	7,921	704,969	9.4340	4.4647	.011236	24,884.556	559.205
90	8,100	729,000	9.4868	4.4814	.011111	25,446.901	565.487
91	8,281	753,571	9.5394	4.4979	.010989	26,015.529	571.770
92	8,464	778,688	9.5917	4.5144	.010870	26,590.441	578.053
93	8,649	804,357	9.6437	4.5307	.010753	27,171.635	584.336
94	8,836	830,584	9.6954	4.5468	.010688	27,759.113	590.619
95	9,025	857,375	9.7468	4.5629	.010526	28,352.874	596.903
96	9,216	884,736	9.7980	4.5789	.010417	28,952.918	603.186
97	9,409	912,673	9.8489	4.5947	.010309	29,559.246	609.469
98	9,604	941,192	9.8995	4.6104	.010204	30,171.856	615.752
99	9,801	970,299	9.9499	4.6261	.010101	30,790.750	622.035
100	10,000	1,000,000	10.0000	4.6416	.010000	31,415.927	628.319

Table 65.—*Squares, cubes, square roots, cube roots and reciprocals*—Continued

N	N²	N³	N^{½}	N^{⅓}	$\frac{1}{N}$
101	10,201	1,030,301	10.0498756	4.6570095	.009900990
102	10,404	1,061,208	10.0995049	4.6723287	.009803922
103	10,609	1,092,727	10.1488916	4.6875482	.009708738
104	10,816	1,124,864	10.1980390	4.7026694	.009615385
105	11,025	1,157,625	10.2469508	4.7176940	.009523810
106	11,236	1,191,016	10.2956301	4.7326235	.009433962
107	11,449	1,225,043	10.3440804	4.7474594	.009345794
108	11,664	1,259,712	10.3923048	4.7622032	.009259259
109	11,881	1,295,029	10.4403065	4.7768562	.009174312
110	12,100	1,331,000	10.4880885	4.7914199	.009090909
111	12,321	1,367,631	10.5356538	4.8058955	.009009009
112	12,544	1,404,928	10.5830052	4.8202845	.008928571
113	12,769	1,442,897	10.6301458	4.8345881	.008849558
114	12,996	1,481,544	10.6770783	4.8488076	.008771930
115	13,225	1,520,875	10.7238053	4.8629442	.008695652
116	13,456	1,560,896	10.7703296	4.8769990	.008620690
117	13,689	1,601,613	10.8166538	4.8909732	.008547009
118	13,924	1,643,032	10.8627805	4.9048681	.008474576
119	14,161	1,685,159	10.9087121	4.9186847	.008403361
120	14,400	1,728,000	10.9544512	4.9324242	.008333333
121	14,641	1,771,561	11.0000000	4.9460874	.008264463
122	14,884	1,815,848	11.0453610	4.9596757	.008196721
123	15,129	1,860,867	11.0905365	4.9731898	.008130081
124	15,376	1,906,624	11.1355287	4.9866310	.008064516
125	15,625	1,953,125	11.1803399	5.0000000	.008000000
126	15,876	2,000,376	11.2249722	5.0132979	.007936508
127	16,129	2,048,383	11.2694277	5.0265257	.007874016
128	16,384	2,097,152	11.3137085	5.0396842	.007812500
129	16,641	2,146,689	11.3578167	5.0527743	.007751938
130	16,900	2,197,000	11.4017543	5.0657970	.007692308
131	17,161	2,248,091	11.4455231	5.0787531	.007633588
132	17,424	2,299,968	11.4891253	5.0916434	.007575758
133	17,689	2,352,637	11.5325626	5.1044687	.007518797
134	17,956	2,406,104	11.5758369	5.1172299	.007462687
135	18,225	2,460,375	11.6189500	5.1299278	.007407407
136	18,496	2,515,456	11.6619038	5.1425682	.007352941
137	18,769	2,571,353	11.7046999	5.1551367	.007299270
138	19,044	2,628,072	11.7473401	5.1676493	.007246377
139	19,321	2,685,619	11.7898261	5.1801015	.007194245
140	19,600	2,744,000	11.8321596	5.1924941	.007142857
141	19,881	2,803,221	11.8743421	5.2048279	.007092199
142	20,164	2,863,288	11.9163753	5.2171034	.007042254
143	20,449	2,924,207	11.9582607	5.2293215	.006993007
144	20,736	2,985,984	12.0000000	5.2414828	.006944444
145	21,025	3,048,625	12.0415946	5.2535879	.006896552
146	21,316	3,112,136	12.0830460	5.2656374	.006849315
147	21,609	3,176,523	12.1243557	5.2776321	.006802721
148	21,904	3,241,792	12.1655251	5.2895725	.006756757
149	22,201	3,307,949	12.2065556	5.3014592	.006711409
150	22,500	3,375,000	12.2474487	5.3132928	.006666667

Table 65.—*Squares, cubes, square roots, cube roots and reciprocals*—Continued

N	N^2	N^3	$N^{\frac{1}{2}}$	$N^{\frac{1}{3}}$	$\dfrac{1}{N}$
151	22,801	3,442,951	12.2882057	5.3250740	.006622517
152	23,104	3,511,808	12.3288280	5.3368083	.006578947
153	23,409	3,581,577	12.3693769	5.3484812	.006535948
154	23,716	3,652,264	12.4096736	5.3601084	.006493506
155	24,025	3,723,875	12.4498996	5.3716854	.006451613
156	24,336	3,796,416	12.4899960	5.3832126	.006410256
157	24,649	3,869,893	12.5299641	5.3946907	.006369427
158	24,964	3,944,312	12.5698051	5.4061202	.006329114
159	25,281	4,019,679	12.6095202	5.4175015	.006289308
160	25,600	4,096,000	12.6491106	5.4288352	.006250000
161	25,921	4,173,281	12.6885775	5.4401218	.006211180
162	26,244	4,251,528	12.7279221	5.4513618	.006172840
163	26,569	4,330,747	12.7671453	5.4625556	.006134969
164	26,896	4,410,944	12.8062485	5.4737037	.006097561
165	27,225	4,492,125	12.8452326	5.4848066	.006060606
166	27,556	4,574,296	12.8840987	5.4958647	.006024096
167	27,889	4,657,463	12.9228480	5.5068784	.005988024
168	28,224	4,741,632	12.9614814	5.5178484	.005952381
169	28,561	4,826,809	13.0000000	5.5287748	.005917160
170	28,900	4,913,000	13.0384048	5.5396583	.005882353
171	29,241	5,000,211	13.0766968	5.5504991	.005847953
172	29,584	5,088,448	13.1148770	5.5612978	.005813953
173	29,929	5,177,717	13.1529464	5.5720546	.005780347
174	30,276	5,268,024	13.1909060	5.5827702	.005747126
175	30,625	5,359,375	13.2287566	5.5934447	.005714286
176	30,976	5,451,776	13.2664992	5.6040787	.005681818
177	31,329	5,545,233	13.3041347	5.6146724	.005649718
178	31,684	5,639,752	13.3416641	5.6252263	.005617978
179	32,041	5,735,339	13.3790882	5.6357408	.005586592
180	32,400	5,832,000	13.4164079	5.6462162	.005555556
181	32,761	5,929,741	13.4536240	5.6566528	.005524862
182	33,124	6,028,568	13.4907376	5.6670511	.005494505
183	33,489	6,128,487	13.5277493	5.6774114	.005464481
184	33,856	6,229,504	13.5646600	5.6877340	.005434783
185	34,225	6,331,625	13.6014705	5.6980192	.005405405
186	34,596	6,434,856	13.6381817	5.7082675	.005376344
187	34,969	6,539,203	13.6747943	5.7184791	.005347594
188	35,344	6,644,672	13.7113092	5.7286543	.005319149
189	35,721	6,751,269	13.7477271	5.7387936	.005291005
190	36,100	6,859,000	13.7840488	5.7488971	.005263158
191	36,481	6,967,871	13.8202750	5.7589652	.005235602
192	36,864	7,077,888	13.8564065	5.7689982	.005208333
193	37,249	7,189,057	13.8924440	5.7789966	.005181347
194	37,636	7,301,384	13.9283883	5.7889604	.005154639
195	38,025	7,414,875	13.9642400	5.7988900	.005128205
196	38,416	7,529,536	14.0000000	5.8087857	.005102041
197	38,809	7,645,373	14.0356688	5.8186479	.005076142
198	39,204	7,762,392	14.0712473	5.8284767	.005050505
199	39,610	7,880,599	14.1067360	5.8382725	.005025126
200	40,000	8,000,000	14.1421356	5.8480355	.005000000

Table 65.—*Squares, cubes, square roots, cube roots and reciprocals*—Continued

N	N^2	N^3	$N^{\frac{1}{2}}$	$N^{\frac{1}{3}}$	$\dfrac{1}{N}$
201	40,401	8,120,601	14.1774469	5.8577660	.004975124
202	40,804	8,242,408	14.2126704	5.8674643	.004950495
203	41,209	8,365,427	14.2478068	5.8771307	.004926108
204	41,616	8,489,664	14.2828569	5.8867653	.004901961
205	42,025	8,615,125	14.3178211	5.8963685	.004878049
206	42,436	8,741,816	14.3527001	5.9059406	.004854369
207	42,849	8,869,743	14.3874946	5.9154817	.004830918
208	43,264	8,998,912	14.4222051	5.9249921	.004807692
209	43,681	9,129,329	14.4568323	5.9344721	.004784689
210	44,100	9,261,000	14.4913767	5.9439220	.004761905
211	44,521	9,393,931	14.5258390	5.9533418	.004739336
212	44,944	9,528,128	14.5602198	5.9627320	.004716981
213	45,369	9,663,597	14.5945195	5.9720926	.004694836
214	45,796	9,800,344	14.6287388	5.9814240	.004672897
215	46,225	9,938,375	14.6628783	5.9907264	.004651163
216	46,656	10,077,696	14.6969385	6.0000000	.004629630
217	47,089	10,218,313	14.7309199	6.0092450	.004608295
218	47,524	10,360,232	14.7648231	6.0184617	.004587156
219	47,961	10,503,459	14.7986486	6.0276502	.004566210
220	48,400	10,648,000	14.8323970	6.0368107	.004545455
221	48,841	10,793,861	14.8660687	6.0459435	.004524887
222	49,284	10,941,048	14.8996644	6.0550489	.004504505
223	49,729	11,089,567	14.9331845	6.0641270	.004484305
224	50,176	11,239,424	14.9666295	6.0731779	.004464286
225	50,625	11,390,625	15.0000000	6.0822020	.004444444
226	51,076	11,543,176	15.0332964	6.0911994	.004424779
227	51,529	11,697,083	15.0665192	6.1001702	.004405286
228	51,984	11,852,352	15.0996689	6.1091147	.004385965
229	52,441	12,008,989	15.1327460	6.1180332	.004366812
230	52,900	12,167,000	15.1657509	6.1269257	.004347826
231	53,361	12,326,391	15.1986842	6.1357924	.004329004
232	53,824	12,487,168	15.2315462	6.1446337	.004310345
233	54,289	12,649,337	15.2643375	6.1534495	.004291845
234	54,756	12,812,904	15.2970585	6.1622401	.004273504
235	55,225	12,977,875	15.3297097	6.1710058	.004255319
236	55,696	13,144,256	15.3622915	6.1797466	.004237288
237	56,169	13,312,053	15.3948043	6.1884628	.004219409
238	56,644	13,481,272	15.4272486	6.1971544	.004201681
239	57,121	13,651,919	15.4596248	6.2058218	.004184100
240	57,600	13,824,000	15.4919334	6.2144650	.004166667
241	58,081	13,997,521	15.5241747	6.2230843	.004149378
242	58,564	14,172,488	15.5563492	6.2316797	.004132231
243	59,049	14,348,907	15.5884573	6.2402515	.004115226
244	59,536	14,526,784	15.6204994	6.2487998	.004098361
245	60,025	14,706,125	15.6524758	6.2573248	.004081633
246	60,516	14,886,936	15.6843871	6.2658266	.004065041
247	61,009	15,069,223	15.7162336	6.2743054	.004048583
248	61,504	15,252,992	15.7480157	6.2827613	.004032258
249	62,001	15,438,249	15.7797338	6.2911946	.004016064
250	62,500	15,625,000	15.8113883	6.2996053	.004000000

Table 65.—*Squares, cubes, square roots, cube roots and reciprocals*—Continued

N	N²	N³	N^{1/2}	N^{1/3}	1/N
251	63,001	15,813,251	15.8429795	6.3079935	.003984064
252	63,504	16,003,008	15.8745079	6.3163596	.003968254
253	64,009	16,194,277	15.9059737	6.3247085	.003952569
254	64,516	16,387,064	15.9373775	6.3330256	.003937008
255	65,025	16,581,875	15.9687194	6.3413257	.003921569
256	65,536	16,777,216	16.0000000	6.3496042	.003906250
257	66,049	16,974,593	16.0312195	6.3578611	.003891051
258	66,564	17,173,512	16.0623784	6.3660968	.003875969
259	67,081	17,373,979	16.0934769	6.3743111	.003861004
260	67,600	17,576,000	16.1245155	6.3825043	.003846154
261	68,121	17,779,581	16.1554944	6.3906765	.003831418
262	68,644	17,984,728	16.1864141	6.3988279	.003816794
263	69,169	18,191,447	16.2172747	6.4069585	.003802281
264	69,696	18,399,744	16.2480768	6.4150687	.003787879
265	70,225	18,609,625	16.2788206	6.4231583	.003773585
266	70,756	18,821,096	16.3095064	6.4312276	.003759398
267	71,289	19,034,163	16.3401346	6.4392767	.003745318
268	71,824	19,248,832	16.3707055	6.4473057	.003731343
269	72,361	19,465,109	16.4012195	6.4553148	.003717472
270	72,900	19,683,000	16.4316767	6.4633041	.003703704
271	73,441	19,902,511	16.4620776	6.4712736	.003690037
272	73,984	20,123,648	16.4924225	6.4792236	.003676471
273	74,529	20,346,417	16.5227116	6.4871541	.003663004
274	75,076	20,570,824	16.5529454	6.4950653	.003649635
275	75,625	20,796,875	16.5831240	6.5029572	.003636364
276	76,176	21,024,576	16.6132477	6.5108300	.003623188
277	76,729	21,253,933	16.6433170	6.5186839	.003610108
278	77,284	21,484,952	16.6733320	6.5265189	.003597122
279	77,841	21,717,639	16.7032931	6.5343351	.003584229
280	78,400	21,952,000	16.7332005	6.5421326	.003571429
281	78,961	22,188,041	16.7630546	6.5499116	.003558719
282	79,524	22,425,768	16.7928556	6.5576722	.003546099
283	80,089	22,665,187	16.8226038	6.5654144	.003533569
284	80,656	22,906,304	16.8522995	6.5731385	.003521127
285	81,225	23,149,125	16.8819430	6.5808443	.003508772
286	81,796	23,393,656	16.9115345	6.5885323	.003496503
287	82,369	23,639,903	16.9410743	6.5962023	.003484321
288	82,944	23,887,872	16.9705627	6.6038545	.003472222
289	83,521	24,137,569	17.0000000	6.6114890	.003460208
290	84,100	24,389,000	17.0293864	6.6191060	.003448276
291	84,681	24,642,171	17.0587221	6.6267054	.003436426
292	85,264	24,897,088	17.0880075	6.6342874	.003424658
293	85,849	25,153,757	17.1172428	6.6418522	.003412969
294	86,436	25,412,184	17.1464282	6.6493998	.003401361
295	87,025	25,672,375	17.1755640	6.6569302	.003389831
296	87,616	25,934,336	17.2046505	6.6644437	.003378378
297	88,209	26,198,073	17.2336879	6.6719403	.003367003
298	88,804	26,463,592	17.2626765	6.6794200	.003355705
299	89,401	26,730,899	17.2916165	6.6868831	.003344482
300	90,000	27,000,000	17.3205081	6.6943295	.003333333

Table 65.—*Squares, cubes, square roots, cube roots and reciprocals*—Continued

N	N²	N³	N^(1/2)	N^(1/3)	1/N
301	90,601	27,270,901	17.3493516	6.7017593	.003322259
302	91,204	27,543,608	17.3781472	6.7091729	.003311258
303	91,809	27,818,127	17.4068952	6.7165700	.003300330
304	92,416	28,094,464	17.4355958	6.7239508	.003289474
305	93,025	28,372,625	17.4642492	6.7313155	.003278689
306	93,636	28,652,616	17.4928557	6.7386641	.003267974
307	94,249	28,934,443	17.5214155	6.7459967	.003257329
308	94,864	29,218,112	17.5499288	6.7533134	.003246753
309	95,481	29,503,629	17.5783958	6.7606143	.003236246
310	96,100	29,791,000	17.6068169	6.7678995	.003225806
311	96,721	30,080,231	17.6351921	6.7751690	.003215434
312	97,344	30,371,328	17.6635217	6.7824229	.003205128
313	97,969	30,664,297	17.6918060	6.7896613	.003194888
314	98,596	30,959,144	17.7200451	6.7968844	.003184713
315	99,225	31,255,875	17.7482393	6.8040921	.003174603
316	99,856	31,554,496	17.7763888	6.8112847	.003164557
317	100,489	31,855,013	17.8044938	6.8184602	.003154574
318	101,124	32,157,432	17.8325545	6.8256242	.003144654
319	101,761	32,461,759	17.8605711	6.8327714	.003134796
320	102,400	32,768,000	17.8885438	6.8399037	.003125000
321	103,041	33,076,161	17.9164729	6.8470213	.003115265
322	103,684	33,386,248	17.9443584	6.8541240	.003105590
323	104,329	33,698,267	17.9722008	6.8612120	.003095975
324	104,976	34,012,224	18.0000000	6.8682855	.003086420
325	105,625	34,328,125	18.0277564	6.8753443	.003076923
326	106,276	34,645,976	18.0554701	6.8823888	.003067485
327	106,929	34,965,783	18.0831413	6.8894188	.003058104
328	107,584	35,287,552	18.1107703	6.8964345	.003048780
329	108,241	35,611,289	18.1383571	6.9034359	.003039514
330	108,900	35,937,000	18.1659021	6.9104232	.003030303
331	109,561	36,264,691	18.1934054	6.9173964	.003021148
332	110,224	36,594,368	18.2208672	6.9243556	.003012048
333	110,889	36,926,037	18.2482876	6.9313008	.003003003
334	111,556	37,259,704	18.2756669	6.9382321	.002994012
335	112,225	37,595,375	18.3030052	6.9451496	.002985075
336	112,896	37,933,056	18.3303028	6.9520533	.002976190
337	113,569	38,272,753	18.3575598	6.9589434	.002967359
338	114,244	38,614,472	18.3847763	6.9658198	.002958580
339	114,921	38,958,219	18.4119526	6.9726826	.002949853
340	115,600	39,304,000	18.4390889	6.9795321	.002941176
341	116,281	39,651,821	18.4661853	6.9863681	.002932551
342	116,964	40,001,688	18.4932420	6.9931906	.002923977
343	117,649	40,353,607	18.5202592	7.0000000	.002915452
344	118,336	40,707,584	18.5472370	7.0067962	.002906977
345	119,025	41,063,625	18.5741756	7.0135791	.002898551
346	119,716	41,421,736	18.6010752	7.0203490	.002890173
347	120,409	41,781,923	18.6279360	7.0271058	.002881844
348	121,104	42,144,192	18.6547581	7.0338497	.002873563
349	121,801	42,508,549	18.6815417	7.0405806	.002865330
350	122,500	42,875,000	18.7082869	7.0472987	.002857143

Table 65.—*Squares, cubes, square roots, cube roots and reciprocals—Continued*

N	N²	N³	N^½	N^⅓	1/N
351	123,201	43,243,551	18.7349940	7.0540041	.002849003
352	123,904	43,614,208	18.7616680	7.0606967	.002840909
353	124,609	43,986,977	18.7882942	7.0673767	.002832861
354	125,316	44,361,864	18.8148877	7.0740440	.002824859
355	126,025	44,738,875	18.8414437	7.0806988	.002816901
356	126,736	45,118,016	18.8679623	7.0873411	.002808989
357	127,449	45,499,293	18.8944436	7.0939709	.002801120
358	128,164	45,882,712	18.9208879	7.1005885	.002793296
359	128,881	46,268,279	18.9472953	7.1071937	.002785515
360	129,600	46,656,000	18.9736660	7.1137866	.002777778
361	130,321	47,045,881	19.0000000	7.1203674	.002770083
362	131,044	47,437,928	19.0262976	7.1269360	.002762431
363	131,769	47,832,147	19.0525589	7.1334925	.002754821
364	132,496	48,228,544	19.0787840	7.1400370	.002747253
365	133,225	48,627,125	19.1049782	7.1465695	.002739726
366	133,956	49,027,896	19.1311265	7.1530901	.002732240
367	134,689	49,430,863	19.1572441	7.1595988	.002724796
368	135,424	49,836,032	19.1833261	7.1660957	.002717391
369	136,161	50,243,409	19.2093727	7.1725809	.002710027
370	136,900	50,653,000	19.2353841	7.1790544	.002702703
371	137,641	51,064,811	19.2613603	7.1855162	.002695418
372	138,384	51,478,848	19.2873015	7.1919663	.002688172
373	139,129	51,895,117	19.3132079	7.1984050	.002680965
374	139,876	52,313,624	19.3390796	7.2048322	.002673797
375	140,625	52,734,375	19.3649167	7.2112479	.002666667
376	141,376	53,157,876	19.3907194	7.2176522	.002659574
377	142,129	53,582,633	19.4164878	7.2240450	.002652520
378	142,884	54,010,152	19.4422221	7.2304268	.002645503
379	143,641	54,439,939	19.4679223	7.2367972	.002638522
380	144,400	54,872,000	19.4935887	7.2431565	.002631579
381	145,161	55,306,341	19.5192213	7.2495045	.002624672
382	145,924	55,742,968	19.5448203	7.2558415	.002617801
383	146,689	56,181,887	19.5703858	7.2621675	.002610966
384	147,456	56,623,104	19.5959179	7.2684824	.002604167
385	148,225	57,066,625	19.6214169	7.2747864	.002597403
386	148,996	57,512,456	19.6468827	7.2810794	.002590674
387	149,769	57,960,603	19.6723156	7.2873617	.002583979
388	150,544	58,411,072	19.6977156	7.2936330	.002577320
389	151,321	58,863,869	19.7230829	7.2998936	.002570694
390	152,100	59,319,000	19.7484177	7.3061436	.002564103
391	152,881	59,776,471	19.7737199	7.3123828	.002557545
392	153,664	60,236,288	19.7989899	7.3186114	.002551020
393	154,449	60,698,457	19.8242276	7.3248295	.002544529
394	155,236	61,162,984	19.8494332	7.3310369	.002538071
395	156,025	61,629,875	19.8746069	7.3372339	.002531646
396	156,816	62,099,186	19.8992487	7.3434205	.002525253
397	157,609	62,570,773	19.9248588	7.3495966	.002518892
398	158,404	63,044,792	19.9499373	7.3557624	.002512563
399	159,201	63,521,199	19.9749844	7.3619178	.002506266
400	160,000	64,000,000	20.0000000	7.3680630	.002500000

Table 65.—*Squares, cubes, square roots, cube roots and reciprocals*—Continued

N	N²	N³	N^(1/2)	N^(1/3)	1/N
401	160,801	64,481,201	20.0249844	7.8741979	.002493766
402	161,604	64,964,808	20.0499377	7.8808227	.002487562
403	162,409	65,450,827	20.0748599	7.8864373	.002481390
404	163,216	65,939,264	20.0997512	7.8925418	.002475248
405	164,025	66,430,125	20.1246118	7.8986363	.002469136
406	164,836	66,923.416	20.1494417	7.4047206	.002463054
407	165,649	67,419,143	20.1742410	7.4107950	.002457002
408	166,464	67,917,312	20.1990099	7.4168595	.002450980
409	167,281	68,417,929	20.2237484	7.4229142	.002444988
410	168,100	68,921,000	20.2484567	7.4289589	.002439024
411	168,921	69,426,531	20.2731349	7.4349988	.002433090
412	169,744	69,934,528	20.2977831	7.4410189	.002427184
413	170,569	70,444,997	20.3224014	7.4470342	.002421308
414	171,396	70,957,944	20.3469899	7.4530399	.002415459
415	172,225	71,473,375	20.3715488	7.4590359	.002409639
416	173,056	71,991,296	20.3960781	7.4650223	.002403846
417	173,889	72,511,713	20.4205779	7.4709991	.002398082
418	174,724	73,034,632	20.4450483	7.4769664	.002392344
419	175,561	73,560,059	20.4694895	7.4829242	.002386635
420	176,400	74,088,000	20.4939015	7.4888724	.002380952
421	177,241	74,618,461	20.5182845	7.4948113	.002375297
422	178,084	75,151,448	20.5426386	7.5007406	.002369668
423	178,929	75,686,967	20.5669688	7.5066607	.002364066
424	179,776	76,225,024	20.5912603	7.5125715	.002358491
425	180,625	76,765,625	20.6155281	7.5184780	.002352941
426	181,476	77,308,776	20.6397674	7.5243652	.002347418
427	182,329	77,854,483	20.6639783	7.5302482	.002341920
428	183,184	78,402,752	20.6881609	7.5361221	.002336449
429	184,041	78,953,589	20.7123152	7.5419867	.002331002
430	184,900	79,507,000	20.7364414	7.5478423	.002325581
431	185,761	80,062,991	20.7605395	7.5536888	.002320186
432	186,624	80,621,568	20.7846097	7.5595263	.002314815
433	187,489	81,182,737	20.8086520	7.5653548	.002309469
434	188,356	81,746,504	20.8326667	7.5711743	.002304147
435	189,225	82,312,875	20.8566536	7.5769849	.002298851
436	190,096	82,881,856	20.8806130	7.5827865	.002293578
437	190,969	83,453,453	20.9045450	7.5885793	.002288330
438	191,844	84,027,672	20.9284495	7.5943633	.002283105
439	192,721	84,604,519	20.9523268	7.6001385	.002277904
440	193,600	85,184,000	20.9761770	7.6059049	.002272727
441	194.481	85,766,121	21.0000000	7.6116626	.002267574
442	195,364	86,350,888	21.0237960	7.6174116	.002262443
443	196,249	86,938,307	21.0475652	7.6231519	.002257336
444	197,136	87,528,384	21.0713075	7.6288837	.002252252
445	198,025	88,121,125	21.0950231	7.6346067	.002247191
446	198,916	88,716,536	21.1187121	7.6403213	.002242152
447	199,809	89,314,623	21.1423745	7.6460272	.002237136
448	200,704	89,915,392	21.1660105	7.6517247	.002232143
449	201,601	90,518,849	21.1896201	7.6574138	.002227171
450	202,500	91,125,000	21.2132084	7.6630948	.002222222

Table 65.—*Squares, cubes, square roots, cube roots and reciprocals*—Continued

N	N^2	N^3	$N^{\frac{1}{2}}$	$N^{\frac{1}{3}}$	$\frac{1}{N}$
451	203,401	91,733,851	21.2367606	7.6687665	.002217295
452	204,304	92,345,408	21.2602916	7.6744303	.002212389
453	205,209	92,959,677	21.2837967	7.6800857	.002207506
454	206,116	93,576,664	21.3072758	7.6857328	.002202643
455	207,025	94,196,375	21.3307290	7.6913717	.002197802
456	207,936	94,818,816	21.3541565	7.6970023	.002192982
457	208,849	95,443,993	21.3775583	7.7026246	.002188184
458	209,764	96,071,912	21.4009346	7.7082388	.002183406
459	210,681	96,702,579	21.4242853	7.7138448	.002178649
460	211,600	97,336,000	21.4476106	7.7194426	.002173913
461	212,521	97,972,181	21.4709106	7.7250325	.002169197
462	213,444	98,611,128	21.4941853	7.7306141	.002164502
463	214,369	99,252,847	21.5174348	7.7361877	.002159827
464	215,296	99,897,344	21.5406592	7.7417532	.002155172
465	216,225	100,544,625	21.5638587	7.7473109	.002150538
466	217,156	101,194,696	21.5870331	7.7528606	.002145923
467	218,089	101,847,563	21.6101828	7.7584023	.002141328
468	219,024	102,503,232	21.6333077	7.7639361	.002136752
469	219,961	103,161,709	21.6564078	7.7694620	.002132196
470	220,900	103,823,000	21.6794834	7.7749801	.002127660
471	221,841	104,487,111	21.7025344	7.7804904	.002123142
472	222,784	105,154,048	21.7255610	7.7859928	.002118644
473	223,729	105,823,817	21.7485632	7.7914875	.002114165
474	224,676	106,496,424	21.7715411	7.7969745	.002109705
475	225,625	107,171,875	21.7944947	7.8024538	.002105263
476	226,576	107,850,176	21.8174242	7.8079254	.002100840
477	227,529	108,531,333	21.8403297	7.8133892	.002096436
478	228,484	109,215,352	21.8632111	7.8188456	.002092050
479	229,441	109,902,239	21.8860686	7.8242942	.002087683
480	230,400	110,592,000	21.9089023	7.8297353	.002083333
481	231,361	111,284,641	21.9317122	7.8351688	.002079002
482	232,324	111,980,168	21.9544984	7.8405949	.002074689
483	233,289	112,678,587	21.9772610	7.8460134	.002070393
484	234,256	113,379,904	22.0000000	7.8514244	.002066116
485	235,225	114,084,125	22.0227155	7.8568281	.002061856
486	236,196	114,791,256	22.0454077	7.8622242	.002057613
487	237,169	115,501,303	22.0680765	7.8676130	.002053388
488	238,144	116,214,272	22.0907220	7.8729944	.002049180
489	239,121	116,930,169	22.1133444	7.8783684	.002044990
490	240,100	117,649,000	22.1359436	7.8837352	.002040816
491	241,081	118,370,771	22.1585198	7.8890946	.002036660
492	242,064	119,095,488	22.1810730	7.8944468	.002032520
493	243,049	119,823,157	22.2036033	7.8997917	.002028398
494	244,036	120,553,784	22.2261108	7.9051294	.002024291
495	245,025	121,287,375	22.2485955	7.9104599	.002020202
496	246,016	122,023,936	22.2710575	7.9157832	.002016129
497	247,009	122,763,473	22.2934968	7.9210994	.002012072
498	248,004	123,505,992	22.3159136	7.9264085	.002008032
499	249,001	124,251,499	22.3383079	7.9317104	.002004008
500	250,000	125,000,000	22.3606798	7.9370053	.002000000

Table 65.—*Squares, cubes, square roots, cube roots and reciprocals*—Continued

N	N²	N³	N$^{\frac{1}{2}}$	N$^{\frac{1}{3}}$	$\frac{1}{N}$
501	251,001	125,751,501	22.8880298	7.9422931	.001996008
502	252,004	126,506,008	22.4053565	7.9475789	.001992032
503	253,009	127,263,527	22.4276615	7.9528477	.001988072
504	254,016	128,024,064	22.4499448	7.9581144	.001984127
505	255,025	128,787,625	22.4722051	7.9633743	.001980198
506	256,086	129,554,216	22.4944488	7.9686271	.001976285
507	257,049	130,323,843	22.5166605	7.9788731	.001972387
508	258,064	131,096,512	22.5388558	7.9791122	.001968504
509	259,061	131,872,229	22.5610288	7.9843444	.001964637
510	260,100	132,651,000	22.5881796	7.9895697	.001960784
511	261,121	133,432,831	22.6058091	7.9947883	.001956947
512	262,144	134,217,728	22.6274170	8.0000000	.001953125
513	263,169	135,005,697	22.6495033	8.0052049	.001949318
514	264,196	135,796,744	22.6715681	8.0104032	.001945525
515	265,225	136,590,875	22.6936114	8.0155946	.001941748
516	266,256	137,388,096	22.7156334	8.0207794	.001937984
517	267,289	138,188,413	22.7376340	8.0259574	.001934236
518	268,324	138,991,832	22.7596134	8 0311287	.001930502
519	269,361	139,798,359	22.7815715	8.0362985	.001926782
520	270,400	140,608,000	22.8035085	8.0414515	.001923077
521	271,441	141,420,761	22.8254244	8.0466030	.001919386
522	272,484	142,236,648	22.8473193	8 0517479	.001915709
523	273,529	143,055,667	22.8691933	8.0568862	.001912046
524	274,576	143,877,824	22.8910463	8.0620180	.001908397
525	275,625	144,703,125	22.9128785	8.0671432	.001904762
526	276,676	145,531,576	22.9346899	8.0722620	.001901141
527	277,729	146,363,183	22.9564806	8.0773743	.001897533
528	278,784	147,197,952	22.9782506	8.0824800	.001893939
529	279,841	148,035,889	23.0000000	8.0875794	.001890359
530	280,900	148,877,000	23.0217289	8.0926723	.001886792
531	281,961	149,721,291	23.0434372	8.0977589	.001883239
532	283,024	150,568,768	23.0651252	8.1028390	.001879699
533	284,089	151,419,437	23.0867928	8.1079128	.001876173
534	285,156	152,273,304	23.1084400	8.1129803	.001872659
535	286,225	153,130,375	23.1300670	8.1180414	.001869159
536	287,296	153,990,656	23.1516738	8.1230962	.001865672
537	288,369	154,854,153	23.1732605	8.1281447	.001862197
538	289,444	155,720,872	23.1948270	8.1331870	.001858736
539	290,521	156,590,819	23.2163735	8.1382280	.001855288
540	291,600	157,464,000	23.2379001	8.1432529	.001851852
541	292,681	158,340,421	23.2594067	8.1482765	.001848429
542	293,764	159,220,088	23.2808935	8.1532989	.001845018
543	294,849	160,103,007	23.3023604	8.1583051	.001841621
544	295,936	160,989,184	23.3238076	8.1633102	.001838235
545	297,025	161,878,625	23.3452351	8.1683092	.001834862
546	298,116	162,771,336	23.3666429	8.1733020	.001831502
547	299,209	163,667,323	23.3880311	8.1782888	.001828154
548	300,304	164,566,592	23.4093998	8.1832695	.001824818
549	301,401	165,469,149	23.4307490	8.1882441	.001821494
550	302,500	166,375,000	23.4520788	8.1932127	.001818182

Table 65.—*Squares, cubes, square roots, cube roots and reciprocals*—Continued

N	N^2	N^3	$N^{\frac{1}{2}}$	$N^{\frac{1}{3}}$	$\dfrac{1}{N}$
551	303,601	167,284,151	23.4733892	8.1981753	.001814882
552	304,704	168,196,608	23.4946802	8.2031319	.001811594
553	305,809	169,112,377	23.5159520	8.2080825	.001808318
554	306,916	170,031,464	23.5372046	8.2130271	.001805054
555	308,025	170,953,875	23.5584380	8.2179657	.001801802
556	309,136	171,879,616	23.5796522	8.2228985	.001798561
557	310,249	172,808,693	23.6008474	8.2278254	.001795332
558	311,364	173,741,112	23.6220236	8.2327463	.001792115
559	312,481	174,676,879	23.6431808	8.2376614	.001788909
560	313,600	175,616,000	23.6643191	8.2425706	.001785714
561	314,721	176,558,481	23.6854386	8.2474740	.001782531
562	315,844	177,504,328	23.7065392	8.2523715	.001779359
563	316,969	178,453,547	23.7276210	8.2572633	.001776199
564	318,096	179,406,144	23.7486842	8.2621492	.001773050
565	319,225	180,362,125	23.7697286	8.2670294	.001769912
566	320,356	181,321,496	23.7907545	8.2719089	.001766784
567	321,489	182,284,263	23.8117618	8.2767726	.001763668
568	322,624	183,250,432	23.8327506	8.2816355	.001760563
569	323,761	184,220,009	23.8537209	8.2864928	.001757469
570	324,900	185,193,000	23.8746728	8.2913444	.001754386
571	326,041	186,169,411	23.8956063	8.2961903	.001751313
572	327,184	187,149,248	23.9165215	8.3010304	.001748252
573	328,329	188,132,517	23.9374184	8.3058651	.001745201
574	329,476	189,119,224	23.9582971	8.3106941	.001742160
575	330,625	190,109,375	23.9791576	8.3155175	.001739130
576	331,776	191,102,976	24.0000000	8.3203353	.001736111
577	332,929	192,100,033	24.0208243	8.3251475	.001733102
578	334,084	193,100,552	24.0416306	8.3299542	.001730104
579	335,241	194,104,539	24.0624188	8.3347553	.001727116
580	336,400	195,112,000	24.0831891	8.3395509	.001724138
581	337,561	196,122,941	24.1039416	8.3443410	.001721170
582	338,724	197,137,368	24.1246762	8.3491256	.001718213
583	339,889	198,155,287	24.1453929	8.3539047	.001715266
584	341,056	199,176,704	24.1660919	8.3586784	.001712329
585	342,225	200,201,625	24.1867732	8.3634466	.001709402
586	343,396	201,230,056	24.2074369	8.3682095	.001706485
587	344,569	202,262,003	24.2280829	8.3729668	.001703578
588	345,744	203,297,472	24.2487113	8.3777188	.001700680
589	346,921	204,336,469	24.2693222	8.3824653	.001697793
590	348,100	205,379,000	24.2899156	8.3872065	.001694915
591	349,281	206,425,071	24.3104916	8.3919423	.001692047
592	350,464	207,474,688	24.3310501	8.3966729	.001689189
593	351,649	208,527,857	24.3515913	8.4013981	.001686341
594	352,836	209,584,584	24.3721152	8.4061180	.001683502
595	354,025	210,644,875	24.3926218	8.4108326	.001680672
596	355,216	211,708,736	24.4131112	8.4155419	.001677852
597	356,409	212,776,173	24.4335834	8.4202460	.001675042
598	357,604	213,847,192	24.4540385	8.4249448	.001672241
599	358,801	214,921,799	24.4744765	8.4296383	.001669449
600	360,000	216,000,000	24.4948974	8.4343267	.001666667

Table 65.—*Squares, cubes, square roots, cube roots and recip-rocals*—Continued

N	N²	N³	N^(1/2)	N^(1/3)	1/N
601	861,201	217,081,801	24.5153013	8.4390098	.001663894
602	862,404	218,167,208	24.5356883	8.4436877	.001661130
603	863,609	219,256,227	24.5560583	8.4483605	.001658375
604	864,816	220,348,864	24.5764115	8.4530281	.001655629
605	866,025	221,445,125	24.5967478	8.4576906	.001652893
606	867,236	222,545,016	24.6170673	8.4623479	.001650165
607	868,449	223,648,543	24.6373700	8.4670001	.001647446
608	869,664	224,755,712	24.6576560	8.4716471	.001644737
609	870,881	225,866,529	24.6779254	8.4762892	.001642036
610	872,100	226,981,000	24.6981781	8.4809261	.001639344
611	873,321	228,099,131	24.7184142	8.4855579	.001636661
612	874,544	229,220,928	24.7386338	8.4901848	.001633987
613	875,769	230,346,397	24.7588368	8.4948065	.001631321
614	876,996	231,475,544	24.7790234	8.4994233	.001628664
615	878,225	232,608,375	24.7991935	8.5040350	.001626016
616	879,456	233,744,896	24.8193473	8.5086417	.001623377
617	880,689	234,885,113	24.8394847	8.5132435	.001620746
618	881,924	236,029,032	24.8596058	8.5178403	.001618123
619	883,161	237,176,659	24.8797106	8.5224321	.001615509
620	884,400	238,328,000	24.8997992	8.5270189	.001612903
621	885,641	239,483,061	24.9198716	8.5316009	.001610306
622	886,884	240,641,848	24.9399278	8.5361780	.001607717
623	888,129	241,804,367	24.9599679	8.5407501	.001605136
624	889,876	242,970,624	24.9799920	8.5453173	.001602564
625	890,625	244,140,625	25.0000000	8.5498797	.001600000
626	891,876	245,314,376	25.0199920	8.5544372	.001597444
627	893,129	246,491,883	25.0399681	8.5589899	.001594896
628	894,384	247,673,152	25.0599282	8.5635377	.001592857
629	895,641	248,858,189	25.0798724	8.5680807	.001589825
630	896,900	250,047,000	25.0998008	8.5726189	.001587302
631	898,161	251,239,591	25.1197134	8.5771523	.001584786
632	899,424	252,435,968	25.1396102	8.5816809	.001582278
633	900,689	253,636,137	25.1594913	8.5862047	.001579779
634	901,956	254,840,104	25.1793566	8.5907238	.001577287
635	903,225	256,047,875	25.1992063	8.5952380	.001574803
636	904,496	257,259,456	25.2190404	8.5997476	.001572327
637	905,769	258,474,853	25.2388589	8.6042525	.001569859
638	907,044	259,694,072	25.2586619	8.6087526	.001567398
639	908,321	260,917,119	25.2784493	8.6132480	.001564945
640	909,600	262,144,000	25.2982213	8.6177388	.001562500
641	910,881	263,374,721	25.3179778	8.6222248	.001560062
642	912,164	264,609,288	25.3377189	8.6267063	.001557632
643	913,449	265,847,707	25.3574447	8.6311830	.001555210
644	914,736	267,089,984	25.3771551	8.6356551	.001552795
645	916,025	268,336,125	25.3968502	8.6401226	.001550388
646	917,316	269,586,136	25.4165801	8.6445855	.001547988
647	918,609	270,840,023	25.4361947	8.6490437	.001545595
648	919,904	272,097,792	25.4558441	8.6534974	.001543210
649	921,201	273,359,449	25.4754784	8.6579465	.001540832
650	922,500	274,625,000	25.4950976	8.6623911	.001538462

Table 65.—*Squares, cubes, square roots, cube roots and reciprocals—Continued*

N	N²	N³	N^½	N^⅓	1/N
651	423,801	275,894,451	25.5147016	8.6668310	.001536098
652	425,104	277,167,808	25.5342907	8.6712665	.001533742
653	426,409	278,445,077	25.5538647	8.6756974	.001531394
654	427,716	279,726,264	25.5734237	8.6801287	.001529052
655	429,025	281,011,375	25.5929678	8.6845456	.001526718
656	430,336	282,300,416	25.6124969	8.6889630	.001524390
657	431,649	283,593,393	25.6320112	8.6933759	.001522070
658	432,964	284,890,312	25.6515107	8.6977843	.001519757
659	434,281	286,191,179	25.6709953	8.7021882	.001517451
660	435,600	287,496,000	25.6904652	8.7065877	.001515152
661	436,921	288,804,781	25.7099203	8.7109827	.001512859
662	438,244	290,117,528	25.7293607	8.7153784	.001510574
663	439,569	291,434,247	25.7487864	8.7197596	.001508296
664	440,896	292,754,944	25.7681975	8.7241414	.001506024
665	442,225	294,079,625	25.7875939	8.7285187	.001503759
666	443,556	295,408,296	25.8069758	8.7328918	.001501502
667	444,889	296,740,963	25.8263431	8.7372604	.001499250
668	446,224	298,077,632	25.8456960	8.7416246	.001497006
669	447,561	299,418,309	25.8650343	8.7459846	.001494768
670	448,900	300,763,000	25.8843582	8.7503401	.001492537
671	450,241	302,111,711	25.9036677	8.7546913	.001490313
672	451,584	303,464,448	25.9229628	8.7590383	.001488095
673	452,929	304,821,217	25.9422435	8.7633809	.001485884
674	454,276	306,182,024	25.9615100	8.7677192	.001488680
675	455,625	307,546,875	25.9807621	8.7720532	.001481481
676	456,976	308,915,776	26.0000000	8.7763830	.001479290
677	458,329	310,288,733	26.0192237	8.7807084	.001477105
678	459,684	311,665,752	26.0384331	8.7850296	.001474926
679	461,041	313,046,839	26.0576284	8.7893466	.001472754
680	462,400	314,432,000	26.0768096	8.7936593	.001470588
681	463,761	315,821,241	26.0959767	8.7979679	.001468429
682	465,124	317,214,568	26.1151297	8.8022721	.001466276
683	466,489	318,611,987	26.1342687	8.8065722	.001464129
684	467,856	320,013,504	26.1533937	8.8108681	.001461988
685	469,225	321,419,125	26.1725047	8.8151598	.001459854
686	470,596	322,828,856	26.1916017	8.8194474	.001457726
687	471,969	324,242,703	26.2106848	8.8237307	.001455604
688	473,344	325,660,672	26.2297541	8.8280099	.001453488
689	474,721	327,082,769	26.2488095	8.8322850	.001451379
690	476,100	328,509,000	26.2678511	8.8365559	.001449275
691	477,481	329,939,371	26.2868789	8.8408227	.001447178
692	478,864	331,373,888	26.3058929	8.8450854	.001445087
693	480,249	332,812,557	26.3248932	8.8493440	.001443001
694	481,636	334,255,384	26.3438797	8.8535985	.001440922
695	483,025	335,702,375	26.3628527	8.8578489	.001438849
696	484,416	337,153,536	26.3818119	8.8620952	.001436782
697	485,809	338,608,873	26.4007576	8.8663375	.001434720
698	487,204	340,068,392	26.4196896	8.8705757	.001432665
699	488,601	341,532,099	26.4386081	8.8748099	.001430615
700	490,000	343,000,000	26.4575131	8.8790400	.001428571

Table 65.—*Squares, cubes, square roots, cube roots and reciprocals*—Continued

N	N²	N³	N^½	N^⅓	$\frac{1}{N}$
701	491,401	344,472,101	26.4764046	8.8832661	.001426534
702	492,804	345,948,408	26.4952826	8.8874882	.001424501
703	494,209	347,428,927	26.5141472	8.8917063	.001422475
704	495,616	348,913,664	26.5329983	8.8959204	.001420455
705	497,025	350,402,625	26.5518361	8.9001304	.001418440
706	498,436	351,895,816	26.5706605	8.9043366	.001416431
707	499,849	353,393,243	26.5894716	8.9085387	.001414427
708	501,264	354,894,912	26.6082694	8.9127369	.001412429
709	502,681	356,400,829	26.6270539	8.9169311	.001410487
710	504,100	357,911,000	26.6458252	8.9211214	.001408451
711	505,521	359,425,431	26.6645833	8.9253078	.001406470
712	506,944	360,944,128	26.6833281	8.9294902	.001404494
713	508,369	362,467,097	26.7020598	8.9336687	.001402525
714	509,796	363,994,344	26.7207784	8.9378433	.001400560
715	511,225	365,525,875	26.7394839	8.9420140	.001398601
716	512,656	367,061,696	26.7581763	8.9461809	.001396648
717	514,089	368,601,813	26.7768557	8.9503438	.001394700
718	515,524	370,146,232	26.7955220	8.9545029	.001392758
719	516,961	371,694,959	26.8141754	8.9586581	.001390821
720	518,400	373,248,000	26.8328157	8.9628095	.001388889
721	519,841	374,805,361	26.8514432	8.9669570	.001386963
722	521,284	376,367,048	26.8700577	8.9711007	.001385042
723	522,729	377,933,067	26.8886593	8.9752406	.001383126
724	524,176	379,503,424	26.9072481	8.9793766	.001381215
725	525,625	381,078,125	26.9258240	8.9835089	.001379310
726	527,076	382,657,176	26.9443872	8.9876373	.001377410
727	528,529	384,240,583	26.9629375	8.9917620	.001375516
728	529,984	385,828,352	26.9814751	8.9958829	.001373626
729	531,441	387,420,489	27.0000000	9.0000000	.001371742
730	532,900	389,017,000	27.0185122	9.0041134	.001369863
731	534,361	390,617,891	27.0370117	9.0082229	.001367989
732	535,824	392,223,168	27.0554985	9.0123288	.001366120
733	537,289	393,832,837	27.0739727	9.0164309	.001364256
734	538,756	395,446,904	27.0924344	9.0205293	.001362398
735	540,225	397,065,375	27.1108834	9.0246239	.001360544
736	541,696	398,688,256	27.1293199	9.0287149	.001358696
737	543,169	400,315,553	27.1477489	9.0328021	.001356852
738	544,644	401,947,272	27.1661554	9.0368857	.001355014
739	546,121	403,583,419	27.1845544	9.0409655	.001353180
740	547,600	405,224,000	27.2029410	9.0450417	.001351351
741	549,081	406,869,021	27.2213152	9.0491142	.001349528
742	550,564	408,518,488	27.2396769	9.0531831	.001347709
743	552,049	410,172,407	27.2580263	9.0572482	.001345895
744	553,536	411,830,784	27.2763634	9.0613098	.001344086
745	555,025	413,493,625	27.2946881	9.0653677	.001342282
746	556,516	415,160,936	27.3130006	9.0694220	.001340483
747	558,009	416,832,723	27.3313007	9.0734726	.001338688
748	559,504	418,508,992	27.3495887	9.0775197	.001336898
749	561,001	420,189,749	27.3678644	9.0815631	.001335113
750	562,500	421,875,000	27.3861279	9.0856030	.001333333

Table 65.—*Squares, cubes, square roots, cube roots and reciprocals*—Continued

N	N^2	N^3	$N^{\frac{1}{2}}$	$N^{\frac{1}{3}}$	$\dfrac{1}{N}$
751	564,001	423,564,751	27.4043792	9.0896392	.001331558
752	565,504	425,259,008	27.4226184	9.0936719	.001329787
753	567,009	426,957,777	27.4408455	9.0977010	.001328021
754	568,516	428,661,064	27.4590604	9.1017265	.001326260
755	570,025	430,368,875	27.4772633	9.1057485	.001324503
756	571,536	432,081,216	27.4954542	9.1097669	.001322751
757	573,049	433,798,093	27.5136330	9.1137818	.001321004
758	574,564	435,519,512	27.5317998	9.1177931	.001319261
759	576,081	437,245,479	27.5499546	9.1218010	.001317523
760	577,600	438,976,000	27.5680975	9.1258053	.001315789
761	579,121	440,711,081	27.5862284	9.1298061	.001314060
762	580,644	442,450,728	27.6043475	9.1338034	.001312336
763	582,169	444,194,947	27.6224546	9.1377971	.001310616
764	583,696	445,943,744	27.6405499	9.1417874	.001308901
765	585,225	447,697,125	27.6586334	9.1457742	.001307190
766	586,756	449,455,096	27.6767050	9.1497576	.001305483
767	588,289	451,217,663	27.6947648	9.1537375	.001303781
768	589,824	452,984,832	27.7128129	9.1577139	.001302083
769	591,361	454,756,609	27.7308492	9.1616869	.001300390
770	592,900	456,533,000	27.7488739	9.1656565	.001298701
771	594,441	458,314,011	27.7668868	9.1696225	.001297017
772	595,984	460,099,648	27.7848880	9.1735852	.001295337
773	597,529	461,889,917	27.8028775	9.1775445	.001293661
774	599,076	463,684,824	27.8208555	9.1815003	.001291990
775	600,625	465,484,375	27.8388218	9.1854527	.001290323
776	602,176	467,288,576	27.8567766	9.1894018	.001288660
777	603,729	469,097,433	27.8747197	9.1933474	.001287001
778	605,284	470,910,952	27.8926514	9.1972897	.001285347
779	606,841	472,729,139	27.9105715	9.2012286	.001283697
780	608,400	474,552,000	27.9284801	9.2051641	.001282051
781	609,961	476,379,541	27.9463772	9.2090962	.001280410
782	611,524	478,211,768	27.9642629	9.2130250	.001278772
783	613,089	480,048,687	27.9821372	9.2169505	.001277189
784	614,656	481,890,304	28.0000000	9.2208726	.001275510
785	616,225	483,736,625	28.0178515	9.2247914	.001273885
786	617,796	485,587,656	28.0356915	9.2287068	.001272265
787	619,369	487,443,403	28.0535203	9.2326189	.001270648
788	620,944	489,303,872	28.0713377	9.2365277	.001269086
789	622,521	491,169,069	28.0891438	9.2404333	.001267427
790	624,100	493,039,000	28.1069386	9.2443355	.001265823
791	625,681	494,913,671	28.1247222	9.2482344	.001264223
792	627,264	496,793,088	28.1424946	9.2521300	.001262626
793	628,849	498,677,257	28.1602557	9.2560224	.001261084
794	630,436	500,566,184	28.1780056	9.2599114	.001259446
795	632,025	502,459,875	28.1957444	9.2637973	.001257862
796	633,616	504,358,336	28.2134720	9.2676798	.001256281
797	635,209	506,261,573	28.2311884	9.2715592	.001254705
798	636,804	508,169,592	28.2488938	9.2754352	.001253183
799	638,401	510,082,399	28.2665881	9.2793081	.001251564
800	640,000	512,000,000	28.2842712	9.2831777	.001250000

Table 65.—*Squares, cubes, square roots, cube roots and reciprocals*—Continued

N	N^2	N^3	$N^{\frac{1}{2}}$	$N^{\frac{1}{3}}$	$\dfrac{1}{N}$
801	641,601	513,922,401	28.3019434	9.2870440	.001248439
802	643,204	515,849,608	28.3196045	9.2909072	.001246883
803	644,809	517,781,627	28.3372546	9.2947671	.001245330
804	646,416	519,718,464	28.3548938	9.2986239	.001243781
805	648,025	521,660,125	28.3725219	9.3024775	.001242236
806	649,636	523,606,616	28.3901391	9.3063278	.001240695
807	651,249	525,557,943	28.4077454	9.3101750	.001239157
808	652,864	527,514,112	28.4253408	9.3140190	.001237624
809	654,481	529,475,129	28.4429253	9.3178599	.001236094
810	656,100	531,441,000	28.4604989	9.3216975	.001234568
811	657,721	533,411,731	28.4780617	9.3255320	.001233046
812	659,344	535,387,328	28.4956137	9.3293634	.001231527
813	660,969	537,367,797	28.5131549	9.3331916	.001230012
814	662,596	539,353,144	28.5306852	9.3370167	.001228501
815	664,225	541,343,375	28.5482048	9.3408386	.001226994
816	665,856	543,338,496	28.5657137	9.3446575	.001225490
817	667,489	545,338,513	28.5832119	9.3484731	.001223990
818	669,124	547,343,432	28.6006993	9.3522857	.001222494
819	670,761	549,353,259	28.6181760	9.3560952	.001221001
820	672,400	551,368,000	28.6356421	9.3599016	.001219512
821	674,041	553,387,661	28.6530976	9.3637049	.001218027
822	675,684	555,412,248	28.6705424	9.3675051	.001216545
823	677,329	557,441,767	28.6879766	9.3713022	.001215067
824	678,976	559,476,224	28.7054002	9.3750963	.001213592
825	680,625	561,515,625	28.7228132	9.3788873	.001212121
826	682,276	563,559,976	28.7402157	9.3826752	.001210654
827	683,929	565,609,283	28.7576077	9.3864600	.001209190
828	685,584	567,663,552	28.7749891	9.3902419	.001207729
829	687,241	569,722,789	28.7923601	9.3940206	.001206273
830	688,900	571,787,000	28.8097206	9.3977964	.001204819
831	690,561	573,856,191	28.8270706	9.4015691	.001203369
832	692,224	575,930,368	28.8444102	9.4053387	.001201923
833	693,889	578,009,537	28.8617394	9.4091054	.001200480
834	695,556	580,093,704	28.8790582	9.4128690	.001199041
835	697,225	582,182,875	28.8963666	9.4166297	.001197605
836	698,896	584,277,056	28.9136646	9.4203873	.001196172
837	700,569	586,376,253	28.9309523	9.4241420	.001194743
838	702,244	588,480,472	28.9482297	9.4278936	.001193317
839	703,921	590,589,719	28.9654967	9.4316423	.001191895
840	705,600	592,704,000	28.9827535	9.4353880	.001190476
841	707,281	594,823,321	29.0000000	9.4391307	.001189061
842	708,964	596,947,688	29.0172363	9.4428704	.001187648
843	710,649	599,077,107	29.0344623	9.4466072	.001186240
844	712,336	601,211,584	29.0516781	9.4503410	.001184834
845	714,025	603,351,125	29.0688837	9.4540719	.001183432
846	715,716	605,495,736	29.0860791	9.4577999	.001182033
847	717,409	607,645,423	29.1032644	9.4615249	.001180638
848	719,104	609,800,192	29.1204396	9.4652470	.001179245
849	720,801	611,960,049	29.1376046	9.4689661	.001177856
850	722,500	614,125,000	29.1547595	9.4726824	.001176471

Table 65.—*Squares, cubes, square roots, cube roots and reciprocals—*Continued

N	N^2	N^3	$N^{\frac{1}{2}}$	$N^{\frac{1}{3}}$	$\frac{1}{N}$
851	724,201	616,295,051	29.1719043	9.4768957	.001175088
852	725,904	618,470,208	29.1890390	9.4801061	.001173709
853	727,609	620,650,477	29.2061637	9.4838136	.001172333
854	729,816	622,835,864	29.2232784	9.4875182	.001170960
855	731,025	625,026,375	29.2403830	9.4912200	.001169591
856	732,736	627,222,016	29.2574777	9.4949188	.001168224
857	734,449	629,422,793	29.2745623	9.4986147	.001166861
858	736,164	631,628,712	29.2916370	9.5023078	.001165501
859	737,881	633,839,779	29.3087018	9.5059980	.001164144
860	739,600	636,056,000	29.3257566	9.5096854	.001162791
861	741,321	638,277,381	29.3428015	9.5133699	.001161440
862	743,044	640,503,928	29.3598365	9.5170515	.001160093
863	744,769	642,735,647	29.3768616	9.5207303	.001158749
864	746,496	644,972,544	29.3938769	9.5244063	.001157407
865	748,225	647,214,625	29.4108823	9.5280794	.001156069
866	749,956	649,461,896	29.4278779	9.5317497	.001154734
867	751,689	651,714,363	29.4448637	9.5354172	.001153403
868	753,424	653,972,082	29.4618397	9.5390818	.001152074
869	755,161	656,234,909	29.4788059	9.5427437	.001150748
870	756,900	658,503,000	29.4957624	9.5464027	.001149425
871	758,641	660,776,311	29.5127091	9.5500589	.001148106
872	760,384	663,054,848	29.5296461	9.5537123	.001146789
873	762,129	665,338,617	29.5465734	9.5573630	.001145475
874	763,876	667,627,624	29.5634910	9.5610108	.001144165
875	765,625	669,921,875	29.5803989	9.5646559	.001142857
876	767,376	672,221,376	29.5972972	9.5682982	.001141553
877	769,129	674,526,133	29.6141858	9.5719877	.001140251
878	770,884	676,836,152	29.6310648	9.5755745	.001138952
879	772,641	679,151,439	29.6479342	9.5792085	.001137656
880	774,400	681,472,000	29.6647939	9.5828397	.001136364
881	776,161	683,797,841	29.6816442	9.5864682	.001135074
882	777,924	686,128,968	29.6984848	9.5900939	.001133787
883	779,689	688,465,387	29.7153159	9.5937169	.001132503
884	781,456	690,807,104	29.7321375	9.5973373	.001131222
885	783,225	693,154,125	29.7489496	9.6009548	.001129944
886	784,996	695,506,456	29.7657521	9.6045696	.001128668
887	786,769	697,864,103	29.7825452	9.6081817	.001127396
888	788,544	700,227,072	29.7993289	9.6117911	.001126126
889	790,321	702,595,369	29.8161030	9.6153977	.001124859
890	792,100	704,969,000	29.8328678	9.6190017	.001123596
891	793,881	707,347,971	29.8496231	9.6226030	.001122334
892	795,664	709,732,288	29.8663690	9.6262016	.001121076
893	797,449	712,121,957	29.8881056	9.6297975	.001119821
894	799,236	714,516,984	29.8998328	9.6333907	.001118568
895	801,025	716,917,375	29.9165506	9.6369812	.001117318
896	802,816	719,323,136	29.9332591	9.6405690	.001116071
897	804,609	721,734,273	29.9499583	9.6441542	.001114827
898	806,404	724,150,792	29.9666481	9.6477367	.001113586
899	808,201	726,572,699	29.9833287	9.6513166	.001112347
900	810,000	729,000,000	30.0000000	9.6548938	.001111111

Table 65.—*Squares, cubes, square roots, cube roots and reciprocals*—Continued

N	N^2	N^3	$N^{\frac{1}{2}}$	$N^{\frac{1}{3}}$	$\dfrac{1}{N}$
901	811,801	731,432,701	30.0166620	9.6584684	.001109878
902	813,604	733,870,808	30.0333148	9.6620403	.001108647
903	815,409	736,314,327	30.0499584	9.6656096	.001107420
904	817,216	738,763,264	30.0665928	9.6691762	.001106195
905	819,025	741,217,625	30.0832179	9.6727403	.001104972
906	820,836	743,677,416	30.0998339	9.6763017	.001103753
907	822,649	746,142,643	30.1164407	9.6798604	.001102536
908	824,464	748,613,312	30.1330383	9.6834166	.001101322
909	826,281	751,089,429	30.1496269	9.6869701	.001100110
910	828,100	753,571,000	30.1662063	9.6905211	.001098901
911	829,921	756,058,031	30.1827765	9.6940694	.001097695
912	831,744	758,550,528	30.1993377	9.6976151	.001096491
913	833,569	761,048,497	30.2158899	9.7011583	.001095290
914	835,396	763,551,944	30.2324329	9.7046989	.001094092
915	837,225	766,060,875	30.2489669	9.7082369	.001092896
916	839,056	768,575,296	30.2654919	9.7117723	.001091708
917	840,889	771,095,213	30.2820079	9.7153051	.001090513
918	842,724	773,620,632	30.2985148	9.7188354	.001089825
919	844,561	776,151,559	30.3150128	9.7223631	.001088139
920	846,400	778,688,000	30.3315018	9.7258883	.001086957
921	848,241	781,229,961	30.3479818	9.7294109	.001085776
922	850,084	783,777,448	30.3644529	9.7329309	.001084599
923	851,929	786,330,467	30.3809151	9.7364484	.001083424
924	853,776	788,889,024	30.3973683	9.7399684	.001082251
925	855,625	791,453,125	30.4138127	9.7434758	.001081081
926	857,476	794,022,776	30.4302481	9.7469857	.001079914
927	859,329	796,597,983	30.4466747	9.7504930	.001078749
928	861,184	799,178,752	30.4630924	9.7589979	.001077586
929	863,041	801,765,089	30.4795013	9.7575002	.001076426
930	864,900	804,357,000	30.4959014	9.7610001	.001075269
931	866,761	806,954,491	30.5122926	9.7644974	.001074114
932	868,624	809,557,568	30.5286750	9.7679922	.001072961
933	870,489	812,166,237	30.5450487	9.7714845	.001071811
934	872,356	814,780,504	30.5614186	9.7749743	.001070664
935	874,225	817,400,875	30.5777697	9.7784616	.001069519
936	876,096	820,025,856	30.5941171	9.7819466	.001068376
937	877,969	822,656,953	30.6104557	9.7854288	.001067236
938	879,844	825,293,672	30.6267857	9.7889087	.001066098
939	881,721	827,936,019	30.6431069	9.7923861	.001064963
940	883,600	830,584,000	30.6594194	9.7958611	.001063830
941	885,481	833,237,621	30.6757233	9.7993336	.001062699
942	887,364	835,896,888	30.6920185	9.8028086	.001061571
943	889,249	838,561,807	30.7083051	9.8062711	.001060445
944	891,136	841,232,384	30.7245830	9.8097362	.001059322
945	893,025	843,908,625	30.7408523	9.8131989	.001058201
946	894,916	846,590,536	30.7571130	9.8166591	.001057082
947	896,809	849,278,123	30.7733651	9.8201169	.001055966
948	898,704	851,971,392	30.7896086	9.8235723	.001054852
949	900,601	854,670,349	30.8058436	9.8270252	.001053741
950	902,500	857,375,000	30.8220700	9.8304757	.001052632

Table 65.—*Squares, cubes, square roots, cube roots and reciprocals*—Continued

N	N²	N³	N^½	N^⅓	$\frac{1}{N}$
951	904,401	860,085,351	30.8382879	9.8339288	.001051525
952	906,304	862,801,408	30.8544972	9.8373695	.001050420
953	908,209	865,523,177	30.8706981	9.8408127	.001049318
954	910,116	868,250,664	30.8868904	9.8442586	.001048218
955	912,025	870,983,875	30.9030743	9.8476920	.001047120
956	913,936	873,722,816	30.9192497	9.8511280	.001046025
957	915,849	876,467,493	30.9354166	9.8545617	.001044932
958	917,764	879,217,912	30.9515751	9.8579929	.001043841
959	919,681	881,974,079	30.9677251	9.8614218	.001042753
960	921,600	884,736,000	30.9838668	9.8648483	.001041667
961	923,521	887,503,681	31.0000000	9.8682724	.001040583
962	925,444	890,277,128	31.0161248	9.8716941	.001039501
963	927,369	893,056,347	31.0322413	9.8751135	.001038422
964	929,296	895,841,344	31.0483494	9.8785305	.001037344
965	931,225	898,632,125	31.0644491	9.8819451	.001036269
966	933,156	901,428,696	31.0805405	9.8853574	.001035197
967	935,089	904,231,063	31.0966236	9.8887673	.001034126
968	937,024	907,039,232	31.1126984	9.8921749	.001033058
969	938,961	909,853,209	31.1287648	9.8955801	.001031992
970	940,900	912,673,000	31.1448230	9.8989830	.001030928
971	942,841	915,498,611	31.1608729	9.9023835	.001029866
972	944,784	918,330,048	31.1769145	9.9057817	.001028807
973	946,729	921,167,317	31.1929479	9.9091776	.001027749
974	948,676	924,010,424	31.2089731	9.9125712	.001026694
975	950,625	926,859,375	31.2249900	9.9159624	.001025641
976	952,576	929,714,176	31.2409987	9.9193513	.001024590
977	954,529	932,574,833	31.2569992	9.9227379	.001023541
978	956,484	935,441,352	31.2729915	9.9261222	.001022495
979	958,441	938,313,739	31.2889757	9.9295042	.001021450
980	960,400	941,192,000	31.3049517	9.9328839	.001020408
981	962,361	944,076,141	31.3209195	9.9362613	.001019368
982	964,324	946,966,168	31.3368792	9.9396368	.001018330
983	966,289	949,862,087	31.3528308	9.9430092	.001017294
984	968,256	952,763,904	31.3687743	9.9463797	.001016260
985	970,225	955,671,625	31.3847097	9.9497479	.001015228
986	972,196	958,585,256	31.4006369	9.9531188	.001014199
987	974,169	961,504,808	31.4165561	9.9564775	.001013171
988	976,144	964,430,272	31.4324673	9.9598389	.001012146
989	978,121	967,361,669	31.4483704	9.9631981	.001011122
990	980,100	970,299,000	31.4642654	9.9665549	.001010101
991	982,081	973,242,271	31.4801525	9.9699095	.001009082
992	984,064	976,191,488	31.4960315	9.9732619	.001008065
993	986,049	979,146,657	31.5119025	9.9766120	.001007049
994	988,036	982,107,784	31.5277655	9.9799599	.001006036
995	990,025	985,074,875	31.5436206	9.9833055	.001005025
996	992,016	988,047,936	31.5594677	9.9866488	.001004016
997	994,009	991,026,973	31.5753068	9.9899900	.001003009
998	996,004	994,011,992	31.5911880	9.9933289	.001002004
999	998,001	997,002,999	31.6069613	9.9966656	.001001001
1000	1,000,000	1,000,000,000	31.6227766	10.0000000	.001000000

Table 66.—*Difference of elevation in feet per mile for various angles of slope*

Angle	0°	1°	2°	3°	4°	5°	6°	7°	8°	9°	10°
′											
0	.....	92.2	184.4	276.7	369.2	461.9	555.0	648.3	742.1	836.3	931.0
1	1.5	93.7	185.9	278.3	370.8	463.5	556.5	649.9	743.6	837.8	932.6
2	3.1	95.2	187.5	279.8	372.3	465.0	558.1	651.4	745.2	839.4	934.2
3	4.6	96.8	189.0	281.3	373.8	466.6	559.6	653.0	746.8	841.0	935.8
4	6.1	98.3	190.5	282.9	375.4	468.1	561.2	654.5	748.3	842.6	937.4
5	7.7	99.8	192.1	284.4	376.9	469.7	562.7	656.1	749.9	844.2	938.9
6	9.2	101.4	193.6	286.0	378.5	471.2	564.3	657.7	751.5	845.7	940.5
7	10.8	102.9	195.1	287.5	380.0	472.8	565.8	659.2	753.0	847.3	942.1
8	12.3	104.4	196.7	289.0	381.6	474.3	567.4	660.8	754.6	848.9	943.7
9	13.8	106.0	198.2	290.6	383.1	475.9	568.9	662.4	746.2	850.5	945.3
10	15.4	107.5	199.8	292.1	384.7	477.4	570.5	663.9	757.7	852.0	946.9
11	16.9	109.1	201.3	293.7	386.2	479.0	572.0	665.5	759.3	853.6	948.5
12	18.4	110.6	202.8	295.2	387.7	480.5	573.6	667.0	760.9	855.2	950.0
13	20.0	112.1	204.4	296.7	389.3	482.1	575.2	668.6	762.4	856.8	951.6
14	21.5	113.7	205.9	298.3	390.8	483.6	576.7	670.2	764.0	858.3	953.2
15	23.0	115.2	207.5	299.8	392.4	485.2	578.3	671.7	765.6	859.9	954.8
16	24.6	116.7	209.0	301.4	393.9	486.7	579.8	673.3	767.1	861.5	956.4
17	26.1	118.3	210.5	302.9	395.5	488.3	581.4	674.8	768.7	863.1	958.0
18	27.6	119.8	212.1	304.4	397.0	489.8	582.9	676.4	770.3	864.7	959.6
19	29.2	121.4	213.6	306.0	398.6	491.3	584.5	678.0	771.8	866.2	961.1
20	30.7	122.9	215.1	307.5	400.1	492.9	586.0	679.5	773.4	867.8	962.7
21	32.3	124.4	216.7	309.1	401.6	494.5	587.6	681.1	775.0	869.4	964.3
22	33.8	126.0	218.2	310.6	403.2	496.0	589.1	682.6	776.6	871.0	965.9
23	35.3	127.5	219.8	312.1	404.7	497.6	590.7	684.2	778.1	872.5	967.5
24	36.9	129.0	221.3	313.7	406.3	499.1	592.2	685.8	779.7	864.1	969.1
25	38.4	130.6	222.8	315.2	407.8	500.7	593.8	687.3	781.3	875.7	970.7
26	39.9	132.1	224.4	316.8	409.4	502.2	595.4	688.9	782.8	877.3	972.2
27	41.5	133.6	225.9	318.3	410.9	503.8	596.9	690.5	784.4	878.8	973.8
28	43.0	135.2	227.5	319.9	412.5	505.3	598.5	692.0	786.0	880.4	975.4
29	44.5	136.7	229.0	321.4	414.0	506.9	600.0	693.6	787.5	882.0	977.0
30	46.1	138.3	230.5	322.9	415.5	508.4	601.6	695.1	789.1	883.6	978.6
31	47.6	139.8	232.1	324.5	417.1	510.0	603.1	696.7	790.7	885.2	980.2
32	49.2	141.3	233.6	326.0	418.6	511.5	604.7	698.3	792.2	886.7	981.8
33	50.7	142.9	235.1	327.6	420.2	513.0	606.3	699.8	793.8	888.3	983.4
34	52.2	144.4	236.7	329.1	421.7	514.6	607.8	701.4	795.4	889.9	985.0
35	53.8	146.0	238.2	330.6	423.3	516.2	609.4	702.9	796.9	891.5	986.5
36	55.3	147.5	239.8	332.2	424.8	517.7	610.9	704.5	798.5	893.1	988.1
37	56.8	149.0	241.3	333.7	426.4	519.3	612.5	706.1	800.1	894.6	989.7
38	58.4	150.6	242.8	335.3	427.9	520.8	614.0	707.6	801.7	896.2	991.3
39	59.9	152.1	244.4	336.8	429.5	522.4	615.5	709.2	803.2	897.8	992.9
40	61.4	153.6	245.9	338.4	431.0	523.9	617.2	710.8	804.8	899.4	994.5
41	63.0	155.2	247.5	339.9	432.5	525.5	618.7	712.3	806.4	901.0	996.1
42	64.5	156.7	249.0	341.4	434.1	527.0	620.3	713.9	808.0	902.5	997.7
43	66.0	158.2	250.5	343.0	435.6	528.6	621.8	715.5	809.5	904.1	999.3
44	67.6	159.8	252.1	344.5	437.2	530.1	623.4	717.0	811.1	905.7	1000.9

Table 66.—*Difference of elevation in feet per mile for various angles of slope*—Continued

Angle	0°	1°	2°	3°	4°	5°	6°	7°	8°	9°	10°
45	69.1	161.3	253.6	346.1	438.7	531.7	624.9	718.6	812.7	907.3	1,002.5
46	70.6	162.9	255.2	347.6	440.3	533.2	626.5	720.2	814.2	908.9	1,004.0
47	72.2	164.4	256.7	349.2	441.8	534.8	628.0	721.7	815.8	910.5	1,005.6
48	73.7	165.9	258.2	350.7	443.4	536.3	629.6	723.3	817.4	912.0	1,007.2
49	75.3	167.5	259.8	352.2	444.9	537.9	631.2	724.8	819.0	913.6	1,008.8
50	76.8	169.0	261.3	353.8	446.5	539.4	632.7	726.4	820.5	915.2	1,010.4
51	78.3	170.6	262.9	355.3	448.0	541.0	634.3	728.0	822.1	916.8	1,012.0
52	79.9	172.1	264.4	356.9	449.6	542.5	635.8	729.5	823.7	918.4	1,013.6
53	81.4	173.6	265.9	358.4	451.1	544.1	637.4	731.1	825.3	919.9	1,015.2
54	82.9	175.2	267.5	360.0	452.7	545.6	638.9	732.7	826.8	921.5	1,016.8
55	84.5	176.7	269.0	361.5	454.2	547.2	640.5	734.2	828.4	923.1	1,018.4
56	86.0	178.2	270.6	363.0	455.8	548.7	642.1	735.8	830.0	924.7	1,020.0
57	87.5	179.8	272.1	364.6	457.3	550.3	643.6	737.4	831.5	926.3	1,021.5
58	89.1	181.3	273.6	366.1	458.8	551.8	645.2	738.9	833.1	927.8	1,023.1
59	90.6	182.8	275.2	367.7	460.4	553.4	646.7	740.5	834.7	929.4	1,024.7

Angle	11°	12°	13°	14°	15°	16°	17°	18°	19°	20°
0	1,026.3	1,122	1,219	1,316	1,415	1,514	1,614	1,716	1,818	1,922
1	1,027.9	1,124	1,221	1,318	1,416	1,516	1,616	1,717	1,820	1,924
2	1,029.5	1,126	1,222	1,320	1,418	1,517	1,618	1,719	1,822	1,925
3	1,031.1	1,127	1,224	1,321	1,420	1,519	1,619	1,721	1,823	1,927
4	1,032.7	1,129	1,225	1,323	1,421	1,521	1,621	1,723	1,825	1,929
5	1,034.3	1,130	1,227	1,325	1,423	1,522	1,623	1,724	1,827	1,931
6	1,035.9	1,132	1,229	1,326	1,425	1,524	1,624	1,726	1,828	1,932
7	1,037.5	1,134	1,230	1,328	1,426	1,525	1,626	1,728	1,830	1,934
8	1,039.1	1,135	1,232	1,330	1,428	1,527	1,628	1,729	1,832	1,936
9	1,040.7	1,137	1,234	1,331	1,430	1,529	1,629	1,731	1,834	1,937
10	1,042.3	1,138	1,235	1,333	1,431	1,531	1,631	1,733	1,835	1,939
11	1,043.8	1,140	1,237	1,334	1,433	1,532	1,633	1,734	1,837	1,941
12	1,045.4	1,142	1,238	1,336	1,435	1,534	1,634	1,736	1,839	1,943
13	1,047.0	1,143	1,240	1,338	1,436	1,535	1,636	1,738	1,840	1,944
14	1,048.6	1,145	1,242	1,339	1,438	1,537	1,638	1,739	1,842	1,946

Table 66.—*Difference of elevation in feet per mile for various angles of slope*—Continued

Angle	11°	12°	13°	14°	15°	16°	17°	18°	19°	20°
15	1,050.2	1,146	1,243	1,341	1,440	1,539	1,639	1,741	1,844	1,948
16	1,051.8	1,148	1,245	1,343	1,441	1,541	1,641	1,743	1,846	1,950
17	1,053.4	1,150	1,247	1,344	1,443	1,542	1,643	1,744	1,847	1,951
18	1,055.0	1,151	1,248	1,346	1,444	1,544	1,644	1,746	1,849	1,953
19	1,056.6	1,153	1,250	1,348	1,446	1,546	1,646	1,748	1,851	1,955
20	1,058.2	1,154	1,251	1,349	1,448	1,547	1,648	1,750	1,853	1,957
21	1,059.8	1,156	1,253	1,351	1,449	1,549	1,649	1,751	1,854	1,958
22	1,061.4	1,158	1,255	1,352	1,451	1,551	1,651	1,753	1,856	1,960
23	1,063.0	1,159	1,256	1,354	1,453	1,552	1,653	1,755	1,858	1,962
24	1,064.6	1,161	1,258	1,356	1,454	1,554	1,655	1,756	1,860	1,964
25	1,066.2	1,163	1,260	1,357	1,456	1,556	1,656	1,758	1,861	1,965
26	1,067.8	1,164	1,261	1,359	1,458	1,557	1,658	1,760	1,863	1,967
27	1,069.4	1,166	1,263	1,361	1,459	1,559	1,660	1,762	1,865	1,969
28	1,071.0	1,167	1,264	1,362	1,461	1,561	1,661	1,763	1,866	1,971
29	1,072.6	1,169	1,266	1,364	1,463	1,562	1,663	1,765	1,868	1,972
30	1,074.2	1,171	1,268	1,366	1,464	1,564	1,665	1,767	1,870	1,974
31	1,075.8	1,172	1,269	1,367	1,466	1,566	1,666	1,768	1,871	1,976
32	1,077.4	1,174	1,271	1,369	1,468	1,567	1,668	1,770	1,873	1,978
33	1,079.0	1,175	1,273	1,370	1,469	1,569	1,670	1,772	1,875	1,979
34	1,080.6	1,177	1,274	1,372	1,471	1,571	1,672	1,773	1,877	1,981
35	1,082.2	1,179	1,276	1,374	1,473	1,572	1,673	1,775	1,878	1,983
36	1,083.8	1,180	1,277	1,375	1,474	1,574	1,675	1,777	1,880	1,985
37	1,085.4	1,182	1,279	1,377	1,476	1,576	1,677	1,779	1,882	1,986
38	1,087.0	1,183	1,281	1,379	1,478	1,577	1,678	1,780	1,884	1,988
39	1,088.6	1,185	1,282	1,380	1,479	1,579	1,680	1,782	1,885	1,990
40	1,090.2	1,187	1,284	1,382	1,481	1,581	1,682	1,784	1,887	1,992
41	1,091.8	1,188	1,286	1,384	1,483	1,582	1,683	1,786	1,889	1,993
42	1,093.4	1,190	1,287	1,385	1,484	1,584	1,685	1,787	1,891	1,995
43	1,095.0	1,192	1,289	1,387	1,486	1,586	1,687	1,789	1,892	1,997
44	1,096.6	1,193	1,290	1,388	1,487	1,587	1,688	1,791	1,894	1,999
45	1,098.2	1,195	1,292	1,390	1,489	1,589	1,690	1,792	1,896	2,000
46	1,099.8	1,196	1,294	1,392	1,491	1,591	1,692	1,794	1,898	2,002
47	1,101.5	1,198	1,295	1,393	1,492	1,592	1,694	1,796	1,899	2,004
48	1,103.1	1,200	1,297	1,395	1,494	1,594	1,695	1,798	1,901	2,006
49	1,104.7	1,201	1,299	1,397	1,496	1,596	1,697	1,799	1,903	2,007
50	1,106.3	1,203	1,300	1,398	1,497	1,597	1,699	1,801	1,904	2,009
51	1,107.9	1,204	1,302	1,400	1,499	1,599	1,700	1,803	1,906	2,011
52	1,109.5	1,206	1,303	1,402	1,501	1,601	1,702	1,804	1,908	2,013
53	1,111.1	1,208	1,305	1,403	1,502	1,602	1,704	1,806	1,910	2,014
54	1,112.7	1,209	1,307	1,405	1,504	1,604	1,705	1,808	1,911	2,016
55	1,114.3	1,211	1,308	1,407	1,506	1,606	1,707	1,809	1,913	2,018
56	1,115.9	1,213	1,310	1,408	1,507	1,607	1,709	1,811	1,915	2,020
57	1,117.5	1,214	1,312	1,410	1,509	1,609	1,711	1,813	1,917	2,021
58	1,119.1	1,216	1,313	1,411	1,511	1,611	1,712	1,815	1,918	2,023
59	1,120.7	1,217	1,315	1,413	1,512	1,612	1,714	1,816	1,920	2,025

Table 67.—*Correction in feet for curvature and refraction*

$$[\,h = 0.574D^2\,]$$

Distance in miles	.0	.1	.2	.3	.4	.5	.6	.7	.8	.9
1	.6	.7	.8	1.0	1.1	1.3	1.5	1.7	1.9	2.1
2	2.3	2.5	2.8	3.0	3.3	3.6	3.9	4.2	4.5	4.8
3	5.2	5.5	5.9	6.2	6.6	7.0	7.4	7.8	8.3	8.7
4	9.2	9.6	10.1	10.6	11.1	11.6	12.1	12.7	13.2	13.8
5	14.3	14.9	15.5	16.1	16.7	17.3	18.0	18.6	19.3	20.0
6	20.7	21.4	22.1	22.8	23.5	24.2	25.0	25.7	26.5	27.3
7	28.1	28.9	30.6	30.6	31.4	32.3	33.2	34.1	35.0	35.9
8	36.7	37.6	38.6	39.5	40.4	41.4	42.4	43.4	44.4	45.5
9	46.5	47.5	48.6	49.7	50.7	51.8	52.9	54.0	55.1	56.3
10	57.4	58.6	59.7	60.9	62.1	63.3	64.5	65.7	67.0	68.2
11	69.5	70.7	71.9	73.2	74.5	75.8	77.1	78.5	79.8	81.2
12	82.7	84.0	85.4	86.8	88.3	89.7	91.1	92.6	94.0	95.5
13	97.0	98.5	100.0	101.5	103.1	104.6	106.2	107.7	109.3	110.9
14	112.5	114.1	115.7	117.4	119.0	120.7	122.4	124.0	125.7	127.4
15	129.1	130.9	132.6	134.3	136.1	137.9	139.7	141.5	143.3	145.1
16	146.9	148.7	150.6	152.5	154.4	156.3	158.2	160.1	162.0	163.9
17	165.8	167.8	169.8	171.7	173.7	175.7	177.7	179.7	181.8	183.8
18	185.9	188.0	190.1	192.2	194.3	196.4	198.5	200.7	202.8	205.0
19	207.1	209.3	211.5	213.7	216.0	218.2	220.4	222.7	224.9	227.2
20	229.5	231.8	234.2	236.5	238.8	241.2	243.5	245.9	248.3	250.7
21	253.1	255.5	257.9	260.4	262.8	265.3	267.7	270.2	272.7	275.2
22	277.7	280.3	282.8	285.4	288.0	290.5	293.1	295.7	298.3	301.0
23	303.6	306.2	308.9	311.5	314.2	316.9	319.6	322.3	325.0	327.8
24	330.5	333.3	336.1	338.9	341.7	344.5	347.3	350.1	352.9	355.8
25	358.6	361.5	364.4	367.3	370.2	373.1	376.0	379.0	381.9	384.9
26	387.9	390.9	393.9	396.9	400.0	403.0	406.0	409.1	412.2	415.3
27	418.3	421.4	424.5	427.7	430.8	434.0	437.1	440.3	443.5	446.7
28	449.9	453.1	456.3	459.6	462.8	466.1	469.4	472.7	476.0	479.3
29	482.6	485.9	489.3	492.6	496.0	499.4	502.8	506.2	509.6	513.0
30	516.5	519.9	523.4	526.8	530.3	533.8	537.3	540.8	544.4	547.9
31	551.5	555.0	558.6	562.2	565.8	569.4	573.0	576.7	580.3	584.0
32	587.6	591.3	595.0	598.7	602.4	606.1	609.9	613.6	617.3	621.1
33	624.9	628.7	632.5	636.3	640.2	644.0	647.9	651.7	655.6	659.5
34	663.4	667.3	671.2	675.1	679.1	683.0	687.0	690.9	694.9	698.9
35	702.9	707.0	711.0	715.1	719.1	723.2	727.3	731.4	735.5	739.6
36	743.7	747.8	752.0	756.1	760.3	764.5	768.7	772.9	777.1	781.3
37	785.6	789.8	794.1	798.4	802.6	806.9	811.3	815.6	819.9	824.2
38	828.6	833.0	837.4	841.8	846.2	850.6	855.0	859.4	863.9	868.3
39	872.8	877.3	881.8	886.3	890.8	895.3	899.9	904.4	909.0	913.5
40	918.1	922.7	927.3	931.9	936.6	941.2	945.9	950.5	955.2	959.9

Table 68.—*Deflections and chords for circular arcs—deflection angle and chord in terms of radius and arc*

Δ=Central angle. $\dfrac{\Delta}{2}$=Deflection angle

L=Length of arc in feet=0.0174533ΔR

R=Radius in feet=57.2958$\dfrac{L}{\Delta}$

R (feet)	1-foot arc	10-foot arc		25-foot arc		50-foot arc		100-foot arc		R (feet)
		$\frac{\Delta}{2}$	Chord	$\frac{\Delta}{2}$	Chord	$\frac{\Delta}{2}$	Chord	$\frac{\Delta}{2}$	Chord	
	'	° '	Feet	° '	Feet	° '	Feet	° '	Feet	
50	34.38	5 43.78	9.98	14 19.43	24.74	28 38.87	47.94	57 17.75	84.15	50
60	28.65	4 46.48	9.98	11 56.20	24.82	23 52.40	48.57	47 44.79	88.82	60
70	24.56	4 5.56	9.99	10 13.88	24.87	20 27.77	48.94	40 55.53	91.71	70
80	21.49	3 34.86	9.99	8 57.15	24.90	17 54.30	49.19	36 48.59	93.62	80
90	19.10	3 10.99	10.00	7 57.46	24.92	15 54.93	49.36	31 49.86	94.93	90
100	17.19	2 51.89	10.00	7 9.72	24.94	14 19.44	49.48	28 38.87	95.89	100
110	15.63	2 36.26	10.00	6 30.65	24.95	13 1.31	49.57	26 2.61	96.59	110
120	14.32	2 23.24	10.00	5 58.10	24.96	11 56.20	49.64	23 52.40	97.13	120
130	13.22	2 12.22	10.00	5 30.55	24.96	11 1.11	49.69	22 2.21	97.55	130
140	12.28	2 2.78	10.00	5 6.94	24.96	10 13.88	49.76	20 27.77	97.89	140
150	11.46	1 54.59	10.00	4 46.48	24.97	9 32.96	49.77	19 5.92	98.16	150
160	10.74	1 47.43	10.00	4 28.57	24.97	8 57.15	49.80	17 54.30	98.38	160
170	10.11	1 41.11	10.00	4 12.78	24.98	8 25.55	49.82	15 51.10	98.57	170
180	9.55	1 35.49	10.00	3 58.73	24.98	7 57.46	49.84	15 54.93	98.72	180
190	9.05	1 30.47	10.00	3 46.17	24.98	7 32.34	49.86	15 4.67	98.85	190
200	8.59	1 25.94	10.00	3 34.86	24.98	7 9.72	49.87	14 19.44	98.96	200
210	8.19	1 21.85	10.00	3 24.63	24.99	6 49.26	49.88	13 38.51	99.06	210
220	7.81	1 18.13	10.00	3 15.33	24.99	6 30.65	49.89	13 1.31	99.14	220
230	7.47	1 14.73	10.00	3 6.83	24.99	6 13.67	49.90	12 27.34	99.22	230
240	7.16	1 11.62	10.00	2 59.05	24.99	5 58.10	49.91	11 56.20	99.28	240
250	6.88	1 8.75	10.00	2 51.89	24.99	5 43.77	49.92	11 27.55	99.33	250
260	6.61	1 6.11	10.00	2 45.28	24.99	5 30.55	49.92	11 1.11	99.38	260
270	6.37	1 3.66	10.00	2 39.15	24.99	5 18.31	49.93	10 36.62	99.43	270
280	6.14	1 1.39	10.00	2 33.47	24.99	5 6.94	49.93	10 13.88	99.47	280
290	5.93	0 59.27	10.00	2 28.18	24.99	4 56.36	49.94	9 52.72	99.51	290
300	5.73	0 57.30	10.00	2 23.24	24.99	4 46.48	49.94	9 32.96	99.54	300
310	5.54	0 55.45	10.00	2 18.62	24.99	4 37.24	49.95	9 14.48	99.57	310
320	5.37	0 53.71	10.00	2 14.29	24.99	4 28.57	49.95	8 57.15	99.59	320
330	5.21	0 52.09	10.00	2 10.22	24.99	4 20.44	49.95	8 40.87	99.62	330
340	5.06	0 50.56	10.00	2 6.39	24.99	4 12.78	49.96	8 25.55	99.64	340
350	4.91	0 49.11	10.00	2 2.78	24.99	4 5.55	49.96	8 11.11	99.66	350
360	4.77	0 47.75	10.00	1 59.37	24.99	3 58.73	49.96	7 57.46	99.68	360
370	4.65	0 46.46	10.00	1 56.14	25.00	3 52.28	49.96	7 44.56	99.69	370
380	4.52	0 45.23	10.00	1 53.08	25.00	3 46.17	49.96	7 32.34	99.72	380
390	4.41	0 44.07	10.00	1 50.18	25.00	3 40.37	49.96	7 20.74	99.73	390
400	4.30	0 42.97	10.00	1 47.43	25.00	3 34.86	49.97	7 9.72	99.74	400

Table 68.—*Deflections and chords for circular arcs—deflection angle and chord in terms of radius and arc*—Continued

R (feet)	1-foot arc	10-foot arc		25-foot arc		50-foot arc		100-foot arc		R (feet)
		$\frac{\Delta}{2}$	Chord	$\frac{\Delta}{2}$	Chord	$\frac{\Delta}{2}$	Chord	$\frac{\Delta}{2}$	Chord	
	′	° ′	Feet	° ′	Feet	° ′	Feet	° ′	Feet	
450	3.82	0 38.20	10.00	1 35.49	25.00	3 10.99	49.98	6 21.97	99.79	450
500	3.44	0 34.38	10.00	1 25.94	25.00	2 51.89	49.98	5 43.77	99.83	500
550	3.13	0 31.25	10.00	1 18.13	25.00	2 36.26	49.98	5 12.52	99.86	550
600	2.86	0 28.65	10.00	1 11.62	25.00	2 23.24	49.99	4 46.48	99.89	600
650	2.64	0 26.44	10.00	1 6.11	25.00	2 12.22	49.99	4 24.44	99.90	650
700	2.46	0 24.56	10.00	1 1.39	25.00	2 2.78	49.99	4 5.55	99.92	700
750	2.29	0 22.92	10.00	0 57.30	25.00	1 54.59	49.99	3 49.18	99.92	750
800	2.15	0 21.49	10.00	0 53.71	25.00	1 47.43	49.99	3 34.86	99.94	800
850	2.02	0 20.22	10.00	0 50.56	25.00	1 41.11	49.99	3 22.22	99.95	850
900	1.91	0 19.10	10.00	0 47.75	25.00	1 35.49	49.99	3 10.99	99.95	900
1,000	1.72	17.19	10.00	42.97	25.00	1 25.94	49.99	2 51.89	99.96	1,000

Table 69.—*Stadia table*

100 (Stadia Intercept)		100	200	300	400	500	600	700	800	900
0°	2′	0.06	0.1	0.2	0.2	0.3	0.3	0.4	0.5	0.5
	4	0.12	0.2	0.3	0.5	0.6	0.7	0.8	0.9	1.0
	6	0.17	0.3	0.5	0.7	0.9	1.0	1.2	1.4	1.6
	8	0.23	0.5	0.7	0.9	1.2	1.4	1.6	1.9	2.1
	10	0.29	0.6	0.9	1.2	1.5	1.7	2.0	2.3	2.6
	12	0.35	0.7	1.0	1.4	1.7	2.1	2.4	2.8	3.1
	14	0.41	0.8	1.2	1.6	2.0	2.4	2.8	3.3	3.7
	16	0.47	0.9	1.4	1.9	2.3	2.8	3.3	3.7	4.2
	18	0.52	1.0	1.6	2.1	2.6	3.1	3.7	4.2	4.7
	20	0.58	1.2	1.7	2.3	2.9	3.5	4.1	4.6	5.2
	22	0.64	1.3	1.9	2.6	3.2	3.8	4.5	5.1	5.8
	24	0.70	1.4	2.1	2.8	3.5	4.2	4.9	5.6	6.3
	26	0.76	1.5	2.3	3.0	3.8	4.5	5.3	6.0	6.8
	28	0.81	1.6	2.4	3.2	4.1	4.9	5.7	6.5	7.3
	30	0.87	1.7	2.6	3.5	4.4	5.2	6.1	7.0	7.8
	32	0.93	1.9	2.8	3.7	4.6	5.6	6.5	7.4	8.4
	34	0.99	2.0	3.0	3.9	4.9	5.9	6.9	7.9	8.9
	36	1.05	2.1	3.1	4.2	5.2	6.3	7.3	8.4	9.4
	38	1.11	2.2	3.3	4.4	5.5	6.6	7.7	8.8	9.9
	40	1.16	2.3	3.5	4.6	5.8	7.0	8.1	9.3	10.5
	42	1.22	2.4	3.7	4.9	6.1	7.3	8.5	9.8	11.0
	44	1.28	2.6	3.8	5.1	6.4	7.7	9.0	10.2	11.5
	46	1.34	2.7	4.0	5.3	6.7	8.0	9.4	10.7	12.0
	48	1.40	2.8	4.2	5.6	7.0	8.4	9.8	11.2	12.5
	50	1.45	2.9	4.4	5.8	7.2	8.7	10.2	11.6	13.1
	52	1.51	3.0	4.5	6.0	7.5	9.1	10.6	12.1	13.6
	54	1.57	3.1	4.7	6.3	7.8	9.4	11.0	12.6	14.1
	56	1.63	3.3	4.9	6.5	8.1	9.8	11.4	13.0	14.6
	58	1.69	3.4	5.0	6.7	8.4	10.1	11.8	13.5	15.2
	60	1.74	3.5	5.2	7.0	8.7	10.5	12.2	14.0	15.7
1°	2′	1.80	3.6	5.4	7.2	9.0	10.8	12.6	14.4	16.2
	4	1.86	3.7	5.6	7.4	9.3	11.2	13.0	14.9	16.7
	6	1.92	3.8	5.8	7.7	9.6	11.5	13.4	15.4	17.3
	8	1.98	4.0	5.9	7.9	9.9	11.9	13.8	15.8	17.8
	10	2.03	4.1	6.1	8.1	10.2	12.2	14.2	16.3	18.3
	12	2.09	4.2	6.3	8.4	10.5	12.6	14.7	16.7	18.8
	14	2.15	4.3	6.5	8.6	10.8	12.9	15.1	17.2	19.4
	16	2.21	4.4	6.6	8.8	11.0	13.3	15.5	17.7	19.9
	18	2.27	4.5	6.8	9.1	11.3	13.6	15.9	18.1	20.4
	20	2.33	4.7	7.0	9.3	11.6	14.0	16.3	18.6	20.9
	22	2.38	4.8	7.2	9.5	11.9	14.3	16.7	19.1	21.5
	24	2.44	4.9	7.3	9.8	12.2	14.7	17.1	19.5	22.0
	26	2.50	5.0	7.5	10.0	12.5	15.0	17.5	20.0	22.5
	28	2.56	5.1	7.7	10.2	12.8	15.3	17.9	20.5	23.0
	30	2.62	5.2	7.8	10.5	13.1	15.7	18.3	20.9	23.5
	32	2.67	5.3	8.0	10.7	13.4	16.0	18.7	21.4	24.1
	34	2.73	5.5	8.2	10.9	13.7	16.4	19.1	21.9	24.6
	36	2.79	5.6	8.4	11.2	14.0	16.7	19.5	22.3	25.1
	38	2.85	5.7	8.5	11.4	14.2	17.1	19.9	22.8	25.6
	40	2.91	5.8	8.7	11.6	14.5	17.4	20.3	23.3	26.2
	42	2.97	5.9	8.9	11.9	14.8	17.8	20.8	23.7	26.7
	44	3.02	6.0	9.1	12.1	15.1	18.1	21.2	24.2	27.2
	46	3.08	6.2	9.2	12.3	15.4	18.5	21.6	24.6	27.7
	48	3.14	6.3	9.4	12.6	15.7	18.8	22.0	25.1	28.3
	50	3.20	6.4	9.6	12.8	16.0	19.2	22.4	25.6	28.8
	52	3.26	6.5	9.8	13.0	16.3	19.5	22.8	26.0	29.3
	54	3.31	6.6	9.9	13.2	16.6	19.9	23.2	26.5	29.8
	56	3.37	6.7	10.1	13.5	16.9	20.2	23.6	27.0	30.3
	58	3.43	6.9	10.3	13.7	17.1	20.6	24.0	27.4	30.9
	60	3.49	7.0	10.5	14.0	17.4	20.9	24.4	27.9	31.4
Horizontal distance		*99.9*	*199.8*	*299.6*	*399.5*	*499.4*	*599.3*	*699.2*	*799.0*	*898.9*

Table 69.—*Stadia table*—Continued

Slant distance		100	200	300	400	500	600	700	800	900
2°	2'	3.55	7.1	10.6	14.2	17.7	21.3	24.8	28.4	31.9
	4	3.60	7.2	10.8	14.4	18.0	21.6	25.2	28.8	32.4
	6	3.66	7.3	11.0	14.6	18.3	22.0	25.6	29.3	33.0
	8	3.72	7.4	11.2	14.9	18.6	22.3	26.0	29.8	33.5
	10	3.78	7.6	11.3	15.1	18.9	22.7	26.4	30.2	34.0
	12	3.84	7.7	11.5	15.3	19.2	23.0	26.9	30.7	34.5
	14	3.90	7.8	11.7	15.6	19.5	23.4	27.3	31.2	35.1
	16	3.95	7.9	11.9	15.8	19.8	23.7	27.7	31.6	35.6
	18	4.01	8.0	12.0	16.0	20.0	24.1	28.1	32.1	36.1
	20	4.07	8.1	12.2	16.3	20.3	24.4	28.5	32.5	36.6
	22	4.13	8.3	12.4	16.5	20.6	24.8	28.9	33.0	37.1
	24	4.18	8.4	12.6	16.7	20.9	25.1	29.3	33.5	37.7
	26	4.24	8.5	12.7	17.0	21.2	25.5	29.7	33.9	38.2
	28	4.30	8.6	12.9	17.2	21.5	25.8	30.1	34.4	38.7
	30	4.36	8.7	13.1	17.4	21.8	26.1	30.5	34.9	39.2
	32	4.42	8.8	13.2	17.7	22.1	26.5	30.9	35.3	39.7
	34	4.47	8.9	13.4	17.9	22.4	26.8	31.3	35.8	40.3
	36	4.53	9.1	13.6	18.1	22.7	27.2	31.7	36.3	40.8
	38	4.59	9.2	13.8	18.4	23.0	27.5	32.1	36.7	41.3
	40	4.65	9.3	13.9	18.6	23.2	27.9	32.5	37.2	41.8
	42	4.71	9.4	14.1	18.8	23.5	28.2	32.9	37.6	42.4
	44	4.76	9.5	14.3	19.1	23.8	28.6	33.3	38.1	42.9
	46	4.82	9.6	14.5	19.3	24.1	28.9	33.8	38.6	43.4
	48	4.88	9.8	14.6	19.5	24.4	29.3	34.2	39.0	43.9
	50	4.94	9.9	14.8	19.8	24.7	29.6	34.6	39.5	44.4
	52	5.00	10.0	15.0	20.0	25.0	30.0	35.0	40.0	45.0
	54	5.05	10.1	15.2	20.2	25.3	30.3	35.4	40.4	45.5
	56	5.11	10.2	15.3	20.4	25.6	30.7	35.8	40.9	46.0
	58	5.17	10.3	15.5	20.7	25.8	31.0	36.2	41.4	46.5
	60	5.23	10.5	15.7	20.9	26.1	31.4	36.6	41.8	47.1
Horizontal dist.		*99.7*	*199.3*	*299.2*	*398.9*	*498.7*	*598.4*	*698.1*	*797.8*	*897.5*
3°	2'	5.28	10.6	15.9	21.1	26.4	31.7	37.0	42.3	47.6
	4	5.34	10.7	16.0	21.4	26.7	32.1	37.4	42.7	48.1
	6	5.40	10.8	16.2	21.6	27.0	32.4	37.8	43.2	48.6
	8	5.46	10.9	16.4	21.8	27.3	32.7	38.2	43.7	49.1
	10	5.52	11.0	16.5	22.1	27.6	33.1	38.6	44.1	49.6
	12	5.57	11.1	16.7	22.3	27.9	33.4	39.0	44.6	50.2
	14	5.63	11.3	16.9	22.5	28.2	33.8	39.4	45.0	50.7
	16	5.69	11.4	17.1	22.8	28.4	34.1	39.8	45.5	51.2
	18	5.75	11.5	17.2	23.0	28.7	34.5	40.2	46.0	51.7
	20	5.80	11.6	17.4	23.2	29.0	34.8	40.6	46.4	52.2
	22	5.86	11.7	17.6	23.4	29.3	35.1	41.0	46.9	52.8
	24	5.92	11.8	17.8	23.7	29.6	35.5	41.4	47.4	53.3
	26	5.98	12.0	17.9	23.9	29.9	35.9	41.8	47.8	53.8
	28	6.04	12.1	18.1	24.1	30.2	36.2	42.2	48.3	54.3
	30	6.09	12.2	18.3	24.4	30.5	36.6	42.6	48.7	54.8
	32	6.15	12.3	18.4	24.6	30.8	36.9	43.0	49.2	55.4
	34	6.21	12.4	18.6	24.8	31.0	37.3	43.5	49.7	55.9
	36	6.27	12.5	18.8	25.1	31.3	37.6	43.9	50.1	56.4
	38	6.32	12.6	19.0	25.3	31.6	37.9	44.3	50.6	56.9
	40	6.38	12.8	19.1	25.5	31.9	38.3	44.7	51.1	57.4
	42	6.44	12.9	19.3	25.8	32.2	38.6	45.1	51.5	58.0
	44	6.50	13.0	19.5	26.0	32.5	39.0	45.5	52.0	58.5
	46	6.55	13.1	19.7	26.2	32.8	39.3	45.9	52.4	59.0
	48	6.61	13.2	19.8	26.4	33.1	39.7	46.3	52.9	59.5
	50	6.67	13.3	20.0	26.7	33.4	40.0	46.7	53.4	60.0
	52	6.73	13.5	20.2	26.9	33.6	40.4	47.1	53.8	60.6
	54	6.78	13.6	20.4	27.1	33.9	40.7	47.5	54.3	61.1
	56	6.84	13.7	20.5	27.4	34.2	41.1	47.9	54.7	61.6
	58	6.90	13.8	20.7	27.6	34.5	41.4	48.3	55.2	62.1
	60	6.96	13.9	20.9	27.8	34.8	41.7	48.7	55.7	62.6
Horizontal dist.		*99.5*	*199.0*	*298.5*	*398.0*	*497.6*	*597.1*	*696.6*	*796.1*	*895.6*

Table 69.—*Stadia table*—Continued

Slant distance		100	200	300	400	500	600	700	800	900
4°	2′	7.02	14.0	21.0	28.1	35.1	42.1	49.1	56.1	63.1
	4	7.07	14.1	21.2	28.3	35.4	42.4	49.5	56.6	63.7
	6	7.13	14.3	21.4	28.5	35.7	42.8	49.9	57.0	64.2
	8	7.19	14.4	21.6	28.8	35.9	43.1	50.3	57.5	64.7
	10	7.25	14.5	21.7	29.0	36.2	43.5	50.7	58.0	65.2
	12	7.30	14.6	21.9	29.2	36.5	43.8	51.1	58.4	65.7
	14	7.36	14.7	22.1	29.4	36.8	44.2	51.5	58.9	66.2
	16	7.42	14.8	22.3	29.7	37.1	44.5	51.9	59.3	66.8
	18	7.48	15.0	22.4	29.9	37.4	44.9	52.3	59.8	67.3
	20	7.53	15.1	22.6	30.2	37.7	45.2	52.7	60.3	67.8
	22	7.59	15.2	22.8	30.4	38.0	45.5	53.1	60.7	68.3
	24	7.65	15.3	22.9	30.6	38.2	45.9	53.5	61.2	68.8
	26	7.71	15.4	23.1	30.8	38.5	46.2	53.9	61.6	69.3
	28	7.76	15.5	23.3	31.1	38.8	46.6	54.3	62.1	69.9
	30	7.82	15.6	23.5	31.3	39.1	46.9	54.7	62.6	70.4
	32	7.88	15.8	23.6	31.5	39.4	47.3	55.1	63.0	70.9
	34	7.94	15.9	23.8	31.7	39.7	47.6	55.5	63.5	71.4
	36	7.99	16.0	24.0	32.0	40.0	48.0	56.0	63.9	71.9
	38	8.05	16.1	24.2	32.2	40.3	48.3	56.4	64.4	72.5
	40	8.11	16.2	24.3	32.4	40.5	48.6	56.8	64.9	73.0
	42	8.17	16.3	24.5	32.7	40.8	49.0	57.2	65.3	73.5
	44	8.22	16.4	24.7	32.9	41.1	49.3	57.6	65.8	74.0
	46	8.28	16.6	24.8	33.1	41.4	49.7	58.0	66.2	74.5
	48	8.34	16.7	25.0	33.4	41.7	50.0	58.4	66.7	75.0
	50	8.40	16.8	25.2	33.6	42.0	50.4	58.8	67.2	75.6
	52	8.45	16.9	25.4	33.8	42.3	50.7	59.2	67.6	76.1
	54	8.51	17.0	25.5	34.0	42.6	51.1	59.6	68.1	76.6
	56	8.57	17.1	25.7	34.3	42.8	51.4	60.0	68.5	77.1
	58	8.63	17.3	25.9	34.5	43.1	51.8	60.4	69.0	77.6
	60	8.68	17.4	26.0	34.7	43.4	52.1	60.8	69.5	78.1
Horizontal dist.		*99.2*	*198.3*	*297.7*	*397.0*	*496.2*	*595.4*	*694.7*	*793.9*	*893.0*
5°	2′	8.74	17.5	26.2	35.0	43.7	52.4	61.2	69.9	78.7
	4	8.80	17.6	26.4	35.2	44.0	52.8	61.6	70.4	79.2
	6	8.85	17.7	26.6	35.4	44.3	53.1	62.0	70.8	79.7
	8	8.91	17.8	26.7	35.6	44.6	53.5	62.4	71.3	80.2
	10	8.97	17.9	26.9	35.9	44.8	53.8	62.8	71.7	80.7
	12	9.03	18.1	27.1	36.1	45.1	54.2	63.2	72.2	81.2
	14	9.08	18.2	27.2	36.3	45.4	54.5	63.6	72.7	81.7
	16	9.14	18.3	27.4	36.6	45.7	54.8	64.0	73.1	82.3
	18	9.20	18.4	27.6	36.8	46.0	55.2	64.4	73.6	82.8
	20	9.25	18.5	27.8	37.0	46.3	55.5	64.8	74.0	83.3
	22	9.31	18.6	27.9	37.2	46.6	55.9	65.2	74.5	83.8
	24	9.37	18.7	28.1	37.5	46.8	56.2	65.6	74.9	84.3
	26	9.43	18.9	28.3	37.7	47.1	56.6	66.0	75.4	84.8
	28	9.48	19.0	28.4	37.9	47.4	56.9	66.4	75.9	85.3
	30	9.54	19.1	28.6	38.2	47.7	57.2	66.8	76.3	85.9
	32	9.60	19.2	28.8	38.4	48.0	57.6	67.2	76.8	86.4
	34	9.65	19.3	29.0	38.6	48.3	57.9	67.6	77.2	86.9
	36	9.71	19.4	29.1	38.8	48.6	58.3	68.0	77.7	87.4
	38	9.77	19.5	29.3	39.1	48.8	58.6	68.4	78.1	87.9
	40	9.83	19.7	29.5	39.3	49.1	59.0	68.8	78.6	88.4
	42	9.88	19.8	29.6	39.5	49.4	59.3	69.2	79.0	88.9
	44	9.94	19.9	29.8	39.8	49.7	59.6	69.6	79.5	89.4
	46	10.00	20.0	30.0	40.0	50.0	60.0	70.0	80.0	90.0
	48	10.05	20.1	30.2	40.2	50.3	60.3	70.4	80.4	90.5
	50	10.11	20.2	30.3	40.4	50.5	60.7	70.8	80.9	91.0
	52	10.17	20.3	30.5	40.7	50.8	61.0	71.2	81.3	91.5
	54	10.22	20.4	30.7	40.9	51.1	61.3	71.6	81.8	92.0
	56	10.28	20.6	30.8	41.1	51.4	61.7	72.0	82.2	92.5
	58	10.33	20.7	31.0	41.4	51.7	62.0	72.4	82.7	93.0
	60	10.40	20.8	31.2	41.6	52.0	62.4	72.8	83.2	93.6
Horizontal dist.		*98.9*	*197.8*	*296.7*	*395.6*	*494.5*	*593.5*	*692.4*	*791.3*	*890.2*

Table 69.—*Stadia table*—Continued

Slant distance	100	200	300	400	500	600	700	800	900
6° 2′	10.45	20.9	31.4	41.8	52.3	62.7	73.2	83.6	94.1
4	10.51	21.0	31.5	42.0	52.5	63.1	73.6	84.1	94.6
6	10.57	21.1	31.7	42.3	52.8	63.4	74.0	84.5	95.1
8	10.62	21.2	31.9	42.5	53.1	63.7	74.4	85.0	95.6
10	10.68	21.4	32.0	42.7	53.4	64.0	74.8	85.4	96.1
12	10.74	21.5	32.2	42.9	53.7	64.4	75.2	85.9	96.6
14	10.79	21.6	32.4	43.2	54.0	64.8	75.5	86.3	97.1
16	10.85	21.7	32.5	43.4	54.2	65.1	75.9	86.8	97.6
18	10.91	21.8	32.7	43.6	54.5	65.4	76.3	87.2	98.2
20	10.96	21.9	32.9	43.8	54.8	65.8	76.7	87.7	98.7
22	11.02	22.0	33.1	44.1	55.1	66.1	77.1	88.2	99.2
24	11.08	22.2	33.2	44.3	55.4	66.5	77.5	88.6	99.7
26	11.13	22.3	33.4	44.5	55.6	66.8	77.9	89.1	100.2
28	11.19	22.4	33.6	44.8	55.9	67.1	78.3	89.5	100.7
30	11.25	22.5	33.7	45.0	56.2	67.5	78.7	90.0	101.2
32	11.30	22.6	33.9	45.2	56.5	67.8	79.1	90.4	101.7
34	11.36	22.7	34.1	45.4	56.8	68.2	79.5	90.9	102.2
36	11.42	22.8	34.2	45.7	57.1	68.5	79.9	91.3	102.7
38	11.47	22.9	34.4	45.9	57.4	68.8	80.3	91.8	103.2
40	11.53	23.1	34.6	46.1	57.6	69.2	80.7	92.2	103.8
42	11.59	23.2	34.8	46.3	57.9	69.5	81.1	92.7	104.3
44	11.64	23.3	34.9	46.6	58.2	69.9	81.5	93.1	104.8
46	11.70	23.4	35.1	46.8	58.5	70.2	81.9	93.6	105.3
48	11.76	23.5	35.3	47.0	58.8	70.5	82.3	94.0	105.8
50	11.81	23.6	35.4	47.2	59.1	70.9	82.7	94.5	106.3
52	11.87	23.7	35.6	47.5	59.3	71.2	83.1	95.0	106.8
54	11.93	23.9	35.8	47.7	59.6	71.6	83.5	95.4	107.3
56	11.98	24.0	35.9	47.9	59.9	71.9	83.9	95.9	107.8
58	12.04	24.1	36.1	48.2	60.2	72.2	84.3	96.3	108.4
60	12.10	24.2	36.3	48.4	60.5	72.6	84.7	96.8	108.9
Horizontal dist.	*98.5*	*197.0*	*295.5*	*394.0*	*492.6*	*591.1*	*689.6*	*788.1*	*886.6*
7° 2′	12.15	24.3	36.5	48.6	60.8	72.9	85.1	97.2	109.4
4	12.21	24.4	36.6	48.8	61.0	73.2	85.5	97.7	109.9
6	12.26	24.5	36.8	49.1	61.3	73.6	85.8	98.1	110.4
8	12.32	24.6	37.0	49.3	61.6	73.9	86.2	98.6	110.9
10	12.38	24.8	37.1	49.5	61.9	74.3	86.6	99.0	111.4
12	12.43	24.9	37.3	49.7	62.2	74.6	87.0	99.5	111.9
14	12.49	25.0	37.5	50.0	62.4	74.9	87.4	99.9	112.4
16	12.55	25.1	37.6	50.2	62.7	75.3	87.8	100.4	112.9
18	12.60	25.2	37.8	50.4	63.0	75.6	88.2	100.8	113.4
20	12.66	25.3	38.0	50.6	63.3	75.9	88.6	101.3	113.9
22	12.71	25.4	38.1	50.9	63.6	76.3	89.0	101.7	114.4
24	12.77	25.5	38.3	51.1	63.8	76.6	89.4	102.2	114.9
26	12.83	25.7	38.5	51.3	64.1	77.0	89.8	102.6	115.4
28	12.88	25.8	38.6	51.5	64.4	77.3	90.2	103.1	115.9
30	12.94	25.9	38.8	51.8	64.7	77.6	90.6	103.5	116.4
32	13.00	26.0	39.0	52.0	65.0	78.0	91.0	104.0	117.0
34	13.05	26.1	39.2	52.2	65.3	78.3	91.4	104.4	117.5
36	13.11	26.2	39.3	52.4	65.5	78.6	91.7	104.9	118.0
38	13.16	26.3	39.5	52.7	65.8	79.0	92.1	105.3	118.5
40	13.22	26.4	39.7	52.9	66.1	79.3	92.5	105.8	119.0
42	13.28	26.6	39.8	53.1	66.4	79.7	92.9	106.2	119.5
44	13.33	26.7	40.0	53.3	66.7	80.0	93.2	106.7	120.0
46	13.39	26.8	40.2	53.6	66.9	80.3	93.7	107.1	120.5
48	13.44	26.9	40.3	53.8	67.2	80.7	94.1	107.6	121.0
50	13.50	27.0	40.5	54.0	67.5	81.0	94.5	108.0	121.5
52	13.56	27.1	40.7	54.2	67.8	81.3	94.9	108.5	122.0
54	13.61	27.2	40.8	54.5	68.1	81.7	95.3	108.9	122.5
56	13.67	27.3	41.0	54.7	68.3	82.0	95.7	109.4	123.0
58	13.73	27.5	41.2	54.9	68.6	82.3	96.1	109.8	123.5
60	13.78	27.6	41.3	55.1	68.9	82.7	96.4	110.3	124.0
Horizontal dist.	*98.1*	*196.1*	*294.2*	*392.2*	*490.3*	*588.4*	*686.4*	*784.5*	*882.6*

Table 69.—*Stadia table*—Continued

Slant distance		100	200	300	400	500	600	700	800	900
8°	5′	13.92	27.8	41.8	55.7	69.6	83.5	97.4	111.4	125.3
	10	14.06	28.1	42.2	56.2	70.3	84.4	98.4	112.5	126.6
	15	14.20	28.4	42.6	56.8	71.0	85.2	99.4	113.6	127.8
	20	14.34	28.7	43.0	57.4	71.7	86.0	100.4	114.7	129.1
	25	14.48	29.0	43.4	57.9	72.4	86.9	101.4	115.8	130.3
	30	14.62	29.2	43.9	58.5	73.1	87.7	102.3	116.9	131.6
	35	14.76	29.5	44.2	59.0	73.7	88.4	103.1	117.8	132.5
	40	14.90	29.8	44.7	59.6	74.5	89.4	104.3	119.2	134.1
	45	15.04	30.1	45.1	60.1	75.2	90.2	105.2	120.3	135.3
	50	15.17	30.3	45.5	60.7	75.9	91.0	106.2	121.4	136.6
	55	15.31	30.6	45.9	61.2	76.6	91.9	107.2	122.5	137.8
	60	15.45	30.9	46.4	61.8	77.3	92.7	108.2	123.6	139.1
Horizontal dist.		*97.5*	*195.1*	*292.7*	*390.2*	*487.8*	*585.3*	*682.9*	*780.4*	*878.0*
9°	5′	15.59	31.2	46.8	62.4	77.9	93.5	109.1	124.7	140.3
	10	15.73	31.5	47.2	62.9	78.6	94.5	110.2	125.9	141.6
	15	15.86	31.7	47.6	63.5	79.3	95.2	111.1	126.9	142.8
	20	16.00	32.0	48.0	64.0	80.0	96.0	112.0	128.0	144.0
	25	16.14	32.3	48.4	64.6	80.7	96.8	113.0	129.0	145.3
	30	16.28	32.6	48.8	65.1	81.4	97.7	113.9	130.2	146.5
	35	16.42	32.8	49.2	65.7	82.1	98.5	114.9	131.3	147.7
	40	16.55	33.1	49.7	66.2	82.8	99.3	115.9	132.4	148.0
	45	16.69	33.4	50.1	66.8	83.5	100.1	116.8	133.5	150.2
	50	16.83	33.7	50.5	67.3	84.4	101.0	117.8	134.6	151.4
	55	16.96	33.9	50.9	67.9	84.8	101.8	118.7	135.7	152.7
	60	17.10	34.2	51.3	68.4	85.5	102.6	119.7	136.8	153.9
Horizontal dist.		*97.0*	*194.0*	*291.0*	*387.9*	*484.9*	*581.9*	*678.9*	*775.9*	*872.9*
10°	5′	17.24	34.5	51.7	68.9	86.2	103.4	120.7	137.9	155.1
	10	17.37	34.7	52.1	69.5	86.9	104.2	121.6	139.0	156.4
	15	17.51	35.0	52.5	70.0	87.6	105.1	122.6	140.1	157.6
	20	17.65	35.3	52.9	70.6	88.2	105.9	123.5	141.2	158.8
	25	17.78	35.6	53.3	71.1	88.9	106.7	124.5	142.3	160.0
	30	17.92	35.8	53.8	71.7	89.6	107.5	125.4	143.3	161.3
	35	18.05	36.1	54.2	72.2	90.3	108.3	126.4	144.4	162.5
	40	18.19	36.4	54.6	72.7	90.9	109.1	127.3	145.5	163.7
	45	18.37	36.6	55.0	73.4	91.8	110.1	128.5	146.9	165.3
	50	18.46	36.9	55.4	73.8	92.3	110.8	129.2	147.7	166.1
	55	18.60	37.2	55.8	74.4	93.0	111.6	130.2	148.8	167.4
	60	18.73	37.5	56.2	74.9	93.7	112.4	131.1	149.8	168.5
Horizontal dist.		*96.4*	*192.7*	*289.1*	*385.4*	*481.8*	*578.2*	*674.5*	*770.9*	*867.7*
11°	5′	18.86	37.7	56.6	75.5	94.3	113.2	132.1	150.9	169.8
	10	19.00	38.0	57.0	76.0	95.0	114.0	133.0	152.0	171.0
	15	19.13	38.3	57.4	76.5	95.7	114.8	133.9	153.1	172.2
	20	19.27	38.5	57.8	77.1	96.3	115.6	134.9	154.1	173.4
	25	19.40	38.8	58.2	77.6	97.0	116.4	135.8	155.2	174.6
	30	19.54	39.1	58.6	78.1	97.7	117.2	136.8	156.3	175.8
	35	19.67	39.3	59.0	78.7	98.4	118.0	137.7	157.4	177.0
	40	19.80	39.6	59.4	79.2	99.0	118.8	138.6	158.4	178.2
	45	19.94	39.9	59.8	79.7	99.7	119.6	139.6	159.5	179.4
	50	20.07	40.1	60.2	80.3	100.4	120.4	140.5	160.6	180.6
	55	20.20	40.4	60.6	80.8	101.0	121.2	141.4	161.6	181.8
	60	20.34	40.7	61.0	81.4	101.7	122.0	142.4	162.7	183.0
Horizontal dist.		*95.7*	*191.3*	*287.0*	*382.7*	*478.4*	*574.1*	*669.7*	*765.4*	*861.1*

Table 69.—*Stadia table*—Continued

Slant distance		100	200	300	400	500	600	700	800	900
12°	5′	20.47	40.9	61.4	81.9	102.3	122.8	143.3	163.8	184.2
	10	20.60	41.2	61.8	82.4	103.0	123.6	144.2	164.8	185.4
	15	20.73	41.5	62.2	82.9	103.7	124.4	145.1	165.9	186.6
	20	20.87	41.7	62.6	83.5	104.3	125.2	146.1	166.9	187.8
	25	21.00	42.0	63.0	84.0	105.0	126.0	147.0	168.0	189.0
	30	21.13	42.3	63.4	84.5	105.7	126.8	147.9	169.0	190.2
	35	21.26	42.5	63.8	85.1	106.3	127.6	148.8	170.1	191.4
	40	21.39	42.8	64.2	85.6	107.0	128.4	149.8	171.2	192.5
	45	21.52	43.1	64.6	86.1	107.6	129.2	150.7	172.2	193.7
	50	21.66	43.3	65.0	86.6	108.3	129.9	151.6	173.2	194.9
	55	21.79	43.6	65.4	87.2	108.9	130.7	152.5	174.3	196.1
	60	21.92	43.8	65.7	87.7	109.6	131.5	153.4	175.3	197.3
Horizontal dist.		*94.9*	*189.9*	*284.8*	*379.8*	*474.7*	*569.6*	*664.6*	*759.5*	*854.5*
13°	5′	22.05	44.1	66.1	88.2	110.2	132.3	154.3	176.3	198.4
	10	22.18	44.4	66.5	88.7	110.9	133.1	155.3	177.4	199.6
	15	22.31	44.6	66.9	89.2	111.6	133.9	156.2	178.5	200.8
	20	22.44	44.9	67.3	89.8	112.2	134.6	157.1	179.5	202.0
	25	22.57	45.1	67.7	90.3	112.8	135.4	158.0	180.6	203.1
	30	22.70	45.4	68.1	90.8	113.5	136.2	158.9	181.6	204.3
	35	22.83	45.7	68.5	91.3	114.1	137.0	159.8	182.6	205.5
	40	22.96	45.9	68.9	91.8	114.8	137.7	160.7	183.7	206.6
	45	23.09	46.2	69.3	92.4	115.4	138.5	161.6	184.7	207.8
	50	23.22	46.4	69.6	92.9	116.1	139.3	162.5	185.7	208.9
	55	23.35	46.7	70.0	93.4	116.7	140.1	163.4	186.8	210.1
	60	23·47	46.9	70.4	93.9	117.4	140.8	164.3	187.8	211.3
Horizontal dist.		*94.2*	*188.3*	*282.4*	*376.6*	*470.7*	*564.9*	*659.0*	*753.2*	*847.3*
14°	5′	23.60	47.2	70.8	94.4	118.0	141.6	165.2	188.8	212.4
	10	23.73	47.5	71.2	94.9	118.6	142.4	166.1	189.8	213.6
	15	23.86	47.7	71.6	95.4	119.3	143.2	167.0	190.9	214.7
	20	23.99	48.0	72.0	95.9	119.9	143.9	167.9	191.9	215.9
	25	24.11	48.2	72.3	96.5	120.6	144.7	168.8	192.9	217.0
	30	24.24	48.5	72.7	97.0	121.2	145.4	169.7	193.9	218.2
	35	24.37	48.7	73.1	97.5	121.8	146.2	170.6	194.9	219.3
	40	24.49	49.0	73.5	98.0	122.5	147.0	171.5	196.0	220.4
	45	24.62	49.2	73.9	98.5	123.1	147.7	172.3	197.0	221.6
	50	24.75	49.5	74.2	99.0	123.7	148.5	173.2	198.0	222.7
	55	24.87	49.7	74.6	99.5	124.4	149.2	174.1	199.0	223.9
	60	25.00	50.0	75.0	100.0	125.0	150.0	175.0	200.0	225.0
Horizontal dist.		*93.3*	*186.6*	*279.9*	*373.2*	*466.5*	*559.8*	*653.1*	*746.4*	*839.7*
15°	5′	25.13	50.3	75.4	100.5	125.6	150.8	175.9	201.0	226.1
	10	25.25	50.5	75.8	101.0	126.3	151.5	176.8	202.0	227.3
	15	25.38	50.8	76.1	101.5	126.9	152.3	177.6	203.0	228.4
	20	25.50	51.0	76.5	102.0	127.5	153.0	178.5	204.0	229.5
	25	25.63	51.3	76.9	102.5	128.1	153.8	179.4	205.0	230.6
	30	25.75	51.5	77.3	103.0	128.8	154.5	180.3	206.0	231.8
	35	25.88	51.8	77.6	103.5	129.4	155.3	181.1	207.0	232.9
	40	26.00	52.0	78.0	104.0	130.0	156.0	182.0	208.0	234.0
	45	26.12	52.2	78.4	104.5	130.6	156.7	182.9	209.0	235.1
	50	26.25	52.5	78.7	105.0	131.2	157.5	183.7	210.0	236.2
	55	26.37	52.7	79.1	105.5	131.9	158.2	184.6	211.0	237.4
	60	26.50	53.0	79.5	106.0	132.5	159.0	185.5	212.0	238.5
Horizontal dist		*93.4*	*184.8*	*277.2*	*369.6*	*462.0*	*554.4*	*646.8*	*739.2*	*831.6*

Table 69.—*Stadia table*—Continued

Slant distance		100	200	300	400	500	600	700	800	900
16°	5′	26.62	53.2	79.9	106.5	133.1	159.7	186.3	213.0	239.6
	10	26.74	53.5	80.2	107.0	133.7	160.5	187.2	213.9	240.7
	15	26.86	53.7	80.6	107.5	134.3	161.2	188.0	214.9	241.8
	20	26.99	54.0	81.0	108.0	134.9	161.9	188.9	215.9	242.9
	25	27.11	54.2	81.3	108.4	135.6	162.7	189.8	216.9	244.0
	30	27.23	54.5	81.7	108.9	136.2	163.4	190.6	217.9	245.1
	35	27.35	54.7	82.1	109.4	136.8	164.1	191.5	218.8	246.2
	40	27.48	55.0	82.4	109.9	137.4	164.9	192.4	219.8	247.3
	45	27.60	55.2	82.8	110.4	138.0	165.6	193.2	220.8	248.4
	50	27.72	55.4	83.2	110.9	138.6	166.3	194.0	221.7	249.5
	55	27.84	55.7	83.5	111.4	139.2	167.0	194.9	222.7	250.6
	60	27.96	55.9	83.9	111.8	139.8	167.8	195.7	223.7	251.6
Horizontal dist.		*91.4*	*183*	*274*	*366*	*457*	*549*	*640*	*732*	*823*
17°	5′	28.08	56.2	84.2	112.3	140.4	168.5	196.6	224.6	252.7
	10	28.20	56.4	84.6	112.8	141.0	169.2	197.4	225.6	253.8
	15	28.32	56.6	85.0	113.3	141.6	169.9	198.2	226.6	254.9
	20	28.44	56.9	85.3	113.8	142.2	170.6	199.1	227.5	256.0
	25	28.56	57.1	85.7	114.2	142.8	171.4	199.9	228.5	257.0
	30	28.68	57.4	86.0	114.7	143.4	172.1	200.8	229.4	258.1
	35	28.80	57.6	86.4	115.2	144.0	172.8	201.6	230.4	259.2
	40	28.92	57.8	86.7	115.7	144.6	173.5	202.4	231.3	260.2
	45	29.04	58.1	87.1	116.1	145.3	174.2	203.2	232.3	261.3
	50	29.15	58.3	87.5	116.6	145.8	174.9	204.1	233.2	262.4
	55	29.27	58.5	87.8	117.1	146.4	175.6	204.9	234.2	263.4
	60	29.39	58.8	88.2	117.6	146.9	176.3	205.7	235.1	264.5
Horizontal dist.		*90.4*	*181*	*271*	*362*	*452*	*543*	*633*	*724*	*814*
18°	5′	29.51	59.0	88.5	118.0	147.5	177.0	206.5	236.1	265.6
	10	29.62	59.2	88.9	118.5	148.1	177.7	207.4	237.0	266.6
	15	29.74	59.5	89.2	119.0	148.7	178.4	208.2	237.9	267.7
	20	29.86	59.7	89.6	119.4	149.3	179.1	209.0	238.9	268.7
	25	29.97	59.9	89.9	119.9	149.9	179.8	209.8	239.8	269.8
	30	30.09	60.2	90.3	120.4	150.5	180.5	210.6	240.7	270.8
	35	30.21	60.4	90.6	120.8	151.0	181.2	211.4	241.7	271.9
	40	30.32	60.6	91.0	121.3	151.6	181.9	212.3	242.6	272.9
	45	30.44	60.9	91.3	121.8	152.2	182.6	213.1	243.5	273.9
	50	30.55	61.1	91.7	122.2	152.8	183.3	213.9	244.4	275.0
	55	30.67	61.3	92.0	122.7	153.3	184.0	214.7	245.4	276.0
	60	30.78	61.6	92.3	123.1	153.9	184.7	215.5	246.3	277.0
Horizontal dist.		*89.4*	*179*	*268*	*358*	*447*	*536*	*626*	*715*	*805*
19°	5′	30.90	61.8	92.7	123.6	154.5	185.4	216.3	247.2	278.1
	10	31.01	62.0	93.0	124.0	155.1	186.1	217.1	248.1	279.1
	15	31.12	62.3	93.4	124.5	155.6	186.8	217.9	249.0	280.1
	20	31.24	62.5	93.7	125.0	156.2	187.4	218.7	249.9	281.2
	25	31.35	62.7	94.1	125.4	156.8	188.1	219.5	250.8	282.2
	30	31.47	62.9	94.4	125.9	157.3	188.8	220.3	251.7	283.2
	35	31.58	63.2	94.7	126.3	157.9	189.5	221.1	252.6	284.2
	40	31.69	63.4	95.1	126.8	158.5	190.1	221.8	253.5	285.2
	45	31.80	63.6	95.4	127.2	159.0	190.8	222.6	254.4	286.2
	50	31.92	63.8	95.7	127.7	159.6	191.5	223.4	255.3	287.2
	55	32.03	64.1	96.1	128.1	160.1	192.2	224.2	256.2	288.3
	60	32.14	64.3	96.4	128.6	160.7	192.8	225.0	257.1	289.3
Horizontal dist.		*88.3*	*177*	*265*	*353*	*442*	*530*	*618*	*706*	*795*

Table 70.—*Average weight, in pounds per cubic foot, of various substances*

SUBSTANCE	WEIGHT		SUBSTANCE	WEIGHT
Soil components:	*Dry*	*Satu-rated*	Masonry and its materials—Continued	
Clay:			Masonry of sandstone or stone of like weight weighs about seven-eights of the above.	
Hard	105	128		
Very soft	60	100		
Silt:			Mortar, hardened	90–115
Compact	115	134	Stone	135–195
Loose	70	106	Stone, quarried, loosely piled	80–110
Sand:			Stone, broken, loose	77–112
Dense	125	140	Stone, broken, rammed	79–121
Loose	75	109	Metals and alloys:	
Gravel:			Brass (copper and zinc)	487–524
Dense	130	143	Bronze (copper and tin)	524–537
Loose	95	122	Copper, cast	537–548
Soil mixtures:			Copper, rolled	548–562
Sandy gravels:			Iron and steel, cast	438–483
Dense	140	149	Average	450
Loose	105	128	Iron and steel, wrought	475–494
Clay-silt-sand-gravel:			Average	481
Dense	150	156	Spelter or zinc	425–450
Loose	100	125	Tin, cast	450–470
Loose dry-processed bentonite	60		Woods, seasoned and dry:	
			Ash	40–53
Masonry and its materials:			Hemlock	25
Brick, best pressed	150		Hickory	37–58
Brick, common hard	125		Oak, white	37–56
Brick, soft inferior	100		Oak, red, black, etc	32–45
Brickwork, pressed brick, fine joints	140		Pine, white	22–31
Brickwork, medium quality	125		Pine, yellow, northern	30–39
Brickwork, coarse, inferior soft bricks	100		Pine, yellow, southern	40–50
Cement, pulverized, loose	72–105		Poplar	22–31
Cement, pressed	115		Spruce	25
Cement, set	168–187		Woods weigh one-fifth to one-half more green than dry; and ordinary building timber, tolerably seasoned, weighs about one-sixth more than dry timber.	
Concrete, structure, 1½-in. max. aggregate	143–152			
Masonry of granite or stone of like weight:				
Well dressed	165			
Well-scabbled rubble, 20% mortar	154			
Roughly scabbled rubble, 25% to 35% mortar	150			
Well-scabbled dry rubble	138			
Roughly scabbled dry rubble	125			

Table 71.—*Convenient equivalents*

LENGTH

1 mil = 0.001 inch.
1 inch = 2.54 centimeters.
1 foot = 12 inches = 0.3048 meter.
1 yard = 36 inches = 3 feet = 0.9144 meter.
1 rod = 198 inches = 16.5 feet = 5.5 yards = 5.0292 meters.
1 mile = 63,360 inches = 5,280 feet = 1,760 yards = 320 rods = 1.60935 kilometers.
1 meter = 100 centimeters = 0.001 kilometer = 39.37 inches = 3.2808 feet.

SURFACE

1 circular mil = $\frac{\pi}{4}$ (0.001)² or 0.0000007854 square inch.

1 square inch = 1,273,240 circular mils = 6.45163 square centimeters.
1 square foot = 144 square inches = 0.092903 square meter.
1 square yard = 1,296 square inches = 9 square feet = 0.83613 square meter.
1 square rod = 39,204 square inches = 272.25 square feet = 30.25 square yards = 25.293
    square meters.
1 acre = 6,272,640 square inches = 43,560 square feet = 4,840 square yards = 160 square
    rods = 208.71 feet square = 0.404687 hectare.
1 square mile = 27,878,400 square feet = 3,097,600 square yards = 102,400 square rods =
    640 acres = 259 hectares.
1 square meter = 10,000 square centimeters = 0.0001 hectare = 0.000001 square kilo-
    meter = 1,550 square inches = 10.7639 square feet.

VOLUME

1 cubic inch = 16.3872 cubic centimeters.
1 U. S. gallon = 231 cubic inches = 3.78543 liters.
1 cubic foot = 1,728 cubic inches = 7.4805 U. S. gallons = 28.317 liters.
1 cubic yard = 46,656 cubic inches = 201.974 U. S. gallons = 27 cubic feet = 0.76456 cubic
    meter.
1 acre-foot = 325,851 U. S. gallons = 43,560 cubic feet = 1,613½ cubic yards = 1233.49 cubic
    meters.
1 cubic meter or kiloliter = 1,000,000 cubic centimeters = 1,000 liters = 61,023.4 cubic
    inches = 35.3145 cubic feet.

WEIGHT

1 U. S. gallon of water weighs 8.34 pounds avoirdupois.
1 cubic foot of water weighs 62.4 pounds avoirdupois.
1 avoirdupois pound = 7,000 grains = 0.4536 kilogram.
1 kilogram = 1,000 grams = 0.001 tonne (metric ton) = 15,432 grains = 2.2046 pounds
    avoirdupois.
1 atmosphere = about $\begin{cases} \text{15 pounds per square inch.} \\ \text{1 ton per square foot.} \\ \text{1 kilogram per square centimeter.} \end{cases}$

HYDRAULICS

1 second-foot = 448.8 U. S. gallons per minute = 26,929.9 gallons per hour = 646,317
    gallons per day.
    = 60 cubic feet per minute = 3,600 cubic feet per hour = 86,400 cubic feet
        per day = 31,536,000 cubic feet per year = 0.000214 cubic mile per year.
    = 0.9917 acre-inch per hour = 0.082645 acre-foot per hour = 1.9835 acre-feet
        per day = 723.9669 acre-feet per year.
    = 50 miner's inches in Idaho, Kansas, Nebraska, New Mexico, North
        Dakota, and South Dakota = 40 miner's inches in Arizona, California,
        Montana, and Oregon = 38.4 miner's inches in Colorado.
    = 0.028317 cubic meter per second = 1.699 cubic meters per minute =
        101.941 cubic meters per hour = 2,446.59 cubic meters per day.
1 acre-foot = 726 second-feet for 1 minute = 12.1 second-feet for 1 hour = 0.5042 second-
    foot for 1 day.
1 cubic meter per minute = 0.5886 second-foot per minute = 4.403 U. S. gallons per
    second = 1.1674 acre-feet per day.
1 second-foot falling 8.81 feet = 1 horsepower.
1 second-foot at 1 foot head = 0.1135 theoretical horsepower = 0.0846 kilowatt.
1 inch depth of water = 53.33 acre-feet per square mile.
1 second-foot for 1 year will cover 1 square mile 1.131 feet or 13.574 inches deep.

Table 71.—*Convenient equivalents*—Continued

SPEED

1 foot per second=0.68 mile per hour=1.097 kilometers per hour.
Acceleration of gravity, g=32.16 feet per second per second.

MISCELLANEOUS

1 horsepower=5,694,120 foot-gallons of water per day=550 foot-pounds per second= 33,000 fcot-pounds per minute=1,980,000 foot-pounds per hour=2,545 B. t. u. per hour=76.0 kilogrammeters per second=4,562 kilogrammeters per minute=746 watts=8.81 second-feet of water falling 1 foot.

1 B. t. u.=778 foot-pounds.

1 pound of bituminous coal contains about 14,100 B. t. u. or 11,000,000 foot-pounds of energy.

Energy in kw.-hr.=1.024 times acre-feet times static head in feet times overall efficiency.

1 foot per year=0.0329 inch per day=0.00137 inch per hour=1×10⁻⁶ centimeters per second. (For use in determining coefficient of permeability in soil.)

INDEX

347

☆ U. S. GOVERNMENT PRINTING OFFICE : 1961 O - 604603

www.ingramcontent.com/pod-product-compliance
Lightning Source LLC
Chambersburg PA
CBHW060541200326
41521CB00007B/440